TISSUE ENGINEERING AND REGENERATIVE MEDICINE

A Nano Approach

TISSUE ENGINEERING AND REGENERATIVE MEDICINE

A Nano Approach

Edited by
Murugan Ramalingam
Pekka Vallittu
Ugo Ripamonti
Wan-Ju Li

CRC Press
Taylor & Francis Group
Boca Raton London New York

CRC Press is an imprint of the
Taylor & Francis Group, an **informa** business

CRC Press
Taylor & Francis Group
6000 Broken Sound Parkway NW, Suite 300
Boca Raton, FL 33487-2742

Printed in the United States of America on acid-free paper
Version Date: 20120719

International Standard Book Number: 978-1-4398-8185-9 (Hardback)

Library of Congress Cataloging-in-Publication Data

Tissue engineering and regenerative medicine : a nano approach / editors, Murugan Ramalingam ... [et al.].
p. ; cm.
Includes bibliographical references and index.
ISBN 978-1-4398-8185-9 (hardback : alk. paper)
I. Ramalingam, Murugan.
[DNLM: 1. Regenerative Medicine--trends. 2. Tissue Engineering--methods. 3. Nanotechnology--methods. QT 37]

612'.028--dc23 2012025561

Visit the Taylor & Francis Web site at
http://www.taylorandfrancis.com

and the CRC Press Web site at
http://www.crcpress.com

Dedication

This book is dedicated to students, researchers and teachers who contributed to the development of tissue engineering and regenerative medicine.

Contents

Preface

Tissue engineering and regenerative medicine, including dentistry, promise to provide solutions based on integrated strategies of life science, engineering, and clinical medicine to current health challenges. Given that the field of tissue engineering and regenerative medicine is developed at a rapid pace and new findings are disclosed constantly, the state of the art of knowledge has to be continually updated. In the past several years, a number of textbooks that aim at covering current topics of tissue engineering and regenerative medicine have been published, but few of them are intended to compile all the aspects of tissue engineering and regenerative medicine and dentistry from fundamental principles to current advances and future trends. We, the editors, decided to edit a new textbook that is distinct from other available with a specific theme focusing on the utilization of nanotechnology, biomaterials science in tissue engineering, and regenerative medicine without forgetting certain important clinical aspects. We invited internationally known basic scientists, engineers, and clinicians to contribute to this book.

This book contains 24 chapters, and the topics cover the areas of nanobiomaterial and scaffold (Chapters 1–11), tissue mechanics (Chapter 12), stem cell (Chapter 13), bone tissue engineering (Chapters 14–16), cartilage tissue engineering (Chapters 17–19), controlled release (Chapters 20–22), and animal science and clinical medicine (Chapters 23 and 24). Readers will find that this book collectively bridges the gap between nanotechnology and tissue engineering and regenerative medicine to facilitate the merger of these two emerging research fields for translating laboratory discovery of tissue engineering and regenerative medicine to clinical applications through the use of nanoscale approaches.

We are grateful to all the contributors for their time and effort to deliver their chapters within a limited time frame. Without their contribution, this book could not have been completed. We also thank the publisher, CRC Press, in particular, Allison Shatkin, Amy Blalock, and Andrea Dale, for their support to publish this book on time. Finally, we hope readers will find it enjoyable to read this book to satisfy their intellectual needs.

Murugan Ramalingam
Pekka Vallittu
Ugo Ripamonti
Wan-Ju Li

Acknowledgments

The editors gratefully acknowledge all the authors who contributed to the success of the book. This book will not succeed without their insights and perspectives. We also appreciate the support and encouragement from the colleagues and students in the following departments: Centre for Stem Cell Research at the Christian Medical College in India, Faculty of Dental Surgery at the University of Strasbourg in France, World Premier International Advanced Institute for Materials Research at the Tohoku University in Japan, Turku Clinical Biomaterials Centre at the University of Turku in Finland, Bone Research Laboratory at the University of Witwatersrand in South Africa, and Musculoskeletal Biology and Regenerative Medicine Laboratory at the University of Wisconsin–Madison in the United States. Other valuable suggestions were made by anonymous referees, whom the editors would like to convey our special thanks. Our publisher, CRC Press, has been extraordinarily supportive and patient with our process. Our special thanks go to the editorial staff at CRC Press, especially Allison Shatkin, Editor of Materials Science and Chemical Engineering, and other supportive staff, who greatly amended the text format, technical corrections, and presentation style. We also wish to thank and formally acknowledge all the publishers and authors, indicated in the text and figures, who granted us permission to use their material in this book. This list is incomplete and we apologize to anyone we omitted.

Editors

Murugan Ramalingam is an Associate Professor and Scientist 'G' at the Centre for Stem Cell Research (CSCR), Department of Biotechnology, Government of India, Christian Medical College Bagayam Campus, Vellore, Tamil Nadu, India. Concurrently, he is an Adjunct Associate Professor at the Tohoku University, Sendai, Miyagi, Japan. Prior to joining the CSCR, he was an Associate Professor of Biomaterials and Tissue Engineering at the Institut National de la Santé et de la Recherche Médicale (INSERM) U977, Faculté de Chirurgie Dentaire, Université de Strasbourg, Strasbourg, France. He has worked at the WPI Advanced Institute for Materials Research, Japan, as an Assistant Professor. He has also worked at the National Institute of Standards and Technology, Gaithersburg, MD, and the National Institutes of Health, Bethesda, MD, under the U.S. National Academies Associateship program. He received his degree in the Doctor of Philosophy (Biomaterials) from the University of Madras, Chennai, India. He has also undergone training in Ethical and Policy Issues on Stem Cells from Harvard University, Cambridge, MA, and in Operations Management from the University of Illinois-Chicago. His current research interests are focused on the development of multiphase biomaterials, through conventional to nanotechnology to biomimetic approaches, cell patterning, stem cell differentiation, and tissue engineering. He is the author of over 125 publications, including peer-reviewed journal papers, conference proceedings, book chapters, authored books, edited books, and patents relevant to biomaterials and tissue engineering. He serves on the editorial boards of multiple biomaterials and tissue engineering-related journals, e.g., as the Editor-in-Chief of the *Journal of Bionanoscience* and the *Journal of Biomaterials and Tissue Engineering*. He is a recipient of Fellowships from the Council of Scientific and Industrial Research (India), Singapore Millenium Foundation (Singapore), the National Research Council (U.S.A), the National Professeur des Universités (France), and the Royal Society of Chemistry (U.K.).

Professor Pekka Vallittu has earned his degrees of certified dental technician in 1988, Doctor of Dental Surgery and Doctor of Philosophy in 1994, received Adjunct Professorship in 1995, and specialized in prosthodontics and stomatognathic physiology (European Prosthodontic Association Recognized Prosthodontist) in 2000. Presently, he is a Full Professorship of Biomaterials Science in the Faculty of Medicine, University of Turku, Turku, Finland, and works as the Dean of the Institute of Dentistry and the Director of Turku Clinical Biomaterials Centre (http://www.biomaterials.utu.fi). He is also an Honorary Professor at the University of Hong Kong, Pokfulam, Hong Kong. His research activity on fiber-reinforced composites started from his hobby of model aircraft and has lasted for more than 25 years; first applications for fiber-reinforced composites were in dentistry, followed by active research to utilize fiber-reinforced composites as biomimetic materials, also in medicine. He has been coordinating several large research projects, for example, Nonmetallic Bioactive Fiber-Reinforced Composite Implant – research project (NEWBONE), which is funded by the European Commission and aimed at developing nonmetallic bioactive orthopedic implants. Pekka Vallittu has over 300 original research papers, 500 other papers and abstracts, and 140 granted international patents based on 17 innovations and has been awarded by Distinguished Scientist Award in Prosthodontics and Implant Research by the International Association of Dental Research.

Ugo Ripamonti is a Professor and the Director of the Bone Research Laboratory, a Research Unit of the South African Medical Research Council and the University of the Witwatersrand, Johannesburg, South Africa, within the Faculty of Health Sciences at the Medical School. He received his degrees of the Doctor of Medicine (1978) and the Doctor of Dental Surgery (1990), *cum laude*, from the University of Milan, Milan, Italy, and the Doctor of Philosophy degree in cell biology (1993) from the University of the Witwatersrand. Dr. Ripamonti is a national and an international scholar focusing on experimentation on tissue engineering of bone, tissue induction, and morphogenesis by transforming growth factor-βs and osteoinductive biomimetic matrices in preclinical and clinical contexts. He has received numerous awards, including the Marshall R. Urist Awarded Lecture at the Sixth International Conference on Bone Morphogenetic Proteins in 2006 in Dubrovnik. Dr. Ripamonti was recognized by the University of the Witwatersrand as the recipient of the prestigious Vice-Chancellor Research Award (1993) and the Distinguished Researchers Award (1995). He has published more than 200 scientific papers, including patents and book chapters for specialists.

Professor Wan-Ju Li is the Principal Investigator of the Musculoskeletal Biology and Regenerative Medicine Laboratory at the University of Wisconsin-Madison. He is also an affiliated Faculty Member in the Cellular and Molecular Biology Program, Stem Cell and Regenerative Medicine Center. His research interests include stem cell, tissue engineering, nanobiomaterial, and skeletal biology. Professor Li is a Member of the International Society for Stem Cell Research, the Orthopaedic Research Society, the Tissue Engineering International and Regenerative Medicine Society, the American Society for Cell Biology, and the American Society for Bone and Mineral Research. He has published 30 papers, 9 book chapters, and more than 60 abstracts. He holds three patents in cartilage, intervertebral disc, and tendon/ligament applications. Professor Li has received the Fellow Award for Research Excellence from the National Institutes of Health and the Young Investigator Research Award from the North American Spine Society. He also serves in the editorial board of the American Journal of Stem Cells, the Journal of Regenerative Medicine and Tissue Engineering, the Formosan Journal of Musculoskeletal Disorders, and the Journal of Biosensors and Bioelectronics.

Contributors

Mauro Alini
AO Research Institute Davos
Davos, Switzerland

John G. Barber
Department of Orthopedics and Rehabilitation
University of Wisconsin-Madison
Madison, Wisconsin

Clark E. Barrett
Department of Materials Science and Metallurgy
University of Cambridge
Cambridge, U.K.

Serena M. Best
Department of Materials Science and Metallurgy
University of Cambridge
Cambridge, U.K.

Patrick T. Brown
Department of Orthopedics and Rehabilitation
University of Wisconsin-Madison
Madison, Wisconsin

Ruth E. Cameron
Department of Materials Science and Metallurgy
University of Cambridge
Cambridge, U.K.

Yoon N. Cho
Department of Bioengineering and Regenerative Medicine
Utah-Inha DDS and Advanced Therapeutics Research Center
Incheon, Republic of South Korea

Marcus T. Cicerone
Polymers Division
National Institute of Standards and Technology
Gaithersburg, Maryland

Meng Deng
Institute for Regenerative Engineering
University of Connecticut Health Center
Farmington, Connecticut

Dilip Dey
Aesthetic and Plastic Surgery Institute
University of California, Irvine
Irvine, California

Ivan Donati
Department of Life Sciences
University of Trieste
Trieste, Italy

David Eglin
AO Research Institute Davos
Davos, Switzerland

Elodie Faure
Department of Materials Science
Polytech Grenoble
Grenoble, France

Matthew B. Fisher
Department of Orthopedic Surgery
University of Pennsylvania
Philadelphia, Pennsylvania

Aziz Ghahary
Department of Surgery
University of British Columbia
Vancouver, British Columbia, Canada

Jyotsnendu Giri
Polymers Division
National Institute of Standards and Technology
Gaithersburg, Maryland

Sibylle Grad
AO Research Institute Davos
Davos, Switzerland

Z. S. Haidar
Department of BioEngineering
University of Utah
Salt Lake City, Utah

Andrew M. Handorf
Department of Orthopedics and Rehabilitation
University of Wisconsin-Madison
Madison, Wisconsin

Ryan Hartwell
Department of Surgery
University of British Columbia
Vancouver, British Columbia, Canada

James J. Hickman
NanoScience Technology Center
University of Central Florida
Orlando, Florida

Leena Hupa
Process Chemistry Centre
Åbo Akademi University
Turku, Finland

Roshan James
Institute for Regenerative Engineering
University of Connecticut Health Center
Farmington, Connecticut

Shaheen Jeeawoody
Department of Bioengineering
Stanford University
Stanford, California

Frank Ko
Department of Materials Engineering
University of British Columbia
Vancouver, British Columbia, Canada

Sangamesh G. Kumbar
Institute for Regenerative Engineering
University of Connecticut Health Center
Farmington, Connecticut

Kurt Lädrach
Inselspital
Bern, Switzerland

Janice H. Lai
Department of Mechanical Engineering
Stanford University
Stanford, California

Cato T. Laurencin
Institute for Regenerative Engineering
University of Connecticut Health Center
Farmington, Connecticut

Victor Leung
Department of Materials Engineering
University of British Columbia
Vancouver, British Columbia, Canada

Wan-Ju Li
Department of Orthopedics and Rehabilitation
University of Wisconsin-Madison
Madison, Wisconsin

Paula Linderbäck
Institute of Dentistry
University of Turku
Turku, Finland

Eleonora Marsich
Department of Life Sciences
University of Trieste
Trieste, Italy

Robert L. Mauck
Department of Orthopedic Surgery
University of Pennsylvania
Philadelphia, Pennsylvania

Daniel G. Miller
Excorp Medical, Inc.
St. Paul, Minnesota

Peter Molnar
Department of Zoology
University of West Hungary
Szombathely, Hungary

Debasish Mondal
Department of Clinical and Experimental Medicine
Linköping University
Linköping, Sweden

Niko Moritz
Institute of Dentistry
University of Turku
Turku, Finland

Shirley Keolebogile Motaung
Department of Orthopedic Surgery
University of California, Davis
Sacramento, California

Timo O. Närhi
Institute of Dentistry
University of Turku
Turku, Finland

Sergio Paoletti
Department of Life Sciences
University of Trieste
Trieste, Italy

Jean-Claude Petit
Faculty of Health Sciences
Medical Research Council/University of the Witwatersrand
Parktown, South Africa

Murugan Ramalingam
National Institute of Health and Medical Research UMR977
University of Strasbourg
Strasbourg, France
and
World Premier International Advanced Institute for Materials Research (WPI-AIMR)
Tohoku University
Sendai, Japan
and
Centre for Stem Cell Research
Christian Medical College Campus
Vellore, India

Anusuya Ramasubramanian
Department of Bioengineering
University of California, Berkeley
Berkeley, California

A. Hari Reddi
Department of Orthopedic Surgery
University of California, Davis
Sacramento, California

Louise Renton
Faculty of Health Sciences
Medical Research Council/University of the Witwatersrand
Parktown, South Africa

George K.B. Sándor
Regea Institute for Regenerative Medicine
University of Tampere
Tampere, Finland
and
Oral and Maxillofacial Surgery
University of Oulu
Oulu, Finland

Koteshwar N. Sridhar
Sri Research for Tissue Engineering Private Limited
Bangalore, India

Martin J. Stoddart
AO Research Institute Davos
Davos, Switzerland

Andrea Travan
Department of Life Sciences
University of Trieste
Trieste, Italy

Tuomas Waltimo
Institute of Preventive Dentistry and Oral Microbiology
University of Basel
Basel, Switzerland

Fan Yang
Department of Bioengineering
Stanford University
Stanford, California

Heejae Yang
Department of Materials Engineering
University of British Columbia
Vancouver, British Columbia, Canada

Nicola U. Zitzmann
School of Dental Medicine
University of Basel
Basel, Switzerland

1 Biostable Composite Biomaterials in Medical Applications

Pekka K. Vallittu

CONTENTS

1.1 INTRODUCTION

Replacement of damaged tissues by medical or dental biomaterials after an injury or disease requires certain properties from biomaterials. There is a trend to utilize non-metallic biomaterials, polymers, ceramics, and composites rather than metals, although metals are durable and can withstand physiological stress well. However, metals and devices made out of metals do not fulfill biomechanical requirements (isoelasticity) of bone and tooth substance, resulting in insufficient (stress-shielding) or overloading situations, and involve potential cytotoxicity arising from metal ion liberation and harmful corrosion products [1,2]. In addition, metallic objects interfere with medical diagnostics by magnetic resonance imaging (MRI) [3–5]. There are a number of polymer-based composites that are often utilized in tissue engineering and biodegradable scaffolds and drug-releasing systems and also as actual reconstructive devices [6,7]. Consequently, medical biomaterials can be classified according to their use as *reconstructive* and *regenerative* biomaterials—the first one having long-term durability, and the latter one to be typically resorbable in nature. One new group of engineering materials that has started to be utilized in reconstructive medical implants is fiber-reinforced composites (FRCs). They are presently biostable, although bioresorbable FRCs are to be developed and known for high strength and bone-like mechanical behavior [8–13]. A number of studies have been conducted to demonstrate the biomechanical suitability and biocompatibility of FRCs to be used as reconstructive medical implant material and prosthodontic and restorative material in dentistry. These aspects are reported and discussed.

1.2 MATERIALS

1.2.1 Resin Matrices

Resins that are used in medical and dental FRCs are thermoplastics, thermosets, or their combinations in the form of semiinterpenetrating polymer networks (semi-IPNs).

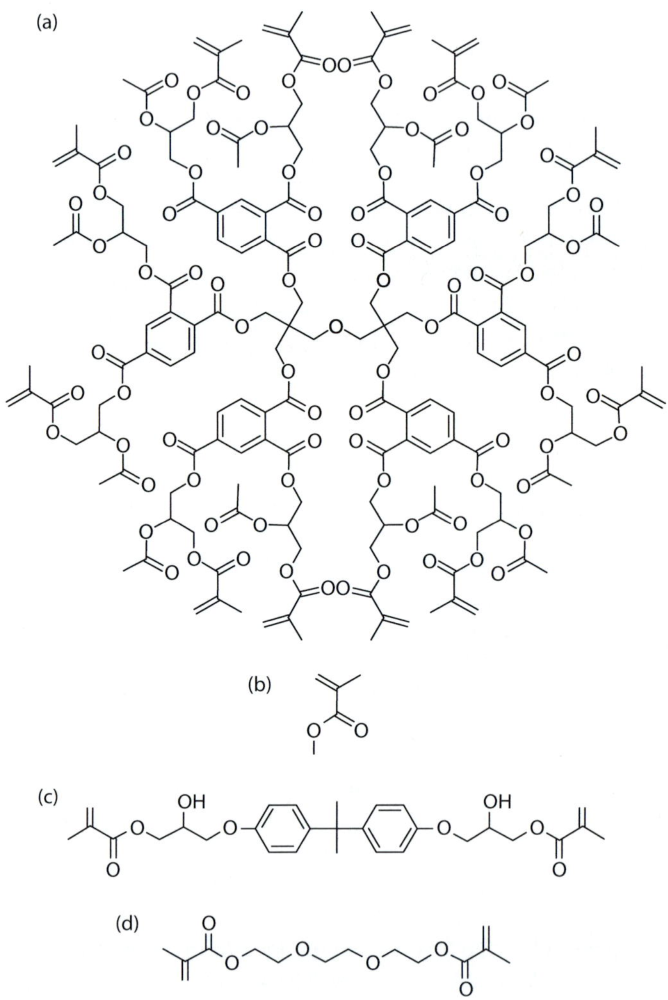

FIGURE 1.1 Chemical structures of monomers used in matrices of composites. (a) Methacrylated dendrimer. (b) Methylmethacrylate. (c) Bis-GMA. (d) TEGDMA.

Examples of thermoplastics are polyethylene (PE), polyetheretherketone (PEEK), polyacetal, and polyurethane. Examples of thermosets that are utilized as biomaterials are epoxies and bis-glycidyl-A-dimethacrylate (Bis-GMA), bis-glycidyl ethylmethacrylate (Bis-EMA), and triethylene glycol dimethacrylate (TEGDMA). Methacrylated dendrimers have also been tested as a resin matrix for FRCs (Figure 1.1).

Thermosets exhibit high cross-linking density, which makes them stiffer and more fragile. Mechanical properties of biostable polymers are presented in Table 1.1.

Resin matrix for the reinforcing fibers is cured in contact to the glass fibers (thermosets and semi-IPN matrices) or melted or dissolved for impregnation of fibers (thermoplastics). By sizing the hydroxyl-covered glass fibers with silane and by using a monomer resin system of thermoset or semi-IPN polymer, the adhesion of polymer matrix to fibers is based on chemical reaction and physical attachment.

The polymerization reaction of monomer systems of thermoset and semi-IPN polymers is based on free radical (vinyl) polymerization. Initiation of the polymerization is made by blue light radiation or by increasing the temperature [16]. Normally, the light-initiated polymerization is based on the activation of initiator camphorquinone

TABLE 1.1
Mechanical Properties of Typical Bulk Biomaterials and Hard Tissues

Material	Modulus (GPa)	Tensile Strength (MPa)
Polymers		
Polyethylene	0.88	35
Polyetheretherketone	8.3	139
Polyacetal	2.1	67
Polyurethane	0.02	35
Polymethylmethacarylate	2.5	59
Polytetrafluoroethylene	0.5	28
Polyethylene terepthalate	2.85	61
Poly(bis-glycidyl-A-dimethacrylate)-poly(triethylene glycol dimethacrylate)	8.0	52
Metals		
Stainless steel	190	586
Cobalt–chromium alloy	210	1085
Titanium alloy	116	965
Hard Tissues		
Cortical bone (longitudinal)	17.7	133
Cortical bone (transversal)	12.8	52
Cancellous bone	0.4	7.4
Enamel	84.3	10
Dentine	11.0	39.3

Source: S. Ramakrishna et al., *Composites and Technology* 61 (2001); E. Asmussen and A. Peutzfeldt. *Dental Materials* 14 (1998) 51.

and activator amine. Heat-induced polymerization is based on activations of benzoyl peroxide initiator by increased temperature. If the monomer system is aimed to be cured without temperature increase or light radiation, an activator of tertiary amine is increased to the initiator system. Examples of these autopolymerizing resins are polymethylmethacarylate (PMMA) and pBis-GMA–TEGDMA bone cements and PMMA denture rebasing and repair resins [17,18]. Autopolymerization and light curing polymerization result in a polymer with a lower degree of monomer conversion than that of the polymer cured by increased temperature. Lower monomer conversion also increases the quantity of leachable residual monomers and thus reduces the biocompatibility of the polymer [19]. Light-cured polymers can be postcured by heat after initial curing, which considerably increases the degree of monomer conversion, reduces the quantity of residual monomers, and improves biocompatibility [20–22]. Optimal postcuring temperature is close to the glass transition temperature where there is enough thermal energy in the system to create free volume, which enables unreacted carbon–carbon double bonds to form free radicals and react with each other.

1.2.2 Fillers

Fillers that have been used in biostable polymer composites are particulate fillers of fibers. Nonresorbable and hydrolytically stable particulate fillers are used in dental filling composites, which have been in clinical use since the 1950s [23]. Since then, after many significant material improvements, restorative composite resins still suffer from two key shortcomings: deficiencies of mechanical strength and high polymerization shrinkage. Resin matrices of filling composites are dimethacrylates (Bis-GMA–TEGDMA, Bis-EMA–TEGDMA, and urethane dimethacrylate), which are polymerized by light initiation. Improvements in silanation and filler composition have led to better adhesion of fillers to the polymer matrix. Filler technology has led to the development of composite resins characterized by containing zirconia or silica nanoparticle filler of approximately 25 nm in size and nanoaggregates of approximately 75 nm in size, which improves polishability and wear resistance of the composite [24]. However, the addition of nanofillers up to 30 wt% to the microfilled composite resin decreased the degree of monomer conversion [25].

Particulate fillers have also been tested to improve the osteoconductivity of bone cements. Biodegradable particulate fillers of hydroxyapatite and bioactive glass (BAG) in the autopolymerizing PMMA matrix were preclinically studied by animal experiments [26]. By histological analysis, it was found that bone contact was observed only when BAG or hydroxyapatite was present as exposed. Fibrous tissue was found on the surface of PMMA. BAG was better able to withstand the detrimental effect of PMMA than hydroxyapatite. Hydroxyapatite has also been incorporated into the PE matrix [2]. Besides biodegradable hydroxyapatite fillers and BAG fillers, tricalcium phosphate has been added to the polymers to improve osteoconductivity. Particulate fillers, regardless of their biostability or biodegradability or their good adhesion to the polymer matrix, have limited influence to the physical properties of the composite. Thus, they cannot be utilized in load-bearing applications where high-bending, shear, or torsional stresses occur.

FRCs are reputed for their high mechanical properties. The action of reinforcing fibers is to arrest propagating cracks in the polymer matrix and reinforce the composite. FRCs is a class of materials in which the basic properties of polymers are given mechanical reinforcement by the addition of fibrous materials such as glass, carbon/graphite, ultra high molecular weight PE, or aramid fibers. The optimum properties of a reinforced resin cannot be obtained, unless there is an effective bond between the two phases. Other factors with considerable influence on the physical properties of FRCs are impregnation of the fibers with the resin matrix, the quantity of reinforcing fibers, and the orientation of the fibers [27]. Fibers that are suitable to be used in medical and dental composites are glass fibers that have been silanized for good adhesion to the polymer matrix. Adhesion has been obtained by covalent bonding of dimethacrylate monomers to the silanated surface of glass fibers during polymerization (Figure 1.2) [28]. Carbon/Graphite fibers have been utilized as medical FRCs with thermoplastic PEEK matrix. Ultra high molecular weight PE fibers lack adequate bonding to the resin matrix, and therefore, their reinforcing effect is modest.

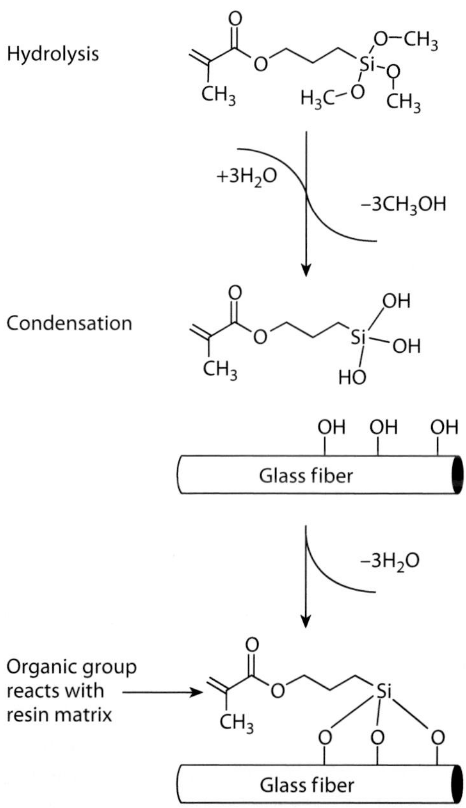

FIGURE 1.2 Schematic of the adhesion of resin matrix to the surface of silanated glass fibers. (Modified from Puska, M. et al., Evaluation of bis-GMA/MMA resin adhesion to silica-coated and silanized titanium, *Adhesion Aspects in Dentistry*, VSP, Leiden, 2009, p. 147.)

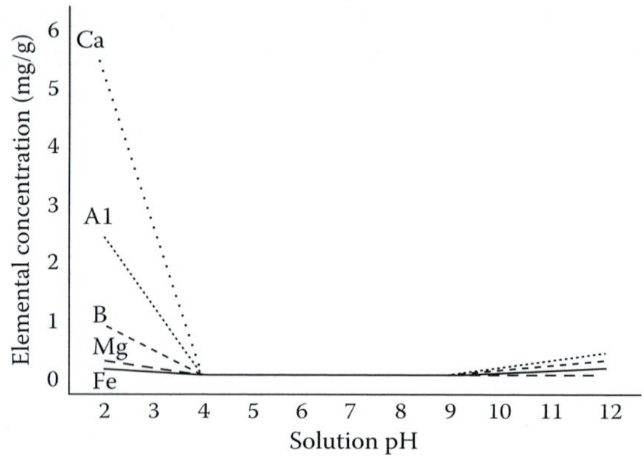

FIGURE 1.3 Stability of E-glass fibers in solutions of different pH values.

1.2.3 Hydrolytic Stability

The long-term success of reconstructive composite materials in biological environments such as in the oral cavity or insides of tissues depends on the hydrolytic stability of the composite. Hydrolytic stability is dependent on the stability of polymer matrix, fillers, and the interface between the fillers and the polymer matrix. In earlier times, dental filling composites failed, because the glass filler particles contained considerable quantities of earth alkali oxides, which were prone to leaching in the presence of water. Once the filler composition was improved to withstand the effects of water, the next weak point was the silane-promoted adhesion between the glass and the polymer matrix. Improvement of the hydrolytic stability of the interface was found to relate to the curing temperature of the silane on the glass surface. Present particulate filling composites and glass FRCs exhibit good long-term hydrolytic stability that is based on the stability of polymer matrix, fillers and fibers, and their interface [10,29–31]. It is known that good-quality and surface-purified E-glass fibers itself exhibit stability in pH values between 4 and 10 (Figure 1.3) [32].

1.3 MATERIAL PROPERTIES

1.3.1 Direction Dependency

Particulate filler composites demonstrate isotropic physical, thermal, and optical properties, whereas FRCs are either isotropic, anisotropic, or orthotropic. Thus, FRCs allow materials designing to obtain biomechanically anisotropic medical and dental devices that correspond to the behavior of bone or tooth, which are known by their anisotropic nature in terms of collagen fibril and osteon orientation (Figure 1.4) [33]. The anisotropic nature is for the optimization of mechanical load bearing of bone structures and to avoid excessive peak stresses. There is an intimate relationship between the forces acting upon bone-and-bone structure. On the topic of bone structure and bone loading, there are descriptions by Bourgery and Bell from 1832 to 1834 [34,35]. It is surprising that the utilization of this information, which is often

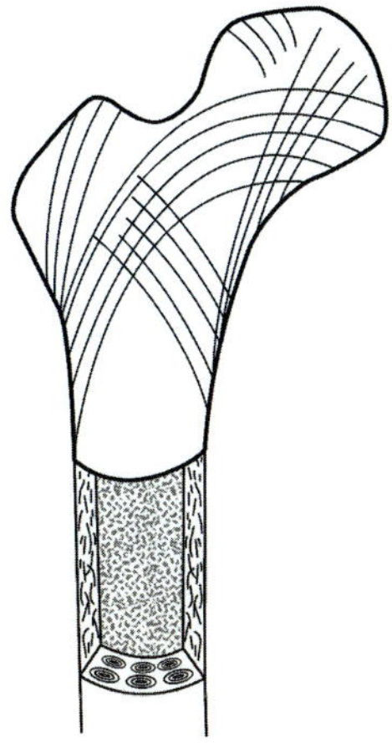

FIGURE 1.4 Anisotropic structure of bone from the level of collagen fibrils to structures of osteons can be mimicked and reconstructed by the use of fiber-reinforced composites.

called Wolff's law [36], has not been made by implant manufacturers. One reason could be in the use of isotropic metallic implants, which from the material basis, do not allow and require tailoring the biomechanical properties of the implant.

Present knowledge of biomechanical properties of FRCs, especially glass FRCs, shows good biomechanical match of FRC implants to that of bone. Studies by finite element modeling (FEM) demonstrate the stress at the bone around FRC implants to be more even than with metal implants [37]. This has also been verified by mechanical tests and animal experiments, which revealed better attachment of an FRC implant to surrounding bone-like material than what can be obtained with titanium implants [9,38]. For the optimization of the load-bearing capacity of an FRC device, the direction of load and stress needs to be known, and the fiber direction should be oriented to carry the stress. Reinforcing the efficiency of fibers is defined as Krenchel's factor—a theoretical maximum reinforcing effect that can be obtained with certain fiber orientation (Figure 1.5) [39]. Thus, the strength of FRCs is largely

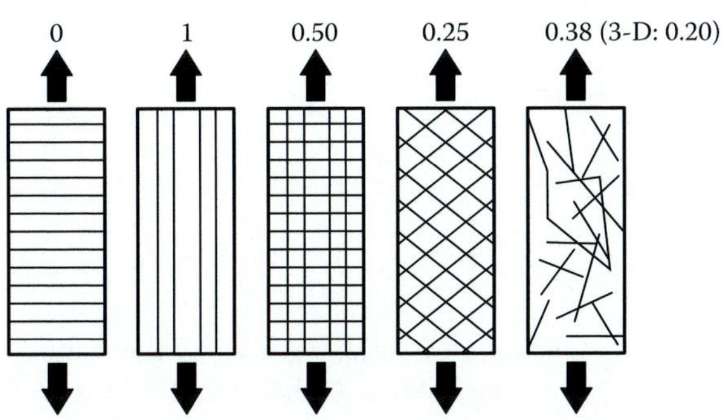

FIGURE 1.5 Reinforcing efficiency (percentage) of fibers against the direction of tensile force (arrows).

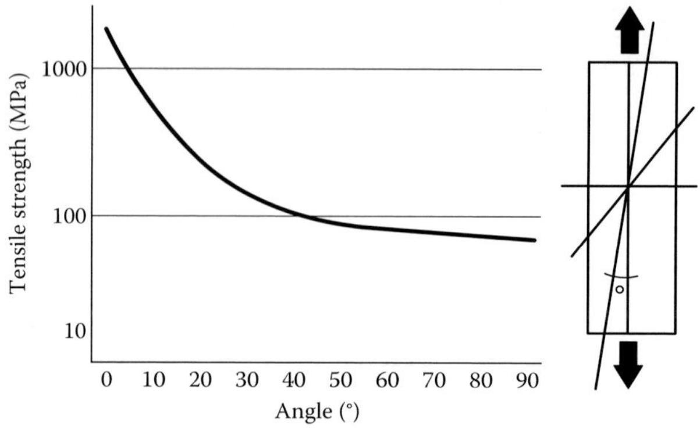

FIGURE 1.6 Alignment of reinforcing fibers against the direction of tensile force (arrow) considerably influence reinforcing effect by fibers.

dependent on the fiber direction, and the efficiency of the fiber reinforcement varies in FRC laminates with different fiber orientations (Figure 1.6) [39]. The continuous unidirectional fibers give the highest strength and modulus of elasticity for the FRCs, but the property is available only in the direction of stress equal to that of the direction of the fibers. The anisotropic behavior of unidirectional FRCs can also be seen in other properties such as thermal expansion or polymerization shrinkage [40,41].

1.3.2 Mechanical Strength

The static strength (ultimate flexural strength) of the FRCs is dependent on the fiber quantity to the level of approximately 70 vol%. A high-quality glass FRC material with high fiber quantity provides high flexural bending properties (with E-glass up to 1250 MPa; Figure 1.7) [42,43]. A positive correlation exists between the water sorption of polymer matrix and the reduction of flexural properties. For instance, the high water sorption of polyamide (nylon) matrix causes reduction of over 50% in the

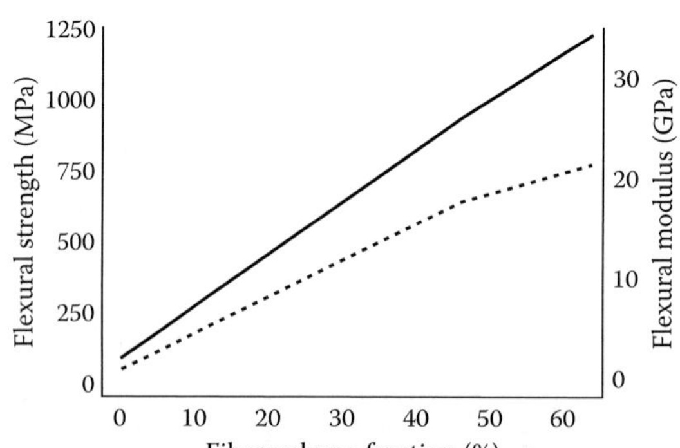

FIGURE 1.7 Flexural strength (solid line) and flexural modulus (dotted line) of unidirectional FRCs plotted against fiber volume fraction.

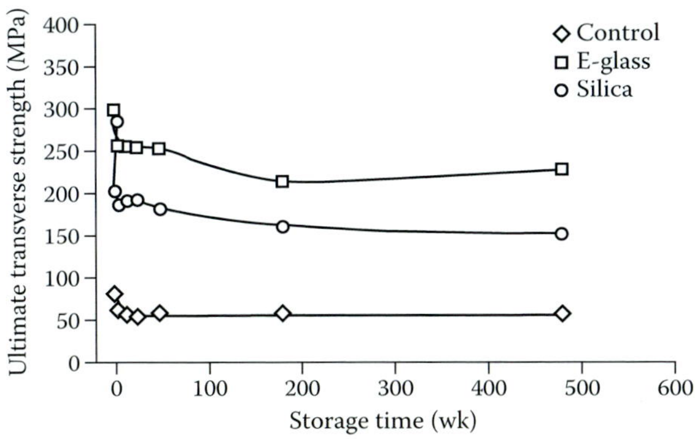

FIGURE 1.8 Flexural strength of unidirectional E-glass–PMMA, silica glass–PMMA FRCs, and plain PMMA (control) plotted against storage time in water at 37°C up to 10 years.

strength of FRCs. The reduction of the flexural properties was reversible (i.e., dehydration of the FRCs recovered the mechanical properties) [44]. Water sorption of the polymer matrix reduces the strength and modulus of elasticity of the FRCs with the semi-IPN polymer matrix of pBis-GMA–PMMA by approximately 15% within 30 days of water storage time at 37°C [44]. No significant reduction of flexural strength and modulus, even in long-term water storage (up to 10 years), occurred (Figure 1.8) [29,30].

Several studies have dealt with the strength of FRCs, which may have shown misleading values of the material strength. Testing of specimens of small dimensions, such as dental root canal posts used for anchoring crowns, can lead to incorrect results in megapascals. It is known that the commonly used mathematical formulas for calculating the flexural strength and modulus of elasticity of test specimens are dependent on the diameter (height) of the specimen and span length of the test setup for a three-point bending test [45]. With a constant span length, thinner specimens reveal higher flexural strength and modulus of elasticity values than those obtained with specimens of the same material but of a larger dimension. Thus, only the strength values of specimens of exactly the same diameter and span length, as well as crosshead speed, in the test setup are comparable. In biomedical applications, there is a need to compare different material properties such as the strength and modulus of elasticity and the load-bearing capacity and structural stiffness of the device. Clinically, the latter is more important.

1.4 BIOCOMPATIBILITY

The biocompatibility of biostable composites relates to the biocompatibility of polymer matrices, reinforcing particulates and fibers, and their combination. Biocompatibility indicates the biological performance of materials in terms of structural and surface compatibility [46,47]. Structural compatibility refers to the biomechanical properties of the implant material, such as the modulus of elasticity and strength, implant design (i.e., stiffness—a product of elastic modulus and seconds moment of area), and optimal load transmission at the implant tissue interface (minimum interfacial strain mismatch and load transfer) [14]. Metals and ceramics have at least 10–20 times higher elastic modulus than hard tissues, and therefore, one of the major problems in orthopedic surgery has been the mismatch of stiffness between the bone and metallic implant. Thus, the bone is insufficiently loaded compared to the implants, and this condition is called stress shielding [14]. Stress shielding affects bone remodeling and healing, leading to increased bone atrophy [48]. Utilization of biomechanical properties of FRCs can overcome the aforementioned biomechanical problems, because FRC materials simultaneously exhibit bone-like modulus of elasticity and high strength. An additional merit of FRCs is that, by controlling the volume fraction and arrangement of the fibers, the properties and designs can be varied and tailored to suit the mechanical and physiological conditions of the host tissues. By arranging fibers, the FRCs can mimic natural anisotropic tissues, which depend on the structural arrangements of its components of collagen, elastin, and hydroxyapatite. By selecting FRCs for implant material, interfacial stress and strain between implant and bone can be transferred more evenly than with titanium implants [37]. The biocompatibility of using FRCs over metals includes the absence of corrosion and fatigue failures and the release of metal ions and metal nanoparticles, which may cause loosening of the implant. In the case of composite implants, tribological properties of composites and fixation devices should be optimized to eliminate the potential risk of debris formation from the composite implants.

The biocompatibility of FRC implants is basically related to the biocompatibility of its major components of polymer matrix and reinforcing fibers. Thermoset polymer FRCs have been made of epoxy polymers and of dimethacrylate polymers. The use of epoxy polymer has been criticized due to the potential toxic and sensitizing (allergic) effects of its monomers, which are present as residuals in the FRCs [49,50]. On the other hand, thermoset polymers made of dimethacrylate monomer systems of Bis-GMA and TEGDMA monomers have shown good biocompatibility after polymerization at elevated temperatures and not in the presence of oxygen (Figure 1.9) [22,51–53].

Thermoset polymers provide the chemical bonding of polymer matrix to glass fibers or other OH-covered substrate through silane coupling agents. Chemical bonding cannot be obtained by the thermal molding of thermoplastic polymers of polymer matrix for FRCs [54–56]. In thermoplastic polymer matrix FRCs, which have been processed by thermal plasticization and molding, only physical attachment of the polymer matrix to the surface of fibers occurs, whereas the polymerization of the monomer system of the thermoset resins in contact to the silanized glass fibers results in covalent and hydrogen bonding of the polymer matrix to the reinforcing fibers. Covalent bonding of polymer matrix to fibers is desired for the long-term stability of FRCs in a biological environment.

Biological testing of glass FRCs by cell cultures and animal testing has shown materials biocompatibility. Cell culture studies by fibroblasts with E-glass fibers without resin matrix have shown no signs of cytotoxicity, as it has been also demonstrated with fibroblasts by agar diffusion cytotoxicity test [57,58]. By using osteoblasts, no signs of toxic reactions of the material can be found either [59]. Rat bone marrow-derived osteoblast-like cells were harvested and cultured on experimental

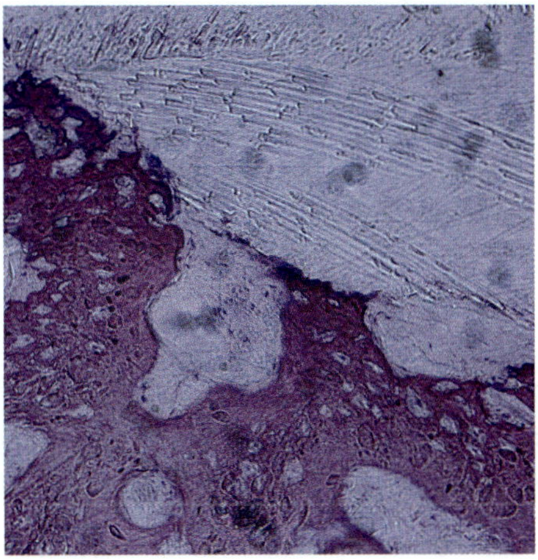

FIGURE 1.9 Histological image of E-glass FRCs showing good biological response of bone to contact of implant.

FRC material plates and on commercially pure titanium plates (control) for 21 days [60]. The materials' surfaces were characterized by roughness testing and scanning electron microscopy. Cell growth and differentiation kinetics were subsequently investigated by evaluating proliferation, alkaline phosphatase (ALP) activity, osteocalcin (OC), and bone sialoprotein (BSP) production. The maximal ALP activities on FRCs and titanium were observed after 3 weeks. The expression of osteoblastic markers (OC, BSP) indicates that the fastest osteogenic differentiation takes place on FRCs after 7 days. In contrast, a slower differentiation process was observed on titanium than on FRCs, as confirmed by the increased mRNA expression of OC and BSP. It was concluded that the proliferation and maturation of osteoblast-like cells on FRCs appears to be comparable to titanium. The presence of BAG enhances cell maturation.

A number of animal experiments have been carried out to show biological responses to E-glass FRCs. In many experimental FRC materials, there have been additional BAG particulates on the surface of the FRC implant. BAGs are synthetic resorbable biocompatible osteoconductive bone substitutes with bone bonding capacity and antibacterial and angiogenesis-promoting properties [61–68]. FRC–BAG implants have been tested by animal tests for head-and-neck applications and for orthopedics and oral implantology. Head-and-neck applications have been preclinically tested with a calvarial critical size defect model in rabbits [51,52]. Between FRC laminates of the FRC implant, there were particulates of BAG for improving osteoconductivity, angiogenesis, and antimicrobial properties. FRCs have the potential for use as load-bearing orthopedic implants if the high strength and elastic modulus of the FRC implant can be matched with local requirements. An animal study was carried out to test the *in vivo* performance of novel FRC implants made of unidirectional glass fibers (E-glass fibers in pBis-GMA–pTEGDMA polymeric matrix) [69]. The implant surface was covered with BAG granules. Control implants were made of surface-roughened titanium. Stress-shielding effects of the implants were predicted by FEM. Surgical stabilization of bone metastasis in the subtrochanteric region of the femur was simulated in rabbits. An oblong subtrochanteric defect of a standardized size (reducing the torsional strength of the bones approximately by 66%) was created, and an intramedullary implant made of titanium or the FRC composite was inserted. The contralateral femur served as the intact control. At 12 weeks of healing, the femurs were harvested and analyzed by radiography, torsional testing, microcomputerized tomography imaging, and hard-tissue histology. The functional recovery was unremarkable in both groups, although the final analysis revealed two healed undisplaced peri-implant fractures in the group of FRC implants. FEM studies demonstrated differences in stress-shielding effects of the titanium and FRC implants, but the expected biological consequences did not become evident during the follow-up time of the animal study. Biomechanical testing of the retrieved femurs showed no significant differences between the groups. The torsional strength of the fixed bones had returned the level of contralateral intact femurs. Both implants showed ongrowth of intramedullary new bone. No adverse tissue reactions were observed.

Oral implant research has also utilized E-glass fibers in a pBis-GMA–pTEGDMA polymeric matrix system. Aside from FRCs' adequate biomechanical strength, the studies have also shown FRCs' biocompatibility in bone to be comparable to that of titanium after 4 and 12 weeks. The addition of BAG to the implant increased osteogenesis and bone maturation [70,71].

1.5 CLINICAL USE

Biostable composite implants have been used as bone plates with screw fixation made of carbon FRCs and PEEK resin matrix. Tests have also been conducted to use glass FRCs/PEEK as intramedullary nail and in spine fusions [14]. Traditional skull and facial bone reconstructions, for example, in cranioplasties are based on using hard tissues (bone grafts), which typically causes donor site morbidity. Metallic implants have been used alone or in combination with bone or hydroxyapatite cement in cranial bone reconstructions. Although these materials are readily available for immediate application, they all require preparation, adaptation, contouring into defect, and later on, resistance to traumatic insults is uncertain. In addition, metallic implants cause problems in medical imaging by X-rays or by MRI. The clinical outcome of treatments of cranioplasties showed that the most often used material is autogenous bone graft, but more than one-third of the treatments are primarily treated with artificial materials. Of all treatments, almost one-fourth are complicated, and infections and material exposure are the most critical complications [72]. Of the artificial materials, hydroxyapatatite cement was associated with the worst results, and bone grafts showed a high grade of partial resorption, which required further surgery for correction. Complications related to the exposure of the implant are related to immunoguided delayed inflammatory reactions that lead to thinning of the skin. Secondary operations for repair are difficult in these cases. Comparison of PMMA and bone grafts has also been undertaken by other researchers [73], demonstrating the exposure of PMMA and secondary infection to be the complications for PMMA.

Because PMMA is one of the most widely used alloplastic material in surgery, it has also been used as a matrix polymer for *ex vivo*–produced biostable composite implants. Because PMMA is polymerized *in situ*, the amount of the exotherm and residual monomers may cause even severe clinical problems, which lead to foreign body reaction [74,75]. To overcome these problems, patient-specific skull bone implants based on patient's preoperative three-dimensional model have started to be produced. The first-generation implants were made of PMMA that has been polymerized *ex vivo* to receive an optimal degree of monomer conversion and, thus, the lowest possible amount of residual monomers [76]. PMMA implants were coated BAG particles. By clinical and radiological follow-ups, it was demonstrated that the implants performed well 24 months postoperatively. Through computerized tomography and positron emission tomography, new bone formation was found on the surface, which had been modified by BAG. No periprosthetic infections have been supported by the *in vitro* and clinical findings of the antimicrobial properties of BAG. Based on the clinical experiences by PMMA implants, further improvements in terms of allowing new bone formation to occur faster and to have thinner margins for the implants were made by introducing glass FRC–BAG implants for a clinical trial. The FRC–BAG implants are made of load-bearing laminate and scaffold, which absorbs blood and patient's own substances (cells, growth factors, and signaling molecules) for bone formation in the scaffold (Figure 1.10). The design of the implant enables the use of nonmetallic patient-specific implants and scaffolds in tissue-engineering approaches in replacing tissues. The FRC–BAG implants have been in preliminary clinical testing since 2007 without any kinds of complications or infections [77], and they show high potential for replacing the present bulk material implants made of metals and bulk polymers. Further developments of nonmetallic biostable implants will likely be made by adding bioactive substances and drugs to the implant to be released locally in the body.

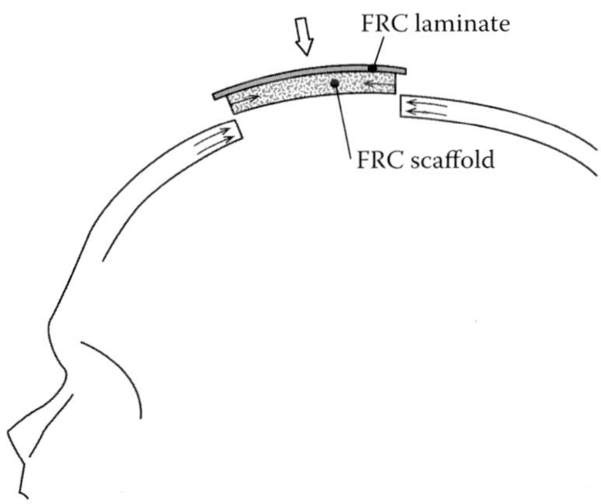

FIGURE 1.10 Schematic of the design of FRC implants for skull bone defects. The implant is having a load-bearing phase (FRC laminate) and a porous phase (FRC scaffold) into which the bone is growing.

1.6 CONCLUSION

Current needs from the biomechanical and medical imaging perspective, as well as the concern for potential harm from released metal nanoparticles and ions, have led to the development of biostable composite implants. In applications where large defects are repaired or the implant will be heavily loaded, the most suitable material alternative seems to be FRCs. FRCs made of glass fibers and dimethacrylate matrix have shown good biocompatibility, and the FRC implants have started to be tested in clinical studies with encouraging preliminary results.

REFERENCES

1. J.B. Park and R.S. Lakes. 1992. *Biomaterials: An Introduction*. New York: Plenum Press.
2. W. Bonfield, M. Grynpas, A.E. Tully, J. Bowman, and J. Abram. Hydroxyapatite reinforced polyethylene—A mechanically compatible implant material for bone replacement. *Biomaterials* 2 (1981) 185.
3. F.G. Shellock. Metallic neurosurgical implants: Evaluation of magnetic field interactions, heating, and artifacts at 1.5-tesla. *Journal of Magnetic Resonance Imaging* 14 (2001) 295.
4. A.M. Sawyer-Glover and F.G. Shellock. Pre-MRI procedure screening: Recommendations and safety considerations for biomedical implants and devices. *Journal of Magnetic Resonance Imaging* 12 (2001) 92.
5. F.G. Shellock. Biomedical implants and devices: Assessment of magnetic field interactions with a 3.0-tesla MR system. *Journal of Magnetic Resonance Imaging* 16 (2002) 721.
6. S.J. Holland, B.J. Tighe, and P.L.J. Goud. Polymers for biodegradable medical devices—Part 1: The potential of polyesters as controlled macromolecular release systems. *Controlled Release* 4 (1986) 155.
7. J.C. Middleton and A.J. Tipton. Synthetic biodegradable polymers as orthopaedic devices. *Biomaterials* 21 (2000) 2335.
8. P.K. Vallittu. The effect of glass fiber reinforcement on the fracture resistance of a provisional fixed partial denture. *Journal of Prosthetic Dentistry* 79 (1998) 125.
9. R.H. Mattila, M. Puska, L.V.J. Lassila, and P.K. Vallittu. Fiber-reinforced composite implant: *In vitro* interfacial failure mechanics and residual monomer analysis. *Journal of Materials Science: Materials in Medicine* 41 (2006) 4321.
10. S. Bouillaguet, A. Schutt, P. Alander, P.K. Vallittu, P. Schwaller, G. Buerki, J. Michler, M. Cattani-Lorente, and I. Krejci. Hydrothermal and mechanical stresses degrade fiber–matrix interfacial bond strength in dental fiber-reinforced composites. *Journal of Biomedical Materials Research, Part B: Applied Biomaterials* 76 (2006) 98.
11. A.M. Ballo, L.V.J. Lassila, T. Närhi, and P.K. Vallittu. *In vitro* mechanical testing of glass fiber-reinforced composite used as dental implant. *Journal of Contemporary Dental Practice* 9 (2008) 41.
12. D.S. Zhao, N. Moritz, P. Laurila, R.H. Mattila, L.V.J. Lassila, N. Strandberg, T. Mäntylä, P.K. Vallittu, and H.T. Aro. Development of a biomechanically optimized multicomponent fiber-reinforced composite implant for load-sharing conditions. *Medical Engineering Physics* 31 (2009) 461.
13. S.L. Evans and P.J. Gregson. Composite technology in load-bearing orthopaedic implants. *Biomaterials* 19 (1998) 1329.
14. S. Ramakrishna, J. Mayer, E. Wintermantel, and K.W. Leon. Biomedical applications of polymer-composite materials: A review. *Composites Science and Technology* 61 (2001) 1189.

15. E. Asmussen and A. Peutzfeldt. Influence of UEDMA BisGMA and TEGDMA on selected mechanical properties of experimental resin composites. *Dental Materials* 14 (1998) 51.
16. S.H. Min, J. Ferracane, and I.B. Lee. Effect of shrinkage strain, modulus and instrument compliance on polymerization shrinkage stress of light-cures composites during the initial curing stage. *Dental Materials* 26 (2010) 1024.
17. C. Grimsrud, R. Raven, A.W. Fothergill, and H.T. Kim. The *in vitro* elution characteristics of antifungal-loaded PMMA bone cement and calcium sulphate bone substitute. *Orthopaedics* 34 (2011) 378.
18. M.A. Puska, A.K. Kokkari, T.O. Närhi, and P.K. Vallittu. Mechanical properties of oligomer-modified acrylic bone cement. *Biomaterials* 24 (2003) 417.
19. P.K. Vallittu, I.E. Ruyter, and S. Buykuilmaz. Polymerization time and temperature affects the residual monomer content of denture base polymers. *European Journal of Oral Sciences* 106 (1998) 588.
20. E.K. Viljanen, M. Skrifvars, and P.K. Vallittu. Degree of conversion of an experimental monomer and methyl methacrylate copolymer for dental applications. *Journal of Applied Polymer Science* 93 (2004) 1908.
21. E.K. Viljanen, L.V.J. Lassila, M. Skrifvars, and P.K. Vallittu. Degree of conversion and flexural properties of a dendrimer/methyl methacrylate copolymer: A statistical modeling. *Dental Materials* 21 (2005) 172.
22. S.M.-R. Tuusa, M.A. Puska, L.V.J. Lassila, and P.K. Vallittu. Residual monomers released from glass fiber-reinforced composite photopolymerized in contact with bone and blood. *Journal of Materials Science: Materials in Medicine* 16 (2005) 15.
23. R.L. Bowen. Properties of a silica-reinforced polymer for dental restorations. *Journal of the American Dental Association, JADA* 66 (1963) 57.
24. S. Ruttermann, C. Wandrey, W.H. Raab, and R. Janda. Novel nanoparticles as fillers for an experimental resi-based restorative material. *Acta Biomaterialia* 4 (2008) 1835.
25. S. Garoushi, L.V.J. Lassila, and P.K. Vallittu. Influence of nanometer scale particulate fillers on some properties of microfilled composite resin. *Journal of Materials Science: Materials in Medicine* 22 (2011) 1645.
26. J.T. Heikkilä, A.J. Aho, I. Kangasniemi, and A. Yli-Urpo. Polymethylmethacrylate composites: Disturbed bone formation at the surface of bioactive glass and hydroxyapatite. *Biomaterials* 17 (1996) 1755.
27. P.K. Vallittu. Flexural properties of acrylic polymers reinforced with unidirectional and woven glass fibers. *Journal of Prosthetic Dentistry* 81 (1999) 318.
28. M. Puska, L.V.J. Lassila, P.K. Vallittu, J. Seppälä, and J. Matinlinna. Evaluation of bis-GMA/MMA resin adhesion to silica-coated and silanized titanium. *Adhesion Aspects in Dentistry*. Leiden: VSP (2009) 147.
29. P.K. Vallittu, I.E. Ruyter, and K. Ekstrand. Effect of water storage on the flexural properties of E-glass and silica fiber acrylic resin composite. *International Journal of Prosthodontics* 11 (1998) 340.
30. P.K. Vallittu. Effect of ten years of *in vitro* aging on the flexural properties of fiber-reinforced resin composites. *International Journal of Prosthodontics* 20 (2007) 43.
31. M. Arksornnukit, H. Takahashi, and N. Nishiyama. Effects of silane coupling agent amount on the mechanical properties and hydrolytic durability of composite resin after hot water storage. *Dental Material Journal* 23 (2004) 31.
32. A. Norström, H. Watson, B. Engström, and J. Rosenholm. Treatment of E-glass fibers with acid, base and silanes. *Colloids and Surfaces—Part A: Physicochemical Engineering Aspects* 194 (2001) 143.
33. S. Hansson. The dental implant meets bone—A clash of two paradigms. *Applied Osseointegration Research* 5 (2006) 5.

34. J.M. Bourgery. Traité complet de l'anatomie de l'homme—Part 1: Osteologie. Paris (1832).
35. C. Bell. *Illustration of Paley's Natural Theologie* (1834).
36. J. Wolff. *Das Gesetz der Transformation der Knochen*. Berlin: A. Hirschwald (1892).
37. A. Shinya, A.M. Ballo, L.V.J. Lassila, T.O. Närhi, and P.K. Vallittu. Stress and strain analysis of the bone–implant interface: A comparison of fiber-reinforced composite and titanium implants utilizing 3-dimensional finite element study. *Journal of Oral Implantology* 37 (2011) 133.
38. R.H. Mattila, P. Laurila, J. Rekola, J. Gunn, L.V. Lassila, T. Mäntylä, A.J. Aho, and P.K. Vallittu. Bone attachment to glass fiber-reinforced composite implant with porous surface. *Acta Biomaterialia* 5 (2009) 1639.
39. H. Krenchel. Fiber reinforcement. Thesis. Technical University of Denmark, Copenhagen (1963).
40. A. Tezvergil, L.V.J. Lassila, and P.K. Vallittu. The effect of fiber orientation on the thermal expansion coefficients of the fiber-reinforced composites. *Dental Materials* 19 (2003) 471.
41. E. Anttila, O. Krintilä, T. Laurila, L.V.J. Lassila, P.K. Vallittu, and R. Hernberg. Polymerization shrinkage and hygroscopic expansion of fiber-reinforced composite using fiber Bragg grating sensors. *Dental Materials* 24 (2008) 1720.
42. P.K. Vallittu. Interpenetrating polymer networks (IPNs) in dental polymers and composites. *Journal of Adhesion Science and Technology* 14 (2009) 961.
43. L.V.J. Lassila, J. Tanner, A.M. Le Bell, K. Narva, and P.K. Vallittu. Flexural properties of fiber reinforced root canal posts. *Dental Materials* 20 (2004) 29.
44. L.V.J. Lassila, T. Nohrström, and P.K. Vallittu. The influence of short-term water storage on the flexural properties of unidirectional glass fiber-reinforced composite. *Biomaterials* 23 (2002) 2221.
45. P. Alander, L.V.J. Lassila, and P.K. Vallittu. The span length and cross-sectional design affect values of strength. *Dental Materials* 21 (2005) 347.
46. D.F. Williams. Consensus and definitions in biomaterials. In: *Advances in Biomaterials*, C. de Putter, K. de Lange, K. Groot, and A.J.C. Lee, eds. Amsterdam: Elsevier Science (1988) 11.
47. E. Wintermantel and J. Mayer. Anisotropic biomaterials strategies and developments for bone implants. In: *Encyclopedic Handbook of Biomaterials and Bioengineering—Part B-1*, D.L. Wise, D.J. Trantolo, D.E. Altobelli, J.D. Yaszemiski, J.D. Gresser, and E.R. Schwarzt, eds. New York: Marcel Dekker (1955) 3.
48. P. Christel, L. Claes, and S.A. Brown. Carbon-reinforced composites in orthopaedic surgery. In: *High-Performance Biomaterials: A Comprehensive Guide to Medical and Pharmaceutical Applications*, M. Szycher, ed. Lancaster, PA: Technomic (1991) 499.
49. C. Morrison, R. Macnair, C. MacDonald et al. *In vitro* biocompatibility testing of polymers for orthopaedic implants using cultured fibroblasts and osteoblasts. *Biomaterials* 16 (1995) 987.
50. G. Peluso, L. Ambrosio, M. Cinquegrani, L. Nicolais, S. Saiello, and G. Tajana. Rat peritonela immune response to carbon fiber-reinforced epoxy composite implants. *Biomaterials* 12 (1991) 231.
51. S.M.-R. Tuusa, M.J. Peltola, T. Tirri, L.V.J. Lassila, and P.K. Vallittu. Frontal bone defect repair with experimental glass fiber-reinforced composite with bioactive glass granule coating. *Journal of Biomedical Material Research—Part B: Applied Biomaterials* 82 (2007) 149.
52. S.M.-R. Tuusa, M.J. Peltola, T. Tirri, M.A. Puska, M. Röyttä, H. Aho, J. Saandholm, L.V. Lassila, and P.K. Vallittu. Reconstruction of critical-size calvarial bone defect in rabbits with glass fiber-reinforced composite with bioactive glass granule coating. *Journal of Biomedical Material Research—Part B: Applied Biomaterials* 84 (2008) 510.

53. A.M. Ballo, A. Kokkari, V.V. Meretoja, L.L. Lassila, P.K. Vallittu, and T.O. Närhi. Osteoblast proliferation and maturation on bioactive fiber-reinforced composite. *Journal of Materials Science: Materials in Medicine* 19 (2008) 3169.
54. J.P. Matinlinna, M. Özcan, L.V.J. Lassila, W. Kalk, and P.K. Vallittu. Effect of the cross-linking silane concentration in a novel silane system on bonding resin-composite cement. *Acta Odontologica Scandinavica* 66 (2008) 250.
55. J.P. Matinlinna, J.E. Dahl, S. Karlsson, L.V.J. Lassila, and P.K. Vallittu. The effect of novel silane system to the flexural properties of E-glass fiber-reinforced composites. *Silanes and Other Coupling Agents* 5 (2009) 107.
56. P.K. Vallittu. Comparison of two different silane compounds used for improving adhesion between fibers and acrylic denture base material. *Journal of Oral Rehabilitation* 20 (1993) 241.
57. M. Väkiparta, M.K. Koskinen, P. Vallittu, T. Närhi, and A. Yli-Urpo. *In vitro* cytotoxicity of E-glass fiber weave preimpregnated with novel biopolymer. *Journal of Materials Science: Materials in Medicine* 15 (2004) 69.
58. P.K. Vallittu and K. Ekstrand. *In vitro* cytotoxicity of fiber–polymethyl methacrylate composite used in dentures. *Journal of Oral Rehabilitation* 26 (1999) 666.
59. M. Hautamäki, V.V. Meretoja, R.H. Mattila, A.J. Aho, and P.K. Vallittu. Osteoblast response to polymethyl methacrylate bioactive glass composite. *Journal of Materials Science: Materials in Medicine* 21 (2010) 1685.
60. A.M. Ballo, E.A. Akca, T. Ozen, L. Lassila, P.K. Vallittu, and T.O. Närhi. Bone tissue responses to glass fiber-reinforced composite implants—A histometric study. *Clinical Oral Implants Research* 20 (2009) 608.
61. R.M. Day. Bioactive glass stimulates the secretion of angiogenetic growth factors and angiogenesis *in vitro*. *Tissue Engineering* 11 (2005) 768.
62. P. Stoor, E. Söderling, and R. Grenman. Interactions between the bioactive glass S53P4 and the atrophic rhinitis–associated microorganism *Klebsiella ozaenae*. *Journal of Biomedical Materials Research—Part B: Applied Biomaterials* 48 (1999) 869.
63. L.L. Hench and J. Wilson. Surface active biomaterials. *Science* 226 (1984) 630.
64. L.L. Hench and H.A. Paschal. Direct chemical bond of bioactive glass–ceramic materials to bone and muscle. *Journal of Biomedical Materials Research* 4 (1973) 25.
65. Ö.H. Andersson and I. Kangasniemi. Calcium phosphate formation at the surface of bioactive glass *in vitro*. *Journal of Biomedical Materials Research* 24 (1991) 1019.
66. Ö.H. Andersson, K.H. Karlsson, and K. Kangasniemi. Calcium phosphate formation at the surface of bioactive glass *in vivo*. *Journal of Noncrystalline Solids* 119 (1990) 290.
67. N.C. Lindfors, P. Hyvönen, M. Nyyssönen et al. Bioactive glass S53P4 as bone graft substitute in treatment of osteomyelitis. *Bone* 47 (2010) 212.
68. A.J. Aho, E. Suominen, A. Alanen, A. Yli-Urpo, J. Knuuti, and H.J. Aho. Remodelling of the tibia after grafting of a large cavity with particulate bioactive glass–hydroxyapatite—Case report on treatment of fibrous dysplasia with 13 years' follow-up. *Acta Orthopaedica Scandinavica* 74 (2003) 766.
69. D.S. Zhao, N. Moritz, P. Laurila, R. Mattila, L.V. Lassila, N. Strandberg, T. Mäntylä, and H. Aro. Development of a multicomponent fiber-reinforced composite implant for load-sharing conditions. *Medical Engineering and Physics* 31 (2009) 461.
70. A.M. Ballo, E.A. Akca, T. Ozen, L.V. Lassila, P.K. Vallittu, and T.O. Närhi. Bone tissue responses to glass fiber-reinforced composite implants—A histometric study. *Clinical Oral Implants Research* 6 (2009) 608.
71. A.M. Ballo, T.O. Närhi, E.A. Akca, T. Ozed, S.M. Syrjänen, L.V. Lkassila, and P.K. Vallittu. Prepolymerized versus *in situ* polymerized fiber-reinforced composite implants—A pilot study. *Journal of Dental Research* 90 (2011) 263.

72. A. Mareira-Gonzalez, I.T. Jackson, T.T. Miyawaki, K. Barakat, and V. DiNick. Clinical outcome in cranioplasty: Critical review in long-term follow-up. *Journal of Craniofacial Surgery* 14 (2003) 114.
73. N. Sahoo, I.D. Roy, A.P. Desai, and V. Gupta. Comparative evaluation of autogenous calvarial bone graft and alloplastic materials for secondary reconstruction of cranial defects. *Journal of Craniofacial Surgery* 21 (2010) 79.
74. E.J. Stelnicki and D.K. Ousterhout. Prevention of thermal injury induced by application of polymethylmethacrylate to the calvarioum. *Journal of Craniofacial Surgery* 7 (1996) 192.
75. P.K. Vallittu. Peak temperature of some prosthetic acrylates on polymerization. *Journal of Oral Rehabilitation* 23 (1996) 776.
76. M.J. Peltola, P.K. Vallittu, V. Vuorinen, A.A. Aho, A. Puntala, and K.M. Aitasalo. 2011. Novel composite implant in craniofacial bone reconstruction. *European Archives of Otorhinolaryngology* 269 (2012) 623.
77. K. Aitasalo, M. Peltola, V. Vuorinen, and P. Vallittu. *Novel Composite Implants in Craniofacial Reconstruction*. North American Skull Base Society, Los Angeles, CA (2009).

2 Bioceramic and Biopolymer Nanocomposite Materials for Use in Orthopedic Applications

Clark E. Barrett, Ruth E. Cameron, and Serena M. Best

CONTENTS

2.1 INTRODUCTION

Bone is an organ capable of self-repair following an injury. However, the loss of significant bone volume due to infection or trauma may result in a permanent defect at an injury site. Current surgical techniques employed to repair large defects include bone grafting and metallic implants. Bone grafting is limited by the quantity of bone available from a possible donor site, making the procedure unsuitable for large defects. Metallic implants are of particular use in load-bearing environments due to their high strength and toughness. However, these devices are far from optimal solutions due to stress shielding (Park and Lakes 1992) and associated bone atrophy, which is often a cause of device loosening and ultimate failure (Wolff 1987). Long-term metallic implant concerns involve potential cytotoxicity arising from heavy metal ion liberation and harmful corrosion products (Williams 1981).

Much research has been undertaken with the aim of optimizing osseointegration on metal surfaces that possess nanometer-scale topographic features. Often, these techniques incorporate hydroxyapatite (HA) and various substituted HA types (Best et al. 2008). Nanomanipulation is not limited to the bone—surface nanopatterning is employed in the cartilage, vascular tissue, bladder, and other tissue engineering applications.

A basic literature search reveals a plethora of investigations into polymer–ceramic nanocomposite (PMNC) development for clinical use which is logical considering the number of degrees of freedom in these systems. PMNCs are defined as polymers containing a ceramic phase that possesses at least one dimension less than 100 nm. Due to the manifold variables that determine PMNC behavior, this chapter considers only a limited number of materials used in PMNC designs. Emphasis is given to the effects of nanoceramic addition on material properties: degradation kinetics, mechanical and biological effects (Sections 2.3.1, 2.3.2, and 2.3.4), and the specific changes manifest by nanoparticles as opposed to conventional microparticle inclusions. Furthermore, modeling attempts to account for modifications to material parameters effected by nanoparticle properties are reported (Section 2.4).

In order to account for nanoparticle effects in these materials, it is first necessary to describe other material properties and modifications wrought by nanoscale phenomena. To this end, some of the most common polymers and ceramics used in osseous tissue regeneration are elucidated in Sections 2.2.1 and 2.2.2 before proceeding to nanoparticle effects on material properties.

2.2 MATERIALS

2.2.1 Polymers

The use of polymers in prostheses is long established and widespread. The choice of polymer that may be used in a biological implant is limited by the requirements that the polymer be nontoxic, biocompatible, and biodegradable and the material's degradation products should have no negative effects on the surrounding tissues and organs of a host organism (Yang et al. 2006). Most available polymeric devices currently on the market do not degrade *in vivo*, e.g., ultra high molecular weight polyethylene, which is used in total hip replacements (Wang 2004). The majority of polymers investigated for osseous tissue regeneration or defect filling pertain to the polyester family to which poly(α-hydroxy acids) [incorporating polylactide (PLA), polyglycolide, and their copolymers] belong (Balasundaram and Webster 2007). A biopolymer's mechanical properties, biological effects, and degradation kinetics are influenced by its hydrophobicity, molecular weight distribution, glass transition temperature, crystallinity, geometry, temperature, and polymer processing technique. Accordingly, a wide range of possibilities exist to manipulate a polymer's properties to suit different requirements. In addition, radiation is often used as a sterilization method prior to clinical use; therefore, the effects of radiation damage have been investigated to ascertain changes to molecular weight, glass transition temperatures, and other properties with demonstrable effects (Yoshioka et al. 1995; Tan et al. 2009).

Poly(α-hydroxy acids) have a long history of the U.S. Food and Drug Administration (FDA)-approved use in medical devices and a large variation in properties between different poly(lactic-co-glycolic acid) (PLGA) forms, making this class of polymer a preferred candidate for a resorbable tissue engineering scaffold polymeric phase (Li et al. 1990; Lu et al. 2000). PLGA and other polyesters degrade by hydrolysis (or partly via enzymatic action *in vivo*), producing lactic acid, which is metabolized by the body via the tricarboxylic acid cycle to be excreted as CO_2 and H_2O (Holland et al. 1986), and glycolic acid, which is excreted in urine.

Polyester degradation in an aqueous environment is classified as bulk heterogeneous degradation in which de-esterification is catalyzed by $H^+_{(aq)}$ ions generated by acid dissociation of degradation products. The acids also diffuse through the material; hence, degradation is a reaction–diffusion problem, depending on material geometry, oligomer diffusion coefficients, and associated boundary conditions, among other factors. Due to autocatalysis, acid accumulates at the center of the material, whereas acids produced closer to the material boundary may diffuse into the surrounding medium, resulting in position- and time-dependent degradation rates with de-esterification proceeding faster at the material center compared to the boundary. Accumulation of acidic degradation products from PLGA has been shown to induce chronic inflammatory responses *in vivo* and a propensity to elicit fibrous tissue encapsulation at the polymer–hose interface (Lickorish et al. 2007); fibrosis at the implant–bone interface can cause device failure at early degradation times (Athanasiou et al. 1996). For more information regarding PLGA degradation, the reader is directed to the work of Therin et al. (1992) and Li et al. (1990).

In this chapter, the D,L-lactide/glycolide molar ratio in PLGA studies is given in parentheses. A plethora of other biodegradable polymers have been extensively investigated, and a brief description of some common types is given in Table 2.1.

The motivations for ceramic nanoparticle addition to polymers include possible improvements to degradation kinetics control offered by polymer–ceramic microcomposites, biological enhancement, and possible nanoparticle stiffening improvements to PMNC mechanical properties affected by nanoparticle addition. PMNCs

TABLE 2.1
Material Properties of Various Biopolymers

Polymer	Crystallinity	Tensile Strength (MPa)	Young's Modulus (GPa)	Approximate Degradation Time
Polyglycolide	Semicrystalline	~70	~6	6–12 months
Polycaprolactone	Semicrystalline	20–25	0.4	~3 years
Poly-L-lactide	Semicrystalline	60–70	3	~3 years
Collagen	Variable	50–100	1	–

are viewed as an evolution of traditional micrometer-sized polymer–ceramic composites. Basic properties of the most commonly used ceramics are described in Sections 2.2.2 and 2.2.2.1.

2.2.2 Ceramics

The concept of the incorporation of HA or another bioceramic to a polymer was introduced by Bonfield et al. (1981) based on the observation that the bone consists of an organic matrix reinforced with a mineral phase. Nanocrystalline HA is similar in composition to the main inorganic constituent of the bone. Consequently, many scaffold designs are based on or have incorporated HA as the primary ceramic phase. Nevertheless, the physical properties and mechanical reliability of pure HA ceramics are poor compared to the bone and, for this reason, is principally used in the form of powders, implant coatings, low-loaded porous implants, and bone cement (Best et al. 2008; Lewandrowski et al. 2000). HA exhibits low biodegradability, with some studies reporting incomplete reabsorption of sintered HA after 9 months *in vivo* (Klein et al. 1983). Numerous investigations have attested to enhanced osteoconductivity and osteoblast metabolism on nanoscale HA (Huang et al. 2004; Pezzatini et al. 2006). Much research has focused on modifying the chemical properties of HA to enhance its osteoconductivity; the interested reader is directed to the summary in the work of Suchanek et al. (1997).

For certain applications, high ceramic degradation rates are required; hence, interest in the tricalcium phosphate (TCP) $Ca_3(PO_4)_2$ class of bioceramic has increased. TCP has four known polymorphs: α, β, γ, and super-alpha ($\bar{\alpha}$). The $\bar{\alpha}$ and γ phases are only observed at high temperatures and pressures (Nurse et al. 1959). The β phase is the most stable form at standard temperature and pressure; the α phase is thermodynamically stable between 1120°C and 1470°C in the absence of impurities (Dorozhkin 2009) and occurs at room temperature by quenching or rapid cooling from its stable state. Both α- and β-TCPs are prepared by the thermal decomposition of calcium-deficient HA, and often, both TCP phases occur with HA after processing (Gibson et al. 1996).

2.2.2.1 Calcium Phosphate Solubility and pH Buffering Reactions

The acidity solubility regimes and apatite reactions with carboxylic acids are fundamental phenomena in composite behavior and affect local pH variant autocatalysis (see Sections 2.2.1 and 2.4). There are only three stable calcium phosphates at standard temperature and pressure: (1) monocalcium phosphate monohydrate [$Ca(H_2PO_4)_2$, pH < 2.51]; (2) brushite ($CaHPO_4 \cdot 2H_2O$, 1.5 < pH < 4.2); and (3) HA (pH > 4.2; Driessens 1982). The pH dependence on solubility is an important factor of any model of these materials' temporal and spatial degradation. The solubility of α- and β-TCP at pH 7.4 is $K_{sp}^{\alpha} = 10^{-25.5}$ and $K_{sp}^{\beta} = 10^{-29.5}$, respectively (Bohner et al. 1997). HA possesses a much lower solubility product at pH 7.4 ($K_{sp}^{HA} = 10^{58.6}$) (Fernández et al. 1999). Dissolved TCP, given a sufficient concentration of calcium and phosphate ions, will reprecipitate as HA, which is the only stable calcium phosphate at standard temperature and pressure, forming an HA bounding layer on TCP crystal surfaces (Klein et al. 1990; cf., Figure 2.1).

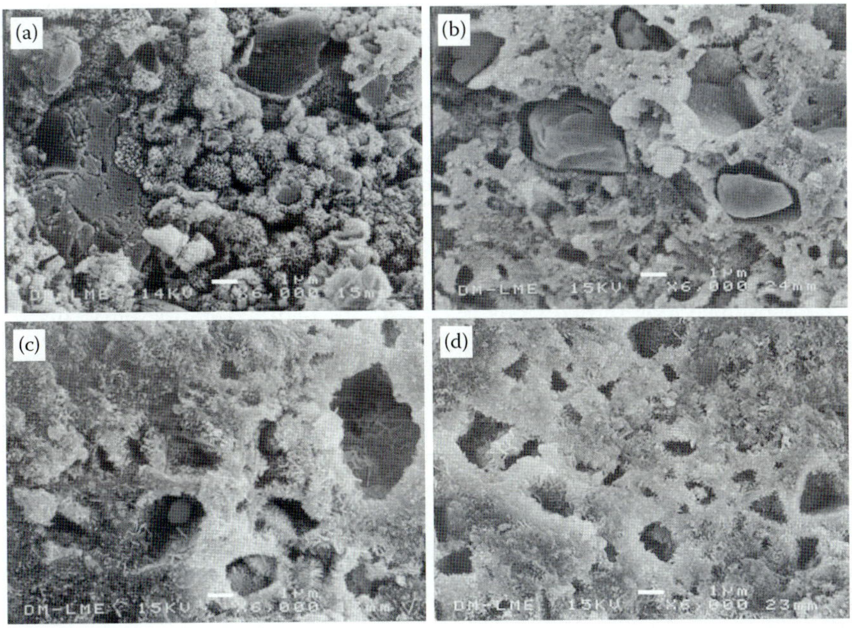

FIGURE 2.1 Evolution of the structure of α-TCP crystals observed using scanning electron microscopy (SEM) after (a) 2 h, (b) 8 h, (c) 64 h, (d) and 360 h in aqueous medium. (From Ginebra, M.P. et al., Effect of the particle size on the micro and nanostructural features of a calcium phosphate cement: A kinetic analysis, *Biomaterials*, 25, 3453, 2003. Reproduced with permission from Elsevier.)

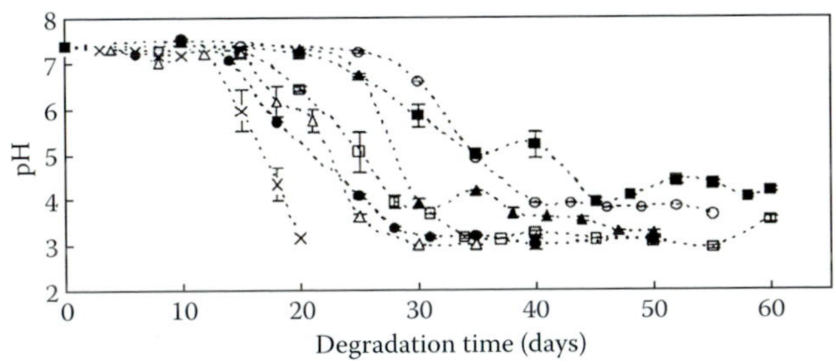

FIGURE 2.2 Variation of pH of solution containing α-TCP–PLGA composite materials as a function of degradation time. ×, Pure PLGA; Δ, 5% TCP; ●, 10% TCP; □, 15% TCP; ▲, 20% TCP; ○, 30% TCP; ■, 40% TCP. (From Ehrenfried, L.M. et al., Mechanical behavior of interpenetrating co-continuous beta-TCP–PDLLA composites, *Bioceramics*, 20, 361, 2008. Reproduced with permission from Elsevier.)

A primary reason for the ceramic addition to PMNCs is to buffer acidity and to modify degradation kinetics in the degrading nanocomposite (see Figure 2.2). The end scission reaction of PLGA autocatalyzed by H^+_{aq} may be written as

$$\text{PLGA} + H_2O \overset{H^+}{\Leftrightarrow} R - \text{COOH} + R - \text{OH}. \tag{2.1}$$

The carboxylic acid has a high degree of acid dissociation, and thus, degradation releases H^+_{aq}, which autocatalyzes further hydrolysis, i.e.,

$$R-COOH \Leftrightarrow R-COO + H^+_{aq}. \tag{2.2}$$

In an aqueous medium, the calcium phosphate dissociates to form

$$Ca_3(PO_4)_2 \Leftrightarrow Ca^{2+}_{aq} + (PO_4^{3-})_{aq}. \tag{2.3}$$

The phosphate ions subsequently react with the hydrons via the buffering reactions, forming monohydrogen phosphate

$$PO_4^{3-} + H^+ \Leftrightarrow HPO_4^{2-} \tag{2.4}$$

dihydrogen phosphate

$$PO_4^{3-} + H^+ \Leftrightarrow H_2PO_4^- \tag{2.5}$$

and phosphoric acid

$$H_2PO_4^- + H^+ \Leftrightarrow H_3PO_4. \tag{2.6}$$

Equations 2.4 through 2.6 elucidate the mechanisms that remove catalytic H^+_{aq} from the solution. TCP and HA dissolution is affected by acidity, physical disintegration into small particles due to preferential dissolution at grain boundaries, and biological factors (Lu et al. 2002). Equations 2.2 through 2.6 are important in nanocomposite material modeling incorporating a ceramic phase (Section 2.4).

2.2.3 Other Constituents

Further considerations to PMNC design include improved coupling between the mineral and organic phases. Numerous investigations have highlighted the critical role played by polymer–ceramic bonding interactions in reducing nanoparticle agglomeration (Jancar and Kucera 1990) or improving mechanical properties (either by reductions in stress concentrations in the polymer matrix or indirectly via T_g manipulation), especially when considering device degradation and performance in an aqueous environment. The interaction strength between the two phases has important ramifications for material behavior, specifically mechanical behavior and glass transition temperature modifications, which are further addressed in Sections 2.3.2 and 2.3.3. Often, only physical adsorption is achieved between the polymeric and ceramic phases. To address this, the ceramic surface may be treated to become more chemically compatible with the polymer matrix (Hong et al. 2005). A common reactive treatment is the incorporation of silanes to the surface of the ceramic phase, which results in increased end-group availability fomenting stronger polymer–ceramic bonding (Jancar and Kucera 1990). Thus, the interaction between the inorganic surface and the matrix polymer can be tuned by selecting appropriate end groups. These considerations, albeit important with regard to nanocomposite behavior, are not considered in detail in this chapter. For a treatise on the silane treatment of fillers, the reader is directed to the work of Plueddemann (1991).

Other PMNC topics that will not be discussed include the addition of growth factors such as bone morphogenic proteins (Wozney 2002) to induce new bone formation and a plethora of other factors, including scaffold geometry, that play important roles in degradation kinetics (Braunecker et al. 2004).

2.3 MATERIAL PROPERTIES

2.3.1 Nanocomposite Degradation Behavior and Acidity Regulation

TCP–PLGA composite acidity buffering capacity generally improves with TCP content (Ehrenfried et al. 2008; Wang 2004). Yang et al. (2009) degraded α-TCP–PLGA(50:50) composites with varying weight ceramic loadings and particle sizes in simulated body fluid (SBF). Differences in degradation kinetics are shown in Figures 2.3 and 2.4. Figure 2.3 clearly demonstrates the reduction in composite degradation rates due to reaction rate modification resulting from the ceramic inhibition of the autocatalytic effect. Moreover, Figure 2.3 shows ceramic crystal-size effects, with 30 wt% nanocomposites exhibiting a delay of approximately 7 days in pH change, onset of polymer mass loss, and time of maximum water absorption compared to

FIGURE 2.3 Morphological state of α-TCP–PLGA(50:50) nanocomposites at different degradation times in phosphate-buffered saline (PBS). (a) and (b) α-TCP (30 wt%) and nanocomposite (20 wt%) after 43 days. (c) Pure PLGA(50:50) at 21 days. (d) Microcomposite (30 wt%) at 36 days. (From Yang, Z. et al., The influence of α-tricalcium phosphate nanoparticles and microparticles on the degradation of poly(D,L-lactide-coglycolide), *Advanced Materials*, 21, 3900, 2009. Reproduced with permission from Wiley.)

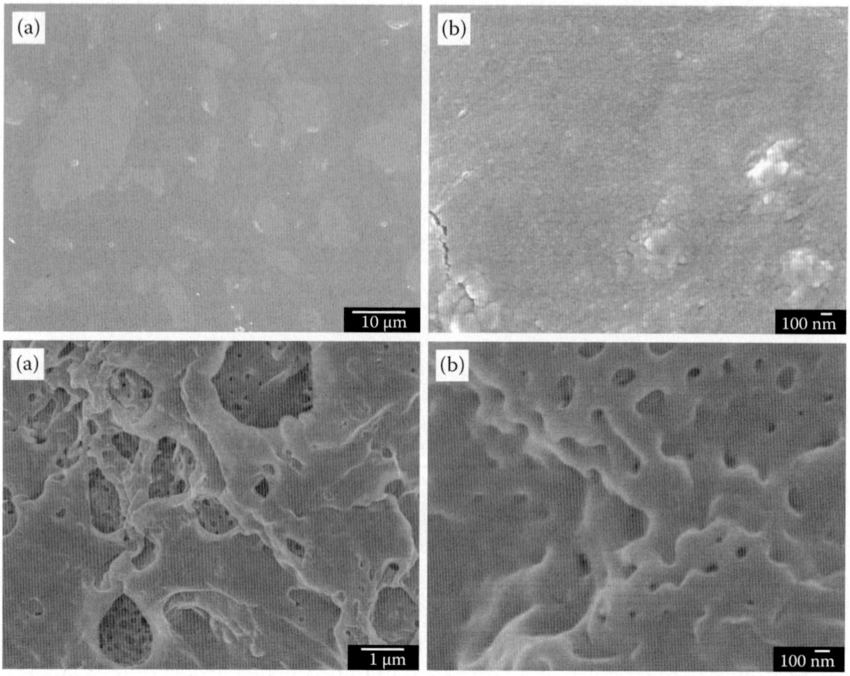

FIGURE 2.4 SEM images of 30 wt% α-TCP–PLGA(50:50) microcomposites and nanocomposites at various degradation times *in vitro* in SBF. (a) Undegraded microcomposite. (b) Microcomposite degraded for 36 days. (c) Undegraded nanocomposite. (d) Nanocomposite degraded for 30 days. (From Yang, Z. et al., The influence of α-tricalcium phosphate nanoparticles and microparticles on the degradation of poly(D,L-lactide-coglycolide), *Advanced Materials*, 21, 3900, 2009. Reproduced with permission from Wiley.)

the equivalent microcomposite. A comparison of surface morphologies for both composite types at different times is shown in Figure 2.4.

The overall pH change in buffering medium during *in vitro* degradation is independent of ceramic particle size, although the rate of pH change is observed to be particle size dependent, with nanoparticles exhibiting a slower pH decrease during degradation. This is important, as slower changes allow acids to be removed by the body, preventing accumulation and cytotoxicity. However, numerous studies have established that, above particular ceramic weight loadings (approximately 30%), little change in degradation kinetics or acid release profile is observed. This convergence has often been attributed to nanoparticle agglomeration, which is now challenged by new material modeling results, as summarized in Section 2.4.

Tang et al. (2008) investigated the water absorption of poly(3-hydroxybutyrate-co-3-hydroxyvalerate)/HA nanocomposites using a standard Fickian diffusion description (Becker et al. 2004; Chuang et al. 2004). The nano-HA used in the study was modified with a silane coupling agent. Water absorption in the materials is greatly affected by HA leaching. Furthermore, water diffusion coefficients have an inverse relationship with nano-HA filler content, although diffusion coefficients remained within the same order of magnitude for all composite types.

2.3.2 Mechanical Behavior

There is much evidence regarding the enhancement of biopolymer mechanical strength by the incorporation of nanoparticulate bioceramics. McManus et al. (2005) reported significantly greater bending moduli in PLA containing 40 and 50 wt% nanophase alumina, titania, and HA than composites with coarser grained ceramics. Liu and Webster (2010) characterized the compressive and tensile properties of composites composed of dispersed and agglomerated particles of 30 wt% HA and titania contained in PLGA(50:50) matrices. Titania nanocomposites exhibited superior elastic and bulk moduli and greater tensile strength at yield than pure PLGA and agglomerated composites, whereas dispersed nano-HA demonstrated inferior moduli but greater ductility than the agglomerated composite. Nano titania and HA composite differences were attributed to different polymer–ceramic bonding strengths.

PMNC nanoparticle effects may be expressed directly as a matrix stiffener or indirectly by changing the polymer phase's thermodynamic state by modifying the material's glass transition temperature, T_g. Wilberforce et al. (2001, 2010) reported large T_g reductions with increasing nano α-TCP loadings in PLGA(50:50), which were attributed to the large interfacial area and poor polymer–ceramic bonding. Results from expanding the study to gauge changes wrought by differing polymer-processing methods are presented in Figure 2.5, which demonstrates T_g reduction due to polymer–ceramic interface effects (variations between polymer-processing routes probably result from polymer thermal degradation during solvent evaporation). Also shown in Figure 2.5 is the dramatic effect of water diffusion and material plasticization, which is detrimental to the composite's mechanical strength—an important factor influencing composite behavior, which is further assessed in Section 2.3.3.

The transition to a rubbery state in polymers results in increased polymer chain mobility and significant changes to material properties. Elastic polymers are much more viscoelastic and present reduced storage moduli for $T > T_g$ than in the glassy state. Consequently, systematic studies and empirical descriptions of MPNC T_g variation with the ceramic interface area, polymer–ceramic bonding, water content and other factors affecting T_g are required.

Neglecting material property changes at temperatures close to T_g, numerous studies have demonstrated that, in general, a polymer's bulk and elastic moduli increase with the incorporation of nanoparticles with the increase being proportional to the nanoparticle loading and increasing for smaller particles. The strengthening mechanism is not understood completely (Hu et al. 2010). Mechanical deformation theory indicates that the high-volume fraction of interfacial regions compared to bulk materials leads to increased deformation by grain-boundary sliding and short-range diffusion-healing events as the grain size is reduced, thus increasing ductility for nanocrystalline ceramics (Narayan et al. 2004). This issue is complicated close to T_g and is exemplified in Figure 2.6, which shows the storage modulus of α-TCP–PLGA(50:50) nanocomposites prepared by injection molding as a function

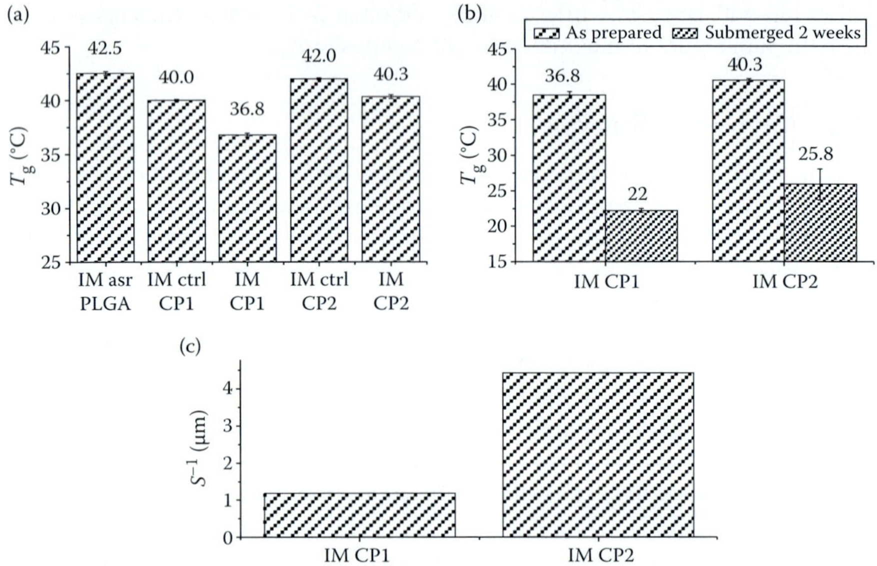

FIGURE 2.5 (a) T_g measured using differential scanning calorimetry of injection-molded samples (IM) processed via solvent evaporation (CP1) or twin screw extrusion (CP2). IM asr PLGA denotes injection-molded PLGA as received from the supplier. IM Ctrl CP1 and IM Ctrl CP2 represent injection-molded PLGA prepared and compacted via solvent evaporation and twin screw extrusion, respectively, which are used as controls. (b) T_g of samples after 2 weeks of degradation in PBS solution. (c) Reciprocal surface area of α-TCP particles per unit volume of composite. (From Wilberforce, S.I.J. et al., The influence of the compounding process and testing conditions on the compressive mechanical properties of poly(D,L-lactide-co-glycolide)/α-tricalcium phosphate nanocomposites, *Journal of Biomedical Materials Research*, 4, 1081, 2001. Reproduced with kind permission from Springer.)

of temperature (Wilberforce et al. 2010). The pure polymer exhibits the largest T_g, which is approximately 50°C, and the lowest T_g occurs in the 30 wt% nanocomposite. The higher ceramic loaded PMNCs demonstrate higher transition temperatures due to particle agglomeration in these samples, which indicates that the phenomenon is attributable to nanoparticle effects. Far from T_g, the storage modulus increases with filler content. However, due to nanoparticle-induced changes in T_g and subsequent deterioration of mechanical strength, at body temperature, any enhancement of PLGA(50:50) properties caused by nanoparticle addition is lost, and the pure polymer possesses superior properties. Consequently, nanoparticle addition to some polymers such as PLGA and resulting T_g reduction might be detrimental to the desired materials' mechanical properties compared to pure polymer or conventional microcomposites. This factor is considered further in Section 2.3.3. Wilberforce et al. (2011) showed that the addition of α-TCP nanoparticles to poly-L-lactide (PLLA), followed by quenching or annealing, reduced the T_g to a minimum of 60°C, making

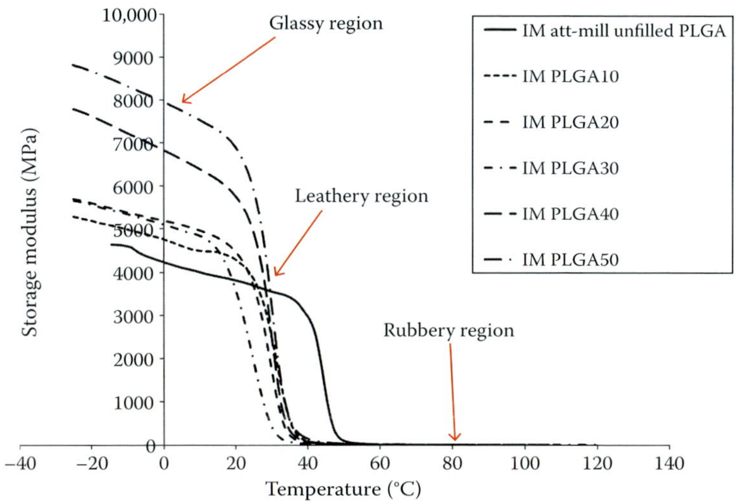

FIGURE 2.6 Storage modulus as a function of temperature measured by dynamical mechanical testing performed at 1 Hz for dry nano α-TCP–PLGA(50:50), with different ceramic weight loadings illustrating the critical role of T_g on mechanical properties of PLGA PMNCs. (From Wilberforce, S.I.J. et al., A dynamic mechanical thermal analysis study of the viscoelastic properties and glass transition temperature behavior of bioresorbable polymer matrix nanocomposites, *Journal of Materials Science: Materials in Medicine*, 21, 3085, 2010. Reproduced with kind permission from Springer.)

rubber transition effects unimportant at body temperature (pure PLLA exhibits a glass transition at approximately 70°C).

2.3.3 Glass Transition Behavior

Reductions in T_g as a function of film thickness have been well documented for polymer nanofilms (Ash et al. 2002). Bulk polymers may be described as having two regions with two different glass transition temperatures for the surface and bulk, T_g^{surf} and T_g^{bulk}, respectively. The surface layer extends to approximately 100 nm from the surface and is hypothesized to represent a liquid-like surface layer with increased polymer mobility compared to a bulk polymer substratum (Keddie et al. 1994). For a freestanding pure polymer thin film of total thickness h, the contribution of the surface and bulk glass transition temperatures to the total material glass transition temperature as a function of film thickness is

$$T_g(h) = T_g^{bulk} + \frac{2\varepsilon}{h}\left(T_g^{surf} - T_g^{bulk}\right) \tag{2.7}$$

where ε represents the surface region thickness. This description assumes a freestanding polymer film that produces the largest change in T_g. Glass transition temperature deviations are observed in supported films whereby the change in T_g depends on the strength of the interaction between the polymer and the substrate. Polymer–substrate interfaces exhibiting attractive interface properties (wetting) and those with weak interactions (nonwetting) manifest T_g increases and decreases, respectively (Ellison et al. 2005). Interfacial polymer–ceramic properties play a critical role in composite mechanical properties (Charvet et al. 2000). Moreover, improvements in polymer–ceramic wettability aid nanoparticle dispersion, as demonstrated by Guo et al. (2008), in which improved nanoparticle dispersion and consequent tensile strength increases were observed for Fe_2O_3 nanoparticle–resin composite treated using the bifunctional coupling agent methacryloxypropyltrimethoxysilane compared to uncoupled nanocomposites.

Bansal et al. (2005) demonstrated an equivalence principle between thermomechanical thin film properties and polymer–nanoparticle surface interactions in PMNCs. The authors calculated three-dimensional average particle separations of silica nanoparticles dispersed in polystyrene (PS) from transmission electron microscopy images and showed a good correlation between T_g as a function of particle separation for the composites and T_g as a function of polymer film thickness for PS–silica nanocomposites (cf., Figure 2.7).

PMNCs exhibit a wide variation in glass transition temperature, which depends on factors such as polymer and nanoparticle composition, solvent extraction, and nanoparticle sizes and loadings (Section 2.3.2). Generally, increasing the polymer's molecular order and greater intermolecular energy raises T_g (Donth 2001). As a general rule, T_g increases for larger polymer cohesive energies and greater molecular order, i.e.,

$$T_g = \frac{\Delta H}{\Delta S} \tag{2.8}$$

where ΔH and ΔS represent the enthalpy and entropy changes wrought by the transition to a rubbery state, respectively. Increasing polymer cross-linking results in greater cohesion energy, which raises ΔH. More importantly, in view of water-induced plasticization (cf., Figure 2.5), composite water absorption increases the system disorder, which can result in a dramatic T_g decrease.

The structural relaxation (aging) of polymers is a closely related phenomenon to T_g surface modification, which is influenced by interface effects with relaxation rate changes caused by surface interactions extending more than 100 nm into the film interior (Priestley et al. 2005). The authors believe these phenomena to be of great importance in designing PMNCs for clinical use, not only in the assessment of device behavior in the laboratory but also in aging-induced changes in properties due to storage time before device implantation in patients. Moreover, these properties must be addressed for all PMNC types and also evaluated as a function of degradation time. Furthermore, nanoparticle-induced changes in composite mechanical properties and aging can be modified by the use of binding agents, which warrants further development.

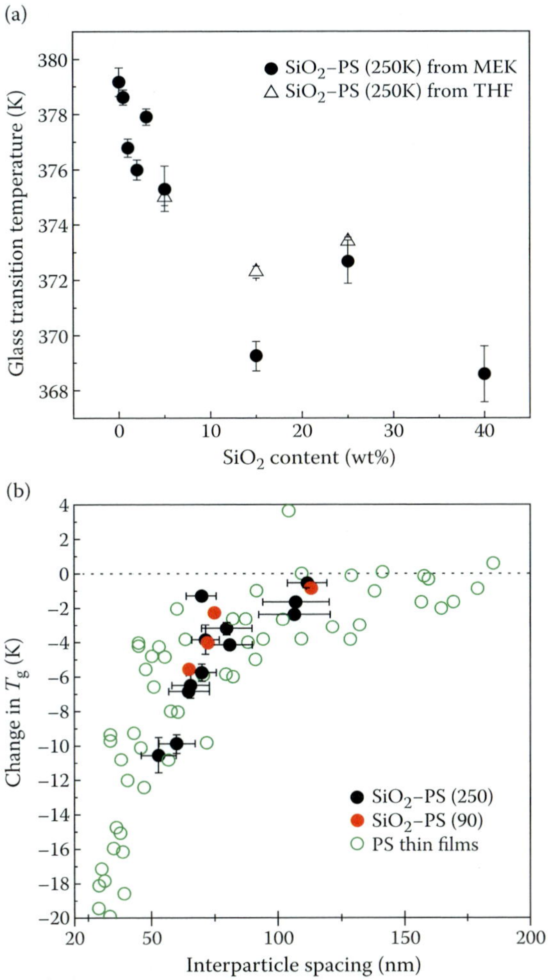

FIGURE 2.7 (a) T_g of SiO_2–PS nanocomposites as a function of silica weight content for composites prepared using the solvents methyl ethyl ketone (MEK) and tetrahydrofuran (THF). Large deviations observed at high weight loadings are caused by particle agglomeration. (b) Comparison between T_g reduction measured for thin PS films as a function of film thickness and T_g reduction of PS–silica nanocomposites as a function of mean interparticle spacing. Numbers in parentheses denote matrix molecular weight. (From Bansal, A. et al., Quantitative equivalence between polymer nanocomposites and thin polymer films, *Nature*, 4, 693, 2005. Courtesy of Nature Publishing Group.)

2.3.4 Bioactivity

Yang et al. (2008) evaluated the *in vitro* bioactivity of nanostructured α-TCP–PLGA(75:25) composites by quantifying the rapid formation of bonelike apatite layers on the material's surface while immersed in SBF (the α phase was chosen due to its greater solubility than β-TCP). Enhanced apatite nucleation and dense lamellar-like apatite formation were observed after 7 and 14 days of immersion, respectively. Acellular indicators of improved bioactivity were corroborated by *in vitro* cell culture studies using human osteoblast-like (HOB) cells, elucidating greater cell numbers on the nanocomposite as the culture proceeded and the observation of a confluent lamellar-like apatite layer beneath a proliferating HOB layer at day 16. Enhanced apatite nucleation was attributed to the nanostructured surface providing large numbers of nucleation sites for apatite precipitation or fast dissolution of nanoparticles near the surface. Both of these mechanisms most likely work in conjunction to create rougher surface topographies. In addition, in a later study, Yang et al. (2009) reported the liberation of residual micron-scale ceramic particles (which are known to cause adverse cell response) from the microcomposite into the buffer medium, which did not occur in the nanocomposites.

Biocompatibility enhancement is not confined to calcium phosphate addition; other nanoceramics such as titania and alumina have demonstrated the inducement of improved cell function (Webster et al. 1999, 2000a). Nanophase titania–PLGA(50:50) composites investigated by Liu et al. (2006) demonstrated that the greatest *in vitro* osteoblast adhesion and optimal indicators of bioactivity such as collagen, alkaline phosphatase activity, intracellular and extracellular protein synthesis, and calcium-containing mineral deposition occurred for composites possessing surface roughness values, which most closely matched that of the natural bone (cf., Figure 2.8). In addition, the material surface roughness as opposed to ceramic surface coverage was shown to dominate the cell response. The surface topographic effect is not limited to bioceramics. Webster and Ejiofor (2004) analyzed osteoblast adhesion on several nanophase metals, which exhibited increased osteoblast adhesion to nanopatterned surfaces relative to conventional surfaces presenting micrometer-sized topographic features. Interestingly, the study observed preferential osteoblast adhesion at grain boundaries.

The compendium of studies performed to assay cell responses to nanometer-sized topographic surface modification clearly demonstrates that nanometer topographies alter cell behavior significantly. In the study of Palin et al. (2005), the authors sought to assess the preponderance of nanoscale topography over extraneous factors such as surface chemistry and surface energy. To this end, nanometer-scale and conventional topographic structures were formed on PLGA by molding and seeded with osteoblasts. The bioactivities of the different surfaces were assessed by quantifying osteoblast proliferation as a function of time, with nanopatterned PLGA demonstrating more than double the initial number of adhered osteoblasts than the conventional surface after 1 day and 5 days in culture.

Enhancing osteoclast activity is also important for nanocomposite clinical performance; rapid bone deposition mediated by osteoblasts and remodeling via osteoclasts is a vital component of wound healing and the incorporation of the orthopedic device. Webster et al. (2001) showed that indicators of osteoclast activity, the number of resorption pits, and tartrate-resistant acid phosphatase increased on nanophase alumina and HA compared to conventional-sized particles. Increased wettability or hydrophilicity is associated with increased protein absorption and, consequently, increased cell adhesion and enhanced function. The study showed increased numbers of absorption pits on nanometer surfaces compared to micro surfaces.

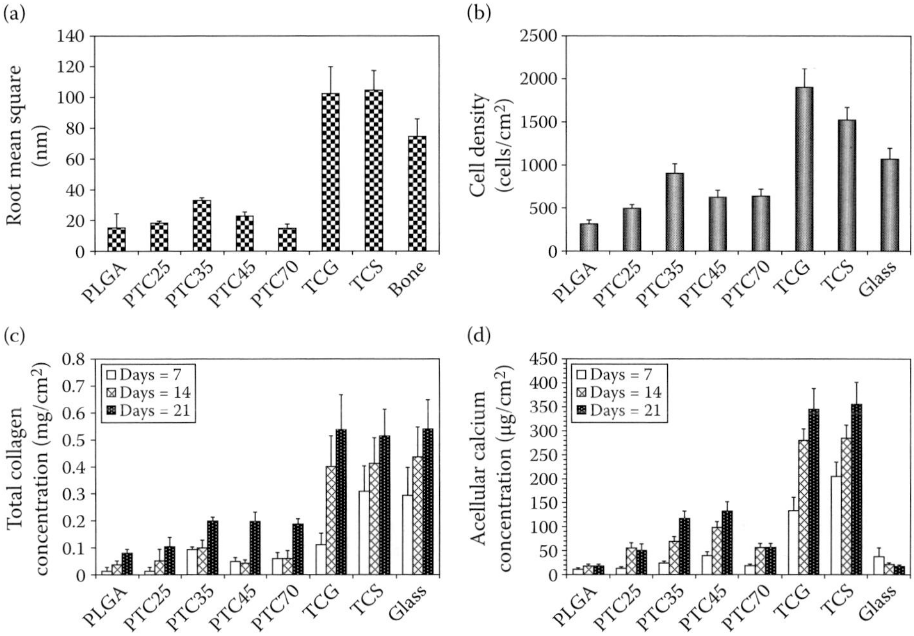

FIGURE 2.8 (a) Surface roughness (root mean square) of PLGA, nanoparticulate titania–PLGA (ultrasonicated at 25%, 35%, 45%, and 70% maximum power), green titania (TCG), sintered titania (TCS), and porcine bone. (b) Cell adhesion to various materials [PLGA, nanoparticulate titania–PLGA (ultrasonicated at 25%, 35%, 45%, and 70% maximum power), TCG, TCS, and glass] after incubation for 4 h. (c) and (d) Total collagen content and acellular calcium concentration for various surfaces analyzed. (From Liu, H. et al., Increased osteoblast functions among nanophase titania/poly(lactide-co-glycolide) composites of the highest nanometer surface roughness, *Journal of Biomedical Materials Research, Part A*, 78, 798, 2006. Reproduced with permission from Wiley.)

Further biological considerations of PMNC performance include fibroblast activity, because excessive fibrosis at the implant–host interface is a known factor in device failure (Vance et al. 2004). Prolonged increased fibroblast numbers at implant interfaces is associated with callous formation and associated soft-tissue formation rather than bone juxtaposition. Thus, a reduction in fibroblast activity (or limited increase in fibroblast activity relative to osteoblasts) might be beneficial to device-mediated wound healing. Promisingly, some studies have shown that the ratio of osteoblast to fibroblast adhesion increased from 1:1 on microparticle alumina surfaces to 3:1 on nanoparticle surfaces (Vance et al. 2004).

Qualitative arguments for the improvement in biological action of nanostructured materials assumed that the improvement was due to mimicking the naturally occurring topographies of natural biological tissues; HA crystals in bone are approximately 50 nm long and 5 nm in diameter. More quantitatively, the enhanced bioactivity observed with nanocomposites results from protein interactions with nanopatterned surfaces (Christenson et al. 2006), whereby cumulative protein absorption has been shown in many studies to be higher on nanograined materials (Webster et al. 1999, 2000a, 2000b). Several proteins that are known to augment osteoblast function have been shown to be deposited much faster and in greater quantity on nanocomposites compared to microcomposites. Dimensionality effects are also observed in cartilage and tendon tissue engineering constructs. Research indicates that chondrocyte and mesenchymal stem cell activities are strongly dependent on scaffold dimensionality (Lu et al. 2002).

Nanoparticulate addition to composites and surface coatings is not the only application of nanotechnology to medical devices. Other techniques have been used to create nano-sized surface features via chemical etching, electron beam lithography, and polymer demixing (Schift et al. 2002; Thapa et al. 2003), and nanopatterned surfaces produced by electrospraying, which has demonstrated the ability to direct osteoblast growth (Thian et al. 2008).

2.4 THEORETICAL APPROACH

There has been significant development of mathematical models designed to predict the rate of drug release from pure biopolymers using a variety of modeling techniques, many of which elucidate the observed phenomena to a high degree of accuracy for materials with simple geometries such as planes, cylinders, and spheres. However, modeling descriptions of PMNC behavior and degradation kinetics are underdeveloped due to the complexity of degradation processes. Many polymer degradation models employing a plethora of techniques, ranging from Monte Carlo statistical analyses of PLGA microspheres to direct analytical solutions of Fick's equations, have been described in the literature. Finite element modeling is now being used as an indispensable modeling tool, and models have been developed to predict degradation mechanics and diffusion mechanisms in pure polymer matrices. Pan et al. (2011) have advanced model complexity to include the effects of ceramic incorporation. For simplicity, the degradation model of TCP–PLGA composites assumes abundant water absorption and a spherical representative unit volume of a TCP particle surrounded by a PLGA matrix. PLGA degradation is primarily a chain end scission. Wang et al. (2008), Han and Pan (2009), and Han et al. (2010) showed that the rate of chain scission $\frac{\mathrm{d}R}{\mathrm{d}t}$ is equal to

$$\frac{\mathrm{d}R}{\mathrm{d}t} = \frac{\mathrm{d}C_{\mathrm{end}}}{\mathrm{d}t} = k_1 C_{\mathrm{e}} + k_2 C_{\mathrm{e}} C_{\mathrm{H^+}} \tag{2.9}$$

where C_{e}, $C_{\mathrm{H^+}}$, k_1, and k_2 represent the ester concentration, hydrogen ion concentration, random polymer scission rate (uncatalyzed reaction), and autocatalysis rate constants, respectively. The concentrations of the ions at each step are calculated by dissociation rate constants. Moreover, the model must account for the differing solubilities of the calcium phosphates. Equilibrium in Equation 2.3 is not reached due to removal of phosphate ions via processes in Equations 2.4 through 2.6. Accordingly, there is an undersaturation leading to TCP dissolution. The rate of TCP dissociation may be described in terms of an ion flux J, which depends on the surface area of the particle and the undersaturation, which is typically represented in terms of a power law, i.e.,

$$J = A_m \sigma^n$$

where σ is the undersaturation, and A and n are material constants (Bohner et al. 1997). Using reaction equilibria constants, mass conservation equations, and the reaction equations, Pan et al. (2011) defined a relative rate of calcium phosphate dissolution to autocatalytic hydrolysis S_{cp} as

$$S_{cp} = \frac{A_d \left(\frac{A_{cp}}{V_{polymer}} \right)}{k_2 \, (C_{H^+})_{pH=7.4} \times (C_{Ca^{2+}})_{eq}} \tag{2.10}$$

where A_d and A_{cp} represent the TCP dissolution rate and the TCP crystal surface area, respectively. Accordingly, due to the surface area term, not only is the autocatalytic reaction rate dependent on the quantity of TCP, but the size of the particles also causes the different degradation kinetics between composites of the same TCP loadings but different morphologies.

Using this approach, Pan et al. (2011) predicted that there exists a saturation limit for the effectiveness of α-TCP as defined in terms of S_{cp}. Above a certain limit, the dissolution of calcium phosphate has no effect on the rate of polymer degradation, as any hydrogen ions produced by degradation react with phosphate ions and are removed from the reaction–diffusion process. In the case of S_{cp} saturation, the controlling parameter on the reaction rate is the noncatalyzed reaction rate constant k_1. For completeness, the degradation kinetics of polymers that degrade by uncatalyzed hydrolysis only are not affected by TCP quantity or form.

Graphical representations of applying this formalism to empirical α-TCP–PLGA(50:50) degradation data are shown in Figures 2.9 and 2.10. Figure 2.9 shows data

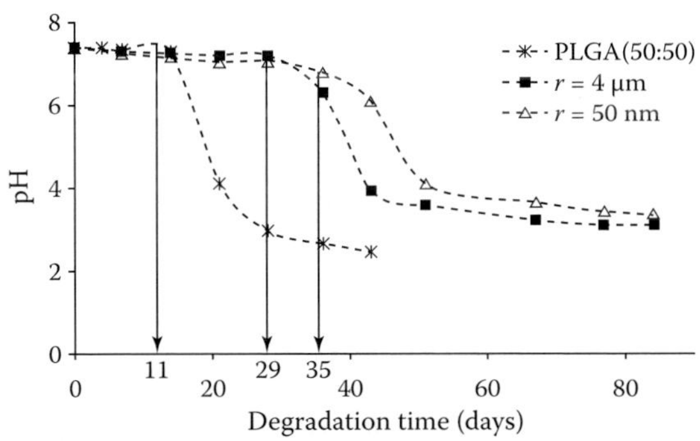

FIGURE 2.9 pH variation as a function of degradation time of buffer medium during degradation of pure PLGA(50:50) and PLGA(50:50) incorporating 30 wt% α-TCP inclusions of 4 μm and 50 nm. Data from Yang et al. (2009), fitted with model results from Pan et al. (2011). (From Pan, J. et al., A model for biodegradation of composite materials made of polyesters and tricalcium phosphates, *Biomaterials*, 32, 2248, 2011. Reproduced with permission from Elsevier.)

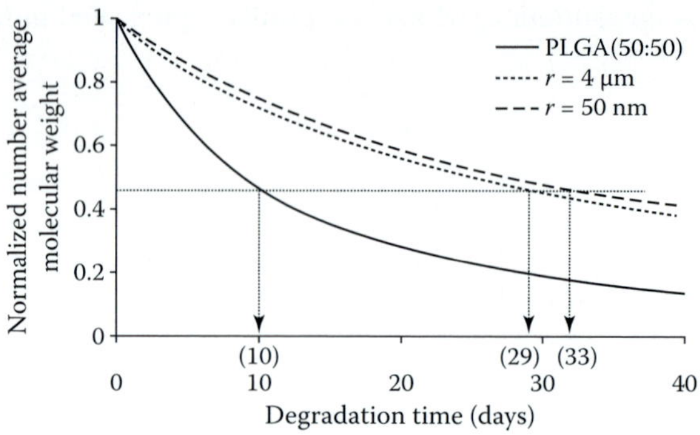

FIGURE 2.10 Predicted half degradation times for pure PLGA(50:50) and α-TCP–PLGA(50:50) composites (30 wt% α-TCP), assuming spherical particles of 4 μm and 50 nm in diameter. (From Pan, J. et al., A model for biodegradation of composite materials made of polyesters and tricalcium phosphates, *Biomaterials*, 32, 2248, 2011. Reproduced with permission from Elsevier.)

from the degradation study of Yang et al. (2009), where the onset of pH decrease was hypothesized to herald the simultaneous start of oligomer loss from degrading composites. These mass loss onset times demonstrated close agreement with simulated polymer molecular mass loss shown in Figure 2.10 using the formalism described above.

The data suggest that, due to TCP dissolution saturation (cf., Figure 2.11), nanosized α-TCP crystals have a small but noticeable effect on degradation rates compared to microparticles with the same weight loading. However, for lower weight loadings, the difference between microcomposites and nanocomposites is more pronounced.

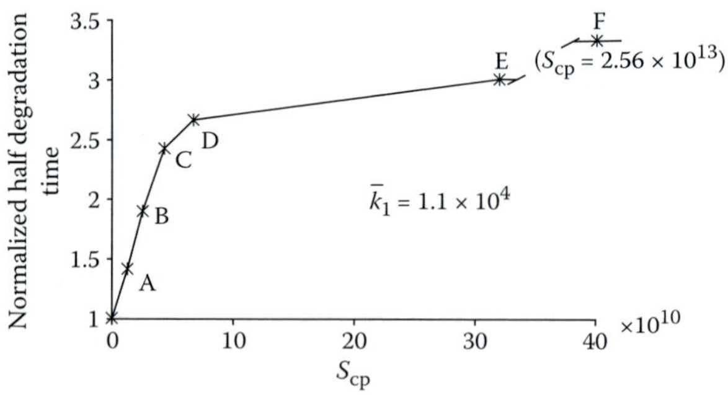

FIGURE 2.11 Predicted normalized degradation times of PLGA(50:50) using noncatalyzed reaction rate constant of 1.1×10^4. Points A through D denote α-TCP–PLGA(50:50) composites containing 30-μm-diameter spherical particles at loadings of 10, 20, 30, and 40 wt%, respectively. Points E and F represent 30 wt% loadings of α-TCP of 4-μm and 50-nm diameter, respectively. N.B. Point F lies off the main axis. (From Pan, J. et al., A model for biodegradation of composite materials made of polyesters and tricalcium phosphates, *Biomaterials*, 32, 2248, 2011. Reproduced with permission from Elsevier.)

Note that these models are still in their infancy, and further phenomena such as water transport, swelling, and polymer phase changes are required to complete a total material description. However, these early results are promising and will aid future material development.

2.5 CONCLUSION

The theoretical modeling of α-TCP–PLGA(50:50) degradation kinetics raises important questions with regard to the cost–benefit returns of using ceramic nanoparticles to improve composite degradation times. Moreover, the decrease in T_g observed due to poor bonding at the interface between TCP and PLGA complicates the assumption that nanoparticle inclusions have a positive effect on mechanical properties. On the other hand, a plethora of evidence indicates that biocompatibility is vastly enhanced using nanoceramic fillers. The specific analysis performed in Section 2.4 and subsequent discussion is only applicable to TCP–PLGA composites; however, the technique is applicable to other PMNC types and should be extended.

Due to the multiplicity of materials, processing conditions, and geometries available to the experimenter, PMNCs present a wealth of different device design directions and possible applications. The authors believe that, to further the goal of employment in clinical applications, many other issues must be addressed. These include the systematic study of coupling agents to improve the interfacial bonding between ceramic and polymer phases. Moreover, there is a dearth of information regarding T_g and the mechanical property evolution of nanocomposites and microcomposites caused by polymer hydration and possible debonding between the organic and inorganic phases. Much more data are required regarding material properties at physiological temperatures. More practically, for materials intended for clinical use, the choice of sterilization technique such as γ irradiation prior to implantation has demonstrable effects on polymer degradation rates (Faisant et al. 2002; Tan et al. 2009). Systematic investigations of sterilization effects on mechanical and biological PMNC properties are required. Furthermore, a fundamental prerequisite for commercial PMNC validity are the effects of material aging and changes in performance between fabrication and implantation.

On the biological level, important issues regard the effects of free nanoparticles released into the host if the polymer degrades before total ceramic dissolution. Moreover, although good evidence exists regarding the *in vitro* indications of superior nanocomposite bioactivity, there is a distinct lack of histological data comparing microparticle and nanoparticle polymer composites.

REFERENCES

B.J. Ash, L.S. Schadler, and R.W. Siegel. Glass transition behavior of alumina/polymethylmethacrylate nanocomposites. *Materials Letters* 55 (2002) 83.

K. Athanasiou, G.G. Niederauer, and C.M. Agrawal. Sterilization, toxicity, biocompatibility and clinical applications of polylactic acid/polyglycolic acid copolymers. *Biomaterials* 17 (1996) 93.

G. Balasundaram and T.J. Webster. Overview of nanopolymers for orthopedic applications. *Macromolecular Bioscience* 7 (2007) 635.

A. Bansal, H. Yang, C. Li, K. Cho et al. Quantitative equivalence between polymer nanocomposites and thin polymer films. *Nature* 4 (2005) 693.

O. Becker, R.J. Varley, and G.P. Simon. Thermal stability and water uptake of high-performance epoxy-layered silicate nanocomposites. *European Polymer Journal* 40 (2004) 187.

S.M. Best, A.E. Porter, E.S. Thian, and J. Huang. Bioceramics: Past, present and for the future. *Journal of the European Ceramic Society* 28 (2008) 1319.

M. Bohner, J. Lemaitre, and T.A. Ring. Kinetics of dissolution of beta-tricalcium phosphate. *Journal of Colloid and Interface Science* 190 (1997) 37.

W. Bonfield, M. Grynpas, A.E. Tully, J. Bowman et al. Hydroxyapatite-reinforced polyethylene—A mechanically compatible implant material for bone replacement. *Biomaterials* 2 (1981) 185.

J. Braunecker, M. Baba, G.E. Milroy, and R.E. Cameron. The effects of molecular weight and porosity on the degradation and drug release from polyglycolide. *International Journal of Pharmaceutics* 282 (2004) 19.

J.L. Charvet, J.A. Cordes, and H. Alexander. Mechanical and fracture behavior of a fiber-reinforced bioabsorbable material for orthopedic applications. *Journal of Materials Science: Materials in Medicine* 11 (2000) 101.

E.M. Christenson, K.S. Anseth, J.J.J.P. van den Beucken, and C.K. Chan. Nanobiomaterial applications in orthopedics. *Journal of Orthopaedic Research* 25 (2006) 11.

T.H. Chuang, T.C.K. Yang, T.Y. Chen, and A.H. Chang. Effects of glass microfillers on the water-transport behavior of polyurethane composites. *International Journal of Polymer Materials* 53 (2004) 553.

E. Donth. *The Glass Transition: Relaxation Dynamics in Liquids and Disordered Materials.* Heidelberg: Springer (2001).

S.V. Dorozhkin. Calcium orthophosphates in nature, biology and medicine. *Materials* 2 (2009) 399.

F.C.M. Driessens. *Mineral Aspects of Dentistry*. Basel: Karger (1982).

L.M. Ehrenfried, D. Farrar, D. Morsley, and R.E. Cameron. Mechanical behavior of interpenetrating co-continuous beta-TCP–PDLLA composites. *Bioceramics* 20 (2008) 361.

C.J. Ellison, M.K. Mundra, and J.M. Torkelson. Impacts of polystyrene molecular weight and modification to the repeat unit structure on the glass transition—Nanoconfinement effect and the cooperativity length scale. *Macromolecules* 38 (2005) 1767.

N. Faisant, J. Siepmann, P. Oury, V. Laffineur et al. The effect of gamma-irradiation on drug release from bioerodible microparticles: A quantitative treatment. *International Journal of Pharmaceutics* 242 (2002) 281.

E. Fernández, F.J. Gil, M.P. Ginebra, F.C. Driessens et al. Calcium phosphate bone cements for clinical applications—Part I: Solution chemistry. *Journal of Materials Science: Materials in Medicine* 10 (1999) 169.

I.R. Gibson, S.M. Best, and W. Bonfield. Phase transformations of tricalcium phosphates using high-temperature X-ray diffraction. *Bioceramics* 9 (1996) 173.

M.P. Ginebra, F.C.M. Driessens, and J.A. Planell. Effect of the particle size on the micro- and nanostructural features of a calcium phosphate cement: A kinetic analysis. *Biomaterials* 25 (2003) 3453.

Z. Guo, K. Lei, Y. Li, H.W. Ng et al. Fabrication and characterization of iron oxide nanoparticles reinforced vinyl–ester resin nanocomposites. *Composite Science and Technology* 68 (2008) 1513.

X. Han and J. Pan. A model for simultaneous crystallisation and biodegradation of biodegradable polymers. *Biomaterials* 30 (2009) 423.

X. Han, J. Pan, F. Buchanan, N. Weir et al. Analysis of degradation data of poly(L-lactide-co-D,L-lactide) and poly(L-lactide) obtained at elevated and physiological temperatures using mathematical models. *Acta Biomaterialia* 6 (2010) 3882.

S.J. Holland, B.J. Tighe, and P.L.J. Gould. Polymers for biodegradable medical devices—Part 1: The potential of polyesters as controlled macromolecular release systems. *Controlled Release* 4 (1986) 155.

Z.K. Hong, P.B. Zhang, C.L. He, X.Y. Qiu et al. Nanocomposite of poly(L-lactide) and surface grafted hydroxyapatite: Mechanical properties and biocompatibility. *Biomaterials* 26 (2005) 6296.

H. Hu, L. Onyebueke, and A. Abatan. Characterizing and modeling mechanical properties of nanocomposites—Review and evaluation. *Journal of Minerals and Materials Characterization and Engineering* 9 (2010) 275.

J. Huang, S. Best, W. Bonfield, R.A. Brooks et al. *In vitro* assessment of the biological response to nano-sized hydroxyapatite. *Journal of Materials Science: Materials in Medicine* 15 (2004) 441.

J. Jancar and J. Kucera. Yield behavior of PP/$CaCO_3$ and PP/$Mg(OH)_2$ composites—Part II: Enhanced interfacial adhesion. *Polymer Engineering and Science* 30 (1990) 714.

J.L. Keddie, R.A.L. Jones, and R.A. Cory. Size-dependent depression of the glass transition temperature in polymer films. *Europhysics Letters* 27 (1994) 59.

C.P. Klein, A.A. Driessens, K. de Groot, and A. van den Hooff. Biodegradation behavior of various calcium phosphate materials in bone tissue. *Journal of Biomedical Materials Research* 17 (1983) 769.

C.P.A.T. Klein, J.M.A. de Blieck-Hogemrst, J.G.C. Wolke, and K. de Groot. Studies of the solubility of different calcium phosphate ceramic particles *in vitro*. *Biomaterials* 11 (1990) 509.

K.U. Lewandrowski, J.D. Gresser, D.L. Wise, R.L. White et al. Osteoconductivity of an injectable and bioresorbable poly(propylene glycol-co-fumaric acid) bone cement. *Biomaterials* 21 (2000) 293.

S.M. Li, H. Garreau, and M. Vert. Structure–property relationships in the case of the degradation of massive poly(α-hydroxy acids) in aqueous media—Part 1: Poly(DL-lactic acid). *Journal of Materials Science: Materials in Medicine* 1 (1990) 198.

D. Lickorish, L. Guan, and J.E. Davies. A three-phase, fully resorbable, polyester/calcium phosphate scaffold for bone tissue engineering: Evolution of scaffold design. *Biomaterials* 28 (2007) 495.

H. Liu, E.B. Slamovich, and T.J. Webster. Increased osteoblast functions among nanophase titania/poly(lactide-co-glycolide) composites of the highest nanometer surface roughness. *Journal of Biomedical Materials Research—Part A* 78 (2006) 798.

H. Liu and T.J. Webster. Mechanical properties of dispersed ceramic nanoparticles in polymer composites for orthopedic applications. *International Journal of Nanomedicine* 5 (2010) 299.

J. Lu, M. Descamps, J. Dejou, G. Koubi et al. The biodegradation mechanism of calcium phosphate biomaterials in bone. *Journal of Biomedical Materials Research* 63 (2002) 408.

L. Lu, S.J. Petera, M.D. Lyman, H.L. Lai et al. *In vitro* and *in vivo* degradation of porous poly(lactic-co-glycolic acid) foams. *Biomaterials* 21 (2000) 1837.

A.J. McManus, R.H. Doremus, R.W. Siegel, and R. Bizios. Evaluation of cytocompatibility and bending modulus of nanoceramic/polymer composites. *Journal of Biomedical Materials Research* 72A (2005) 98.

R.J. Narayan, P.N. Kumta, C. Sfeir, D.H. Lee et al. Nanostructured ceramics in medical devices: Applications and prospects. *Journal of the Minerals, Metals and Materials Society* 56 (2004) 38.

R.W. Nurse, J.H. Welsh, and W.J. Gutt. High-temperature phase equilibria in the system dicalcium silicate–tricalcium phosphate. *Journal of the Chemical Society* 220 (1959) 1077.

E. Palin, H. Liu, and T.J. Webster. Mimicking the nanofeatures of bone increases bone-forming cell adhesion and proliferation. *Nanotechnology* 16 (2005) 1828.

J. Pan, X. Han, W. Niu, and R.E. Cameron. A model for biodegradation of composite materials made of polyesters and tricalcium phosphates. *Biomaterials* 32 (2011) 2248.

J.B. Park and R.S. Lakes. *Biomaterials: An Introduction*. New York: Plenum Press (1992).

S. Pezzatini, R. Solito, L. Morbidelli, S. Lamponi et al. The effect of hydroxyapatite nanocrystals on microvascular endothelial cell viability and functions. *Journal of Biomedical Materials Research, Part A* 76 (2006) 656.

E.P. Plueddemann. *Silane Coupling Agents*, 3rd edition. New York: Plenum Press (1991).

R.D. Priestley, C.J. Ellison, L.J. Broadbelt, and J.M. Torkelson. Structural relaxation of polymer glasses at surfaces, interfaces, and in between. *Science* 308 (2005) 456.

H. Schift, L.J. Heyderman, C. Padeste, and J. Gobrecht. Chemical nanopatterning using hot embossing lithography. *Microelectron Engineering* 61 (2002) 423.

W. Suchanek, M. Yashima, M. Kakihana, and M. Yoshimura. Hydroxyapatite ceramics with selected sintering additives. *Biomaterials* 18 (1997) 923.

H.Y. Tan, E. Widjaja, F. Boey, and S.C.J. Loo. Spectroscopy techniques for analyzing the hydrolysis of PLGA and PLLA. *Journal of Biomedical Materials Research—Part B: Applied Biomaterials* 91 (2009) 433.

C.Y. Tang, D.Z. Chen, T.M. Yue, K.C. Chan et al. Water absorption and solubility of PHBHV/HA nanocomposites. *Composites Science and Technology* 68 (2008) 1927.

A. Thapa, D.C. Miller, T.J. Webster, and K.M. Haberstroh. Nanostructured polymers enhance bladder smooth muscle cell function. *Biomaterials* 24 (2003) 2915.

M. Therin, P. Christel, S. Li, H. Garreau, and M. Vert. *In vivo* degradation of massive poly(α-hydroxy acids): Validation of *in vitro* findings. *Biomaterials* 13 (1992) 594.

E.S. Thian, Z. Ahmad, J. Huang, and M.J. Edirisinghe. The role of electrosprayed apatite nanocrystals in guiding osteoblast behavior. *Biomaterials* 29 (2008) 1833.

R.J. Vance, D.C. Miller, A. Thapa, K.M. Haberstroh et al. Decreased fibroblast cell density on chemically degraded poly(lactic-co-glycolic) acid, polyurethane, and polycaprolactone. *Biomaterials* 25 (2004) 2095.

M. Wang. Developing bioactive composite materials for tissue replacement. *Biomaterials* 24 (2004) 2133.

Y. Wang, J. Pan, X. Han, C. Sinka et al. A phenomenological model for the degradation of biodegradable polymers. *Biomaterials* 29 (2008) 339.

T.J. Webster and J.U. Ejiofor. Increased osteoblast adhesion on nanophase metals: Ti, Ti6Al4V, and CoCrMo. *Biomaterials* 25 (2004) 4731.

T.J. Webster, C. Ergun, R.H. Doremus, R.W. Siegel et al. Enhanced functions of osteoblasts on nanophase ceramics. *Biomaterials* 21 (2000a) 1803.

T.J. Webster, C. Ergun, R.H. Doremus, R.W. Siegel et al. Specific proteins mediate enhanced osteoblast adhesion on nanophase ceramics. *Journal of Biomedical Materials Research* 51 (2000b) 475.

T.J. Webster, C.D. Ergun, R.W. Siegel, and R. Bizios. Enhanced functions of osteoclast-like cells on nanophase ceramics. *Biomaterials* 22 (2001) 1327.

T.J. Webster, R.W. Siegel, and R. Bizios. Osteoblasts adhesion on nanophase ceramics. *Biomaterials* 20 (1999) 1221.

S.I.J. Wilberforce, S.M. Best, and R.E. Cameron. A dynamic mechanical thermal analysis study of the viscoelastic properties and glass transition temperature behavior of bioresorbable polymer matrix nanocomposites. *Journal of Materials Science: Materials in Medicine* 21 (2010) 3085.

S.I.J. Wilberforce, C.E. Finlayson, S.M. Best, and R.E. Cameron. The influence of the compounding process and testing conditions on the compressive mechanical properties of poly(D,L-lactide-co-glycolide)/-tricalcium phosphate nanocomposites J. Mechanical Behaviour of Biomedical Materials 4 (2001) 1081.

S.I.J. Wilberforce, C.E. Finlayson, S.M. Best, and R.E. Cameron. A comparative study of the thermal and dynamic mechanical behavior of quenched and annealed bioresorbable poly-L-lactide/α-tricalcium phosphate nanocomposites. *Acta Biomaterialia* 7 (2011) 2176.

D.F. Williams. *Fundamental Aspects of Biocompatibility*. Boca Raton, FL: CRC Press (1981).

J. Wolff. *Das Gesetz der Transformation der Knochen*. Berlin: Herschwald (1987).

J.M. Wozney. Overview of bone morphogenetic proteins. *Spine* 27 (2002) S2.

F. Yang, W. Cui, Z. Xiong, L. Liu et al. Poly(L,L-lactide-co-glycolide)/tricalcium phosphate composite scaffold and its various changes during degradation *in vitro*. *Polymer Degradation and Stability* 91 (2006) 3065.

Z. Yang, E.S. Thian, R.A. Brooks, N. Rushton et al. Apatite formation on alpha-tricalcium phosphate/poly(D,L-lactide-co-glycolide) nanocomposite. *Bioceramics* 20 (1) (2008) 459.

Z. Yang, S.M. Best, and R.E. Cameron. The influence of α-tricalcium phosphate nanoparticles and microparticles on the degradation of poly(D,L-lactide-coglycolide). *Advanced Materials* 21 (2009) 3900.

S. Yoshioka, Y. Aso, and S. Kojima. Drug release from poly(DL-lactide) microspheres controlled by γ-irradiation. *Journal of Controlled Release* 37 (1995) 263.

3 Tailoring of Bioactive Glasses

Leena Hupa

CONTENTS

3.1 INTRODUCTION

The discovery of bioactive glasses in the late 1960s opened an entirely new area of using glasses as materials that promote tissue growth in medicine (Hench et al. 1972). The development of bioactive glasses was a result of the purposeful development of a novel material that would satisfy several critical requirements of a prosthetic material. One of the most important requirements was to find a material that would not form interfacial scar tissue around the implant but instead bond to the living tissue (Hench 2006). The approach was to tailor the new prosthetic material of components that are naturally found in bones in the human body. The first bioactive glasses were tailored of large amounts of calcium oxide and phosphorus pentoxide (P_2O_5) in a sodium oxide (Na_2O)–silicon dioxide (SiO_2) matrix (Hench 2006). From the point of view of glass technology, the choice of the composition was simple but genius—the oxide compositions chosen are not chemically durable; instead, they will dissolve gradually, first by releasing ions of sodium, calcium, and phosphate from the glass surface. Sodium ions are abundant in the human body, whereas the ions of calcium and phosphate that are released from the glass precipitate as hydroxyapatite as the interfacial solution becomes supersaturated. Some glasses within the Na_2O–CaO–P_2O_5–SiO_2 system formed a strong adherent bond to bone. This bone-bonding capability distinguishes bioactive glasses from bioinert materials that form a nonadherent fibrous capsule at the interface of the host tissue. One of the glasses, 45S5 Bioglass, formed a strong bond with the bone within a few days. It was found that 45S5 also bonded to the soft tissue (Wilson et al. 1981). Later, materials showing this kind of tissue-bonding ability have been defined as bioactive materials (Williams 1999).

The bonding to the bone and soft tissue of the first glasses and glass–ceramics within the system Na_2O–CaO–SiO_2 was used to estimate the influence of oxide composition on the *in vitro* reactivity (Hench 1996). All the glasses contained a constant level of 6 wt% P_2O_5. Too high a content of SiO_2 gave no bone bonding due to the low reactivity of the glass. The glasses formed a fibrous tissue at the implant–tissue interface. Correspondingly, too low a content of CaO gave glasses that dissolved too rapidly to support hydroxyapatite precipitation. However, within the range of around 40–60 wt% SiO_2, 10–50 wt% CaO, and 0–40 wt% Na_2O, the glasses and glass–ceramics containing 6 wt% P_2O_5 bonded to the bone. Glasses within a more narrow oxide range around the 45S5 Bioglass composition (in wt%: 45 SiO_2, 24.5 Na_2O, 24.5 CaO, and 6 P_2O_5) bonded to the soft and to the bone. These early studies of bioactive glasses provided the limits still used today in tailoring compositions to various prosthetic applications. Since then, many bioactive glasses based on the composition of 45S5 have been studied *in vitro* and *in vivo*.

The high tissue-bonding ability of the bioactive glasses was achieved by compositions with a low SiO_2 content and a high CaO content. From the point of view of glass manufacturing, such compositions have a low glass forming ability, and they show a strong tendency to crystallize during forming processes at elevated temperatures. Accordingly, the first bioactive glasses have been studied as small monoliths or powdered fractions. Andersson et al. (1988, 1990) studied compositions within the oxide range (wt%) 15–30 Na_2O, 10–25 CaO, 0–3 boron trioxide (B_2O_3), 0–8 P_2O_5, and 45–65.5 SiO_2. The goal was to develop mathematical models that describe the properties of the bioactive glasses as functions of the oxide composition. These studies gave the first mathematical model of the influence of glass composition on the bone bonding *in vivo* (Andersson et al. 1990). One important result was that aluminum oxide (Al_2O_3) inhibited bone bonding if the content exceeded approximately 1.5 wt%. In the glass series studied, one alumina-free composition, S53P4, also known as BonAlive (in wt%: 53 SiO_2, 23 Na_2O, 20 CaO, and 4 P_2O_5), was successfully used in several *in vivo* studies that dealt with nonload-bearing applications (Heikkilä et al. 1993, 1996; Peltola et al. 2000). Both 45S5 Bioglass and S53P4 have received approval from the U.S. Food and Drug Administration (FDA) to be used in certain clinical applications.

The strong crystallization tendency of the bioactive glasses was suppressed when introducing potassium oxide (K_2O), magnesium oxide (MgO), and B_2O_3 into the compositions (Brink 1997; Brink et al. 1997a). From this series of experimental glasses, one composition, denoted as 13-93, has been employed in several *in vivo* studies (Brink et al. 1997b; Ylänen et al. 1999; Pirhonen et al. 2006). Glass 13-93 (in wt%: 6 Na_2O, 12 K_2O, 5 MgO, 20 CaO, and 4 P_2O_5) tolerates repeated hot working with conventional forming methods.

Today, bioactive glasses are clinically mainly used as bone grafting substitutes that induce and support bone regeneration in defects and cavities (Heikkilä et al., 2011; Lindfors et al. 2010a,b). Recent development of bioactive glasses has been focused on usage in bone tissue engineering scaffolds and implants together with bioresorbable or biostable polymers (Kellomäki et al. 2000; Ballo et al. 2007, 2008; Blaker et al. 2011; Gerhardt and Boccaccini 2010). The ions dissolving from the bioactive glass are essential not only for bone regeneration but also for bone tissue vascularization (Hoppe et al. 2011).

In principle, the role of the ions on biological processes is known. However, the dissolution kinetics of the glasses in the physiological environments must be properly known in order to tailor optimal compositions for various products. The manufacture of the glasses into desired product calls for a thorough understanding of the physical properties and their composition dependence. Accordingly, tailoring glass compositions to various medical applications requires interdisciplinary collaboration among the medical, biological, and materials sciences.

Below, the properties of melt-derived, silicate-based bioactive glasses are discussed. Currently, only a few bioactive glasses within a narrow composition range are commercially available. However, several *in vitro* and *in vivo* studies have been conducted on glasses within wider ranges of compositions, and these studies give guidelines for developing novel optimal compositions for various purposes. In tailoring novel compositions, controlled reaction kinetics that promotes the desired biological responses is the most important property to achieve. The glass composition should enable efficient and controlled forming into a desired product. Some requirements to be taken into account when tailoring bioactive glasses to bone graft materials are reviewed here. In addition, the requirements of tissue engineering, scaffold production, and continuous fibers used in polymer composites are discussed here.

3.2 INFLUENCE OF OXIDE COMPOSITION ON GLASS PROPERTIES

Glass, a noncrystalline solid, is an excellent solvent of almost every chemical element. Each element affects a certain property of the glass in a manner specific to that element. Oftentimes, the property values vary fairly smoothly with composition across a certain composition range. Within limited areas, the properties of bioactive glasses can be presented by linear functions of composition, whereas higher order terms must be considered for broader composition fields. The property values of a known glass composition can be calculated when the composition–property relationships are known. Accordingly, glasses can be tailored for any particular application merely by adjusting the oxide composition.

Two different sets of properties have to be considered when choosing the glass composition. The first set includes the properties of the glass in its final application, and the second set deals with the properties that affect the manufacture of the glass. In the final applications of bioactive glasses, the reactivity of the glass in the physiological environment and the influence of the physiological environment on tissue regeneration are of most crucial importance. The chemical reactions of the glasses not only affect tissue growth but also change the mechanical properties of the implant as the glass dissolves and new tissue grows. However, the brittle nature of glasses restricts their usage in load-bearing applications. In addition, because the practical strength of glasses is determined mainly by surface flaws, strength cannot be used as a criterion in the choice of glass composition for medical implant. Certain product forms of glass are fixed at high-temperature processes. The most important properties within the temperature range of glassworking are composition dependent.

Thus, these high-temperature properties are important parameters in the tailoring of glass compositions.

3.2.1 Important Properties in Glass Manufacture

The most important glass forming properties are viscosity and its change with temperature, the liquidus temperature that gives the temperature at which the glass melt starts to crystallize upon cooling, and the crystallization characteristics. Only when manufacturing crushed and powdered fractions of glass do exact values of the forming properties become of lesser importance. Figure 3.1 illustrates the measured viscosity–temperature points for bioactive glasses 45S5, S53P4, and 13-93 within the temperature range of glass forming (Vedel et al. 2008). The bioactivity of glass 13-93 has been verified *in vivo* (Brink et al. 1997a; Ylänen et al. 1999). Typical viscosity ranges utilized in the sintering of porous bodies from powdered glass or in the drawing of continuous fibers from melt are highlighted in Figure 3.1. The viscosity of 45S5 could be measured within a limited high temperature range and only at one low temperature; in the intermediate range, the glass proceeded to crystallize. Glass S53P4 is equally prone to crystallizing at the intermediate temperatures, whereas the low-temperature viscosity values suggest that viscous flow sintering into porous bodies may be possible for this composition. Glass 13-93 was originally developed to enable versatile hot working (Brink 1997; Brink et al. 1997a). The addition of K_2O and MgO increased the viscosities at all temperatures but also narrowed the crystallization range. Empirical models that describe the viscosity–temperature relationship as functions of the oxide compositions of bioactive glasses have been presented by Karlsson and Rönnlöf (1998) and Vedel et al. (2008). These models can be utilized in tailoring compositions within fairly wide composition ranges in the

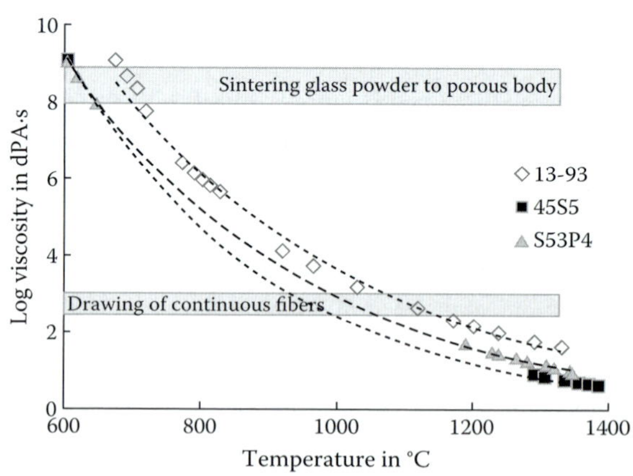

FIGURE 3.1 Measured viscosity–temperature values for bioactive glasses 45S5, S53P4, and 13-93. A line is drawn to guide the eye. (Values are adapted from Vedel, E. et al., *Glass Technology: European Journal of Glass Science and Technology—Part A*, 49, 251–9, 2008.)

system Na_2O–K_2O–MgO–CaO–B_2O_3–P_2O_5–SiO_2. The viscosity models are valid within the compositional range (wt%): Na_2O (5–25), K_2O (0–15), MgO (0–6), CaO (15–25), B_2O_3 (0–4), P_2O_5 (0–4), and SiO_2 (50–65). As shown in Figure 3.1, the prediction over the whole temperature range from 600°C to 1400°C is not meaningful because of the strong crystallization tendency. However, a suitable temperature for the sintering of porous bodies at around 10^8 dPa·s can be estimated from the oxide composition (x, in wt%) of the glass using Equation 3.1 (Vedel et al. 2008)

$$T_{\text{sintering}}\ (°\text{C}) = 322.895 - 4.346 \cdot x_{\text{Na}_2\text{O}} + 2.643 \cdot x_{\text{MgO}} + 4.720 \cdot x_{\text{CaO}} + 4.202 \cdot x_{\text{P}_2\text{O}_5} + 5.201 \cdot x_{\text{SiO}_2} \quad (3.1)$$

Figure 3.1 shows that the viscosity values of the glasses 13-93 and S53P4 could be measured within the temperature range of interest for sintering of porous bodies, whereas 45S5 crystallized at around 600°C. Glass 13-93 has been successfully sintered into porous glassy implants (Ylänen et al. 1999, 2000). Detailed knowledge of the crystallization kinetics of glasses may be utilized to manufacture glass–ceramics that have higher mechanical strength than the parent glasses. The powdered glass 45S5 has been found to be a feasible material for the sintering of porous glass–ceramic tissue engineering scaffolds via the foam replica technique (Chen et al. 2006; Bretcanu et al. 2007, 2008; Boccaccini et al. 2007). The final phase composition and the crystal size of the scaffolds are controlled by the parameters of the heat treatment. The choice of these parameters is based on the crystallization characteristics of the glasses in thermal treatments.

The temperature at which the crystallization commences depends strongly on the composition of the primary crystals forming in the thermal treatment (Arstila et al. 2008a). Most silicate glasses that show bioactivity form either sodium calcium silicate or calcium silicate ($CaSiO_3$) as the primary crystal phase in thermal treatments. $Na_2Ca_2Si_3O_9$, $Na_2CaSi_3O_8$, and $Na_2CaSi_2O_6$ have been suggested for the sodium calcium silicate crystals in glasses 45S5 and S53P4 (Rizkalla et al. 1996; Chen et al. 2006; Lefebvre et al. 2007, 2008; Arstila et al. 2008a; Bretcanu et al. 2009; Fagerlund and Hupa 2010). The discrepancy in the crystal phase composition can be explained partially by the solid solubility of the three oxides. Fokin and Zanotto (2007) reported that the crystallization of a stoichiometric glass $Na_2Ca_2Si_3O_8$ did not commence with the nucleation of the stoichiometric composition but approached that composition during crystallization. In addition, the particle size has been found to affect the crystallization mechanisms. Massera et al. (2012) reported that fine-grained 45S5 showed surface crystallization, whereas liquid-to-liquid phase separation preceded the three-dimensional (3-D) crystallization of a coarse-grained fraction of 45S5. For the S53P4 surface, crystallization was observed independent of the glass fraction. This suggests that small changes in the glass composition or the particle size may affect the final phase morphology formed in the thermal treatment.

Glass 13-93 crystallizes with $CaSiO_3$ as the primary phase. In general, the sodium calcium silicate crystals form at 50°C–200°C lower temperatures than the $CaSiO_3$ crystals (Arstila et al. 2008a). The slow crystal growth rate for the $CaSiO_3$-type glasses verifies their suitability for conventional glass forming operations (Arstila et al. 2008b; Fagerlund et al. 2011). Models for calculating glass crystallization temperature and the composition of the primary phase formed in heating glass have been suggested by Arstila et al. (2008). Equation 3.2 can be used to approximate the temperature at which crystallization starts in the thermal treatment of glasses. The validity range of the equation is the same as for Equation 3.1.

$$T_{\text{crystallization}}\ (^\circ\text{C}) = 191.70 - 9.51 \cdot x_{\text{Na}_2\text{O}} + 5.91 \cdot x_{\text{MgO}} + 5.93 \cdot x_{\text{CaO}} + 15.29 \cdot x_{\text{B}_2\text{O}_3} + 12.92 \cdot x_{\text{P}_2\text{O}_5} + 9.77 \cdot x_{\text{SiO}_2} \quad (3.2)$$

Equations 3.1 and 3.2 can be utilized to do first estimations of the suitability of the composition for sintering into a porous implant without crystallization. However, because these equations do not take into account the crystal growth rate, the thermal treatment times have to be experimentally verified. Although the activation energy value of crystallization of glasses 13-93 and 1-98 (in wt%: 53 SiO_2, 6 Na_2O, 11 K_2O, 5 MgO, 22 CaO, 1 B_2O_3, and 2 P_2O_5) are of the same order, the nucleation and crystal growths rates are one magnitude higher in 1-98 than in 13-93 (Fagerlund et al. 2012). This suggests that 13-93 is better suited for sintering processes in which long thermal treatment is desired. For example, in the manufacture of porous scaffolds using the foam replica technique, extended heat treatment is needed to burn out the organic sacrificial template.

Several different techniques have been utilized to manufacture porous scaffolds of microspheres, particles, or fibers of glass 13-93 (Ylänen et al. 2000; Fu et al. 2008; Brow et al. 2008; Meszaros et al. 2011; Doiphode et al. 2011; Liu et al. 2011). Common to all these is the densification and consolidation to the final product at around 700°C. The same temperature range was used to sinter porous implants of 1-98 (Itälä et al. 2002). Amorphous scaffolds have been manufactured using the foam replica fabrication method of two B_2O_3-containing glasses 0106 and 1806 (Mantsos et al. 2009; Erol et al. 2012). Hold times of several hours close to 700°C could be utilized to fabricate scaffolds with adequate porosity, pore size, and mechanical properties. The oxide compositions of 0106 and 1806 have been tailored using Equations 3.1 and 3.2 to allow the manufacture of porous implants via viscous flow sintering (Vedel et al. 2009).

The high-temperature viscosity range in Figure 3.1 is of interest when molding monoliths or drawing continuous fibers of glass melts. The utilization of monoliths of bioactive glasses in prosthetic applications is limited by the brittle nature of glasses. Thus, the high temperature range is of interest mainly in manufacturing of continuous fibers that can be used as reinforcement of polymers and to induce bioactivity to polymer composites. The viscosity range for fiber drawing is quite narrow—typically within 500–1000 dPa·s. When measuring the high-temperature viscosity of bioactive glasses 45S5 and S53P4, the glasses crystallized at clearly lower values than 500 dPa·s, whereas 13-93 fulfilled the viscosity requirement of fiber drawing (Figure 3.1). In addition, the liquidus temperature, that is, the temperature at which crystallization commences in the cooling of a molten glass, should be considered when tailoring suitable composition glass for fibers. The temperature span between the liquidus and the fiber drawing viscosity should be high enough to guarantee the production of fibers with no crystalline inclusions. The information available on the effect of the composition on the liquidus for glasses showing bioactivity is sparse. The liquidus temperatures have been from the endothermic peaks, indicating the melting of crystals in the thermal analysis of bioactive glasses (Arstila et al. 2008b; Bretcanu et al. 2009; Fagerlund et al. 2012). The offset of the peak suggests liquidus values around 1200°C for 45S5, S53P4, and 13-93. Remembering that all melt forming operations must start at higher temperatures than the liquidus questions the suitability of 13-93 for industrial fiber drawing. However, continuous fibers have been successfully drawn from 13-93 and 1-98 in laboratory-scale conditions (Pirhonen et al. 2006; Alm et al. 2010). The nucleation and crystal growth rates close to the liquidus of these compositions are likely slow enough to facilitate fiber drawing (Fagerlund et al. 2012).

The bioactive glass fibers aimed for the reinforcement of polymers have small diameters, typically 10–20 μm, and a very high surface-to-volume ratio. In order to avoid too rapid dissolution, the thin fibers need to be made from glasses of low bioactivity. Such slowly dissolving fibers may be feasible alternatives for reinforcement of biodegradable polymers. The bioactivity of the glass is known to decrease with increasing silica content. Simultaneously, the stability of the melt against crystallization increases. The progressive increase of boron oxide in a glass containing 65 wt% SiO_2 was found to decrease the liquidus (Massera et al. 2011). These glasses were assumed to be suitable for slowly dissolving reinforcing fibers with limited osteoconductive capability in biopolymer composites.

3.2.2 Important Glass Properties in Physiological Environment

The special feature of glass compositions showing bioactivity relies on their controlled surface reactions in the physiological environment. Traditionally, these reactions have been discussed as 11 successive but partly overlapping steps (Hench 1991). The first five steps take place at the glass surface and lead to the formation of a layered structure of silica and hydroxyapatite on the glass surface. These first reactions take place also in *in vitro* conditions and are thus often used to study and compare the bioactivity of different glass compositions. The following six steps are related with the interactions of the layered structure with biological moieties, such as blood proteins, growth factors, collagens, macrophages, and stem cells. Finally, these reaction steps lead to the proliferation and growth of bone.

The *in vivo* reactions depend primarily on the bulk and surface compositions of the glass. In addition, the surface area and roughness must be considered. Factors depending on the biological environment, such as the implantation site and the rate of the solution exchange around the implant, also affect the reaction rates. The product form has also been found to affect the reactivity—the bone formation and osteoinduction *in vivo* depended not only on the glass composition but also on the size of the granules of the bioactive glasses (Lindfors and Aho 2003). Composite fabrics made of thin fibers of 1-98 and polylactide dissolved rapidly *in vivo* without supporting osteogenesis observed for the same glass used as porous implants (Alm et al.

2010). On the contrary, long-term clinical follow-up studies of bioactive glass S53P4 as filler in bone defects showed a good remodeling of the bone but also remnants of glass particles within the tissue still after several years of implantation (Lindfors et al. 2010a,b; Heikkilä et al. 2011). These observations imply that, in each particular application, the composition of the glass should be chosen to support the desired tissue remodeling effect. In addition, the influence of the product forms on the dissolution and reaction rates should be considered when tailoring glasses for certain clinical applications. The increasing trend to use the bioactive glasses together with bioresorbable polymers may put special requirements to the glass composition. The interactions of the glasses and polymers are not properly established.

The first bioactive glasses were developed to bond to the bone without the formation of a fibrous tissue at the interface. The glass compositions were studied mainly according to their ability to support the formation of chemical bonding with bone tissue. Considering that the main applications of bioactive glasses still are found as synthetic bone graft substitutes for bone cavity filling, the bone bonding and bone tissue growth stimulating properties are essential properties to be considered also in the future. Interestingly, fine-grained bioactive glass S53P4 was found to create interfacial environments that prevented the colonization of microorganisms (Stoor et al. 1998). The dissolution of certain ions from glasses is essential not only for the osteogenesis and antibacterial activity but also for the angiogenesis of the new bone tissue that forms in the implantation of bioactive glass in the bone (Hoppe et al. 2011). Accordingly, the various biological interactions of bioactive glasses are key questions for the future development of bioactive glasses. The response of the glass depends on its chemical durability in the biological environment. Because chemical durability is strongly dependent on composition, the glasses can, in principle, be tailored for the desired biological performance. This can be done by adjusting the bulk glass composition; surface doping with specific ions that give antibacterial effects (e.g., silver) has also been studied (Balagna et al. 2011).

3.2.2.1 Tailoring Glasses for Bone Bonding

The bone bonding of glasses is facilitated by the layered silica and hydroxyapatite structure formed on glasses in physiological solutions. This surface structure forms only in glasses of limited composition, around 45–55 wt% SiO_2. One important prerequisite of hydroxyapatite precipitations is the selective alkali and alkaline earth ion dissolution leading to a silica-rich gel with acid hydroxyl groups, such as $\equiv SiOH$ and $=Si(OH)_2$ in the glass surface. The hydroxide groups provide an ideal environment for the precipitation of the hydroxyapatite crystals with a unit cell size that is compatible with the biological apatite (Karlsson 1999). The gel structure is thought to provide suitable bonding sites for collagen fibers too. Collagen, the main organic component in the bone, is bonded to apatite via hydroxyl groups in the surfaces.

The influence of glass composition on the formation of a layered structure of silica and hydroxyapatite has been studied both *in vivo* and *in vitro*. Formation of the layered structure takes place not only in the physiological environment but also in buffered solutions such as the simulated body fluid (SBF) developed by Kokubo et al. (1990). The effect of glass composition on the bioactivity has been studied along two approaches. In the first approach, the goal has been to establish the effect of changes in some component of the original 45S5 Bioglass on the bioactivity. The goal has often been to adjust the forming properties of the glass. In several studies, doping with some specific ions has been found to enhance bone tissue regeneration. For example, replacing calcium oxide with strontium oxide enhanced bone formation (Gentleman et al. 2010). In the second approach, the goal has been to produce empirical models describing the effect of composition on the bioactivity. The *in vivo* models of bioactivity as functions of the oxide composition of glass are based on relative values describing the layer formation on glass cones, the tissue attachment, and the bone bonding of the cones (Andersson et al. 1990; Brink et al. 1997a). The models can be utilized to get a first estimation of the bioactivity of a glass composition.

When tailoring novel glass compositions, *in vitro* studies are usually utilized to compare and measure the bioactivity. The compositional dependence of *in vitro* bioactivity may be described as the changes taking place in the glass surface and in the composition of the immersion solution as functions of the incubation time. Figure 3.2 shows the pH of SBF versus the thickness of silica gel formed in bioactive glasses after 72-h incubation in SBF. And the primary crystals of either sodium calcium silicate (NCS) of calcium silicate (CS) formed in thermal treatment of the same glasses. The figure is based on the works of Zhang et al. (2009) and Arstila et al (2008). In general, the glasses forming calcium silicate in the thermal treatment increased the pH of SBF less and had higher chemical durability than the glasses forming sodium calcium silicates. The compositions giving low pH values dissolved less and gave thinner silica gel in the glass surface. The slow dissolution and the good melt forming properties of the calcium silicate-type glasses make them interesting alternatives for products with a high surface area such as thin-walled porous structures thin fibers.

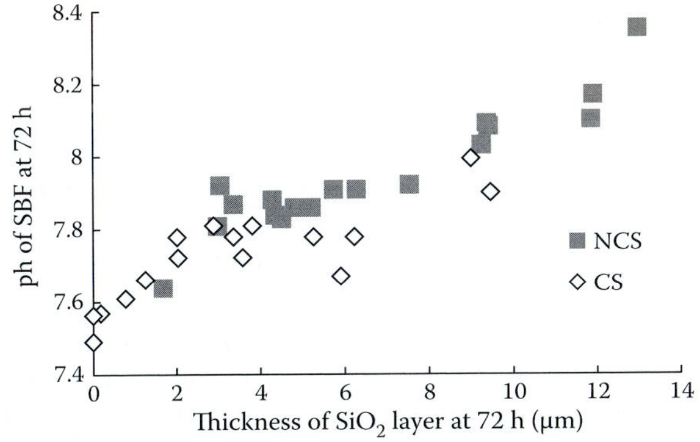

FIGURE 3.2 pH of SBF versus the thickness of silica gel formed in bioactive glasses after 72-h incubation in SBF, and the primary crystals of either sodium calcium silicate (NCS) or calcium silicate (CS) formed in thermal treatment of the same glasses. (From Zhang et al. [2009] and Arstila et al. [2008].)

Some compositions did not develop any layered structure on the glasses, but they slightly increased the pH of the solution, thus indicating a slow but gradual dissolution. The relatively wide ranges of the silica gel layer thickness enable tailoring of compositions with different degrees of bioactivity for both sodium calcium silicate and calcium silicate glass types. In general, the thicker the silica layer, the thicker the hydroxyapatite layer formed within and on the silica gel. The dissolution of the glasses and the formation of the layered structure can be estimated from the oxide composition of the glasses with the models suggested by Zhang et al. (2009).

The layered structure formed only on monoliths and large-grained fractions of the glasses (Zhang et al. 2008). This depended both on the large surface area and restricted fluid flow between the bulk solution and the interfacial solution. The dissolution of the glass may alter due to locally high pH values of the interfacial solution.

3.2.2.2 Tailoring of Glasses for Controlled Dissolution of Ions

The dissolution tendency is strongly dependent on the glass composition, especially the silica content. Traditionally, the dissolution of bioactive glasses has been related with the capability to form the silica gel and hydroxyapatite crystalline layer in the surface, providing the bonding to the bone and soft tissue. Because the calcium and phosphate ions dissolve from the glass implanted in the bone, the interfacial solution becomes supersaturated, and hydroxyapatite precipitates inside and on the silica gel. This silica gel not only acts as a nucleation site for the hydroxyapatite but also provides silicon ions essential for metabolic processes and the formation of cartilage and bone (Carlisle 1970). Recent microbiological research has shown that the ionic dissolution products stimulate the gene expression of human osteoblasts and thus enhance bone tissue growth, repair, and remodeling (Xynos et al. 2000a,b, 2001; Hench 2009). These findings have led to the research and development of 3-D substrates based on bioactive glasses that may be used as scaffolds for tissue regeneration *in vitro* and *in vivo*. Several ions have been found to affect the bone growth and vascularization of the remodeled bone tissue. Hoppe et al. (2011) have reviewed the effect of ionic dissolution products from bioactive glasses on the osteogenesis and angiogenesis of bone tissue. The ions of Ca, P, Si, Sr, Zn, Cu, Mg, and B are considered of importance for bone remodeling. Ions dissolving from bioactive glasses may also exhibit antibacterial effects against clinically important bacterial species (Stoor et al. 1998; Munukka et al. 2008; Leppäranta et al. 2008). The antibacterial effects are not properly understood, but the ion dissolution of the glasses in the physiological environment is assumed decisive.

The critical concentrations of the ions dissolving from the bioactive glasses on the gene stimulating effects may be analyzed using different static or dynamic cell culture methods. The results can be used to tailor glass compositions for controlled ion release kinetics. Several factors affect the ion release from the glass, e.g., oxide composition, surface condition, surface area-to-volume ratio, the pH of the interfacial solution, and the flow of solution. The ion release kinetics in physiological solutions is a key factor when choosing the glass composition to any particular application. Most ion dissolution studies have been conducted in static conditions. Dissolution kinetics has also been measured by using continuous measurement arrangements (Fagerlund et al. 2010a,b). The initial dissolution of Si and Ca ions from bioactive

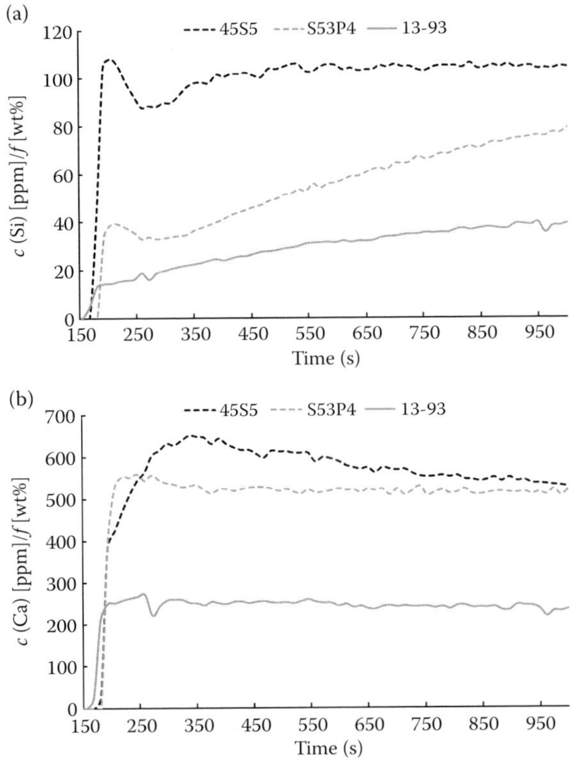

FIGURE 3.3 Initial dissolution of (a) silicon and (b) calcium from bioactive glasses 45S5, S53P4, and 13-93 in a 0.2-ml/min Tris buffer solution at 40°C. The concentrations given are normalized to the element concentration per mass fraction of oxide in glass. (Redrawn from Fagerlund, S. et al., *Proceedings of the International Congress on Glass (ICG 2010)*, Bahia, Brasil. www.advidrio.org.br/icg2010/bookOfAbstracts.)

glasses 45S5, S53P4, and 13-93 in Tris buffer solution is given in Figure 3.3. Fresh solution at 40°C was fed through the samples at a flow rate of 0.2 ml/min. Clear differences in the dissolution profiles of the two ions from the glasses were measured *in situ*. The method can be used not only to compare the dissolution profiles of different glasses qualitatively and quantitatively but also to determine activation energies for glass dissolution in different *in vitro* solutions (Fagerlund et al. 2010a). The method gives simultaneously the dissolution profile of all ions released from the glasses. When combined with information of critical concentrations needed to stimulate biological processes, the dissolution profile measurement provides a powerful tool for developing and testing novel bioactive and biocompatible glasses *in vitro*.

3.3 SUMMARY

When tailoring new bioactive glasses to meet certain clinical requirements, the properties important to the biological performance of the glass have to be established. In general, bioactive glasses have the unique ability to enhance and support repair, growth, and remodeling of the bone tissue. These biological processes are stimulated and activated partially by the ions dissolving from the bioactive glasses. Thus, controlled ion dissolution is the main property to be considered when developing composition for novel glasses. Ideally, the bioactive glass product has a highly porous 3-D structure and a surface morphology that support bone tissue ingrowth and ongrowth, and vascularization. However, the high-temperature forming processes giving bioactive glasses in the desired product forms are restricted by the strong tendency of the glass to crystallize during the processing. Accordingly, the working properties also have to be considered when tailoring novel glasses.

Oxide composition controls the dissolution reactions, osseointegration, and high-temperature processing properties of bioactive glasses. During the past years, several models describing the relationship between the glass properties and the oxide compositions have been established. These models are valuable when tailoring novel bioactive glasses into products that may have desired biological performance. The biological performance depends also on the surface area and fluid flow around the implanted bioactive glass. Detailed dissolution profiles of the various ions released from the glasses in the biological fluids will help generate fundamental understanding of the ion dissolution kinetics. The main clinical uses of bioactive glasses in bone tissue engineering suggest that strength should also be considered. However, the brittle nature of glasses limits the use of bioactive glasses in load-bearing applications. Using the glass together with bioresorbable polymers in composites gives a possibility of utilizing the bone tissue stimulating properties of the glass also in some moderately loaded applications. The interaction of the dissolution reactions of the glass and degradation reactions of the biopolymers is not fully understood. This means that more work needs to be done to optimize the glass compositions that are used together with biopolymers in composites.

REFERENCES

Alm, J.J., Frantzén, J.P.A., Moritz, N., Lankinen, P., Tukiainen, M., Kellomäki, M., and Aro, H.T. (2010). *In vivo* testing of a biodegradable woven fabric made of bioactive glass fibers and PLGA80—A pilot study in the rabbit. *Journal of Biomedical Materials Research—Part B: Applied Biomaterials* 93B: 573–80.

Andersson, Ö.H., Karlsson, K.H., Kangasniemi, K., and Yli-Urpo, A. (1988). Models for physical properties and bioactivity of phosphate opal glasses. *Glastechnische Berichte* 61: 300–5.

Andersson, Ö.H., Liu, G., Karlsson, K.H., Niemi, L., Miettinen, J., and Juhanoja, J. (1990). *In vivo* behavior of glasses in the SiO_2–Na_2O–CaO–P_2O_5–Al_2O_3–B_2O_3 system. *Journal of Materials Science: Materials in Medicine* 1: 219–27.

Arstila, H., Tukiainen, M., Taipale, S., Kellomäki, M., and Hupa, L. (2008b). Liquidus temperatures of bioactive glasses. *Advanced Materials Research (Durnten-Zurich, Switzerland)* 39–40 (Glass): 287–92.

Arstila, H., Vedel, E., Hupa, L, and Hupa, M. (2008a). Predicting physical and chemical properties of bioactive glasses from chemical composition—Part 2: Devitrification characteristics. *Glass Technology: European Journal of Glass Science and Technology—Part A* 49: 260–5.

Balagna, C., Vitale-Brovarone, C., Miola, M., Verne, E., Canuto, R.A., Saracino, S., Muzio, G., Fucale, G., and Maina, G. (2011). Biocompatibility and antibacterial effect of silver-doped 3-D glass–ceramic scaffolds for bone grafting. *Journal of Biomaterials Applications* 25: 595–617.

Ballo, A.M., Kokkari, A.K., Meretoja, V.V., Lassila, L.L., Vallittu, P.K., and Närhi, T.O. (2008). Osteoblast proliferation and maturation on bioactive fiber-reinforced composite surface. *Journal of Materials Science: Materials in Medicine* 19: 3169–77.

Ballo, A.M., Lassila, L.V., Vallittu, P.K., and Närhi, T.O. (2007). Load-bearing capacity of bone-anchored fiber-reinforced composite device. *Journal of Materials Science: Materials in Medicine* 18: 2025–31.

Blaker, J.J., Nazhat, S.N., Maquet, V., and Boccaccini, A.R. (2011). Long-term *in vitro* degradation of PDLLA/Bioglass bone scaffolds in acellular simulated body fluid. *Acta Biomaterialia* 7: 829–40.

Boccaccini, A.R., Chen, Q., Lefebvre, L., Gremillard, L., and Chevalier, J. (2007). Sintering, crystallization and biodegradation behavior of Bioglass-derived glass–ceramics. *Faraday Discussions* 136: 27–44.

Bretcanu, O., Chatzistavrou, X., Paraskevopoluos, K., Condradt, R., Thompson, I., and Boccaccini, A.R. (2009). Sintering and crystallization of 45S5 Bioglass® powder. *Journal of the European Ceramic Society* 29: 3299–306.

Bretcanu, O., Chen, Q., Misra, S.K., Boccaccini, A.R., Roy, I., Verne, E., and Brovarone, C.V. (2007). Biodegradable polymer coated 45S5 Bioglass-derived glass–ceramic scaffolds for bone tissue engineering. *Glass Technology: European Journal of Glass Science and Technology—Part A* 48: 227–34.

Bretcanu, O., Samaille, C., and Boccaccini, A.R. (2008). Simple methods to fabricate Bioglass-derived glass–ceramic scaffolds exhibiting porosity gradient. *Journal of Materials Science* 43: 4127–34.

Brink, M. (1997). The influence of alkali and alkaline earths on the working range for bioactive glasses. *Journal of Biomedical Materials Research* 36: 109–17.

Brink, M., Laine, P., Narva, K., and Yli-Urpo, A. (1997b). Implantation of bioactive and inert glass fibers in rats—Soft tissue response and short-term reactions of the glass. *Bioceramics* 10: 61–4.

Brink, M., Turunen, T., Happonen, R.P., and Yli-Urpo, A. (1997a). Compositional dependence of bioactivity of glasses in the system Na_2O–K_2O–MgO–CaO–B_2O_3–P_2O_5–SiO_2. *Journal of Biomedical Materials Research* 37: 114–21.

Brow, R.F., Day, D.E., Day, T.E., Jung, S., Rahaman, M.N., and Fu, Q. (2008). Growth and differentiation of osteoblastic cells on 13-93 bioactive glass fibers and scaffolds. *Acta Biomaterialia* 4: 387–96.

Carlisle, E.M. (1970). Silicon: A possible factor in bone calcification. *Science (Washington, DC, United States)* 167: 279–80.

Chen, Q.Z., Thompson, I.D., and Boccaccini, A.R. (2006). 45S5 Bioglass-derived glass–ceramic scaffolds for bone tissue engineering. *Biomaterials* 27: 2414–25.

Doiphode, N.D., Huang, T., Leu, M.C., Rahaman, M.N., and Day, D.E. (2011). Freeze extrusion fabrication of 13-93 bioactive glass scaffolds for bone repair. *Journal of Materials Science: Materials in Medicine* 22: 515–23.

Erol, M.M., Mouriño, V., Newby, P., Chatzistavrou, X., Hupa, L., and Boccaccini, A.R. (2012). Copper releasing boron containing bioactive glass-based scaffolds coated with alginate for bone tissue engineering. *Acta Biomaterialia* 8: 792–801.

Fagerlund, S., Ek, P., Hupa, M., and Hupa, L. (2010a). On determining chemical durability of glasses. *Glass Technology: European Journal of Glass Science and Technology—Part A* 51: 235–40.

Fagerlund, S., Ek, P., Hupa, L., and Hupa, M. (2010b). Dissolution kinetics of bioactive glasses by continuous measurement. *Proceedings of the International Congress on Glass (ICG 2010)*, Bahia, Brasil. http://www.abividro.org.br/icg2010/bookOfAbstracts/.

Fagerlund, S. and Hupa, L. (2010). Crystallization of 45S5 during isothermal heat treatment. *Ceramic Materials* 62: 34–54.

Fagerlund, S., Massera, J., Hupa, L., and Hupa, M. (2012). T–T–T behavior of bioactive glasses 1-98 and 13-93. *Journal of the European Ceramic Society*, 32: 2731–8.

Fokin, V.M. and Zanotto, E.D. (2007). Continuous compositional changes of crystal and liquid during crystallization of a sodium calcium silicate glass. *Journal of Non-Crystalline Solids* 353: 2459–68.

Fu, Q., Rahaman, M.N., Bal, B.S., Brow, R.F., and Day, D.E. (2008). Mechanical and *in vitro* performance of 13-93 bioactive glass scaffolds prepared by a polymer foam replication technique. *Acta Biomaterialia* 4: 1854–64.

Gentleman, E., Fredholm, Y.C., Jell, G., Lotfibakhshaiesh, N., O'Donell, M.D., Hill, R.G., and Stevens, M.M. (2010). The effects of strontium-substituted bioactive glasses on osteoblasts and osteoclasts *in vitro*. *Biomaterials* 31: 3949–56.

Gerhardt, L.C. and Boccaccini, A.R. (2010). Bioactive glass and glass–ceramic scaffolds for bone tissue engineering. *Materials* 3: 3867–910.

Heikkilä, J.T., Aho, A.J., Yli-Urpo, A., Andersson, Ö.H., Aho, H.J., and Happonen, R.P. (1993). Bioactive glass versus hydroxyapatite in reconstruction of osteochondral defects in the rabbit. *Acta Orthopaedic Scandinavica* 64: 678–82.

Heikkilä, J.T., Kukkonen, J., Aho, A.J., Moisander, S., Kyyrönen, T., and Mattila, K. (2011). Bioactive glass granules: A suitable bone substitute material in the operative treatment of depressed lateral tibial plateau fractures: A prospective, randomized 1-year follow-up study. *Journal of Materials Science: Materials in Medicine* 22: 1073–80.

Heikkilä, J.T., Salonen, H.R., Yli-Urpo, A., and Aho, A.J. (1996). Long-term behavior of bioactive glass cone and granules in rabbit bone. *Bioceramics* 9: 123–6.

Hench, L.L. (1991). Bioceramics: From concept to clinic. *Journal of the American Ceramic Society* 74: 1487–1510.

Hench, L.L. (1996). Ceramics, glasses and glass–ceramics. In *Biomaterials Science: An Introduction to Materials in Medicine*, Eds Ratner, B.D., Hoffman, A.S., Shoen, F.J., and Lemons, J.E. San Diego, CA: Academic Press, 73–84.

Hench, L.L. (2006). The story of Bioglass. *Journal of Materials Science: Materials in Medicine* 17: 967–78.

Hench, L.L. (2009). Genetic design of bioactive glass. *Journal of the European Ceramic Society* 29: 1257–65.

Hench, L.L., Splinter, R.J., Allen, W.C., and Greenlee Jr., T.K. (1972). Bonding mechanisms at the interface of ceramic prosthetic materials. *Journal of Biomedical Materials Research Symposium (Interscience, New York)* 2: 117–41.

Hoppe, A., Güldal, N.S., and Boccaccini, A.R. (2011). A review of the biological response to the ionic dissolution products from bioactive glasses and glass–ceramics. *Biomaterials* 32: 2757–74.

Itälä, A., Ylänen, H.O., Yrjans, J., Heino, T., Hentunen, T., Hupa, M., and Aro, H.T. (2002). Characterization of microrough bioactive glass surface: Surface reactions and osteoblast responses *in vitro*. *Journal of Biomedical Materials Research* 62: 404–11.

Karlsson, K.H. (1999). Bone implants—A challenge to materials science. *Annales Chirurgiae et Gynaecologiae* 88: 226–35.

Karlsson, K.H. and Rönnlöf, M. (1998). Property–composition relationships for potentially bioactive glasses. *Glass Science and Technology (Frankfurt/Main)* 71: 141–5.

Kellomäki, M., Niiranen, H., Puumanen, K., Ashammakhi, N., Waris, T., and Törmälä, P. (2000). Bioabsorbable scaffolds for guided bone regeneration and generation. *Biomaterials* 21: 2495–505.

Kokubo, T., Kushitani, H., Sakka, S., Kitsugi, T., and Yamamuro, T. (1990). Solutions able to reproduce *in vivo* surface-structure changes in bioactive glass–ceramic A–W. *Journal of Biomedical Materials Research* 24: 2161–75.

Lefebvre, L., Chevalier, J., Gremillard, L., Zenati, R., Thollet, G., Bernache-Assolant, D., and Govin, A. (2007). Structural transformations of bioactive glass 45S5 with thermal treatments. *Acta Materialia* 55: 3305–13.

Lefebvre, L., Gremillard, L., Chevalier, J., Zenati, R., and Bernache-Assolant, D. (2008). Sintering behaviour of 45S5 bioactive glass. *Acta Biomaterialia* 4: 1894–903.

Leppäranta, O., Vaahtio, M., Peltola, T., Zhang, D., Hupa, L., Hupa, M., Ylänen, H., Salonen, J.I., Viljanen, M.K., and Eerola, E. (2008). Antibacterial effect of bioactive glasses on clinically important anaerobic bacteria *in vitro*. *Journal of Materials Science: Materials in Medicine* 19: 547–51.

Lindfors, N.C. and Aho, A.J. (2003). Granule size and composition of bioactive glasses affect osteoconduction in rabbit. *Journal of Materials Science: Materials in Medicine* 14: 365–72.

Lindfors, N.C., Hyvönen, P., Nyyssönen, M., Kirjavainen, M., Kankare, J., Gullichsen, E., and Salo, J. (2010a). Bioactive glass S53P4 as bone graft substitute in treatment of osteomyelitis. *Bone (Amsterdam, Netherlands)* 47: 212–8.

Lindfors, N.C., Koski, I., Heikkilä, J.T., Mattila, K., and Aho, A.J. (2010b). A prospective randomized 14-year follow-up study of bioactive glass and autogenous bone as bone graft substitutes in benign bone tumors. *Journal of Biomedical Materials Research—Part B: Applied Biomaterials* 94B: 157–64.

Liu, X., Rahaman, M.N., and Fu, Q. (2011). Oriented bioactive glass (13-93) scaffolds with controllable pore size by unidirectional freezing of camphene-based suspensions: Microstructure and mechanical response. *Acta Biomaterialia* 7: 406–16.

Mantsos, T., Chatzistavrou, X., Roether, J.A., Hupa, L., Arstila, H., and Boccaccini, A.R. (2009). Non-crystalline composite tissue engineering scaffolds using boron-containing bioactive glass and poly(D,L-lactic acid) coatings. *Biomedical Materials (Bristol, United Kingdom)* 4: 055002/1–12.

Massera, J., Claireaux, C., Lehtonen, T., Tuominen, J., Hupa, L., and Hupa, M. (2011). Control of the thermal properties of slow bioresorbable glasses by boron addition. *Journal of Non-Crystalline Solids* 357: 3623–30.

Massera, J., Fagerlund, S., Hupa, L., and Hupa, M. (2012). Crystallization mechanism of the bioactive glasses 45S5 and S53P4. *Journal of the American Ceramic Society* 95: 607–13.

Meszaros, R., Zhao, R., Travitzky, N., Fey, T., Greil, P., and Wondraczek, L. (2011). Three-dimensional printing of a bioactive glass. *Glass Technology: European Journal of Glass Science and Technology—Part A* 52: 111–6.

Munukka, E., Leppäranta, O., Korkeamäki, M., Vaahtio, M., Peltola,T., Zhang, D., Hupa, L., Ylänen, H., Salonen, J., Viljanen, M.K., and Eerola, E. (2008). Bactericidal effects of bioactive glasses on clinically important aerobic bacteria. *Journal of Materials Science: Materials in Medicine* 19: 27–32.

Peltola, M.J., Suonpää, J.T.K., Andersson, Ö.H., Määttänen, H.S., Aitasalo, K.M.J., Yli-Urpo, A., and Laippala, P.J. (2000). *In vitro* model for frontal sinus obliteration with bioactive glass S53P4. *Journal of Biomedical Materials Research* 53: 161–6.

Pirhonen, E., Niiranen, H., Niemelä, T., Brink, M., and Törmälä, P. (2006). Manufacturing, mechanical characterization, and *in vitro* performance of bioactive glass 13-93 fibers. *Journal of Biomedical Materials Research—Part B: Applied Biomaterials* 77B: 227–33.

Rizkalla, A.S., Jones, D.W., Clark, D.B., and Hall, G.C. (1996). Crystallization of experimental bioactive glass compositions. *Journal of Biomedical Materials Research* 32: 119–24.

Stoor, P., Söderling, E., and Salonen, J. (1998). Antibacterial effects of a bioactive glass paste on oral microorganisms. *Acta Odontologica Scandinavica* 48: 161–5.

Vedel, E., Arstila, H., Ylänen, H., Hupa, L., and Hupa, M. (2008). Predicting physical and chemical properties of bioactive glasses from chemical composition—Part 1. Viscosity characteristics. *Glass Technology: European Journal of Glass Science and Technology—Part A* 49: 251–9.

Vedel, E., Zhang, D., Arstila, H., Hupa, L., and Hupa, M. (2009). Predicting physical and chemical properties of bioactive glasses from chemical composition—Part IV: Tailoring compositions with desired properties. *Glass Technology: European Journal of Glass Science and Technology—Part A* 50: 9–16.

Williams, D.F. (1999). *The Williams Dictionary of Biomaterials*. Liverpool, U.K.: Liverpool University Press.

Wilson, J., Pigott, G.H., Shoen, F.J., and Hench, L.L. (1981). Toxicology and biocompatibility of bioglasses. *Journal of Biomedical Materials Research* 15: 805–17.

Xynos, I.D., Edgar, A.J., Buttery, L.D.K., Hench, L.L., and Polak, J.M. (2000b). *Biochemical and Biophysical Research Communications* 276: 461–5.

Xynos, I.D., Edgar, A.J., Buttery, L.D.K., Hench, L.L., and Polak, J.M. (2001). Gene-expression profiling of human osteoblasts following treatment with the ionic products of Bioglass 45S5 dissolution. *Journal of Biomedical Materials Research* 55: 151–7.

Xynos, I.D., Hukkanen, M.V.J., Batten, J.J., Buttery, L.D., Hench, L.L., and Polak, J.M. (2000a). Bioglass 45S5 stimulates osteoblast turnover and enhances bone formation *in vitro*: Implications and applications for bone tissue engineering. *Calcified Tissue International* 67: 321–9.

Yllänen, H.O., Helminen, T., Helminen, A., Rantakokko, J., Karlsson, K.H., and Aro, H.T. (1999). Porous bioactive glass matrix in reconstruction of articular osteochondral defects. *Annales Chirurgiae et Gynaecologiae* 88: 237–45.

Yllänen, H., Karlsson, K.H., Itälä, A., and Aro, H.T. (2000). Effect of immersion in SBF on porous bioactive bodies made by sintering bioactive glass microspheres. *Journal of Non-Crystalline Solids* 275: 107–15.

Zhang, D., Hupa, M., and Hupa, L. (2008). *In situ* pH within particle beds of bioactive glasses. *Acta Biomaterialia* 4: 1498–1505.

Zhang, D., Vedel, E., Hupa, L., and Aro, H.T. (2009). Predicting physical and chemical properties of bioactive glasses from chemical composition—Part III: *In vitro* reactivity of glasses. *Glass Technology: European Journal of Glass Science and Technology—Part A* 50: 1–8.

4 Bioactive Ceramic Coatings for Metallic Implants

Niko Moritz, Paula Linderbäck, and Timo O. Närhi

CONTENTS

4.1 INTRODUCTION

Stainless steel, Co–Cr alloys, titanium, and titanium alloys are common implant materials in reconstructive medicine and trauma surgery. The success of these biomaterials is attributed to their reliable mechanical performance, which is essential in load-bearing applications. In general, metallic biomaterials are biologically almost inert and, therefore, have good biocompatibility. They can easily be manufactured in complex shapes by standard machining procedures and sterilized by conventional methods.

However, metal implants have intrinsic drawbacks. The most notable is the mismatch of the mechanical properties of load-bearing metallic implants and bone, which leads to the physiological underloading of the bone, known as "stress shielding," and increases the possibility of an adverse remodeling of the periprosthetic bone. Another well-known problem is related to debris formation and corrosion (Archibeck et al. 2000; Hallab et al. 2004), which originate in articulating surfaces, modular component interfaces, fixation surfaces, and devices. Debris formation results in periprosthetic bone loss and increases the likelihood of the aseptic loosening of orthopedic implants. Due to bone loss, the revision surgery is significantly more complex and has an increased risk of complications. In addition, metallic implants produce considerable artifacts in modern diagnostic imaging systems such as computed tomography and magnetic resonance imaging. However, although alternative materials (e.g., composites) are being developed (Zhao et al. 2009), it is unlikely that these materials will soon replace metals that have a history of reliable clinical performance.

Stainless steel 316 is a common clinically used implant material. It contains a considerable amount of nickel. Nickel is also present in some Co–Cr alloys. Therefore, concerns exist over the possibility of nickel allergy due to the use of these biomaterials. Consequently, efforts have been made to develop nickel-free materials such as Co–Cr–Mo alloys and nickel-free stainless steels (Niinomi 2008).

Titanium and titanium alloys have been the focus of research and development efforts. These materials are widely used in orthopedic and dental applications. Although inferior to Co–Cr alloys, titanium alloys have good load-bearing capacity, which justifies their use in clinical applications (e.g., implants for cementless total hip arthroplasty). In dentistry, titanium is commonly used in the form of commercially pure titanium grade 4, which contains less than 1% impurities, and an $\alpha + \beta$-type Ti–6Al–4V alloy, which contains 6% aluminum and 4% vanadium (Dohan Ehrenfest et al. 2010). However, because the strength of commercially pure titanium is not sufficient for most load-bearing applications, Ti–6Al–4V alloy is the material of choice for orthopedic applications. Concerns exist over the toxicity of aluminum and vanadium; therefore, extensive research has been done to develop titanium alloys free of aluminum and vanadium, e.g., Ti–6Al–7Nb and Ti–15Zr–4Nb–2Ta–0.2Pd alloys (Niinomi 2008). Another topic of interest has been the development of titanium alloys for better biomechanical compatibility to reduce the effect of "stress shielding." To achieve this goal, new β-type alloys were introduced. These alloys have low Young's modulus, high strength, and good corrosion resistance. For example, Ti–12Mo–6Zr–2Fe (TMZF) alloy has a Young's modulus of 74–85 GPa (Wang 1996), and Ti–29Nb–13Ta–4.6Zr (TNTZ) alloy has a Young's modulus of 60 GPa (Niinomi 2008) compared to that of 110 GPa for Ti–6Al–4V alloy. Because the Young's modulus of the human cortical bone is in the range of 7–25 GPa, it is expected that implants made of β-type titanium alloys would provide a more physiologically favorable load sharing than the commonly used $\alpha + \beta$-type titanium alloys.

In addition to load-bearing capacity, titanium and titanium alloys retain excellent biocompatibility. Titanium is a highly reactive metal and is covered by a thin (2–10 nm) surface oxide layer (except for certain artificial conditions). It is the chemical stability of this oxide layer, mainly titanium dioxide (TiO_2), that provides the superior corrosion resistance of implants made of titanium and its alloys. It has also been shown that titanium dioxide has bioactive properties—it can form chemical bonds with the bone (Li and de Groot 1993) and, possibly, soft tissues (Areva et al. 2004; Rossi et al. 2007). Therefore, surface modification of titanium has the potential for improving the integration of implants in host tissues.

The nickel–titanium (NiTi, Nitinol) shape memory alloys are promising materials for surgical implants in reconstructive medicine because of their unique shape memory effect and superelasticity (Morgan 2004). Within a given temperature range, shape memory alloys can be strained several times more than conventional metal alloys without being plastically deformed.

NiTi has been proven safe and biocompatible in many *in vitro* and *in vivo* studies (Ryhänen et al. 1999; Rhalmi et al. 1999; Kapanen et al. 2001; Kujala et al. 2004). The good biocompatibility is due to the oxide surface film of NiTi, which is mainly titanium oxide, and prevents the nickel from corroding and leaching in the biological environment. However, some adverse effects such as the inferior osteogenesis process, lower osteonectin synthesis activity, and higher cell death rate have also been reported (Jia et al 1999; Shih et al. 2000; Es-Souni et al. 2002). These problems are attributed to an increased cytotoxicity due to poor corrosion resistance. Thus, in some applications, the titania surface layer on NiTi materials is not thick enough, or the adhesion of the TiO_2 layer to the implant is not sufficient to protect the NiTi implant from corrosion. In addition, one important issue is that the nickel ions released from the alloys can cause allergic reactions in nickel-hypersensitive patients.

4.2 INCORPORATION OF IMPLANTS WITH THE BONE

When a biomaterial is implanted into a living body, it always elicits certain responses from the tissues (Williams 1987, 1992), because no material is biologically inert. Implants made of nontoxic materials are either incorporated into the host tissue or substituted by the host tissue.

Biologically inactive materials (e.g., metals) cause fibrous tissue capsule formation around the implant. This is sometimes referred to as "morphological fixation" (Cao and Hench 1996). The thickness of the capsule depends on the chemical reactivity of the implant material. For more stable materials such as Al_2O_3, the capsule is thin, whereas for more reactive materials, the capsule comprises thicker interfacial layers of fibrous tissue. Porous implants with porosity in the 100–500 μm range or implants with significant surface roughness are incorporated through tissue ingrowths called "biological fixation" (Cao and Hench 1996). Until the introduction of the concept of bioactive materials, surface reactivity was considered an unwanted property for an implant material. However, some surface-reactive (biologically active and bioactive) materials have shown the ability to form an interfacial chemical bond with surrounding tissues through a cascade of chemical and biochemical reactions, causing "bioactive fixation" of the implant (Cao and Hench 1996). Bioactive materials can be biostable (i.e., synthetic hydroxyapatite) or bioresorbable (i.e., bioactive glasses and glass–ceramics).

Two classes of bioactive materials were suggested by Cao and Hench (1996). "Class A" biomaterial elicits extracellular and intracellular responses that lead to osteoproduction or enhancement in proliferation of bone cells. "Class B" biomaterials elicit only extracellular responses, thus providing only suitable surfaces for the attachment of cells or osteoconduction. Bioactive glasses belong to class A, whereas synthetic hydroxyapatites (HAs) are class B materials. The rapid bone production induced by class A bioactive materials could be attributed to the capability of these materials to create a local chemical environment that influences the cell cycle, leading to the terminal differentiation of osteoprogenitor cells into bone cells. It has been suggested that bioactive glasses release Si and Ca ions at the rate needed to maintain differentiation of the cells (Hench 1994).

4.2.1 Surface Modifications for the Biological Fixation of Metallic Implants

Different methods exist for improving the incorporation of metallic implants with the bone, including mechanical, chemical, biochemical, and combined techniques. Mechanical surface modifications, common for orthopedic and dental implants, are aim at increasing the roughness of the surface by grit or abrasive blasting. It is generally accepted that surface roughness below 100 μm affects cell adhesion, morphology, and orientation, whereas surface roughness above 100 μm is important for the biological fixation of the implant (Pilliar 1998). In total hip arthroplasty, the stability of the fixation of the femoral and the acetabular components is often amended by the modification of the implant surface. A common strategy is to roughen the surface or to introduce surface pores; this way, the implants are incorporated by biological rather than morphological fixation (Figure 4.1).

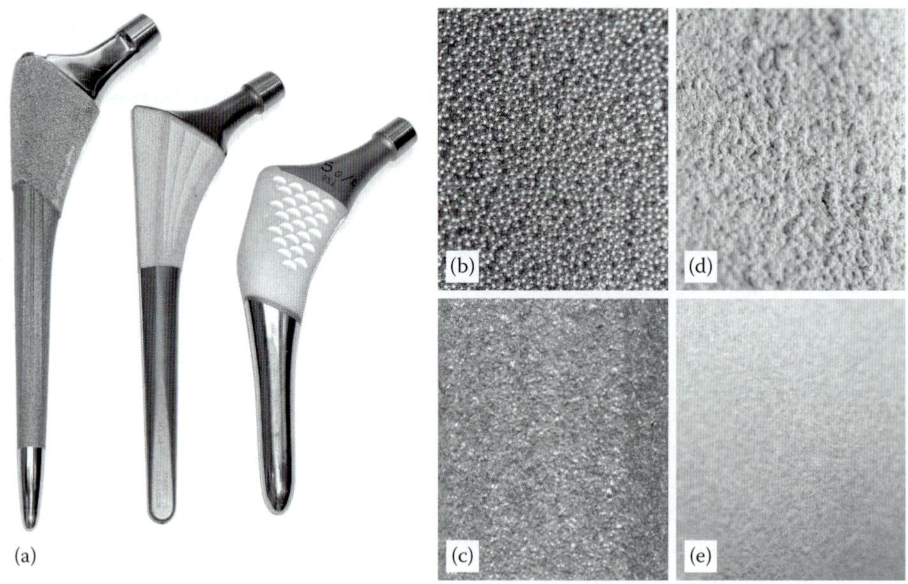

FIGURE 4.1 Different designs and surface structures of femoral stems for cementless total hip arthroplasty. (a) Left: Synergy-tapered hip system, forged Ti alloy (Smith & Nephew, London, U.K.). Center: Symax, forged Ti–6A1–4V alloy, electrolytic conversion of the surface of the distal part of a stem (Stryker). Right: ABG II (Stryke). (b) Sintered bead porous coating (Smith & Nephew). (c) Roughened surface (Smith & Nephew). (d) Titanium plasma-sprayed surface with a biomimetic HA coating (Stryker). (e) Plasma-sprayed 50-μm-thick HA coating (Stryker).

The majority of currently marketed oral implants permit bone ingrowth into minor surface irregularities, which ensures biomechanical bonding to bone. These moderately rough surfaces (Ra between 1.0 and 2.0 μm) have shown stronger bone responses than smoother or rougher surfaces (Albrektsson and Wennerberg 2004).

In the early 1970s, scientists introduced a concept of porous metal biomaterials that allowed bone ingrowth into porous interfaces. The porosity was reported to influence cell adhesion and morphology, proliferation, and differentiation (Levine 2008; Bowers et al. 1992). The porous orthopedic devices are most often surface blasted and acid etched, and with sintered Co–Cr beads or diffusion-bonded fiber metal meshes. Titanium is cancellous-like structured or plasma sprayed onto various solids, for example, onto Co–Cr and stainless steel. Buser et al. (1991) studied cylindrical titanium implants with different surface modifications (electron-polished, sand-blasted, acid-pickled HF/HNO_3 with medium or high grit, titanium plasma sprayed, and HA coated). They observed that the bone–implant interface integration in miniature pig femur metaphyses and tibia was significantly improved with increasing surface roughness (Buser et al. 1991). More recently, the desire for increasingly porous surfaces for enhanced bone ingrowth paired with low-surface-friction characteristics and availability for a coating has led to the development of novel highly porous metal interfaces. These have porosity in the range of 100–600 μm and an overall porosity of 60%–70% (Lombardi et al. 2007; Frenkel et al. 2004). In one study with implantable chambers in dogs, high-porosity titanium interfaces were compared to titanium and cobalt chrome beads and then revealed enhanced bone ingrowth, which may contribute to an enhanced long-term implant survival (Frenkel et al. 2004). An open-cell tantalum construct has recently been presented in orthopedic surgery. The mechanical properties show a low modulus of elasticity (~3 GPa) while allowing the support of physiological load and bone ingrowth (Zardiackas et al. 2001). Porous tantalum was thus indicated as bioactive *in vitro* and *in vivo* and showed enhanced bone ingrowth in a canine study, indicating higher shear fixation strength than reference implants with lower volumetric porosity.

4.2.2 Calcium Phosphate Bioactive Coatings

Another strategy for enhancing the integration of implants in the bone is to apply a coating of a bioactive material. This achieves bioactive fixation usually in combination with biological fixation. Calcium phosphates (CaPs), e.g., synthetic HAs, have a composition similar to the bone mineral. CaPs are the most commonly used coating materials. There exist a number of different methods of depositing CaP on metal implants, including plasma spraying, sputter deposition, sol–gel coating, electrophoretic deposition, and biomimetic precipitation (Ellingsen et al. 2006; Junker et al. 2009).

Plasma spraying is the method of choice for coating orthopedic implants (Campbell 2003). There are, however, certain concerns over the performance and quality of these coatings and the need to apply the coatings. Plasma-sprayed HA coatings may exhibit low mechanical strength, chemical inhomogeneity, and high porosity (which can also be an advantage), as well as a partial loss of crystallinity (Overgaard et al. 1996; Cheang and Khor 1996). These problems originate in the high production temperature and excessive thickness of the coating. In the process of plasma spraying, the HA particles are quenched with a high cooling rate. This leads to the dehydroxylation of the HA and the reduction of the crystallinity of the coating. Although small amounts of the amorphous phase could promote the integration of the implant in bone, the high content of the amorphous phase can cause excessive dissolution. In animal experiments, the HA coatings were shown to enhance the integration of implants with the bone (Søballe 1993a). Furthermore, studies that involve radiostereometric analysis demonstrated a decrease in the migration of HA-coated implants (Søballe 1993b; Kärrholm et al. 1994). Later studies demonstrated good clinical results for HA-coated implants (Theis and Ball 2003; Donnelly et al. 1997; Singh et al. 2004).

Nevertheless, little data exist on the medium- and long-term performance of these implants. There are concerns over the delamination of the coatings, which may create third body polyethylene wear, resulting in peri-implant bone loss and loosening of the implant (Bloebaum et al. 1994). A study of the Danish Arthroplasty Registry indicated that, in the medium term, HA-coated implants were not associated with a reduced risk of revision compared to noncoated implants (Paulsen et al. 2007). Another study implied that, for HA-coated cups in primary total hip arthroplasty, HA coating was a risk factor for cup revision (Lazarinis et al. 2010). A recent meta-analysis of clinical studies of the HA-coated femoral stems in primary total hip arthroplasty demonstrated no clear benefit of HA/porous coating over porous coating (Gandhi et al. 2009). It should be noted that a bioactive coating can only contribute to the implant fixation during the first few months; however, it cannot improve a poor implant design (Paulsen et al. 2007).

Dental implants in clinical practice are coated by plasma spraying, sputter deposition, or, recently, the sol–gel method (Junker et al. 2009). Similar to reconstructive orthopedic surgery, in dentistry, there are concerns over the delamination or partial degradation of the HA coatings causing failure of the implants, as discussed in the review of Ellingsen et al. (2006). There are no adequate data on the long-term clinical performance of coated implants in comparison with noncoated implants (Ellingsen et al. 2006; Lee et al. 2000). Recently, a radio frequency (RF) magnetron sputtering technique has been developed and investigated for the deposition of CaP coatings on titanium oral implants. This method offers several advantages over other coating techniques, including the deposition of thin (1–2 μm) and homogeneous coatings that preserve the original implant topography (Vercaigne et al. 2000). RF magnetron sputtering of CaP coatings has been shown to be an economically feasible process and is currently used commercially for coating oral implants with CaP coatings. However, HA-coated implants are not commonly used.

4.2.3 Bioactive Glass Coatings

Bonding of certain compositions of silicate glasses to bone was first demonstrated by Hench as early as 1969 (Hench et al. 1971). These bioactive glasses were developed in the Na_2O–CaO–P_2O_5–SiO_2 system. The bioactive glasses in this system are similar to ordinary soda lime glass but contain a lower amount of SiO_2. According to Hench (1991), a bioactive glass should contain 45–60 wt% SiO_2, the

Na_2O content should be high, and the Ca/P ratio should also be high. The latter is achieved by keeping the phosphate content low. These types of glasses are considered "short," because their viscosity is changed dramatically over a narrow temperature range. This leads to the tendency of these glasses to crystallize. Due to the low SiO_2 content, these glasses are easily phase separated with SiO_2- and P_2O_5-rich phases (Brink 1997a), because the three-dimensional (3-D) silica network cannot be formed.

In an attempt to extend the working range of bioactivity, Brink (1997b) studied glasses in the $Na_2O–K_2O–MgO–CaO–B_2O_3–P_2O_5–SiO_2$ system. It was possible to extend the working range of the glasses by lowering the amount of alkali ions, as this led to improvement in the viscosity–temperature dependence. The partial replacement of Na_2O with K_2O and the addition of a small amount of MgO, B_2O_3, and P_2O_5 had a positive effect on the working range (Brink 1997a,b). The work led to the development of glass that could be handled outside the melting furnace without the risk of crystallization, allowing the manufacturing of glass in the form of fibers, blown shapes, and microspheres (Brink 1997b). Another advantage of glass with a broad viscosity range is the possibility of repeated heat treatment of the glass. Thus, microspheres manufactured from bioactive glass can be sintered into porous implants of different sizes and shapes (Itälä et al. 2001; Ylänen et al. 2000).

Several methods have been applied to obtain a bioactive glass coating on the load-bearing part of the implants (Hench and Andersson 1993b). Numerous attempts have been made to create bioactive glass coatings on alumina (Greenspan and Hench 1976; Griss et al. 1976; Hench et al. 1977), stainless steel (Piotrowski et al. 1975; Schepers et al. 1989), Co–Cr–Mo alloy (Ducheyne et al. 1984; Lacefield and Hench 1986; Pajamäki et al. 1995), and titanium alloys (Kitshugi et al. 1996; West et al. 1990). A number of coating techniques are available, such as enameling, glazing, flame spray coating, rapid immersion coating, and ablating/sputtering. Each of these techniques has its limitations for producing bioactive glass coating on medical implants.

Conventional methods with time-consuming heat treatments often lead to the diffusion of unwanted ions through the coating, resulting in decreased bioactivity of the glasses. In addition, different expansion coefficients between bioactive glasses and titanium result in a detachment of the coating from the substrate (Hench and Andersson 1993b). The application of a double glass coating has been suggested to solve this problem (Andersson et al. 1995). Moreover, repeated heating of the glass during the coating procedure may result in a phase separation of the amorphous material and decreased bioactivity of the glass (Pietl Filho et al. 1996). In addition, the extensive oxidation of the titanium surface resulting from the heat treatment contributes to poor attachment of the coating (Hautaniemi 1993).

Recently, several attempts have been made to fabricate coatings on titanium and Ti–6Al–4V with glasses in the $SiO_2–CaO–MgO–Na_2O–K_2O–P_2O_5$ system with the silica content ranging from 44 to 70 wt% (Bloyer et al. 1999; Saiz et al. 2002). Coatings fabricated of glass with 55–60 wt% silica content were able to precipitate CaP from simulated body fluid (SBF). Coatings made of glasses having silica contents more than 60 wt% were inactive, even after 2 months in SBF. The coatings of 50–70 μm thickness were deposited on substrates from the suspension of glass powder heat treated at 800°C–820°C in a vacuum for 30 s. This processing led to the partial crystallization of the glass, but a mismatch of the thermal expansion coefficients initiated the delamination of the coating in SBF. The application of a ground layer prepared from inert glass in the same system having 68 wt% silica and a thermal expansion coefficient close to that of Ti–6Al–4V provided good adhesion to the substrate. This was achieved due to the formation of a thin (~150 nm) titanium silicide (Ti_5Si_3) adhesive interfacial layer between the glass and the coating (Oku et al. 2001). This ground coating can be used in combination with more surface-reactive glass coatings (Gomez-Vega et al. 2001), with embedded HA and/or bioactive glass particles (Gomez-Vega et al. 2000a) or a sol–gel-derived silica coating (Gomez-Vega et al. 2000b).

Attempts were made to create coatings of bioactive glass on titanium surfaces by a CO_2 laser treatment using 1-98 glass composition suitable for repeated heat treatments (Moritz et al. 2004a). With respect to the bioactive glass coatings, as heat treatment occurs locally and rapidly, it mostly affects the glass layer. Thus, the oxidation of the substrate is significantly less than in conventional heat treatment. Moreover, a glass coating produced by this method is not uniform but consists of separate glass "drops" attached to the surface of titanium. Thus, the mechanical properties of the coating could be superior to the coatings produced by conventional methods.

When implanted in the femoral condyles of the rabbit, these implants demonstrated a superior performance compared with the control titanium implants and the implants treated with NaOH (Moritz et al. 2004b). The rationale for using bioactive glass as a coating material is that, unlike HA, bioactive glass is an osteopromotive class A biomaterial with a high resorption rate. It is expected that the coating will only have a short-term effect and will soon be replaced with the peri-implant bone. However, the delamination of bioactive glass coatings may still be a problem for implant stability.

4.2.4 Chemical Modification of Titanium Surfaces

It was suggested that a surface-reactive titania gel can be formed on the surface of a titanium metal via the chemical treatment of the surface with NaOH and subsequent heat treatments (Kim et al. 1996, 2000; Kokubo et al. 2000; Takadama et al. 2001; Uchida et al. 2003). The chemical treatment leads to the formation of an alkali titanate hydrogel layer. This alkali titanate hydrogel, even without heat treatment, can form an apatite in SBF but is unusable for practical considerations (Kim et al. 1996). As a result of heat treatment, the hydrogel layer is densified to an amorphous or crystalline alkali titanate layer. When subjected to SBF, the surface is rehydrated and transformed into titanate hydrogel through the release of alkali ions into the solution; at the same time, the pH of the fluid in the vicinity of the surface increases, giving an increase in the ionic activity product of apatite. Thus, hydrated titania induce apatite nucleation and growth. Moreover, this treatment allows for the creation of compositional gradients from the apatite layer into the titanium metal (Kim et al. 1996). A possible application of these materials was suggested for bone–implant interfaces. NaOH-treated and, subsequently, heat-treated macroporous titanium allows bone ingrowth into pores and provides combined biological and bioactive fixation.

Another well-established method of improving the biocompatibility and bioactivity of titanium is treatment with a mild hydrogen peroxide (H_2O_2) solution, which leads to the formation of Ti–peroxy gel. This method was originally developed to model the inflammatory response to titanium (Tengvall et al. 1989a,b) but can also be used to oxidize titanium surfaces (Wälivaara et al. 1994). It was also suggested that H_2O_2 treatment leads to the formation of a double-layer oxide structure, with the porous outer layer capable of precipitating apatite; *in vivo*, this could result in the integration of titanium implants into the host bone (Pan et al. 1998). The *in vitro* bioactivity of titanium can be enhanced by using higher concentrations of H_2O_2 and subsequent heat treatment (Wang et al. 2001).

Moreover, biochemical modification of the surface of the implant can be achieved by attaching specific peptides that could promote the adhesion of specific cells and affect the rate of their differentiation and the healing rate (Xiao et al. 2001).

4.2.5 Sol–Gel-Derived Titania Coating

The sol–gel method provides a means of producing ceramic materials from chemical precursors at low temperatures. Unlike traditional methods used to make ceramics, the sol–gel method allows a great degree of control over the purity and homogeneity of the resulting ceramic material. Moreover, selecting appropriate parameters of the chemical reactions can control the structure of the material.

A sol is a colloidal suspension that consists of small particles (1–1000 nm). The gelation of the sol leads to the formation of a wet gel that is an interconnected solid 3-D matrix within an interconnected liquid phase. Gels can be formed as polymeric gels that are covalently linked, gelatin gels that are formed by the entanglement of chains, or particulate gels that are kept together by van der Waals forces. Upon drying, a gel becomes a dry gel (xerogel) that can be processed (densified) further by the application of heat to form a porous ceramic. Sol–gel-derived materials usually have high porosity and surface area in the range of 100–1000 m^2/g. Further heat treatment that causes sintering of the material leads to a dense ceramic (Brinker and Scherer 1990).

Typical precursors in the sol–gel process are alkoxides, acetylacetonates, and inorganic salts dissolved in alcohol or water. The precursors undergo two steps: (1) hydrolysis and (2) polymerization by condensation. In hydrolysis, the hydroxyl ion becomes attached to the center atom such as Si or Ti. Depending on the amount of water and catalyst present, all of the alkoxy groups or only part of them are substituted by hydroxyl groups. Two partly hydrolyzed molecules can join together by condensation reaction and free water or alcohol. This reaction goes on through polymerization and leads to the formation of the larger molecules and, eventually, large 3-D structures.

Changes in the pH of the solution alter the relative rates of hydrolysis and condensation reactions, leading to the formation of either weakly branched (low pH) or particulate sols (high pH). The water/alkoxide molar ratio has a significant effect on the microstructure (e.g., pore sizes and surface area) of the xerogels. When the ratio is low, the alcohol condensation is dominating, the gelation time is longer, and the resulting gel has a finer pore structure than the gel obtained from a sol with a higher water content.

The sol–gel method provides the possibility of producing materials with the desired pore structure of the outermost layer. Furthermore, sol–gel-derived materials have higher OH contents than traditional glasses. This, although a disadvantage for the fabrication of optical grade materials, is usually seen as a benefit in terms of bioactivity.

Film deposition by the sol–gel method is considered one of the most important capabilities offered by the method. The films are prepared from the fluid sol prior to gelation by dipping, sputtering, or spraying. Although some differences between the formation of bulk gels and films exist, the physical and chemical processes that determine polymer growth and gel formation are the same (Brinker and Scherer 1990).

In addition, the sol–gel method allows for the fabrication of noncrystalline solids outside the range of normal glass formation and at much lower temperatures than the traditional heat treatment methods (Hench and Andersson 1993a). Alkoxide-based bioactive sol–gel glasses have a broader compositional range than the melt-derived glasses; sol–gel-derived glasses that contain 88 mol% of SiO_2 can form an apatite layer in SBF compared to a 60 mol% limit for melt-derived glasses (Cao and Hench 1996). The bioactivity of these glasses is dependent on their compositional and structural properties. Compared to conventional melt-derived bioactive glasses, for sol–gel bioactive glasses, the possibility of controlling their compositional and chemical properties leads to a higher rate of material resorption than can be matched with the rate of bone formation (Hamadouche et al. 2001).

There exist several advantages of sol–gel synthesis when the film deposition is considered one of the most important preparation steps. The sol–gel film can be deposited onto an implant, for example, by dip coating, spinning, or spraying. By appropriate chemical modifications of precursors, the control maybe achieved over the rates of hydrolysis and condensation and colloid particle and pore size, and the porosity and pore wall chemistry of the final material. The chemical conditions are mild, and extreme pH conditions can be easily avoided, which give the possibility of entrapping biological molecules such as proteins, enzymes, or even whole cells into the matrix and still retain their functionalities (Wright 2001).

4.2.6 Bioactivity of Sol–Gel-Derived Ceramic Coatings

It was experimentally demonstrated, both *in vitro* and *in vivo*, that the application of a sol–gel-derived titania coating to a titanium substrate induces HA formation (Figure 4.2), bone (Li and de Groot 1993), and connective tissue bonding (Areva et al. 2004). The incorporation of an experimental implant coated with sol–gel-derived titania is shown in Figure 4.3.

After the discovery of the bioactive properties of titania (Li and de Groot 1993), extensive research on the sol–gel-derived titania coatings have been conducted (Jokinen et al. 1998; Peltola 2000; Peltola et al. 1998, 2000). It has been suggested that it is the nanoscale topography of the outermost surface layer of the film rather than the porosity, pore size, or surface area that influence the ability of titania to induce apatate formation. It has been pointed out that nanoscale topography is related to the charge and charge density of the surface (Jokinen et al. 1998). Furthermore,

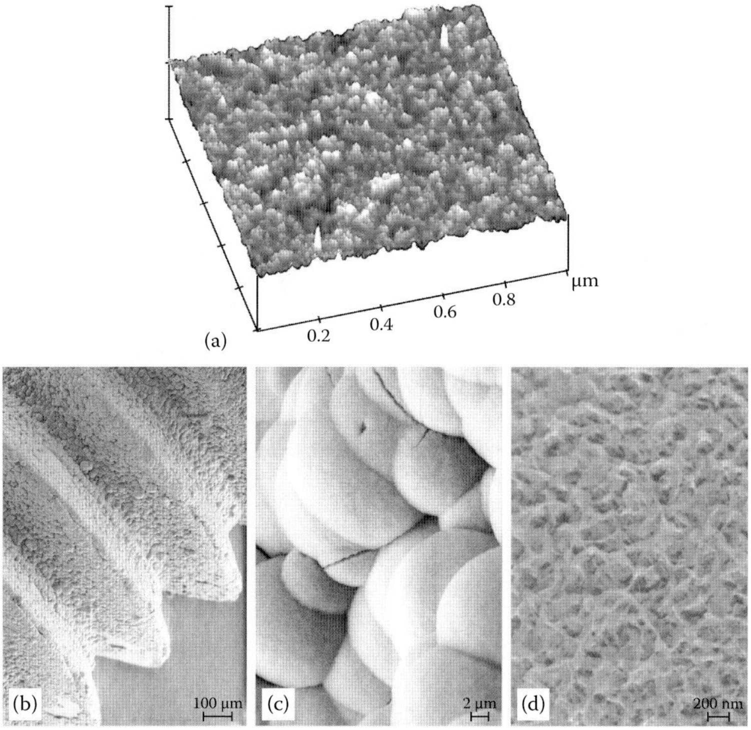

FIGURE 4.2 Sol–gel-derived TiO_2 coating. (a) AFM height field of the nanoscale structure of sol–gel-derived TiO_2 coating. (b) CaP formation *in vitro* on the surface of TiO_2-coated grooved implant. (c) Closer view of the CaP precipitate. (d) Fine structure of the CaP precipitate at the nanoscale.

it has been shown that the nanoscale topography of the surface of the films can be modified as a function of heat treatment temperature, the aging time of the sol, and a number of coating layers (Peltola et al. 1998). In addition, it has been demonstrated that suitable nanoscale dimensions promote CaP formation on a titania surface. This surface topography was studied in terms of the distribution of distances between peaks in line profiles obtained by atomic force microscopy (AFM) scanning of the surface. This distribution of distances between peaks approximates the pore size distribution, and the optimal range for apatite formation is reported to be between 5 and 50 nm, particularly between 5 and 20 nm, with the peak height being >1 nm. These dimensions were suggested to correspond to the size of CaP crystals in the bone (Peltola 2000; Peltola et al. 2000; Figure 4.2).

Many studies have been conducted on sol–gel-derived materials that are used to manufacture integrated optical circuits and other optical devices. The optical or electrochromic properties of the coatings have been a major concern (Araújo et al. 1994; Fabes et al. 1990; Pelli et al. 1994, 1996; Shaw and King 1990; Taylor et al. 1996; Wilson and Merwin 1988). The main reason for the use of sol–gel-derived materials in this field is the possibility of producing films with a controlled composition

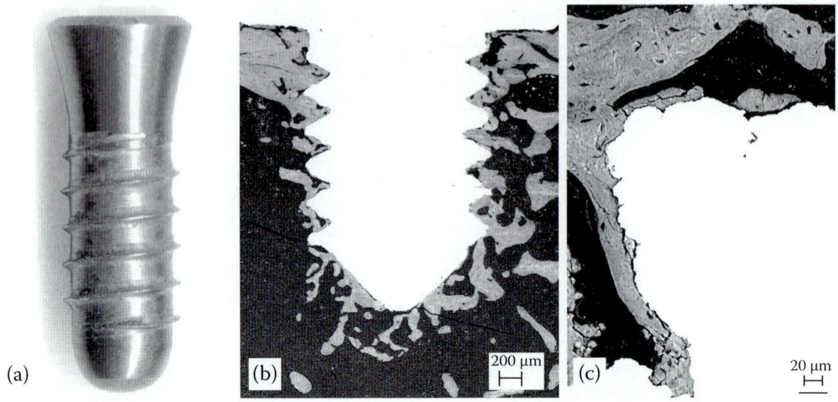

FIGURE 4.3 TiO_2-coated implants used in *in vivo* experiments. (a) Sol–gel-derived nanoporous TiO_2 thin film on the smooth coronal part ITI Straumann implant. (Adapted from Rossi, S. et al., Peri-implant tissue response to TiO_2 surface modified implants, *Clinical Oral Implants Research*, 19 (4), 348–355, 2008.) (b) New bone formation around an experimental titanium implant with sol–gel-derived TiO_2 coating implanted in rat tibia. (c) Close-up view of the bone attachment to the surface of the implant.

and almost free of contamination (Pelli et al. 1996). Oftentimes, the requirement of coatings on such devices is that different areas of the same coating possess different properties. Thus, it was suggested that localized adsorbed laser irradiation could provide an alternative heating method to furnace firing (Taylor et al. 1996).

Silica and titania absorb wavelengths in the ultraviolet (UV) and infrared (IR) spectral regions, thus excimer lasers (UV, 193–354 nm) or CO_2 lasers (IR, 10.6 μm), can be used in the processing of these materials. Compared to excimer lasers, CO_2 lasers present a preferable technique because of their broader availability and simplicity of operation (Righini and Pelli 1997). However, the processing temperature cannot be measured directly. Therefore, models for temperature calculations for laser-irradiated sol–gel-derived materials have been suggested (Chia et al. 1990; Taylor et al. 1995).

The bioactivity of sol–gel-derived TiO_2 coatings was studied *in vitro* (Moritz et al. 2003, 2005) and *in vivo* (Rossi et al. 2007). Results of the *in vitro* studies suggested that the anatase or rutile crystalline phase, alone or both phases in the same coating, is required for the *in vitro* bioactivity of the titania coatings. However, the nanoscale topography of the surface could be the key factor, irrespective of the crystal phase. Results indicate the possibility of controlling bioactivity by altering the laser power used through the anatase/rutile crystallinity enhancement.

An experimental animal study performed in rats involved subcutaneous placement of the laser-processed and control implants. Results indicated that, 3 days after the implantation, a soft connective tissue with possible attachment to the implant surface was formed to both anatase-structured sol–gel-derived TiO_2 coatings produced by heat treatment at 500°C and rutile-structured sol–gel-derived TiO_2 coatings produced by CO_2 laser treatment with furnace preheat treatment at 500°C. The study indicates that a rutile crystal phase produced by CO_2 laser treatment does not inhibit the bioactivity of the coatings and can be used for soft-tissue attachment equally well as anatase produced by conventional heat treatment (Rossi et al. 2007).

To further enhance the bioactivity of TiO_2, coating ions or molecules, such as Zn^{2+}, Ca^{2+}, MgO, or Sr_2, could be added to a surface oxide or matrix networks. Such elements reportedly change the biomimetic apatite layer composition in SBF and enhance hard tissue healing by increasing the number of CaP nucleation sites (Arcos and Vallet-Regí 2010; Chen et al. 2010; Kokubo et al. 2003).

4.3 CONTROLLED DRUG DELIVERY BY BIOMATERIALS

There exist several goals in controlled drug delivery. One of them is to target the drug of interest to a particular place, tissue, organ, or cell type of the body. Other goals may be to control the release rate and optimize the effective treatment duration and drug concentration in circulation (Langer and Peppas 2003). Various types of controlled release systems currently exist, ranging from implants that release drugs via surface erosion mechanisms with duration up to 5 years to osmotically driven pills that may release drugs during months at a constant release rate (Langer 1998). Common examples of drugs or molecules released from materials include, for example, antibiotics to combat biomaterial-related infections; antiinflammatory drugs to modulate wound healing; growth factors to upregulate or downregulate specific cell or tissue growth; cell binding peptides and proteins such as fibronectin, fibrinogen, vitronectin, and laminin to attract and bind specific cells to surfaces; anticancer drugs; anticontraceptives for the prevention of pregnancy; osteoporosis drugs to slow down the bone remodeling rate; gene vectors; and statins.

Controlled drug release occurs mainly by one of three mechanisms. First, in diffusion-controlled devices, the drug is generally uniformly distributed through the carrier material (such as polymer), or the drug is surrounded by a barrier membrane. The drug then diffuses out through the carrier polymer or through the membrane. Second, in case of a chemical reaction–mediated release mechanism, the drug is released when a carrier material is degraded by water (hydrolysis) or chemical reactions, or alternatively, the drug can be attached to the host material by a covalent bond that is hydrolytically cleaved or released by enzymatic cleavage. Third, solvent activation and transport applies either an osmotically driven process, which subsequently pushes the drug out from the matrix, or the drug is released from a hydrogel by swelling and increased diffusion constant of the drug (Langer and Peppas 2003).

4.3.1 Pharmacological Substrates on Implant Surface

There is a clear goal to modify bone-anchored implant surfaces to decrease complications such as infections, increase the implant survival, and minimize the healing time, with the best possible biomechanical outcome. If these goals could be reached, it would be possible to treat even less suitable patient cases with a surgical procedure and likely enhance the life quality for many patients. Furthermore, illnesses such as diabetes or osteoporosis increase the risk of implant failure. As the risk for late loosening of implants is likely coupled to early fixation (Ryd 1992), even minor improvements in early implant healing might provide an enhanced long-term outcome. One means of obtaining this may be to immobilize pharmacological agents onto the implant surface for local implant improvement.

4.3.2 Bisphosphonates

There exist several pharmacological agents that have been reported to stimulate osteoblast activation and bone generation. Bisphosphonates (BPs) bound to metallic implants via surface-deposited HA, calcium ions, or through a poly (D,L)-lactide matrix increased bone formation around implants. BP is a geminal (structurally similar) of pyrophosphate, which is a natural metabolite of bone and binds strongly to the bone mineral apatite through the P–R1CR2–P backbone structure (cf., pyrophosphate P–O–P), which allows a great number of possible variations either by changing the two lateral chains on the carbons (R1 and R2) or by esterification of phosphate groups.

The two structurally different BPs (one *N*-BP and one without nitrogen) act via different mechanisms—the former via the inhibition of farnesyl diphosphate during protein prenylation and the latter by disturbance and affected cell ATP synthesis and metabolism (Fleisch 2002; Rodan and Fleisch 1996). They are used for the treatment of different bone resorbing diseases such as osteoporosis, tumor-associated osteolysis, and hyperparathyroidism (Fleisch 1997, 1998; Coleman 2001). Currently, many different compounds of the BP family are commercially available for the treatment of bone diseases. Each BP has its own chemical, physiochemical, and biological characteristics and thus vary greatly in efficacy (Fleisch 1998). The length of the aliphatic carbon chain and the addition of a hydroxyl group to the central carbon atom at position 1 increases the potency of BPs by enhancing their affinity to the bone. Furthermore, derivatives with an amino group at the end of the side chain, the length of the side chain, and a nitrogen-containing ring structure further increase the activity by regulating the potency of bone resorption (Fleisch 2001).

BPs act via two separate mechanisms. First, BPs act through the inhibition of bone resorption when given at high doses (Rodan and Fleisch 1996). The overall effects can be divided into three levels: (1) the reduction of bone turnover; (2) osteoclast inhibition at the cellular level; and (3) actions at the molecular level. At the tissue level, the actions of all BPs are similar—both new bone formation and old bone reduction is decreased, although the latter process dominates. Thus, a decrease in bone turnover leads to a decrease in bone loss. The effect is due to a decrease in the generation of new bone remodeling units and a decrease in the depth of the erosion cavities (Ott 1993; Storm et al. 1993). At the cellular level, the main BP target cell is the osteoclast, and its inactivation decreases bone resorption. BPs reduce the bone resorption directly and/or indirectly through the inhibition of osteoclast recruitment to bone surfaces, the inhibition of osteoclast activity on the bone surface, and the shortening of the osteoclast life span (Rodan and Fleisch 1996). BPs generate changes in osteoclast cytoskeleton and in the ruffled border of active osteoclasts, indicating that, under certain conditions, BPs can enter osteoclasts but not other cells in the bone and bone marrow (Felix et al. 1984; Murakami et al. 1995; Sato and Grasser 1990; Sato et al. 1991). Cellular uptake of BP drugs requires fluid-phase endocytosis and is enhanced by Ca^{2+} ions, whereas transfer from endocytic vesicles into the cytosol requires endosomal acidification (Thompson et al. 2006).

Second, BPs act by shortening the life span of osteoclasts, possibly through apoptosis. The possible action at the molecular level also includes action on cell surface receptors and BP uptake by cells, where it interacts with an enzyme or other molecules, affecting the cellular metabolism by inhibiting protein tyrosine phosphatase (Rodan and Fleisch 1996).

4.3.3 Local Delivery of Bisphosphonates

BPs have been given orally or locally in order to improve the fixation of orthopedic implants. Oral treatment with clondronate, risedronate, alendronate, and/or zoledronate has shown improved fixation of pins, screws, and knee protestheses (Hilding and Aspenberg 2006, 2007; Moroni et al. 2007; Yamasaki et al. 2007). Various coatings have been developed to immobilize BP to implants, with the aim of enhancing the fixation to the surrounding new bone. The most common way of applying BP to implant surfaces is adsorption to HA, as BPs possess high affinity to calcium minerals (Meraw and Reeve 1999; Denissen et al. 1994, 2000; Kajiwara et al. 2005; McLeod et al. 2006; Meraw et al. 1999; Peter et al. 2005, 2006; Gao et al. 2009). However, the lack of thicker HA coating on clinical practice has led researchers to find other approaches. Tengvall et al. (2004) bound alendronate successfully to a metallic implant surface through a thin 50-nm fibrinogen layer. Immobilization of fibrinogen to the implant surface was done via amine group binding to activated carboxyls. The so-fabricated fibrinogen network was approximately 50 nm thick and allowed *N*-BP (e.g., pamidronate) to be chemically immobilized to this matrix. Another approach for binding BPs to implant surfaces through the TiO_2-derived sol–gel technique and preadsorbed thin (50-nm) CaP crystal layer was reported to enhance mechanical screw fixation in rat tibia after 4 weeks (Linderbäck et al. 2010).

4.3.4 Strontium Ranelate and Orthopedic Implants

A relatively new antiosteoporotic drug, i.e., strontium ranelate (Protelos), is proposed to induce simultaneously, in contrast to other antiosteoporotic drugs (such as BPs), both an antiresorption and bone-forming effect (Ammann 2006; Ammann et al. 2004, 2007; Blake and Fogelman 2006; Bonnelye et al. 2008; Canalis et al. 1996; Marie et al. 1985). Strontium is a trace metal in the human body in its physiologically stable, divalent cationic form (Sr^{2+}). The ion has a relatively high affinity to the bone and may replace Ca in apatite crystals by ion exchange (Morohashi et al. 1995; Boivin and Meunier 2003). Strontium ranelate contains two Sr^{2+} ions and ranelic acid.

The effect of strontium ranelate is not fully understood, but it is suggested to activate Ca sensing receptors on osteoblasts (Quarles 1997). Moreover, the beneficial action of strontium ranelate on bone tissue quality (density) may be the result of a modification of the bone matrix, as it has been shown that strontium ranelate stimulates the synthesis of bone collagen *in vitro* (Canalis et al. 1996; Maïmoun et al. 2010). Several *in vitro* and *in vivo* studies suggest that strontium increases the number of osteoblasts, decreases the number and activity of osteoclasts, decreases bone resorption, and maintains a high *de novo* bone formation in osteoporotic rats, thereby preventing bone loss (Ammann 2006; Ammann et al. 2007; Bonnelye et al. 2008; Canalis et al. 1996; Marie et al. 1993). Strontium ranelate has also shown increased lumbar spine and hip bone mineral density and reduced the incidences of vertebral and nonvertebral fractures during long-term treatment of postmenopausal osteoporotic women (Meunier et al. 2004; Reginster et al. 2008; Roux et al. 2006; Seeman et al. 2006). A couple of studies show strontium ranelate–related improvement of implant osseointegration and indicate the improvement of bone microarchitecture around the implant (Maïmoun et al. 2010). The *in vivo* kinetics of bone turnover upon the administration of strontium ranelate is largely unknown, particularly around bone-anchored metal implants.

4.3.5 Other Pharmacological Agents in Orthopedics

Parathyroid hormone was observed to increase the number of osteoblasts and positively affect the proliferation and differentiation of osteoprogenitor cells (Aspenberg et al. 2008; Nishida et al. 1994; Skripitz and Aspenberg 2001; Skripitz et al. 2005). However, in order to obtain an anabolic effect, it should be given intermittently and, thus, is yet not applicable for controlled drug delivery through coating technologies.

Bone morphometric proteins (BMPs), especially BMP-2 and MP-7, induce osteoblast differentiation and, according to several studies, accelerate fracture healing (Grgurevic et al. 2011; Li et al. 2010; Luo et al. 2011; Stenport et al. 2003). Furthermore, some studies have observed that locally delivered Simvastatin stimulates BMP and thus improves fracture healing (Du et al. 2009; Pauly et al. 2009). In one study, researchers reported that BMP-2 significantly enhanced bone mineralization upon implantation in rat calvarial bone in the hydrogel matrix (Patterson et al. 2010).

Recently, a rat study has reported positive effects of a novel pharmacological agent (antisclerostin) that affected the Wnt-signaling pathway and the GSK3β complex (ten Dijke et al. 2008). Sclerostin is a product of the SOST gene, and mutation in this gene results in high bone mass in humans (Balemans et al. 2001). Sclerostin is reported to antagonize Wnt signaling at GSK3β level and is therefore a negative regulator for bone formation. Systemically administered sclerostin-neutralizing antibody was thus reported to successfully promote bone healing (Agholme et al. 2010).

4.4 CONCLUSION

The success of metallic implants (stainless steel, Co–Cr, titanium, and tantalum) is attributed to high strength, ductility, and satisfactory resistance to wear and corrosion. However, these materials possess certain intrinsic limitations such as the mismatch of their mechanical properties with the bone and susceptibility to debris formation and corrosion, which originate in articulating surfaces, modular component interfaces, fixation surfaces, and devices. In addition, although metals are biocompatible, they are also biologically inert. Therefore, to achieve predictable performance and long-term stability of the implants, there is a need not only for the optimization of the material properties and the design of the implants but also for the optimization of the structure and chemistry of the implant surfaces. Engineering of the implant surfaces provides strategies for achieving a favorable biological reaction, leading to bone healing and osteointegration of the implants. Because the risk for late loosening of the implant is lined to early fixation, even minor improvements in early implant healing might provide an enhanced long-term outcome. Therefore, the enhanced biocompatibility through surface engineering leads to shorter healing periods and fewer implant-related complications. In addition, a desired response of the peri-implant bone can be attained when implant surfaces are used as carriers for biomolecules.

REFERENCES

Agholme, F., Li, X., Isaksson, H., Ke, H.Z., and Aspenberg, P. Sclerostin antibody treatment enhances metaphyseal bone healing in rats. *J Bone Miner Res* 2010;25(11):2412–2418.

Albrektsson, T., and Wennerberg, A. Oral implant surfaces—Part 1: Review focusing on topographic and chemical properties of different surfaces and *in vivo* response to them. *Int J Prosthodont* 2004;17(5):536–543.

Ammann, P. Strontium ranelate: A physiological approach for an improved bone quality. *Bone* 2006;38(2 Suppl 1):15–18.

Ammann, P., Badoud, I., Barraud, S., Dayer, R., and Rizzoli, R. Strontium ranelate treatment improves trabecular and cortical intrinsic bone tissue quality: A determinant of bone strength. *J Bone Miner Res* 2007;22(9):1419–1425.

Ammann, P., Shen, V., Robin, B., Mauras, Y., Bonjour, J.P., and Rizzoli, R. Strontium ranelate improves bone resistance by increasing bone mass and improving architecture in intact female rats. *J Bone Miner Res* 2004;19(12):2012–2020.

Andersson, Ö.H., Karlsson, K.H., Hero, H., Vedel, E., Yli-Urpo, A., Pajamäki, K.J.J., and Lindholm, T.S. Bioactive glass coatings for Co–Cr–Mo alloy. *J Mater Sci: Mater Med* 1995;6:242–247.

Araújo, F.G., Chia, T., and Hench, L.L. Laser densification of chanel waveguides in gel–silica substrates. *J Sol–Gel Sci Technol* 1994;2:729–735.

Archibeck, M.J., Jacobs, J.J., Roebuck, K.A., and Glant, T.T. The basic science of periprosthetic osteolysis. *J Bone Jt Surg* 2000;82-A:1478–1489.

Arcos, D., and Vallet-Regí. M. Sol–gel silica-based biomaterials and bone tissue regeneration. *Acta Biomater* 2010;6(8):2874–2888.

Areva, S., Paldan, H., Peltola, T., Närhi, T., Jokinen, M., and Lindén, M. Use of sol–gel-derived titania coating for direct soft tissue attachment. *J Biomed Mater Res—Part A* 2004;70:169–178.

Aspenberg, P., Wermelin, K., Tengwall, P., and Fahlgren, A. Additive effects of PTH and bisphosphonates on the bone healing response to metaphyseal implants in rats. *Acta Orthop* 2008;79(1):111–115.

Balemans, W., Ebeling, M., Patel, N., Van Hul, E., Olson, P., Dioszegi, M. et al. Increased bone density in sclerosteosis is due to the deficiency of a novel secreted protein (SOST). *Hum Mol Genet* 2001;10(5):537–543.

Blake, G.M., and Fogelman, I. Strontium ranelate: A novel treatment for postmenopausal osteoporosis: A review of safety and efficacy. *Clin Interv Aging* 2006;1(4):367–375.

Bloebaum, R.D., Beeks, D., Dorr, L.D., Savory, C.G., DuPont, J.A., and Hofmann, A.A. Complications with hydroxyapatite particulate separation in total hip arthroplasty. *Clin Orthop Relat Res* 1994;298:19–26.

Bloyer, D.R., Gomez-Vega, J.M., Saiz, E., McNaney, J.M., Cannon, R.M., and Tomsia, A.P. Fabrication and characterization of a bioactive glass coating on titanium implant alloys. *Acta Mater* 1999;47:4221–4224.

Boivin, G., and Meunier, P.J. The mineralization of bone tissue: A forgotten dimension in osteoporosis research. *Osteoporosis Int* 2003;14(Suppl 3):S19–S24.

Bonnelye, E., Chabadel, A., Saltel, F., and Jurdic, P. Dual effect of strontium ranelate: Stimulation of osteoblast differentiation and inhibition of osteoclast formation and resorption *in vitro*. *Bone* 2008;42(1):129–138.

Bowers, K.T., Keller, J.C., Randolph, B.A., Wick, D.G., and Michaels, C.M. Optimization of surface micromorphology for enhanced osteoblast responses *in vitro*. *Int J Oral Maxillofac Implants* 1992;7(3):302–310.

Brink, M. Bioactive glasses with a large working range. Academic Dissertation. Åbo Akademi University, Finland, 1997a.

Brink, M. The influence of alkali and alkali earths on the working range for bioactive glasses. *J Biomed Mater Res—Part A* 1997b;36:109–117.

Brinker, C.J., and Scherer, G.W. *Sol–Gel Science: The Physics and Chemistry of Sol–Gel Processing*. Academic Press Inc., San Diego, CA, 1990.

Buser, D., Schenk, R.K., Steinemann, S., Fiorellini, J.P., Fox, C.H., and Stich, H. Influence of surface characteristics on bone integration of titanium implants: A histomorphometric study in miniature pigs. *J Biomed Mater Res—Part A* 1991;25(7):889–902.

Campbell, A.A. Bioceramics for implant coatings. *Mater Today* 2003;11:26–30.

Canalis, E., Hott, M., Deloffre, P., Tsouderos, Y., and Marie, P.J. The divalent strontium salt S12911 enhances bone cell replication and bone formation *in vitro*. *Bone* 1996;18(6):517–523.

Cao, W., and Hench, L.L. Bioactive materials. *Ceram Int* 1996;22:493–507.

Cheang, P., and Khor, K.A. Addressing processing problems associated with plasma spraying of hydroxyapatite coatings. *Biomaterials* 1996;17(5):537–544.

Chen, X., Liao, X., Huang, Z., You, P., Chen, C., Kang, Y. et al. Synthesis and characterization of novel multiphase bioactive glass–ceramics in the CaO–MgO–SiO(2) system. *J Biomed Mater Res—Part B* 2010;93(1):194–202.

Chia, T., Hench, L.L., Qin, C., and Hsieh, C.K. Laser densification modeling. *Mater Res Soc Symp Proc* 1990;180:819–824.

Coleman, R.E. Metastatic bone disease: Clinical features, pathophysiology and treatment strategies. *Cancer Treat Rev* 2001;27(3):165–176.

Denissen, H., Martinetti, R., van Lingen, A., and van den Hooff, A. Normal osteoconduction and repair in and around submerged highly bisphosphonate-complexed hydroxyapatite implants in rat tibiae. *J Periodontol* 2000;71(2):272–278.

Denissen, H., van Beek, E., Löwik, C., Papapoulos, S., and van den Hooff, A. Ceramic hydroxyapatite implants for the release of bisphosphonate. *Bone Miner* 1994;25(2):123–134.

Dohan Ehrenfest, D.M., Coelho, P.G., Kang, B.-S., Sul, Y.-T., and Albrektsson, T. Classification of osseointegrated implant surfaces: Materials, chemistry and topography. *Trends Biotechnol* 2010;28(4):198–206.

Donnelly, W.J., Kobayashi, A., Freeman, M.A., Chin, T.W., Yeo, H., West, M. et al. Radiological and survival comparison of four methods of fixation of a proximal femoral stem. *J Bone Jt Surg, Br Vol* 1997;79(3):351–360.

Du, Z., Chen, J., Yan, F., and Xiao, Y. Effects of Simvastatin on bone healing around titanium implants in osteoporotic rats. *Clin Oral Implants Res* 2009;20(2):145–150.

Ducheyne, P., Martens, M., and Burssens, A. Materials, clinical and morphological evaluation of custom-made bioactive-glass–coated canine hip prostheses. *J Biomed Mater Res—Part A* 1984;18:1017–1030.

Ellingsen, J.E., Thomsen, P., and Lyngstadaas, S.P. Advances in dental implant materials and tissue regeneration. *Periodontology 2000* 2006;41:136–156.

Es-Souni, M., Es-Souni, M., and Fischer-Brandies, H. On the properties of two binary NiTi shape memory alloys: Effects of surface finish on the corrosion behavior and *in vitro* biocompatibility. *Biomaterials* 2002;23(14):2887–2894.

Fabes, B.D., Taylor, D.J., Weisenbach, L., Stuppi, M.M., Klein, D.L., Raymond, L.J., Zelinski, B.J.J., and Birnie, D.P. III. *SPIE Proc* 1990;1328:319–328.

Felix, R., Gunether, H.L., and Fleisch, H. The subcellular distribution of [14C]dichloromethylenebisphosphonate and [14C]1-hydroxyethylidene-1,1-bisphosphonate in cultured calvaria cells. *Calcif Tissue Int* 1984;36(1):108–113.

Fleisch, H. Bisphosphonates: Mechanisms of action. *Endocr Rev* 1998;19(1):80–100.

Fleisch, H. Bisphosphonates: Mechanisms of action. *Expert Opin Ther Pat* 2001;11(9):1371–1381.

Fleisch, H. Development of bisphosphonates. *Breast Cancer Res* 2002;4(1):30–34.

Fleisch, H.A. Bisphosphonates: Preclinical aspects and use in osteoporosis. *Ann Med* 1997;29(1):55–62.

Frenkel, S.R., Jaffe, W.L., Dimaano, F., Iesaka, K., and Hua, T. Bone response to a novel highly porous surface in a canine implantable chamber. *J Biomed Mater Res—Part B* 2004;71(2):387–391. Epub 2004/09/25.

Gandhi, R., Davey, J.R., and Mahomed, N.N. Hydroxyapatite coated femoral stems in primary total hip arthroplasty: A meta-analysis. *J Arthroplasty* 2009;24(1):38–42.

Gao, Y., Zou, S., Liu, X., Bao, C., and Hu, J. The effect of surface immobilized bisphosphonates on the fixation of hydroxyapatite-coated titanium implants in ovariectomized rats. *Biomaterials* 2009;30(9):1790–1796.

Gomez-Vega, J.M., Hozumi, E., Saiz, E., Tomsia, A.P., Sugimura, H., and Takai, O. Bioactive glass–mesoporous silica coatings on Ti6Al4V through enameling and triblock–copolymer-templated sol–gel processing. *J Biomed Mater Res* 2001;56:382–389.

Gomez-Vega, J.M., Saiz, E., Tomsia, A.P., Marshall, G.W., and Marshall, S.J. Bioactive glass coatings with hydroxyapatite and Bioglass particles on Ti-based implants—Part I: Processing. *Biomaterials* 2000a;21:105–111.

Gomez-Vega, J.M., Saiz, E., Tomsia, A.P., Oku, T., Suganuma, K., Marshall, G.W., and Marshall, S.J. Novel bioactive glass functionally graded coatings on Ti6Al4V. *Adv Mater* 2000b;12:894–898.

Greenspan, D.C., and Hench, L.L. Chemical and mechanical properties of bioglass-coated alumina. *J Biomed Mater Res* 1976;10:503–509.

Grgurevic, L., Macek, B., Mercep, M., Jelic, M., Smoljanovic, T., Erjavec, I. et al. Bone morphogenetic protein (BMP)1-3 enhances bone repair. *Biochem Biophys Res Commun* 2011;408(1):25–31.

Griss, P., Greenspan, D.C., Heimke, G., Krempien, B., Buchinger, R., Hench, L.L., and Jentschura, G. Evaluation of a bioglass-coated Al2O3 total hip prosthesis in sheep. *J Biomed Mater Res* 1976;10:511–518.

Hallab, N.D., Jacobs, J.J., and Katz, J.L. Orthopaedic Applications. In *Biomaterials Science: An Introduction to Materials in Medicine*, Rutner, B.D., et al. editors. Elsevier Academic Press, 2004, pp. 527–556.

Hamadouche, M., Meunier, A., Greenspan, D.C., Blanchat, C., Zhong, J., La Torre, G.P., and Sedel L. Long-term *in vivo* bioactivity and degradability of bulk sol–gel bioactive glasses. *J Biomed Mater Res* 2001;54:560–566.

Hautaniemi, J. Oxidation and porcelain veneering of Ti and Pd alloys. Academic dissertation. University of Turku, Finland, 1993.

Hench, L.L., ed. Bioactive implants. *Transactions of the Annual Meeting of the Society for Biomaterials in Conjunction with the Interna*, 1985.

Hench, L.L. Bioceramics: From concept to clinic. *J Am Ceram Soc* 1991;74:1487–1510.

Hench, L.L., Bioactive ceramics: Theory and clinical applications. In *Bioceramics*, Vol. 7, Anderson, et al. editors. Butterworth-Heinemann, Oxford, 1994, pp. 3–14.

Hench, L.L., and Andersson, Ö.H. Bioactive glass coatings. In *An Introduction to Bioceramics*, Hench, L.L., and Wilson, J., editors. World Scientific, 1993a, p. 239.

Hench, L.L., and Andersson, Ö.H. Bioactive glasses. In *An Introduction to Bioceramics*, Hench LL and Wilson J, editors. World Scientific, 1993b, pp. 41–62.

Hench, L.L., Pantano, C.G., Buscemi, P.J., and Greenspan, D.C. Analysis of bioglass fixation of hip prostheses. *J Biomed Mater Res, Part A* 1977;11:267–282.

Hench, L.L., and Polak, J.M. Third-generation biomedical materials. *Science* 2002;295(5557): 1014–1017.

Hench, L.L., Splinter, R.J., Allen, W.C., and Greenlee, T.K. Bonding mechanisms at the interface of ceramic prosthetic materials. *J Biomed Mater Res* 1971;5:117–141.

Hench, L.L., Wheeler, D., and Greenspan, D.C. Molecular control of bioactivity in sol–gel glasses. *J Sol–Gel Sci Technol* 1998;13:245–250.

Hench, L.L., Xynos, I.D., and Polak, J.M. Bioactive glasses for *in situ* tissue regeneration. *J Biomater Sci, Polym Ed* 2004;15(4):543–562.

Hilding, M., and Aspenberg, P. Postoperative clodronate decreases prosthetic migration: 4-year follow-up of a randomized radiostereometric study of 50 total knee patients. *Acta Orthop* 2006;77(6):912–916.

Hilding, M., and Aspenberg, P. Local peroperative treatment with a bisphosphonate improves the fixation of total knee prostheses: A randomized, double-blind radiostereometric study of 50 patients. *Acta Orthop* 2007;78(6):795–799.

Itälä, A., Nordström, E., Ylänen, H., Aro, H., and Hupa, M. Creation of microrough surface on sintered bioactive glass microspheres. *J Biomed Mater Res* 2001;56:282–288.

Jia, W., Beatty, M.W., Reinhardt, R.A., Petro, T.M., Cohen, D.M., Maze, C.R. et al. Nickel release from orthodontic arch wires and cellular immune response to various nickel concentrations. *J Biomed Mater Res* 1999;48(4):488–495.

Jokinen, M., Pätsi, M., Rahiala, H., Peltola, T., Ritala, M., and Rosenholm, J.B. Influence of sol and surface properties on *in vitro* bioactivity of sol–gel-derived TiO_2 and TiO_2–SiO_2 films deposed by dip-coating method. *J Biomed Mater Res* 1998;42:295–302.

Junker, R., Dimakis, A., Thoneick, M., and Jansen, J.A. Effects of implant surface coatings and composition on bone integration: A systematic review. *Clin Oral Implants Res* 2009;20(Suppl 4):185–206.

Kajiwara, H., Yamaza, T., Yoshinari, M., Goto, T., Iyama, S., Atsuta, I. et al. The bisphosphonate pamidronate on the surface of titanium stimulates bone formation around tibial implants in rats. *Biomaterials* 2005;26(6):581–587.

Kapanen, A., Ryhänen, J., Danilov, A., and Tuukkanen, J. Effect of nickel-titanium shape memory metal alloy on bone formation. *Biomaterials* 2001;22(18):2475–2480.

Kärrholm, J., Malchau, H., Snorrason, F., and Herberts, P. Micromotion of femoral stems in total hip arthroplasty: A randomized study of cemented, hydroxyapatite-coated, and porous-coated stems with Roentgen stereophotogrammetric analysis. *J Bone Jt Surg, Am Vol* 1994;76(11):1692–1705.

Kim, H.-M., Kokubo, T., Fujibayashi, S., Nishiguchi, S., and Nakamura, T. Bioactive macroporous titanium surface layer on titanium substrate. *J Biomed Mater Res* 2000;52:553–557.

Kim, H.-M., Miyaji, F., Kokubo, T., and Nakamura, T. Preparation of bioactive Ti and its alloys via simple chemical surface treatment. *J Biomed Mater Res* 1996;32:409–417.

Kitshugi, T., Nakamura, T., Oka, M., Senaha, Y., Goto, T., and Shibuya, T. Bone-bonding behavior of plasma-sprayed coatings of Bioglass, AW–glass ceramic, and tricalcium phosphate on titanium alloy. *J Biomed Mater Res* 1996;30:261–269.

Kokubo, T., Kim, H.M., and Kawashita, M. Novel bioactive materials with different mechanical properties. *Biomaterials* 2003;24(13):2161–2175.

Kokubo, T., Kim, H.M., Nishiguchi, S., and Nakamura, T. *In vivo* apatite formation induced on titanium metal and its alloys by chemical treatment. Bioceramics 13 *Key Eng Mater* 2000;192–195:3–6.

Kokubo, T., Kushitani, H., Sakka, S., Kitsugi, T., and Yamamuro, T. Solutions able to reproduce *in vivo* surface-structure changes in bioactive glass–ceramic A–W. *J Biomed Mater Res* 1990;24:721–731.

Kujala, S., Pajala, A., Kallioinen, M., Pramila, A., Tuukkanen, J., and Ryhänen, J. Biocompatibility and strength properties of nitinol shape memory alloy suture in rabbit tendon. *Biomaterials* 2004;25(2):353–358.

Lacefield, W.R., and Hench, L.L. The bonding of Bioglass to a cobalt–chromium surgical implant alloy. *Biomaterials* 1986;7:104–108.

Langer, R. Drug delivery and targeting. *Nature* 1998;392(Suppl 6679):5–10.

Langer, R., and Peppas, N.A. Advances in biomaterials, drug delivery, and bionanotechnology. *AIChE J* 2003;49(12):2990–3006.

Lazarinis, S., Kärrholm, J., and Hailer, N.P. Increased risk of revision of acetabular cups coated with hydroxyapatite. *Acta Orthop* 2010;81(1):53–59.

Lee, J.J., Rouhfar, L., and Beirne, O.R. Survival of hydroxyapatite coated implants: A meta-analytic review. *J Oral Maxillofac Surg* 2000;58:1372–1379.

Levine, B. A new era in porous metals: Applications in orthopaedics. *Adv Eng Mater* 2008;10(9):788–792.

Li, J., Li, Y., Ma, S., Gao, Y., Zuo, Y., and Hu, J. Enhancement of bone formation by BMP-7 transduced MSCs on biomimetic nanohydroxyapatite/polyamide composite scaffolds in repair of mandibular defects. *J Biomed Mater Res—Part A* 2010;95(4):973–981.

Li, P., and de Groot, K. Calcium phosphate formation within sol–gel prepared titania *in vitro* and *in vivo*. *J Biomed Mater Res* 1993;27:1495–1500.

Li, P., Ohtsuki, C., Kokubo, T., Nakanishi, K., Soga, N., Nakamura, T., and Yamamuro, T. Process of formation of bone-like apatite layer on silica gel. *J Mater Sci: Mater Med* 1993;4:127–131.

Linderbäck, P., Areva, S., Aspenberg, P., and Tengvall, P. Sol–gel derived titania coating with immobilised bisphosphonate enhances screw fixation in rat tibia. *J Biomed Mater Res—Part A* 2010;94(2):389–395.

Lombardi, A.V., Berasi, C.C., and Berend, K.R. Evolution of tibial fixation in total knee arthroplasty. *J Arthroplasty* 2007;22(4 Suppl 1):25–29. Epub 2007/08/25.

Luo, T., Zhang, W., Shi, B., Cheng, X., and Zhang, Y. Enhanced bone regeneration around dental implant with bone morphogenetic protein 2 gene and vascular endothelial growth factor protein delivery. *Clin Oral Implants Res* 2011;23(4):467–473.

Maïmoun, L., Brennan, T.C., Badoud, I., Dubois-Ferriere, V., Rizzoli, R., and Ammann, P. Strontium ranelate improves implant osseointegration. *Bone* 2010;46(5):1436–1441.

Marie, P.J., Garba, M.T., Hott, M., and Miravet, L. Effect of low doses of stable strontium on bone metabolism in rats. *Miner Electrolyte Metab* 1985;11(1):5–13.

Marie, P.J., Hott, M., Modrowski, D., De Pollak, C., Guillemain, J., Deloffre. P. et al. An uncoupling agent containing strontium prevents bone loss by depressing bone resorption and maintaining bone formation in estrogen-deficient rats. *J Bone Miner Res* 1993;8(5):607–615. Epub 1993/05/01.

McLeod, K., Anderson, G.I., Dutta, N.K., Smart, R.S.C., Voelcker, N.H., Sekel, R. et al. Adsorption of bisphosphonate onto hydroxyapatite using a novel co-precipitation technique for bone growth enhancement. *J Biomed Mater Res—Part A* 2006;79(2):271–281.

Meraw, S.J., and Reeve, C.M. Qualitative analysis of peripheral peri-implant bone and influence of alendronate sodium on early bone regeneration. *J Periodontol* 1999;70(10):1228–1233.

Meraw, S.J., Reeve, C.M., and Wollan, P.C. Use of alendronate in peri-implant defect regeneration. *J Periodontol* 1999;70(2):151–158.

Meunier, P.J., Roux, C., Seeman, E., Ortolani, S., Badurski, J.E., Spector, T.D. et al. The effects of strontium ranelate on the risk of vertebral fracture in women with postmenopausal osteoporosis. *N Engl J Med* 2004;350(5):459–468. Epub 2004/01/30.

Morgan, N.B. Medical shape memory alloy applications—The market and its products. *Mater Sci Eng, A* 2004;378(1–2):16–23.

Moritz, N., Areva, S., Wolke, J., and Peltola T. TF-XRD examination of surface-reactive TiO_2 coatings produced by heat treatment and CO_2 laser treatment. *Biomaterials* 2005;26(21):4460–4467.

Moritz, N., Jokinen, M., Peltola, T., Areva, S., and Yli-Urpo, A. Local induction of calcium phosphate formation on TiO_2 coatings on titanium via surface treatment with a CO_2 laser. *J Biomed Mater Res—Part A* 2003;65(1):9–16.

Moritz, N., Rossi, S., Vedel, E., Tirri, T., Ylänen, H., Aro, H., and Närhi, T. Implants coated with bioactive glass by CO_2 laser: An *in vivo* study. *J Mater Sci: Mater Med* 2004a;15(7):795–802.

Moritz, N., Vedel, E., Ylänen, H., Jokinen, M., Hupa, M., and Yli-Urpo, A. Characterization of bioactive glass coatings on titanium substrates produced using a CO_2 laser. *J Mater Sci: Mater Med* 2004b;15(7):787–794.

Morohashi, T., Sano, T., Harai, K., and Yamada, S. Effects of strontium on calcium metabolism in rats—Part II: Strontium prevents the increased rate of bone turnover in ovariectomized rats. *Jpn J Pharmacol* 1995;68(2):153–159.

Moroni, A., Faldini, C., Hoang-Kim, A., Pegreffi, F., and Giannini, S. Alendronate improves screw fixation in osteoporotic bone. *J Bone Jt Surg, Ser A* 2007;89(1):96–101.

Murakami, H., Takahashi, N., Sasaki, T., Udagawa, N., Tanaka, S., Nakamura, I. et al. A possible mechanism of the specific action of bisphosphonates on osteoclasts: Tiludronate preferentially affects polarized osteoclasts having ruffled borders. *Bone* 1995;17(2): 137–144.

Niinomi, M. Metallic biomaterials. *J Artif Organs* 2008;11:105–110.

Nishida, S., Yamaguchi, A., Tanizawa, T., Endo, N., Mashiba, T., Uchiyama, Y. et al. Increased bone formation by intermittent parathyroid hormone administration is due to the stimulation of proliferation and differentiation of osteoprogenitor cells in bone marrow. *Bone* 1994;15(6):717–723.

Oku, T., Suganuma, K., Wallenberg, L.R., Tomsia, A.P., Gomez-Vega, J.M., and Saiz, E. Structural characterization of the metal/glass interface in bioactive glass coatings on Ti–6Al–4V. *J Mater Sci: Mater Med* 2001;12:413–417.

Ott, S.M. Clinical effects of bisphosphonates in involutional osteoporosis. *J Bone Miner Res* 1993;8(Suppl 2):S597–S606.

Overgaard, S., Søballe, K., Josephsen, K., Hansen, E.S., and Bünger, C. Role of different loading conditions on resorption of hydroxyapatite coating evaluated by histomorphometric and stereological methods. *J Orthop Res* 1996;14(6):888–894.

Pajamäki, K.J., Lindholm, T.S., Andersson, Ö.H., Karlsson, K.H., Vedel, E., Yli-Urpo, A., and Happonen, R.P. Bioactive glass and glass–ceramic-coated hip endoprosthesis: Experimental study in rabbit. *J Mater Sci: Mater Med* 1995;6:14–18.

Pan, J., Liao, H., Leygraf, C., Thierry, D., and Li, J. Variayion of oxide films on titanium induced by osteoblast-like cell culture and influence of an H2O2 pretreatment. *J Biomed Mater Res* 1998;40:244–256.

Patterson, J., Siew, R., Herring, S.W., Lin, A.S., Guldberg, R., and Stayton, P.S. Hyaluronic acid hydrogels with controlled degradation properties for oriented bone regeneration. *Biomaterials* 2010;31(26):6772–6781.

Paulsen, A., Pedersen, A.B., Johnsen, S.P., Riis, A., Lucht, U., and Overgaard, S. Effect of hydroxyapatite coating on risk of revision after primary total hip arthroplasty in younger patients: Findings from the Danish hip arthroplasty registry. *Acta Orthop* 2007;7(5):622–628.

Pauly, S., Luttosch, F., Morawski, M., Haas, N.P., Schmidmaier, G., and Wildemann, B. Simvastatin locally applied from a biodegradable coating of osteosynthetic implants improves fracture healing comparable to BMP-2 application. *Bone* 2009;45(3):505–511.

Pelli, S., Reghini, G.C., Scaglione, A., Ascoli, C., Frediani, C., Martucci, A., and Guglielmi, M. Characterization of laser-written sol–gel strip waveguides. *SPIE Proc* 1994;2288:573–579.

Pelli, S., Reghini, G.C., Scaglione, A., Guglielmi, M., and Martucci, A. Direct writing of ridge optical waveguides in silica–titania glass sol–gel films. *Opt Mater* 1996;5:119–126.

Peltola, T. Nanoscale dimensions and *in vitro* calcium phosphate formation: Studies on sol–gel-derived materials and bioactive glass. Academic dissertation, Ser. A Tom. 263. Astronomica-Chemica-Physica-Mathematica, University of Turku, Turku, 2000.

Peltola, T., Jokinen, M., Rahiala, H., Pätsi, M., Heikkilä, J., Kangasniemi, I., and Yli-Urpo, A. Effect of aging time of sol on structure and *in vitro* calcium phosphate formation of sol–gel-derived titania films. *J Biomed Mater Res* 2000;51:200–208.

Peltola, T., Pätsi, M., Rahiala, H., Kangasniemi, I., and Yli-Urpo, A. Calcium phosphate induction by sol–gel-derived titania coatings on titanium substrates *in vitro*. *J Biomed Mater Res* 1998;41:504–510.

Peter, B., Gauthier, O., Laïb, S., Bujoli, B., Guicheux, J., Janvier, P. et al. Local delivery of bisphosphonate from coated orthopedic implants increases implants mechanical stability in osteoporotic rats. *J Biomed Mater Res—Part A* 2006;76(1):133–143.

Peter, B., Pioletti, D.P., Laïb, S., Bujoli, B., Pilet, P., Janvier, P. et al. Calcium phosphate drug delivery system: Influence of local zoledronate release on bone implant osteointegration. *Bone* 2005;36(1):52–60.

Pietl Filho, O., LaTorre, G.P., and Hench, L.L. Effect of crystallization on apatite-layer formation of bioactive glass 45S5. *J Biomed Mater Res* 1996;30:509–514.

Pilliar, R.M. Overview of surface variability of metallic endosseous dental implants: Textured and porous surface structured designs. *Implant Dent* 1998;7:305–312.

Piotrowski, G., Hench, L.L., Allen, W.C., and Miller, G.J. Mechanical studies of the bone Bioglass interfacial bond. *J Biomed Mater Res* 1975;9:47–61.

Pors Nielsen, S. The biological role of strontium. *Bone* 2004;35(3):583–588.

Quarles, L.D. Cation sensing receptors in bone: A novel paradigm for regulating bone remodeling? *J Bone Miner Res* 1997;12(12):1971–1974.

Reginster, J.Y., Felsenberg, D., Boonen, S., Diez-Perez, A., Rizzoli, R., Brandi, M.L. et al. Effects of long-term strontium ranelate treatment on the risk of nonvertebral and vertebral fractures in postmenopausal osteoporosis: Results of a five-year, randomized, placebo-controlled trial. *Arthritis Rheum* 2008;58(6):1687–1695. Epub 2008/06/03.

Rhalmi, O., Assad, T., and Rivard, Y. Hard, soft tissue and in vitro cell response to porous nickel-titanium: A biocompatibility evaluation. *Biomed Mater Eng* 1999;9(3):151.

Righini, G.C., and Pelli, S. Sol–gel glass waveguides. *J Sol–Gel Sci Technol* 1997;8:991–997.

Rodan, G.A., and Fleisch, H.A. Bisphosphonates: Mechanisms of action. *J Clin Invest* 1996;97(12):2692–2696.

Rossi. S., Moritz, N., Tirri, T., Peltola, T., Areva, S., Jokinen, M., Happonen, R.P., and Närhi, T. Comparison between sol–gel-derived anatase- and rutile-structured TiO_2 coatings in soft-tissue environment. *J Biomed Mater Res—Part A* 2007;82(4):965–974.

Rossi, S., Tirri, T., Paldan, H., Kuntsi-Vaattovaara, H., Tulamo, R., and Närhi, T. Peri-implant tissue response to TiO_2 surface modified implants. *Clin Oral Implants Res* 2008;19(4):348–355.

Roux, C., Reginster, J.Y., Fechtenbaum, J., Kolta, S., Sawicki, A., Tulassay, Z. et al. Vertebral fracture risk reduction with strontium ranelate in women with postmenopausal osteoporosis is independent of baseline risk factors. *J Bone Miner Res* 2006;21(4):536–542.

Ryd, L. Roentgen stereophotogrammetric analysis of prosthetic fixation in the hip and knee joint. *Clin Orthop Relat Res* 1992;276:56–65.

Ryhänen, J., Kallioinen, M., Tuukkanen, J., Lehenkari, P., Junila, J., Niemelä, E., Sandvik, P. et al. Bone modeling and cell–material interface responses induced by nickel–titanium shape memory alloy after periosteal implantation. *Biomaterials* 1999;20(14):1309–1317.

Saiz, E., Goldman, M., Gomez-Vega, J.M., Tomsia, A.P., Marshall, G.W., and Marshall, S.J. *In vitro* behavior of silicate glass coatings on Ti6Al4V. *Biomaterials* 2002;23:3749–3756.

Sato, M., and Grasser, W. Effects of bisphosphonates on isolated rat osteoclasts as examined by reflected light microscopy. *J Bone Miner Res* 1990;5(1):31–40.

Sato, M., Grasser, W., Endo, N., Akins, R., Simmons, H., Thompson, D.D. et al. Bisphosphonate action: Alendronate localization in rat bone and effects on osteoclast ultrastructure. *J Clin Invest* 1991;88(6):2095–2105.

Schepers, E., Ducheyne, P., and De Clerq, M. Interfacial analysis of fiber-reinforced bioactive glass dental root implants. *J Biomed Mater Res* 1989;23:735–752.

Seeman, E., Vellas, B., Benhamou, C., Aquino, J.P., Semler, J., Kaufman, J.M. et al. Strontium ranelate reduces the risk of vertebral and nonvertebral fractures in women 80 years of age and older. *J Bone Miner Res* 2006;21(7):1113–1120.

Shaw, D.J., and King, T.A. Densification of sol–gel silica glass by laser irradiation. *SPIE Proc* 1990;132:474–481.

Shih, C., Lin, S., Chen, Y., Su, Y., Lai, S., Wu, G.J. et al. The cytotoxicity of corrosion products of nitinol stent wire on cultured smooth muscle cells. *J Biomed Mater Res* 2000;52(2):395–403.

Singh, S., Trikha, S.P., and Edge, A.J. Hydroxyapatite ceramic-coated femoral stems in young patients: A prospective 10-year study. *J Bone Jt Surg, Br Vol* 2004;86(8):1118–1123.

Skripitz, R., and Aspenberg, P. Parathyroid hormone (1-34) increases attachment of PMMA cement to bone. *J Orthop Sci* 2001;6(6):540–544.

Skripitz, R., Bohling, S., Ruther, W., and Aspenberg, P. Stimulation of implant fixation by parathyroid hormone (1-34)—A histomorphometric comparison of PMMA cement and stainless steel. *J Orthop Res* 2005;23(6):1266–1270.

Søballe, K., Hansen, E.S., Brockstedt-Rasmussen, H., and Bunger, C. Hydroxyapatite coating converts fibrous tissue to bone around loaded implants. *J Bone Jt Surg, Br Vol* 1993a;75(2):270–278.

Søballe, K., Toksvig-Larsen, S., Gelineck, J., Fruensgaard, S., Hansen, E.S., Ryd, L. et al. Migration of hydroxyapatite coated femoral prostheses: A Roentgen stereophotogrammetric study. *J Bone Jt Surg, Br Vol* 1993b;75(5):681–687.

Stenport, V.F., Johansson, C., Heo, S.J., Aspenberg, P., and Albrektsson, T. Titanium implants and BMP-7 in bone: An experimental model in the rabbit. *J Mater Sci: Mater Med* 2003;14(3):247–254.

Storm, T., Steiniche, T., Thamsborg, G., and Melsen, F. Changes in bone histomorphometry after long-term treatment with intermittent, cyclic etidronate for postmenopausal osteoporosis. *J Bone Miner Res* 1993;8(2):199–208.

Takadama, H., Kim, H.-M., Kokubo, T., and Nakamura, T. TEM–EDX study of mechanism of bonelike apatite formation on bioactive titanium metal in simulated body fluid. *J Biomed Mater Res* 2001;57:441–448.

Taylor, D.J., Birnie, D.P. III, and Fabes, B.D. Temperature calculation for laser-irradiated sol–gel films on oxide substrates. *J Mater Res* 1995;10:1429–1434.

Taylor, D.J., Cronin, J.P., Allard, L.F. Jr., and Birnie, D.P. III. Microstructure of laser-fired, sol–gel-derived tungsten oxide films. *Chem Mater* 1996;8:1396–1401.

ten Dijke, P., Krause, C., de Gorter, D.J., Löwik, C.W., and van Bezooijen, R.L. Osteocyte-derived sclerostin inhibits bone formation: Its role in bone morphogenetic protein and Wnt signaling. *J Bone Jt Surg, Am Vol* 2008;90(Suppl 1):31–35.

Tengvall, P., Elwing, H., Sjoqvist, L., Lundstrom, I., and Bjursten, L.M. Interaction between hydrogen peroxide and titanium: A possible role in the biocompatibility of titanium. *Biomaterials* 1989a;10:118–120.

Tengvall, P., Lundstrom, I., Sjoqvist, L., Elwing, H., and Bjursten, L.M. Titanium–hydrogen peroxide interaction: Model studies of the influence of the inflammatory response on titanium implants. *Biomaterials* 1989b;10:166–175.

Tengvall, P., Skoglund, B., Askendal, A., and Aspenberg, P. Surface immobilized bisphosphonate improves stainless-steel screw fixation in rats. *Biomaterials* 2004;25(11):2133–2138.

Theis, J.C., and Ball, C. Medium-term results of cementless hydroxyapatite-coated primary total hip arthroplasty: A clinical and radiological review. *J Orthop Surg (Hong Kong)* 2003;11(2):1591–665.

Thompson, K., Rogers, M.J., Coxon, F.P., and Crockett, J.C. Cytosolic entry of bisphosphonate drugs requires acidification of vesicles after fluid-phase endocytosis. *Mol Pharmacol* 2006;69(5):1624–1632.

Uchida, M., Kim, H.-M., Kokubo, T., Fujibayashi, S., and Nakamura, T. Effect of water treatment on the apatite-forming ability of NaOH-treated titanium metal. *J Biomed Mater Res* 2003;63:522–533.

Vercaigne, S., Wolke, J.G.C., Naert, I., and Jansen, J.A. A histological evaluation of TiO_2-gritblasted and Ca–P magnetron sputter coated implants placed into the trabecular bone of the goat—Part 2. *Clin Oral Implants Res* 2000;11:314–324.

Wälivaara, B., Aronsson, B.-O., Rodahl, M., Lausmaa, J., and Tengvall, P. Titanium with different oxides: *In vitro* studies of protein adsorption and contact activation. *Biomaterials* 1994;15:827–834.

Wang, K. The use of titanium for medical applications in the USA. *Mater Sci Eng, A* 1996;213:134–137.

Wang, X.-X., Hayakawa, S., Tsuru, K., and Osaka, A. A comparative study of *in vitro* apatite deposition on heat-, H2O2-, and NaOH-treated titanium surfaces. *J Biomed Mater Res* 2001;54:172–178.

Wermelin, K., Aspenberg, P., Linderbäck, P., and Tengvall, P. Bisphosphonate coating on titanium screws increases mechanical fixation in rat tibia after 2 weeks. *J Biomed Mater Res—Part A* 2008;86(1):220–227.

West, J.K., Clark, A.E., Hall, M.B., and Turner, G.E. *In vivo* bone-bonding study of Bioglass-coated titanium alloy. In *Handbook of Bioactive Ceramics*—Vol. I, Yamamuro, T, et al. editors. CRC Press, Boca Raton, FL, 1990, pp. 161–166.

Williams, D.F. Definitions in biomaterials. *Proceedings of a Consensus Conference of the European Society for Biomaterials Chester*, England, 1986. Elsevier, Amsterdam, 1987.

Williams, D.F. Second consensus conference on definitions in biomaterials. In *Biomaterial–Tissue Interfaces*, Doherty, P.J. et al. editors. Elsevier, Amsterdam. *Adv Biomater* 1988;10:99–104.

Wilson, J., and Merwin, G.E. Biomaterials for facial bone augmentation: Comparative studies. *J Biomed Mater Res* 1988;22:159–177.

Wright, J.D. NAJMS. *Sol–Gel Materials Chemistry and Applications*, Phillips, D., editor. CRC Press, Boca Raton, FL, 2001.

Xiao, S.-J., Kenausis, G., and Textor, M. 2001. Biochemical modification of titanium surfaces. In *Titanium in Medicine*, Brunette, D.M. et al. editors. Springer-Verlag, Berlin, 2001, pp. 417–455.

Yamasaki, S., Masuhara, K., Yamaguchi, K., Nakai, T., Fuji, T., and Seino, Y. Risedronate reduces postoperative bone resorption after cementless total hip arthroplasty. *Osteoporosis Int* 2007;18(7):1009–1015.

Ylänen, H., Karlsson, K.H., Itälä, A., and Aro, H. Effect of immersion in SBF on porous bioactive bodies made by sintering bioactive glass microspheres. *J Non-Cryst Solids* 2000;275:107–205.

Zardiackas, L.D., Parsell, D.E., Dillon, L.D., Mitchell, D.W., Nunnery, L.A., and Poggie, R. Structure, metallurgy, and mechanical properties of a porous tantalum foam. *J Biomed Mater Res* 2001;58(2):180–187. Epub 2001/03/10.

Zhao, D.S., Moritz, N., Laurila, P., Mattila, R., Lassila, L.V.J., Strandberg, N., Mäntylä, T., Vallittu, P.K., and Aro, H.T. Development of a multicomponent fiber-reinforced composite implant for load-sharing conditions. *Med Eng Phys* 2009;31(4):461–469.

5 Induction of Bone Formation by Calcium Phosphate–Based Biomimetic Macroporous Constructs

Ugo Ripamonti

CONTENTS

5.1 INDUCTION OF BONE FORMATION AND THE TISSUE ENGINEERING PARADIGM

The induction of bone formation starts by erecting scaffolds of insoluble signals or substrata controlling the expression and synthesis of the soluble osteogenic molecular signals of the transforming growth factor-β (TGF-β) supergene family.[1–6] There is no bone formation without the osteogenic soluble molecular signals of the TGF-β supergene family.[3,4]

The induction of bone formation requires three key components:[4–6] (1) soluble osteogenic molecular signals; (2) responding stem cells; and (3) insoluble signals or substrata. The latter is the scaffold upon which differentiating and transforming mesenchymal stem cells erect "Bone: Formation by Autoinduction."[7,8] Last century research has thought us a fundamental truth in bone tissue engineering and regenerative medicine at large,[8–10] insoluble signals or substrata, when recombined or reconstituted with soluble osteogenic molecular signals, trigger the ripple-like cascade of tissue induction and morphogenesis.[1,2,4–6]

The induction of bone formation by recombining or reconstituting osteogenic soluble molecular signals with insoluble signals or substrata has been pivotal for setting the rules of the tissue engineering paradigm and regenerative medicine at large;

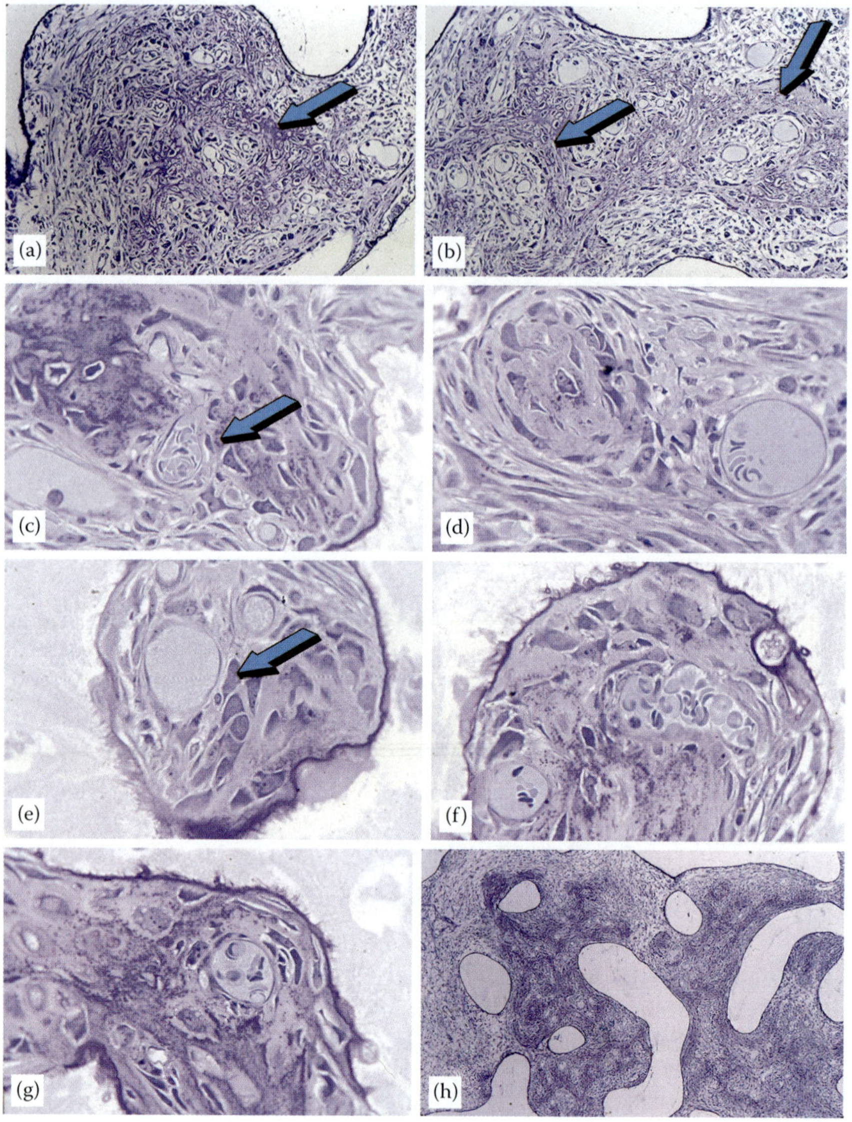

FIGURE 5.1 Tissue induction and morphogenesis by combinatorial molecular protocols whereby highly purified osteogenic protein fractions purified greater than 50,000-fold by sequential chromatography on hydroxyapatite Ultrogel adsorption and heparin-Sepharose affinity chromatography followed by gel filtration chromatography on tandem S-200 Sephacryl columns were reconstituted with discs of coral-derived macroporous constructs and implanted in the subcutaneous space of Long–Evans rats;[23] tissue specimens harvested on days 7 and 12 after heterotopic implantation were processed for undecalcified histology and embedded in historesin.[23] (a) and (b) Lower power microphotographs on day 7 highlighting the pronounced vascular invasion within the macroporous spaces surrounded by differentiating mesenchymal condensations populated by contiguous osteoblasts facing the invading central capillaries of Trueta's definition,[24] the "osteogenetic vessels" (blue arrows).

ultimately, thus, the tissue engineering paradigm is the induction of tissue morphogenesis by combinatorial molecular protocols.[1,4,5,9–12]

The induction of bone formation has evolved as a prototype of the tissue engineering paradigm continuously developing the concept and further expanding the depths and the breadth of the novel developmental, biological, and surgical concepts.[4,7,8,13–17] The induction of bone formation initiates by the invocation of osteogenic soluble molecular signals that, when combined with insoluble signals or substrata, trigger the ripple-like cascade of bone differentiation by induction (Figure 5.1).[4,7,8,18–21]

The osteogenic soluble molecular signals of the TGF-β supergene family, the BMPs/OPs, and, uniquely in primates, the three mammalian TGF-β isoforms[29–31] induce endochondral bone formation as a recapitulation of embryonic development.[20,21,32–35] More importantly, for tissue engineering in clinical contexts, neither the solubilized proteins nor the insoluble signals or residue were active;[9,10] the reconstitution of the soluble and insoluble components of the bone matrix, however, restored the osteogenic activity of the extraskeletal heterotopic bioassay.[4,6,9–11] This operational reconstitution of the soluble molecular signal with an insoluble signal or substratum was a key experiment that provided a bioassay for *bona fide* initiators of endochondral bone differentiation in heterotopic extraskeletal sites of animal models (Figure 5.1), including nonhuman primates (Figures 5.2 and 5.3).[5,6,20,21,33,36]

5.2 BIOMIMETISM, BIOMIMETIC MATRICES, AND THE INDUCTION OF BONE FORMATION

Can we initiate the induction of bone formation without the exogenous application of the osteogenic soluble molecular signals of the TGF-β supergene family? Can we engineer functional bioinspired biomimetic matrices that *per se* direct cell attachment and differentiation with expression of selected mRNA species of the TGF-β supergene family secreted and embedded within specific geometric cues of the biomimetic matrices, that is, a solid-sate matrix that, on its own right, expresses

FIGURE 5.1 (Continued) The invading sprouting capillaries are not only "osteogenetic" but also "morphogenetic," as proposed by Aristotle,[25,26] since are constructing the tissue patterning and morphogenesis of the osteonic bone surrounding each morphogenetic osteogenetic vessel. (c)–(g) High-power views of the morphological and molecular intimate relationships between the invading capillaries and the differentiating osteoblastic-like cells (arrows) of the bone forming compartment within the macroporous spaces. Arrows point to the morphological connection of osteoblastic-like cells with the invading capillaries. Based on the morphological data presented by Trueta, introducing the osteogenetic vessels,[24] Reddi's group has suggested that mesenchymal stem cells/osteoprogenitors are in very close contact with sprouting capillaries and may see and respond to specific laminin and type IV collagen sequence motifs and set into motion the ripple-like cascade of cell differentiation, osteoblast-like synthesis, and osteoid secretion, culminating in the induction of bone formation.[27,28] (h) Woven bone formation within the macroporous spaces on day 11; note the induction of bone formation around central capillaries. (From Ripamonti, U. et al., *Matrix*, 12(3), 202–12, 1992. With permission.)

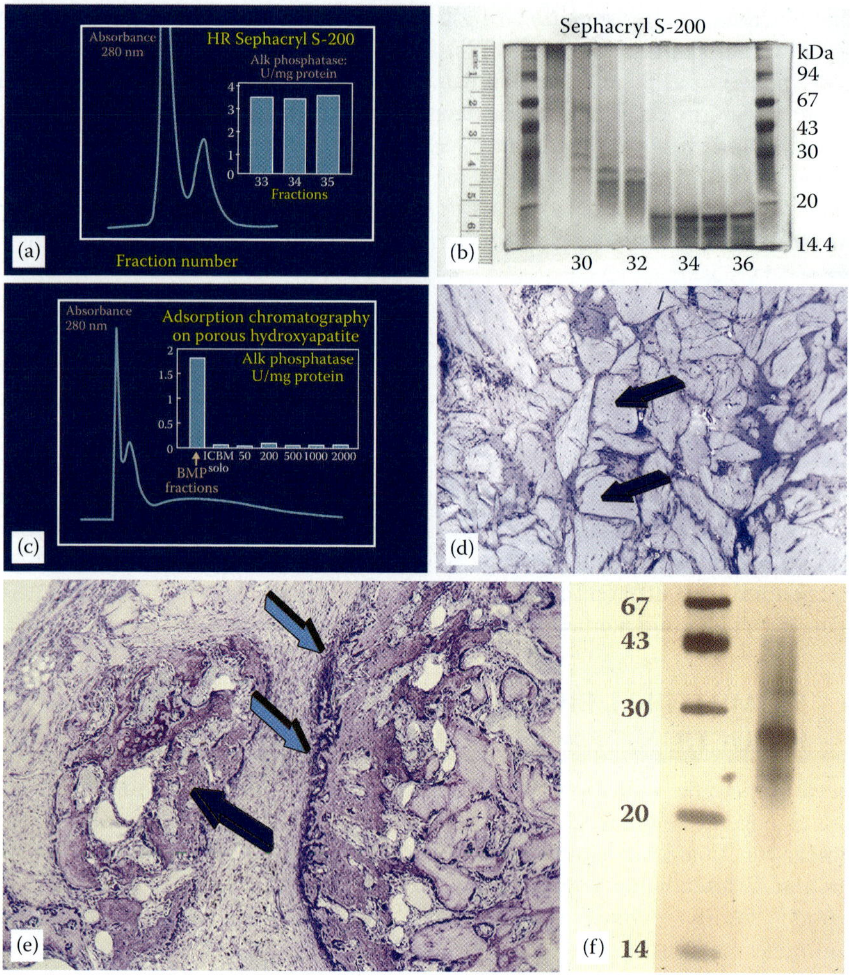

FIGURE 5.2 Reconstituting molecular signals by adsorption chromatography.[36] Adsorption chromatography of highly purified osteogenic proteins isolated and purified from baboon bone matrices onto macroporous discs of coral-derived hydroxyapatite replicas.[36] (a) Sephacryl S-200 gel filtration profile of baboon protein fractions purified after adsorption chromatography on Ultrogel hydroxyapatite matrix and affinity chromatography on heparin-Sepharose gel; the osteogenic activity is isolated in a single peak (a) after molecular sieve gel filtration on S-200 Sephacryl columns. (a, inset) Alkaline phosphatase activity on day 11 of implants of rat insoluble collagenous matrix as carrier reconstituted with protein fractions separated from high-molecular-weight components in a single peak (a) with absent high-molecular-weight contaminants on sodium dodecyl sulfate polyacrylamide gel electrophoresis (b); the alkaline phosphatase activity of the supernatant after homogenization of implants was determined with 0.1 M *p*-nitrophenylphosphate (Sigma) as substrate at pH 9.3 and 37°C for 30 min.[8] (c) Chromatographic adsorption profile of purified osteogenic fractions of the S-200 profile (a) onto macroporous coral-derived hydroxyapatite replicas.

the osteogenic soluble molecular signals endowed with the striking prerogative of initiating osteogenesis in angiogenesis?[6,15–17,21,32,33]

Several biomaterials have been implanted in extraskeletal intramuscular and/or subcutaneous sites in a variety of experimental animal models.[37] Heughebaert et al.[38] physiochemically characterized deposits associated with hydroxyapatite bioceramics implanted in heterotopic sites of hamsters.[38] The authors, based on the characterization of newly deposited material associated with the implanted hydroxyapatite bioceramics, suggested that it could have been tempting to speculate to call this material "bonelike material" and to "claim that the bioceramic promoted new bone formation," that is, osteoinductive, although the published studies did not give the histological and morphological evidence and characteristics of the true bone.[38]

The morphogenesis of bone in macroporous calcium phosphate bioceramics when implanted in heterotopic intramuscular sites was first reported after the implantation of coral-derived macroporous calcium carbonate/calcium phosphate constructs in the *rectus abdominis* muscle of adult nonhuman primates of the species *Papio ursinus* (Figures 5.4 and 5.5).[39,40] The most basic tissue engineering paradigm is the reconstitution of soluble molecular signals with insoluble signals or substrata, which triggers the induction of tissue morphogenesis.[4,9–12] The tissue engineering paradigm is fundamentally condensed by the fascinating phenomenon of "Bone: Formation by

FIGURE 5.2 (Continued) Fractions 33, 34, and 35 in 4 M Gdn–HCl after Sephacryl S-200 with biological activity (a, inset) in the rodent subcutaneous bioassay[36] were pooled and exchanged with 5 mM HCl acid and loaded, in a final concentration of 4.2 mg in 4 ml of 0.5 mM HCl, onto a chromatographic column containing discs of macroporous coral-derived hydroxyapatite replicas;[36] pooled highly osteogenic protein fractions were loaded into a Pharmacia chromatography column 25 mm in diameter packed with discs of coral-derived hydroxyapatite replicas 25 mm in diameter fitting by friction fit into the column. Adsorption was monitored at absorbance of 280 nm. After loading, the column was run in close circuit, recirculating the hydroxyapatite eluate. The column was left to dry, and the hydroxyapatite eluate (20 mL) was collected and bioassayed for residual osteogenic activity. (d) Hydroxyapatite-unbound material showed the lack of biological activity when reconstituted with the collagenous matrix (dark blue arrows) and implanted in the rat subcutaneous assay. Before adsorption chromatography, aliquots of the loading material were analyzed electrophoretically and further bioassayed in the subcutaneous space of the rat. (c, inset) Alkaline phosphatase activity on day 11 of implants of rat insoluble inactive collagenous matrix reconstituted with pooled osteogenic fractions; collagenous matrix without osteogenic proteins (ICBM *solo* as control, and collagenous matrix reconstituted with 50, 200, 500, 1000, and 2000 μL of hydroxyapatite-unbound eluate; values are the result of four to eight observations per group. (e) Pronounced osteogenic activity of pooled osteogenic fractions after gel filtration chromatography and just before adsorption chromatography (c); pronounced induction of bone formation on day 11 with newly formed woven bone (dark blue arrow) populated by contiguous osteoblasts (light blue arrows). (f) Electrophoretic profile of the pooled highly purified osteogenic fractions before (b) and after (c) adsorption chromatography on coral-derived macroporous constructs; lane b shows bands with an apparent molecular weight of 32/34 kDa with prominent biological activity in the rodent (e) and nonhuman primate bioassays (Figure 5.3). (Ripamonti, U. et al., *Biochem Biophys Res Commun*, 193(2), 509–17, 1993. With permission.)

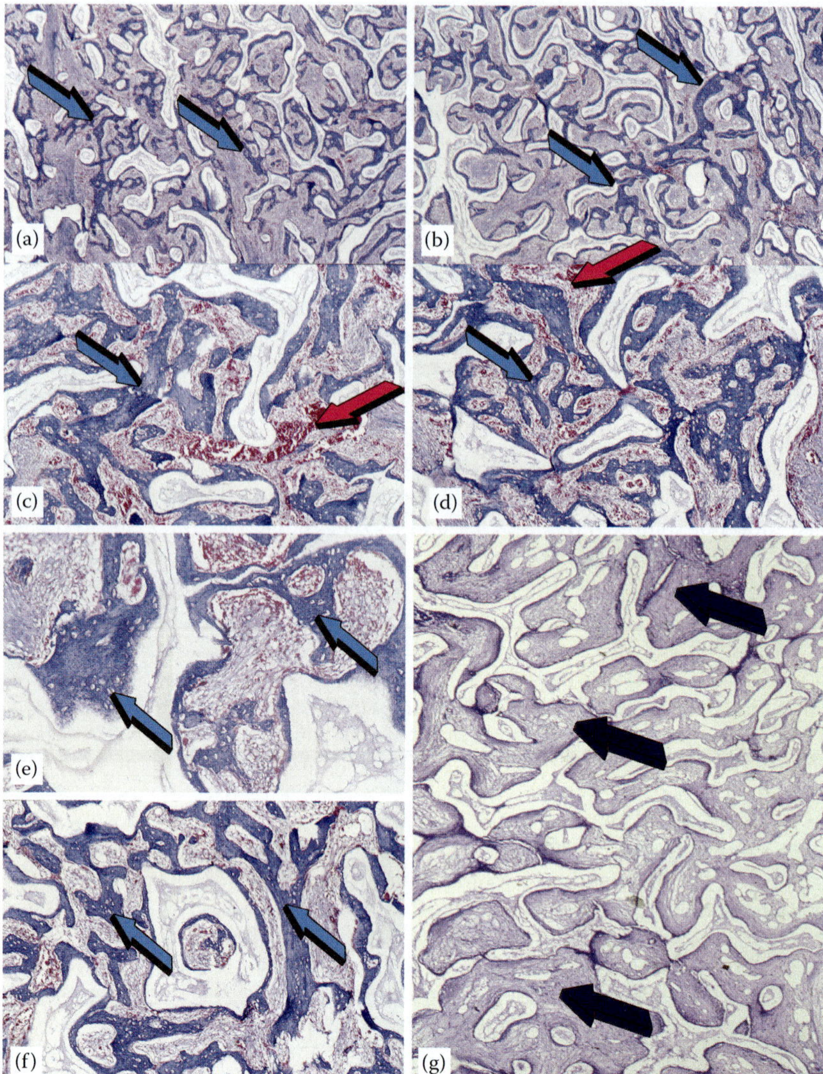

FIGURE 5.3 Induction of bone formation on day 30 by the prefabricated osteogenic constructs of coral-derived macroporous hydroxyapatite replicas after the adsorption chromatography of highly purified osteogenic proteins extracted and purified from the Chacma baboon *Papio ursinus*.[36] (a) and (b) Low-power microphotographs of the chromatographed macroporous hydroxyapatite replicas showing extensive bone differentiation by induction (light blue arrows) within the macroporous spaces. (c) and (d) Middle-power view of the newly induced woven bone (light blue arrows) together with prominent vascular invasion (red/magenta arrows) within the macroporous spaces. (e) and (f) High-power view of reconstituted macroporous samples after adsorption chromatography showing the induction of woven bone formation (light blue arrows) within the macroporous spaces. (g) Untreated coral-derived construct implanted in the rectus abdominis muscle without chromatographic adsorption of osteogenic proteins—lack of bone formation across the porous spaces (dark blue arrows).

Autoinduction."[7] The novel tissue engineering paradigm is to fabricate functionalized biomimetic biomaterial matrices that, *per se*, without the exogenous application of the osteogenic soluble molecular signals of the TGF-β supergene family, initiate the cascade of bone differentiation by induction even when implanted in heterotopic extraskeletal sites and without the exogenous applications of the osteogenic proteins of the TGF-β supergene family (Figures 5.4 and 5.5).[16,39–41]

We have modified the basic tissue engineering paradigm by implanting biomimetic matrices into the *rectus abdominis* muscle of adult nonhuman primates *P. ursinus*. There is a direct spatial and temporal relationship of morphological and molecular events, which emphasizes the biomimetism of the remodeling cycle of the osteonic corticocancellous bone *versus* the induction of bone formation by *smart* macroporous biomimetic calcium phosphate-based constructs (Figures 5.6 and 5.7).[16,17,32,44,45]

Osteoclastogenesis is a critical cellular and molecular event for the induction of bone formation by macroporous calcium phosphate-based biomaterials. Pretreatment of coral-derived macroporous surfaces by doses of the bisphosphonate zoledronate Zometa, which inhibits osteoclastic differentiation and function, blocks the bone induction cascade (Figure 5.7e and f).[43,44]

Carved concavities within biomimetic matrices induce the induction of bone differentiation (Figure 5.6). The lacunae, pits, and concavities cut by osteoclastogenesis are the driving morphogenetic cues that induce bone formation in a *continuum* of sequential phases of resorption/dissolution and bone formation (Figure 5.6f). The concavity ("shape of life")[32,33] biomimetizes the remodeling cycle of the primate corticocancellous osteonic bone (Figure 5.7a–d).[16,17,32] The activation phase results in osteoclastogenesis with the resorption of the mineral phase of the corticocancellous bone, leaving a trench with cross-sectional geometric cues of concavities after cyclic episodes of osteoclastogenesis.[16,47–49]

Concavities, as induced by osteoclastogenesis, precipitate the formation phase, whereas calcium ions (Ca^{2+}), released by osteoclasts during the resorption of the implanted matrices, stimulate angiogenesis and myoblastic/myoendothelial and/or pericytic/endothelial stem cell proliferation and differentiation toward osteoblastic cell lines.[43,50–55] Ultimately, which resident mesenchymal stem cells are capable of transformation and differentiation into actively secreting osteoblastic-like cells within the selected microenvironment at the concavity's interface?[56] The *rectus abdominis* striated muscle is obviously endowed with few, if not several, stem cell "niches" that provide a continuous stream of stem cells that attach and differentiate within the concavities of the nanotopographically functionalized surfaces.

Specific paravascular pericytic stem cell "niches" are present within the vascular and paravascular spaces; possibly, as first postulated by Keith,[57] endothelial cells are the first cell lines capable of transformation and differentiation into osteoblastic precursors; endothelial cells in single layers forming the primary structure of the invading capillaries may provide a continuous flow of stem cells from the invading "osteogenetic vessels" (Figure 5.4b–e);[24] the invading capillary is the functional and anatomical structure that will eventually construct the osteonic bone and, thus, the remodeling pattern of the skeleton.[16,24,27,47–49] It is tempting to suggest that the observed significant neovascular network formation with invading capillaries

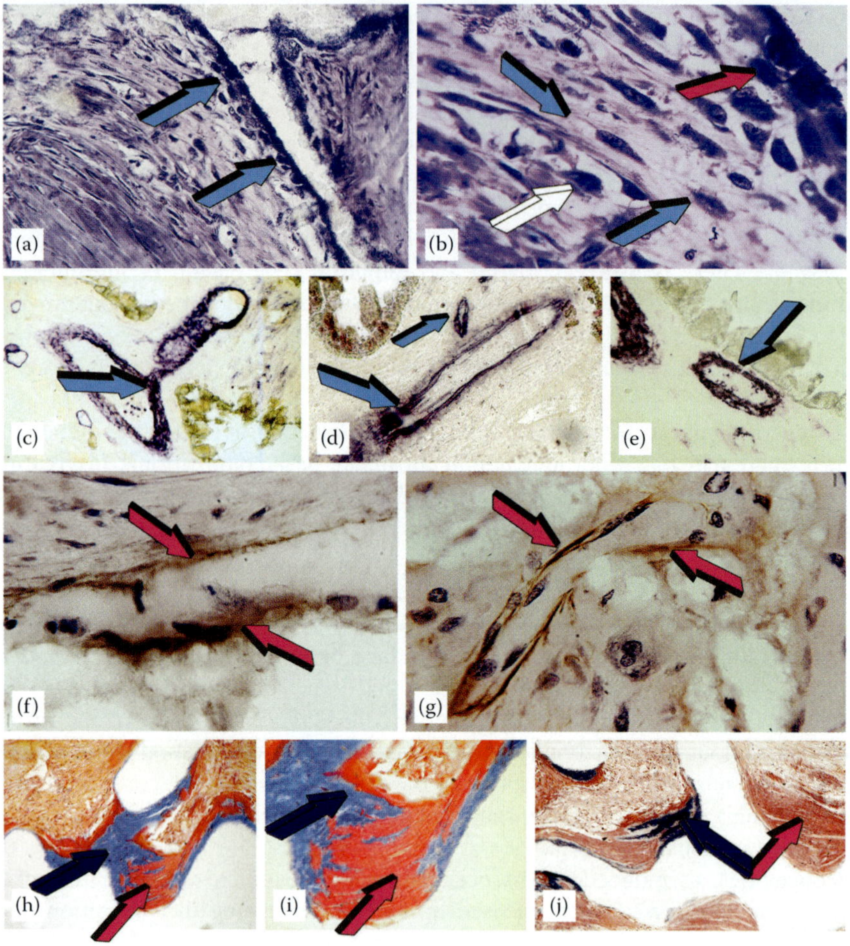

FIGURE 5.4 Stem cell differentiating events at the coral-derived hydroxyapatite mesenchymal interface 60 and 90 days after heterotopic implantation in the *rectus abdominis* muscle of adult nonhuman primates *P. ursinus*.[42] (a) and (b) Osteoblastic-like cell differentiation against the implanted substratum (light blue arrows). Arrows indicate large nucleated hyperchromatic cells differentiating against the coral-derived macroporous surface. (b) High-power view highlighting the differentiating hyperchromatic cells at the interface (magenta arrow) and cellular trafficking from the invading "osteogenetic" capillaries with large hyperchromatic nuclei (light blue arrows) from the invading capillary (white arrow) to the osteoblastic-like compartment (magenta arrow); arrows indicate a stream of differentiating and migrating cells from the vascular-activated endothelial compartment (white arrow) to the osteoblastic-like compartment trafficking through the extracellular permissive matrix. (c)–(g) Osteogenetic vessels of Trueta's definition[24] displaying alkaline phosphatase activity [light blue arrow in (c)–(e)] within the invading sprouting capillaries within the macroporous spaces as well as laminin immunostaining [magenta arrows in (f) and (g)] of the invading capillaries. Note the intimate relationship of the endothelial cells with the paravascular spaces and pericytic stem cells in close relationship with the basement membrane of the invading capillaries with laminin staining [magenta arrows in (f) and (g)].

sprouting within the macroporous spaces is further regulated by mechanical forces[58] as imparted by the geometric configuration of the macroporous substrata.

The conceptual framework of the osteonic bone and of the induction of bone formation would not be possible without the knowledge of the binding and segregation of angiogenic and osteogenic proteins to the invading sprouting capillaries that provide the framework of the supramolecular assembly of the extracellular matrix of bone.[32,33] Angiogenic and bone morphogenetic proteins bind to type IV collagen of the invading capillaries;[59–62] morphogenetic progression is sustained by the continuous recruitment of mesenchymal stem cells together with capillary invasion within concavities. Our systematic studies in the *rectus abdominis* muscle of adult Chacma baboons *P. ursinus* have clearly identified that the striated *rectus abdominis* muscle of *P. ursinus* is endowed with stem cell "niches" that provide a continuous flow of differentiating stem cells with osteoblastic-like transformation when attached to the osteophilic microenvironment of the concavities carved within the biomimetic matrices.[16,56]

Novel stem cell "niches" have recently been identified, including a myogenic endothelial stem cell "niche" within a human primate striated muscle.[63] Myoendothelial cells possibly derive from a common somatic precursor coexpressing myogenic and endothelial cell markers, that is, myoendothelial stem cells, which may indeed contribute to postnatal tissue morphogenesis.[63] More importantly, myoendothelial cells may respond to BMPs/OPs and angiogenic proteins previously bound to collagen type IV and other extracellular matrix components of the basement membrane of the invading capillaries; this may further enhance the ripple-like cascade of bone differentiation by induction within the macroporous spaces of the biomimetic matrices, mimicking the corticocancellous remodeling cycle of the primate osteonic bone.[16,56] Additional stem cell niches have been reported within the paravascular space,[64] which provide nonterminally differentiated vessel-resident stem cells, possibly endowed with osteogenic differentiation. Current data have shown that the perivascular space harbors a multilineage progenitor cell "niche" defined as regenerative units as sources of cytokines;[65] furthermore, it has been recently and provocatively speculated that all mesenchymal stem cells are paravascular pericytes,[66] with additional trophic activities to structure a regenerative microenvironment.[66,67]

FIGURE 5.4 (Continued) Laminin sequence motifs bind osteogenic proteins of the TGF-β supergene family including TGF-β_1, OP-1/BMP-7, and BMP-3[33] and may present it in an immobilized form to responding migrating cells. (h)–(j) Generation of collagenous condensations at the interface of coral-derived macroporous hydroxyapatite replica—mesenchymal condensations (magenta arrows) predate the induction of bone formation with mineralization (dark blue arrows) of the newly formed matrix developing within concavities of the coral-derived constructs.[16,26,33,39,40] The concavity is thus the geometric signal that sets into motion mesenchymal stem cell recruitment, differentiation, proliferation, collagenous synthesis, and the generation of collagenous condensations aligned against the macroporous surface, osteoblastic-like cell differentiation and synthesis, and secretion and embedding of the osteogenic gene products of the TGF-β supergene family within the concavities of the macroporous constructs.[16,41,43] (Ripamonti, U. et al., *Matrix*, 13, 491–502, 1993. With permission.)

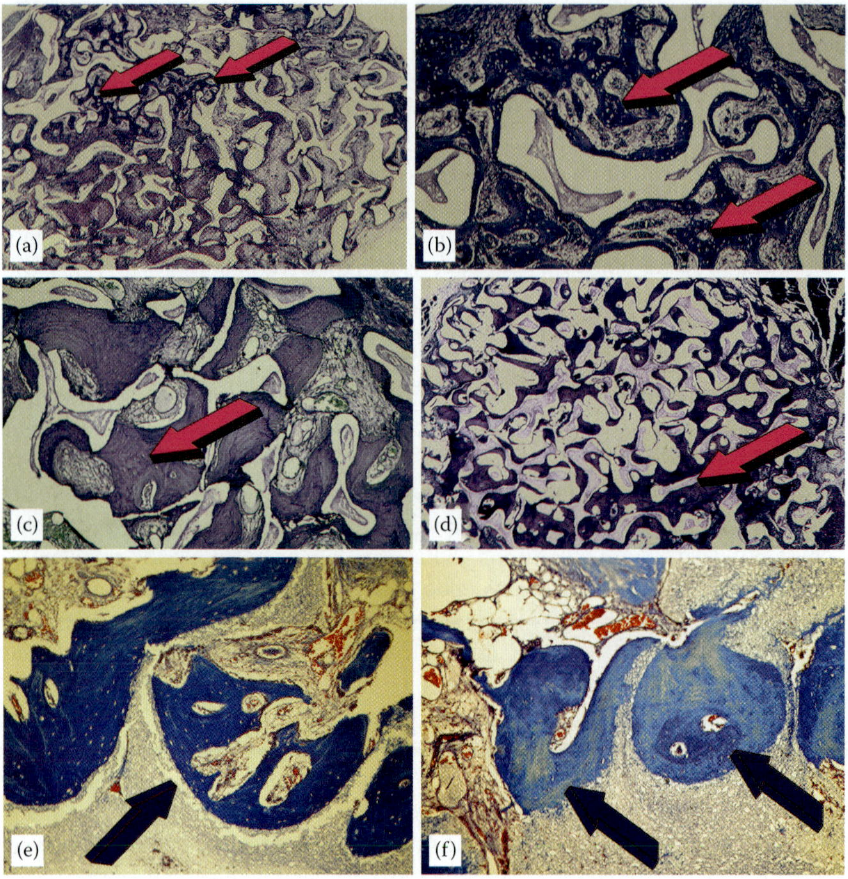

FIGURE 5.5 Microphotographic digital images of the "spontaneous" and "intrinsic" induction of bone formation by coral-derived macroporous constructs implanted in heterotopic sites of the *rectus abdominis* muscle of adult nonhuman primates *P. ursinus*.[40] (a) and (b) Induction of bone formation (magenta arrows) by the macroporous construct tightly attached to the macroporous surface (a) also protruding within the macroporous spaces (b) [magenta arrows in (a) and (b)] on day 90 after implantation. (c) Remodeling of the newly formed bone with the induction of solid trabeculae (magenta arrow) 6 months after heterotopic implantation in *P. ursinus*. (d) Further remodeling of the newly formed bone (magenta arrow) 9 months after heterotopic implantation in the rectus abdominis muscle with generation of bone marrow across the macroporous spaces. (e) and (f) The "concavity motif" (dark blue arrows) assembled in coral-derived macroporous constructs initiates the bone induction cascade tightly attached to the macroporous surface as seen 90 days after implantation of the coral-derived construct in the rectus abdominis muscle. (Ripamonti, U., *J Bone Jt Surg Am*, 73(5), 692–703, 1991. With permission.)

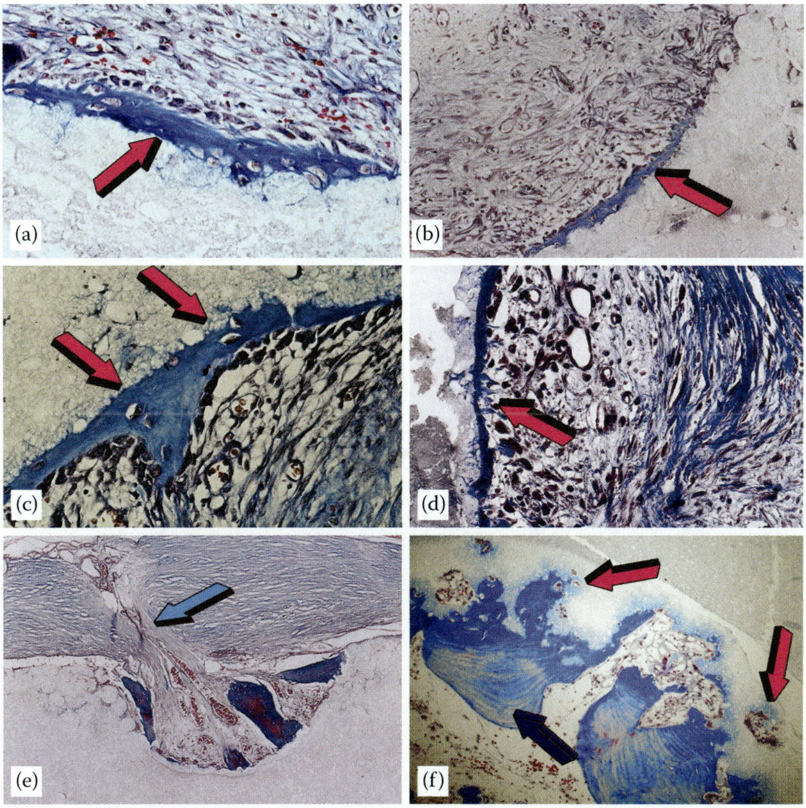

FIGURE 5.6 Concavity ("shape of life")[32] and the induction of bone formation. The induction of bone biomimetizes the remodeling cycle of the corticocancellous osteonic bone of the primate skeleton.[16,17,32,44] (a) and (b) Spontaneous intrinsic induction of bone formation on day 30 (magenta arrows) within concavities prepared in solid discs of highly crystalline hydroxyapatite.[41] (Ripamonti, U. et al., *S Afr J Sci*, 95, 335–43, 1999. With permission.) (c) and (d) Newly formed bone (magenta arrows) on day 30 tightly attached to the macroporous substratum of the concavity with embedded osteocytes and surfaced by contiguous osteoblasts. Note that the invading sprouting capillaries are always intimately connected with the newly formed bone. (e) Pronounced vascular invasion (light blue arrow) targeting the newly formed bone within a concavity of highly crystalline hydroxyapatite harvested on day 90.[41] (f) Concavity motif as cut by osteoclasts (magenta arrows) induces newly formed bone in a *continuum* of matrix resorption and dissolution and bone formation by induction.[46] (Ripamonti, U. et al., *J Cell Mol Med*, 12, 2609–21, 2008. With permission.) Solid discs of biphasic hydroxyapatite/β-tricalcium phosphate (20/80 HA/β-TCP presintering ratio) were implanted in the rectus abdominis muscle of *P. ursinus* and harvested on days 90 and 365 after heterotopic implantation.[46] Dark blue arrow in (e) indicates newly formed lamellar bone within the large concavity of the substratum 365 days after implantation; magenta arrows indicate the *continuum* of resorption, molecular and morphological induction of lacunae, and concavities by osteoclastogenesis in a *continuum* of bone formation by induction within the lacunae and concavities of the resorbed substrata.[46]

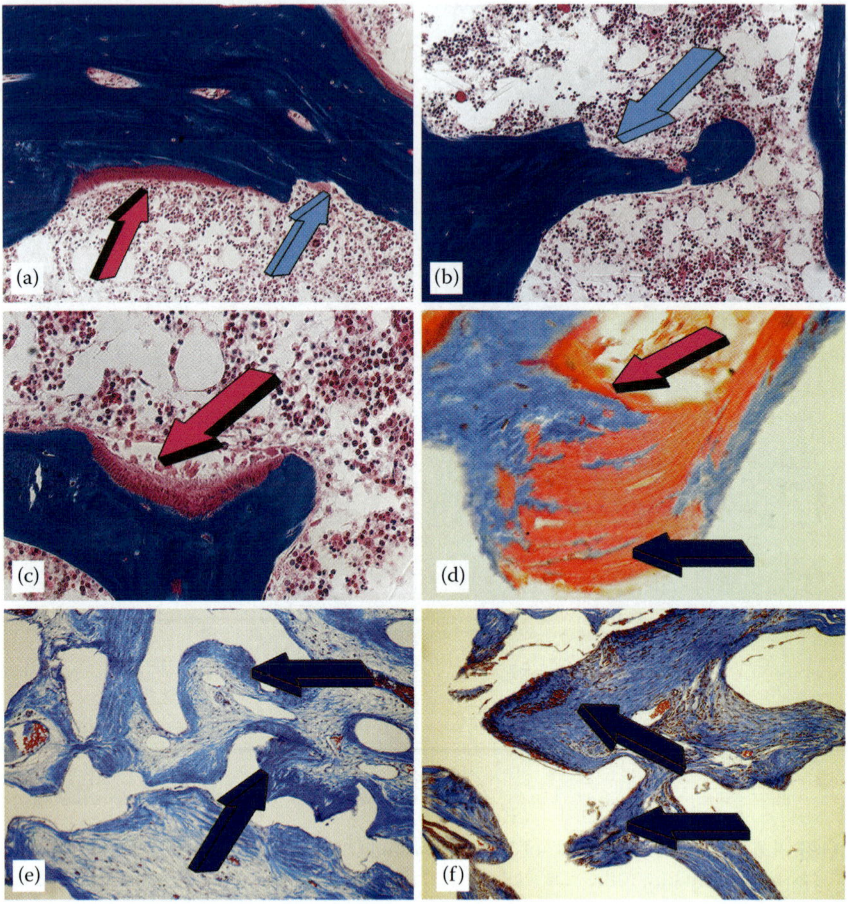

FIGURE 5.7 Repetitive concavities, as prepared within calcium phosphate-based macroporous constructs, biomimetize the remodeling cycle of the primate corticocancellous bone. (a) and (b) Osteoclastogenesis along trabeculae of mineralized bone result in the formation of lacunae and resorptive pits configuring concavities of various dimensions and radii of curvature (light blue arrows); the formation phase is initiated within concavities as prepared by osteoclasts and results in the induction of bone formation with osteoid synthesis and deposition [magenta arrows in (a) and (c)]. (d) Concavities, as prepared and assembled within the calcium phosphate-based macroporous constructs, biomimetize the remodeling cycle and initiate the induction of bone formation after osteoclastogenesis and nanotopographical substrate modifications (d). (e) and (f) Coral-derived calcium carbonate/hydroxyapatite constructs were thus treated with the osteoclastic inhibitors zoledronate Zometa and implanted in the rectus abdominis muscle of *P. ursinus*; harvested specimens showed collagenous condensations at the interface (dark blue arrows) with the lack of bone formation by induction (e) and (f); nanotopographical surface modifications as cut and prepared by osteoclasts after priming the macroporous surfaces are essential for stem cell differentiation into osteoblastic-like cells.[43] (Ripamonti, U. et al., *Biomaterials*, 31, 6400–10, 2010. With permission.)

Further stem cell differentiation within the conducive microenvironment of the concavities results in the expression and synthesis of selected osteogenic gene products of the TGF-β supergene family; expressed proteins are later embedded within the concavities of the substratum, resulting in the induction of bone formation as a secondary response.[16,68]

Based on the sequential time studies using coral-derived macroporous calcium phosphate-based constructs,[39,40,43,56] highly sintered crystalline hydroxyapatite[41] or biphasic hydroxyapatite/β-tricalcium phosphate constructs at various ratios by weight,[46,68] we are now in the position to unfold the *conundrum* of the induction of bone formation by macroporous calcium phosphate-based constructs.

Upon implantation into the *rectus abdominis* muscle, the implanted specimens will rest within the severed *rectus abdominis* muscle and on the fascia of the striated muscle. A fibrin clot will form, surrounding the specimens and also penetrating the macroporous spaces; the surgical pouch of the *rectus abdominis* muscle contains few stem cell niches, the myoblastic/myoendothelial stem cell niche,[63] and paravascular/pericytic niches[64] including pericytic stem cells at large identified now as the multifaceted multiorgan-related mesenchymal stem cells.[65,66]

The lacunae and concavities cut by osteoclastogenesis within the biomimetic matrices are the driving morphogenetic cues for the differentiation of myoblastic stem cells to induce bone formation in a *continuum* of sequential phases of resorption/dissolution and induction of bone formation (Figure 5.6f).[16,46] The concavities assembled in the biomimetic matrices biomimetizes the remodeling cycle of the primate corticocancellous bone and are endowed with multifunctional pleiotropic self-assembly capacities initiating angiogenesis and bone formation by induction.[16,47–49] Indeed, the concavities, as prepared within the biomimetic matrices, biomimetize nature's archaic but constantly functional designs initiating the induction of bone formation (Figure 5.7).

Pluripotent stem cells resting in cell niches of the striated *rectus abdominis* muscle have provided the cue into the spontaneous and/or intrinsic induction of bone formation by macroporous calcium phosphate-based constructs; the nanotopography and, in general, the topography and geometry of the biomimetic extracellular matrix substrata regulate cell migration, attachment, and the induction of selected cellular phenotypes constructing osteogenesis in angiogenesis.[43,69–71] Osteoclastic cells prime the macroporous surfaces releasing Ca^{2+}, initiating the ripple-like cascade of bone differentiation by induction.[43,50–55] The multiple short- and long-term studies in the *rectus abdominis* muscle of the Chacma baboon *P. ursinus* have shown that the concavities cut within the extracellular matrix substrata provide a morphogenetic microenvironment in which myoblastic/myoendothelial and/or endothelial/pericytic paravascular stem cells migrate and attach to the biomimetic substratum, eventually differentiating into osteoblastic-like cells under the direct influence of osteoclastic nanotopographical geometric modifications coupled with released Ca^{2+} together with angiogenesis and capillary sprouting always intimately connected with differentiating stem cells (Figures 5.4 and 5.5). Our latest data have shown that priming by osteoclastic cells results in micro- and macro-patterned self-initiating topographies;[43,45] nanotopographical geometric alterations differentiate resident stem cells into the osteoblastic phenotype with expression of the osteogenic soluble molecular signals of the TGF-β supergene family. By inhibiting osteoclastic activity within the macroporous coral-derived calcium phosphate surfaces, zoledronate-treated macroporous constructs showed limited, if any, bone formation as a result of osteoclastic inhibition (Figure 5.7e and f).[43] In agreement to the morphological observations of the lack of bone induction, molecular analyses by quantitative reverse transcriptase polymerase chain reaction showed a significant reduction of *OP-1* gene expression.[43,45] These critical observations have *par force* indicated that the spontaneous and/or intrinsic induction of bone formation by macroporous calcium phosphate-based biomatrices is initiated by secreted and embedded osteogenic gene products of the BMPs/OPs family within the biomimetic concavities of the substratum.[43,45]

The above should now be used to manipulate differentiating cellular morphologies through the printing of nanotopographical geometric configurations to effectively induce significant amounts of bone formation in clinical contexts. We now know that stem cells feel and respond to nanotopographies of the substratum to which they attach, migrate, and differentiate,[72,73] and feel their way in organogenesis.[74]

5.3 SURGICAL PERSPECTIVES OF SELF-INDUCING FUNCTIONAL BIOMIMETIC MATRICES

A number of important studies have conclusively shown that nanostructured functionalized surfaces with nanotopographical modifications of macroporous biomaterial surfaces are endowed with the striking characteristic of inducing the osteogenic phenotype both *in vitro* and *in vivo*.[71,75–77] Our laboratories have further showed that osteoclastogenesis along the macroporous surfaces of coral-derived constructs is a prerequisite for bone induction to occur.[43] Indeed, nanotopography sensing by osteoclasts[78] is critical for osteoclastic activity; this has indicated that the spontaneous induction of bone formation by macroporous constructs is the result of osteoclastic activity with high Ca^{2+} release, promoting angiogenesis and cell differentiation, resulting in the induction of bone formation.[43]

An important challenge for the future of tissue engineering in clinical contexts is the development and testing in human patients of biomimetic matrices that *per se* direct cell differentiation with the expression of selected mRNA species of the TGF-β supergene family; in these functionalized matrices, the induction of bone formation is initiated, even in the absence of endogenously applied osteogenic proteins of the TGF-β supergene family. Biomimetic matrices of highly crystalline or biphasic hydroxyapatite/β-tricalcium phosphate constructed with a series of repetitive concavities offer a geometric configuration that vividly reproduce and biomimetize the remodeling process of the primate osteonic bone.[16,17,32] Osteoinductive bioceramics prepared with selected surface microporosity, either primed by osteoclastic surface modifications[43] or by specific physiochemical structural characteristics,[22,37] induce stem cell differentiation, BMPs/OPs synthesis, expression, and bone formation by induction as a secondary response. Implantation in a large bone defect in sheep unequivocally demonstrated the concept that osteoinductive bioceramics are equally efficient in bone repair as autologous bone grafts.[79]

The geometric assembly of a series of repetitive concavities within macroporous biomimetic matrices mimic Nature archaic repetitive and tested effective biological designs; the assembly of concavities within macroporous constructs is learning from

Nature how to design and functionalize "smart" intelligent biomimetic matrices for tissue induction and morphogenesis. Research data in the nonhuman primate *P. ursinus* have shown that the role of geometrically designed biomimetic matrices in regenerative medicine is guided by the biomimetism of the extracellular matrix.[16] The geometric design functionalizes the biomimetic matrix and sets into motion the ripple-like cascade of tissue induction and morphogenesis, initiating the induction of bone formation as a secondary response. The future of bone tissue engineering is thus to construct biomimetic matrices to predictably induce "Bone: Formation by Autoinduction."[7]

ACKNOWLEDGMENTS

The "concavity," "The shape of life" and the "geometric induction of bone formation" have been the source of a fascinating intellectual stimulation particularly for the vision of changing the paradigms of bone tissue engineering and material science for skeletal reconstruction in clinical contexts. We thank the Medical Research Council of South Africa, the University of the Witwatersrand, Johannesburg, the National Research Foundation, and *ad hoc* grants to the Bone Research Laboratory for the support of the continuous and systematic studies in nonhuman primates since the late 1980s; several students, laboratory technologists, scientists, and visiting professors who have significantly contributed to the several discoveries highlighted in the chapter, in particular, Barbara van den Heever, Louise Renton, Manolis Heliotis, and Carlo Ferretti; and the Materials Science and Manufacturing Technology Group, Council for Scientific and Industrial Research (CSIR), Pretoria, South Africa, for the long-standing collaborative research and for the preparation of the biomimetic matrices for heterotopic and orthotopic implantation. Coral-derived constructs have first shown and documented the "spontaneous" and "intrinsic" osteoinduction by macroporous calcium phosphate-based biomaterials. We also thank Edwin Clayton Shors and Mike Ponticiello of Interpore International and Biomet, respectively, for the continuous supply of coral-derived calcium carbonate/calcium phosphate constructs for intramuscular implantation in *P. ursinus*, Karuna Bhavanbhai of the Bone Research Unit for the expert bibliographical assistance and the final assembly of the manuscript, and M.R. Urist, A.H. Reddi, and "Bone: Formation by Autoinduction"[7] for continuously providing the stimulus to create, discover, and write.

REFERENCES

1. Reddi, A.H. (1994). Symbiosis of biotechnology and biomaterials: Applications in tissue engineering of bone and cartilage. *J Cell Biochem*, 56(2), 192–5.
2. Reddi, A.H. (1998). Role of morphogenetic proteins in skeletal tissue engineering and regeneration. *Nat Biotechnol*, 16(3), 247–52.
3. Wozney, J.M., Rosen, V., Celeste, A.J., Mitsock, L.M., Whitters, M.J., Kriz, R.W., Hewick, R.M., and Wang, E.A. (1988). Novel regulators of bone formation: Molecular clones and activities. *Science*, 242, 1528–34.
4. Reddi, A.H. (2000). Morphogenesis and tissue engineering of bone and cartilage: Inductive signals, stem cells, and biomimetic biomaterials. *Tissue Eng*, 6(4), 351–9.
5. Ripamonti, U., van den Heever, B., Crooks, J., Tucker, M.M., Sampath, T.K., Rueger, D.C., and Reddi, A.H. (2000). Long-term evaluation of bone formation by osteogenic

protein 1 in the baboon and relative efficacy of bone-derived bone morphogenetic proteins delivered by irradiated xenogeneic collagenous matrices. *J Bone Miner Res*, 15, 1798–809.

6. Ripamonti, U., Ramoshebi, L.N., Patton, J., Matsaba, T., Teare, J., and Renton, L. (2004). Soluble signals and insoluble substrata: Novel molecular cues instructing the induction of bone. In *The Skeleton*, E.J. Massaro and J.M. Rogers (Eds.), Chapter 15. Humana Press, Totowa, New Jersey, pp. 217–27.
7. Urist, M.R. (1965). Bone: Formation by autoinduction. *Science*, 150(698), 893–9.
8. Reddi, A.H. and Huggins, C. (1972). Biochemical sequences in the transformation of normal fibroblasts in adolescent rats. *Proc Natl Acad Sci USA*, 69(6), 1601–5.
9. Sampath, T.K. and Reddi, A.H. (1981). Dissociative extraction and reconstitution of extracellular matrix components involved in local bone differentiation. *Proc Natl Acad Sci USA*, 78(12), 7599–603.
10. Sampath, T.K. and Reddi, A.H. (1983). Homology of bone-inductive proteins from human, monkey, bovine, and rat extracellular matrix. *Proc Natl Acad Sci USA*, 80(21), 6591–5.
11. Ripamonti, U. and Reddi, A.H. (1995). Bone morphogenetic proteins: Applications in plastic and reconstructive surgery. *Adv Plast Reconstr Surg*, 11, 47–65.
12. Khouri, R.K., Koudsi, B., and Reddi, H. (1991). Tissue transformation into bone *in vivo*: A potential practical application. *JAMA*, 266(14), 1953–5.
13. Viola, J., Lal, B., and Grad, O. (2003). NSF: ABT report on the emergence of tissue engineering as a research field. http://www./nsf.gov/pub/2005/msf0450/start.htm.
14. Reddi, A.H. (2005). BMPs: From bone morphogenetic proteins to body morphogenetic proteins. *Cytokine Growth Factor Rev*, 16(3), 249–50.
15. Ripamonti, U., Heliotis, M., and Ferretti, C. (2007). Bone morphogenetic proteins and the induction of bone formation: From laboratory to patients. *Oral Maxillofac Surg Clin North Am*, 19(4), 575–89, vii.
16. Ripamonti, U. (2009). Biomimetism, biomimetic matrices and the induction of bone formation. *J Cell Mol Med*, 13(9B), 2953–72.
17. Ripamonti, U. and Roden, L. (2010). Biomimetics for the induction of bone formation. *Expert Rev Med Devices*, 7(4), 469–79.
18. Urist, M.R., Silverman, B.F., Büring, K., Dubuc, F.L., and Rosenberg, J.M. (1967). The bone induction principle. *Clin Orthop Relat Res*, 53, 243–83.
19. Reddi, A.H. (1981). Cell biology and biochemistry of endochondral bone development. *Coll Relat Res*, 1(2), 209–6.
20. Ripamonti, U., Ramoshebi, L.N., Matsaba, T., Tasker, J., Crooks, J., and Teare, J. (2001). Bone induction by BMPs/OPs and related family members in primates. *J Bone Jt Surg Am*, 83-A Suppl 1(Pt 2), S116–27.
21. Ripamonti, U., Ferretti, C., and Heliotis, M. (2006). Soluble and insoluble signals and the induction of bone formation: Molecular therapeutics recapitulating development. *J Anat*, 209, 447–68.
22. Habibovic, P., Sees, T.M., Van Den Doel, M.A., Van Blitterswijk, C.A., and De Groot, K. (2006). Osteoinduction by biomaterials—Physicochemical and structural influences. *J Biomed Mater Res A*, 77(4), 747–62.
23. Ripamonti, U., Ma, S., and Reddi, A.H. (1992). The critical role of geometry of porous hydroxyapatite delivery system in induction of bone by osteogenin, a bone morphogenetic protein. *Matrix*, 12(3), 202–12.
24. Trueta, J. (1963). The role of the vessels in osteogenesis. *J Bone Jt Surg Br*, 45, 402–18.
25. Lanza, D. and Vegetti, M. (1971). *Opere biologiche: Di Aristotele. A cura di Diego Lanza e Mario Vegetti Torino*. UTET, Torino.
26. Crivellato, E., Nico, B., and Ribatti, D. (2007). Contribution of endothelial cells to organogenesis: A modern reappraisal of an old aristotelian concept. *J Anat*, 211, 415–27.

27. Reddi, A.H. (1997). Bone morphogenesis and modeling: Soluble signals sculpt osteosomes in the solid state. *Cell*, 89, 159–61.
28. Vukicevic, S., Luyten, F.P., Kleinman, H.K., and Reddi, A.H. (1990). Differentiation of canalicular cell processes in bone cells by basement membrane matrix components: Regulation by discrete domains of laminin. *Cell*, 63, 437–45.
29. Ripamonti, U., Duneas, N., van den Heever, B., Bosch, C., and Crooks, J. (1997). Recombinant transforming growth factor-β_1 induces endochondral bone in the baboon and synergizes with recombinant osteogenic protein-1 (bone morphogenetic protein-7) to initiate rapid bone formation. *J Bone Miner Res*, 12, 1584–95.
30. Ripamonti, U., Crooks, J., Matsaba, T., and Tasker, J. (2000). Induction of endochondral bone formation by recombinant human transforming growth factor-β_2 in the baboon (*Papio ursinus*). *Growth Factors*, 17(4), 269–85.
31. Ripamonti, U., Ramoshebi, L.N., Teare, J., Renton, L., and Ferretti, C. (2008). The induction of endochondral bone formation by transforming growth factor-β_3: Experimental studies in the nonhuman primate *Papio ursinus*. *J Cell Mol Med*, 12(3), 1029–48.
32. Ripamonti, U. (2004). Soluble, insoluble and geometric signals sculpt the architecture of mineralized tissues. *J Cell Mol Med*, 8(2), 169–80.
33. Ripamonti, U. (2006). Soluble osteogenic molecular signals and the induction of bone formation. *Biomaterials*, 27, 807–22.
34. Ripamonti, U. (2003). Osteogenic proteins of the transforming growth factor-β superfamily. In *Encyclopedia of Hormones*, H.L. Henry and A.W. Norman (Eds.). Academic Press, San Diego, California, pp. 80–6.
35. Ripamonti, U., Ferretti, C., Teare, J., and Blann, L. (2009). Transforming growth factor-β isoforms and the induction of bone formation: Implications for reconstructive craniofacial surgery. *J Craniofac Surg*, 20, 1544–55.
36. Ripamonti, U., Yeates, L., and van den Heever, B. (1993). Initiation of heterotopic osteogenesis in primates after chromatographic adsorption of osteogenin, a bone morphogenetic protein, onto porous hydroxyapatite. *Biochem Biophys Res Commun*, 193(2), 509–17.
37. Barradas, A.M., Yuan, H., van Blitterswijk, C.A., and Habibovic, P. (2011). Osteoinductive biomaterials: Current knowledge of properties, experimental models and biological mechanisms. *Eur Cell Mater*, 21, 407–29; discussion 429.
38. Heughebaert, M., Legeros, R.Z., Gineste, M., Guilhem, A., and Bonel, G. (1988). Physicochemical characterization of deposits associated with HA ceramics implanted in nonosseous sites. *J Biomed Mater Res*, 22(3 Suppl), 257–68.
39. Ripamonti, U. (1990). Inductive bone matrix and porous hydroxyapatite composites in rodents and nonhuman primates. In *Handbook of Bioactive Ceramics—Volume II: Calcium Phosphate and Hydroxylapatite Ceramics*, T. Yamamuro, J. Wilson-Hench, and L.L. Hench (Eds.). CRC Press, Boca Raton, FL, pp. 245–53.
40. Ripamonti, U. (1991). The morphogenesis of bone in replicas of porous hydroxyapatite obtained from conversion of calcium carbonate exoskeletons of coral. *J Bone Jt Surg Am*, 73(5), 692–703.
41. Ripamonti, U., Crooks, J., and Kirkbride, A.N. (1999). Sintered porous hydroxyapatites with intrinsic osteoinductive activity: Geometric induction of bone formation. *S Afr J Sci*, 95, 335–43.
42. Ripamonti, U., van den Heever, B., and van Wyk, J. (1993). Expression of the osteogenic phenotype in porous hydroxyapatite implanted extraskeletally in baboons. *Matrix*, 13, 491–502.
43. Ripamonti, U., Klar, R.M., Renton, L.F., and Ferretti, C. (2010). Synergistic induction of bone formation by hOP-1, hTGF-β_3 and inhibition by zoledronate in macroporous coral-derived hydroxyapatites. *Biomaterials*, 31, 6400–10.
44. Ripamonti, U., Herbst, N.N., and Ramoshebi, L.N. (2005). Bone morphogenetic proteins in craniofacial and periodontal tissue engineering: Experimental studies in the nonhuman primate *Papio ursinus*. *Cytokine Growth Factor Rev*, 16(3), 357–68.

45. Ripamonti, U. and Klar, R.M. (2010). Redundancy of osteogenic molecular signals initiating the induction of bone formation. Science in Africa: http://www.scienceinafrica.co.za/2010/october/bone_regeneration_full.htm.
46. Ripamonti, U., Richter, P.W., Nilen, R.W., and Renton, L. (2008). The induction of bone formation by smart biphasic hydroxyapatite tricalcium phosphate biomimetic matrices in the non-human primate *Papio ursinus*. *J Cell Mol Med*, 12, 2609–21.
47. Parfitt, A.M. (1994). Osteonal and hemi-osteonal remodeling: The spatial and temporal framework for signal traffic in adult human bone. *J Cell Biochem*, 55, 273–86.
48. Manolagas, S.C. and Jilka, R.L. (1995). Bone marrow, cytokines, and bone remodeling. Emerging insights into the pathophysiology of osteoporosis. *N Engl J Med*, 332(5), 305–11.
49. Parfitt, A.M., Mundy, G.R., Roodman, G.D., Hughes, D.E., and Boyce, B.F. (1996). A new model for the regulation of bone resorption, with particular reference to the effects of bisphosphonates. *J Bone Miner Res*, 11, 150–9.
50. Kohn, E., Alessandro, R., Spoonster, J., Wersto, R.P., and Liotta, L.A. (1995). Angiogenesis: Role of calcium-mediated signal transduction. *Proc Natl Acad Sci USA*, 92, 1307–11.
51. Munaron, L. (2006). Intracellular calcium, endothelial cells and angiogenesis. *Recent Pat Anti-Cancer Drug Discovery*, 1, 105–19.
52. Wu, Q., Shao, H., Eton, D., Li, J., Yang, B. et al. (2009). Extracellular calcium increases CXCR4 expression on bone marrow-derived cells and enhances proangiogenesis therapy. *J Cell Mol Med*, 13, 1–10.
53. Zayzafoon, M. (2006). Calcium/calmodulin signaling controls osteoblast growth and differentiation. *J Cell Biochem*, 97, 56–70.
54. Ohgushi, H., Dohi, Y., Tamai, S., and Tabata, S. (1993). Osteogenic differentiation of marrow stromal stem cells in porous hydroxyapatite ceramics. *J Biomed Mater Res*, 27, 1401–7.
55. Kanatani, M., Sugimoto, T., and Fukase, T. (1991). Effect of elevated extracellular calcium on the proliferation of osteoblastic MC3T3-E1 cells: Its direct and indirect effects via monocytes. *Biochem Biophys Res Commun*, 181, 1425–30.
56. Ripamonti, U., Crooks, J., Khoali, L., and Roden, L. (2009). The induction of bone formation by coral-derived calcium carbonate/hydroxyapatite constructs. *Biomaterials*, 30(7), 1428–39.
57. Keith, A. (1927). Concerning the origin and nature of osteoblasts. *Proc R Soc Med*, 21(2), 301–8.
58. Boerckel, J.D.,Uhriig, B.A., Willett, N.J., Huebsch, N., and Guldberg, R.E. (2011). Mechanical regulation of vascular growth and tissue regeneration *in vivo*. *Proc Natl Acad Sci USA*, 108, E674–80.
59. Vlodavsky, I., Folkman, J., Sullivan, R. et al. (1987). Endothelial cell-derived basic fibroblast growth factor: Synthesis and deposition into subendothelial extracellular matrix. *Proc Natl Acad Sci USA*, 84, 2292–6.
60. Folkman, J., Klagsbrun, M., Sasse, J. et al. (1988). A heparin binding angiogenic protein—basic fibroblast growth factor—is stored within basement membranes. *Am J Pathol*, 130, 393–400.
61. Paralkar, V.M., Nandedkar, A.K.N., Pointer, R.H., Kleinman, H.K., and Reddi, A.H. (1990). Interaction of osteogenin, a heparin binding bone morphogenetic protein, with type IV collagen. *J Biol Chem*, 265, 17281–4.
62. Paralkar, V.M., Vukicevic, S., and Reddi, A.H. (1991). Transforming growth factor β type 1 binds to collagen type IV of basement membrane matrix: Implications for development. *Dev Biol*, 143, 303–9.
63. Zheng, B., Cao, B., Crisan, M., Sun, B., Li, G., Logar, A., Yap, S., Pollett, J.B., Drowley, L., Cassino, T., Gharaibeh, B., Deasy, B.M., Huard, J., and Peault, B. (2007). Prospective identification of myogenic endothelial cells in human skeletal muscle. *Nat Biotechnol*, 25, 1025–34.

64. Kovacic, J.C. and Boehm, M. (2009). Resident vascular progenitor cells: An emerging role for non-terminally differentiated vessel-resident cells in vascular biology. *Stem Cell Res*, 2, 2–15.
65. Chen, C.-W., Montelatici, E., Crisan, M., Corselli, M., Huard, J., Lazzari, L. et al. (2009). Perivascular multi-lineage progenitor cells in human organs: Regenerative units, cytokine sources or both? *Cytokine Growth Factor Rev*, 20, 429–34.
66. Crisan, M., Yap, S., Casteilla, L., Chen, C.-W., Corselli, M., Park, T.S., Andriolo, G., Sun, B., Zheng, B., Zhang, L., Norotte, C., Teng, P.N., Traas, J., Schugar, R., Deasy, B.M., Badylak, S., Buhring, H.J., Giacobino, J.P., Lazzari, L., Huard, J., and Peault, B. (2008). A perivascular origin for mesenchymal stem cells in multiple human organs. *Cell Stem Cell*, 3, 301–13.
67. Caplan, A.I. (2008). All MSCs are pericytes? *Cell Stem Cell*, 3(3), 229–30.
68. Ripamonti, U., Richter, P.W., and Thomas, M.E. (2007). Self-inducing shape memory geometric cues embedded within smart hydroxyapatite-based biomimetic matrices. *Plast Reconstr Surg*, 120(7), 1796–807.
69. Singhvi, R., Kumar, A., Lopez, G.P., Stephanopoulos, G.N., Wang, D.I., Whitesides, G.M., and Ingber, D.E. (1994). Engineering cell shape and function. *Science*, 264(5159), 696–8.
70. Chou, L., Firth, J.D., Uitto, V.-J., and Brunette, D.M. (1995). Substratum surface topography alters cell shape and regulates fibronectin mRNA level, mRNA stability, secretion and assembly in human fibroblasts. *J Cell Sci*, 108, 1563–73.
71. Dalby, M.J., Gadegaard, N., Tare, R., Andar, A., Riehle, M.O., Herzyk, P., Wilkinson, C.D., and Oreffo, R.O. (2007). The control of human mesenchymal cell differentiation using nanoscale symmetry and disorder. *Nat Mater*, 6(12), 997–1003.
72. Fu, J., Wang, Y.K., Yang, M.T., Desai, R.A., Yu, X., Liu, Z., and Chen, C.S. (2010). Mechanical regulation of cell function with geometrically modulated elastomeric substrates. *Nat Methods*, 7(9), 733–6.
73. Buxboim, A. and Discher, D.E. (2010). Stem cells feel the difference. *Nat Methods*, 7(9), 695–7.
74. Discher, D.E., Janmey, P., and Wang, Y.-L. (2005). Tissue cells feel and respond to the stiffness of their substrate. *Science*, 310, 1139–43.
75. You, M.H., Kwak, M.K., Kim, D.H., Kim, K., Levchenko, A., Kim, D.Y., and Suh, K.Y. (2010). Synergistically enhanced osteogenic differentiation of human mesenchymal stem cells by culture on nanostructured surfaces with induction media. *Biomacromolecules*, 11(7), 1856–62.
76. Wilkinson, A., Hewitt, R.N., McNamara, L.E., McCloy, D., Dominic Meek, R.M., and Dalby, M.J. (2011). Biomimetic microtopography to enhance osteogenesis *in vitro*. *Acta Biomater*, 7, 2919–25.
77. McNamara, L.E., McMurray, R.J., Biggs, M.J., Kantawong, F., Oreffo, R.O., and Dalby, M.J. (2010). Nanotopographical control of stem cell differentiation. *J Tissue Eng*, 120623.
78. Geblinger, D., Addadi, L., and Geiger, B. (2010). Nano-topography sensing by osteoclasts. *J Cell Sci*, 123, 1503–10.
79. Yuan, H., Fernandes, H., Habibovic, P., De Boer, J., Barradas, A.M., De Ruiter, A., Walsh, W.R., van Blitterswijk, C.A., and De Bruijn, J.D. (2010). Osteoinductive ceramics as a synthetic alternative to autologous bone grafting. *Proc Natl Acad Sci USA*, 107, 13614–9.

6 Bioinspired Titanium Implants

The Concavity—The Shape of Life

Ugo Ripamonti, Louise Renton, and Jean-Claude Petit

CONTENTS

6.1 SMART BIOACTIVE BIOMIMETIC MATRICES INDUCE THE INDUCTION OF BONE FORMATION

Regenerative medicine is the great multidisciplinary challenge of molecular, cellular, and evolutionary biology requiring the integration of tissue biology, tissue engineering, developmental biology, and experimental surgery to explore how to trigger *de novo* tissue induction and morphogenesis, with the ultimate goal of inducing *de novo* and *ex novo* tissue induction and organogenesis in man.[1–7] The emerging novel frontier of tissue engineering, and regenerative medicine at large, is to construct biomimetic matrices that mimic Nature's hierarchical structural assemblages and mechanisms of simplicity and elegance that are conserved throughout genera and species.[7–9] The novel and exciting concept of bone tissue engineering is to develop *smart* and functional bioinspired biomimetic matrices that *per se* initiate the ripple-like cascade of bone differentiation by induction without the osteogenic soluble molecular signals of the transforming growth factor-β (TGF-β) supergene family.[8–10] The driving force of the intrinsic induction of bone formation by bioactive bioinspired biomimetic matrices is the shape of the implanted scaffold. The language of shape is the language of geometry; the language of geometry is the language of a sequence of repetitive concavities that biomimetize the remodeling cycle of primate osteonic bone.[7,8] The concavity *per se*, as fabricated into calcium phosphate macroporous constructs or generated by osteoclastogenesis during the remodeling cycle of the primate osteonic bone, is the geometric signal that initiates the induction of bone formation throughout the corticocancellous bone of the primate skeleton.[8,11–15] The concavity as cut into the biomimetic matrices biomimetizes the ancestral repetitive and multimillion-year-old tested designs and nanotopographies of nature; the concavity is thus the geometric signal that initiates the induction of bone formation.

A novel geometric titanium dental implant was devised by incorporating a series of repetitive concavities on the titanium surface coated by a multimicron layer of sintered highly crystalline hydroxyapatite (Figure 6.1). Devices were implanted in

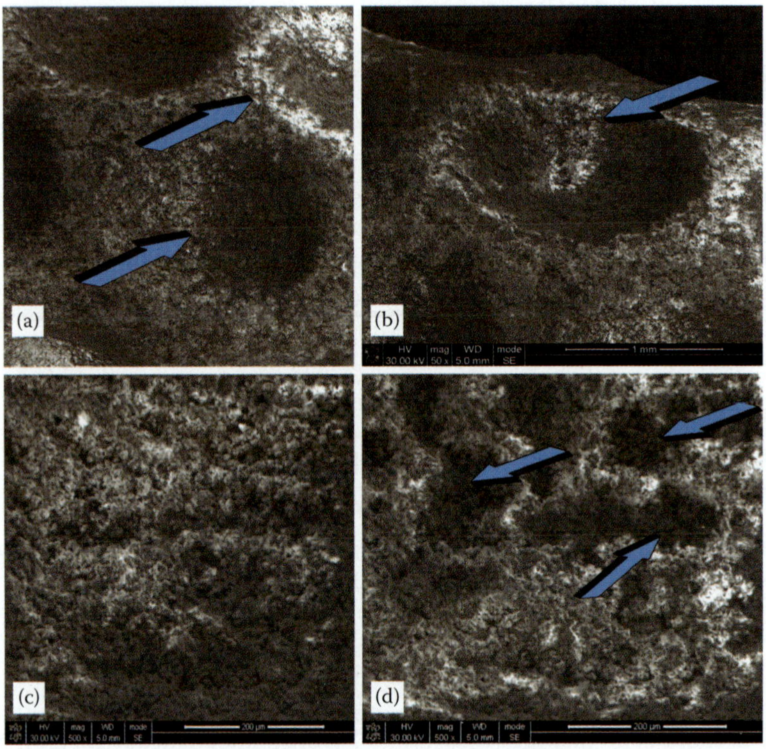

FIGURE 6.1 Scanning electron micrographs (SEMs) of geometric titanium implants coated with highly crystalline sintered hydroxyapatite. (a) Concavity motif as carved within the titanium constructs; concavities are highlighted by blue arrows. (b) Detail of a concavity (blue arrow) across the coated titanium surface; concavities are 1000 μm in diameter and 400 μm in depth. (c) and (d) SEM analysis of the nanotopographical landscape of the sintered hydroxyapatite coating showing extensive microporosity. (d) SEM nanotopographical image indicating topographical surface alterations consistent with lacunae, pits, and microconcavities (blue arrows) of corticocancellous bone as carved and excavated by osteoclastogenesis *in vivo*. Original magnifications: a, ×32; b, ×50; c, ×500; d, ×500.

edentulous mandibular ridges and in the anterior surface of the tibia of six non-human primates *Papio ursinus*. Histological undecalcified sections were analyzed histomorphometrically to determine the measurement of bone in contact (BIC) along the retrieved implants. On day 90, the ratio of BIC to the surface of geometric implants was greater than on the planar control surfaces of the titanium implants in both mandibular and tibial sites. This is indicative that the concavities are osteogenic due to the inherent characteristics of the generated microenvironment, the concavities providing a unique environment to foster stem cell differentiation and the induction of bone formation.

Understanding the attainment of tissue form and function or morphogenesis has been a grand challenge since the Aristotelian vision of the body plan forming around the invading vasculature, thus, "morphogenetic" vessels complementary to the "osteogenetic" vessels of Trueta's definition;[16] the morphogenetic vessels of Aristotelian definition[17,18] further provide the anatomical and molecular basis of pattern formation during tissue morphogenesis.

Tissue morphogenesis, the genesis of form and function,[1] has also determined the continuous never-ending search for morphogens, first defined by Turing as "forms generating substances."[19] The realization that morphogens are the propelling soluble molecular force behind tissue morphogenesis has indicated that only "Morpheus unbound"[20] can trigger the ripple-like cascade of pattern formation and the attainment of tissue form and function, ultimately resulting in tissue induction and morphogenesis.[1,20–23]

Through the centuries, another great challenge for tissue biologists and morphologists alike has been the search for the postulated morphogens that were hypothesized but not yet experimentally shown to exist as the soluble molecular signals responsible for pattern formation, tissue induction, and morphogenesis.[6,7,19,21] As it happens frequently by serendipity, many different yet converging fields were set to study the fascinating phenomenon of bone: formation by autoinduction[21] and to try to mechanistically unravel the ripple-like cascade of the induction of bone formation.[1,6,23–26] Unexpectedly, studies on the induction of bone formation heralded firstly the vision and secondly the realization that several morphogens indeed exist with marked phylogenetically ancient pleiotropic activities.[6–8,23]

The aim of this chapter is to convey a concise perspective on bone formation by autoinduction.[21] The rationale of linking bone formation by induction to nanotopography and titanium surface characteristics is based on the discovery that functionalized nanotopographical surface characteristics are endowed with the striking capacity to induce the osteogenic phenotype in responding mesenchymal stem cells, ultimately leading to the differentiation of bone.[28–35]

6.2 TISSUE INDUCTION AND MORPHOGENESIS: SOLUBLE AND INSOLUBLE SIGNALS AND THE TISSUE ENGINEERING PARADIGM

The above critical studies on nanotopographical control of stem cell differentiation[32–36] have stemmed from early reports that have indicated an important role of nanotopography of the substratum on cell differentiation, contact guidance and cell alignment, synthesis, and gene expression.[37–39] These fundamental studies have indicated the molecular cross-talk and multiple interactions between stem cells, differentiated cells, and progenitors alike, with the substrata surface topography and geometry.[37,38] The results have provided a quantum leap forward to the identification of the critical importance of surface nanotopography and geometry on the induction of the cellular phenotype,[39,40] with particular reference to the induction of the osteogenic phenotype.[8,30]

Recent cardinal studies have investigated the mechanistic control of stem cell differentiation by substratum's topography together with forces that control stem cell differentiation.[41] These studies have shown the mechanical modulation of cell function by geometrically different substrata;[42] geometrical configurations by regulating substrata's rigidity modulate cell signaling and the cellular phenotype.[41] *In vitro*, differentiating osteogenic cells demonstrate higher traction forces and cell spreading when regulated by the rigidity of the substratum.[41–43]

An important experiment that led to the design of geometric titanium dental implants was an *in vitro* study of preosteoblastic cell lines seeded onto coral-derived macroporous calcium carbonate/calcium phosphate constructs; these studies steamed the preparation of *ex vivo* bioreactors preseeded with mouse-derived fibroblasts (NIH3T3) and preosteoblast (MC3T3-E1) cell lines. Note that cells grown onto the coral-derived substratum grow and align within concavities of the substratum for further differentiation and assembly into the osteogenic phenotype (Figure 6.2).

Within the concavity microenvironment of the titanium surfaces coated with highly crystalline sintered hydroxyapatite, several signals must exist that divert differentiation of pericytic mesenchymal stem cells and/or progenitor osteoblastic-like cells into secreting differentiated osteoblastic cells. In recent studies,[30] we have shown that the osteoclastic priming of calcium phosphate-based macroporous constructs is a critical step for the release of calcium ions that are required for the induction of angiogenesis and of cell differentiation into the osteoblastic phenotype.[44–49] Differentiating cells must, however, attach to the calcium phosphate-based construct to secrete and embed the osteogenic gene products to eventually initiate the induction of bone formation.[30] It is tempting to suggest that the osteogenic differentiation of locally available stem cells and/or progenitors is further regulated by the rigidity of the implanted substratum.[42] We have now learned that "stem cells feel the difference"[43] of the composition and physical status of the substrata, dramatically setting the importance of the insoluble signal in controlling cell attachment and cell differentiation, ultimately initiating the induction of bone formation *per se* without the invocation of the soluble osteogenic molecular signals of the TGF-β supergene family.[8,30,31]

The emergence of the research field of tissue engineering has been prominently supported and established by the grand studies on the bone induction principle,[21,24] to which many scientists worldwide contributed significantly to unravel the mechanisms behind the induction of bone formation.[6,23] The term tissue engineering, as we know it today, has been traced to Y.C. Fung of the University of California at San Diego, who submitted an unsuccessful proposal to the U.S. National Science

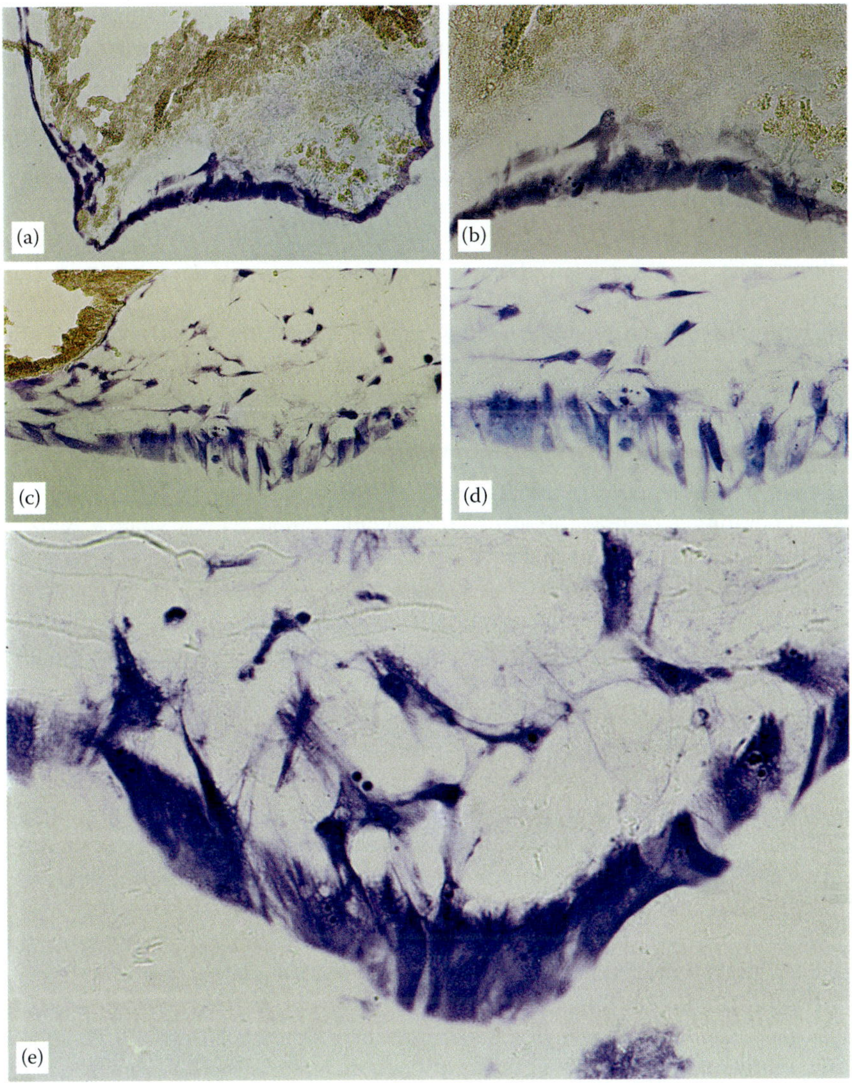

FIGURE 6.2 *In vitro* bioreactor of coral-derived macroporous calcium carbonate/calcium phosphate hydroxyapatite replicas seeded with (a) and (b) mouse-derived fibroblasts (NIH3T3) and (c)–(e) preosteoblastic cell lines (MC3T3-E1). Of significant importance is that cells grow, polarize, and align within the concavities of the coral-derived substratum, mimicking the remodeling cycle of the corticocancellous osteonic bone (undecalcified sections cut at 3 μm).

Foundation for an engineering center.[50] Incisive studies on the induction of bone formation unexpectedly heralded that the "bone induction principle"[24] is propelled by what was eventually known as the tissue engineering paradigm, whereas soluble molecular signals are reconstituted or recombined with insoluble signals or substrata to trigger the ripple-like cascade of bone differentiation by induction.[2–7,21,27,51–53]

The recombination studies (i.e., combining soluble molecular signals with insoluble signals or substrata) implied several marvelous interdisciplinary discoveries summarized by the pleiotropic activity of the bone induction principle,[1–8,24] where 1) there are soluble osteogenic molecular signals that initiate and trigger the induction of bone formation; 2) the signals are physiologically recombined with insoluble signals or substrata to deliver the biological activity of the soluble molecular signals; 3) soluble molecular signals are combined with extracellular matrix components and thus need to be extracted and identified; 4) several extracellular matrices, not only the bone matrix, must host a plethora of molecular signals in solution or tightly bound to the extracellular matrix components; 5) the physiological recombination of the hypothesized soluble molecular signals with insoluble signals or extracellular matrix substrata implies the underlying molecular mechanism, where insoluble signals *par force* are acting as biomimetic matrices to finely tune and deliver the biological activity of the soluble molecular signals; and 6) insoluble signals or substrata are required, when combined to soluble molecular signals, to trigger the ripple-like cascade of bone formation by induction.[6–8]

The operational reconstitution of an insoluble signal with a soluble molecular signal was a key experiment that further propelled the bone induction principle[24] in preclinical and clinical contexts[6,27,51–53] and set the tissue engineering paradigm of diverse tissues and organs of the mammalian body,[51–53] including the construction of a prefabricated osteoinductive flap for bone tissue engineering of a large mandibular defect in a human (Figure 6.3).[54]

A significant step forward in bone tissue engineering and regenerative medicine at large were the grand studies of Sampath and Reddi, who identified the elusive osteogenic proteins to be part of the extracellular matrix of bone.[51,52] The bone, however, is in the solid state;[4] yet, the bone is in both the soluble and insoluble states, in which there is a *continuum* between the insoluble and soluble states of the matrix that is regulated by signals in solution interacting with the insoluble signals of the extracellular matrix.[4–6]

The novel concept of the induction of bone formation and, thus, bone tissue engineering is the induction of bone by the implantation of *smart*, bioinspired, self-inducing functionalized biomimetic matrices that *per se*, without the addition of exogenously applied osteogenic soluble molecular signals of the TGF-β supergene family, initiate the ripple-like cascade of bone differentiation by induction (Figure 6.4).[7,8] The induction of bone formation starts by erecting scaffolds of insoluble signals or substrata controlling the expression and synthesis of the soluble osteogenic molecular signals of the TGF-β supergene family. There is no bone formation without the osteogenic soluble molecular signals of the TGF-β supergene family.[8] The insoluble signal or the carrier substratum provides an exciting and novel concept of bone tissue engineering for the development of functional biomimetic matrices that *per se* initiate the induction of bone formation, that is, a matrix that induces differentiation of resident myoblastic/pericytic and/or endothelial/paravascular stem cells into osteoblastic-like cells capable of expressing, secreting, and embedding the osteogenic gene products into the biomimetic matrices, initiating bone formation as a secondary response (Figure 6.4).[7–10]

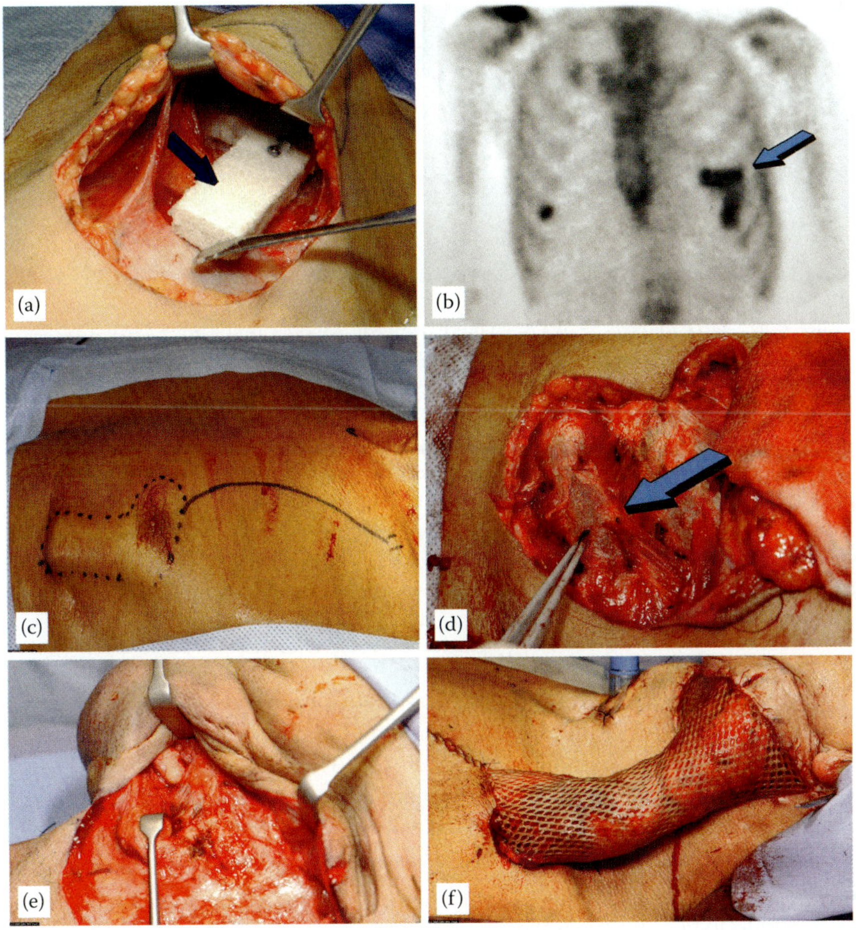

FIGURE 6.3 *In vivo* bioreactor of a prefabricated coral-derived macroporous hydroxyapatite construct into a vascularized pedicled bone flap in a human chest. (a) Surgical insertion of L-shaped blocks of macroporous coral-derived hydroxyapatite as used for the preparation of the *in vitro* bioreactor (Figure 6.2); the macroporous biomimetic matrices were preloaded with doses of recombinant human osteogenic protein-1 (hOP-1).[54] (b) Skeletal scintigraphy demonstrates osteogenesis on the L-shaped prefabricated heterotopic chest implant (arrow). (c) Preparation of the chest for the pectoralis major myo-osseous-hydroxyapatite flap pedicled on the thoracoacromial artery. (d) Surgical debridement of the newly formed heterotopic mandible (arrow). (e) Preparation of the recipient mandibular bed. (f) Placement of the pedicled flap in the left mandible.[54] (From Heliotis, M. et al., *Int J Oral Maxillofac Surg*, 35(3), 265–9, 2006. With permission.)

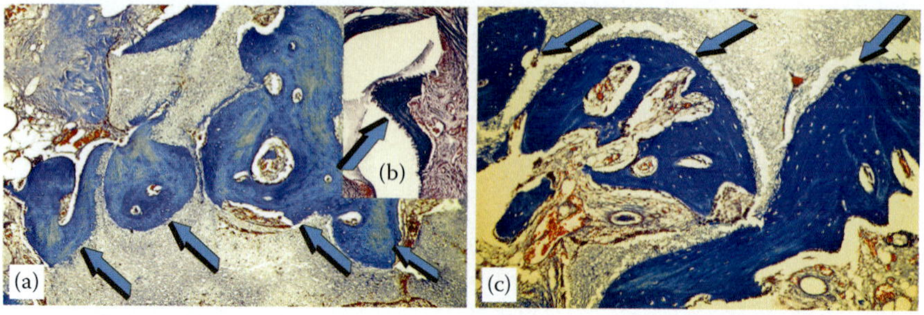

FIGURE 6.4 Modified tissue engineering paradigm—induction of bone formation by macroporous calcium phosphate-based constructs without the exogenous applications of the osteogenic soluble molecular signals of the TGF-β supergene family. Coral-derived macroporous constructs were implanted in the rectus abdominis muscle of adult nonhuman primates of the species *P. ursinus*.[28,56] Harvested specimens on day 90 were processed for histological analyses and cut at 3–6 μm and stained with toluidine blue in 6% ethanol. (a)–(c) Bone forms without the exogenous application of osteogenic soluble molecular signals within a series of repetitive concavities of the coral-derived macroporous constructs (blue arrows); blue arrow (b, inset) points to the induction of bone formation within a concavity (arrow) of granular/particulate coral-derived construct implanted in the rectus abdominis of an adult Chacma baboon *P. ursinus*.[63] Original magnifications: a–c, ×75.

It is clear that the formulation of smart bioinspired and functionalized biomimetic matrices that *per se* initiate the induction of bone formation proposes a modified tissue engineering paradigm in which the osteogenic molecular signals are not added to the implanted substrata but are rather expressed and secreted post-implantation by differentiating myoblastic/pericytic stem cells residing within the *smart* biomimetic concavities; these, as such, regulate the induction of bone formation.[22,30,31]

The spontaneous and/or intrinsic induction of bone formation of various calcium phosphate-based macroporous constructs when implanted in heterotopic intramuscular sites of the nonhuman primate *P. ursinus* has unequivocally shown that the driving force of the intrinsic induction of bone formation by bioactive bio-inspired biomimetic matrices is the shape of the implanted scaffold,[8,55] that is, a sequence of repetitive concavities that biomimetize the remodeling cycle of the primate corticocancellous bone.[8,9] There is a direct spatial and temporal relationship of morphological and molecular events that emphasize the similarity between the remodeling cycles of the corticocancellous bone and the geometric induction of bone formation, that is, the induction of bone by smart concavities assembled in biomimetic matrices of highly crystalline hydroxyapatites.[8,10,15] The basic multicellular unit of the corticocancellous bone excavates a trench across the surface, leaving in its wake with some geometrical latitude, a hemiosteon rather than an osteon,[11–13] that is, a trench with cross-sectional geometric cues of concavities at different stages of osteoclastogenesis, eventually leading to osteogenesis.[8,10,14] The concavities are thus regulators of bone induction and deposition during the remodeling cycle of the corticocancellous osteonic bone and biomimetize the induction of bone formation when assembled into calcium phosphate-based macroporous constructs (Figure 6.4).[8,9]

6.3 PLASMA-SPRAYED TITANIUM IMPLANTS WITH GEOMETRIC CUES OF REPETITIVE CONCAVITIES ENHANCE THE INDUCTION OF BONE FORMATION AND OSTEOINTEGRATION IN A NONHUMAN PRIMATE MODEL

Since the late 1980s, we have learned that macroporous calcium phosphate-based constructs spontaneously initiate the induction of bone formation[28,56] *via* the geometric induction of bone formation.[8,55] In a series of experiments in *P. ursinus*, we have exploited the geometric motif of the concavity not carved in macroporous calcium phosphate-based constructs but rather in solid titanium dental implants coated with highly crystalline sintered hydroxyapatite. We then wanted to determine the possible influence of the macro surface topography on the rate of osteointegration on plasma-sprayed geometrically modified titanium dental implants (Figure 6.5). Titanium implants were fabricated with a series of repetitive concavities of 1000 μm in diameter and 400 μm in depth; titanium implants were plasma sprayed with highly crystalline hydroxyapatite (Figures 6.1 and 6.5e). A total of six nonhuman primates Chacma baboon *P. ursinus* were used for the study. Mandibular molars and premolars were extracted to prepare edentulous mandibular ridges for later implantation. Each animal received implants in both mandibular ($n = 20$) and tibial ($n = 14$) sites.

One animal had to be euthanized 5 days after surgical implantation, which included the implantation of the hydroxyapatite-coated titanium implants in heterotopic sites of the *rectus abdominis* muscle. Harvested samples were processed for SEM analyses (Figure 6.6). Carbon-coated samples (5 nm in thickness) were examined on an FEI Nova Nanolab SEM (FEI Company, Hillsboro, OR), at 30 kV. Geometric implants showed dramatically how the geometric design of the concavity is the "shape of life."[14] When implanted in the rectus abdominis muscle known to harbor several paravascular/pericytic stem cell niches,[57] concavities induced multiple cellular attachments, cell differentiation, and cellular trafficking across the modified surfaces and the margins of the exposed concavities (Figure 6.6e and f). This morphological and spatial organization of the invading myoblastic/pericytic stem cells attached to the borders of the concavities is responsible for the induction of bone across concavities as prepared in discs of highly crystalline hydroxyapatite when implanted in heterotopic sites of the *rectus abdominis* muscle of *P. ursinus* (Figure 6.7).

Animals were euthanized on days 30 and 90 after implantation. After euthanasia, the titanium implants together with surrounding bone were harvested, fixed in 10% buffered formalin, dehydrated through graded alcohols, and embedded in resin. Tissue samples were treated according to the method of Donath and Breuner.[58] Longitudinal sections were precision sawed and ground to an approximate final thickness of 80 μm. Undecalcified sections were ground and polished to 11 μm and stained with a modified Goldner's trichrome. All sample preparation was performed

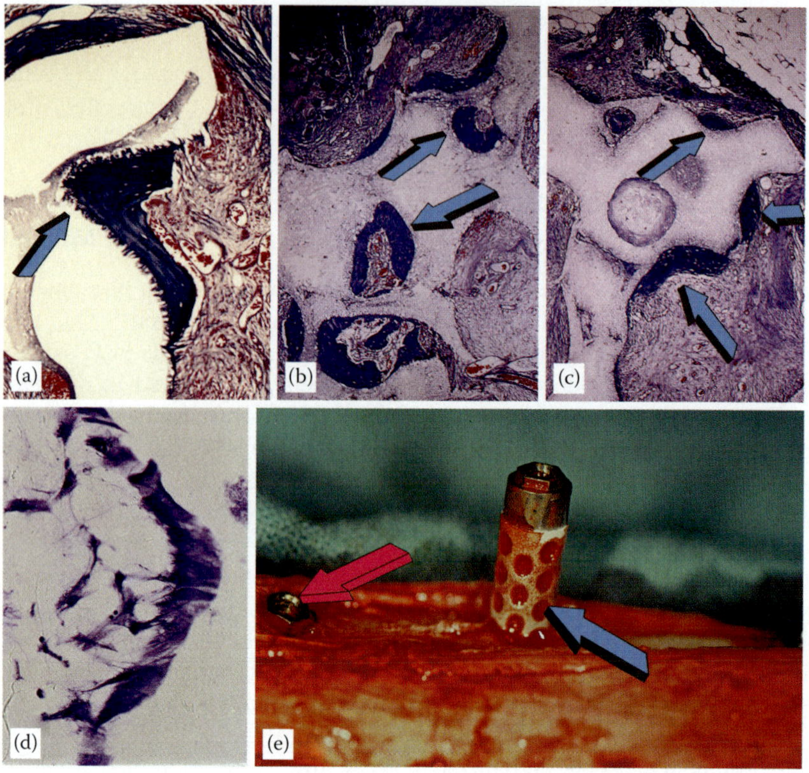

FIGURE 6.5 Conceptualization and genesis of the geometric titanium implant with concavities prepared across the implant later coated by sintered highly crystalline hydroxyapatite. (a) and (d) were instrumental for the preparation of the geometric titanium dental implant with concavities. (a) Bone forms within a concavity of a specimen of granular/particulate coral-derived hydroxyapatite implanted in the *rectus abdominis* muscle of a nonhuman primate *P. ursinus*; bone never formed in the particulate granular specimens, with the exception of the minute amount of the bone forming within a concavity [arrow in (a)] of the particulate coral-derived hydroxyapatite.[63] (b) and (c) Intrinsic spontaneous induction of bone formation across the macroporous spaces of sintered macroporous constructs implanted in the *rectus abdominis* muscle of nonhuman primates *P. ursinus*;[55] bone has formed exclusively within concavities of the macroporous sintered constructs. (c) Histological section with the induction of bone formation within adjacent concavities (blue arrows) suggested the sequential osteointegration of a dental implant with the titanium threads inserted into bone. (d) *In vitro* organization, alignment, and orientation of preosteoblastic cells within concavities of the coral-derived biomimetic construct; (a) and (d) were instrumental for the conceptualization and fabrication of self-inducing geometric titanium implants. (e) Clinical macrophotograph of geometric titanium implants implanted in the exposed tibial bone (magenta arrow) of the nonhuman primate *P. ursinus*. The blue arrow points to the concavities that are highly suited for the adsorption of plasma and plasma protein products that set into motion the osteoinductive capacity of the concavity as a vital microenvironment for the spontaneous initiation of bone formation (the concavity is the shape of life).[14] Original magnifications: a, ×75; b, ×45; c, ×45; d, ×1700. (From Ripamonti, U., *J Cell Mol Med*, 8(2), 169–80, 2004. With permission.)

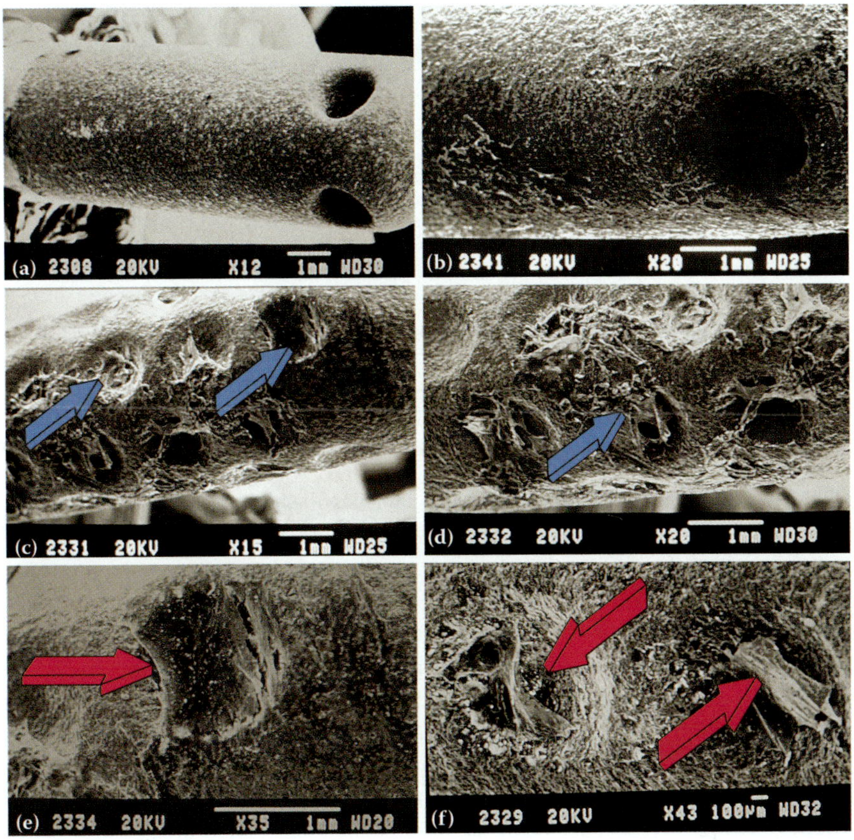

FIGURE 6.6 SEM images of hydroxyapatite-coated titanium implants harvested 5 days after implantation in the *rectus abdominis* muscle of an adult nonhuman primate *P. ursinus*. (a) and (b) Planar coated implants with minor cellular activity along the planar hydroxyapatite surfaces. (c) and (d) Myoblastic/pericytic cellular trafficking, attachment, and differentiation across two geometric implants with a series of repetitive concavities. (e) and (f) Note the important cellular attachment and differentiation encompassing the exposed concavities (blue arrows) and the establishment of tractional forces across the margins of the exposed concavities with cellular and mesenchymal tissue attachment across the concavities (magenta arrows). The resulting tissue patterning across the concavities as established by cellular tractional forces from the exposed margins resulted 85 days later the induction of bridging bone across the exposed concavities as illustrated in Figure 6.7.

using the EXAKT precision cutting and grinding system (EXAKT Apparatebau, Nordestedt, Hamburg, Germany).[58]

Digital images of the sections at both low and intermediate magnification were captured with an Olympus AnalySIS system mounted on an Olympus Provis research microscope, and linear measurements were obtained using the system software.[30]

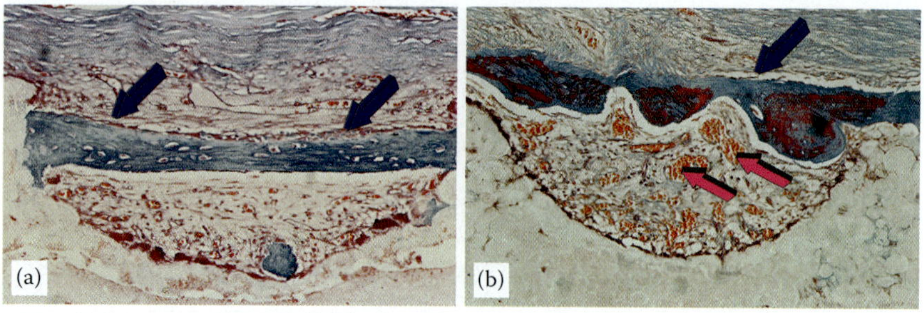

FIGURE 6.7 Tractional forces, tissue patterning, and morphogenesis of newly formed bone (arrows) across concavities prepared in highly crystalline substrata of sintered hydroxyapatite.[55] (a) Newly induced bone surfaced by contiguous osteoblasts had formed across a concavity when implanted in the *rectus abdominis* muscle of *P. ursinus*. (b) Substantial bone induction across a concavity with newly formed bone (blue arrows) containing a rich vascular space populated by invading capillaries (magenta arrows) in contact with both the newly formed bone and the sintered hydroxyapatite substratum. Original magnifications: a and b, ×175.

Measurements of BIC on both geometric versus planar control implants were made, and the bone/implant osteointegration ratio (in percentage) was determined. Results are graphically represented in Figure 6.8a and b (30 and 90 days, respectively).

On day 30 (Figure 6.8a), there was no difference between the geometric and the standard planar implants; on day 90, the ratio of BIC to surface within the geometric implants was greater than on the standard planar implants in both mandibular and tibial sites (Figure 6.8b). The data strongly indicate that geometrically constructed titanium implants are more osteogenic than standard planar implants, the concavities providing a unique microenvironment to foster bone differentiation, remodeling, and osteointegration (Figure 6.9). At implantation (Figures 6.5e and 6.9c), the

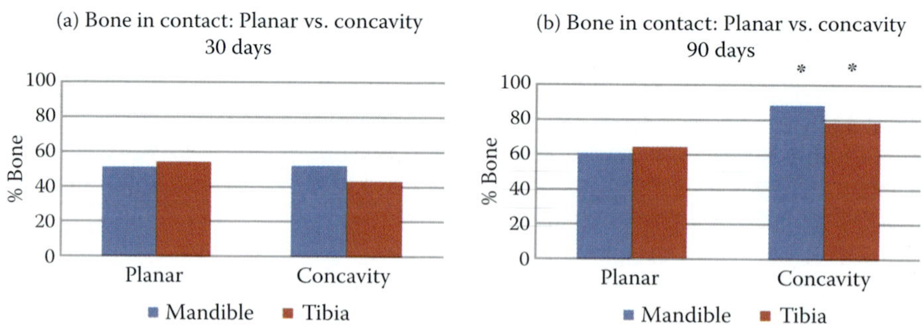

FIGURE 6.8 BIC on both geometric and planar control implants coated with crystalline hydroxyapatite. (a) No major differences were found on day 30 between planar versus concavity geometric surfaces. (b) On day 90, there was a statistically significant difference between the two topographical configurations (*$p < 0.05$ on day 90). There were no differences between the intratibial and intramandibular sites of implantation.

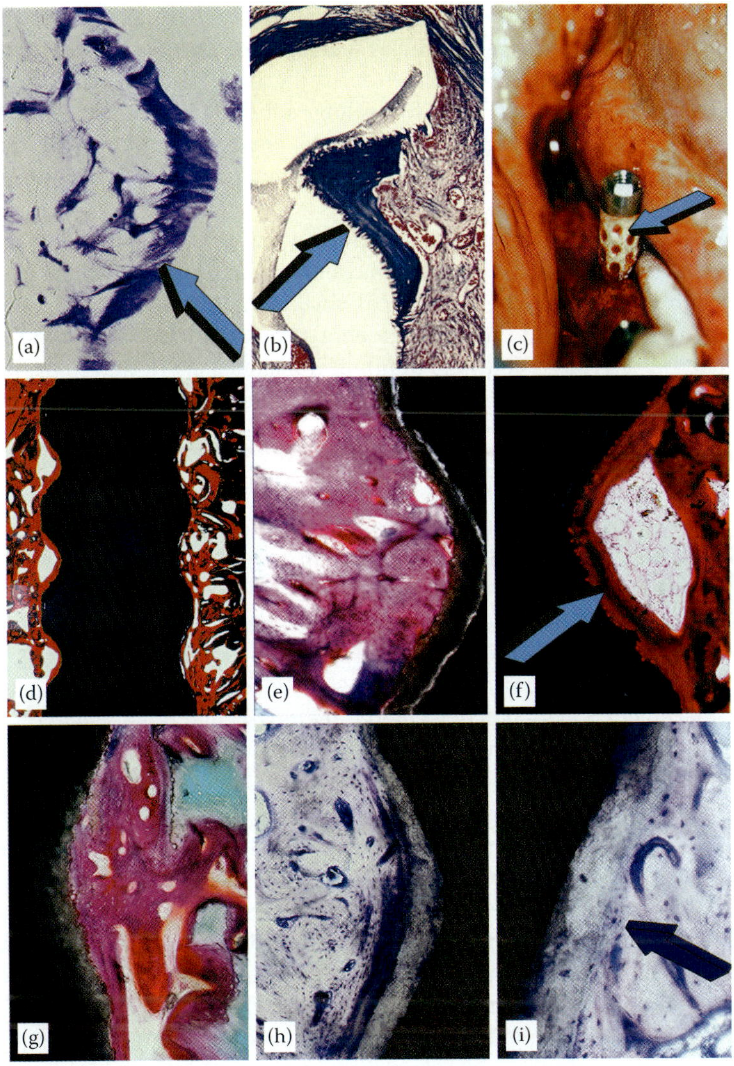

FIGURE 6.9 Cellular attachment, polarization, orientation, assembly (a), and the induction of bone formation within concavities of macroporous calcium carbonate/hydroxyapatite coral-derived constructs (a) *in vitro* and (b) *in vivo* (blue arrows). (c) Generation of hydroxyapatite-coated titanium implants with a series of repetitive concavities on the plasma-sprayed hydroxyapatite surface (blue arrow). (d) Complete osteointegration with the surrounding mandibular bone 90 days after the insertion of the dental implant within the edentulous ridge of an adult nonhuman primate *P. ursinus*.[29] (From Ripamonti, U., Tissue engineering of bone by novel substrata instructing gene expression during the *de novo* bone formation, *Science in Africa*: http://www.scienceinafrica.co.za/2002/march/bone2.htm, 2002. With permission.) (e–i) High-power views of osteointegrated concavities with the newly formed mandibular bone. Note the intimate relationship of the newly formed bone with the plasma-sprayed substratum with the induction of marrow [blue arrow in (f)] and optimal osteointegration of the newly formed bone with the plasma-sprayed hydroxyapatite [dark blue arrow in (i)].

traumatic release of blood and the resulting clotting factors such as fibronectin and fibrinogen pool in the concavities together with platelets that are known to retain several molecular signals including platelet-derived growth factor and TGF-β proteins. Fibrinogen is known to be important for cellular adhesion via membrane-bound integrin receptors.[59,60] These receptors translate the environmental cues to responding stem cells and/or osteoblastic precursor cells and determine the final commitment of the responding differentiating cells.

Several studies have indicated that mesenchymal stem cells sense and respond in a predetermined manner to several noncellular stimuli; these mechanical stimuli include roughness and nanotopographical surface modifications/alterations of the prefabricated substrata and the surface energy and stiffness of the material.[41–43,61] These properties are intrinsic to the titanium implants used in this study. Similarly, another property is cell stretching, whether in terms of permissible space for growth and expansion or as a result of cytoskeletal tension due to microgravitational forces, as the attached cells are likely to be "suspended" from the margins of the implanted concavities (Figure 6.6e and f). The geometric microenvironment is thus influential in determining the final fate of the responding stem cells, in context toward a more "stress-resistant" lineage, that is, an osteoblast that will ultimately form bone (Figure 6.5) with osteointegration.[62]

Geometric plasma-sprayed titanium implants showed substantial osteointegration, as evaluated histomorphometrically (Figure 6.8) and morphologically (Figures 6.9 and 6.10). Morphological analyses showed pronounced integration with newly formed mineralized bone—newly formed mineralized and remodeled bone showed an exquisite and intimate osteointegration with the plasma-coated layer of the hydroxyapatite enveloping the implants within the mandibular bone (Figures 6.9i and 6.10g).

Titanium implants harvested from one animal euthanized 31 months after implantation in the *rectus abdominis* muscle, which is a nonbony site, showed the induction of bone formation within the concavities (Bone Research Laboratory, unpublished data). This unique finding has set the formulation of functionalized osteoinductive hydroxyapatite-coated titanium implants that *per se* without the exogenous application of the soluble osteogenic molecular signals of the TGF-β supergene family initiate the ripple-like cascade of bone differentiation by induction.

Within the limits of the study without data on loaded implants in edentulous ridges of the nonhuman primate *P. ursinus*, the reported data confirm the critical role of implant geometry and of the geometric cue of the concavity created across the titanium surface to enhance the rate of osteointegration of hydroxyapatite-coated titanium implanted in orthotopic bony sites of the nonhuman primate *P. ursinus*.

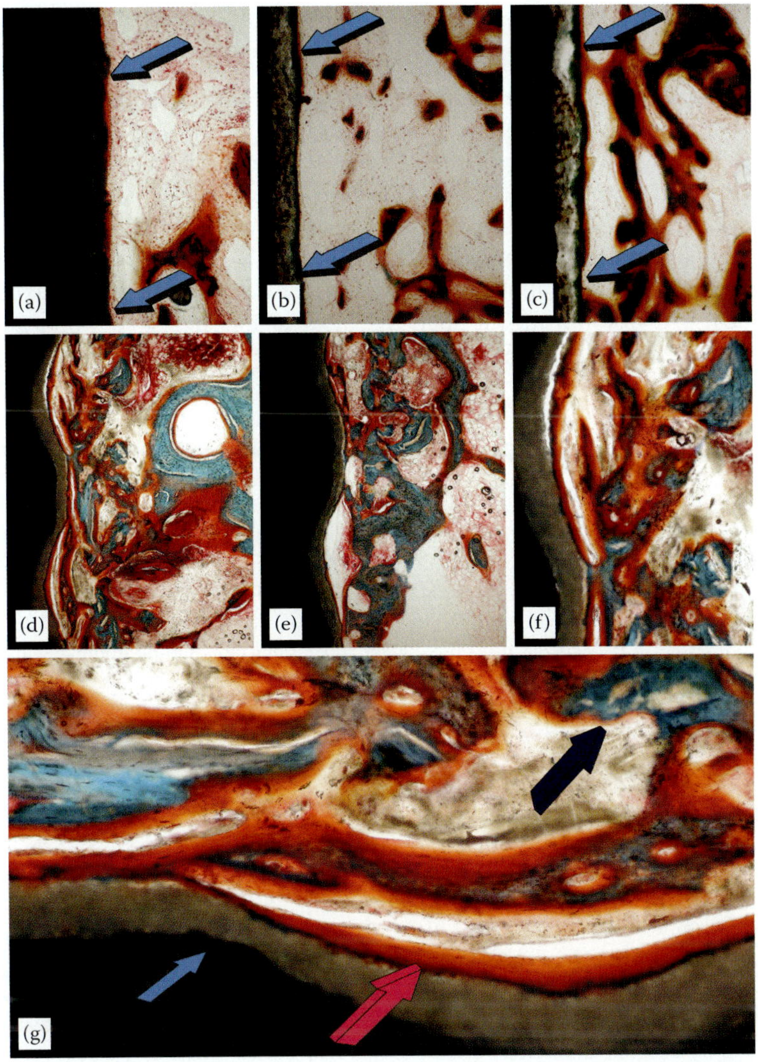

FIGURE 6.10 Details of osteointegration of the plasma-sprayed hydroxyapatite titanium implants in the tibia of the Chacma baboon *P. ursinus* and harvested 30 days after implantation. (a)–(c) Planar control implants showing thin layers of newly formed bone (arrows) against the hydroxyapatite coating with minimal bone forming activity distally located to the inserted implants. (d)–(f) Pronounced induction of bone formation and osteoid synthesis along concavities of the geometric implants. (g) Detail of tissue generation within the microenvironment of a concavity on day 30 after implantation in the tibial bone. The light blue arrow indicates the plasma-sprayed hydroxyapatite/titanium interface; the magenta arrow indicates the newly formed osteoid matrix within the geometric concavity of the implant, and the dark blue arrow indicates the osteoid synthesis on newly formed and mineralized bone.

ACKNOWLEDGMENTS

The "geometric induction of bone formation" has been constantly supported by the Medical Research Council of South Africa, the University of the Witwatersrand, Johannesburg, the National Research Foundation, and by *ad hoc* grants to the Bone Research Laboratory. We thank the Materials Science and Manufacturing Technology Group, Council for Scientific and Industrial Research (CSIR), Pretoria, South Africa, for the long-standing collaborative research and for the preparation of the titanium implants; the late Dr. Michael Thomas, PhD (CSIR) for the construction of the plasma-sprayed titanium implants; Karuna Bhavanbhai for the expert bibliographical analysis and for the final assembly of the presented work; and Prof. Alexander Ziegler for the scanning electron microscope images.

REFERENCES

1. Reddi, A.H. (1984). Extracellular matrix and development. In *Extracellular Matrix Biochemistry*, K.A. Piez and A.H. Reddi (Eds.). Elsevier, New York, pp. 375–412.
2. Reddi, A.H. (1994). Symbiosis of biotechnology and biomaterials: Applications in tissue engineering of bone and cartilage. *J Cell Biochem*, 56(2), 192–5.
3. Khouri, R.K., Koudsi, B., and Reddi, A.H. (1991). Tissue transformation into bone *in vivo*: A potential practical application. *JAMA*, 266(14), 1953–5.
4. Reddi, A.H. (1997). Bone morphogenesis and modeling: Soluble signals sculpt osteosomes in the solid state. *Cell*, 89(2), 159–61.
5. Reddi, A.H. (1998). Role of morphogenetic proteins in skeletal tissue engineering and regeneration. *Nat Biotechnol*, 16(3), 247–52.
6. Reddi, A.H. (2000). Morphogenesis and tissue engineering of bone and cartilage: Inductive signals, stem cells, and biomimetic biomaterials. *Tissue Eng*, 6(4), 351–9.
7. Ripamonti, U. (2006). Soluble osteogenic molecular signals and the induction of bone formation. *Biomaterials*, 27(6), 807–22.
8. Ripamonti, U. (2009). Biomimetism, biomimetic matrices and the induction of bone formation. *J Cell Mol Med*, 13(9B), 2953–72.
9. Ripamonti, U. and Roden, L.C. (2010). Biomimetics for the induction of bone formation. *Exp Rev Med Devices*, 7(4), 469–79.
10. Ripamonti, U., Heliotis, M., and Ferretti, C. (2007). Bone morphogenetic proteins and the induction of bone formation: From laboratory to patients. *Oral Maxillofac Surg Clin North Am*, 19(4), 575–89, vii.
11. Parfitt, A.M. (1994). Osteonal and hemiosteonal remodeling: The spatial and temporal framework for signal traffic in adult human bone. *J Cell Biochem*, 55(3), 273–86.
12. Parfitt, A.M., Mundy, G.R., Roodman, G.D., Hughes, D.E., and Boyce, B.F. (1996). A new model for the regulation of bone resorption, with particular reference to the effects of bisphosphonates. *J Bone Miner Res*, 11(2), 150–9.
13. Manolagas, S.C. and Jilka, R.L. (1995). Bone marrow, cytokines, and bone remodeling: Emerging insights into the pathophysiology of osteoporosis. *N Engl J Med*, 332(5), 305–11.
14. Ripamonti, U. (2004). Soluble, insoluble and geometric signals sculpt the architecture of mineralized tissues. *J Cell Mol Med*, 8(2), 169–80.
15. Ripamonti, U., Herbst, N.N., and Ramoshebi, L.N. (2005). Bone morphogenetic proteins in craniofacial and periodontal tissue engineering: Experimental studies in the non-human primate *Papio ursinus*. *Cytokine Growth Factor Rev*, 16(3), 357–68.
16. Trueta, J. (1963). The role of the vessels in osteogenesis. *J Bone Jt Surg*, 45-B, 402–18.

17. Lanza, D. and Vegetti, M. (1971). *Opere biologiche: Di Aristotele. A cura di Diego Lanza e Mario Vegetti Torino*. UTET, Torino.
18. Crivellato, E., Nico, B., and Ribatti, D. (2007). Contribution of endothelial cells to organogenesis: A modern reappraisal of an old aristotelian concept. *J Anat*, 211(4), 415–27.
19. Turing, A.M. (1952). The chemical basis of morphogenesis. *Philos Trans R Soc London*, 237, 37.
20. Lander, A.D. (2007). Morpheus unbound: Reimagining the morphogen gradient. *Cell*, 128(2), 245–56.
21. Urist, M.R. (1965). Bone: Formation by autoinduction. *Science*, 150(698), 893–9.
22. Ripamonti, U., Crooks, J., Khoali, L., and Roden, L. (2009). The induction of bone formation by coral-derived calcium carbonate/hydroxyapatite constructs. *Biomaterials*, 30(7), 1428–39.
23. Ripamonti, U., Ferretti, C., and Heliotis, M. (2006). Soluble and insoluble signals and the induction of bone formation: Molecular therapeutics recapitulating development. *J Anat*, 209(4), 447–68.
24. Urist, M.R., Silverman, B.F., Büring, K., Dubuc, F.L., and Rosenberg, J.M. (1967). The bone induction principle. *Clin Orthop Relat Res*, 53, 243–83.
25. Reddi, A.H. and Huggins, C. (1972). Biochemical sequences in the transformation of normal fibroblasts in adolescent rats. *Proc Natl Acad Sci USA*, 69(6), 1601–5.
26. Reddi, A.H. (1981). Cell biology and biochemistry of endochondral bone development. *Coll Relat Res*, 1(2), 209–26.
27. Ripamonti, U. and Reddi, A.H. (1995). Bone morphogenetic proteins: Applications in plastic and reconstructive surgery. *Adv Plastic Reconstr Surg*, 11, 47–65.
28. Ripamonti, U. (1991). The morphogenesis of bone in replicas of porous hydroxyapatite obtained from conversion of calcium carbonate exoskeletons of coral. *J Bone Jt Surg Am*, 73(5), 692–703.
29. Ripamonti, U. (2002). Tissue engineering of bone by novel substrata instructing gene expression during the *de novo* bone formation. *Science in Africa*. http://www.scienceinafrica.co.za/2002/march/bone2.htm.
30. Ripamonti, U., Klar, R.M., Renton, L.F., and Ferretti, C. (2010). Synergistic induction of bone formation by hOP-1, hTGF-β_3 and inhibition by zoledronate in macroporous coral-derived hydroxyapatites. *Biomaterials*, 31(25), 6400–10.
31. Ripamonti, U. and Klar, R.M. (2010). Redundancy of osteogenic molecular signals initiating the induction of bone formation. *Science in Africa*. http://www.scienceinafrica.co.za/2010/october/bone_regeneration_full.htm.
32. Dalby, M.J., Gadegaard, N., Tare, R., Andar, A., Riehle, M.O., Herzyk, P., Wilkinson, C.D., and Oreffo, R.O. (2007). The control of human mesenchymal cell differentiation using nanoscale symmetry and disorder. *Nat Mater*, 6(12), 997–1003.
33. Verrier, S., Peroglio, M., Voisard, C., Lechmann, B., and Alini, M. (2010). The osteogenic differentiation of human osteoprogenitor cells on anodic plasma chemical-treated Ti6AI7Nb. *Biomaterials*, 32(3), 672–80. doi:10.1016/j.biomaterials.2010.09028.
34. McNamara, L.E., McMurray, R.J., Biggs, M.J., Kantawong, F., Oreffo, R.O., and Dalby, M.J. (2010). Nanotopographical control of stem cell differentiation. *J Tissue Eng*, 120623.
35. Wilkinson, A., Hewitt, R.N., McNamara, L.E., McCloy, D., Dominic Meek, R.M., and Dalby, M.J. (2011). Biomimetic microtopography to enhance osteogenesis *in vitro*. *Acta Biomater*, 7(7), 2919–25.
36. Vlacic-Zischke, J., Hamlet, S.M., Friis, T., Tonetti, M.S., and Ivanovski, S. (2011). The influence of surface microroughness and hydrophilicity of titanium on the upregulation of TGFβ/BMP signalling in osteoblasts. *Biomaterials*, 32, 665–71.

37. Brunette, D.M. (1988). The effects of implant surface topography on the behavior of cells. *Int J Oral Maxillofac Implants*, 3(4), 231–46.
38. Singhvi, R., Kumar, A., Lopez, G.P., Stephanopoulos, G.N., Wang, D.I., Whitesides, G.M., and Ingber, D.E. (1994). Engineering cell shape and function. *Science*, 264(5159), 696–8.
39. Chou, L., Firth, J.D., Uitto, V.-J., and Brunette, D.M. (1995). Substratum surface topography alters cell shape and regulates fibronectin mRNA level, mRNA stability, secretion and assembly in human fibroblasts. *J Cell Sci*, 108.
40. Tirrell, M., Kokkoli, E., and Biesalski, M. (2002). The role of surface science in bioengineered materials. *Surf Sci*, 500, 61–83.
41. Discher, D.E., Mooney, D.J., and Zandstra, P.W. (2009). Growth factors, matrices, and forces combine and control stem cells. *Science*, 26, 324(5935), 1673–7.
42. Fu, J., Wang, Y.K., Yang, M.T., Desai, R.A., Yu, X., Liu, Z., and Chen, C.S. (2010). Mechanical regulation of cell function with geometrically modulated elastomeric substrates. *Nat Methods*, 7(9), 733–6.
43. Buxboim, A. and Discher, D.E. (2010). Stem cells feel the difference. *Nat Methods*, 7(9), 695–7.
44. Kohn, E., Alessandro, R., Spoonster, J., Wersto, R.P., and Liotta, L.A. (1995). Angiogenesis: Role of calcium-mediated signal transduction. *Proc Natl Acad Sci USA*, 92, 1307–11.
45. Munaron, L. (2006). Intracellular calcium, endothelial cells and angiogenesis. *Recent Pat Anti-Cancer Drug Discovery*, 1(1), 105–19.
46. Wu, Q., Shao, H., Eton, D., Li, J., Yang, B. et al. (2009). Extracellular calcium increases CXCR4 expression on bone marrow-derived cells and enhances proangiogenesis therapy. *J Cell Mol Med*, 13, 1–10.
47. Zayzafoon, M. (2006). Calcium/calmodulin signaling controls osteoblast growth and differentiation. *J Cell Biochem*, 97(1), 56–70.
48. Ohgushi, H., Dohi, Y., Tamai, S., and Tabata, S. (1993). Osteogenic differentiation of marrow stromal stem cells in porous hydroxyapatite ceramics. *J Biomed Mater Res*, 27(11), 1401–7.
49. Kanatani, M., Sugimoto, T., Kano, J., Fukase, M., and Fujita, T. (1991). Effect of 17 beta-estradiol on the proliferation of osteoblastic mc3t3-e1 cells via human monocytes. *Biochem Biophys Res Commun*, 178(3), 866–70.
50. Viola, J., Lal, B., and Grad, O. (2003). NSF: ABT report on the emergence of tissue engineering as a research field. http://www./nsf.gov/pub/2005/msf0450/start.htm.
51. Sampath, T.K. and Reddi, A.H. (1981). Dissociative extraction and reconstitution of extracellular matrix components involved in local bone differentiation. *Proc Natl Acad Sci USA*, 78(12), 7599–603.
52. Sampath, T.K. and Reddi, A.H. (1983). Homology of bone-inductive proteins from human, monkey, bovine, and rat extracellular matrix. *Proc Natl Acad Sci USA*, 80(21), 6591–5.
53. Ripamonti, U., Ramoshebi, L.N., Patton, J., Matsaba, T., Teare, J., and Renton, L. (2004). Soluble signals and insoluble substrata: Novel molecular cues instructing the induction of bone. In *The Skeleton*, Chapter 15, E.J. Massaro and J.M. Rogers (Eds.). Totowa, New Jersey: Humana Press, pp. 217–27.
54. Heliotis, M., Lavery, K.M., Ripamonti, U., Tsiridis, E., and Di Silvio, L. (2006). Transformation of a prefabricated hydroxyapatite/osteogenic protein-1 implant into a vascularised pedicled bone flap in the human chest. *Int J Oral Maxillofac Surg*, 35(3), 265–9.
55. Ripamonti, U., Crooks, J., and Kirkbride, A.N. (1999). Sintered porous hydroxyapatites with intrinsic osteoinductive activity: Geometric induction of bone formation. *S Afr J Sci*, 95, 335–43.

56. Ripamonti, U., van den Heever, B., and van Wyk, J. (1993). Expression of the osteogenic phenotype in porous hydroxyapatite implanted extraskeletally in baboons. *Matrix*, 13(6), 491–502.
57. Kovacic, J.C. and Boehm, M. (2009). Resident vascular progenitor cells: An emerging role for non-terminally differentiated vessel-resident cells in vascular biology. *Stem Cell Res*, 2(1), 2–15.
58. Donath, K. and Breuner, G. (1982). A method for the study of undecalcified bones and teeth with attached soft tissues. The Sage–Schliff (sawing and grinding) technique. *J Oral Pathol*, 11(4), 318–26.
59. Mosesson, M.W. (1983). Fibrinogen heterogeneity. *Ann NY Acad Sci*, 408, 97–113.
60. Gronowicz, G. and McCarthy, M.B. (1996). Response of human osteoblasts to implant materials: Integrin-mediated adhesion. *J Orthop Res*, 14(6), 878–87.
61. Breuls, R.G., Jiya, T.U., and Smit, T.H. (2008). Scaffold stiffness influences cell behavior: Opportunities for skeletal tissue engineering. *Open Orthop J*, 2, 103–9.
62. McBeath, R., Pirone, D.M., Nelson, C.M., Bhadriraju, K., and Chen, C.S. (2004). Cell shape, cytoskeletal tension, and RhoA regulate stem cell lineage commitment. *Dev Cell*, 6(4), 483–95.
63. van Eeden, S.P. and Ripamonti, U. (1994). Bone differentiation in porous hydroxyapatite in baboons is regulated by the geometry of the substratum: Implications for reconstructive craniofacial surgery. *Plast Reconstr Surg*, 93(4), 959–66.

7 Bioactive Surface Modifications for Dental and Orthopedic Implants

Ivan Donati, Eleonora Marsich,
Andrea Travan, and Sergio Paoletti

CONTENTS

7.1 TISSUE–MATERIAL INTERFACE IN DENTAL AND ORTHOPEDIC IMPLANTS

Until the 1970s, the central challenge in orthopedic surgery was the development of "bioinert" permanent implants that can functionally replace a damaged bone tissue and provide a safe and durable mechanical load transfer between bone and orthopedic graft; the loosening of this mechanical interlock was, in fact, a major cause of implant failure and was often related to inadequate interface interactions based on a nonphysiological implant material behavior (Sakka and Coulthard 2011). The use of morphological fixation methods such as press-fitting, bone cementation, and mechanical interlocking by using screw-shaped and rough endo-osseous implants was commonly employed to increase the implant stability (DiMaio 2002; Holt et al. 2011; Javed et al. 2011; Varini et al. 2008).

In the last three decades, with the development of tissue engineering and regenerative medicine techniques, in orthopedic implantation, the new concept of "bioactive" biomaterial has arisen, that is, a material, permanent or biodegradable, that can induce specific and tailored biological responses by intimate interactions with the surrounding living tissue (Beutner et al. 2010; Porter et al. 2009). In particular, in orthopedics, three main properties are sought for an ideal bone substitute with improved performances: 1) osteoconductivity, bone growth into the inner portion of a bone device through pores, channels, or pipes; 2) osteoinductivity, a process by which osteogenesis is induced; and 3) osseointegration, direct, stable biological/chemical bond of implant to bone with no fibrous tissue interface (Albrektsson and Johansson 2001; Branemark 1983; Hulbert et al. 1973). The numerous orthopedic and dental implants reported in the literature and some of those currently used in clinical practice aim at combining, more or less successfully, the above-mentioned properties.

All the approaches described above are led by the basic notion that the success or failure of biomaterials for joint and dental implants or for any other bone graft depends on events at the molecular level that determine subsequent feedback of cells and tissues.

The biocompatibility and the biological response of a tissue to an implanted biomaterial (i.e., cell differentiation, tissue healing, integration, and regeneration) are deeply influenced and governed by the events of cell contact and adhesion on material surface. Cell attachment, spreading, and migration are the first phase of cell/material interplay, and its nature and quality will influence all the following biological pathways. The primary interactions between biomaterial and host occur on a molecular scale and involve the adsorption and reactions of biomolecules, water, and inorganic ions from the biological fluids on biomaterial surface, as well as the dissolution of atomic, ionic, or molecular fragments from the biomaterial. It is well described that implanted materials are immediately coated with proteins from blood and interstitial fluids (defined as surface biofouling), and it is through this adsorbed layer that the cells sense foreign surfaces. Protein absorption is an energy-driven process that involves noncovalent interactions such as electrostatic, hydrophobic, hydrogen bonding, and van der Waals interactions (Andrade and Hlady 1991; Thull 2002). Hence, cells initially respond to adsorbed proteins rather than to the surface itself.

The fouling proteins and molecules on material surface are the active elements by which osteoblasts anchor onto implant via different membrane adhesion receptors such as integrins, cell adhesion molecules, and gap junctions (Anselme 2000). These membrane adhesion receptors regulate multiple and different gene expression pathways such as cell migration, differentiation, and proliferation by a signal transduction induced by cell–substrate interactions. In particular, integrins are considered the main cellular molecules involved in host response to bone-replacing materials (García 2005; Siebers et al. 2005). Integrins are the primary cell adhesion receptors to extracellular matrix (ECM) components, including fibronectin, laminin, and type I collagen (Hynes 2002). They are a widely expressed family of glycosylated transmembrane receptors consisting of two subunits: 1) α chains and 2) β chains. Eighteen α chains and eight β chains have been identified to dimerize into 24 distinct receptors that are expressed in different cell types. Moreover, a cell type expresses several different integrin receptors. As outlined in Figure 7.1, when an integrin is engaged by interaction with an extracellular ligand, the formation and activation of focal adhesions takes place. They are supramolecular complexes consisting of integrins and cytoskeletal proteins, such as talin, vinculin, and α-actin, which are physically and functionally linked to signaling molecules, including focal adhesion kinase (FAK), src, and paxillin. Focal adhesions activate signaling pathways, such as extracellular

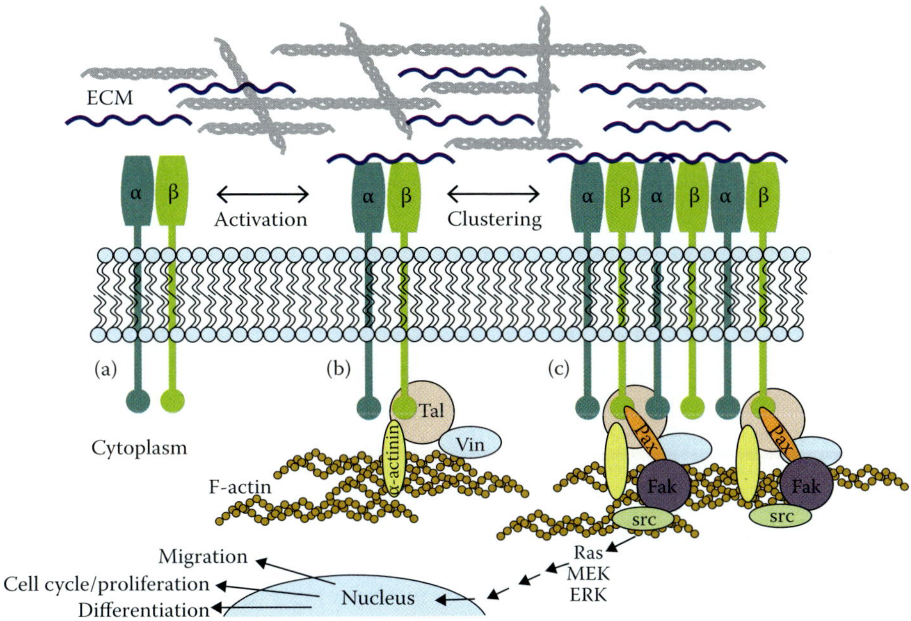

FIGURE 7.1 Schematic of integrin-mediated signal transduction. (a) In their inactive state, integrins do not bind to ligands and do not transduce signals. (b) Integrins are activated upon interaction with a ligand. (c) Clusterization of activated integrins promotes signal transduction. Tal, talin; Vin, vinculin; Pax, paxillin; F-actin, filamentous actin.

signal-regulated kinase (ERK) and mitogen-activated protein kinase/ERK kinase, which regulate transcription factor activity and direct cell cycle progression and differentiation (Byron 2011; Campbell and Humphries 2011; Zambuzzi et al. 2011).

The topography, roughness, chemistry, wettability, charge, stiffness, and energy of the material surface influence nature, composition, density, distribution, and structural rearrangement of the adsorbed proteins and, consequently, cell behavior upon contact (Anselme et al. 2010b; Ponche et al. 2010; Saranya et al. 2011). Much experimental evidence, mostly *in vitro*, is reported in the literature about the influence of the chemical–physical features of material surface on bone tissue response. For example, it has been demonstrated that MG63 osteoblast-like cells adhere and proliferate on smooth surfaces of polycarbonate membranes more than on rough ones, but the phenotype expression of the MG63 cells appears more influenced on rough surfaces than on smooth ones (Lee et al. 2004b). Saldaña et al. (2007) reported that the thermal oxidation of Ti–6Al–4V, resulting in the formation of an outer ceramic layer, improves primary osteoblast behavior such as cell attachment, spreading, and cytoskeleton organization, as well as cell proliferation, viability, and collagen secretion. More recently, Kubo et al. (2009) described the development of TiO_2 microhybrid and nanohybrid surfaces in titanium surfaces, consisting of nanoscale nodules within microscale pits. A study that aims at comparing surface topographies with different nanonodule dimensions revealed significant effects on protein adsorption, on osteoblast behavior, such as attachment, spread, adhesion, proliferation, and differentiation, and on the strength of bone–implant integration *in vivo*. Overall, the strategies for modifying the nature of tissue/material interface frequently involve changes to the surface, affecting protein adsorption, and, thus, cell–substrate interactions and tissue response. Recent methods and technologies for controlling these biological interactions mainly at the nanometer scale on the surface of implants are reviewed in more detail thereafter.

7.2 SURFACE REACTIVITY OF TITANIUM ALLOYS FOR IMPLANT APPLICATIONS

Metals represent the gold standard in implantology due to their mechanical properties and biocompatibility. At present, most common metallic materials for orthopedics and dentistry include stainless steel, cobalt-based alloys, and titanium-based alloys (Navarro et al. 2008). Titanium alloys are generally preferred to stainless steel and cobalt alloys because of their superior biocompatibility, lower elastic modulus, and higher corrosion resistance. Table 7.1 sums up the main characteristics and use of the three metallic alloys.

In biomaterials, titanium alloys are mainly used for implants such as hip and dental prosthesis, screws, nails, and fixation plates. A brief overview about titanium

TABLE 7.1
Comparison between Different Metallic Implant Materials

	Titanium and Ti-Base Alloys	Stainless Steel	Cobalt-Base Alloys
Designation	ASTM F-67 (ISO 5832/II) ASTM F-136 (ISO 5832/II) ASTM F-1295 (Cast and wrought)	ASTM F-138 ("316 LDVM")	ASTM F-75 ASTM F-799 ASTM F-1537 (Cast and wrought)
Principal alloying elements (wt%)	Ti(bal.) Al(6) V(4) Nb(7)	Fe(bal.) Cr(17–20) Ni(12–14) Mo(2–4)	Co(bal.) Cr(19–30) Mo(0–10) Ni(0–37)
Advantages	Biocompatibility Corrosion resistance Low modulus Fatigue strength	Cost Availability Processing	Wear resistance Corrosion resistance Fatigue strength
Disadvantages	Wear resistance Shear strength	Long-term behavior High modulus	High modulus Low biocompatibility
Primary utilizations	Total hip replacement (THR) Screws Dental prosthesis Nails Pacemakers	Temporary devices (fracture plates, screws, hip nails, THR stems)	Dentistry castings Prosthesis stems Load-bearing components in THR

Source: Adapted from Long, M. and Rack, H.J., *Biomaterials* 19, 1621–1639, 1998.

metallurgy, corrosion behavior, mechanical and wear properties, and biocompatibility is described in the next paragraphs in order to understand the advantages of this biomaterial and the possibilities of further enhancing its performance by means of surface modifications.

Titanium is a transition metal that enables the formation of solid solutions with various substitutional elements having a size factor of ±20%. It displays two crystalline microstructures at different temperatures: 1) the α structure (hexagonal close packed) and 2) the β structure (body centered cubic). The type of microstructure influences the mechanical properties (e.g., elastic modulus, fatigue, and wear response) and the corrosion resistance. Alloying elements are used to stabilize α or β phases according to the desired characteristics of the final construct. α-Stabilizers such as Al, O, N, and C form alloys with good corrosion resistance but low strength at ambient temperature. β-Stabilizers such as V, Mo, Nb, and Ta allow the formation of stable and metastable alloys with high strength and hardenability. The choice of proper alloying elements makes it possible to obtain different microstructures (α, β, or α + β) that modulate the properties of the final construct (Carman et al. 2011).

The mechanical behavior of metallic biomaterials with respect to both bulk and surface properties is of primary importance. Compared to the bone, the high elastic modulus and the variable fatigue resistance of metallic biomaterials may eventually lead to prosthesis failure through loosening or fracture. In fact, the difference in stiffness and flexibility (which depend on the elastic modulus) between the natural bone and the implant material may cause insufficient load of the tissue, resulting in bone resorption and loosening of the prosthetic device; this phenomenon is known as "stress shielding." The elastic modulus of the natural bone is approximately 10 times lower than Co–Cr–Mo and stainless steel alloys, and four to five times lower than the Ti–6Al–4V alloy. A stiffness reduction through the modification of composition and microstructure of such alloys promotes load redistribution, minimizing stress shielding. For example, elements such as Nb and Zr in titanium lower the alloy moduli by 20%–25% (Banerjee et al. 2004).

Another critical aspect to take into account for implantable orthopedic and dental biomaterials is the fatigue resistance; in fact, cyclic loading is applied to implants during body motion, causing surface deformation and stress concentration zones produced by notches or microstructural inhomogeneities. The fatigue life of titanium alloys depends mostly on the microstructure obtained after thermomechanical processing (α/β structures, grain size, morphology, and distribution). Improved fatigue resistance may be achieved by the modification of the surface through mechanical or physical/chemical methods (Niinomi and Akahori 2010). For instance, a technique called shot peening—a cold working process applied by bombardment of the surface—aims at introducing residual stresses to hamper crack initiation and propagation. Fatigue properties can also be varied by altering the content of interstitial elements such as O, C, N, and H (Long et al. 1999).

A key feature of titanium-based materials is the presence of a natural oxide layer (TiO_2) that forms spontaneously on the titanium surface upon exposure to oxygen (passivation) and protects the metal from corrosion; in fact, when fresh titanium is exposed to air, it reacts rapidly with atmospheric oxygen to form a surface oxide that is typically a few nanometers thick. For oxidation temperatures less than 200°C, the oxide appears amorphous and glasslike, whereas thicker oxides formed at elevated temperatures or by electrochemical oxidation are more crystalline (Kasemo and Lausmaa 1994).

Considering the typical applications of metal implants, the surface oxide of such materials is subjected to wear due to the variable and cyclic loading of the material within the body. The wear behavior is an important aspect that is related to surface properties and determines the long-term fate of the prosthesis; in fact, friction with artificial materials is generally higher than in natural joints and tissues. Failure generally occurs due to the excessive wear of the components that accumulate, producing an adverse cellular response that leads to inflammation, release of damaging enzymes, bone cell lysis, osteolysis, infection, and, eventually, implant loosening (Hallab and Jacobs 2009; Jasty 1993).

In the case of titanium alloys, wear phenomena are related to the mechanical properties of the oxide layer (Buchanan et al. 1987; Mitsuo 2008); when normal or shear stresses are high enough to induce the breakdown of the passive layer, the oxide will be disrupted. The exposed metal surface may then reform a passive layer, leading to the continuous removal (material disruption) and reformation (oxidation) of the oxide, which results in the gradual consumption of the alloy, surface roughness increase, and release of hard oxide debris, which acts as third body abrasive components (Figure 7.2; Barbour et al. 1999). The mechanical response of the surface layer is related to the chemical composition of the bulk material, which affects the kinetics of repassivation and the shear resistance of the oxide.

Titanium alloys are considered to have poor wear resistance when tribological and chemical events take place at the implant surface (Yildiz et al. 2009); this fact has resulted in the development of surface treatments to enhance the hardness and the abrasive wear resistance of the alloy. Various procedures including physical vapor deposition (PVD)/chemical vapor deposition (CVD) coating (TiN and TiC), thermal oxidation, ion implantation (N), laser alloying with TiC, thermal treatments (nitriding and diffusion hardening), and oxygen diffusion hardening are used to address this issue, which remains of primary importance for orthopedic and dental applications (Bloyce et al. 1998; Hanawa 2001; Streicher et al. 1991).

The oxide layer protects the material from electrochemical processes (corrosion) due to contact with electrolyte solutions such as biological fluids, thus preventing

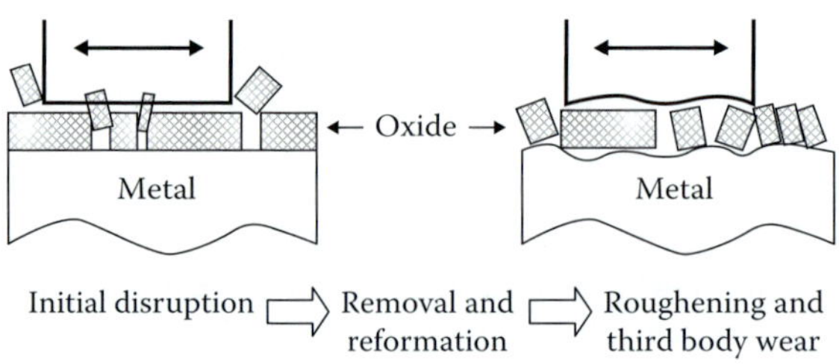

FIGURE 7.2 Schematic of oxidative/abrasive wear of metal surface. (Adapted from Long, M. and Rack, H.J., *Biomaterials* 19, 1621–1639, 1998.)

a massive release of metal ions (e.g., ionic forms of Ni, Cr, Al, and V) that could be harmful to biological systems. This protective layer is the basis of the generally accepted biocompatibility of titanium alloys. In fact, the biocompatibility of a metallic alloy is strictly related to both its corrosion resistance and the biocompatibility of its corrosion products (Olmedo et al. 2009); the release of metal ions from the implant can elicit local adverse tissue reactions (Cremasco et al. 2011; Steinemann 1996). Alloying elements contribute to the corrosion resistance and thus influence the biocompatibility of titanium alloys—it was reported that alloying elements such as Nb and Zr contribute to the formation of the protective oxide and are incorporated into the passive layer, whereas Al and V can be more easily released into the environment as dissolved metal ions (Velich et al. 2006). This potential risk represents one of the major drawbacks of metallic implants such as alloys of titanium or stainless steel.

When a passivated metal surface is exposed to air, the typical oxide layer formed is not a perfect lattice, but it displays defects such as steps, vacancies, and impurities, and it tends to bind molecules and atoms (e.g., water, hydrocarbons, and hydrogen) from the surrounding as a monomolecular layer (Figure 7.3; Kasemo and Lausmaa 1994).

If the material is in contact with a biological environment, various parameters (e.g., chemical constituents, pH, and ionic strength) influence the interactions at the surface. In general, the first event that takes place is water adsorption, followed by the adsorption of plasma and ECM proteins, and, ultimately, cell colonization. Once implanted into the human body, the surface oxide is subjected to several molecular processes that can be summarized as follows (Figure 7.4):

1. Oxygen, hydrogen, and metal atom diffusion through the oxide layer
2. Incorporation of elements (e.g., Ca and P) from the bioliquid into the oxide layer

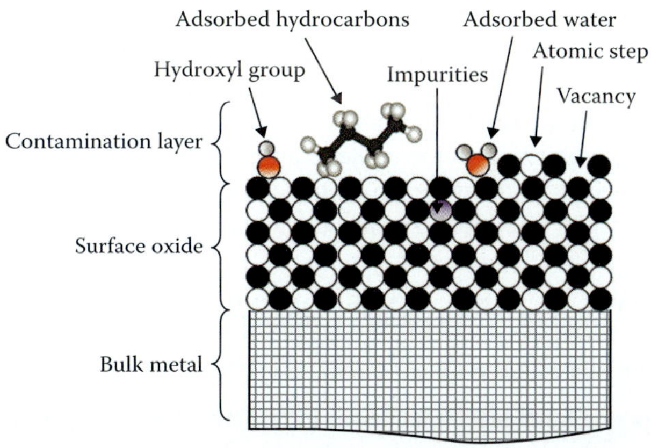

FIGURE 7.3 Schematic of a "real" metal surface after passivation upon exposure to ambient air. (Adapted from Kasemo, B. and Lausmaa, J., *Environ. Health Perspect.* 102 Suppl 5, 41–45, 1994.)

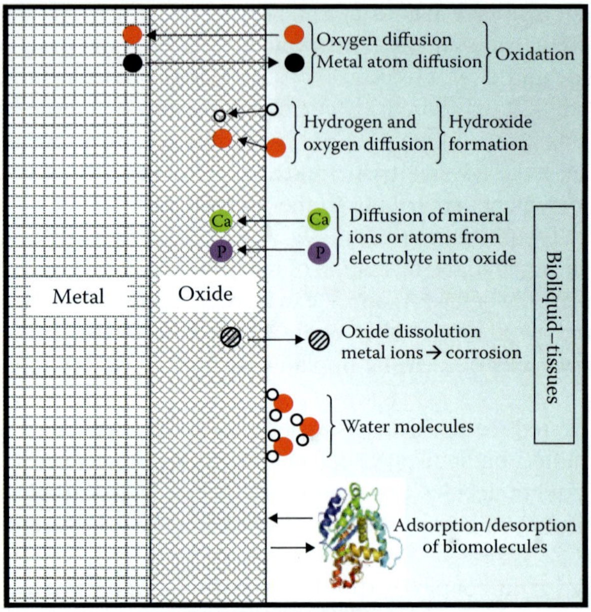

FIGURE 7.4 Schematic of some molecular processes that take place at the interface between the metallic biomaterial and the surrounding biological environment. (Adapted from Kasemo, B. and Lausmaa, J., *Environ. Health Perspect.* 102 Suppl 5, 41–45, 1994.)

3. Release of metal ions due to corrosion
4. Adsorption/desorption of water and biomolecules (proteins)
5. Cell/bacteria colonization

All these mechanisms must be taken into account to understand and foresee the events that take place when a metallic biomaterial is implanted in a biological tissue. In general, the "unmodified" surface of titanium alloys is considered to be biocompatible but bioinert in the sense that it is only passively integrated within the bone without eliciting specific responses in host tissues.

The following sections provide some examples of recent techniques of surface modification aimed at modulating the biological response of titanium-based biomaterials, with particular attention to innovative nanoscale approaches.

7.3 BIOACTIVE MODIFICATION OF SURFACES

7.3.1 Surface Topography

The surface topography of an implant is an important parameter that affects osseointegration. For instance, an increase in the surface roughness of a biomaterial can improve its biological performance. Compared to a smooth surface, roughness at the microscale and nanoscale can enhance osteoblast activity, such as the adhesiveness, proliferation, and production of specific markers of differentiation (Montanaro et al. 2008). The importance of surface topography on cell adhesion and proliferation has been explored by various research groups on bone-derived cells (Anselme and Bigerelle 2006). Surface topography parameters such as roughness amplitude and organization (e.g., autocorrelation length, the number of peaks/lengths, and order) influence cell response (Anselme et al. 2000a,b), in particular, long-term adhesion (Anselme and Bigerelle 2005; Bigerelle and Anselme 2005) and, in combination with surface chemistry, proliferation. Indeed, a synergistic effect of nanostructures and microstructures on cells has been reported and demonstrated the importance of achieving a tight control over both nano roughness and microroughness (Anselme 2011).

Nanotopography and surface chemistry have a strong impact on the interaction with cells and tissues through differential adsorption of the proteins on the surface of the material (Anselme 2011). Nanotopography is also known to alter surface energy via a mechanism similar (although more pronounced) to the one encountered for microstructured surfaces, known as the lotus effect (Bhushan et al. 2006; Martines et al. 2005). The responses of cells to nanotopography are basically mediated by the transduction of the signal to the nucleus, with a modification of the gene expression. The physical structuring of nanofeatures, size, shape, and presence of columns/island/peaks does show an effect of cell behavior [for a recent review, see the work of Anselme et al. (2010a)]. However, it is, at present, very difficult to rationalize the various results reported in the literature, because it is, in several cases, difficult to discriminate between the relative influences of modifications to nanotopography and to surface energy induced by the treatment.

Bioactive nanotopographies can be generated by physical methods such as plasma spray, PVD, and thermal oxidation or by chemical methods such as anodic electrochemical oxidation, acid/base treatments, and CVD (Variola et al. 2011). Indeed, several reviews dealing with topography at the nanoscale can be found in the literature (Anselme et al. 2010a; Blättler et al. 2006).

7.3.2 Biomimetic Signals

Novel nanomolecular engineering strategies on biomaterials aim at developing interactive interfaces presenting gradients of multiple adhesive signals (García 2005). Biomimetic strategies exploit cell adhesive interaction between cells and ECM to engineer bioinspired surfaces promoting osteoblast adhesion and differentiation, bone formation, and osseointegration. The complex interplay, based on a biochemical and mechanical signaling, between cells and ECM is fundamental for the development, maintenance, resorption, and remodeling of osseous tissue (Damsky 1999; De Arcangelis and Georges-Labouesse 2000; Fisher et al. 1993; García and Reyes 2005). Indeed, an important factor for implant osseointegration is given by the possibility of engaging osteogenic cell receptors (García 2005; LeBaron and Athanasiou 2000). The adhesion, organization, differentiation, and matrix mineralization of osteoblasts and osteoprogenitor cells, guided by biochemical signals conveyed from the extracellular environment, are key features for a bone implant to be successful.

A strategy that is widely proposed in the literature for improving cell/biomaterial interactions focuses on the introduction, on the surface of the biomaterial, of factors that promote the adhesion and survival of selected cell types involved in wound healing or tissue regenerative process. These signals include (Shekaran and García 2011):

1. Insoluble molecules highly entangled within the ECM (collagens, fibronectin, vitronectin, osteocalcin, bone sialoprotein, biglycan, decorin, and polysaccharides)
2. Hormones (e.g., parathyroid hormone), growth factors [e.g., bone morphogenetic proteins (BMPs)], cytokines, and locally released chemokines
3. Cell–cell receptors (cadherins, immunoglobulin-like superfamily adhesion molecules, and ephrins)

ECM polymers, ECM-derived peptides, and growth factor-derived peptides have been proposed for the biomimetic functionalization of orthopedic and dental biomaterials [for a comprehensive review, see the work of Shekaran and García (2011)].

Full-length ECM biomolecules, e.g., collagen (Morra et al. 2005; Rammelt et al. 2006; Svehla et al. 2005), fibrin (Kim et al. 2007; Perka et al. 2000), and hyaluronic acid (Gristina 1994; Kurkalli et al. 2010; Solchaga et al. 1999), have been proposed as a potential biomimetic signal for bone healing in view of their important regulatory role *in vivo*. However, some drawbacks have limited the widespread use of ECM molecules for the bioactivation of orthopedic implants. First, the isolation of ECM molecules requires demanding extraction procedures and leads to batch-to-batch variations in physical–chemical properties that might strongly affect the biological outcome. In addition, a reliable and reproducible purification procedure still represents a challenge for biomolecules, even when they are produced through biotechnological approaches. All these aspects are reflected in the high costs of the commercial ECM molecules. Finally, potential immunogenicity of ECM molecules should be seriously taken into account together with the proteolytic degradation, which is increased in case of inflammation and infection and hampers long-lasting applications (Shekaran and García 2011). It is also important to mention the difficulties encountered in the coating of biomaterials with ECM full molecules in a highly controllable manner (Shakesheff et al. 1998). Moreover, for the stochastic characteristic of the process, only a part of ECM proteins display, once adsorbed on the biomaterial's surface, the proper orientation for cell adhesion to occur (Elbert and Hubbell 2001; Hersel et al. 2003).

For the hurdles represented by the use of natural ECM molecules, the use of synthetic ECM-derived peptides has emerged as a cost-effective and tunable alternative. This approach is based on the concept that just few short peptide sequences, within the whole polymer chain of ECM molecules, trigger receptor-mediated cell adhesion and spreading. Short peptide sequences offer several advantages over full ECM molecules, i.e., higher stability toward sterilization conditions, heat treatment, and pH variation (Ito et al. 1991). Moreover, they can be produced in large quantities at a reasonable cost and with a high purity. This aspect allows modulating the amount of bioactive signals on the surface of a biomaterial to compensate for their lack of activity compared to the complete ECM molecule (Hersel et al. 2003; Ruoslahti 1996). However, conformation is a nontrivial aspect influencing cell recognition, even at this level.

The biomimetic tripeptide arginine–glycine–aspartate (RGD) is probably the most widely used signal exploited for surface biofunctionalization. RGD is an adhesive peptide sequence that was proved to promote cell adhesion in 1984 (Pierschbacher and Ruoslahti 1984), and it is found in fibronectin, vitronectin, bone sialoprotein, tenascin, and osteopontin (Hersel et al. 2003; Pytela et al. 1987). RGD has been shown to bind to approximately half of the 24 integrins known to date, such as α3β1, α5β1, α8β1, αIIbβ1, αvβ1, αvβ3, αvβ5, αvβ6, and αvβ8 (Pfaff 1997), although for some of these integrins, the interaction is modulated by other peptide sequences (Redick et al. 2000).

A large number of RDG-containing peptides, varying in specific amino acid sequence and length, have been tested to convey bioactive features to the surface of the biomaterial [for a comprehensive review, see the work of Perlin et al. (2008)]. The results obtained from competition studies against ECM molecules showed that the affinity in solution toward integrin receptors increased along the sequence GRGDSP > RGDS > RGD–NH_2 > RGD. However, the immobilization of the RGD on the surface of the biomaterial leads to a regain of its bioactive properties, very likely due to minimization of entropy loss in the ligand–receptor interaction upon anchoring of the tripeptide (Mammen et al. 1998).

One aspect in the RGD-dependent adhesion is the importance of the residues flanking the bioactive sequence. Varying the peptide sequence that contains the RGD signal can alter the conformation of the latter, and this is reflected in a different affinity for the integrin receptor. The replacement of a serine with a valine in GRGDxY was reported to promote higher levels of adhesion (Lin et al. 1992). This effect is amplified when cyclic peptides are considered, which showed greater affinity (Porté-Durrieu et al. 2004; van der Pluijm et al. 1996) and specific selectivity for integrins (Pfaff et al. 1994) [reviewed in Perlin et al. (2008)].

Additional aspects influencing the biological role of RGD are the presence of a spacer between the ligand and the surface of the biomaterial and its multiple presence (polyvalency) (clustering effect; Figure 7.5) on the biomaterial.

The importance of distancing the RGD ligand from the anchoring surface has been underlined by several authors (Beer et al. 1992; Craig et al. 1995; Kantlehner et al. 2000). This concept takes the move from the consideration that the RGD

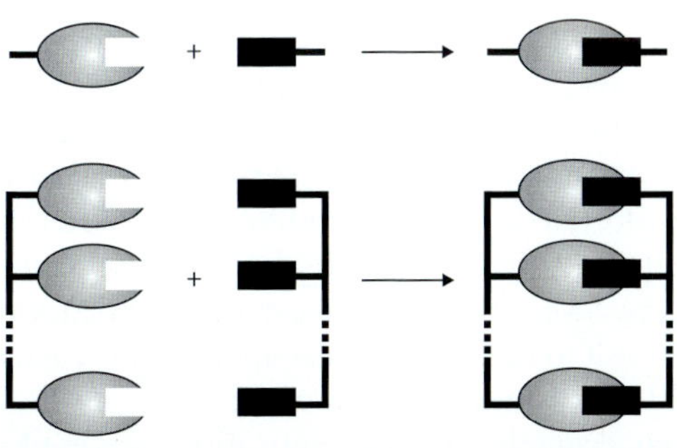

$$K^{Mono} = e^{-\frac{\Delta G^{Mono}}{RT}}$$

$$K^{Poly} = (K^{Mono})^{\alpha N} = e^{-\frac{\alpha N \Delta G^{Mono}}{RT}}$$

FIGURE 7.5 Schematic of the concept of polyvalency: Comparison between the constant of the monovalent process and a polyvalent (*N*-valent) process. α indicates the cooperativity of the process. (Adapted from Mammen, M. et al., *Angew. Chem., Int. Ed.* 37, 2754–2794, 1998.)

sequence is placed in exposed loops in ECM proteins (Pierschbacher and Ruoslahti 1984). Therefore, it has been proven that the RGD needs to extend from the surface to reach the binding site inside the globular head of the integrin (Beer et al. 1992). In particular, it was found that the most efficient binding of an RGDF peptide to integrins was achieved by adding a spacer composed of an oligopeptide containing 13 glycines, which extend approximately 11–32 Å from the surface of the material [poly(acrylonitrile)] (Beer et al. 1992). At variance, the use of very long spacers prevents the adhesion of the cells to the surface of the engineered material (Beer et al. 1992; Drumheller and Hubbell 1994). This is basically caused by the negative entropic contribution arising from the conformational freedom of long and flexible spacers, which opposes ligand binding (Mammen et al. 1998). [The entropy of a Gaussian polymer chain increases upon increasing the number of repeating units (Treloar 1958).] Despite the considerations reported above on the importance of the spacer, good cell adhesion results have also been achieved with RGD peptides anchored on glass without any spacer (Massia and Hubbell 1991). These conflicting results underline the effect of the specific mechanical and physical–chemical properties of the surface, which go beyond the ligand–receptor binding event. In fact, it is known that cells prefer a tight binding to more rigid surfaces (Hersel et al. 2003; Lo et al. 2000).

Ligand surface density and clustering appear to be essential to elicit RGD-mediated cell responses by integrins. The presentation of RGD strongly impacts processes such as cell attachment strength, cytoskeletal reorganization, cell spreading, and migration (Cavalcanti-Adam et al. 2007; Huang et al. 2009; Lee et al. 2004a). The clustering of RGD-based ligands increased cell adhesion and motility (Maheshwari et al. 2000), with the distance among clusters being more important than the overall RGD concentration for osteoblast proliferation and differentiation (Lee et al. 2004a). Distinct patterns can trigger the formation of cytoskeletal structures such as focal adhesions and actin stress fibers (Huang et al. 2009). The clustering of the RGD ligand on the surface of a biomaterial induces the clustering of the integrin receptors (Kornberg et al. 1991), which is required for integrin-mediated signaling and for engaging interactions with cytoskeletal and signaling proteins (Miyamoto et al. 1995a,b). The effect of RGD spacing and concentration on biological events such as cell adhesion, spreading, and proliferation has been widely explored, and several conclusions, even conflicting, have been drawn (Hersel et al. 2003; Perlin et al. 2008). As an example, the importance of the spacing among the clusters has been debated in the literature, and although some results have reported a concentration of 10 fmol cm^{-2} as optimal for cell adhesion (Massia and Hubbell 1991), others have concluded that a 100-fold higher concentration is needed (Herbert et al. 1997). These data underline the fact that the presence of RGD sequence in a proper concentration undoubtedly improves at least one aspect of cell behavior, but the specific conditions for the best result to be achieved rest upon factors such as the specific sequence, spacer used, and the chemical nature of the substrate (Perlin et al. 2008). Indeed, to date, numerous examples of RGD activations reported in the literature [see the work of Hersel et al. (2003)] allow us to draw only semiquantitative conclusions on the effect of inserting an RGD-containing peptide on a specific material. This strongly limits the possibility of reviewing the wide amount of data reported to achieve a well-established general knowledge allowing for the *in silico* design of the ideal RGD modification pattern, regardless of the chemical nature of the surface of biomaterial. Moreover, the biological effect of the RGD sequence is strongly related to the cell type considered. Although it is a generally accepted that cell migration, adhesion, and spreading show a bell-shaped curve with the maximum effect obtained at intermediate RGD concentrations (Sawyer et al. 2005; Schense and Hubbell 2000; Yang et al. 2001), cells adapt to different chemistry and physical–chemical properties (Sinha and Tuan 1996) of the materials upregulating integrin expression and varying the expression pattern (Ross et al. 1993).

In order to highlight the effect of the RGD peptides on the biological systems, the synthetic peptide is typically anchored on nonfouling/nonadhesive supports such as poly(ethylene glycol) (PEG) or alginate. This has the advantage of engineering a surface of a material with a very high control of the pattern of the bioactive signal and, hence, with a tight control over cell behavior and tissue repair. However, this approach prevents one fundamental and physiological step of the biomaterial–tissue interactions, that is, protein adsorption. Whether nonfouling surfaces could represent the ideal biomaterial surface or just an oversimplification of the formidable task of biomaterial integration is still debated.

The approach of surface modification with the bioactive RGD sequence has been proven to lead to interesting results *in vitro*, but its limits, especially for bone repair, have recently been highlighted (García and Reyes 2005; Huang et al. 2009). According to Bellis (2011), "Simply stated, integrin-mediated signaling mechanisms induced by native RGD-containing matrix proteins cannot be completely recapitulated by the isolated RGD motif." Three main reasons to explain this behavior have been proposed. First, the biological activity of the short RGD-containing peptides used is, in any case, lower than a complete ECM molecule; osteoblastic markers are produced in a 50-fold higher concentration on fibronectin-coated substrates than on RGD-decorated polymers (Yang et al. 2001). The same peptide sequence, that is, GRGDSP, is 1000-fold more effective in fibronectin than as an isolated signal on the surface of the material. Second, RGD peptides are required but not sufficient for a specific and efficient binding to integrin receptors. This effect can be traced back to the presence, in the complete protein, of other sequences that activate synergistically the integrin-dependent signaling (García et al. 2002). Third, differentiation in osteogenic cells requires signals from non-RGD-binding integrins. These considerations led to the development of the so-called second-generation bioadhesive surfaces, which differ from the first generation for the contemporary presence of RGD-based and non-RGD-based ligands for integrins (Figure 7.6; García 2006; García and Reyes 2005).

In 1999, a synergistic effect on osteoblast adhesion and differentiation was found in the presence of two different adhesion peptides, i.e., RGD and FHRRIKA, a sequence found in fibronectin with cell adhesive functions (Aota et al. 1994; Rezania and Healy 1999). Moreover, enhanced cell adhesion and metabolic activity of primary rat calvarian osteoblasts (Benoit and Anseth 2005) and human osteoblast-like cells (Kim et al. 2002) was detected *in vitro* on surfaces decorated with both RGD and PHSRN domains. The rationale of these findings is that the binding of the integrin $\alpha5\beta1$ to fibronectin requires both the PHSRN and the RGD motifs; although each domain binds independently to the integrin, their combination leads to a synergistic

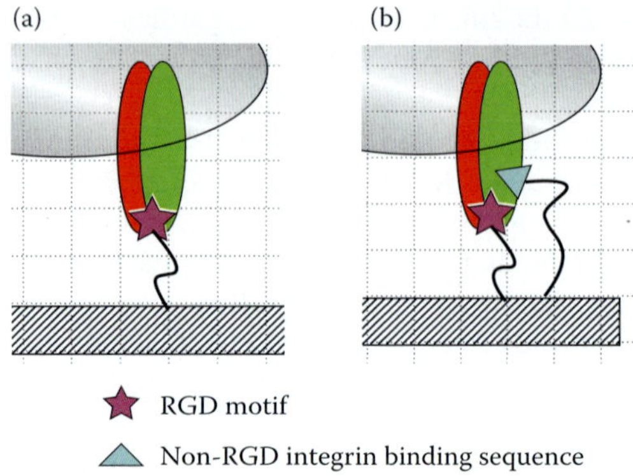

FIGURE 7.6 Comparison between (a) first- and (b) second-generation bioadhesive surfaces in engaging the integrins on the cell surface.

effect (Akiyama et al. 1995; Redick et al. 2000). Although the complexity of the surface modification is increased, the use of coupled synergistic signals is expected to regulate cell behavior in more complex ways and more closely model the naturally occurring interactions between cells and ECM (Bellis 2011).

Although the positive role of the RGD-containing bioadhesive peptides on different cell functions has been largely demonstrated *in vitro*, conflicting results have been obtained *in vivo* (Bellis 2011). In fact, some reports have shown that the presence of RGD on the surface of a biomaterial enhances tissue response (Eid et al. 2001; Ferris et al. 1999), whereas other researchers have found the opposite (Hennessy et al. 2008; Petrie et al. 2008; Schliephake et al. 2002). When the biomaterial decorated with the RGD peptide on its surface is implanted *in vivo*, it is exposed to blood and biological fluids and undergoes all the phases of the tissue–biomaterial interactions, in particular, protein adsorption (Allen et al. 2006; Keselowsky et al. 2003; Roach et al. 2005). This event leads to a net modification of the chemistry of the surface with marked effects on the biological responses such as cell adhesion, spreading, and proliferation, depending on the conformational modification of the protein upon adsorption on the surface (Hautanen et al. 1989). It should also be noted that, *in vivo*, the overall result is modulated by the adsorption of nonintegrin-binding molecules such as albumin, which normally inhibits cell adhesion, or growth factors, which regulate adhesion-related integrin signaling (Plopper et al. 1995).

Some researchers have shown that the presence of synthetic RGD peptides has adverse effects on mesenchymal stem cell adhesion and biomaterial osseointegration (Hennessy et al. 2008). These results have been interpreted in terms of the complex opposing factors of the adsorption of ECM proteins and of competition by the synthetic RGD signals. In particular, it has been reported that a surface coating based on hydroxyapatite (HA) enhances the adsorption of proteins such as fibronectin and vitronectin in a conformation that supports mesenchymal stem cell adhesion (Kilpadi et al. 2001). When synthetic RGD is present on the surface containing HA, it competes for the cell integrins, reducing the overall number of cell–protein interaction, where the synergy with additional signal is guaranteed. These recent results sprout new insight into the use of proper *in vitro* assays to predict the *in vivo* relevance of the RGD synthetic signals.

Among the non-RGD-binding integrins critical for many cellular activities, the collagen-binding α2β1 has a particular relevance, as it is involved in controlling the cell adhesion, migration, proliferation, and differentiation of osteoblast. The hexapeptide GFOGER is found in the type I collagen, and it represents the major recognition site for α2β1 integrin binding. GFOGER-functionalized surfaces supported FAK activation, expression of osteoblast-specific gene, and matrix mineralization (Reyes and García 2004). Moreover, the GFOGER peptide was reported to promote the osteoblast differentiation of MC3T3-E1 and primary bone marrow stromal cells *in vitro* and to enhance bone repair *in vivo* in critical-sized rat femur defect models (Reyes and García 2004; Reyes et al. 2007; Wojtowicz et al. 2010).

Several other peptides have been proposed for the coating of the surface of biomaterials for bone regeneration [for a review, see the work of Shekaran and García (2011)]. Among these, the 24-amino-acid peptide SKIPKASSVPTELSAISTLYLDDD, which was derived from the BMP sequence, is worth mentioning; it has been reported to enhance ectopic bone formation and accelerate bone healing in rat tibial unicortical defect (Saito et al. 2004, 2005) and a rabbit radial unicortical defect (Saito et al. 2006). However, in several of these cases, the use of materials with different physical–chemical properties hampers the comparative examination of the specific effects of the bioactive signals used.

7.3.3 Biomimetic Titanium Surfaces

This section provides evidence of the enhanced biological activity achieved with titanium surface modifications by presenting an overview of *in vitro* studies on cell response and relevant examples of *in vivo* biological and biomechanical performances of implants.

During the last years, many *in vitro* studies have been performed on titanium surfaces modified by various methods; the biological outcomes of these studies have been compared by Variola et al. (2011) in a detailed review on this topic. Overall, these studies pointed out the importance of surface modification to enhance cell activities (e.g., adhesion, proliferation, and differentiation) and modulate protein adsorption in order to improve the osteoconductivity of the material. In the next section, some of these works and other recent studies are briefly discussed in order to illustrate the variety of strategies employed for a biomimetic functionalization of titanium.

As to the effect of surface topography, Variola et al. (2008) tailored the surface properties of Ti–6Al–4V by controlled chemical oxidation, creating surfaces having both a microtexture and a nanotexture; these surfaces showed the presence of both α- and β-phase grains of a network of nanopits and were proved to selectively promote the growth of osteoblasts while inhibiting that of fibroblasts.

As discussed in Section 7.3.1, techniques based on the surface linking of peptides or ECM proteins appear as an effective way of stimulating bone regeneration (Morra 2006). Such molecules can be introduced by simple adsorption methods or by covalent linkage by means of spacers such as silanes and phosphonates. In fact, a widely explored biomimetic strategy for modifying titanium surface is based on the introduction of bioadhesive motifs derived from ECM components to increase matrix formation in the peri-implant zone (Schwartz et al. 2003). Fibronectin was recently coimmobilized with heparin on titanium for blood-contacting implant applications with the objective of inhibiting thrombosis and simultaneously promoting endothelialization (Li et al. 2010). This biomimetic coating was applied through the silanization of the surface to provide reactive amino group for conjugation with carboxyl group on both heparin and fibronectin molecules; the heparin/fibronectin mixture covalently immobilized on the surface improved the anticoagulant activity and caused a favorable blood compatibility.

Another promising approach to endow titanium with bioactive properties is the use of polysaccharides due to the versatility of these polymers; they possess unique biological properties and are prone to chemical and physical manipulations to stimulate specific biological response (Marcon et al. 2005). The conjugation of such macromolecules on surfaces can occur both by covalent bonding and by exploiting electrostatic charges of polyanions and polycations such as in the case of layer-by-layer (LbL) depositions (Cai et al. 2005; Travan et al. 2010b). In such a perspective, Pasqui et al. (2007) functionalized the oxidized titanium surface with a modified carboxymethyl cellulose in order to increase osseointegration; the polysaccharide was modified through the introduction of phosphate groups, which stimulated osteoblasts growth. The coating was realized by the acidic etching of the metal surface, followed by dipping into the polysaccharide solution; *in vitro* biological tests showed increased adhesion and proliferation of MG-63 cells. Cai et al. (2005) exploited the LbL self-assembly technique for the adsorption of chitosan and gelatin on titanium. The rationale for selecting these two components was that chitosan is the analogue of glycosaminoglycans (including hyaluronic acid) and gelatin is directly derived from collagen. The film growth was initialized by the deposition of one layer of positively charged poly(ethyleneimine) (PEI), followed by the alternate deposition of gelatin and exploiting electrostatic interactions. This multilayer coating enhanced biocompatibility by promoting the *in vitro* proliferation and viability of CAL-72 osteoblasts.

Titanium has been modified with ECM-derived bioadhesive peptides. Recently, covalent grafting of RGD complexes has been proposed by Bly et al. (2007) on Ti–6Al–4V; the coating method is based on the self-assembly of alkane phosphonic acid/RGD at physiological conditions. This RGD coating reacts and bonds with the oxide layer on titanium via a cysteine amino acid and alkyl phosphonate link (Figure 7.7a). Biological *in vitro* tests showed that the coating accelerates the initial spreading of human osteosarcoma cells that display increased interfacial shear strength.

Dettin et al. (2009) immobilized two peptides on titanium to promote osteoblast adhesion by means of two different mechanisms: 1) interaction with the RGD motif $(GRGDSP)_4K$ via cell membrane integrin receptor and 2) interaction between heparin sulfate proteoglycans of the cell membrane and the (352–360) heparin binding sequence of the human vibronectin protein. To this end, titanium surfaces were amino-functionalized by 3-aminopropyltriethoxylsilane (APTES), followed by a glutaraldehyde reaction with the terminal amino group of APTES and the amino

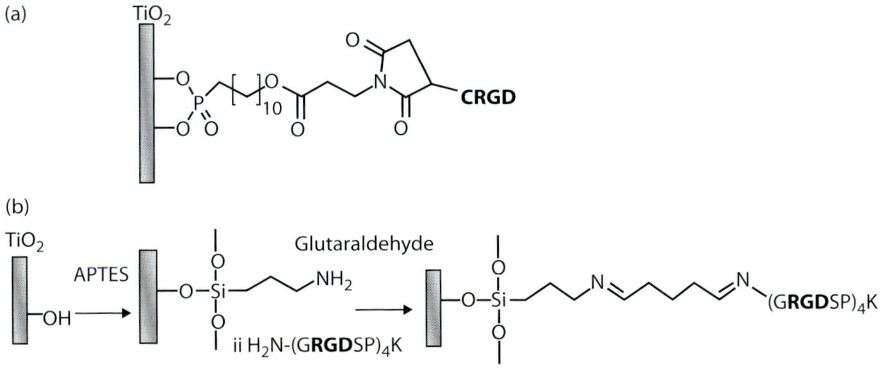

FIGURE 7.7 Examples of biomimetic functionalization of the titanium surface by the introduction of RGD peptide sequences. (a) Adapted from Bly, R.A. et al., *Mater. Sci. Eng., C* 27, 83–89, 2007; (b) Adapted from Dettin, M. et al., *J. Biomed. Mater. Res., Part A* 90A, 35–45, 2009.

group of the peptide through the Schiff base formation (Figure 7.7b). *In vitro* adhesion assays with osteoblasts derived from adult Sprague–Dawley rat bone marrow demonstrated that both biomimetic surfaces significantly increased the amount of cell adhesion; notably, the RGD-containing peptide also promotes the adhesion of endothelial cells, at variance with (352–360) HVP. Another method was proposed by Petrie et al. (2008), who covered clinical-grade titanium implants with a nonfouling oligo (ethylene glycol)-substituted polymer coating functionalized with controlled densities of bioadhesive ligands of different specificity for target integrin receptors: 1) the recombinant fragment $FNIII_{7-10}$ (which presents the RGD motif) and 2) a linear RGD oligopeptide (GRGDSPC). This study indicates that such bioactive coating regulates osteoblast differentiation of marrow-derived progenitor cells and enhances bone tissue healing in rat tibia.

In vivo tests provide further information about the complex interaction of the material with living tissues. Recently, *in vivo* studies were performed by Kerner et al. (2010), who implanted surface-modified titanium implants in rabbit femora for 4 weeks; the metal surface was modified by anionic polymers bearing carboxylate and/or sulfonate groups that affected bone formation around the implants. Morra et al. (2005) covalently linked fibrillar type I porcine collagen to acid-etched titanium implants and performed *in vivo* tests in rabbit; such biomimetic coating increased the rate of peri-implant bone formation, resulting in significantly higher "bone-to-implant contact" both in the femur and tibia after 2 weeks.

HA is a bioceramic with osteoinductive properties due to a chemical composition that mimics the inorganic constituents of the bone, which stimulates the proliferation of osteoblasts (Darimont et al. 2002). Many studies have pointed out the possibility of coating titanium implants with HA, as recently reviewed by Kim and Ramaswamy (2009). Several techniques are used, including chemical coprecipitation (Huang et al. 2010), electrophoretic deposition (Xiao et al. 2009b), hydrothermal nucleation (Calixto de Andrade et al. 2002), and cathodic deposition with *in situ* (nano) HA formation from an electrolyte-containing calcium and phosphate ions (Zhang et al. 2006). Overall, the incorporation of HA on the titanium surface represents another effective means of improving the osseointegration of titanium implants.

This overview has shown a variety of different approaches for titanium modifications that arise from the crossover nanotechnology, material science, and biology, with the goal of creating biomimetic surfaces; future developments in the applications of the infinitesimally small to biomedical surface science will lead to deeper understanding of implant–tissue interface phenomena, leading to a new generation of "smart" biomaterials.

7.4 ANTIMICROBIAL TITANIUM SURFACES

Regardless of the degree of osseointegration, the long-term success of orthopedic and dental implants may be compromised by bacterial infections. In fact, periprosthetic infections represent one of the most critical complications that can ultimately lead to the failure of the implant procedure. The prevention of implant-associated infections has been one of the main challenges in orthopedic and dental surgery (Francolini and Donelli 2010). This challenge is further complicated by the concern over the development of antibiotic resistance as a result of using traditional antibiotics for infection prophylaxis (Trebse et al. 2005). Orthopedic implant surfaces provide the bacterial growth with ideal substrates. Common pathogens such as the *Staphylococcus* species *Staphylococcus aureus* and *S. epidermidis* and *Pseudomonas aeruginosa* can be acquired at the time of surgery or at a later stage (via a hematogenous route), and they can steadily adhere to the fouling layer formed on surfaces of the implants. This often leads to the formation of complex biofilms as a result of a local altered immune host response unable to eliminate, even low microorganism counts (Gristina 1994). Biofilms are defensive barrier structures that dramatically decrease the susceptibility of bacteria to antimicrobial agents such as antibiotics and antiseptics (Jabbouri and Sadovskaya 2010). This drug resistance is due to manifold molecular events that include inactivation and blocking of diffusion of drugs by biofilm matrix components (Cerca et al. 2005; Suci et al. 1994), reduced metabolic activity and growth rate of bacteria (Duguid et al. 1992; Evans et al. 1990), high rate of genetic material exchange and hypermutations, which promote the development of genetic survival mechanism in the microorganisms (Zhang and Mah 2008), and the presence of multidrug-resistant bacterial persister cells (Lewis 2005).

Because the first crucial step that actuates the complex pathway of peri-implant colonization is always the contact and the colonization of bacteria on the device surface, at present, the prevention of bacterial adhesion and survival on implant surface is considered the ideal strategy for combating the progression of an implant-associated infection. Thus, an effective approach for minimizing implant failure would be to modify the surface of the implant to make it appealing for bone-forming cells and antiinfective at the same time (Hu et al. 2010).

The various technologies pursued in the orthopedic field to confer antimicrobial properties to device surfaces can be schematically grouped into three lines of defense: 1) antiadhesiveness strategies; 2) systems for the release of active antibacterial compounds; and 3) bactericidal surfaces active by contact (Figure 7.8).

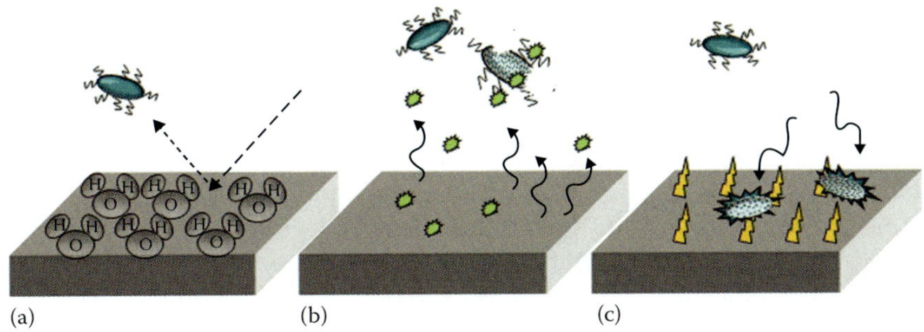

FIGURE 7.8 Antibacterial surfaces. (a) Antiadhesive. (b) Active by the release of antibiotic compounds. (c) Active by contact with bacterial cells.

Considering antimicrobial coatings for titanium-based biomaterials, several methods are explored by means of nanoengineered surfaces with antibiotics, antibacterial nanoparticles, peptides, and antibacterial polymers; whatever the type of coating, one essential requirement is that it should not hamper tissue integration.

Antiadhesiveness approaches are mainly based on the concept that, when a bacterium approaches a surface that can be potentially colonized, a transient attachment takes place. As described by the Derjaguin, Landau, Verwey, and Overbeek (DLVO) theory, the process implies weak attractive thermodynamic forces (Hermansson 1999). Among them, surface free energy, hydrophobic interactions, surface potential, and conductivity are of paramount importance. Bacteria deposition mainly occurs in correspondence to a secondary energy minimum when the charge repulsions are overcome. The knowledge of the basic thermodynamic principles defined by the DLVO theory has led to design surfaces that impede or limit bacterial adhesion due to their physical–chemical characteristics (Liu and Zhao 2011).

Titanium surfaces have been modified by means of physical–chemical techniques such as plasma ion implantation, plasma deposition, photocatalysis, and anodic oxidation (Ahn et al. 2011; Chun et al. 2007; Nurhaerani et al. 2006; Shirai et al. 2011; Suketa et al. 2005). For example, ultraviolet irradiation can lead to a "spontaneous" wettability increase on titanium dioxide (Yu et al. 2003).

Some surface modifications are intended to avoid stable bacterial adhesion on the materials by rendering surfaces more hydrophilic using polyelectrolytes. Polymer coatings such as those based on the hydrophilic poly(methacrylic acid) (PMA) and protein-resistant PEG are the most commonly used. Such hydrophilic polymers that can establish strong interactions with water molecules via hydrogen bonding are supposed to reduce protein adsorption and curtail bacterial adhesion (Ki et al. 1998).

The antibiofouling polymer PEG has been immobilized on biomaterial surfaces to form "brush-like" molecular structures that prevent proteins and cell adsorption (Ito et al. 2007). Recently, Khoo et al. (2009) have developed a PEGylated-peptide infection-resistant coating assembled on titanium through noncovalent binding. In this study, a short peptide (HKH tripeptide motif) was adsorbed on the titanium surface via electrostatic interactions, forming a densely packed peptide layer with an average height of 0.5 nm. This binding agent was used as a nanolinker to create a

PEGylated coating that efficiently blocks the adsorption of fibronectin and reduces the extent of *S. aureus* attachment and *in vitro* biofilm formation (Figure 7.9). Following this rationale, Ito et al. (2007) immobilized a photoreactive PEG on titanium to reduce surface interaction with proteins and cells.

The main drawback correlated with surface modifications aimed at reducing bacterial adhesion is the possible associated reduction/modification of protein absorption process, with ensuing possible impairment of the biological properties. Orthopedic and dental implant surfaces should ideally be designed to prevent bacterial adhesion, and at the same time, they should preserve and/or promote the correct answer of target tissue cells. Some of the strategies developed to combine bacteria antiadhesiveness and biological properties are summarized in the following paragraphs.

Shibata et al. (2004) reported that titanium anodized by discharging in NaCl solutions acquires antibacterial activity against oral bacteria and, at the same time, displays enhanced cell growth through a larger adsorption of fibronectin compared with the pure titanium control (Shibata et al. 2004).

An antiadhesion surface can also be achieved by changing the crystalline structure of the surface oxide layer. It has been shown that a crystalline anatase-type titanium oxide layer can reduce bacterial attachment significantly without negatively affecting the cell metabolic activity. Moreover, this anatase-type oxide layer can stimulate the precipitation of apatite in simulated body fluids, suggesting good osteoconductive ability (Del Curto et al. 2005).

Some strategies are based on the hydrophilicity associated with the higher surface energy of a material-coating polypeptide, which leads to a reduction in bacterial adhesion by increasing repulsion forces between the modified implant surface and

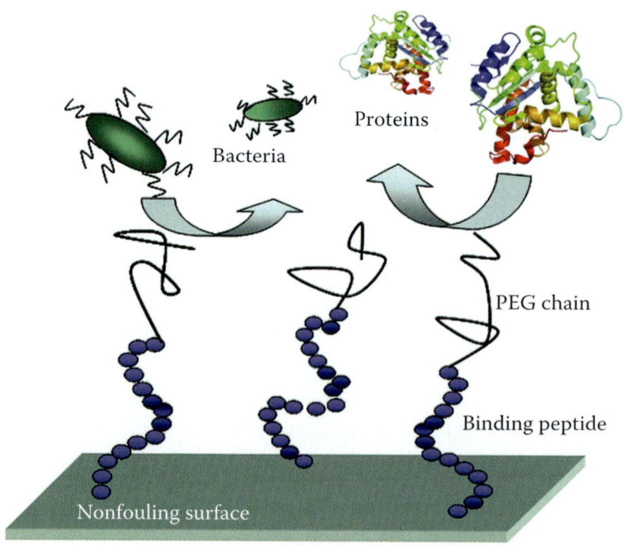

FIGURE 7.9 Schematic of nonfouling coating on the target surface.

bacteria. Additionally, osteoblast adhesion is increased with a local immobilization of growth factors, cell-binding proteins, or peptides, such as RGD peptide (Harris et al. 2004; Hu et al. 2010; Maddikeri et al. 2008). Zhang et al. (2008) have developed titanium implants that can simultaneously inhibit bacterial adhesion and promote osteoblast functions. In this work, titanium surfaces were modified with PMA, followed by the immobilization of silk sericin. Similarly, Chua et al. (2008) functionalized the titanium surface with hyaluronic acid/chitosan polyelectrolyte multilayers (PEMs) and RGD for promoting osteoblast functions and inhibiting bacterial adhesion. Titanium functionalized with the polysaccharides was performed through an LbL technique using a precursor layer of PEI—this surface modification allowed for creating a bioactive multilayer where chitosan exerted antibacterial activity toward *S. aureus*, whereas hyaluronic acid–RGD improved osteoblast adhesion and proliferation *in vitro*.

Chitosan is widely studied in relation to its antimicrobial activity. Recently, its use has been proposed by Hu et al. (2010), who functionalized titanium with carboxymethyl chitosan through covalent grafting with dopamine; this coating was shown to decrease *S. aureus* adhesion *in vitro*. Furthermore, the addition of hyaluronic acid or vascular endothelial growth factor to the chitosan component can concurrently promote osteoblast functions.

Another attempt to minimize bacterial adhesion is to modify implant material surfaces with bioactive polymers bearing sulfonate and carboxylate groups; Kerner et al. (2010) studied the bone tissue response to titanium implant surfaces modified with such functional groups; this coating inhibited bacteria growth and the functionalized implants supported new bone formation in rabbit femoral condyle.

Among the antibacterial systems based on the use of leachable antimicrobial compounds, the controlled delivery of antibiotics at the implant site is the most common way of preventing implant-associated biofilms. Administrating antibiotics directly at the site of implantation allows for maintaining the drug in high local concentration but below the systemic toxicity level (Wu and Grainger 2006).

The use of bone cements loaded with antibiotics such as with gentamicin or vancomycin for fixation of joint prostheses is studied and has been used since the 1970s, and it is considered a standard of care in many countries as prophylaxis and as a therapeutic system. To date, the use of antibiotic-loaded bone cement as an implant coating has been the primary and only practical method of delivering local antibiotics in the clinical setting of total joint replacement (Dunbar 2009; Jiranek et al. 2006). However, in the last years, the routine use of antibiotic-impregnated cement during primary or uninfected revision total hip arthroplasty has become the object of controversial opinions because of issues regarding drug resistance, detrimental effects on the mechanical or structural characteristics of PMA, antibacterial efficacy, and costs (Jiranek et al. 2006; Parvizi et al. 2008). For this reason, researchers are trying to develop coatings loaded with leachable antibiotics directly on the titanium surface. In order to achieve a controlled release of the drug, Popat et al. (2007) fabricated titanium nanotubular surfaces loaded with gentamicin by an anodization technique. The presence of the antibiotic into the nanotubes is effective in minimizing initial bacterial adhesion without adverse influence on the good cytocompatibility of the nanotubes.

Antibiotic local delivery systems are often based on polymers, e.g., poly(D,L-lactide) (Schmidmaier et al. 2006), poly(lactic-co-glycolic acid) (Neut et al. 2011), and chitosan (Norowski et al. 2011). Drug-release coatings based on PEMs created using the LbL technology platform are considered a simple, tunable, economical, and scalable process for the incorporation of antibiotics (Moskowitz et al. 2010). Chuang et al. (2008) described degradable PEMs based on poly(β-amino esters) with controlled release of gentamicin. Nguyen et al. (2007) incorporated into LbL films micelles encapsulating a hydrophobic bactericide, which allowed release of the drug over a period of several weeks.

Calcium phosphate coatings, which are known to be biocompatible and osteoconductive, have been extensively studied as potential vectors of antibacterial drugs from the titanium surface. HA coatings are able to form strong chemical bonds with the bone and are modifiable with surface antibiotics by the immersion of the coating in the antibiotic solution or by means of precipitation processes. Different methods of antibiotic loading into HA coatings, such as biomimetic techniques (Pan et al. 2011) and physical absorption, have been described (Avés et al. 2009). The kinetics of drug release is influenced by the chemical nature of the antibiotics in relation to the mineral. For instance, acid antibiotics can be easily incorporated in the carbonated HA coatings because of the calcium-chelating properties of the carboxylate groups (Stigter et al. 2004).

Besides antibiotic-releasing coatings, there is interest in the covalent bonding of drugs to the implant surface to realize long-lasting antibacterial systems with a reduced risk of systemic antibiotic toxicity. Antoci et al. (2007) reported the inhibition of *S. epidermidis* biofilm formation by vancomycin-modified titanium alloy; in this study, self-protecting Ti–6Al–4V alloy pins were prepared by covalent bonding of bis(ethylene glycol) linkers with vancomycin to the oxidized aminopropylated Ti–6Al–4V alloy surface. Chen et al. and Wickstrom (2010) described the preparation of self-protecting bactericidal titanium surface formed by the covalent bonding of daptomycin bisphosphonates; this antibiotic was covalently bonded via a flexible hydrophilic spacer, and it could prevent *S. aureus* colonization on Ti–6Al–4V foils.

Although clinically applied, antibiotic-releasing systems appear to have some drawbacks: 1) the loss of bacterial susceptibility to antibiotics following the development of drug resistance mechanisms; 2) the challenge of producing antibiotic-loaded coatings that ensure the delivery of antibiotics at effective concentrations for an adequately long time; and 3) the potential local toxicity of some antibiotics against bone cells (Antoci et al. 2007; Naal et al. 2008; Salzmann et al. 2007).

In this perspective, nonantibiotic bactericidal organic and inorganic compounds have been considered good alternatives, mainly because they do not display the risk of inducing the formation of antibiotic-resistant strains. The former ones include, for instance, chlorhexidine (CHX)—an antibacterial agent widely used in dental applications for the treatment of periodontal infections. The interaction and absorption of CHX with smooth and rough types of titanium surfaces have been studied by Kozlovsky et al. (2006). They demonstrated that, at some concentrations, CHX displays good persistence at the titanium surface and bacterial growth inhibition activity. Complex coatings containing CHX in combination with other components have been studied in order to modulate the drug release kinetic. Harris et al.

(2006) described cytocompatible titanium coatings based on poly(D,L-lactide) and poly(terephtalate) loaded with CHX diacetate. Both materials showed to elute CHX over a maximum period of 96 h *in vitro*, lowering the amount of bacteria in comparison with the coatings without CHX.

Quaternary ammonium salts are extensively used as antimicrobial agents and disinfectants. The incorporation of these compounds in dental adhesives and composite resins provides antibacterial properties that can be beneficial in eliminating the detrimental effect caused by bacterial microleakage (Beyth et al. 2010; Chai et al. 2011; Li et al. 2009a,b; Xiao et al. 2009a). Recently, a noncytotoxic and antibacterial coating on titanium surface based on quaternized chitosan has been described (Peng et al. 2011). To improve the biocorrosion resistance and to confer the bactericidal function on its surface, stainless steel material have been coated with inorganic–organic hybrid systems, consisting of inner compact nanostructured titanium oxide multilayers and outer dense poly(vinyl-*N*-hexylpyridinium) quaternized brushes. The material showed excellent antibacterial efficiency of the grafted polycations and significant enhancement of biocorrosion resistance (Yuan et al. 2009).

Among the various inorganic antimicrobial agents such as copper, gold, zinc, and fluorine, the most well known and studied is silver in its different chemical forms (nanoparticles, ions, and salts). Silver as an antimicrobial agent is gaining increasing appeal for medical applications because of a broad antibacterial spectrum spanning from Gram-positive to Gram-negative bacteria and to some drug-resistant bacteria at very low parts per billion (ppb) concentrations. Silver-based medical products, ranging from topical ointments and bandages for wound healing to coated stents, have been proved to be effective in retarding and preventing bacterial infections. Silver, especially in the form of nanoparticles, is widely explored for the preparation of biomaterials with antibacterial surfaces (Travan et al. 2009, 2010a).

On account of a lower risk of drug resistance, silver may be applied *in vivo* for a relatively long period of time. Due to these advantages, as a bactericide, silver has also been introduced on the titanium surface to enhance the bactericidal ability. These systems can be developed both as biocide-releasing systems (Ag^+ and nano-Ag) and as contact-active materials. Silver ions and silver nanoparticles have been implanted on titanium by various techniques such as plasma immersion ion implantation (Cao et al. 2011), adsorption (Flores et al. 2010), surface silanization (Juan et al. 2010), and chemical reduction (Niño-Martínez et al. 2008). Necula et al. (2011) studied the antibacterial activity of porous TiO_2–Ag composite layers against methicillin-resistant *S. aureus* (MRSA) *in vitro*. The coating was produced by the plasma electrolytic oxidation of Ti–6Al–7Nb medical alloy in an electrolyte solution bearing Ag nanoparticles. Following oxidation, the surface of the titanium substrate was converted into a TiO_2 layer that incorporates Ag particles. The biological tests revealed a complete killing of MRSA bacteria within 24 h. Another technique used to endow the titanium surface with silver antimicrobial properties is PVD. This method is based on the vaporization of the metals in an inert argon atmosphere to obtain a surface layer with good adhesiveness and wear resistance, making it particularly appealing for orthopedic and dental applications. Ewald et al. (2006) have recently described the development of an antimicrobial titanium/silver hard coating via PVD process with a silver content in the range of 0.7%–9%. *In vitro* biological tests revealed that such coatings released silver ions (0.5–2.3 ppb) when immersed in phosphate-buffered saline and showed significant antimicrobial potency against *S. epidermidis* and *Klebsiella pneumoniae* strains. At the same time, no cytotoxic effect of the coatings was found on osteoblast and epithelial cells. HA-based composite coatings loaded with silver have been developed both on titanium and stainless steel in order to combine the biological properties of the mineral with the bactericidal activity of the metal (Bai et al. 2010; Chen et al. 2008; Lu et al. 2011; O'Sullivan et al. 2010; Qu et al. 2011; Sandukas et al. 2011; Shimazaki et al. 2010). The widespread use of silver nanoparticles and silver ion-based products is partially justified by the fact that, until a few years ago and even nowadays, silver is considered to have low toxicity *in vivo* and minimal risk with respect to clinical exposure by dermal application, inhalation, ingestion, or through the urological or hematogenous routes. Silver ions are not known as a systemic toxic agent for humans, except at extreme doses. However, because of the large use of silver nanoparticles in consumer products in recent years, the concern about their potential toxicity arose; in fact, the peculiar physical, chemical, and biological properties of nano-sized materials should be considered for applications in biomaterials (Teow et al. 2011). Although titanium orthopedic devices coated with silver can be found in the clinical market, such as the spinal systems commercialized by Bey-Om-Sin, more comprehensive studies are needed to clarify their biocompatibility to tissues adjacent to the implants (Hoet et al. 2004; Oberdörster et al. 2005).

The nanostructured topography of the titanium surface was reported to affect bacterial adhesion, even without adding antibiotics or exogenous antimicrobial agents, pointing out the role of TiO_2 morphology in determining the outcome of infections. Pucket et al. (2010) examined the adhesion of several bacterial strains on conventional Ti, nanorough Ti produced by electron beam evaporation, and nanotubular and nanotextured Ti produced by anodization processes; this *in vitro* study found that the nanorough Ti surfaces decreased bacterial adherence, whereas nanotubular and nanotextured Ti resulted in an increase in bacteria attachment.

In conclusion, infection associated with orthopedic and dentistry implants represents a major concern in clinical research; for this reason, several strategies for obtaining antimicrobial surfaces on titanium alloys have been proposed. Chemical, physical, and topographical modifications at the microscale/nanoscale have allowed a considerable progress during the last years, although *in vitro* characterizations of such surfaces are not sufficient to rule out the risks of cytotoxicity, impaired integration within the tissues, or reduced activity upon implantation; in fact, it must be noted that *in vivo* information on titanium antibacterial coatings are still scarce.

REFERENCES

Ahn, S.J., Han, J.S., Lim, B.S., and Lim, Y.J. 2011. Comparison of ultraviolet light-induced photocatalytic bactericidal effect on modified titanium implant surfaces. *Int. J. Oral Maxillofac. Implants* 26:39–44.

Akiyama, S.K., Aota, S.I., and Yamada, K.M. 1995. Function and receptor specificity of a minimal 20 kilodalton cell adhesive fragment of fibronectin. *Cell Adhes. Commun.* 3:13–25.

Albrektsson, T. and Johansson, C. 2001. Osteoinduction, osteoconduction and osseointegration. *Eur. Spine J.* 10 Suppl 2:S96–S101.

Allen, L.T., Tosetto, M., Miller, I.S. et al. 2006. Surface-induced changes in protein adsorption and implications for cellular phenotypic responses to surface interaction. *Biomaterials* 27:3096–3108.

Andrade, J.D. and Hlady, V. 1991. Vroman effects, techniques, and philosophies. *J. Biomater. Sci., Polym. Ed.* 2:161–172.

Anselme, K. 2000. Osteoblast adhesion on biomaterials. *Biomaterials* 21:667–681.

Anselme, K. 2011. Biomaterials and interface with bone. *Osteoporosis Int.* 22:2037–2042.

Anselme, K. and Bigerelle, M. 2005. Topography effects of pure titanium substrates on human osteoblast long-term adhesion. *Acta Biomater.* 1:211–222.

Anselme, K. and Bigerelle, M. 2006. Modeling approach in cell/material interactions studies. *Biomaterials* 27:1187–1199.

Anselme, K., Bigerelle, M., Noel, B. et al. 2000a. Qualitative and quantitative study of human osteoblast adhesion on materials with various surface roughnesses. *J. Biomed. Mater. Res.* 49:155–166.

Anselme, K., Davidson, P., Popa, A.M., Giazzon, M., Liley, M., and Ploux, L. 2010a. The interaction of cells and bacteria with surfaces structured at the nanometre scale. *Acta Biomater.* 6:3824–3846.

Anselme, K., Linez, P., Bigerelle, M. et al. 2000b. The relative influence of the topography and chemistry of TiAl6V4 surfaces on osteoblastic cell behavior. *Biomaterials* 21:1567–1577.

Anselme, K., Ponche, A., and Bigerelle, M. 2010b. Relative influence of surface topography and surface chemistry on cell response to bone implant materials—Part 2: Biological aspects. *Proc. Inst. Mech. Eng. H* 224:1487–1507.

Antoci Jr., V., Adams, C.S., Hickok, N.J., Shapiro, I.M., and Parvizi, J. 2007. Antibiotics for local delivery systems cause skeletal cell toxicity *in vitro*. *Clin. Orthop. Relat. Res.* 462:200–206.

Aota, S., Nomizu, M., and Yamada, K.M. 1994. The short amino acid sequence Pro–His–Ser–Arg–Asn in human fibronectin enhances cell-adhesive function. *J. Biol. Chem.* 269:24756–24761.

Avés, E.P., Estévez, G.F., Sader, M.S. et al. 2009. Hydroxyapatite coating by sol–gel on Ti–6Al–4V alloy as drug carrier. *J. Mater. Sci.: Mater. Med.* 20:543–547.

Bai, X., More, K., Rouleau, C.M., and Rabiei, A. 2010. Functionally graded hydroxyapatite coatings doped with antibacterial components. *Acta Biomater.* 6:2264–2273.

Banerjee, R., Nag, S., Stechschulte, J., and Fraser, H.L. 2004. Strengthening mechanisms in Ti–Nb–Zr–Ta and Ti–Mo–Zr–Fe orthopaedic alloys. *Biomaterials* 25:3413–3419.

Barbour, P.S., Stone, M.H., and Fisher, J. 1999. A study of the wear resistance of three types of clinically applied UHMWPE for total replacement hip prostheses. *Biomaterials* 20:2101–2106.

Beer, J.H., Springer, K.T., and Coller, B.S. 1992. Immobilized Arg–Gly–Asp (RGD) peptides of varying lengths as structural probes of the platelet glycoprotein IIb/IIIa receptor. *Blood* 79:117–128.

Bellis, S.L. 2011. Advantages of RGD peptides for directing cell association with biomaterials. *Biomaterials* 32:4205–4210.

Benoit, D.S.W. and Anseth, K.S. 2005. The effect on osteoblast function of colocalized RGD and PHSRN epitopes on PEG surfaces. *Biomaterials* 26:5209–5220.

Beutner, R., Michael, J., Schwenzer, B., and Scharnweber, D. 2010. Biological nanofunctionalization of titanium-based biomaterial surfaces: A flexible toolbox. *J. R. Soc., Interface* 7 Suppl 1:S93–S105.

Beyth, N., Yudovin-Farber, I., Perez-Davidi, M., Domb, A.J., and Weiss, E.I. 2010. Polyethyleneimine nanoparticles incorporated into resin composite cause cell death and trigger biofilm stress *in vivo*. *Proc. Natl. Acad. Sci. USA* 107:22038–22043.

Bhushan, B., Tokachichu, D.R., Keener, M.T., and Lee, S.C. 2006. Nanoscale adhesion, friction and wear studies of biomolecules on silicon-based surfaces. *Acta Biomater.* 2:39–49.

Bigerelle, M. and Anselme, K. 2005. Bootstrap analysis of the relation between initial adhesive events and long-term cellular functions of human osteoblasts cultured on biocompatible metallic substrates. *Acta Biomater.* 1:499–510.

Blättler, T., Huwiler, C., Ochsner, M. et al. 2006. Nanopatterns with biological functions. *J. Nanosci. Nanotechnol.* 6:2237–2264.

Bloyce, A., Qi, P.Y., Dong, H., and Bell, T. 1998. Surface modification of titanium alloys for combined improvements in corrosion and wear resistance. *Surf. Coat. Technol.* 107:125–132.

Bly, R.A., Cao, Y., Moore, W.A., and Soboyejo, W.O. 2007. Investigation of the effects of alkane phosphonic acid/RGD coatings on cell spreading and the interfacial strength between human osteosarcoma cells and Ti–6Al–4V. *Mater. Sci. Eng., C* 27:83–89.

Branemark, P.I. 1983. Osseointegration and its experimental background. *J. Prosthet. Dent.* 50:399–410.

Buchanan, R.A., Rigney, E.D., Jr., and Williams, J.M. 1987. Wear-accelerated corrosion of Ti–6Al–4V and nitrogen-ion-implanted Ti–6Al–4V: Mechanisms and influence of fixed-stress magnitude. *J. Biomed. Mater. Res.* 21:367–377.

Byron, A. 2011. Analyzing the anatomy of integrin adhesions. *Sci. Signal.* 4:jc3.

Cai, K., Rechtenbach, A., Hao, J., Bossert, J., and Jandt, K.D. 2005. Polysaccharide–protein surface modification of titanium via a layer-by-layer technique: Characterization and cell behavior aspects. *Biomaterials* 26:5960–5971.

Calixto de Andrade, M., Tavares Filgueiras, M.R., and Ogasawara, T. 2002. Hydrothermal nucleation of hydroxyapatite on titanium surface. *J. Eur. Ceram. Soc.* 22:505–510.

Campbell, I.D. and Humphries, M.J. 2011. Integrin structure, activation, and interactions. *Cold Spring Harb. Perspect. Biol.* 3:1–14

Cao, H., Liu, X., Meng, F., and Chu, P.K. 2011. Biological actions of silver nanoparticles embedded in titanium controlled by microgalvanic effects. *Biomaterials* 32:693–705.

Carman, A., Zhang, L.C., Ivasishin, O.M., Savvakin, D.G., Matviychuk, M.V., and Pereloma, E.V. 2011. Role of alloying elements in microstructure evolution and alloying elements behaviour during sintering of a near-β titanium alloy. *Mater. Sci. Eng. A Struct. Mater.* 528:1686–1693.

Cavalcanti-Adam, E.A., Volberg, T., Micoulet, A., Kessler, H., Geiger, B., and Spatz, J.P. 2007. Cell spreading and focal adhesion dynamics are regulated by spacing of integrin ligands. *Biophys. J.* 92:2964–2974.

Cerca, N., Martins, S., Cerca, F. et al. 2005. Comparative assessment of antibiotic susceptibility of coagulase-negative staphylococci in biofilm versus planktonic culture as assessed by bacterial enumeration or rapid XTT colorimetry. *J. Antimicrob. Chemother.* 56:331–336.

Chai, Z., Li, F., Fang, M. et al. 2011. The bonding property and cytotoxicity of a dental adhesive incorporating a new antibacterial monomer. *J. Oral Rehabil.* 38:849–856.

Chen, C.P. and Wickstrom, E. 2010. Self-protecting bactericidal titanium alloy surface formed by covalent bonding of daptomycin bisphosphonates. *Bioconjug. Chem.* 21:1978–1986.

Chen, Y., Zheng, X., Xie, Y., Ding, C., Ruan, H., and Fan, C. 2008. Antibacterial and cytotoxic properties of plasma sprayed silver-containing HA coatings. *J. Mater. Sci.: Mater. Med.* 19:3603–3609.

Chua, P.H., Neoh, K.G., Kang, E.T., and Wang, W. 2008. Surface functionalization of titanium with hyaluronic acid/chitosan polyelectrolyte multilayers and RGD for promoting osteoblast functions and inhibiting bacterial adhesion. *Biomaterials* 29:1412–1421.

Chuang, H.F., Smith, R.C., and Hammond, P.T. 2008. Polyelectrolyte multilayers for tunable release of antibiotics. *Biomacromolecules* 9:1660–1668.

Chun, M.J., Shim, E., Kho, E.H. et al. 2007. Surface modification of orthodontic wires with photocatalytic titanium oxide for its antiadherent and antibacterial properties. *Angle Orthod.* 77:483–488.

Craig, W.S., Cheng, S., Mullen, D.G., Blevitt, J., and Pierschbacher, M.D. 1995. Concept and progress in the development of RGD-containing peptide pharmaceuticals. *Biopolymers* 37:157–175.

Cremasco, A., Messias, A.D., Esposito, A.R., Duek, E.A.d.R., and Caram, R. 2011. Effects of alloying elements on the cytotoxic response of titanium alloys. *Mater. Sci. Eng., C* 31:833–839.

Damsky, C.H. 1999. Extracellular matrix–integrin interactions in osteoblast function and tissue remodeling. *Bone* 25:95–96.

Darimont, G.L., Cloots, R., Heinen, E., Seidel, L., and Legrand, R. 2002. *In vivo* behavior of hydroxyapatite coatings on titanium implants: A quantitative study in the rabbit. *Biomaterials* 23:2569–2575.

De Arcangelis, A. and Georges-Labouesse, E. 2000. Integrin and ECM functions: Roles in vertebrate development. *Trends Genet.* 16:389–395.

Del Curto, B., Brunella, M.F., Giordano, C. et al. 2005. Decreased bacterial adhesion to surface-treated titanium. *Int. J. Artif. Organs* 28:718–730.

Dettin, M., Bagno, A., Gambaretto, R. et al. 2009. Covalent surface modification of titanium oxide with different adhesive peptides: Surface characterization and osteoblast-like cell adhesion. *J. Biomed. Mater. Res.—Part A* 90A:35–45.

DiMaio, F.R. 2002. The science of bone cement: A historical review. *Orthopedics* 25:1399–1407.

Drumheller, P.D. and Hubbell, J.A. 1994. Polymer networks with grafted cell adhesion peptides for highly biospecific cell adhesive substrates. *Anal. Biochem.* 222:380–388.

Duguid, I.G., Evans, E., Brown, M.R., and Gilbert, P. 1992. Growth rate–independent killing by ciprofloxacin of biofilm-derived *Staphylococcus epidermidis*; evidence for cell-cycle dependency. *J. Antimicrob. Chemother.* 30:791–802.

Dunbar, M.J. 2009. Antibiotic bone cements: Their use in routine primary total joint arthroplasty is justified. *Orthopedics* 32:

Eid, K., Chen, E., Griffith, L., and Glowacki, J. 2001. Effect of RGD coating on osteocompatibility of PLGA-polymer disks in a rat tibial wound. *J. Biomed. Mater. Res.* 57:224–231.

Elbert, D.L. and Hubbell, J.A. 2001. Conjugate addition reactions combined with free-radical cross-linking for the design of materials for tissue engineering. *Biomacromolecules* 2:430–441.

Evans, D.J., Allison, D.G., Brown, M.R., and Gilbert, P. 1990. Effect of growth rate on resistance of gram-negative biofilms to cetrimide. *J. Antimicrob. Chemother.* 26:473–478.

Ewald, A., Glückermann, S.K., Thull, R., and Gbureck, U. 2006. Antimicrobial titanium/silver PVD coatings on titanium. *Biomed. Eng. Online* 5:22.

Ferris, D.M., Moodie, G.D., Dimond, P.M., Giorani, C.W.D., Ehrlich, M.G., and Valentini, R.F. 1999. RGD-coated titanium implants stimulate increased bone formation *in vivo*. *Biomaterials* 20:2323–2331.

Fisher, J.E., Caulfield, M.P., Sato, M. et al. 1993. Inhibition of osteoclastic bone resorption *in vivo* by echistatin, an "arginyl–glycyl–aspartyl" (RGD)-containing protein. *Endocrinology* 132:1411–1413.

Flores, C.Y., Diaz, C., Rubert, A. et al. 2010. Spontaneous adsorption of silver nanoparticles on Ti/TiO_2 surfaces: Antibacterial effect on *Pseudomonas aeruginosa*. *J. Colloid Interface Sci.* 350:402–408.

Francolini, I. and Donelli, G. 2010. Prevention and control of biofilm-based medical device-related infections. *FEMS Immunol. Med. Microbiol.* 59:227–238.

García, A.J. 2005. Get a grip: Integrins in cell–biomaterial interactions. *Biomaterials* 26:7525–7529.

García, A.J. 2006. Interfaces to control cell–biomaterial adhesive interactions. *Polym. Regener. Med.* 203:171–190.

García, A.J. and Reyes, C.D. 2005. Bioadhesive surfaces to promote osteoblast differentiation and bone formation. *J. Dent. Res.* 84:407–413.

García, A.J., Schwarzbauer, J.E., and Boettiger, D. 2002. Distinct activation states of α5β1 integrin show differential binding to RGD and synergy domains of fibronectin. *Biochemistry* 41:9063–9069.

Gristina, A.G. 1994. Implant failure and the immuno-incompetent fibroinflammatory zone. *Clin. Orthop. Relat. Res.* 298:106–118.

Hallab, N.J. and Jacobs, J.J. 2009. Biologic effects of implant debris. *Bull. NYU Hosp. Jt. Dis.* 67:182–188.

Hanawa, T. 2001. Metallic biomaterials in body fluid and their surface modification. In *Structural Biomaterials for 21st Century*, Niinomi, N., Obake, T., Taleff, E.M., Lesuer, D.R., and Lippards, H.E. (Eds.). TMS, Warrendale, PA, pp. 145–156.

Harris, L.G., Mead, L., Müller-Oberländer, E., and Richards, R.G. 2006. Bacteria and cell cytocompatibility studies on coated medical grade titanium surfaces. *J. Biomed. Mater. Res.—Part A* 78A:50–58.

Harris, L.G., Tosatti, S., Wieland, M., Textor, M., and Richards, R.G. 2004. *Staphylococcus aureus* adhesion to titanium oxide surfaces coated with nonfunctionalized and peptide-functionalized poly(L-lysine)-grafted-poly(ethylene glycol) copolymers. *Biomaterials* 25:4135–4148.

Hautanen, A., Gailit, J., Mann, D.M., and Ruoslahti, E. 1989. Effects of modifications of the RGD sequence and its context on recognition by the fibronectin receptor. *J. Biol. Chem.* 264:1437–1442.

Hennessy, K.M., Clem, W.C., Phipps, M.C., Sawyer, A.A., Shaikh, F.M., and Bellis, S.L. 2008. The effect of RGD peptides on osseointegration of hydroxyapatite biomaterials. *Biomaterials* 29:3075–3083.

Herbert, C.B., McLernon, T.L., Hypolite, C.L. et al. 1997. Micropatterning gradients and controlling surface densities of photoactivatable biomolecules on self-assembled monolayers of oligo(ethylene glycol) alkanethiolates. *Chem. Biol.* 4:731–737.

Hermansson, M. 1999. The DLVO theory in microbial adhesion. *Colloids Surf., B* 14:105–119.

Hersel, U., Dahmen, C., and Kessler, H. 2003. RGD modified polymers: Biomaterials for stimulated cell adhesion and beyond. *Biomaterials* 24:4385–4415.

Hoet, P.H., Nemmar, A., and Nemery, B. 2004. Health impact of nanomaterials? *Nat. Biotechnol.* 22:19.

Holt, G., Hook, S., and Hubble, M. 2011. Revision total hip arthroplasty: The femoral side using cemented implants. *Int. Orthop.* 35:267–273.

Hu, X., Neoh, K.G., Shi, Z., Kang, E.T., Poh, C., and Wang, W. 2010. An *in vitro* assessment of titanium functionalized with polysaccharides conjugated with vascular endothelial growth factor for enhanced osseointegration and inhibition of bacterial adhesion. *Biomaterials* 31:8854–8863.

Huang, J., Best, S.M., Bonfield, W., and Buckland, T. 2010. Development and characterization of titanium-containing hydroxyapatite for medical applications. *Acta Biomater.* 6:241–249.

Huang, J., Grater, S.V., Corbellini, F. et al. 2009. Impact of order and disorder in RGD nanopatterns on cell adhesion. *Nano Lett.* 9:1111–1116.

Hulbert, S.F., Cooke, F.W., Klawitter, J.J. et al. 1973. Attachment of prostheses to the musculoskeletal system by tissue ingrowth and mechanical interlocking. *J. Biomed. Mater. Res.* 7:1–23.

Hynes, R.O. 2002. Integrins: Bidirectional, allosteric signaling machines. *Cell* 110:673–687.

Ito, Y., Hasuda, H., Sakuragi, M., and Tsuzuki, S. 2007. Surface modification of plastic, glass and titanium by photoimmobilization of polyethylene glycol for antibiofouling. *Acta Biomater.* 3:1024–1032.

Ito, Y., Kajihara, M., and Imanishi, Y. 1991. Materials for enhancing cell adhesion by immobilization of cell-adhesive peptide. *J. Biomed. Mater. Res.* 25:1325–1337.

Jabbouri, S. and Sadovskaya, I. 2010. Characteristics of the biofilm matrix and its role as a possible target for the detection and eradication of *Staphylococcus epidermidis* associated with medical implant infections. *FEMS Immunol. Med. Microbiol.* 59:280–291.

Jasty, M. 1993. Clinical reviews: Particulate debris and failure of total hip replacements. *J. Appl. Biomater.* 4:273–276.

Javed, F., Almas, K., Crespi, R., and Romanos, G.E. 2011. Implant surface morphology and primary stability: Is there a connection? *Implant. Dent.* 20:40–46.

Jiranek, W.A., Hanssen, A.D., and Greenwald, A.S. 2006. Antibiotic-loaded bone cement for infection prophylaxis in total joint replacement. *J. Bone Jt. Surg. Am.* 88:2487–2500.

Juan, L., Zhimin, Z., Anchun, M., Lei, L., and Jingchao, Z. 2010. Deposition of silver nanoparticles on titanium surface for antibacterial effect. *Int. J. Nanomed.* 5:261–267.

Kantlehner, M., Schaffner, P., Finsinger, D. et al. 2000. Surface coating with cyclic RGD peptides stimulates osteoblast adhesion and proliferation as well as bone formation. *ChemBioChem* 1:107–114.

Kasemo, B. and Lausmaa, J. 1994. Material–tissue interfaces: The role of surface properties and processes. *Environ. Health Perspect.* 102 Suppl 5:41–45.

Kerner, S., Migonney, V., Pavon-Djavid, G., Helary, G., Sedel, L., and Anagnostou, F. 2010. Bone tissue response to titanium implant surfaces modified with carboxylate and sulfonate groups. *J. Mater. Sci.: Mater. Med.* 21:707–715.

Keselowsky, B.G., Collard, D.M., and García, A.J. 2003. Surface chemistry modulates fibronectin conformation and directs integrin binding and specificity to control cell adhesion. *J. Biomed. Mater. Res., Part A* 66A:247–259.

Khoo, X., Hamilton, P., O'Toole, G.A., Snyder, B.D., Kenan, D.J., and Grinstaff, M.W. 2009. Directed assembly of PEGylated-peptide coatings for infection-resistant titanium metal. *J. Am. Chem. Soc.* 131:10992–10997.

Ki, D.P., Young, S.K., Dong, K.H. et al. 1998. Bacterial adhesion on PEG modified polyurethane surfaces. *Biomaterials* 19:851–859.

Kilpadi, K.L., Chang, P.L., and Bellis, S.L. 2001. Hydroxylapatite binds more serum proteins, purified integrins, and osteoblast precursor cells than titanium or steel. *J. Biomed. Mater. Res.* 57:258–267.

Kim, K.H. and Ramaswamy, N. 2009. Electrochemical surface modification of titanium in dentistry. *Dent. Mater. J.* 28:20–36.

Kim, S.J., Jang, J.D., and Lee, S.K. 2007. Treatment of long tubular bone defect of rabbit using autologous cultured osteoblasts mixed with fibrin. *Cytotechnology* 54:115–120.

Kim, T.I., Jang, J.H., Lee, Y.M. et al. 2002. Design and biological activity of synthetic oligopeptides with Pro–His–Ser–Arg–Asn (PHSRN) and Arg–Gly–Asp (RGD) motifs for human osteoblast-like cell (MG-63) adhesion. *Biotechnol. Lett.* 24:2029–2033.

Kornberg, L.J., Earp, H.S., Turner, C.E., Prockop, C., and Juliano, R.L. 1991. Signal transduction by integrins: Increased protein tyrosine phosphorylation caused by clustering of beta 1 integrins. *Proc. Natl. Acad. Sci. USA* 88:8392–8396.

Kozlovsky, A., Artzi, Z., Moses, O., Kamin-Belsky, N., and Greenstein, R.B. 2006. Interaction of chlorhexidine with smooth and rough types of titanium surfaces. *J. Periodontol.* 77:1194–1200.

Kubo, K., Tsukimura, N., Iwasa, F. et al. 2009. Cellular behavior on TiO_2 nanonodular structures in a micro-to-nanoscale hierarchy model. *Biomaterials* 30:5319–5329.

Kurkalli, B.G., Gurevitch, O., Sosnik, A., Cohn, D., and Slavin, S. 2010. Repair of bone defect using bone marrow cells and demineralized bone matrix supplemented with polymeric materials. *Curr. Stem Cell Res. Ther.* 5:49–56.

LeBaron, R.G. and Athanasiou, K.A. 2000. Extracellular matrix cell adhesion peptides: Functional applications in orthopedic materials. *Tissue Eng.* 6:85–103.

Lee, K.Y., Alsberg, E., Hsiong, S. et al. 2004a. Nanoscale adhesion ligand organization regulates osteoblast proliferation and differentiation. *Nano Lett.* 4:1501–1506.

Lee, S.J., Choi, J.S., Park, K.S., Khang, G., Lee, Y.M., and Lee, H.B. 2004b. Response of MG63 osteoblast-like cells onto polycarbonate membrane surfaces with different micropore sizes. *Biomaterials* 25:4699–4707.

Lewis, K. 2005. Persister cells and the riddle of biofilm survival. *Biochemistry (Moscow)* 70:267–274.

Li, F., Chai, Z.G., Sun, M.N. et al. 2009a. Antibiofilm effect of dental adhesive with cationic monomer. *J. Dent. Res.* 88:372–376.

Li, F., Chen, J., Chai, Z. et al. 2009b. Effects of a dental adhesive incorporating antibacterial monomer on the growth, adherence and membrane integrity of *Streptococcus mutans*. *J. Dent.* 37:289–296.

Li, G., Zhang, F., Liao, Y., Yang, P., and Huang, N. 2010. Coimmobilization of heparin/fibronectin mixture on titanium surfaces and their blood compatibility. *Colloids Surf., B* 81:255–262.

Lin, H.B., García-Echeverría, C., Asakura, S., Sun, W., Mosher, D.F., and Cooper, S.L. 1992. Endothelial cell adhesion on polyurethanes containing covalently attached RGD-peptides. *Biomaterials* 13:905–914.

Liu, C. and Zhao, Q. 2011. Influence of surface-energy components of Ni–P–$TiO_{(2)}$-PTFE nanocomposite coatings on bacterial adhesion. *Langmuir* 27:9512–9519.

Lo, C.M., Wang, H.B., Dembo, M., and Wang, Y.L. 2000. Cell movement is guided by the rigidity of the substrate. *Biophys. J.* 79:144–152.

Long, M., Crooks, R., and Rack, H.J. 1999. High-cycle fatigue performance of solution-treated metastable-β titanium alloys. *Acta Mater.* 47:661–669.

Long, M. and Rack, H.J. 1998. Titanium alloys in total joint replacement—A materials science perspective. *Biomaterials* 19:1621–1639.

Lu, X., Zhang, B., Wang, Y. et al. 2011. Nano-Ag-loaded hydroxyapatite coatings on titanium surfaces by electrochemical deposition. *J. R. Soc., Interface* 8:529–539.

Maddikeri, R.R., Tosatti, S., Schuler, M. et al. 2008. Reduced medical infection related bacterial strains adhesion on bioactive RGD modified titanium surfaces: A first step toward cell selective surfaces. *J. Biomed. Mater. Res., Part A* 84A:425–435.

Maheshwari, G., Brown, G., Lauffenburger, D.A., Wells, A., and Griffith, L.G. 2000. Cell adhesion and motility depend on nanoscale RGD clustering. *J. Cell Sci.* 113:1677–1686.

Mammen, M., Choi, S.K., and Whitesides, G.M. 1998. Polyvalent interactions in biological systems: Implications for design and use of multivalent ligands and inhibitors. *Angew. Chem., Int. Ed.* 37:2754–2794.

Marcon, P., Marsich, E., Vetere, A. et al. 2005. The role of galectin-1 in the interaction between chondrocytes and a lactose-modified chitosan. *Biomaterials* 26:4975–4984.

Martines, E., Seunarine, K., Morgan, H., Gadegaard, N., Wilkinson, C.D., and Riehle, M.O. 2005. Superhydrophobicity and superhydrophilicity of regular nanopatterns. *Nano Lett.* 5:2097–2103.

Massia, S.P. and Hubbell, J.A. 1991. An RGD spacing of 440 nm is sufficient for integrin alpha V beta 3-mediated fibroblast spreading and 140 nm for focal contact and stress fiber formation. *J. Cell Biol.* 114:1089–1100.

Mitsuo, N. 2008. Mechanical biocompatibilities of titanium alloys for biomedical applications. *J. Mech. Behav. Biomed. Mater.* 1:30–42.

Miyamoto, S., Akiyama, S.K., and Yamada, K.M. 1995a. Synergistic roles for receptor occupancy and aggregation in integrin transmembrane function. *Science* 267:883–885.

Miyamoto, S., Teramoto, H., Coso, O.A. et al. 1995b. Integrin function: Molecular hierarchies of cytoskeletal and signaling molecules. *J. Cell Biol.* 131:791–805.

Montanaro, L., Campoccia, D., and Arciola, C.R. 2008. Nanostructured materials for inhibition of bacterial adhesion in orthopedic implants: A minireview. *Int. J. Artif. Organs* 31:771–776.

Morra, M. 2006. Biochemical modification of titanium surfaces: Peptides and ECM proteins. *Eur. Cells Mater.* 12:1–15.

Morra, M., Cassinelli, C., Meda, L., Fini, M., Giavaresi, G., and Giardino, R. 2005. Surface analysis and effects on interfacial bone microhardness of collagen-coated titanium implants: A rabbit model. *Int. J. Oral Maxillofac. Implants* 20:23–30.

Moskowitz, J.S., Blaisse, M.R., Samuel, R.E. et al. 2010. The effectiveness of the controlled release of gentamicin from polyelectrolyte multilayers in the treatment of *Staphylococcus aureus* infection in a rabbit bone model. *Biomaterials* 31:6019–6030.

Naal, F.D., Salzmann, G.M., von Knoch F. et al. 2008. The effects of clindamycin on human osteoblasts *in vitro*. *Arch. Orthop. Trauma Surg.* 128:317–323.

Navarro, M., Michiardi, A., Castano, O., and Planell, J.A. 2008. Biomaterials in orthopaedics. *J. R. Soc., Interface* 5:1137–1158.

Necula, B.S., Apachitei, I., Tichelaar, F.D., Fratila-Apachitei, L.E., and Duszczyk, J. 2011. An electron microscopical study on the growth of TiO_2–Ag antibacterial coatings on Ti6Al7Nb biomedical alloy. *Acta Biomater.* 7:2751–2757.

Neut, D., Dijkstra, R.J., Thompson, J.I., van der Mei, H.C., and Busscher, H.J. 2011. Antibacterial efficacy of a new gentamicin-coating for cementless prostheses compared to gentamicin-loaded bone cement. *J. Orthop. Res.* 29:1654–1661.

Nguyen, P.M., Zacharia, N.S., Verploegen, E., and Hammond, P.T. 2007. Extended release antibacterial layer-by-layer films incorporating linear–dendritic block copolymer micelles. *Chem. Mater.* 19:5524–5530.

Niinomi, M. and Akahori, T. 2010. Improvement of the fatigue life of titanium alloys for biomedical devices through microstructural control. *Expert Rev. Med. Devices* 7:481–488.

Niño-Martínez, N., Martínez-Castañón, G.A., Aragon-Piña, A., Martínez-Gutierrez, F., Martínez-Mendoza, J.R., and Ruiz, F. 2008. Characterization of silver nanoparticles synthesized on titanium dioxide fine particles. *Nanotechnology* 19:065711.

Norowski, P.A., Courtney, H.S., Babu, J., Haggard, W.O., and Bumgardner, J.D. 2011. Chitosan coatings deliver antimicrobials from titanium implants: A preliminary study. *Implant. Dent.* 20:56–67.

Nurhaerani, Arita, K., Shinonaga, Y., and Nishino, M. 2006. Plasma-based fluorine ion implantation into dental materials for inhibition of bacterial adhesion. *Dent. Mater. J.* 25:684–692.

O'Sullivan, C., O'Hare, P., O'Leary, N.D. et al. 2010. Deposition of substituted apatites with anticolonizing properties onto titanium surfaces using a novel blasting process. *J. Biomed. Mater. Res.—Part B* 95B:141–149.

Oberdörster, G., Maynard, A., Donaldson, K. et al. 2005. Principles for characterizing the potential human health effects from exposure to nanomaterials: Elements of a screening strategy. *Part. Fibre Toxicol.* 2:8.

Olmedo, D.G., Tasat, D.R., Duffo, G., Guglielmotti, M.B., and Cabrini, R.L. 2009. The issue of corrosion in dental implants: A review. *Acta Odontol. Latinoam.* 22:3–9.

Pan, C.J., Dong, Y.X., Zhang, Y.Y., Nie, Y.D., Zhao, C.H., and Wang, Y.L. 2011. Enhancing the antibacterial activity of biomimetic HA coatings by incorporation of norvancomycin. *J. Orthop. Sci.* 16:105–113.

Parvizi, J., Saleh, K.J., Ragland, P.S., Pour, A.E., and Mont, M.A. 2008. Efficacy of antibiotic-impregnated cement in total hip replacement. *Acta Orthop*. 79:335–341.

Pasqui, D., Rossi, A., Di Cintio, F., and Barbucci, R. 2007. Functionalized titanium oxide surfaces with phosphated carboxymethyl cellulose: Characterization and bonelike cell behavior. *Biomacromolecules* 8:3965–3972.

Peng, Z.X., Tu, B., Shen, Y. et al. 2011. Quaternized chitosan inhibits icaA transcription and biofilm formation by *Staphylococcus* on a titanium surface. *Antimicrob. Agents Chemother*. 55:860–866.

Perka, C., Schultz, O., Spitzer, R.S., Lindenhayn, K., Burmester, G.R., and Sittinger, M. 2000. Segmental bone repair by tissue-engineered periosteal cell transplants with bioresorbable fleece and fibrin scaffolds in rabbits. *Biomaterials* 21:1145–1153.

Perlin, L., MacNeil, S., and Rimmer, S. 2008. Production and performance of biomaterials containing RGD peptides. *Soft Matter* 4:2331–2349.

Petrie, T.A., Raynor, J.E., Reyes, C.D., Burns, K.L., Collard, D.M., and García, A.J. 2008. The effect of integrin-specific bioactive coatings on tissue healing and implant osseointegration. *Biomaterials* 29:2849–2857.

Pfaff, M. 1997. Recognition sites of RGD-dependent integrins. In *Integrin–Ligand Interaction*, Eble, J.A. (Ed.). Springer-Verlag, Heidelberg, pp. 101–121.

Pfaff, M., Tangemann, K., Müller, B. et al. 1994. Selective recognition of cyclic RGD peptides of NMR defined conformation by alpha IIb beta 3, alpha V beta 3, and alpha 5 beta 1 integrins. *J. Biol. Chem*. 269:20233–20238.

Pierschbacher, M.D. and Ruoslahti, E. 1984. Cell attachment activity of fibronectin can be duplicated by small synthetic fragments of the molecule. *Nature* 309:30–33.

Plopper, G.E., McNamee, H.P., Dike, L.E., Bojanowski, K., and Ingber, D.E. 1995. Convergence of integrin and growth factor receptor signaling pathways within the focal adhesion complex. *Mol. Biol. Cell* 6:1349–1365.

Ponche, A., Bigerelle, M., and Anselme, K. 2010. Relative influence of surface topography and surface chemistry on cell response to bone implant materials—Part 1: Physicochemical effects. *Proc. Inst. Mech. Eng. H*. 224:1471–1486.

Popat, K.C., Eltgroth, M., LaTempa, T.J., Grimes, C.A., and Desai, T.A. 2007. Decreased *Staphylococcus epidermis* adhesion and increased osteoblast functionality on antibiotic-loaded titania nanotubes. *Biomaterials* 28:4880–4888.

Porté-Durrieu, M.C., Guillemot, F., Pallu, S. et al. 2004. Cyclo-(DfKRG) peptide grafting onto Ti–6Al–4V: Physical characterization and interest towards human osteoprogenitor cells adhesion. *Biomaterials* 25:4837–4846.

Porter, J.R., Ruckh, T.T., and Popat, K.C. 2009. Bone tissue engineering: A review in bone biomimetics and drug delivery strategies. *Biotechnol. Prog*. 25:1539–1560.

Puckett, S.D., Taylor, E., Raimondo, T., and Webster, T.J. 2010. The relationship between the nanostructure of titanium surfaces and bacterial attachment. *Biomaterials* 31: 706–713.

Pytela, R., Pierschbacher, M.D., Argraves, S., Suzuki, S., and Ruoslahti, E. 1987. Arginine–glycine–aspartic acid adhesion receptors. *Methods Enzymol*. 144:475–489.

Qu, J., Lu, X., Li, D. et al. 2011. Silver/hydroxyapatite composite coatings on porous titanium surfaces by sol–gel method. *J. Biomed. Mater. Res., Part B* 97B:40–48.

Rammelt, S., Illert, T., Bierbaum, S., Scharnweber, D., Zwipp, H., and Schneiders, W. 2006. Coating of titanium implants with collagen, RGD peptide and chondroitin sulfate. *Biomaterials* 27:5561–5571.

Redick, S.D., Settles, D.L., Briscoe, G., and Erickson, H.P. 2000. Defining fibronectin's cell adhesion synergy site by site-directed mutagenesis. *J. Cell Biol*. 149:521–527.

Reyes, C.D. and García, A.J. 2004. $\alpha2\beta1$ Integrin-specific collagen-mimetic surfaces supporting osteoblastic differentiation. *J. Biomed. Mater. Res., Part A* 69A:591–600.

Reyes, C.D., Petrie, T.A., Burns, K.L., Schwartz, Z., and Garcia, A.J. 2007. Biomolecular surface coating to enhance orthopaedic tissue healing and integration. *Biomaterials* 28:3228–3235.

Rezania, A. and Healy, K.E. 1999. Biomimetic peptide surfaces that regulate adhesion, spreading, cytoskeletal organization, and mineralization of the matrix deposited by osteoblast-like cells. *Biotechnol. Prog.* 15:19–32.

Roach, P., Farrar, D., and Perry, C.C. 2005. Interpretation of protein adsorption: Surface-induced conformational changes. *J. Am. Chem. Soc.* 127:8168–8173.

Ross, F.P., Chappel, J., Alvarez, J.I. et al. 1993. Interactions between the bone matrix proteins osteopontin and bone sialoprotein and the osteoclast integrin alpha v beta 3 potentiate bone resorption. *J. Biol. Chem.* 268:9901–9907.

Ruoslahti, E. 1996. RGD and other recognition sequences for integrins. *Annu. Rev. Cell Dev. Biol.* 12:697–715.

Saito, A., Suzuki, Y., Kitamura, M. et al. 2006. Repair of 20-mm long rabbit radial bone defects using BMP-derived peptide combined with an alpha-tricalcium phosphate scaffold. *J. Biomed. Mater. Res., Part A* 77A:700–706.

Saito, A., Suzuki, Y., Ogata, S., Ohtsuki, C., and Tanihara, M. 2004. Prolonged ectopic calcification induced by BMP-2-derived synthetic peptide. *J. Biomed. Mater. Res., Part A* 70A:115–121.

Saito, A., Suzuki, Y., Ogata, S., Ohtsuki, C., and Tanihara, M. 2005. Accelerated bone repair with the use of a synthetic BMP-2-derived peptide and bone-marrow stromal cells. *J. Biomed. Mater. Res., Part A* 72A:77–82.

Sakka, S. and Coulthard, P. 2011. Implant failure: Etiology and complications. *Med. Oral Patol. Oral Cir. Bucal.* 16:e42–e44.

Saldaña, L., Barranco, V., Gonzalez-Carrasco, J.L., Rodriguez, M., Munuera, L., and Vilaboa, N. 2007. Thermal oxidation enhances early interactions between human osteoblasts and alumina blasted Ti6Al4V alloy. *J. Biomed. Mater. Res.—Part A* 81A:334–346.

Salzmann, G.M., Naal, F.D., von, K.F. et al. 2007. Effects of cefuroxime on human osteoblasts *in vitro*. *J. Biomed. Mater. Res.—Part A* 82A:462–468.

Sandukas, S., Yamamoto, A., and Rabiei, A. 2011. Osteoblast adhesion to functionally graded hydroxyapatite coatings doped with silver. *J. Biomed. Mater. Res.—Part A* 97A:490–497.

Saranya, N., Saravanan, S., Moorthi, A., Ramyakrishna, B., and Selvamurugan, N. 2011. Enhanced osteoblast adhesion on polymeric nano-scaffolds for bone tissue engineering. *J. Biomed. Nanotechnol.* 7:238–244.

Sawyer, A.A., Weeks, D.M., Kelpke, S.S., McCracken, M.S., and Bellis, S.L. 2005. The effect of the addition of a polyglutamate motif to RGD on peptide tethering to hydroxyapatite and the promotion of mesenchymal stem cell adhesion. *Biomaterials* 26:7046–7056.

Schense, J.C. and Hubbell, J.A. 2000. Three-dimensional migration of neurites is mediated by adhesion site density and affinity. *J. Biol. Chem.* 275:6813–6818.

Schliephake, H., Scharnweber, D., Dard, M. et al. 2002. Effect of RGD peptide coating of titanium implants on periimplant bone formation in the alveolar crest. *Clin. Oral Implants Res.* 13:312–319.

Schmidmaier, G., Lucke, M., Wildemann, B., Haas, N.P., and Raschke, M. 2006. Prophylaxis and treatment of implant-related infections by antibiotic-coated implants: A review. *Injury* 37 Suppl 2:S105–S112.

Schwartz, J., Avaltroni, M.J., Danahy, M.P. et al. 2003. Cell attachment and spreading on metal implant materials. *Mater. Sci. Eng., C* 23:395–400.

Shakesheff, K.M., Cannizzaro, S.M., and Langer, R. 1998. Creating biomimetic micro-environments with synthetic polymer–peptide hybrid molecules. *J. Biomater. Sci., Polym. Ed.* 9:507–518.

Shekaran, A. and García, A.J. 2011. Extracellular matrix-mimetic adhesive biomaterials for bone repair. *J. Biomed. Mater. Res., Part A* 96A:261–272.

Shibata, Y., Kawai, H., Yamamoto, H., Igarashi, T., and Miyazaki, T. 2004. Antibacterial titanium plate anodized by being discharged in NaCl solution exhibits cell compatibility. *J. Dent. Res.* 83:115–119.

Shimazaki, T., Miyamoto, H., Ando, Y. et al. 2010. *In vivo* antibacterial and silver-releasing properties of novel thermal sprayed silver-containing hydroxyapatite coating. *J. Biomed. Mater. Res.—Part B* 92B:386–389.

Shirai, T., Shimizu, T., Ohtani, K., Zen, Y., Takaya, M., and Tsuchiya, H. 2011. Antibacterial iodine-supported titanium implants. *Acta Biomater.* 7:1928–1933.

Siebers, M.C., ter Brugge, P.J., Walboomers, X.F., and Jansen, J.A. 2005. Integrins as linker proteins between osteoblasts and bone replacing materials. A critical review. *Biomaterials* 26:137–146.

Sinha, R.K. and Tuan, R.S. 1996. Regulation of human osteoblast integrin expression by orthopedic implant materials. *Bone* 18:451–457.

Solchaga, L.A., Dennis, J.E., Goldberg, V.M., and Caplan, A.I. 1999. Hyaluronic acid-based polymers as cell carriers for tissue-engineered repair of bone and cartilage. *J. Orthop. Res.* 17:205–213.

Steinemann, S.G. 1996. Metal implants and surface reactions. *Injury* 27 Suppl 3:SC16–SC22.

Stigter, M., Bezemer, J., de Groot, K., and Layrolle, P. 2004. Incorporation of different antibiotics into carbonated hydroxyapatite coatings on titanium implants, release and antibiotic efficacy. *J. Controlled Release* 99:127–137.

Streicher, R.M., Weber, H., Schon, R., and Semlitsch, M. 1991. New surface modification for Ti–6Al–7Nb alloy: Oxygen diffusion hardening (ODH). *Biomaterials* 12:125–129.

Suci, P.A., Mittelman, M.W., Yu, F.P., and Geesey, G.G. 1994. Investigation of ciprofloxacin penetration into *Pseudomonas aeruginosa* biofilms. *Antimicrob. Agents Chemother.* 38:2125–2133.

Suketa, N., Sawase, T., Kitaura, H. et al. 2005. An antibacterial surface on dental implants, based on the photocatalytic bactericidal effect. *Clin. Implant Dent. Relat. Res.* 7:105–111.

Svehla, M., Morberg, P., Bruce, W., and Walsh, W.R. 2005. No effect of a type I collagen gel coating in uncemented implant fixation. *J. Biomed. Mater. Res.—Part B* 74B:423–428.

Teow, Y., Asharani, P.V., Hande, M.P., and Valiyaveettil, S. 2011. Health impact and safety of engineered nanomaterials. *Chem. Commun. (Cambridge UK)* 47:7025–7038.

Thull, R. 2002. Physicochemical principles of tissue material interactions. *Biomol. Eng.* 19:43–50.

Travan, A., Donati, I., Marsich, E. et al. 2010b. Surface modification and polysaccharide deposition on BisGMA/TEGDMA thermoset. *Biomacromolecules* 11:583–592.

Travan, A., Marsich, E., Donati, I. et al. 2010a. Silver–polysaccharide nanocomposite antimicrobial coatings for methacrylic thermosets. *Acta Biomater.* 7:337–346.

Travan, A., Pelillo, C., Donati, I. et al. 2009. Noncytotoxic silver nanoparticle–polysaccharide nanocomposites with antimicrobial activity. *Biomacromolecules* 10:1429–1435.

Trebse, R., Pisot, V., and Trampuz, A. 2005. Treatment of infected retained implants. *J. Bone Jt. Surg., Br.* 87:249–256.

Treloar, L.R.G. 1958. *The Physics of Rubber Elasticity*, Second edition.

van der Pluijm, G., Vloedgraven, H.J.M., Ivanov, B. et al. 1996. Bone sialoprotein peptides are potent inhibitors of breast cancer cell adhesion to bone. *Cancer Res.* 56:1948–1955.

Varini, E., Cristofolini, L., Traina, F., Viceconti, M., and Toni, A. 2008. Can the rasp be used to predict intraoperatively the primary stability that can be achieved by press-fitting the stem in cementless hip arthroplasty? *Clin. Biomech. (Bristol, Avon)* 23:408–414.

Variola, F., Brunski, J.B., Orsini, G., Tambasco de Oliveira, P., Wazen, R., and Nanci, A. 2011. Nanoscale surface modifications of medically relevant metals: State-of-the art and perspectives. *Nanoscale* 3:335–353.

Variola, F., Yi, J.H., Richert, L., Wuest, J.D., Rosei, F., and Nanci, A. 2008. Tailoring the surface properties of Ti6Al4V by controlled chemical oxidation. *Biomaterials* 29:1285–1298.

Velich, N., Kádár, B., Kiss, G. et al. 2006. Effect of human organism on the oxide layer formed on titanium osteosynthesis plates: A surface analytical study. *J. Craniofac. Surg.* 17:1144–1149.

Wojtowicz, A.M., Shekaran, A., Oest, M.E. et al. 2010. Coating of biomaterial scaffolds with the collagen-mimetic peptide GFOGER for bone defect repair. *Biomaterials* 31:2574–2582.

Wu, P. and Grainger, D.W. 2006. Drug/device combinations for local drug therapies and infection prophylaxis. *Biomaterials* 27:2450–2467.

Xiao, F.J., Zhang, Y., and Yun, L.J. 2009b. Electrophoretic deposition of titanium/silicon-substituted hydroxyapatite composite coating and its interaction with bovine serum albumin. *Trans. Nonferrous. Met. Soc. China* 19:125–130.

Xiao, Y.H., Ma, S., Chen, J.H., Chai, Z.G., Li, F., and Wang, Y.J. 2009a. Antibacterial activity and bonding ability of an adhesive incorporating an antibacterial monomer DMAE-CB. *J. Biomed. Mater. Res.—Part B* 90B:813–817.

Yang, X.B., Roach, H.I., Clarke, N.M.P. et al. 2001. Human osteoprogenitor growth and differentiation on synthetic biodegradable structures after surface modification. *Bone* 29:523–531.

Yildiz, F., Yetim, A.F., Alsaran, A., and Efeoglu, I. 2009. Wear and corrosion behaviour of various surface treated medical grade titanium alloy in bio-simulated environment. *Wear* 267:695–701.

Yu, J.C., Ho, W., Lin, J., Yip, H., and Wong, P.K. 2003. Photocatalytic activity, antibacterial effect, and photoinduced hydrophilicity of TiO_2 films coated on a stainless steel substrate. *Environ. Sci. Technol.* 37:2296–2301.

Yuan, S.J., Pehkonen, S.O., Ting, Y.P., Neoh, K.G., and Kang, E.T. 2009. Inorganic–organic hybrid coatings on stainless steel by layer-by-layer deposition and surface-initiated atom transfer–radical polymerization for combating biocorrosion. *ACS Appl. Mater. Interfaces* 1:640–652.

Zambuzzi, W.F., Ferreira, C.V., Granjeiro, J.M., and Aoyama, H. 2011. Biological behavior of pre-osteoblasts on natural hydroxyapatite: A study of signaling molecules from attachment to differentiation. *J. Biomed. Mater. Res.—Part A* 97A:193–200.

Zhang, F., Zhang, Z., Zhu, X., Kang, E.T., and Neoh, K.G. 2008. Silk-functionalized titanium surfaces for enhancing osteoblast functions and reducing bacterial adhesion. *Biomaterials* 29:4751–4759.

Zhang, L. and Mah, T.F. 2008. Involvement of a novel efflux system in biofilm-specific resistance to antibiotics. *J. Bacteriol.* 190:4447–4452.

Zhang, Y.Y., Tao, J., Pang, Y.C., Wang, W., and Wang, T. 2006. Electrochemical deposition of hydroxyapatite coatings on titanium. *Trans. Nonferrous Met. Soc. China* 16:633–637.

8 Oral Implants
Biological Background and Current Concepts in Peri-Implantitis Treatment

Nicola U. Zitzmann and Tuomas Waltimo

CONTENTS

8.1 INTRODUCTION

The regeneration of soft and hard tissues around dental implants naturally happens following implant placement, and is also the goal in treatment of peri-implantitis as a *restitutio ad integrum*. A thorough understanding of the morphology of the tissues surrounding dental implants and differences compared to natural teeth is essential in compiling an adequate treatment strategy for diseased sites. Furthermore, options for future developments should be based on this knowledge. Current therapeutic concepts have been derived mainly from periodontitis treatment and focus on the removal of the biofilm as the major disease etiology, combined with the debridement and decontamination of the involved surface areas. Although nonsurgical approaches are less successful in peri-implantitis therapy due to the limited access to the undercuts of the screw-shaped implants and the circumferential defect morphologies, they serve as the mandatory initial treatment phase. After this nonsurgical pretreatment, surgical interventions aim at a resolution of the infection and are either conducted as a resective or a regenerative approach.

8.2 MORPHOLOGIC BACKGROUND OF PERIODONTAL AND PERI-IMPLANT TISSUES

Similar to the gingiva around natural teeth, the peri-implant mucosa (PiM) surrounding an implant comprises an epithelial and a connective tissue (CT) compartment attached on the transmucosal portion of the dental implant (Berglundh et al. 1991; Buser et al. 1989; Listgarten et al. 1991; Schroeder et al. 1981). Different from the periodontal ligament (PDL) interposed between the root and the alveolar bone, the enossal implant portion is in direct contact with the bone tissue under light microscopy (Figure 8.1a,b). The transmucosal implant portion is either part of the implant body in those implants placed on the mucosa level or connected as a separate abutment in implants placed at the level of the bone (Figure 8.2). The latter are also called two-piece or two-stage implants, which are intended for submerged healing after implant installation. Implants combining the enossal and the transmucosal portion are also called one-piece or one-stage implants, which are intended for nonsubmerged healing. Obviously, the type of soft tissue healing can be varied in one- and two-piece implants and is selected according to the individual needs (e.g., preference is given to submerged healing when augmentation procedures are applied simultaneously).

The so-called barrier epithelium facing dental implants is about 2 mm long and adheres to the implant surface via a basal lamina and hemidesmosomes, and is similar to the attachment of the junctional epithelium (JE) on the tooth surface (Brunette 1988; Gould et al. 1984; James and Schultz 1974; Listgarten and Lai 1975). By analyzing the attachment at implants made of titanium alloy (Ti–6Al–4V) placed in rats with transmission electron microscopy, it was, however, observed that dense hemidesmosome formations were restricted to the lower region close to the CT. Furthermore, the secretion of laminin-1, which contributes to epidermal cell adhesion, was lower in PiM than in the gingiva around teeth (Ikeda et al. 2000). After the topical application of horseradish peroxidase in the sulcus region, penetration in the CT compartment was observed in PiM, whereas the internal basement lamina of the gingiva inhibited peroxidase penetration and protected the underlying CT (Ikeda et al. 2002). It was assumed that the weaker epithelial adherence around implants may not function sufficiently as a tight biological seal against the invasion of foreign material.

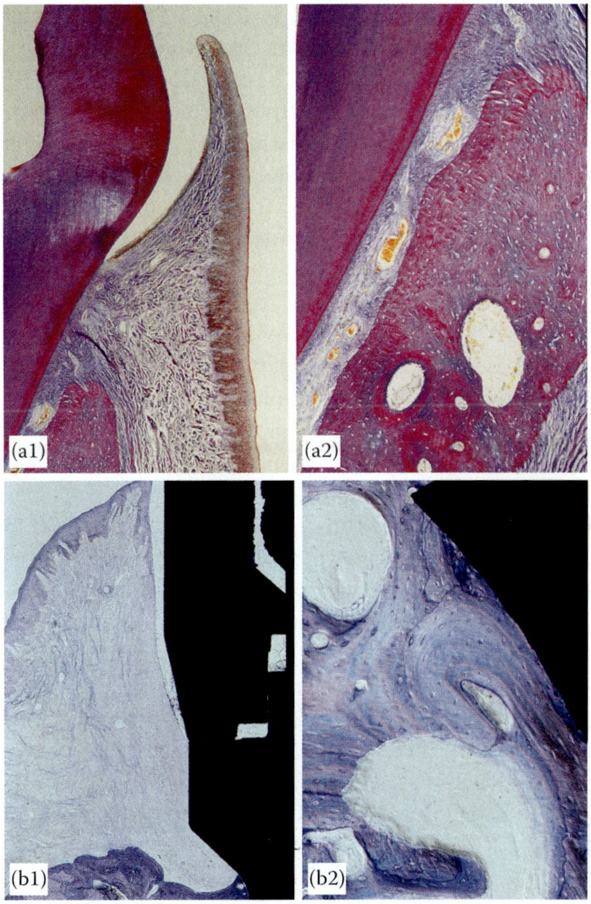

FIGURE 8.1 Histologic sections from a tooth site (dog) illustrating fiber arrangement in an area of CT attachment (a1) and PDL (a2). Histologic sections from an implant site (dog) indicating fiber arrangement in transmucosal compartment (b1) and BIC along the enossal part (b2).

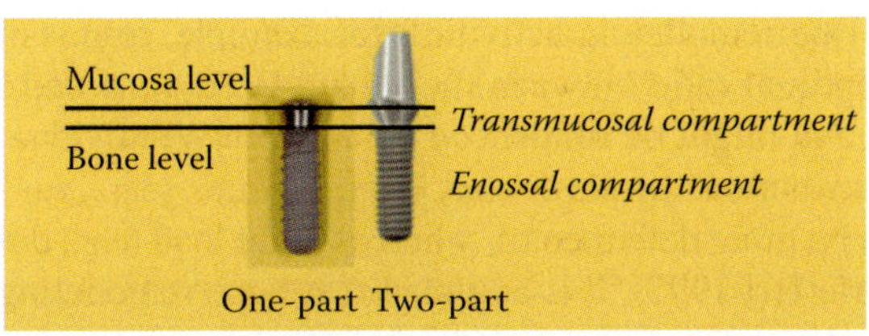

FIGURE 8.2 Schematic illustration of one- and two-part implant types placed on the level of mucosal margin or at the level of the bone.

The CT of the interface portion of PiM is about 1.5 mm high and contains more collagen but fewer fibroblasts and vascular units than the corresponding zone in the gingiva, that is, the supraalveolar CT (Berglundh et al. 1991, 1994). In the gingiva at the teeth, the collagen fibers are embedded within the root cementum and are arranged in groups or bundles with distinct orientations such as dentogingival, dentoperiosteal, circular, and transseptal fibers (Feneis 1952; Page et al. 1974). The implant lacks a lining cementum and, hence, the collagen fibers in PiM are invested in the peri-implant bone and run a course parallel to the implant surface (Abrahamsson et al. 1996; Berglundh et al. 1991; Listgarten et al. 1992). The presence of circular fibers in a zone of PiM close to the implant has also been demonstrated (Buser et al. 1989; Ruggeri et al. 1992), as well as fibers with an orientation perpendicular to the transmucosal surface (Berglundh et al. 2007b). Some authors interpreted the observation of fibers perpendicular to the implant surface as a "functional orientation" (Glauser et al. 2005; Nevins et al. 2008; Schüpbach and Glauser 2007). This assumption, however, does not allow the erroneous conclusion of a fiber anchorage, which is not present around implant surfaces but limited to the periodontium.

Within the PDL, which is approximately 200 μm wide, 28,000 fiber bundles are arranged per square millimeter surrounding several capillary plexi. Fiber fringes are embedded within the acellular extrinsic fiber cement, which is produced by fibroblasts (2–2.5 μm/a) for life (Bosshardt and Schroeder 1996). These fibers, which are anchored in the 110–200 μm wide cementum layer, enter either the PDL or, in the supraalveolar portion, the CT compartment. Note that, in case of periodontal bone loss (as physiologic remodeling or as a result of an inflammatory process), the same fibers entering the PDL space and facilitating the functional connection to the alveolar bone become part of the CT attachment, functioning as a protective seal. In contrast, the enossal implant portion is placed in the hard tissue and is completely encompassed by bone tissue. After placement, the implant is initially fixed by mechanical stabilization; the following remodeling process with bone resorption and apposition leads to a direct contact between living cells of the bone tissue and the implant surface, called "osseointegration" (at the level of light microscopy; Figure 8.1b). This principle of osseointegration has been shown to provide the best long-term results compared to CT integration (Meffert 1996). Note that data related to the percentage of the entire bone compartment in contact with the enossal implant portion [bone-to-implant contact (BIC)] consider only the percentage of mineralized bone and not the bone marrow compartments. Although mineralized bone is primarily responsible for mechanical stability, bone marrow compartments represent the adaptive area bearing remodeling activities, for example, responsible for functional adaptation. The optimal ratio between the mineralized bone and the bone marrow is not yet defined and might be influenced by a patient's individual situation and by the local load distribution. In particular, lateral forces (e.g., during bruxism with grinding) seem to be more detrimental, whereas axial load (e.g., during clenching) is better tolerated (Meffert 1997). It is assumed that bone remodeling following surgical trauma is completed 3–6 months after implant placement, provided that no additional interference and/or trauma occurs, for example, from abutment connection in two-stage implants placed with the implant shoulder at the level of the bone margin (Lang and Berglundh 2011). In particular, in two-stage implants (e.g., *ad modum*

Brånemark) in the work of Adell et al. (1981) with submerged healing protocols, a physiologic bone remodeling of ≤1.5 mm after the first year of loading is anticipated. This remodeling procedure is related to traumatic drilling, resorption of thin cortical bone crests with insufficient vascularization, the abutment connection procedure, and/or the establishment of a certain distance between the abutment–fixture junction and the marginal bone level (biological width; Berglundh and Lindhe 1996). In one-piece implants, in which the transmucosal portion continues with the endosseous implant part, the amount of remodeling was found to be less pronounced (Hermann et al. 1997).

8.3 BIOLOGICAL IMPLICATIONS OF IMPLANT SURFACE MODIFICATIONS

8.3.1 Enossal Implant Portion

For the enossal implant portion, obvious advantages were demonstrated with regard to bone formation and the degree of osseointegration at implants with rough compared to turned surfaces; for example, see the work of Abrahamsson et al. (2001, 2004) and Buser et al. (1991). A better stabilization of the blot clot followed by improved fibrin adherence was observed during the first days after implant placement, and these processes are deemed to be responsible for a more rapid establishment of mineralized bone in contact to the titanium surface (Abrahamsson et al. 2004; Berglundh et al. 2003). A specific cell activation from the rough surface structure is also suggested. Further modifications, for example, with fluoride addition (OsseoSpeed, Astra Tech, Waltham, MA), were found to facilitate more intensive BIC in the initial phase 2 weeks after implant insertion, whereas after 6 weeks, differences between OsseoSpeed and the nonmodified TiOblast implant were no longer evident (Berglundh et al. 2007a).

For dental implants made from commercially pure (c.p.) titanium, surface treatment is subtractive using mechanical or chemical methods, additive, or a combination of both procedures (Table 8.1).

TABLE 8.1
Common Implant Types and Their Surface Treatment

Implant Type	Surface
Straumann	SLA: Sand blasted, large grit, and acid etched
	SLActive: Sand blasted, acid etched, and hydroxylated
Nobel Biocare	TiUnite: Anodic oxidation, increased TiO_2 layer by eletrolytic gas release
BIOMET 3i	Osseotite: Dual acid etched
	Nanotite: Induced buildup of single CaP crystals on acid-etched osseotite surface
Thommen	Sand blasted and thermal acid etched
FRIALIT	FRIADENT plus: Sand blasted and thermal acid etched
Astra Tech	OsseoSpeed: TiO_2 sand blasted and acid etched; modified with fluoride ions

Subtractive methods for surface modifications (mechanical or chemical ablation) comprise:

1. Machined/Turned (anisotropic properties)
2. Sand blasting with aluminum oxide (Al_2O_3) or titanium dioxide (TiO_2; 1–2 μm)
3. Acid etched (with increased surface hydrophily and isotropic properties)
4. Anodic oxidation (Ca/P ions embedded and isotropic properties)
5. Hydroxylation (with increased surface hydrophily)

Additive coatings of the enossal implant surface comprise:

1. Titanium plasma sprayed (TPS; 1–4 μm), hydroxyapatite (HA) sintered, isotropic properties
2. Calcium phosphate nanocrystals
3. Fluoride ions

Surface roughness is frequently estimated by the average height deviation measured from a mean plane and described as S_a values (Albrektsson and Wennerberg 2004). These surface characteristics have been graduated as follows (S_a value in micrometers):

1. Smooth (experimental implants): 0.0–0.4
2. Minimally rough (machined/*turned*): 0.5–1.0
3. Moderately rough (e.g., sand blasted, acid etched, and/or anodic oxidation): 1.0–2.0
4. Rough (e.g., HA and TPS): >2.0

8.3.2 Transmucosal Implant Compartment

For the transmucosal portion, different surface treatments have been established, aiming at an ideal surface free energy that ideally increases wettability for an improved epithelial and CT attachment. Furthermore, salivary proteins and bacteria should be rejected, whereas epithelial cell adhesion should be encouraged, for example, by human epidermal growth factors (Lange et al. 2004). It has been documented that, after implant placement in a nonsubmerged manner, a healing time of 4–6 weeks is required for the maturation of the soft tissue complex (Berglundh et al. 2007b).

Different surface treatments have been introduced for an enhanced CT adherence and improved contact guidance with specific microstructures, which comprise:

1. Additive processes
2. Subtractive processes (e.g., laser-structured microchannels)

Furthermore, different biological coatings are the subject of current investigations:

1. Laminin (for epithelial cell adhesion)
2. Fibronection (for fibroblast adhesion)

3. Growth factors
4. Nanoporous TiO_2 layer
5. Silver nanoparticles (colloidal dispersion of stabilized Ag nanoparticles deposited as sol–gel)

Current research is performed by investigating the antibacterial effect of silver nanoparticles incorporated into diamond-like carbon coatings, which revealed an effective reduction of *Streptococcus sanguinis* (Schwarz et al. 2011; Vig Slenters et al. 2008). A critical aspect of all these coatings is its limited stability so that long-lasting effects in the prevention of peri-implant diseases may not be expected. Further research aims at the development of implant surface modifications in the transmucosal compartment, which could be activated if peri-implant diseases are established (Welch et al. 2010). Bactericidal effects have been documented for hydrophily radicals (–OH) and reactive oxygen species generated after photocatalysis of TiO_2 nanoparticles by irradiation with ultraviolet light. Applying such TiO_2 coatings on the surface of implant abutments for the treatment of peri-implant diseases would, however, require the option of repetitive light activation to ensure a long-term effect.

8.3.2.1 Surface Roughness in Transmucosal Implant Portion

By comparing the CT compartment surrounding moderately rough surfaces (acid etched and oxidized) with minimally rough or smooth transmucosal surfaces, different histometric dimensions were observed. On one hand, it was assumed that the limited epithelial downgrowth with a longer CT attachment around moderately rough surfaces would provide better protective mechanisms (Glauser et al. 2005; Heinrich et al. 2008; Schüpbach and Glauser 2007). The presence of a rough surface within the transmucosal portion of an implant may, on the other hand, constitute a risk for enhanced plaque formation. In a study in man, it was demonstrated that, in the early phase (24 h) of plaque accumulation, titanium discs with a smooth surface ($S_a \leq 0.088$ μm) allowed less plaque to accumulate and maturate than discs with rough surfaces ($S_a = 2.142$ μm) (Rimondini et al. 1997). When implant abutments with rough ($S_a = 0.45$ μm) or smooth ($S_a = 0.22$ μm) surfaces were exposed to long-standing plaque accumulation over 6 months in dogs, no difference was observed in the dimension and composition of the plaque-associated infiltrated connective tissue (ICT) (Zitzmann et al. 2002). Rougher surfaces are found in the enossal implant portion, and if once exposed to the oral cavity, they may promote adherence of the biofilm. Deteriorated clinical long-term results have been reported for very rough surfaces, which had been produced by additive layering mechanisms, such as TPS and HA (Renvert et al. 2011; Rosenberg et al. 2004). According to a review investigating the role of implant surface roughness as confounding factor for peri-implantitis, more implant failures were observed with rough than with minimal or moderately rough surfaces particularly in patients with a history of periodontal disease (Quirynen et al. 2007). It must be noted that, when the enossal implant portion is exposed to oral cavity and biofilm adhesion occurs, peri-implantitis is already established, and its resolution by appropriate therapy is essential (see Section 8.6).

8.3.2.2 Geometric Designs and Abutment Materials in Transmucosal Implant Compartment

In addition to different coatings, varying geometric designs (e.g., straight, divergent, convergent, and platform switching) were experimentally established in the transmucosal portion, aiming at a stabilization of the CT compartment (Kim et al. 2010; Pongnarisorn et al. 2007). Larger infiltrates were observed in the CT portion around implants with a circumferential groove than around straight transmucosal implant portions (Pongnarisorn et al. 2007). Furthermore, concave transmucosal profiles with a microgrooved surface were associated with longer CT attachments and less bone resorption than flared or straight transmucosal implant portions with an anodic oxidized surface in the early healing phase (Kim et al. 2010). Particularly for two-piece implants placed on the level of the bone margin, the concept of platform switching was clinically introduced. Hence, an abutment diameter smaller than the implant shoulder is selected so that the connection of the enossal and the transmucosal portion and a potential microgap in this area is shifted in a region apart from the bone margin. It must be noted that, as soon as such niches and undercuts in the area of the abutment–implant connection are exposed to the oral cavity, the patient's personal and professional oral hygiene is even more difficult than with straight, continuous implant abutment designs. Furthermore, these niches may interfere with axial probing as the most essential diagnostic tool (see Section 8.5).

Although the titanium material in the enossal region is continuing up to the mucosal margin in one-part implants, in two-part implants, the selected abutment material located in the transmucosal region may influence the cell adhesion and the dimensions of the peri-implant soft tissue complex. After a 6-month period, the soft tissue attachment around conventional c.p. titanium abutments or highly sintered Al_2O_3 abutments revealed favorable integrity compared to surrounding abutments made of gold or porcelain fused to a gold alloy (Abrahamsson et al. 1998).

8.4 ETIOLOGY AND RISK FACTORS OF PERIODONTITIS AND PERI-IMPLANTITIS

8.4.1 Definition of Peri-Implantitis and Biofilm as Major Etiologic Factor

In accordance with the classification of periodontal disease around the teeth, peri-implant disease includes *peri-implant mucositis*, which corresponds to gingivitis, and *peri-implantitis*, which corresponds to periodontitis (Zitzmann and Berglundh 2008). According to the consensus report from the First European Workshop on Periodontology (EWOP), the two peri-implant disease entities are defined as follows (Albrektsson and Isidor 1994): "While peri-implant mucositis is a reversible inflammatory reaction in the soft tissues surrounding a functioning implant, peri-implantitis describes inflammatory reactions associated with loss of supporting bone around an implant in function." Peri-implant infection may take years to develop, as is the case with periodontitis. In the infectious disease described by peri-implantitis, there is a clear causal relationship to bacterial biofilm, which has been verified in animal and human studies (Berglundh et al. 1992; Pontoriero et al. 1994; Zitzmann et al. 2001). The colonization around dental implants follows the same steps as around natural teeth (Leonhardt et al. 1992, 1993). The colonizing bacteria bind to the pellicle comprising saliva and sulcus fluid components. In particular, subgingivally (around the implant "submucosally"), a stable biofilm is established, whereas single pathogens can also be present as mobile "swimmers." In the subgingival/submucosal region, bacterial colonies are protected from mechanical disturbance but are still outside the tissue and are therefore almost inaccessible for the immune response and systemic antibacterial therapies. Persisting biofilm accumulation in the area of the gingival/mucosal margin leads to a continuation of the inflammatory process with an increase in the vascularization, the establishment of an inflammatory infiltrate in the ICT, and the reduction of the proportion of collagen fibers.

According to a recent review dealing with the biofilm characteristics in peri-implantitis sites, the disease was described as a mixed anaerobic infection (Mombelli and Decaillet 2011). In most cases, the composition of the flora was similar to the subgingival flora of chronic periodontitis (ChP) and dominated by Gram-negative, strict, or facultative anaerobic species such as *Aggregatibacter actinomycetemcomitans*, *Porphyromonas gingivalis*, *Treponema denticola*, and/or *Tannerella forsythensis*. Peri-implant infections were occasionally linked to a different microbiota, including high numbers of peptostreptococci (i.e., *Peptostreptococcus micra*) or staphylococci (particularly, *Staphylococcus aureus* and *S. epidermidis*). Peptostreptococci are commensal organisms in humans that can cause abscesses and necrotizing soft tissue infections. *S. aureus* and *S. epidermidis* are well-established pathogens implicated in infections of implanted medical devices crossing the epidermal barrier. Longitudinal observations have shown that *S. aureus* may colonize implants early after placement and may persist long term. A checkerboard DNA-DNA hybridization assay encompassing 79 different microorganisms was used to study bacterial counts in 34 cases with peri-implantitis (Persson et al. 2010). The most prevalent bacteria were *Fusobacterium nucleatum* sp., *Staphylococcus* sp., *A. actinomycetemcomitans*, *Helicobacter pylori*, and *T. forsythensis*. In addition to these microbiota, *Candida albicans* was also frequently found (Mombelli and Decaillet 2011).

The loss of marginal/crestal supporting bone indicates the conversion from peri-implant mucositis into peri-implantitis. In periodontitis and peri-implantitis, the tissue destruction is initiated by the activated individual local host response mediated by cytokines (Burt 2005; Gomez et al. 2009). In several experimental studies, ligatures were placed in a submarginal position, and undisturbed plaque formation was allowed to elapse. This local minor trauma with an opening of the gingival/mucosal pocket allowed biofilm to form also in a profound subgingival/submucosal position (Berglundh et al. 2010; Heijl et al. 1976). The resulting acute inflammatory reaction involved a shift to a microflora with Gram-negative anaerobic rods, early tissue breakdown, and bone loss (Lang et al. 1993; Leonhardt et al. 1992). According to a recent review, the tissue reaction following ligature removal is different around implants and teeth (Berglundh et al. 2010). The experimental peri-implantitis lesion extended apical of the pocket epithelium and reached close to the bone crest, whereas a protective CT capsule separated the periodontitis lesion from the bone 1 month after ligature removal (described as the "self-limiting" process; Lindhe and Ericsson

1978). Around implants, the tissue destruction either rested and changed from an active destructive process into a chronic lesion or continued until the apical implant portion was reached and osseointegration was completely lost, resulting in implant mobility 3 months after ligature removal (Marinello et al. 1995). If left untreated over a longer period (1 year), a progression of experimentally induced peri-implant infection tends to be found in most sites, with a possible influence of local factors (Zitzmann et al. 2004). At varying times after ligature removal, experimental peri-implantitis sites, in contrast to experimental periodontitis, exhibited signs of acute inflammation and large amounts of osteoclasts, which lined the surface of the bone crest (Berglundh et al. 2010). The ICT in PiM showed a wider spread with larger proportions of immunocompetent cells than in the gingival ICT. The apical portion of the ICT was uncovered and in direct contact with the biofilm residing on the implant surface in the pocket area. These findings revealed an open/unclosed pocket, with the epithelium incapable of covering the ICT and the CT unable to protect the bone and to encapsulate the ICT (Albouy et al. 2009; Berglundh et al. 2010).

The experimental peri-implantitis model with subsequent spontaneous progression was applied to evaluate the influence of implant surface characteristics on the disease progression (see Section 8.4.1; Berglundh et al. 2007c). Comparing implants with experimental polished with moderately rough surface implants, the area of plaque, the size of the inflammatory lesion in the CT, and bone loss were larger in sand-blasted and acid-etched (SLA) than in polished implant sites (Berglundh et al. 2007c). Using commercially available implants, it was observed that the spontaneous progression of peri-implantitis during a 6-month period after ligature removal differed among the implant types with more destructive processes around rougher implant surfaces (Albouy et al. 2008, 2009).

Clinically, peri-implantitis is often associated with crater-like defect morphologies surrounding the implant, whereas in periodontal tissues, intrabony defects specifically localized in the interproximal regions are frequently found. The circular defect contour reveals a circular spread of the inflammatory infiltrate within the CT of PiM, which results from present microbiota, and is possibly related to the missing fiber anchorage within the compartment of the CT attachment. Clinically, peri-implantitis is often associated with pus release from the pocket area following probing and/or palpation. Pus contains mainly polymorphonuclear granulocytes and fibrin and is indicative for a process of tissue destruction. Clinical observations suggest that submucosal calculus deposits on implants with minimally or moderately rough surfaces are less common than subgingival calculus formation at the teeth. This phenomenon may be related to the observation that peri-implant disease, in general, progresses faster than periodontitis, and several years are required for calculus formation at the teeth or implants in the pocket area.

8.4.2 Risk Factors

The progression of gingivitis into periodontitis is influenced by several individual factors described as host susceptibility. The primary cause is the presence of a pathogenic biofilm (see Section 8.4.1). Additional acquired cofactors are immune suppression, diabetes mellitus, human immunodeficiency virus infection, tobacco use, alcohol consumption, or stress (Heitz-Mayfield 2008; Lindhe and Meyle 2008). Furthermore, genetic cofactors may affect a patient's predisposition to periodontitis. In peri-implantitis, several factors are similarly associated with an increased risk of developing the disease (Heitz-Mayfield 2008), including insufficient personal oral hygiene, history of periodontal disease, and tobacco use (Strietzel et al. 2007). Further correlations have been suggested between peri-implantitis and diabetes mellitus or alcohol consumption, whereas a possible association with genetic factors and rough implant surfaces is currently not proven. A recent study with an observation period of 48 ± 26 months addressed the question of whether periodontally compromised patients have an increased susceptibility to peri-implant infections (De Boever et al. 2009). Particularly, patients with a history of aggressive periodontal disease (AgP) appeared to have an increased risk of developing peri-implantitis and of experiencing implant failures, with 15.3% compared to 3.6% in patients with a history of ChP and 3.1% in the healthy control. Among those with AgP, smokers had a reduced implant success rate of 63%, whereas success was 78% among former smokers and 86% among nonsmokers. In the group of ChP and healthy patients, the smoking status had only a minor influence on the success rates (De Boever et al. 2009). In patients with periodontal diseases, treatment to establish healthy periodontia is essential prior to implant placement. Deep periodontal pockets serve as reservoirs for pathogenic microbiota, which may be transmitted to the peri-implant regions (Koka et al. 1993; Leonhardt et al. 1993; Sbordone et al. 1999; Sumida et al. 2002; van Winkelhoff et al. 2000). Furthermore, *S. aureus* at tooth sites was found to be predictive for its presence at implant sites (Renvert et al. 2008a).

8.4.3 Iatrogenic Factors

Iatrogenic factors are thought to be cofactors affecting the development of peri-implantitis as soon as accessibility for plaque control is insufficient and biofilm establishment is promoted (Figure 8.3; Heitz-Mayfield 2008). These iatrogenic factors comprise excess cement, overhanging restoration margins, poor fit of restorative components, closed interproximal regions, implant malpositioning with implants too close together, and inadequate vertical position/depth of the implant shoulder (Lang and Berglundh 2011). The latter has most detrimental effects in single-tooth implants, particularly when the transmucosal implant component is in a position

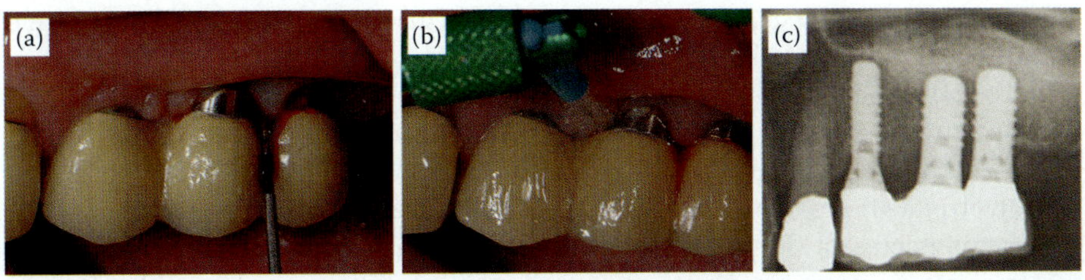

FIGURE 8.3 (a) Iatrogenic cofactors promoting biofilm adherence and spread. Clinical signs of inflammation with bleeding on probing—increased PPD. (b) Closed interproximal areas inhibit proper cleaning. (c) Radiographic loss of supporting bone.

that is too far apical from the interproximal bone septum of the adjacent teeth. This situation entails a long epithelial and CT attachment, which, in turn, may develop into a deep pocket once a submucosal biofilm has become established and the attachment mechanisms are destroyed. As soon as the barrier epithelium is converted into a pocket epithelium and the CT attachment is detached from the implant abutment surface, the inflammatory tissue destruction progresses and possibly includes bone resorption.

In particular, when using cemented restorations placed immediately after implant insertion (immediate loading protocols), the lack of a protecting seal allows cement excess to penetrate in the submucosal compartment, which, otherwise, requires 4 to 6 weeks for sufficient maturation after implant placement (see Section 8.3.2). These cement remnants possibly act as retentive surfaces that promote biofilm formation.

8.5 DIAGNOSIS

Correct diagnosis of peri-implant infection is critical for appropriate management of peri-implant disease. Detection of inflammation in the PiM requires the use of periodontal probing to identify bleeding and/or suppuration following such mechanical testing, that is, approximately 20 s after probing, when quadrant measuring has been completed (Etter et al. 2002; Lang et al. 1994). Additional clinical signs of infection may include pus release and/or suppuration following palpation, increased probing pocket depths (PPD) relative to the baseline, redness, and peri-implant mucosal swelling or hyperplasia. For the assessment of peri-implantitis, the detection of marginal bone loss in radiographs is another prerequisite (Figure 8.3). Ideally, periapical radiographs are used with a film holder and the X-ray beam perpendicular to the implant axis. In this context, it is important to distinguish between "physiologic" bone remodeling, which occurs early after implant installation (see Section 8.2), and the loss of supporting bone, which may be detected at implants during function, that is, after the process of osseointegration is completed (Lang and Berglundh 2011). Recording of baseline data is, therefore, essential and should include radiographs after the insertion of the implant restoration and PPD measurements after reestablishing the barrier (junctional) epithelium 1 to 2 weeks after prosthesis installation. Depending on the contours of the implant restoration, an axial access to the sulcus area may be inhibited by the discrepancy to a smaller diameter of the implant or implant abutment. In this case, the clinician may dispense with assessing four to six sites (i.e., mesiobuccal, midbuccal, distobuccal, distolingual, midlingual, and mesiolingual) and instead selects at least one site for appropriate axial measurement (Figure 8.4). Such a site is frequently found on the lingual aspect and is then used as representative to evaluate the most decisive symptom of bleeding on probing (BoP+). The use of a curved scaled Nabers probe (originally intended for measuring furcation involvement) possibly facilitates appropriate access when an overcontour of the restoration is present (Figure 8.4). Due to the sudden circular spread of the inflammatory process around implants, the assessment of at least one site with adequate access is deemed to provide reliable diagnosis, whereas probing around the teeth requires a circular procedure as if using a sewing machine in order to ensure that localized infrabony defects are detected.

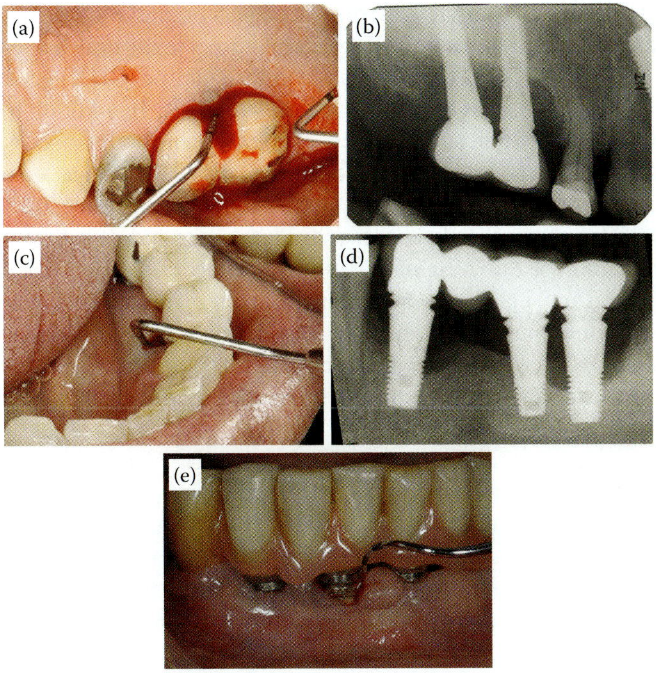

FIGURE 8.4 (a) Increased PPD (9 mm) around implant *regio* 16. (b) Corresponding radiograph *regio* 15-16. (c) When overbulks inhibit proper access, at least one site mainly on the lingual aspect should be used for axial probing. (d) Corresponding radiograph *regio* 34-36-37. (e) Use of a curved probe intended for the measurement of furcation involvement often facilitates adequate probing to diagnose implant health status.

8.6 TREATMENT OF PERI-IMPLANT DISEASES

The aims of treatment of peri-implant diseases are the resolution of the inflammation, the ideal regeneration of lost tissues with reosseointegration (i.e., new BIC), and the prevention of recurrence of the disease. Because adequate debridement is considered an important part in the treatment of peri-implant infections, the issue of removal of mineralized deposits must be further explored. Macroscopic or microscopic differences in the design of implants and/or implant components may require the application of different methods for mechanical debridement and detoxification of the surface.

8.6.1 Initial Nonsurgical Treatment for Infection Control

As soon as peri-implant diseases are diagnosed, the primary therapeutic goal is to control the infection by biofilm removal, plaque control, and antiseptic measures. Peri-implant mucositis requires already appropriate therapy to deal with the inflammation and to prevent it from developing into peri-implantitis. The initial nonsurgical treatment comprises mechanical instrumentation to remove bacterial biofilm and calculus, surface polishing, establishing adequate personal oral hygiene, and antiseptic measures with the local application of chlorhexidine (CHX) gel and/or mouth rinses (0.2% CHX gluconate solution; Lang et al. 1997). In addition to its antibacterial effects, CHX facilitates a reduction of *C. albicans* counts (see Section 8.4.1; Rohrer et al. 2010). According to a recent *in vitro* study, the application of antimicrobial agents only (0.125% CHX digluconate, 0.75% povidone-iodine, 0.1% octenidine, and 0.1%–0.25% polyhexamethylene biguanide) caused a delayed regrowth but did not eliminate adherent *Str. sanguinis* cells completely (Astasov-Frauenhoffer et al. 2011). Thus, these antiseptics may be used as adjunctive measures in order to improve mechanical biofilm control. Furthermore, depending on the restoration design and the patient's manual skills, the feasibility of personal hygiene may require additional adjustments, for example, by opening interproximal spaces and the removal of overhangs (Figure 8.3).

According to a recent review, nonsurgical mechanical cleaning is successful (with reduced numbers of sites with BoP+) in the treatment of peri-implant mucositis and can be supplemented by antimicrobial mouth rinses (Renvert et al. 2008b). These measures are, in most instances, not sufficient to resolve peri-implantitis, although the nonsurgical approach should be applied as initial therapy for a 3 to 4 week period to reduce the signs of acute inflammation (Renvert et al. 2008b, 2009). When increased PPD (≥6 mm) and BoP+ persists after initial therapy, surgical treatment is required in a staged approach to get access to the implant surface for adequate debridement and decontamination and to reduce increased PPD (Lindhe and Meyle 2008). Ideally, the reduction of PPD is accompanied by bone regeneration and the reestablishment of the BIC ("reosseointegration"). By reducing PPD, a more aerobic environment can be established, whereas the anaerobic species are suppressed or eliminated.

In periodontal defects, nonsurgical treatment with scaling and root planing has been shown to be successful also in sites with PPD of up to 7–9 mm, with better long-term results in single-rooted teeth than in molars with furcation involvement (Lindhe et al. 1982). In peri-implantitis sites, however, the limited access due to overcontoured restorations (discrepancy between implant diameter and crown contour), combined with narrow circular defect morphologies and retentive rough implant surfaces with undercuts in the screw regions, often inhibits proper cleaning during a nonsurgical approach, and a surgical intervention usually becomes inevitable in the second stage (see Section 8.6.2).

The application of antibiotics locally and/or systemically at the end of the initial treatment phase has been described in the literature (Lang et al. 2000; Mombelli and Lang 1992; Mombelli et al. 2001). With local applications, drug release should be sustained over several days, whereas systemic antibiotic intake has additional limitations due to the reduced vascular supply of the peri-implant tissues, possibly resulting in low local concentrations of the medication. The benefits of both local and systemic applications should be tested in comparative studies with and without the use of antibiotics (Klinge et al. 2002). To prevent any risk of antibiotic resistance, their current use should be restricted to high-risk patients and surgical interventions with prolonged bone exposure (see Section 8.6.2).

8.6.2 Surface Decontamination in Nonsurgical and Surgical Treatment

Decontamination of the exposed implant surface is a primary goal and can be performed by mechanical, chemical, and/or physical methods. Various measures for implant surface cleaning have been investigated in animal models (e.g., saline, pumice, chemical agents, ultrasonic, sandblasting, airpolishing, and laser), but none of these methods revealed clear benefits [for a review, see the work of Claffey et al. (2008)]. In particular, for HA-coated implant surfaces, the application of citric acid (40%, pH 1, for 30 s) was recommended for effective removal of lipopolysaccharides (Meffert 1996). This effect was related to the demineralization of the HA surface layer, while CHX gluconate and stannous fluoride bound to the HA surface, causing endotoxin remnants (Zablotsky et al. 1992). Plastic curettes (carbon) and ultrasonic tips made of polyetheretherketone have been introduced in order to avoid alterations of the titanium implant surface (e.g., coating removal, decreasing surface roughness, or scratching; Espedito Di Lauro et al. 2003; Speelman et al. 1992). Their efficiency in calculus removal, however, seems to be limited (Ramaglia et al. 2006). Hence, the use of more blunt than sharp steel curettes is still preferable and should be combined with ultrasonic and air-powder polishing (Figure 8.5a,b). The application of air powder has to be clearly directed at the exposed implant surface and should not be slipped in sites with increased PPD without flap opening to avoid embolism. Air powder can be used with sodium bicarbonate powder or with less abrasive resorbable amino acid glycine powder (Deppe et al. 2007; Schwarz et al. 2009). Laser decontamination (CO_2 or Er:YAG) has been used for implant surface cleaning and detoxification, but the clinical evidence to show, whether this improves treatment outcomes, is still lacking (Deppe et al. 2007; Renvert et al. 2008b; Schwarz et al. 2004, 2006). *In vitro* studies revealed that the viability of *Str. sanguinis* (an early colonizer) and *P. gingivalis* (frequently associated with peri-implantitis) adherent to titanium or zirconia surfaces was effectively reduced both with CO_2 laser (100 J/cm^2) and diode laser (150 J/cm^2), while planktonic cells of both species were more resistant to laser irradiations (Hauser-Gerspach et al. 2010a). Furthermore, gaseous ozone (OZOTOP unit, TIP TOP TIPS Sàrl. Rolle, Switzerland; 140 ppm; 2 L/min) treatment showed selective efficacy to reduce adherent bacteria on titanium and zirconia without affecting the adhesion and proliferation of osteoblastic cells. Although

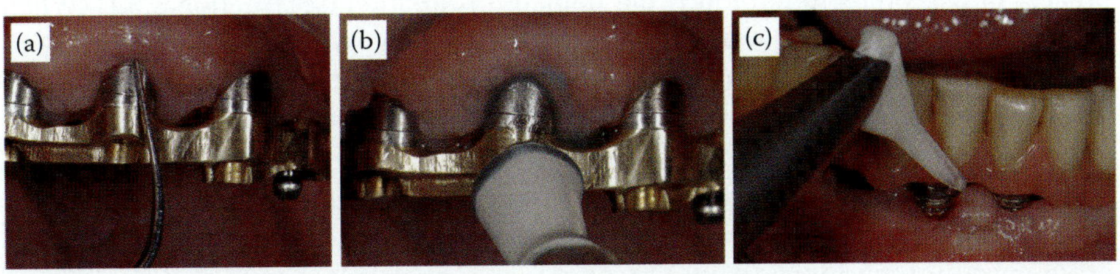

FIGURE 8.5 (a) Use of conventional ultrasonic tips. (b) Polishing of relatively soft titanium abutment and gold surface of the bar to eliminate any scratching. (c) Use of airpolishing (Perio flow with a flexible nozzle).

P. gingivalis was eliminated by ozone from all surfaces within 24 s to below the detection limit (≥99.94% reduction), *Str. sanguinis* was more resistant and showed the highest reduction on zirconia discs (>90% reduction). Titanium (polished) and zirconia (acid etched or polished) had a lower colonization potential than titanium with an SLA surface (Hauser-Gerspach et al. 2011).

Photosensitization in combination with soft laser has been introduced as a physical method for surface decontamination. In an experimental study applying this method at titanium platelets with different surface structures, anaerobic bacteria were found to be reduced, and smears did not reveal any bacterial growth in cultures (Haas et al. 1997). Photosensitization with a diode soft laser was used in a clinical study during surgery. Bacterial samples taken before and after laser application revealed a significant reduction of *P. gingivalis*, *Prevotella intermedia*, and *A. actinomycetemcomitans* on different implant surfaces (machined, plasma flame sprayed, etched, and HA coated), but not a complete elimination of these pathogens (Dörtbudak et al. 2001).

8.6.3 Surgical Peri-Implantitis Therapy

Depending on the disease severity and the esthetic demands, the surgical peri-implantitis therapy is conducted either as a resective or a regenerative intervention. Both measures can also be applied successively, particularly when advanced infections have to be eliminated in a first approach and any interference with augmentation materials should be avoided.

The surgical intervention should improve access for adequate removal of granulation tissue and surface cleaning. The inflammatory lesion in the CT should be excised similar to an internal gingivectomy procedure, while a band of keratinized mucosa has to be maintained. Sharp bone contours may need to be recontoured, particularly in narrow crevice-like defects, in order to ensure sufficient access for the surface cleaning. Reosseointegration with new BIC was observed following surgical peri-implantitis therapy using cotton pellets soaked in saline to clean the exposed titanium surface. It was observed that moderately rough surfaces (SLA) allowed more reosseointegration than experimentally smooth surfaces (Persson et al. 2001). These results indicate that reosseointegration requires not only adequate surface cleaning but also sufficient stability of the blood coagulum similar to the initial process of osseointegration following implant installation (see Section 8.3.1). When using xenogenic grafting material (Bio-Oss, Geistlich, Wolhusen, Switzerland) to fill the peri-implant defects, results were slightly improved around SLA versus TPS surfaces with less bleeding on probing and greater reduction of pocket probing depth after 1 year (Roccuzzo et al. 2011).

When the resective approach is indicated for pocket elimination, soft tissue management comprises either excision (but maintaining 3–4 mm of keratinized mucosa) or an apically repositioned flap. Osteotomy may be required to flatten sharp bone contours for the establishment of a positive bone architecture (Lang et al. 2000). Furthermore, flattening of contaminated rough implant surfaces has been described in the literature ("implantoplasty") using rotary instruments (diamonds, carbide drills, arkansas, and rubber polishers), and beneficial short-term clinical outcomes have been documented (Romeo et al. 2005). Reosseointegration at smoothened surfaces is, however, less likely, as aforementioned (Persson et al. 2001). As a consequence, the implantoplasty is only indicated in resective approaches and should be restricted to the area intended to be exposed following healing and maturation of the peri-implant tissues. Such surface modification should only be performed following surface decontamination in order to avoid a spread of contaminated particles and subsequent foreign body reactions.

In most clinical studies, a regenerative treatment approach is used with the addition of bone substitutes with or without barrier membranes (Claffey et al. 2008). It must be noted that this regenerative approach is primarily used to fill the bone defect but does not necessarily promote the resolution of the infection and may interfere with the healing process. Deppe et al. (2007) compared resective and regenerative therapies [50% tricalcium phosphate, 50% autogenous bone, and expanded polytetrafluoroethylene (ePTFE) membrane] and applied the latter only when restorations were retrievable and could be removed for 4 months of submerged healing. Although, clinically, a resolution was observed with both treatment modalities, the radiographic analysis indicated greater bone fill after augmentation procedures (Deppe et al. 2007). Haas et al. (2000) used autogenous bone in combination with nonresorbable ePTFE membranes following decontamination with toluidine blue and laser irradiation (Haas et al. 2000). Despite the submerged healing protocol with restorations removed, all membranes were prematurely exposed (after 10 days on the average) and occasionally required immediate membrane removal. With an initial defect depth of 5.5 mm, the radiographic analysis revealed a bone gain of 2 mm (±1.9) after an observation period of 9.5 months. In order to avoid detrimental effects from membrane exposures, Behneke et al. (2000) used autogenous corticocancellous bone blocks stabilized with miniscrews, while a fibrin gel was applied for stabilizing additional bone chips (Behneke et al. 2000). Hand curettes were used to remove granulation tissue, and an air powder abrasive instrument (Air-Flow SI, EMS, Nyon, Switzerland) with sodium carbonate solution was applied for 30 s, with the spray vector directed perpendicular to the implant surface. Sites were then rinsed with sterile physiologic saline. After transmucosal healing, the bone blocks showed a stable defect fill according to the radiographic analysis after periods of up to 3 years (Behneke et al. 2000).

The effect of enamel matrix derivates (e.g., Emdogain) applied on titanium surfaces was investigated in *in vitro* studies, and an increased osteoblast spread and proliferation was observed compared to uncoated titanium discs (Miron et al. 2009; Qu et al. 2010). Although clinical data are lacking until now, the use of these amelogenins has been suggested for the treatment of peri-implantitis (Qu et al. 2010).

Almost all human studies reporting about surgical peri-implantitis therapy used antibiotics, although the additional benefit has not yet been demonstrated (Claffey et al. 2008). Penicillin–amoxicillin in combination with clavulan acid (e.g., Augmentin) is the primary choice efficient also against streptococci and staphylococci. A prescription of amoxicillin (875 mg) combined with clavulan acid (125 mg) twice a day for 6 days was used in surgical peri-implantitis treatment in combination with xenogenic augmentation material, and healing without complications was observed in all patients (Roccuzzo et al. 2011). When the presence of *A. actinomycetemcomitans* is evident, the combination of amoxicillin and metronidazole (van Winkelhoff cocktail) has to be considered due to the synergistic effect between the two substances and/or their metabolites (Pavicic et al. 1994a,b; Walter et al. 2011).

Obviously, the removal of the implant restoration is, if feasible, beneficial to improve accessibility during the surgical procedure. Temporary restoration removal and submerged healing appears to be superior, particularly when regenerative measures with bone grafting materials are planned (Deppe et al. 2007). In this respect, short-span screw-retained implant restorations are preferable to cement-retained reconstructions (Zitzmann et al. 2008, 2010).

Because no particular therapeutic protocol described in the literature has turned out to be superior, meticulous biofilm destruction, calculus removal, surface decontamination, plaque control, and pocket reduction are still the major steps in the treatment of peri-implant diseases. According to the systematic treatment of periodontitis, the following strategy is recommended (Lang et al. 1997; Mombelli 1999; Mombelli and Lang 1998; Roos-Jansåker et al. 2003).

Initial phase (3 to 4 weeks):

1. Instruct oral hygiene in person and facilitate oral hygiene, also in the interproximal region, that is, establish adequate interproximal spaces and remove overcontouring.
2. Mechanically remove bacterial biofilm and accessible calculus, which acts as a plaque retentive factor (manual instrumentation or ultrasonic and polishing), topical antibacterial therapy with chlorhexidine gel, and mouth rinses (0.2%) for 3 to 4 weeks.

If signs of inflammation persist during reevaluation (PPD ≥6 mm with BoP+):

1. Resective approach—Perform surgical intervention with access flap, excise granulation tissue, use mechanical instrumentation to remove soft and hard debris around the implant surface, treat the surface antimicrobially (Betadine or chlorhexidine, potentially laser treatment), recontour bone margins, and perform apical flap repositioning or mucosal resection to reduce the pocket depth (warning: keep part of the keratinized mucosa and avoid complete removal); consider implantoplasty in the area proposed for exposure.
2. Regenerative approach—Consider using regenerative therapies mainly in a staged approach as an alternative to the resective procedure, particularly in the anterior region for esthetic reasons.

If oral hygiene following peri-implantitis therapy cannot be ensured in advance (e.g., due to long implant compartments or hollow cylinders exposed to the oral cavity), removal of the implant has to be considered (Figure 8.6). Patients have to be informed about the burden of staged augmentation and reimplantation, and alternative reconstructive treatment options should be considered instead.

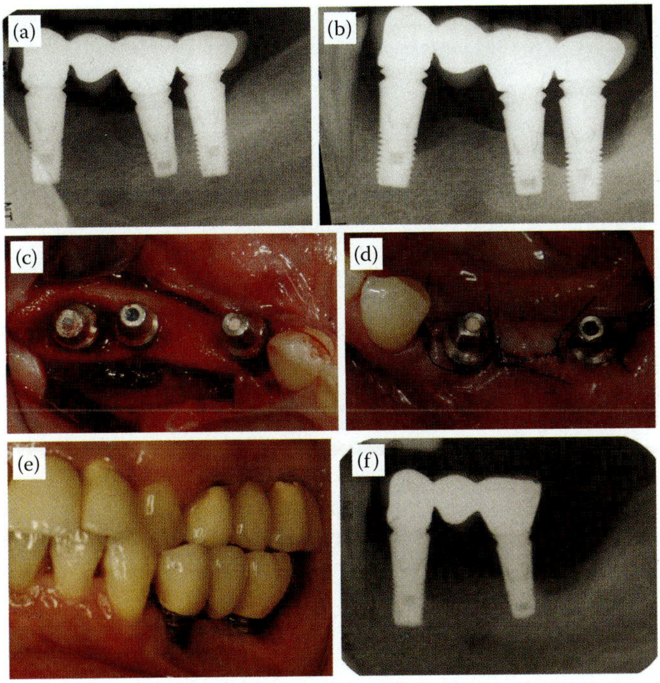

FIGURE 8.6 (a) Radiographic view of peri-implantitis sites 3 years after implant placement *alio loco* (the arrow indicates the misfit of restoration and cement remnants; the same patient as in Figure 8.4). (b) Radiograph 2 years later (the patient initially refused peri-implantitis treatment). (c) Surgical intervention after initial therapy. (d) Removal of posterior implant positioned too close to adjacent implant site and representing the largest bone defect. (e) Healing after 3 months. (f) Corresponding radiograph.

8.7 HEALING AFTER PERIODONTAL AND PERI-IMPLANTITIS THERAPY

In periodontal therapy, healing ideally comprises the regeneration of alveolar bone, a reestablished PDL, and the apposition of new cementum with inserting periodontal fibers. Healing can also be associated with a long JE, representing a new epithelial attachment. Both new epithelial and CT attachments reduce PPD. In periapical radiographs, a corticalization with a clear contrast of the remineralized marginal bone is frequently visible. In peri-implantitis therapy, the clinical signs of healing include reduced peri-implant pockets without BoP+, while radiographs similarly reveal a remineralization with clear marginal/crestal bone contours (Figure 8.7). Where bone substitutes have been used in a regenerative approach, the radiographic view does not reveal whether reosseointegration has occurred or whether CT has become interposed between the bone and the implant surface. In the soft tissue compartment, a new epithelial and CT attachment is established with sparse vascularization of the scar-like tissue. Due to the scarred character of the peri-implant soft tissues with

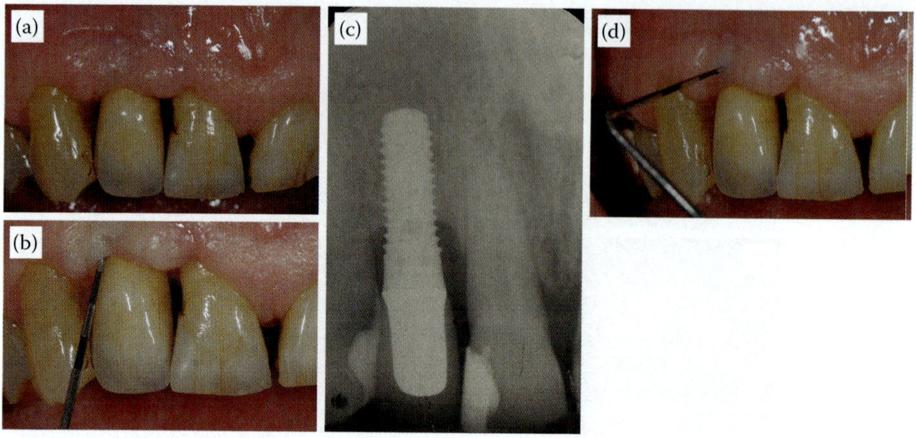

FIGURE 8.7 (a) and (b) two small openings as remains of former fistulae reside in scarred peri-implant tissues. (c) Clear radiographic mineralization of the bone margin. (d) Clinical and radiographic situation in a patient 10 months after successful surgical peri-implantitis therapy—PPD reduced to 2 mm circularly.

reduced vascular supply, the healing potential, for example, to close the opening of a former fistula (Figure 8.7), is very limited, and further surgical interventions to resolve these openings should be avoided.

Long-term results after surgical peri-implantitis therapy are sparse (Claffey et al. 2008). In a human study, surgical treatment with cleaning (hydrogen peroxide followed by saline) in combination with systemic antibiotics showed successful healing in 58% of the sites over a 5-year observation period. Implants (27%) had to be removed, and 44% of the patients required retreatment, with smoking being a risk factor for nonsuccessful outcomes (Leonhardt et al. 2003). Using resective or regenerative treatment methods, 18% of the implants were lost after varying observation periods (3 years on the average, ranging 5–59 months), with higher failure rates following regenerative therapy (25% versus 12%; Deppe et al. 2007).

8.8 CONCLUSIONS AND CONSEQUENCES FOR FUTURE SURFACE DEVELOPMENTS

Several morphological differences exist between periodontal and peri-implant tissues, such as the reduced vascularization and fewer fibroblasts, entailing a lower healing potential. As a result of bacterial pathogens within the biofilm, an inflammatory cell lesion is established in the CT and shows a sudden spread. This ICT may promote bone loss, which proceeds mainly in a circumferential way surrounding the implant and advances to a greater extent than those around the teeth.

The fiber anchorage with distinct fiber anchorage is exclusively reserved for natural teeth with a periodontium but is not feasible around dental implants, unless tissue engineering facilitates different surface characteristics and biological features.

A dense CT attachment within the transmucosal tissue portion would be beneficial also around implants and/or implant abutments. Furthermore, a sufficient epithelial attachment is required to ensure soft tissue integrity and a protective seal. Future surface developments of dental implants aim at a reduction of biofilm adherence in the supramucosal and transmucosal compartment. Along the enossal implant portion, the surface structure allows for adequate osseointegration following implant placement, as already established, and should ideally prevent the progression of peri-implant diseases in the long term. Implant surfaces should enable efficient decontamination if once exposed to the biofilm, followed by reosseointegration with new bone tissue in contact with the implant surface. In the future, tissue engineering and regenerative medicine may offer a broad scope of opportunities to improve the integrity of dental implants and to regenerate soft and hard tissues if peri-implant disease has occurred.

REFERENCES

Abrahamsson, I., T. Berglundh, P.-O. Glantz, and J. Lindhe. 1998. The mucosal attachment at different abutments: An experimental study in dogs. *J Clin Periodontol* 25 (9):721–7.

Abrahamsson, I., T. Berglundh, E. Linder, N.P. Lang, and J. Lindhe. 2004. Early bone formation adjacent to rough and turned endosseous implant surfaces: An experimental study in the dog. *Clin Oral Implants Res* 15 (4):381–92.

Abrahamsson, I., T. Berglundh, J. Wennstrom, and J. Lindhe. 1996. The peri-implant hard and soft tissues at different implant systems: A comparative study in the dog. *Clin Oral Implants Res* 7 (3):212–9.

Abrahamsson, I., N.U. Zitzmann, T. Berglundh, A. Wennerberg, and J. Lindhe. 2001. Bone and soft tissue integration to titanium implants with different surface topography: An experimental study in the dog. *Int J Oral Maxillofac Implants* 16 (3):323–32.

Adell, R., U. Lekholm, B. Rockler, and P.-I. Brånemark. 1981. A 15-year study of osseointegrated implants in the treatment of the edentulous jaw. *Int J Oral Surg* 10 (6):387–416.

Albouy, J.P., I. Abrahamsson, L.G. Persson, and T. Berglundh. 2008. Spontaneous progression of peri-implantitis at different types of implants: An experimental study in dogs—Part I: Clinical and radiographic observations. *Clin Oral Implants Res* 19 (10):997–1002.

Albouy, J.P., I. Abrahamsson, L.G. Persson, and T. Berglundh. 2009. Spontaneous progression of ligatured induced peri-implantitis at implants with different surface characteristics: An experimental study in dogs—Part II: Histological observations. *Clin Oral Implants Res* 20 (4):366–71.

Albrektsson, T. and F. Isidor. 1994. Consensus report of session IV. In *Proceedings of the First European Workshop on Periodontology*, edited by N.P. Lang and T. Karring. London: Quintessence.

Albrektsson, T. and A. Wennerberg. 2004. Oral implant surfaces—Part 1: Review focusing on topographic and chemical properties of different surfaces and in vivo responses to them. *Int J Prosthodont* 17 (5):536–43.

Astasov-Frauenhoffer, M., O. Braissant, I. Hauser-Gerspach, A.U. Daniels, D. Wirz, R. Weiger, and T. Waltimo. 2011. Quantification of vital adherent *Streptococcus sanguinis* cells on protein-coated titanium after disinfectant treatment. *J Mater Sci: Mater Med* 22 (9):2045–51.

Behneke, A., N. Behneke, and B. d'Hoedt. 2000. Treatment of peri-implantitis defects with autogenous bone grafts: Six-month to 3-year results of a prospective study in 17 patients. *Int J Oral Maxillofac Implants* 15 (1):125–38.

Berglundh, T., I. Abrahamsson, J.P. Albouy, and J. Lindhe. 2007a. Bone healing at implants with a fluoride-modified surface: An experimental study in dogs. *Clin Oral Implants Res* 18 (2):147–52.
Berglundh, T., I. Abrahamsson, N.P. Lang, and J. Lindhe. 2003. *De novo* alveolar bone formation adjacent to endosseous implants. *Clin Oral Implants Res* 14 (3):251–62.
Berglundh, T., I. Abrahamsson, M. Welander, N.P. Lang, and J. Lindhe. 2007b. Morphogenesis of the peri-implant mucosa: An experimental study in dogs. *Clin Oral Implants Res* 18 (1):1–8.
Berglundh, T., K. Gotfredsen, N.U. Zitzmann, N.P. Lang, and J. Lindhe. 2007c. Spontaneous progression of ligature induced peri-implantitis at implants with different surface roughness: An experimental study in dogs. *Clin Oral Implants Res* 18 (5):655–661.
Berglundh, T. and J. Lindhe. 1996. Dimension of the periimplant mucosa: Biological width revisited. *J Clin Periodontol* 23 (10):971–3.
Berglundh, T., J. Lindhe, I. Ericsson, C.P. Marinello, B. Liljenberg, and P. Thomsen. 1991. The soft tissue barrier at implants and teeth. *Clin Oral Implants Res* 2 (2):81–90.
Berglundh, T., J. Lindhe, K. Jonsson, and I. Ericsson. 1994. The topography of the vascular systems in the periodontal and peri-implant tissues in the dog. *J Clin Periodontol* 21 (3):189–93.
Berglundh, T., J. Lindhe, C. Marinello, I. Ericsson, and B. Liljenberg. 1992. Soft tissue reaction to *de novo* plaque formation on implants and teeth: An experimental study in the dog. *Clin Oral Implants Res* 3 (1):1–8.
Berglundh, T., N.U. Zitzmann, and M. Donati. 2010. Are peri-implantitis lesions different from periodontitis lesions? *J Clin Periodontol* 38 (Suppl 11):188–202.
Bosshardt, D.D. and H.E. Schroeder. 1996. Cementogenesis reviewed: A comparison between human premolars and rodent molars. *Anat Rec* 245 (2):267–92.
Brunette, D.M. 1988. The effects of implant surface topography on the behavior of cells. *Int J Oral Maxillofac Implants* 3 (4):231–46.
Burt, B. 2005. Position paper: Epidemiology of periodontal diseases. *J Periodontol* 76 (8):1406–19.
Buser, D., R.K. Schenk, S. Steinemann, J.P. Fiorellini, C.H. Fox, and H. Stich. 1991. Influence of surface characteristics on bone integration of titanium implants: A histomorphometric study in miniature pigs. *J Biomed Mater Res* 25 (7):889–902.
Buser, D., H. Stich, G. Krekeler, and A. Schroeder. 1989. Faserstrukturen der periimplantären Mukosa bei Titanimplantaten. *Z Zahnärztl Implantol* 5:15–23.
Claffey, N., E. Clarke, I. Polyzois, and S. Renvert. 2008. Surgical treatment of peri-implantitis. *J Clin Periodontol* 35 (Suppl 8):316–32.
De Boever, A.L., M. Quirynen, W. Coucke, G. Theuniers, and J.A. De Boever. 2009. Clinical and radiographic study of implant treatment outcome in periodontally susceptible and nonsusceptible patients: A prospective long-term study. *Clin Oral Implants Res* 20 (12):1341–50.
Deppe, H., H.H. Horch, and A. Neff. 2007. Conventional versus CO_2 laser-assisted treatment of peri-implant defects with the concomitant use of pure-phase beta-tricalcium phosphate: A 5-year clinical report. *Int J Oral Maxillofac Implants* 22 (1):79–86.
Dörtbudak, O., R. Haas, T. Bernhart, and G. Mailath-Pokorny. 2001. Lethal photosensitization for decontamination of implant surfaces in the treatment of peri-implantitis. *Clin Oral Implants Res* 12 (2):104–8.
Espedito Di Lauro, A., F. Morgese, A. Squillace, and L. Ramaglia. 2003. *In vitro* effects on rough implant surfaces of different instrumentations used in the surgical therapy of peri-implantitis. *Minerva Stomatol* 52 (1–2):1–7.
Etter, T.H., I. Hakanson, N.P. Lang, P.M. Trejo, and R.G. Caffesse. 2002. Healing after standardized clinical probing of the periimplant soft tissue seal: A histomorphometric study in dogs. *Clin Oral Implants Res* 13 (6):571–80.

Feneis, H. 1952. Anatomy and physiology of the normal gingiva. *Dtsch Zahnärztl Z* 7 (8):467–76.

Glauser, R., P. Schüpbach, J. Gottlow, and C.H. Hämmerle. 2005. Peri-implant soft tissue barrier at experimental one-piece mini-implants with different surface topography in humans: A light-microscopic overview and histometric analysis. *Clin Implant Dent Relat Res* 7 (Suppl 1):S44–51.

Gomez, R.S., W.O. Dutra, and P.R. Moreira. 2009. Epigenetics and periodontal disease: Future perspectives. *Inflamm Res* 58 (10):625–9.

Gould, T.R., L. Westbury, and D.M. Brunette. 1984. Ultrastructural study of the attachment of human gingiva to titanium *in vivo*. *J Prosthet Dent* 52 (3):418–20.

Haas, R., M. Baron, O. Dörtbudak, and G. Watzek. 2000. Lethal photosensitization, autogenous bone, and e-PTFE membrane for the treatment of peri-implantitis: Preliminary results. *Int J Oral Maxillofac Implants* 15 (3):374–82.

Haas, R., O. Dörtbudak, N. Mensdorff-Pouilly, and G. Mailath. 1997. Elimination of bacteria on different implant surfaces through photosensitization and soft laser: An *in vitro* study. *Clin Oral Implants Res* 8 (4):249–54.

Hauser-Gerspach, I., S. Stübinger, and J. Meyer. 2010a. Bactericidal effects of different laser systems on bacteria adhered to dental implant surfaces: An *in vitro* study comparing zirconia with titanium. *Clin Oral Implants Res* 21 (3):277–83.

Hauser-Gerspach, I., J. Vadaszan, I. Deronjic, C. Gass, J. Meyer, M. Dard, T. Waltimo, S. Stübinger, and C. Mauth. 2011. Influence of gaseous ozone in peri-implantitis: Bactericidal efficacy and cellular response—An *in vitro* study using titanium and zirconia. *Clin Oral Invest*. doi:10.1007/s00784-011-0603-2.

Heijl, L., B.R. Rifkin, and H.A. Zander. 1976. Conversion of chronic gingivitis to periodontitis in squirrel monkeys. *J Periodontol* 47 (12):710–6.

Heinrich, A., K. Dengler, T. Koerner, C. Haczek, H. Deppe, and B. Stritzker. 2008. Laser-modified titanium implants for improved cell adhesion. *Lasers Med Sci* 23 (1):55–8.

Heitz-Mayfield, L.J. 2008. Peri-implant diseases: Diagnosis and risk factors. *J Clin Periodontol* 35 (Suppl 8):292–304.

Hermann, J.S., D.L. Cochran, P.V. Nummikoski, and D. Buser. 1997. Crestal bone changes around titanium implants: A radiographic evaluation of unloaded nonsubmerged and submerged implants in the canine mandible. *J Periodontol* 68 (11):1117–30.

Ikeda, H., M. Shiraiwa, T. Yamaza, M. Yoshinari, M.A. Kido, Y. Ayukawa, T. Inoue, K. Koyano, and T. Tanaka. 2002. Difference in penetration of horseradish peroxidase tracer as a foreign substance into the peri-implant or junctional epithelium of rat gingivae. *Clin Oral Implants Res* 13 (3):243–51.

Ikeda, H., T. Yamaza, M. Yoshinari, Y. Ohsaki, Y. Ayukawa, M.A. Kido, T. Inoue, M. Shimono, K. Koyano, and T. Tanaka. 2000. Ultrastructural and immunoelectron microscopic studies of the peri-implant epithelium-implant (Ti–6Al–4V) interface of rat maxilla. *J Periodontol* 71 (6):961–73.

James, R.A. and R.L. Schultz. 1974. Hemidesmosomes and the adhesion of junctional epithelial cells to metal implants—A preliminary report. *Oral Implantol* 4 (3):294–302.

Kim, S., K.C. Oh, D.H. Han, S.J. Heo, I.C. Ryu, J.H. Kwon, and C.H. Han. 2010. Influence of transmucosal designs of three one-piece implant systems on early tissue responses: A histometric study in beagle dogs. *Int J Oral Maxillofac Implants* 25 (2):309–14.

Klinge, B., A. Gustafsson, and T. Berglundh. 2002. A systematic review of the effect of anti-infective therapy in the treatment of peri-implantitis. *J Clin Periodontol* 29 (Suppl 3):213–25.

Koka, S., M.E. Razzoog, T.J. Bloem, and S. Syed. 1993. Microbial colonization of dental implants in partially edentulous subjects. *J Prosthet Dent* 70 (2):141–4.

Lang, N.P. and T. Berglundh. 2011. Periimplant diseases: Where are we now?—Consensus of the Seventh European Workshop on Periodontology. *J Clin Periodontol* 38 (Suppl 11):178–81.

Lange, K., M. Herold, L. Scheideler, J. Geis-Gerstorfer, H.P. Wendel, and G. Gauglitz. 2004. Investigation of initial pellicle formation on modified TiO_2 surfaces by reflectometric interference spectroscopy (RIfS) in a model system. *Dent Mater* 20 (9):814–22.

Lang, N.P., U. Brägger, D. Walther, B. Beamer, and K.S. Kornman. 1993. Ligature-induced peri-implant infection in cynomolgus monkeys—Part I: Clinical and radiographic findings. *Clin Oral Implants Res* 4 (1):2–11.

Lang, N.P., A. Mombelli, M.S. Tonetti, U. Brägger, and C.H. Hämmerle. 1997. Clinical trials on therapies for peri-implant infections. *Ann Periodontol* 2 (1):343–56.

Lang, N.P., A.C. Wetzel, H. Stich, and R.G. Caffesse. 1994. Histologic probe penetration in healthy and inflamed peri-implant tissues. *Clin Oral Implants Res* 5 (4):191–201.

Lang, N.P., T.G. Wilson, and E.F. Corbet. 2000. Biological complications with dental implants: Their prevention, diagnosis and treatment. *Clin Oral Implants Res* 11 (Suppl 1):146–55.

Leonhardt, A., B. Adolfsson, U. Lekholm, M. Wikström, and G. Dahlén. 1993. A longitudinal microbiological study on osseointegrated titanium implants in partially edentulous patients. *Clin Oral Implants Res* 4 (3):113–20.

Leonhardt, A., T. Berglundh, I. Ericsson, and G. Dahlén. 1992. Putative periodontal pathogens on titanium implants and teeth in experimental gingivitis and periodontitis in beagle dogs. *Clin Oral Implants Res* 3 (3):112–9.

Leonhardt, A., G. Dahlén, and S. Renvert. 2003. Five-year clinical, microbiological, and radiological outcome following treatment of peri-implantitis in man. *J Periodontol* 74 (10):1415–22.

Lindhe, J. and I. Ericsson. 1978. Effect of ligature placement and dental plaque on periodontal tissue breakdown in the dog. *J Periodontol* 49 (7):343–50.

Lindhe, J. and J. Meyle. 2008. Peri-implant diseases: Consensus Report of the Sixth European Workshop on Periodontology. *J Clin Periodontol* 35 (Suppl 8):282–5.

Lindhe, J., E. Westfelt, S. Nyman, S.S. Socransky, L. Heijl, and G. Bratthall. 1982. Healing following surgical/nonsurgical treatment of periodontal disease: A clinical study. *J Clin Periodontol* 9 (2):115–28.

Listgarten, M.A., D. Buser, S.G. Steinemann, K. Donath, N.P. Lang, and H.P. Weber. 1992. Light and transmission electron microscopy of the intact interfaces between nonsubmerged titanium-coated epoxy resin implants and bone or gingiva. *J Dent Res* 71 (2):364–71.

Listgarten, M.A. and C.H. Lai. 1975. Ultrastructure of the intact interface between an endosseous epoxy resin dental implant and the host tissues. *J Biol Buccale* 3 (1):13–28.

Listgarten, M.A., N.P. Lang, H.E. Schroeder, and A. Schroeder. 1991. Periodontal tissues and their counterparts around endosseous implants. *Clin Oral Implants Res* 2 (3):1–19.

Marinello, C.P., T. Berglundh, I. Ericsson, B. Klinge, P.O. Glantz, and J. Lindhe. 1995. Resolution of ligature-induced peri-implantitis lesions in the dog. *J Clin Periodontol* 22 (6):475–9.

Meffert, R.M. 1996. Periodontitis vs. peri-implantitis: The same disease? The same treatment? *Crit Rev Oral Biol Med* 7 (3):278–91.

Meffert, R.M. 1997. Issues related to single-tooth implants. *J Am Dent Assoc* 128 (10):1383–90.

Miron, R.J., C.J. Oates, A. Molenberg, M. Dard, and D.W. Hamilton. 2009. The effect of enamel matrix proteins on the spreading, proliferation and differentiation of osteoblasts cultured on titanium surfaces. *Biomaterials* 31 (3):449–60.

Mombelli, A. 1999. Prevention and therapy of periimplant infections. In *Proceedings of the 3rd European Workshop on Periodontology*, edited by N.P. Lang, T. Karring, and J. Lindhe. Berlin: Quintessence.

Mombelli, A. and F. Decaillet. 2011. The characteristics of biofilms in peri-implant disease. *J Clin Periodontol* 38 (Suppl 11):203–13.

Mombelli, A., A. Feloutzis, U. Brägger, and N.P. Lang. 2001. Treatment of peri-implantitis by local delivery of tetracycline: Clinical, microbiological and radiological results. *Clin Oral Implants Res* 12 (4):287–94.

Mombelli, A. and N.P. Lang. 1992. Antimicrobial treatment of peri-implant infections. *Clin Oral Implants Res* 3 (4):162–8.

Mombelli, A. and N.P. Lang. 1998. The diagnosis and treatment of peri-implantitis. *Periodontology 2000* 17:63–76.

Nevins, M., M.L. Nevins, M. Camelo, J.L. Boyesen, and D.M. Kim. 2008. Human histologic evidence of a connective tissue attachment to a dental implant. *Int J Periodont Restor Dent* 28 (2):111–21.

Page, R.C., W.F. Ammons, L.R. Schectman, and L.A. Dillingham. 1974. Collagen fibre bundles of the normal marginal gingiva in the marmoset. *Arch Oral Biol* 19 (11):1039–43.

Pavicic, M.J., A.J. van Winkelhoff, N.H. Douque, R.W. Steures, and J. de Graaff. 1994a. Microbiological and clinical effects of metronidazole and amoxicillin in *Actinobacillus actinomycetemcomitans*-associated periodontitis. A 2-year evaluation. *J Clin Periodontol* 21 (2):107–12.

Pavicic, M.J., A.J. van Winkelhoff, Y.A. Pavicic-Temming, and J. de Graaff. 1994b. Amoxicillin causes an enhanced uptake of metronidazole in *Actinobacillus actinomycetemcomitans*: A mechanism of synergy. *J Antimicrob Chemother* 34 (6):1047–50.

Persson, G.R., E. Samuelsson, C. Lindahl, and S. Renvert. 2010. Mechanical nonsurgical treatment of peri-implantitis: A single-blinded randomized longitudinal clinical study—Part II: Microbiological results. *J Clin Periodontol* 37 (6):563–73.

Persson, L.G., T. Berglundh, J. Lindhe, and L. Sennerby. 2001. Reosseointegration after treatment of peri-implantitis at different implant surfaces: An experimental study in the dog. *Clin Oral Implants Res* 12 (6):595–603.

Pongnarisorn, N.J., E. Gemmell, A.E. Tan, P.J. Henry, R.I. Marshall, and G.J. Seymour. 2007. Inflammation associated with implants with different surface types. *Clin Oral Implants Res* 18 (1):114–25.

Pontoriero, R., M.P. Tonetti, G. Carnevale, A. Mombelli, S.R. Nyman, and N.P. Lang. 1994. Experimentally induced peri-implant mucositis: A clinical study in humans. *Clin Oral Implants Res* 5 (4):254–9.

Qu, Z., O. Andrukhov, M. Laky, C. Ulm, M. Matejka, M. Dard, and X. Rausch-Fan. 2010. Effect of enamel matrix derivative on proliferation and differentiation of osteoblast cells grown on the titanium implant surface. *Oral Surg, Oral Med, Oral Pathol, Oral Radiol, Endod* 111 (4):517–22.

Quirynen, M., M. Abarca, N. Van Assche, M. Nevins, and D. van Steenberghe. 2007. Impact of supportive periodontal therapy and implant surface roughness on implant outcome in patients with a history of periodontitis. *J Clin Periodontol* 34 (9):805–15.

Ramaglia, L., A.E. di Lauro, F. Morgese, and A. Squillace. 2006. Profilometric and standard error of the mean analysis of rough implant surfaces treated with different instrumentations. *Implant Dent* 15 (1):77–82.

Renvert, S., C. Lindahl, H. Renvert, and G.R. Persson. 2008a. Clinical and microbiological analysis of subjects treated with Brånemark or AstraTech implants: A 7-year follow-up study. *Clin Oral Implants Res* 19 (4):342–7.

Renvert, S., I. Polyzois, and N. Claffey. 2011. How do implant surface characteristics influence peri-implant disease? *J Clin Periodontol* 38 (Suppl 11):214–22.

Renvert, S., A.-M. Roos-Jansåker, and N. Claffey. 2008b. Non-surgical treatment of peri-implant mucositis and peri-implantitis: A literature review. *J Clin Periodontol* 35 (Suppl 8):305–15.

Renvert, S., E. Samuelsson, C. Lindahl, and G.R. Persson. 2009. Mechanical nonsurgical treatment of peri-implantitis: A double-blind randomized longitudinal clinical study—Part I: Clinical results. *J Clin Periodontol* 36 (7):604–9.

Rimondini, L., S. Fare, E. Brambilla, A. Felloni, C. Consonni, F. Brossa, and A. Carrassi. 1997. The effect of surface roughness on early *in vivo* plaque colonization on titanium. *J Periodontol* 68 (6):556–62.

Roccuzzo, M., F. Bonino, L. Bonino, and P. Dalmasso. 2011. Surgical therapy of peri-implantitis lesions by means of a bovine-derived xenograft: Comparative results of a prospective study on two different implant surfaces. *J Clin Periodontol* 38 (8):738–45.

Rohrer N, A.F. Widmer, T. Waltimo, E.M. Kulik, R. Weiger, E. Filipuzzi-Jenny, and C. Walter. 2010. Antimicrobial efficacy of 3 oral antiseptics containing octenidine, polyhexamethylene biguanide, or Citroxx: Can chlorhexidine be replaced? *Infect Control Hosp Epidemiol* 31 (7):733–9.

Romeo, E., M. Ghisolfi, N. Murgolo, M. Chiapasco, D. Lops, and G. Vogel. 2005. Therapy of peri-implantitis with resective surgery: A 3-year clinical trial on rough screw-shaped oral implants—Part I: Clinical outcome. *Clin Oral Implants Res* 16 (1):9–18.

Roos-Jansåker, A.M., S. Renvert, and J. Egelberg. 2003. Treatment of peri-implant infections: A literature review. *J Clin Periodontol* 30 (6):467–85.

Rosenberg, E.S., H.D. Dent, S.C. Cho, N. Elian, Z.N. Jalbout, S. Froum, and C.I. Evian. 2004. A comparison of characteristics of implant failure and survival in periodontally compromised and periodontally healthy patients: A clinical report. *Int J Oral Maxillofac Implants* 19 (6):873–9.

Ruggeri, A., M. Franchi, N. Marini, P. Trisi, and A. Piatelli. 1992. Supracrestal circular collagen fiber network around osseointegrated nonsubmerged titanium implants. *Clin Oral Implants Res* 3 (4):169–75.

Sbordone, L., A. Barone, R.N. Ciaglia, L. Ramaglia, and V.J. Iacono. 1999. Longitudinal study of dental implants in a periodontally compromised population. *J Periodontol* 70 (11):1322–9.

Schroeder, A., E. van der Zypen, H. Stich, and F. Sutter. 1981. The reactions of bone, connective tissue, and epithelium to endosteal implants with titanium-sprayed surfaces. *J Maxillofac Surg* 9 (1):15–25.

Schüpbach, P. and R. Glauser. 2007. The defense architecture of the human periimplant mucosa: A histological study. *J Prosthet Dent* 97 (Suppl 6):S15–25.

Schwarz, F., K. Bieling, A. Sculean, M. Herten, and J. Becker. 2004. Treatment of periimplantitis with laser or ultrasound: A review of the literature. *Schweiz Monatsschr Zahnmed* 114 (12):1228–35.

Schwarz, F., D. Ferrari, K. Popovski, B. Hartig, and J. Becker. 2009. Influence of different air-abrasive powders on cell viability at biologically contaminated titanium dental implants surfaces. *J Biomed Mater Res, Part B* 88 (1):83–91.

Schwarz, F., E. Nuesry, K. Bieling, M. Herten, and J. Becker. 2006. Influence of an erbium, chromium-doped yttrium, scandium, gallium, and garnet (Er,Cr:YSGG) laser on the reestablishment of the biocompatibility of contaminated titanium implant surfaces. *J Periodontol* 77 (11):1820–7.

Schwarz, F.P., I. Hauser-Gerspach, T. Waltimo, and B. Stritzker. 2011. Antibacterial properties of silver containing diamond like carbon coatings produced by ion induced polymer densification. *Surf Coat Technol* 205:4850–4.

Speelman, J.A., B. Collaert, and B. Klinge. 1992. Evaluation of different methods to clean titanium abutments: A scanning electron microscopic study. *Clin Oral Implants Res* 3 (3):120–7.

Strietzel, F.P., P.A. Reichart, A. Kale, M. Kulkarni, B. Wegner, and I. Kuchler. 2007. Smoking interferes with the prognosis of dental implant treatment: A systematic review and meta-analysis. *J Clin Periodontol* 34 (6):523–44.

Sumida, S., K. Ishihara, M. Kishi, and K. Okuda. 2002. Transmission of periodontal disease-associated bacteria from teeth to osseointegrated implant regions. *Int J Oral Maxillofac Implants* 17 (5):696–702.

van Winkelhoff, A.J., R.J. Goene, C. Benschop, and T. Folmer. 2000. Early colonization of dental implants by putative periodontal pathogens in partially edentulous patients. *Clin Oral Implants Res* 11 (6):511–20.

Vig Slenters, T., I. Hauser-Gerspach, A.U. Daniels, and K.M. Fromm. 2008. Silver coordination compounds as light-stable, nanostructured and antibacterial coatings for dental implant and restorative materials. *J Mater Chem* 18:5359–5362.

Walter, C., E.M. Kulik, R. Weiger, N.U. Zitzmann, and T. Waltimo. 2011. Additive or synergistic antimicrobial effects of amoxicillin and metronidazole on whole plaque samples: A preliminary report. *J Int Acad Periodontol* 13 (1):11–6.

Welch, K., Y. Cai, H. Engqvist, and M. Strømme. 2010. Dental adhesives with bioactive and on-demand bactericidal properties. *Dent Mater* 26 (5):491–9.

Zablotsky, M., R. Meffert, O. Mills, A. Burgess, and D. Lancaster. 1992. The macroscopic, microscopic and spectrometric effects of various chemotherapeutic agents on the plasma-sprayed hydroxyapatite-coated implant surface. *Clin Oral Implants Res* 3 (4):189–98.

Zitzmann, N.U., I. Abrahamsson, T. Berglundh, and J. Lindhe. 2002. Soft tissue reactions to plaque formation at implant abutments with different surface topography: An experimental study in dogs. *J Clin Periodontol* 29 (5):456–61.

Zitzmann, N.U. and T. Berglundh. 2008. Definition and prevalence of peri-implant diseases. *J Clin Periodontol* 35 (Suppl 8):286–91.

Zitzmann, N.U., T. Berglundh, I. Ericsson, and J. Lindhe. 2004. Spontaneous progression of experimentally induced peri-implantitis. *J Clin Periodontol* 31 (10):845–9.

Zitzmann, N.U., T. Berglundh, C.P. Marinello, and J. Lindhe. 2001. Experimental peri-implant mucositis in man. *J Clin Periodontol* 28 (6):517–23.

Zitzmann, N.U., G. Krastl, H. Hecker, C. Walter, T. Waltimo, and R. Weiger. 2010. Strategic considerations in treatment planning: Deciding when to treat, extract or replace a questionable tooth. *J Prosthet Dent* 104 (2):80–91.

Zitzmann, N.U., M.D. Margolin, A. Filippi, R. Weiger, and G. Krastl. 2008. Patient assessment and diagnosis in implant treatment. *Aust Dent J* 53 (Suppl 1):S3–10.

9 Soft Tissue Attachment on Implant Surface

Timo O. Närhi

CONTENTS

9.1 BACKGROUND

Medical devices that penetrate the skin or mucosa are frequently used in patient care. Some of the devices, such as intravascular or peritoneal dialysis catheters and glucose sensors, on one hand, penetrate the skin without any attachment to the surrounding tissues. Bone-anchored percutaneous devices, on the other hand, are stabilized through the osseointegrated part of an implant. External fixation pins and bone-anchored hearing aids are both commonly used implants with skin-penetrating components. Oral implants possess a similar situation, but the part of the implant that penetrates periodontal tissues is exposed to an aqueous oral environment.

In a subcutaneous environment, a connective tissue capsule normally forms around a bioinert implant within a few days. The thickness of the capsule depends on the implant material and the stage of healing (Andersson 1996). In a percutaneous environment, epithelial lining partly separates the implanted devices that penetrate the skin or oral mucosa from the components of skin and connective tissue (Winter 1974). A firm bond between an implant surface and soft tissue components would ensure a stable tissue seal and improve the treatment results.

The increasing popularity of oral implant treatments also increases the demands for a good esthetic treatment outcome (Albrektsson et al. 2003). One of the most important factors hindering the long-term treatment results is the retraction of gingival tissues, which expose the titanium surface to the oral cavity. A firm bond between the implant surface and gingival tissues with the presence of sufficient supporting bone structure is a prerequisite for maintaining sound gingival anatomy. No methods have been developed to ensure firm gingival attachment on oral implant surfaces.

This chapter focuses on the attachment of soft tissues on bone-anchored medical devices, although main emphasis is put on oral implants.

9.2 SOFT TISSUES AROUND ORAL IMPLANTS AND PERCUTANEOUS DEVICES

The epithelial surface of the skin is covered by a keratinized stratified epithelium. Hard palate and attached gingiva in the oral cavity are also covered with keratinized epithelium, while nonkeratinized epithelium covers the buccal mucosa and the floor of the mouth. The epithelium that attaches on the tooth surface is called junctional epithelium. Junctional epithelium has a unique structure compared to other stratified epithelia in the human body. The skin and oral epithelia are key factors for the establishment of firm soft tissue attachment on implant surfaces.

9.2.1 Structure of Peri-Implant Gingival Tissues

Soft tissues around oral implants have many similar features with the gingival tissues around healthy natural teeth. Peri-implant soft tissues comprise marginal gingiva, which is covered by keratinized epithelium in the intraoral area. Attachment to the implant surface begins with peri-implant sulcus, which is covered by the sulcular epithelium. The sulcular epithelium possesses similar morphological and functional characteristics as the gingival sulcus around natural teeth (Mackenzie and Tonetti 1995; Liljenberg et al. 1996).

Beneath is a zone of junctional epithelium. The junctional epithelium of a tooth has a unique ability to attach directly on the tooth enamel. On the tooth surface, the junctional epithelium stops at the cemento–enamel junction, which is approximately 2 mm long. Apically to the junctional epithelium, there is a direct connective tissue attachment raised from collagen fibers, which prevents further epithelial migration. The junctional epithelium and connective tissue contact can also be detected on oral implants (Figure 9.1). The mechanism for the termination of the epithelial migration along the surface of oral implants is yet unknown. It has been suggested that there is an interaction between the titanium implant and connective tissue (Berglundh et al. 1991). There are several electron microscopy studies detailing the ultrastructural tissue response and adaptation of the soft tissues to transgingival or percutaneous

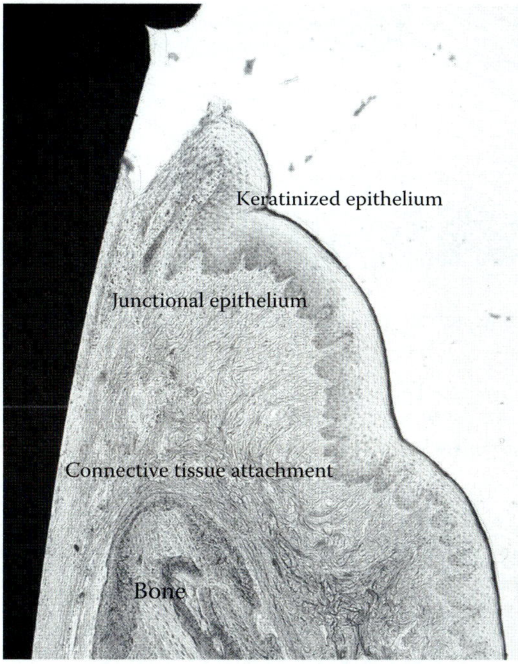

FIGURE 9.1 Keratinized epithelium, junctional epithelium, and connective tissue adjacent to the oral implant surface.

implants. In these studies, both *in vitro* and *in vivo* ultrastructural demonstrations of epithelial cell adhesion organelles against titanium alloy, that is, hemidesmosomes and a basal lamina between junctional epithelium and implant, have been clearly presented (McKinney et al. 1985; Steflik et al. 1991a,b; Bjursteen et al. 1995; Jansen et al. 1985, 1990; Chehroudi et al. 1991; Shimono et al. 2003; Holgers et al. 1995; Kasemo and Lausmaa 1988).

The biggest structural differences between a natural tooth and an implant start below the junctional epithelium. In a natural tooth connective tissue, attachment is achieved by collagen fibers, which run perpendicular to the tooth surface and attach directly on the root cement. This cement layer is missing from oral implants and implant abutments. Periodontal ligament, which mediate tooth attachment to the bone, does not exist on oral implants. Periodontal ligament provides vascular supply to the marginal gingiva. The lack of this vascular supply may weaken the defense system of peri-implant soft tissues compared with a natural tooth.

9.2.2 Structure of the Skin Around Percutaneous Devices

Anatomically, the normal skin structure, which is composed of the epidermis and dermis, can be detected around percutaneous implant devices. The placement of percutaneous implant initiates the wound healing process. Uneventful healing seals the injured area by firm skin adhesion. Epidermal cells attach first on implant surface

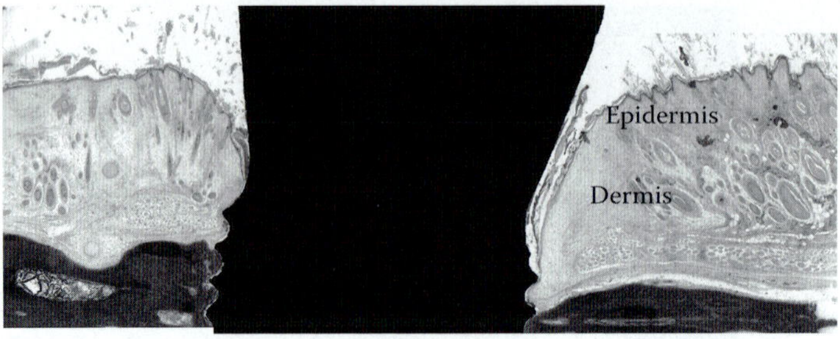

FIGURE 9.2 Left: Uneventful healing of skin. Right: Migration of epidermis over the dermis (marsupialization).

and allow healing of dermal contact. Mature dermal tissue prevents keratinocytes from migrating along the implant surface. Although immediate contact with percutaneous implants has been described, no real biological fixation occurs between an implant and skin. In disturbed healing, keratinocytes migrate on the surface of the dermis, causing marsupialization to the tissue penetrating part of an implant (Figure 9.2; von Recum 1984).

9.3 WOUND HEALING AND ESTABLISHMENT OF SOFT TISSUE ATTACHMENT

The formation of peri-implant structures begins at the state of surgery. The healing process is similar to normal wound healing, in which the initial inflammatory response is activated by injury. Acute inflammation is followed by the chronic inflammation stage and, eventually, the formation of granulation tissue. In normal wound healing, two injured tissue surfaces integrate with each other. In case of implanted devices, the healing process occurs between the injured tissue and the implant surface.

9.3.1 Blood Coagulation on the Implant Surface

Blood components are involved in the first phase of inflammatory reaction. Blood coagulation is a humoral response, and it can be influenced by many mechanisms that affect intrinsic or extrinsic pathways of coagulation system. In healthy conditions, blood begins to coagulate almost instantly after injury. Coagulation is followed by the formation of a fibrin clot, which fills the niche between soft tissues and implant surface. Therefore, the hematocompatibility of an implant surface plays an important role in blood interactions. On one hand, it intrinsically affects clotting through several surface-mediated reactions. Extrinsic pathway, on the other hand, becomes activated by factors released from surrounding tissues. These pathways merge before a common pathway in which a fibrinogen monomer turns into a stable fibrin polymer. Protein adsorption onto the implant surface activates platelets to adhere, aggregate, and finally form a platelet thrombus, which is stabilized by the fibrin (Hanson and Harker 1996).

9.3.2 Acute Inflammation

Acute inflammation is characterized by the exudation and emigration of leukocytes (predominantly neutrophils) from blood vessels to the site of injury. Inflammatory agents regulate the extent of leukocyte emigration. Neutrophils become activated after entering the site of injury, where, together with macrophages, they start to phagocytize microorganisms. In a normal wound healing process, neutrophils phagocytize foreign materials, but in case of large implants, this does not happen. After implantation, the material surface will be coated by serum proteins, for which neutrophils and macrophages can attach and start to release growth factors (Andersson 1996).

9.3.3 Chronic Inflammation

Neovascularization begins during chronic inflammation, which is normally of short duration. The presence of macrophages, monocytes, and lymphocytes is typical for this stage of healing. Several growth factors are released from activated macrophages during the chronic inflammation. Platelet-derived growth factors, fibroblast growth factors, transforming growth factors β and α, interleukin-1 (IL1), tumor necrosis growth factors, and vascular endothelial growth factor are all regulating the proliferation of fibroblasts and regeneration of blood vessels and epithelium (Singer and Clark 1999).

9.3.4 Formation of the Granulation Tissue

Upon activation, fibroblasts and endothelial cells begin to form a granulation tissue. It starts to appear around day 3 after implantation. Fibroblasts proliferate in the granulation tissue and begin to produce new extracellular matrix to support cell ingrowth. The structural molecules of the extracellular matrix, such as fibronectin and fibrin, assist in the cell migration. Newly formed blood vessels transfer oxygen to maintain cell metabolism. The newly formed extracellular matrix is gradually replaced with a collagenous matrix. Collagen amount increases during the maturation of the granulation tissue, after which it will turn from a fibroblast-rich granulation tissue to a more or less acellular tissue (Singer and Clark 1999).

9.3.5 Formation of Epithelial Attachment

The epithelium starts to migrate from the wound edge over the granulation tissue and along the implant surface. During the migration, the originally keratinized epithelium undergoes several structural changes before it attaches to the implant surface. At the same time, connective tissue fibroblasts form attachment to the implant surface just above the bone level.

The implant surface structure and composition may either promote or disturb the morphological and functional modifications that the cells undertake. A smooth titanium surface is known to give good support for the epithelial cell adhesion (Misiek et al. 1984; Räisänen et al. 2000). Epithelial cells appear to attach directly on the titanium surface via basal lamina, and they have been reported to be capable of forming hemidesmosomes comparable with the adhesion on tooth surfaces (Rompen et al.

2006). Hemidesmosomes are very small structures on the inner basal surface of epithelial cells. They are comparable with desmosomes, which link two cells together, while hemidesmosomes attach one cell to the extracellular matrix. Hemidesmosomes use integrin receptors on cell surfaces, which transmit information between cells and surrounding environment. Integrins have an important role in the regulation of cell shape, motility, and cell cycle. They are needed in several steps of wound healing. They bind cell surface and extracellular matrix components through ligands such as fibronectin, collagen, and laminin. Laminin 332, for example, is needed for hemidesmosome formation, and it can be detected from the epithelial cell basement membranes attaching to neighboring cells or the tooth surface (Figure 9.3). Therefore, the adhesion of ligands on the implant surface would be crucial for the formation of firm tissue attachment. However, according to the recent literature, direct attachment between the epithelium and the implant surface is questionable (Larjava et al. 2011).

It is not fully understood what dictates the length of epithelial migration on the oral implant surface, but it seems that mature connective tissue prevents epithelial migration more efficiently than the granulation tissue. The rapidity of connective tissue maturation may play a significant role in the regulation of epithelial migration. Thus, the adhesion and stability of the fibrin cloth to the implant surface is critical at the initial stage of the healing process (Lowenguth et al. 1993). It has also been speculated that interaction between soft tissues and the implant surface may have an effect on epithelial downgrowth (Berglundh et al. 1991).

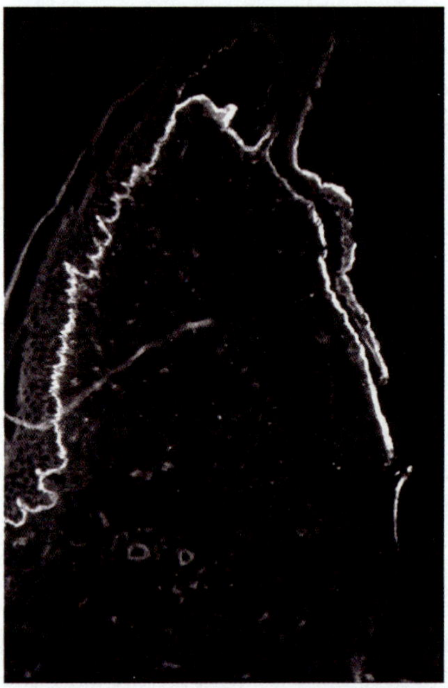

FIGURE 9.3 Immunohistological staining of laminin 332 indicating the interfaces between the epithelium and connective tissue, and the epithelium and tooth surface. (Courtesy of Dr. Jaana Willberg, Institute of Dentistry, University of Turku, Finland.)

9.3.6 Formation of Connective Tissue Attachment

As discussed above, soft tissue adhesion starts upon implant placement or abutment connection similar to normal wound healing. The implant surface should facilitate protein adsorption from an extracellular matrix that mediates fibroblast attachment on the surface. After maturation, the connective tissue area is poor from cells and resembles more or less a scar tissue, with the abundance of collagen fibers and reduced vascularization. Two zones of connective tissues have been described. On one hand, the fiber-rich inner zone is in direct contact with the implant surface and is postulated to function as a protective barrier for the peri-implant bone. Connective tissue components are separated from the implant surface by a 20-nm-thick proteoglycan film. No direct attachment between fibroblasts and the abutment surface has been found *in vivo* so far. The outer zone, on the other hand, contains more cells and randomly oriented collagen fibers and has more blood vessels (Rompen et al. 2006).

Human (Liljenberg et al. 1996) and experimental animal studies (Berglundh et al. 1991; Gotfredsen et al. 1991) have shown that collagen fibers tend to run parallel to the implant surface. However, some earlier studies have also reported about perpendicular fiber orientation toward the implant surface (Schroeder et al. 1981; Deporter et al. 1988). The quality and quantity of soft tissues seem to play a major role in the magnitude and orientation of collagen fibers. As stated earlier, the lack of cement prevents collagen fibers from attaching directly into the abutment or implant surface. Therefore, direct fiber attachment does not exist between connective tissues and implant surfaces, but the attachment can be more or less mechanical in nature (Ericsson and Lindhe 1993; Gray et al. 2005).

9.4 INFLUENCE OF BIOMATERIAL SURFACE PROPERTIES ON TISSUE ATTACHMENT

Biomaterial surface quality can be measured by a combination of physical, chemical, and mechanical properties and its surface structure (Smith et al. 1991; Wennerberg et al. 1995). It is generally accepted that early surface events that occur rapidly upon the implantation of a biomaterial into biological fluids determine the subsequent responses. These involve wetting by physiological liquids, followed by the adsorption of proteins and cells to the biomaterial surface (Davies 1996, 1998).

Cellular behaviors such as adhesion, morphological change, functional alteration, and proliferation are greatly affected by surface properties such as hydrophilicity, roughness, charge, free energy, and morphology (Oshida et al. 1994; Cho et al. 1996; Lampin et al. 1997).

9.4.1 Surface Free Energy

The measurement of the wettability of a material is a predictive index of cytocompatibility (Mekayarajjananonth and Winkler 1999). Surface free energy (SFE) is important for wettability (Baier et al. 1968). Surface energy is the energy at the surface of a solid, which is greater than its interior energy. The outer atoms are not equally attracted to each other as in the inner layer of atoms. The energy is greatest on the outermost atomic layer, because the unsaturated bonds generate surface energies (Anusavice 1996). SFE can be determined by measuring the contact angles formed by a range of liquids on a given surface using several diverse approaches (Combe at al. 2004). When a liquid is placed on a lower energy surface, the contact angle will be higher compared with a higher energy surface (Baier et al. 1968).

It is well known that the surface chemistry, surface energy, and surface topography govern the biological response to an implant material (Smith et al. 1991). Hallab et al. (2001) demonstrated that SFE was a more important surface characteristic than surface roughness for cellular adhesion strength and proliferation and that the SFE components of various tested materials were shown to be related to cellular adhesion strength. It has been reported that, despite the great number of parameters interfering with cellular adhesion and spreading, the solid SFE apparently is a dominant factor in cellular attachment and remains so, even if the solid surface has been covered by a protein layer (Schakenraad et al. 1988).

9.4.2 Surface Texture

The most commonly used dental implant roughness parameter is R_a. It has been recognized that, in order to achieve a stable soft tissue seal around titanium implants, a certain threshold roughness (around R_a of 0.2 μm) is necessary. An increase of the surface roughness of the transmucosal area above 0.2 μm will facilitate early plaque formation (Teughels et al. 2006). Therefore, an ideal transmucosal implant surface should be capable of simultaneously minimizing bacterial adhesion and allow epithelial and connective tissue attachment.

The contact guidance observed as the cells' preference to grow along grooves is a well-known phenomenon. Microgrooves have been shown to guide cell alignment and migration. Actin microfilaments in the lamellipodia and the interplay between the lamellipodium and extracellular matrix molecules have been suggested to be the determining factors in the establishment of contact guidance (Britch and Allen 1980).

Although microgrooves can enhance connective tissue attachment on implant surfaces, they are not commonly used in percutaneous devices. Rough surfaces possess increased risk for microbial adhesion, which is the main reason that smooth surfaces have been preferred in soft tissue penetrating parts of medical devices. However, the development of antimicrobial surfaces may reveal surface texturing for optimized cell alignment.

9.4.3 Surface Chemistry

Considering the effect of implant surface composition, the application of calcium phosphate (CaP) coatings has been shown to benefit the bone response during the initial healing period after implant insertion (Narayanan et al. 2008). Various animal studies have demonstrated that CaP coatings or implant surface treatments that promote CaP deposition increase the bone–implant contact compared to noncoated implants (Vercaigne et al. 2000; Mohammadi et al. 2004). Hydroxyapatite and CaP coatings are already used in many bone-anchored devices. The incorporation of fluoride on implant surfaces has also been reported to improve bone response (Ellingsen et al. 2004). Besides, for improved bone healing, the presence of calcium is beneficial for the wound healing process in general, because it is needed in many steps during the coagulation cascade. Relatively little efforts have been made to chemically modify implant surfaces for direct connective tissue or epithelium integration. Fibronectin coatings have been used to improve fibroblast attachment on medical implants. However, the coatings are subject to desorption from serum protein competition *in vivo*, and biological activities of fibronectin coatings cannot be confirmed. Silanization of implant surfaces has been suggested to solve the problem (Olbrich et al. 1996), but the activity of the coatings after storage and sterilization cannot be verified.

Similar to surface roughening, chemical modifications may create surfaces that not only improve cell adhesion but also increase microbial attachment and subsequent biofilm formation.

9.4.4 Bioactive Surfaces

According to Williams (1999), biomaterial is, on one hand, a material that is designed to elicit or modulate biological activity. In the engineering of medical devices, this usually refers to direct (biochemical) bone bonding and/or osteoinduction *in vivo*. Bioactive material, on the other hand, is intended to interface with biological systems to evaluate, treat, augment, or replace any tissue, organ, or function of the body. In addition, the term is often used in reference to bonelike mineral formation in simulated body fluid. Several methods of fabricating bioactive surfaces on medical implants are available. Recently sol–gel-derived bioactive titania (TiO_2) coatings have been found to enhance fibroblast adhesion and connective tissue attachment on implant surfaces (Jokinen et al. 1998). As stated earlier, cells recognize and attach to biomacromolecules that adsorb nonspecifically onto synthetic surfaces via integrin receptors. It has been shown that titania coatings are able to selectively adsorb certain proteins, which can mediate fibroblast attachment (Areva et al. 2004; Paldan et al. 2008) This soft-tissue bonding property could be unique; to date, no commercial materials that can provide implant integration in the soft tissue exist. The selective protein adsorption property of titania coatings is still a poorly understood phenomenon. The ability to tailor the surface nanoporosity for selective protein adsorption in various tissue environments could lead to significant breakthroughs in the bioengineering of medical implants. The smooth surface of coating with a high-energy negative surface charge and facilitated adsorption of proteins on sol–gel-derived titanium coatings (Peltola et al. 1998, 2000; Rossi et al. 2007) are probably the reasons that explain the improved cell adhesion and proliferation on sol–gel-derived surfaces compared with commercially pure (cp) titanium.

In addition, other reasons can explain good soft tissue reactions. One reason is *rugophilia*. Macrophages are known to prefer rough surfaces (Andersson 1988; Brunette 1988). Because the sol–gel-derived titania-modified surfaces are smooth, there are neither attracting signals for macrophages nor IL1 formation. This may prevent capsule formation, and the amount of new fibroblasts and collagen is reduced. Another reason is *surface energy*. Capsule forms normally around implants with a low-energy surface. The cells do not attach to a low-energy surface but, rather, adhere to each other, which often results in a capsule formation (Brunette 1988). In case of nanoporous titania coatings, there is a large number of specific high-energy OH groups in titania gel containing anatase and rutile structures—the same groups that are also demonstrated to induce apatite formation in simulated body fluid (Uchida et al. 2003). Anatase and rutile are also predominant structures of the sol–gel-derived titania surface (Moritz et al. 2005). Compared to amorphous titanium, which is a predominant structure in cp titanium, anatase- and rutile-containing surfaces allow the adsorption of the physiological fluids and proteins and soft tissues to come very close or attach on it directly much more likely than on the amorphous structure. In addition, due to the high-energy surface and stronger van der Waals bonds, the attachment of biomolecules and the adsorption of proteins such as fibronectin will exponentially increase the adhesion and spreading of fibroblasts, which leads to good tissue integration (Kasemo and Lausmaa 1988). High surface energy has also been achieved with traditional sand-blasted and acid-etched (SLA) surfaces after storage in isotonic NaCl solution protected by N_2 filling (Rupp et al. 2005). These surfaces are strongly hydrophilic (SLAactive), and they have possessed superior wettability compared to traditional SLA surface. Although the surfaces can accelerate cell attachment, no data are available about their effect on fibroblast or epithelial cell behavior.

Under culture conditions, focal integrin adhesions can be disrupted by trypsin–ethylenediaminetetraacetic acid (Wilson et al. 2005), and rounded cells will detach from the surface upon agitation. It has been shown that human gingival fibroblasts are much more resistant to such a treatment on sol–gel-derived TiO_2 coatings than on a noncoated titanium surface. Possible mechanisms to trypsin resistance might include a higher number of focal adhesions per cell or morphological protection against protease cleavage (Meretoja et al. 2010). It has been shown that the contact between cell membranes and the titania coating observed in transmission electron microscopy is as close as that between two tissue cells. The lack of abnormal features in the cultured cells suggests that fibroblasts do not react to titanium oxide as foreign materials. An organotypic *in vitro* cell culture model (Oksanen and Hormia 2002) that mimics the dentoepithelial junction could demonstrate that the epithelium can also attach directly on sol–gel-derived TiO_2 coating (Figure 9.4).

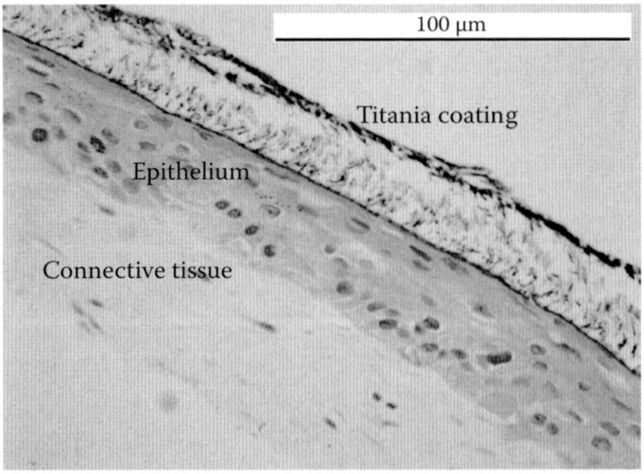

FIGURE 9.4 Epithelial attachment on titania coating in a tissue culture that mimics a dentoepithelial junction. (Courtesy of Dr. Jaana Willberg, Institute of Dentistry, University of Turku, Finland.)

Development in materials science is very fast. New materials are usually tested in a laboratory environment, which rarely meet clinical conditions. Many materials and new surface treatments, for example, are tested in cell culture models. These studies give important information on cell behavior on materials surfaces. However, *in vivo* soft tissues attach on surfaces after complicated biological processes. Human gingival fibroblasts are frequently used in implant materials research, but in real life, fibroblasts attach on surfaces via proteoglycan layers. This situation cannot be imitated in laboratory conditions.

9.4.5 Engineering of Peri-Implant Gingival Structures

Periodontal ligament fibroblasts are capable of regenerating ligament structures (Ivanovski et al. 2006). The use of periodontal ligament progenitor cells has attracted researchers to create tissue-engineered ligaments on titanium implant surfaces. The lack of suitable carriers is a major shortcoming in tissue engineering in general, and so far, tissue engineering attempts have not resulted in major breakthroughs in oral implantology. Lin and coworkers (2011) presented a new method using Matrigel as a carrier for periodontal ligament cells on titanium implants. They demonstrated that periodontal ligament-derived progenitor cells can be used to regenerate periodontal ligament tissues around titanium implants *in situ* in rats. The engineered tissues showed the formation of a cementum-like structure on titanium implants, with Sharpey's fibers running parallel toward the implant surface. It remains to be seen if the experiment can be repeated with human cells and under loaded conditions in the oral cavity. However, fast developments in tissue engineering may provide methods of regenerating periodontal structures around oral implants in the near future.

9.5 BIOFILM ON THE IMPLANT SURFACE

The lack of direct attachment of soft tissues may facilitate the migration of bacteria along the implant surface, having the potential for increasing the risk for peri-implantitis, implant failure, and, finally, a loss of oral implant (Meffert 1988; Esposito et al. 1998; Mombelli and Lang 1998; Hardt et al. 2002). This situation is analogical with percutaneous medical devices, where microbial migration and subsequent infection around the implant is a frequent problem. This is especially important for intravenous implants that are subjected to bloodstream infections.

Oral cavity is a challenging environment for implants. Microbial colonization on oral implants follows the same pattern as on tooth surfaces (Subramani et al. 2009). In both cases, microbial colonization is determined by the formation of a acellular salivary pellicle. Thus, in clinical environments, bacterial adhesion and subsequent biofilm development are strongly determined by the material's ability to adsorb fluids from their surroundings.

Primary adhesion between bacteria and implant devices is mediated by nonspecific interactions (Dunne 2002). Before microbial adhesion, the implant surface is conditioned by the adsorption of water and proteins, which alters its original surface properties.

The colonization of surfaces by pathogens is the first step in pathogenesis. Bacterial cells start to colonize on the pellicle surface within 4 h. *Actinomyces* species are frequently detected among the early colonizers on tooth and on implant surfaces, although large proportions of initial colonizers are streptococci (Rosan and Lamont 2000). *Actinomyces* can activate T cells, which, in turn, can induce osteoclast activation and bone resorption (Taubman and Kawai 2001). *Fusobacterium nucleatum* colonizes in a later stage during biofilm formation, but it is often related with peri-implantitis and cultivated from the surfaces of implants that have lost their osseointegration (Mombelli et al. 1987).

In general, surface roughness seems to be more important to the bacterial adhesion on dental materials, including implant surfaces, than the surface coating (Bürgers et al. 2010). An *in vivo* study indicated that the reduction of surface roughness below 0.2 μm has no major effect on supragingival or subgingival microbial composition on implant surfaces (Bollen et al. 1996). However, biofilm formation is different in various parts of the oral cavity (Hanning et al. 1999; Carlen et al. 2003; Groessner-Schneider et al. 2004).

Antimicrobial implant surfaces may decrease the risk of implant infections, but the effect of antimicrobial agents is limited to the early phases of implant wound healing. The sustained release of antimicrobial agents is of great concern for the development of antibiotic-resistant bacterial strains.

9.6 CONCLUSION

Soft tissue attachment to an implant surface is established during the wound healing process. The healing process around the percutaneous or transgingival part of an implant follows, in principle, the same pattern as in normal wound healing. However, implant wound healing always takes place between living tissues and biomaterial surface. Soft tissue healing around oral implants is closely related with the healing of periodontal tissues after surgical intervention. Surface characteristics of an implant, such as topography, chemistry, and wettability, have a large impact on the healing process. A direct bond between the implant and soft tissues could guarantee a firm seal for preventing bacterial migration into implant and soft tissue interphase. Despite the promising results, firm soft tissue attachment has not yet been validated in long-term clinical trials. In terms of oral implants, the ultimate goal should create an attachment mechanism that resembles periodontal ligament around the natural tooth. Advancements in tissue engineering might make this possible in the future.

REFERENCES

Albrektsson, T., Berglundh, T., and Lindhe, J. (2003). Osseointegration: Historic background and current concepts. *Clinical Periodontology and Implant Dentistry*, Lindhe, J., Karring, T., and Lang, N.P., editors. Oxford: Blackwell Munksgaard, pp. 809–820.

Andersson, J.M. (1988). Inflammatory response to implants. *ASAIO Transactions/American Society of Artificial Internal Organs* **34**: 101–107.

Andersson, J.M. (1996). Inflammation, wound healing, and the foreign body reaction. *Biomaterials Science: An Introduction to Materials in Medicine*, Ratner, B.D., Hoffman, A.S., Schoen, F.J., and Lemons, J.E., editors. San Diego, CA: Academic Press, pp. 165–173.

Anusavice, K.J. (1996). *Phillips' Science of Dental Materials*, 10th edition. Philadelphia, PA: WB Saunders, pp. 25–31.

Areva, S., Paldan, H., Peltola, T., Närhi, T., Jokinen, M., and Lindén, M. (2004). The use of sol–gel-derived titania coating for direct soft tissue attachment. *Journal of Biomedical Materials Research, Part A* **70A**: 169–178.

Baier, R.E., Shafrin, E.G., and Zisman, W.A. (1968). Adhesion: Mechanisms that assist or impede it. *Science* **162**: 1360–1368.

Berglundh, T., Lindhe, J., Ericsson, I., Marinello, C.P., Liljenberg, B., and Thomsen, P. (1991). The soft tissue barrier at the implants and teeth. *Clinical Oral Implants Research* **2**: 81–90.

Bjursteen, L.M., Emanuelsson, L., Ericson, L.E., Thomsen, P., Lausmaa, J., Mattsson, L., Rolander, U., and Kasemo, B. (1995). Method for ultrastructural studies of the intact tissue metal interface. *Biomaterials* **16**: 83–90.

Bollen, C.M., Papaioanno, W., van Eldere, J., Schepers, E., Quirynen, M., van Steenberghe, D. (1996). The influence of abutment surface roughness on plaque accumulation and peri-implant mucositis. *Clinical Oral Implants Research* **7**: 201–211.

Britch, M. and Allen, T.D. (1980). The modulation of cellular contractility and adhesion by trypsin and EGTA. *Experimental Cell Research* **125**: 221–231.

Brunette, D.M. (1988). The effect of implant surface topography on the behavior of cells. *International Journal of Oral and Maxillofacial Implants* **3**: 231–246.

Bürgers, R., Hahnel, S., Reichert, T.E., Rosentritt, M., Behr, M., Gerlach, T., Handel, G., and Gosau, M. (2010). Adhesion of *Candida albicans* to various dental implant surfaces and the influence of salivary pellicle proteins. *Acta Biomater* **6**: 2307–2313.

Carlén, A., Rüdinger, S.G., Loggner, I., and Olsson. J. (2003). Bacteria-binding plasma proteins in pellicles formed on hydroxyapatite *in vitro* and on teeth *in vivo*. *Oral Microbiology and Immunology* **18**: 203–207.

Chehroudi, B., Gould, T.R.L., and Brunette, D.M. (1991). A light and electron microscopic study of the effects of surface topography on the behavior of cells attached to titanium-coated percutaneous implants. *Journal of Biomedical Materials Research* **25**: 387–405.

Cho, C.S., Kobayashi, A., Goto, M., Park, K.H., and Akaide, T. (1996). Difference in adhesion and proliferation of fibroblast between Langmuir–Blodgett films and cast surfaces of poly (gamma-benzyl L-glutamate)/poly(ethylene oxide) diblock copolymer. *Journal of Biomedical Materials Research* **32**: 425–432.

Combe, E.C., Owen, B.A., and Hodges, J.S. (2004). A protocol for determining the surface free energy of dental materials. *Dental Materials* **20**: 262–268.

Davies, J.E. (1996). *In vitro* modeling of the bone/implant interface. *Anatomical Record* **245**: 426–445.

Davies, J.E. (1998). The importance and measurement of surface charge species in cell behavior at the biomaterial interface. *Surface Characterisation of Biomaterials: Progress in Biomedical Engineering*, Ratner, B.D., editor. Amsterdam: Elsevier, pp. 219–234.

Deporter, D.A., Watson, P.A., Pilliar, R.M., Howley, T.P., and Winslow, J. (1988). A histological evaluation of a functional endosseous, porous-surfaced, titanium alloy dental implant system in the dog. *Journal of Dental Research* **67**: 1190–1195.

Dunne, M.W. Jr. (2002). Bacterial adhesion: Seen any good biofilms lately? *Clinical Microbiology Reviews* **15**: 155–166.

Ellingsen, J.E., Johansson, C.B., Wennerberg, A., and Holmén, A. (2004). Improved retention and bone-to implant contact with fluoride-modified titanium implants. *International Journal of Oral and Maxillofacial Implants* **19**: 659–666.

Ericsson, I. and Lindhe, J. (1993). Probing depth at implants and teeth: An experimental study in the dogs. *Journal of Clinical Periodontology* **20**: 623–627.

Esposito, M., Hirsch, J.M., Lekholm, U., and Thomsen, P. (1998). Biological factors contributing to failures of osseointegrated oral implants—Part I: Success criteria and epidemiology. *European Journal of Oral Sciences* **106**: 527–551

Gotfredsen, K., Rostrup, E., Hjörting-Hansen, E., Stoltze, K., and Budtz-Jörgensen, E. (1991). Histological and histomorphometrical evaluation of tissue reactions adjacent to endosteal implants in monkeys. *Clinical Oral Implant Research* **2**: 30–37.

Gray, J.L., Vemino, A.R., and Towle, H.J. (2005). A comparison of the clinical and histologic crestal bone level measurements adjacent to dental implants in the baboon. *International Journal of Periodontics and Restorative Dentistry* **25**: 623–628.

Groessner-Schreiber, B., Hannig, M., Dück, A., Griepentrog, M., and Wenderoth, D.F. (2004). Do different implant surfaces exposed in the oral cavity of humans show different biofilm compositions and activities? *European Journal of Oral Sciencies* **112**: 516–522.

Hallab, N., Bundy, K., O'Connor, K., Moses, R.L., and Jacobs, J.J. (2001). Evaluation of metallic and polymeric biomaterial surface energy and surface roughness characteristics for directed cell adhesion. *Tissue Engineering* **71**: 55–71.

Hanning, M. (1999). Transmission electron microscopy of early plaque formation on dental materials in vivo. *European Journal of Oral Sciences* **107**: 55–64.

Hanson, S.R. and Harker, L.A. (1996). Blood coagulation and blood–materials interactions. *Biomaterials Science: An Introduction to Materials in Medicine*, Ratner, B.D., Hoffman, A.S., Schoen, F.J., and Lemons, J.E., editors. Academic Press, San Diego pp. 193–200.

Hardt, C.R.E., Gröndahl, K., Lekholm, U., and Wennström, J.L. (2002). Outcome of implant therapy in relation to experienced loss of periodontal bone support. *Clinical Oral Implants Research* **13**: 488–494.

Holgers, K.M., Thomsen, A., Tjellsström, A., and Ericson, L.E. (1995). Electron microscopic observations on soft tissue around clinical long-term percutaneous titanium implants. *Biomaterials* **16**: 83–90.

Ivanovski, S., Gronthos, S., Shi, S., and Bartold, P.M. (2006). Stem cells in the periodontal ligament. *Oral Diseases* **12**: 358–363.

Jansen, J.A., de Wijn, J.R., Wolters-Lutgerhorst, J.M.L., and van Mullem, P.J. (1985). Ultrastructural study of epithelial cell attachment to implant materials. *Journal of Dental Research* **64**: 891–896.

Jansen, J.A., van der Waerden, J.P.C.M., van der Lubbe, H.B.M., and de Groot, K. (1990). Tissue response to percutaneous implants in rabbits. *Journal of Biomedical Materials Research* **24**: 295–307.

Jokinen, M., Pätsi, M., Rahiala, H., Peltola, T., Ritala, M., and Rosenholm, J.B. (1998). Influence of sol and surface properties on *in vitro* bioactivity of sol–gel-derived TiO_2 and TiO_2–SiO_2 films deposed by dip-coating method. *Journal of Biomedical Materials Research* **42**: 295–302.

Kasemo, B. and Lausmaa, J. (1988). Biomaterial and implant surfaces: A surface science approach. *International Journal of Oral and Maxillofacial Implants* **3**: 247–259.

Lampin, M., Warocquier, C., Legris, C., Degrange, M., and Sigot-Luizard, M.F. (1997). Correlation between substratum roughness and wettability, cell adhesion, and cell migration. *Journal of Biomedical Materials Research* **36**: 99–108.

Larjava, H., Koivisto, L., Häkkinen, L., and Heino, J. (2011). Epithelial integrins with special reference to oral epithelia. *Journal of Dental Research.* **90**: 1367–1376.

Liljenberg, B., Gualini, F., Berglundh, T., Tonetti, M., and Lindhe, J. (1996). Some characteristics of the ridge mucosa before and after implant installation. A prospective study in humans. *Journal of Clinical Periodontology* **23**: 1008–1013.

Lin, Y., Galluci, G.O., Buser, D., Bosshardt, D., Belser, U.C., and Yelick, P.C. (2011). Bioengineered periodontal tissue formed on titanium implants. *Journal of Dental Research* **90**: 251–256.

Lowenguth, R.A., Polson, A.M., and Caton, J.G. (1993). Oriented cell and fiber attachment systems *in vivo*. *Journal of Periodontology* **64**: 330–342.

Mackenzie, I.C. and Tonetti, M.S. (1995). Formation of normal gingival epithelial phenotypes around osseo-integrated oral implants in humans. *Journal of Periodontology* **66**: 933–943.

McKinney, R.V., Steflik, D.E., and Koth, D.L. (1985). Evidence for a junctional epithelial attachment to ceramic dental implants: A transmission electron microscopic study. *Journal of Periodontology* **10**: 579–591.

Meffert, R. (1988). The soft tissue interface in dental implantology. *International Journal of Oral Implantology* **5**: 55–58.

Mekayarajjananonth, T. and Winkler, S. (1999). Contact angle measurement on dental implant biomaterials. *Journal of Oral Implantology* **25**: 230–236.

Meretoja, V.V., Rossi, S., Peltola, T., Pelliniemi, L.J., and Närhi, T.O. (2010) Adhesion and proliferation of human fibroblasts on sol–gel coated titania. *Journal of Biomedical Materials Research A* **95**: 269–275.

Misiek, D.J., Kent, J.N., and Carr, R.F. (1984). Soft tissue responses to hydroxylapatite particles of different shapes. *Journal of Oral and Maxillofacial Surgery* **42**: 150–160.

Mohammadi, S., Esposito, M., Hall, J., Emanuelsson, L., Krozer, A., and Thomsen, P. (2004). Long-term bone response to titanium implants coated with thin radio frequent magnetron-sputtered hydroxyapatite in rabbits. *International Journal of Oral and Maxillofacial Implants* **19**: 498–509.

Mombelli, A. and Lang, N.P. (1998). The diagnosis and treatment of peri-implantitis. *Periodontology* **17**: 63–76.

Mombelli, A., van Oosten, M., Schurch, E., and Lang, N.P. (1987). The microbiota associated with successful or failing osseointegrated titanium implants. *Oral Microbiology and Immunology* **2**: 145–151.

Moritz, N., Areva, S., Wolke, J., and Peltola, T. (2005). TF-XRD examination of surface-reactive TiO_2 coatings produced by heat treatment and CO_2 laser treatment. *Biomaterials* **26**: 4460–4467.

Narayanan, R., Seshadri, S.K., Kwon, T.Y., and Kim, K.H. (2008). Calcium phosphate-based coatings on titanium and its alloys. *Journal of Biomedical Materials Research—Part B: Applied Biomaterials* **85B**: 279–299.

Oksanen, J. and Hormia, M. (2002). An organotypic *in vitro* model that mimics the dento-epithelial junction. *Journal of Periodontology* **73**: 86–93.

Olbrich, K.C., Andersen, T.T., Blumenstock, F.A., and Bizios, R. (1996). Surfaces modified with covalently immobilized adhesive peptides affect fibroblast population motility. *Biomaterials* **17**: 759–764.

Oshida, Y., Hashem, A., Nishihara, T., and Yapchulay, M.V. (1994). Fractal dimension analysis of mandibular bones: Toward a morphological compatibility of implants. *Biomedical Materials Engineering* **4**: 397–407.

Paldan, H., Areva, S., Tirri, T., Peltola, T., Lindholm, T.C., Lassila, L., Pelliniemi, L.J., Happonen, R.P., and Närhi, T.O. (2008). Soft tissue attachment on sol–gel-treated titanium implants *in vivo*. *Journal of Materials Science: Materials in Medicine* **19**: 1283–1290.

Peltola, T., Jokinen, M., Rahiala, H., Pätsi, M., Heikkilä, J., Kangasniemi, I., and Yli-Urpo, A. (2000). Effect of aging time of sol on structure and *in vitro* calcium phosphate formation of sol–gel-derived titania films. *Journal of Biomedical Materials Research* **51**: 200–208.

Peltola, T., Pätsi, M., Rahiala, H., Kangasniemi, I., and Yli-Urpo, A. (1998). Calcium phosphate induction by sol–gel-derived titania coatings on titania substrates *in vitro*. *Journal of Biomedical Materials Research* **41**: 504–510.

Räisänen, L., Könönen, M., Juhanoja, J., Varpavaara, P., Hautaniemi, J., Kivilahti, J., and Hormia, M. (2000). Expression of cell adhesion complexes in epithelial cells seeded on biomaterial surfaces. *Journal of Biomedical Materials Research* **49**: 79–87.

Rompen, E., Domken, O., Degidi, M., Pontes, A.E., and Piattelli, A. (2006). The effect of material characteristics, of surface topography and of implant components and connections on soft tissue integration: A literature review. *Clinical Oral Implants Research* **17** (Suppl. 2): 55–67.

Rosan, B. and Lamont, R.J. (2000). Dental plaque formation. *Microbes and Infection* **2**: 1599–1607.

Rossi, S., Moritz, N., Tirri, T., Peltola, T., Areva, S., Jokinen, M., Happonen, R.-P., and Närhi, T. (2007). Comparison between sol–gel-derived anatase- and rutile-structured TiO_2 coatings in soft tissue environment. *Journal of Biomedical Materials Research, Part A* **77A**, accepted for publication.

Rupp, F., Scheideler, L., Olshanska, N., de Wild, M., Wieland, M., and Geis-Gerstorfer, J. (2005). Enhancing surface free energy and hydrophilicity through chemical modification of microstructured titanium implant surfaces. *Journal of Biomedical Materials Research—Part A* **76A**: 323–334.

Schakenraad, J.M., Busscher, H.J., Wildevuur, C.R., and Arends, J. (1988). Thermodynamic aspects of cell spreading on solid substrata. *Cell Biophysics* **13**: 75–91.

Schroeder, A., van der Zypen, E., Stich, H., and Sutter, F. (1981). The reactions of bone, connective tissue, and epithelium to endosteal implants with titanium-sprayed surfaces. *Journal of Oral and Maxillofacial Surgery* **9**: 15–25.

Shimono, M., Ishikawa, T., Enokiua, Y., Muramatsu, T., Matsuzaka, K., Inoue, T., Abiko, Y., Yamaza, T., Kido, M., Tanaka, T., and Hashimoto, S. (2003). Biological characteristics of the junctional epithelium. *Journal of Electron Microscopy (Tokyo)* **52**: 627–639.

Singer, A.J. and Clark, R.A.F. (1999). Cutaneous wound healing. *New England Journal of Medicine* **341**: 738–746.

Smith, D.C., Pilliar, R.M., and Chernecky, R. (1991). Dental implant materials—Part I: Some effects of preparative procedures on surface topography. *Journal of Biomedical Materials Research* **25**: 1045–1068.

Steflik, D.E., McKinney, R.V., Parr, G.R., and Koth, D.L. (1991a). What we know about the interface between implants and soft tissue. *Oral and Maxillofacial Surgery Clinics of North America* **3**: 775–793.

Steflik, D.E., McKinney, R.V., Sisk, A.L., Parr, G.R., and Marshall, B.L. (1991b). Dental implants retrieved from humans: A diagnostic light microscopic review of the findings in seven cases failure. *International Journal of Oral and Maxillofacial Implants* **6**: 147–153.

Subramani, K., Jung, R.E., Molenberg, A., and Hammerle, C.H. (2009). Biofilm on dental implants: A review of the literature. *International Journal of Oral and Maxillofacial Implants* **24**: 616–626.

Taubman, M.A. and Kawai, T. (2001). Involvement of T-lymphocytes in periodontal disease and in direct and indirect induction of bone resorption. *Critical Reviews in Oral Biology and Medicine* **12**: 125–135.

Teughels, W., Van Assche, N., Sliepen, I., and Quirynen, M. (2006). Effect of material characteristics and/or surface topography on biofilm development. *Clinical Oral Implants Research* **17** (Suppl. 2): 68–81.

Uchida, M., Kim, H.-M., Kokubo, T., Fujibayashi, S., and Nakamura, T. (2003). Structural dependence of apatite formation on titania gels in a simulated body fluid. *Journal of Biomedical Materials Research, Part A* **64A**: 164–170.

Vercaigne, S., Wolke, J.G.C., Naert, I., and Jansen, J.A. (2000). A histological evaluation of TiO_2-gritblasted and Ca–P magnetron sputter-coated implants placed into the trabecular bone of the goat—Part 2. *Clinical Oral Implants Research* **11**: 314–324.

von Recum, A.F. (1984). Applications and failure modes of percutaneous devices: A review. *Journal of Biomedical Materials Research* **18**: 323–336.

Wennerberg, A., Albrektsson, T., Andersson, B., and Krol, J.J. (1995). A histomorphometric and removal torque study of screw-shaped titanium implants with three different surface topographies. *Clinical Oral Implants Research* **6**: 24–30.

Williams, D.F. (1999). *The Williams Dictionary of Biomaterials*. Liverpool University Press, Liverpool: 343.

Wilson, C.J., Clegg, R.E., Leavesley, D.I., and Pearcy, M.J. (2005). Mediation of biomaterial–cell interactions by adsorbed proteins: A review. *Tissue Engineering* **11**: 1–18.

Winter, G.D. (1974). Transcutaneous implants: Reactions of the skin–implant interface. *Journal of Biomedical Materials Research Symposium* **15**: 99–113.

10 Surface Chemical Determination of the Placement, Growth, and Differentiation of Cells for Drug Screening, Toxin Detection, and Lab-on-a-Chip Applications

Peter Molnar, Murugan Ramalingam, and James J. Hickman

CONTENTS

10.1 INTRODUCTION

The determination and engineering of the behavior of living cells became a crucial task in several emerging fields of tissue engineering and drug screening [1]. The building of complex tissue analogs, the creation of tissue engineered organs, the preparation of stem cells for therapy, the development of high-throughput functional drug screening platforms, or the integration of whole-cell biosensors with silicon-based electronics all require the manipulation of cells and the determination of their placement, interactions, differentiation, and integration [2–5]. Unfortunately, conventional engineering methods may not be suitable to deal with living cells, primarily because the cells are following a self-organizing internal program that we have only limited knowledge about and is usually not our design. Consequently, traditional cell engineering methods do not fulfill their intended purpose, in particular, mimicking the native cellular environment. Recently, new technologies have been emerging, with the aim of engineering the environment of living cells to control their internal program through extracellular cues [4,6–10]. These extracellular cues could be soluble factors introduced in the culture medium or surface-bound topographical, chemical, or physical signals on the culture surface or incorporated into tissue engineered three-dimensional (3-D) scaffolds. At the same time, new technologies were also developed to create spatial and (in a lesser extent) temporal patterns of these signals [4,6,7,11–17]. Contact interactions are major determinants of cell attachment, proliferation, growth, and differentiation. Cell adhesion and attachment molecules play essential roles in embryonic development, integration of cells into tissues and organs, cell migration, axon guidance, immune response, and tumor development [18–20].

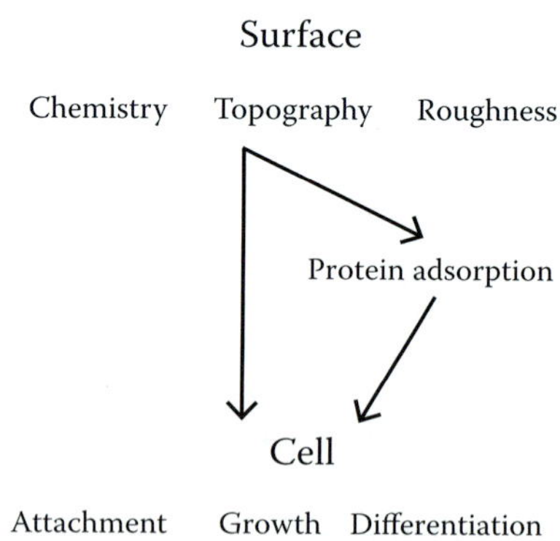

FIGURE 10.1 Complex interaction of surfaces with proteins and cells. Surface properties such as chemistry, topography, and roughness determine cell attachment, proliferation, growth, and differentiation. At the same time, surface properties are major determinants of protein adsorption; the adsorbed protein layer further modifies surface properties and affect cell behavior.

In contrast to soluble factors, contact signals can be engineered and implemented on two-dimensional (2-D) surfaces and in 3-D scaffolds with high spatial resolution and used to guide the attachment, proliferation, growth, differentiation, and integration of a wide variety of cell types. Surface modification became an essential step in any tissue engineering application, because the surface of the engineered material forms the primary interface connecting the scaffold to the living cells or tissues and determines their interactions [1,21] (Figure 10.1).

10.2 SURFACES TO GUIDE CELL ATTACHMENT

Surface treatment is an essential step in any cell culture-based tissue engineering application [22,23]. Most cell types, in particular, anchorage-dependent cells, do not attach to nonbiological untreated surfaces. Several solutions were worked out for this problem during the long history of cell culture—commercial companies are now offering "easy grow" flasks, plates, and other plasticwares and a wide selection of extracellular matrix proteins and other materials for "do-it-yourself" surface treatment. Unfortunately, surface modification procedures were mostly developed by experience on a "trial-and-error" basis; they depend not only on the engineered material but also on the cell type and, in some extent, on the preference and skill of the person who handles the experimental methods. The most widely used methods of enhanceing cell attachment include: 1) the determination of surface topography and roughness [24,25]; 2) the adsorption of an extracellular matrix and other biomimetic molecules to the surface [26]; and 3) the covalent modification of the surface to determine wettability or to attach cell adhesion receptor substrates [27].

In living tissues, cell attachment is mediated through very specific receptor-mediated interactions between the cell and the extracellular matrix or the neighboring cells. Cell adhesion/attachment receptors not only provide cohesion between cells but also have very important signaling functions—they determine the survival, proliferation, growth, and differentiation of cells [16,20,28,29]. Thus, careful selection of contact signals could determine not only the attachment but also the fate of cultured cells. Most important cell adhesion molecules are integrins, immunoglobulin superfamily members, cadherins, and selectins [30]. Each of these molecules has a different function and recognizes different ligands. The role of nonspecific (nonreceptor-mediated) attachment in live tissues is less known. Many surface treatments in *in vitro* tissue engineering applications are utilizing nonspecific interactions to enhance cell attachment. It is also not clear whether contact interactions with nonspecific surfaces could activate signal transduction pathways. Some evidence in the literature indicates that surface topography or contact with a surface-modified nonspecific surface could induce focal contact formation in attached cells—a phenomenon that was considered to require receptor activation [31–33]. Another aspect of surface modification is that surface properties of the materials determine their protein adsorption profile [34–36]. In most cell culture applications, the growth medium contains high concentrations of proteins, and the adsorption of these proteins is a fast process, which usually precedes cell attachment; thus, cells are attaching to the adsorbed protein layer and not to the original surface. Protein adsorption is selective; chemical properties of the surface selectively determine the adsorption rates and conformation of the different proteins.

10.2.1 Surface Modification through the Adsorption of Extracellular Proteins

Based on a long period of experimentation, different surface treatment protocols were optimized for different cell types. The most frequently used extracellular matrix proteins or mimetics are poly-D-lysine (neurons, stem cells), laminin (neurons, stem cells), fibronectin (endothelium), vitronectin (muscle) gelatin (cardiac), Matrigel (cardiac, epithelial, and stem cells), and collagen (bone, cartilage) [37–45]. Still, there is no established theory that would explain the surface preference of the different cells; moreover, only a few systematic studies have been published concerning the effect of surface composition on the behavior of cultured cells [40,43]. The popularity of protein adsorption surface treatment lays in its simplicity, but although it is a seemingly easy technique, there are several potential problems that could affect the reliability and reproducibility of this method such as: 1) the lack of control over the conformation of the adsorbed protein; 2) the lack of reliable surface characterization method and quality control; 3) the variability of the source and composition of available extracellular proteins; 4) the lack of flexibility in determining surface properties of biomaterials; and 5) uncontrolled degradation [1,27]. In order to overcome these drawbacks, covalent modification of the surfaces was introduced.

10.2.2 Covalent Surface Modifications for Cell Attachment

Self-assembled monolayers (SAMs) are frequently used for the modification of surfaces, and they are a good example for the covalent surface modification of biomaterials. They are composed of organic molecules that form strong interactions or covalent bonds with reactive groups on an exposed surface, resulting in a one-molecule-thick modifying layer that determines the physical and chemical properties of the surface [46,47]. Originally, SAMs were characterized on a gold surface [48], but for tissue engineering applications, SAMs on other surfaces might have higher significance, such as on silicon dioxide or glass [49,50], biodegradable polymers [51], and other polymers such as Teflon [52] or poly(dimethylsiloxane) (PDMS) [53]. A large variety of functional groups or combination of functional groups can be located on the terminus opposite the attachment point of a SAM [54], and the chemical composition can be manipulated to systematically vary the surface free energy [49]. Moreover, SAMs could also be functionalized with covalently bound proteins using cross-linkers, which make them an extremely flexible test system to study contact interaction. Fields suggested the use of SAMs to create bioactive surfaces with covalent protein functionalization and controlled protein folding [55,56]. The patterning of SAMs is possible using photolithography, which is an already established industrial technique; thus, the determination of cell attachment is compatible with advanced fabrication methods, which is important in lab-on-a-chip applications [57,58].

10.2.3 Cell Attachment Studies with SAMs

Modifying surfaces with SAMs is an extremely flexible method [54]. A large number of functional groups have been tried out with a wide selection of cell types to determine the surface requirements of cell attachment, proliferation, and differentiation. There is almost no surface that is favored by cells or completely cell repellent in all experimental conditions. The most consistently positive surfaces are usually lightly hydrophilic and contain amino or carboxyl groups, whereas the negative surfaces are very hydrophilic (polyethylene glycol or PEG) or hydrophobic (CF3, CH3) [36,49,59,60]. Moreover, positive surface charges usually enhance cell attachment [61].

Kleinfeld et al. were among the first researchers who performed systematic studies to select SAMs that promote the attachment and growth of neuronal cells [58]. They found that the presence of an amine group on the surface is a common factor in promoting cell adhesion. By systematically varying the aminosilane chain length and structure of the functional group and by measuring the viability and morphological parameters of cultured neurons, they successfully optimized surface properties for neuronal applications. They identified diethylenetriamine-propyl silane (DETA-P) as a cell attachment–promoting surface. Later systematic studies were performed with other cell types such as epithelial cells [62], endothelial cells [27,63–67], fibroblasts [36,68–70], osteoblasts [71,72], liver cells [73], cardiac myocytes [74], muscle cells [3,75], or other neurons [57,76].

In tissue engineering applications, the prevention, not always the promotion, of cell attachment is the goal. The most commonly used cell attachment–preventing surface is PEG or, in most of our experiments, tridecafluoro-1,1,2,2-tetrahydrooctyl-1-trichlorosilane (13F) [27,77]. Besides supplying a cell-resistant background in patterning experiments, PEG has several very important uses in tissue engineering applications such as controling the density of other functional groups on the surface, creating gradients, controling conformation of proteins, and preventing their aggregation in protein immobilization experiments. Moreover, PEG is also resistant to protein adsorption; thus, it can be used in cell patterning experiments where the cell attachment promoting foreground is created by that method.

In our experiments, *N*-1[3-(trimethoxysilyl)propyl]diethylenetriamine (DETA) seemed to be an appropriate surface for the attachment of most of our primary cells and cell lines, except for primary cardiac myocytes. In comparison with standard/widely used surface treatment protocols, we observed only minor differences in the morphology and physiology of cells cultured on DETA (Figure 10.2a and b) [78].

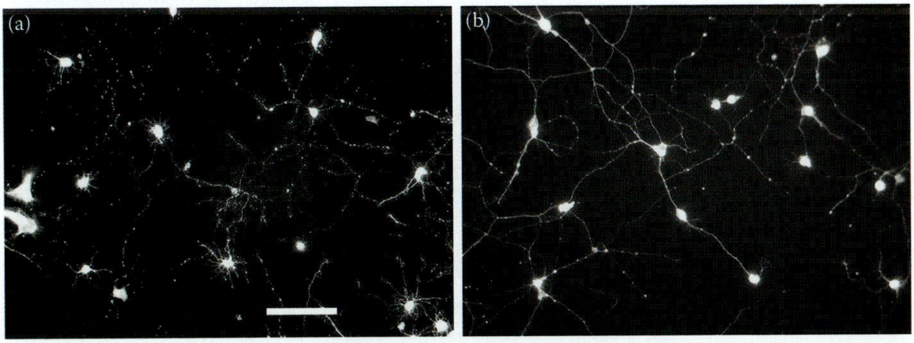

FIGURE 10.2 Purified rat embryonic motoneurons plated on DETA (a) and ornithine/laminin (b) surfaces. Day 6 cultures—epifluorescent pictures. Scale bar = 100 μm. (From Das, M.P. et al., *Biotechnology Progress*, 19, 1756–1761, 2003.)

The SAM surface was remarkably stable; we had different primary cultures and cell lines surviving for several months on these surfaces.

10.2.4 Dynamic Surfaces

One of the major problems with surface-chemical determination of cell behavior is that, in contrast to the dynamic biological environment, the engineered surface signals are static. It is very difficult to create a dynamic system that changes together with the development and growth of the cells. It is even much more difficult to create a defined feedback loop between the cells and their extracellular environment. The versatility of the self-assembled monolayer technique enables the engineering of switchable surfaces where changes in environmental parameters such as temperature, pH, magnetic field, or electrical potential trigger drastic alteration in surface properties [79–82]. Shimizu et al. used thermosensitive polymers to create cardiac myocyte sheets for cardiac tissue engineering [83]. By decreasing the temperature, the cultured cardiac myocyte sheet was released from the underlying thermosensitive polymer layer and could be used to create a multilayer tissue engineered construct.

10.3 SURFACES WITH SPECIFIC SIGNALING FUNCTIONS

Contact signals are major determinants of not only attachment but also the survival, growth, and differentiation of cells. For instance, we have shown that C2C12 skeletal muscle cells do attach to nonspecific surfaces such as DETA and SAMs but do not differentiate without the appropriate contact signals [3]. Using vitronectin (an integrin receptor substrate) on the surface, the differentiation, alignment, and fusion of myocytes could be induced. The patterning of vitronectin using photolithography could result in directed myofiber formation on the predetermined areas (Figure 10.3).

Cell type-specific surfaces have high significance in tissue engineering. The selection and plating of a specific cell type from a mixture of cells is a common problem. One possible solution is to create cell type—specific surfaces that promote

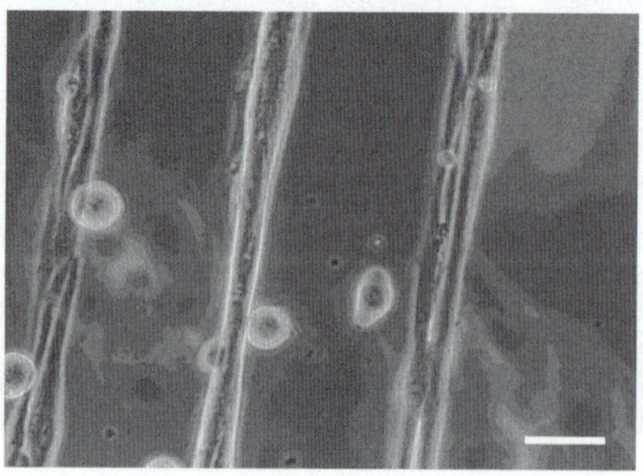

FIGURE 10.3 C2C12 skeletal muscle cells on vitronectin surface patterns. Scale bar = 100 μm.

the adhesion of one type of cell and prevent the adhesion of another type of cell. For instance, Quirk et al. demonstrated that covalent surface functionalization with a galactose derivative enhanced the selective attachment of hepatocytes [84]. Contact signals can also be used to guide the growth or differentiation of cells [85]. In many stem cell applications, differentiation of the cells is achieved by the overexpression of a specific promoter. Shen et al. showed that successful gene transfer could be achieved by using covalently surface-bound DNA, which opens new horizons for tissue engineering with human stem cells [86].

In general, surface modification with SAMs is a more complex process than protein adsorption and requires more equipment, time, and expertise but offers much more flexibility and control over the physicochemical properties of the surface. It is more reliable and quality control is possible using advanced surface characterization methods such as X-ray photoelectron spectroscopy and contact angle measurements.

10.4 PATTERNING OF SAMs

The patterning of SAMs is one of the established ways for positioning cells or proteins into desired regions in order to control cell shape, growth, and function. A wide variety of techniques have been developed in the last two decades for the patterning of SAMs, including photolithography and microcontact printing (μCP). These techniques are primarily applied to the production of micropatterned or nanopatterned surfaces suitable for controlled cell adhesion or protein adsorption. Each method has its own merits and demerits; thus, the selection of the right method purely depends upon the specific purpose of the experiment.

10.4.1 Patterning by Photolithography

Photolithography was the first well-established technique used in the patterning of SAMs in order to regulate cell growth and function. This technique involves three key components: 1) a light source; 2) a photoresist; and 3) a photomask. The light source provides the energy required for the exposure of photoresist or the ablation of SAM over the selected regions of the substrate. Ultraviolet (UV) light has been the source often used for patterning. Photoresist is a photosensitive material, typically made from organic compounds, whose molecular chains are capable of reorganizing or cross-linking upon exposure to energy. A photomask is a solid substrate usually made of quartz and coated with a thin layer (~1 μm) of chrome with a desired geometrical pattern. In the late 1980s, Kleinfeld et al. first demonstrated that neuron cells could be spatially cultured onto patterned SAMs, made of aminosilane and alkylsilane compounds [58]. A schematic of the processing steps involved in the chemical patterning of a silica glass substrate by conventional photolithography (photoresist based) is shown in Figure 10.4 [58]. The substrate used for patterning is first cleaned by covering them with freshly prepared and suitable cleaning agents (such as H_2O_2 and H_2SO_4). The surface-cleaned substrate is then subjected to spin coating with the photoresist in a clean room facility to avoid contamination (Figure 10.4, step 1). The photoresist-coated substrate is exposed to UV light through the photomask (Figure

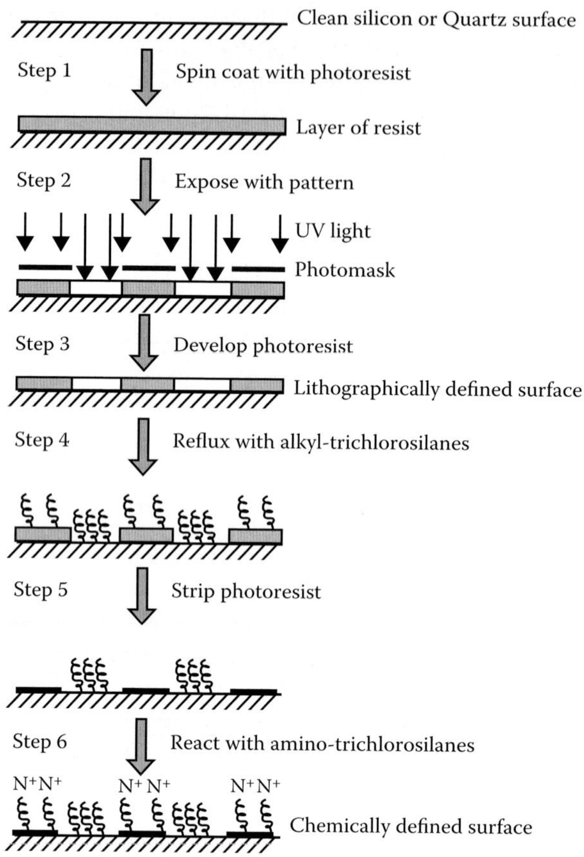

FIGURE 10.4 Schematic of processing steps involved in chemically patterning silicon or silicon dioxide (quartz) substrate. (Adapted with permission from Kleinfeld, D. et al., *Journal of Neuroscience*, 8, 4098–4120, 1988.)

10.4, step 2). Note that the light rays are transmitted only through the quartz background, but not through the chrome pattern on the mask. This chemically alters the solubility of the photoresist in certain developer solutions due to the molecular chain rearrangement. The light-exposed part of the photoresist is then solubilized in a developer solution, which yields a photoresist pattern that corresponds to the image of the pattern designed on the mask (Figure 10.4, step 3). Consequently, the material of interest (e.g., alkylchlorosilane—a cell inhibiting compound) to be patterned is applied on the photoresist pattern (Figure 10.4, step 4), and then, the photoresist is carefully removed (Figure 10.4, step 5). The open areas (the area other than the cell inhibiting compound) are backfilled with another material of interest, e.g., alkylaminosilane, which is a cell promoting compound (Figure 10.4, step 6). Finally, the chemically defined substrate pattern made up of cell promoting and inhibiting compounds, with biphasic regions, is obtained.

The formation of patterned SAMs using different aminosilanes has also been reported for use in neuronal cell studies. Stenger et al. created a DETA/13F patterned

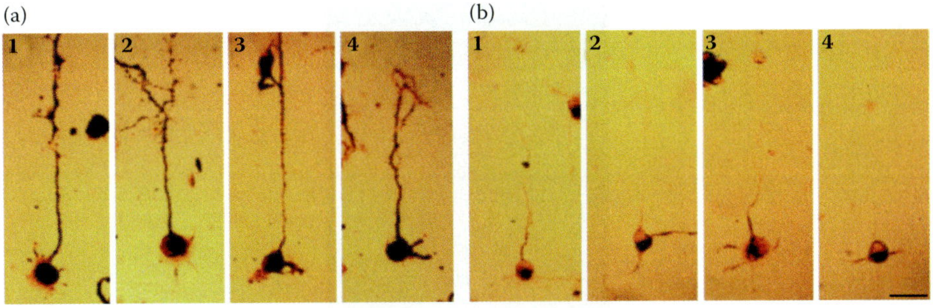

FIGURE 10.5 Immunocytochemical staining of neurons after 3 days in culture on a DETA/13F patterned substrate. Cells are stained using antibodies against (a) MAP5 or (b) MAP2. Scale bar = 50 μm. (Adapted with permission from Stenger, D.A. et al., *Journal of Neuroscience Methods*, 82, 167–173, 1998.)

SAMs without utilizing a photoresist in order to determine their suitability to direct axonal and dendritic process extensions at a single hippocampal neuron level [57]. First, SAMs of DETA were prepared on a glass coverslip, placed in contact with a photomask, and then irradiated with a deep UV light (193 nm) using an Ar/F excimer laser customized with a beam homogenizer. The photomasks used in this study were made from four different patterns over different dimensions. The irradiated surfaces (DETA-derivatized SAMs) were subsequently rederivatized with 13F, which resulted in a coplanar monolayer of permissive and nonpermissive regions. The ability of the patterned system to control neuronal polarity at the single neuron level was evaluated *in vitro* by using embryonic hippocampal cell cultures. To study the overall cellular processes, the cells on the patterned surfaces were stained using antibodies against microtubule associated proteins 5 and 2 (MAP5 and MAP2) (see Figure 10.5). In each case, the dominant process, cell body, and multiple short processes are immunopositive for MAP5 (Figure 10.5a), as expected for rapidly growing neuronal cultures, while MAP2 was more preferentially localized in the somatodendritic domains (Figure 10.5b).

To characterize the polarity of axonal and dendritic processes at the single-cell scale, the cultured neuron was stained with antibodies to a neurofilament (NF150; green) and MAP2 (red). Figure 10.6 shows the double immunostaining of a representative hippocampal neuron after 4 days in culture, which illustrates the localization of NF150, but not MAP2, to the single dominant neurite that is consistent with its identification as the axon. Of the cells with soma localized at the DETA nodes, approximately 76% had axons on the continuous DETA patterns, while dendrites were localized on lines with 13F speed pumps, which emanated from the central DETA node. These results suggested that geometrically designed SAM patterns have had an impact on the neural polarity of *in vitro* cultured hippocampal cells.

10.4.2 Patterning by μCP

μCP is a well-known technique that allows the transfer of patterns onto biomaterial substrates with high spatial resolution suitable for cell studies. This microfabrication technique is one of the most recognized techniques in bioengineering because of its

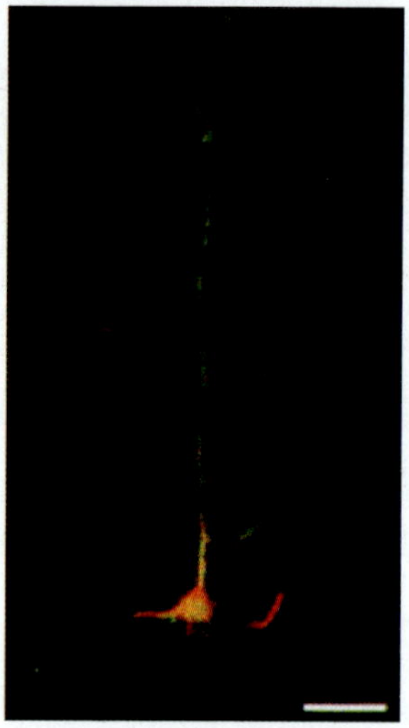

FIGURE 10.6 Double immunostaining of a representative hippocampal neuron after 4 days in culture on a DETA/13F patterned substrate. Neuron was stained with antibodies to NF150 (green) and MAP2 (red) to illustrate induced polarity of axonal and somatodendritic processes, respectively. Scale bar = 20 μm. (Adapted with permission from Stenger, D.A. et al., *Journal of Neuroscience Methods*, 82, 167–173, 1998.)

simplicity, flexibility, and ability to pattern many biomaterials with feature sizes down to 1 μm without using any expensive equipment. In addition, this technique can be extended to pattern a nonplanar (3-D) surface that it is not feasible with the conventional photolithography technique. μCP was initially developed by Singhvi et al. and it was introduced to pattern SAMs of alkanethiolates onto a gold surface to control cell behavior and for engineering cell shape and function by modulating the surface characteristics [87]. The major processing steps involved in the pattern formation by using μCP are schematically shown in Figure 10.7a and b [88]. This method has been based on the utilization of an elastomeric stamp to print a pattern of the material of interest (e.g., cell compatible compounds) on a solid substrate. Three major components are associated with this technique: 1) the master; 2) the stamp; and 3) the ink. The "master" is a solid substrate (e.g., silicon), often created by standard photolithography with a desired geometry with high-resolution features that are specific to a particular application. The "stamp" is a soft elastomeric material, frequently created by casting an elastomeric material (e.g., PDMS) over the predesigned master. The "ink" is a functional material chosen to be patterned onto a substrate material. In the patterning process, the stamp is first inked with a solution made from the materials of interest (e.g., cell-compatible compounds), and the stamp is then brought into contact with the surface of the substrate material. For a period of time, the stamp and the substrate remain undisturbed, which ultimately yields a geometrical pattern of the stamp on the substrate material. The result of the stamping process is the formation of patterns on the surface of the substrate material in regions where the stamp has come into contact with the substrate. The unstamped regions are then backfilled with noncell-compatible compounds in order to ensure that the resultant patterned surfaces have both cell-adhesive and nonadhesive

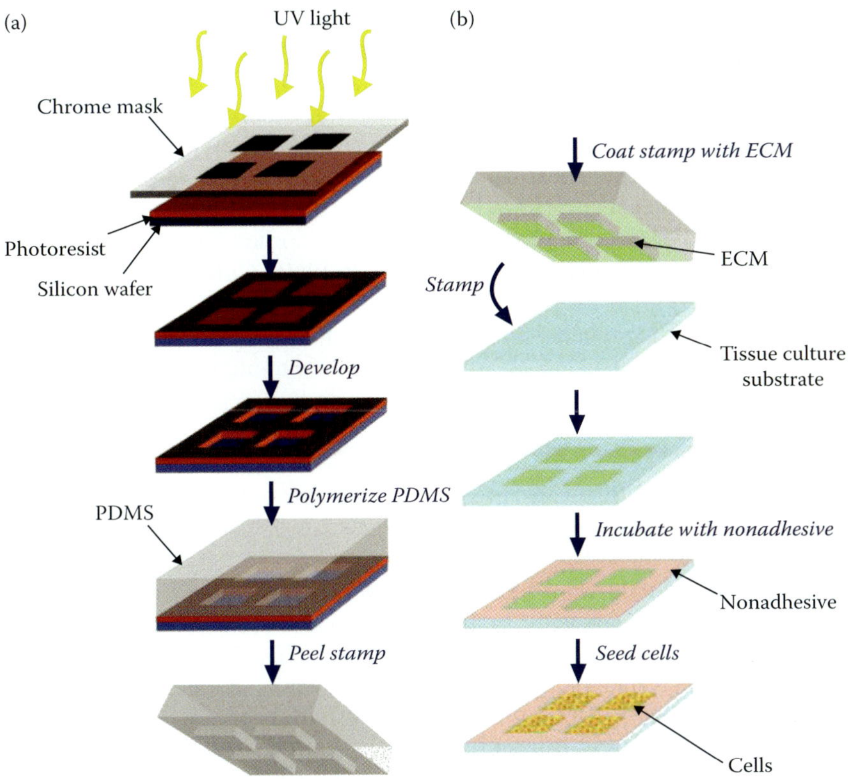

FIGURE 10.7 Schematic of the micropatterning process. PDMS stamps are generated by photolithography (a), and using these stamps, ECM proteins are printed onto tissue culture substrates, upon which cells are seeded (b). (Adapted with permission from Wendy, F.L. and Christopher, S.C., *Materials Today*, 8, 28–35, 2005.)

regions, which directs the cells to grow only on the cell-adhesive regions. The resulting patterned surface can be used to study the fundamental aspects of cell behavior that can be eventually applied to engineering physiologically functional tissues. This kind of patterned substrates will have a tremendous potential in controlling cell fate and function in order to understand the basic mechanism underlying the tissue organization.

10.4.2.1 Topographical Patterning

Topographical patterning refers to the use of substrates with textured patterns by modulating pattern shape and size. Patterning of texture is a physical modification of the surface. Topographical features of the substrate have long been known to play a critical role in dictating the cell behavior. For example, fibroblasts sense the substrate's topography by using filopodia (a cell's main sensory element), which are present as thin extensions on the cell lamellae or emanating directly from the cell body and strongly respond to them [89]. The success of any biomaterial depends on the interaction of the host cells with its surface. Therefore, designing a biomaterial's topography that is suitable to enhance cellular responses is preferred and essential for the organization of the cells in a controlled fashion for many cell-based biomedical and biotechnological applications and is also important in developmental biology in order to study cell behavior, cell–substrate interactions, and cell–cell communications. Harrison was the first researcher who attempted to study the behavior of attached cells to surface structures in the early 1910s [90]; the author compared the effect of nerve cells grown on spider webs and planar coverslips. Later, Weiss studied cell behaviors, in particular, the orientation of neural cells, using milled grooves on mica surfaces, which resulted in the concept of contact guidance [91]. In the early 1960s, Curtis and Varde proposed that cells responded to the microscale topographical environments and their behaviors can be controlled by modulating the surface topography [92]. The results of this study revealed that the cells are sensitive to the degree of curvature of the surface where they are cultured and they are capable of aligning on a cylindrical glass with a diameter less than 100 μm. Yang et al. reported that nanoscale fiber topography of poly(L-lactic acid) promotes neural cell adhesion and differentiation better than its microscale counterparts, which leads to a faster formation of neurite outgrowth and other cellular processes [93]. All of these examples clearly indicate the impact of surface topography in order to regulate cell behavior. Recently, lithographic techniques such as imprint lithography have been developed and have shown a great promise in patterning surfaces with the desired topography. In the following section, we further focus on nanoimprint lithography (NIL), which can have the ability to pattern surfaces with features less than 100 nm in resolution in a convenient, rapid, and inexpensive manner.

10.4.2.2 Nanoimprint Lithography

NIL is a simple and potent technique for patterning surfaces with topographical features that are less than 100 nm in size. This technique is not only simple but also cost effective, because it does not require any expensive instrumentation or sophisticated clean room facilities as conventional lithography (e.g., photolithography), and thus, it is more suitable for biological applications. In addition, it is possible to control the spatial distribution of chemical species on the structured surface. This technique can be applied to pattern 2-D or 3-D nanotopography of different geometries and on a wide range of biomaterial substrates suitable for cell engineering.

This versatile patterning technique, in principle, replicates topographical patterns by means of applied pressure and temperature, in which a rigid master (e.g., silicon) with nanostructure features is imprinted (pressed) onto a polymer resist [e.g., poly(methyl methacrylate) or PMMA], which results in a relief replica of the master on the substrate's surface. Two basic steps are involved in this technique, as shown in Figure 10.8. First is the imprint step, in which a master (also called a "mold") with custom-made nanotopography is pressed onto a polymer resist (usually in the form of a thin film), layered on a substrate material, followed by the removal of the master. This step duplicates the nanotopography of the master on the polymer resist. During the process, the polymer resist is heated to a temperature above its glass transition (T_g), because the resist becomes a viscous liquid at this temperature, which facilitates the polymer to flow and easily mold into the shape of the master. For example, in order to transfer the pattern from a rigid master onto a PMMA resist requires a temperature of around 110°C. PMMA has proven to be an excellent material as a resist

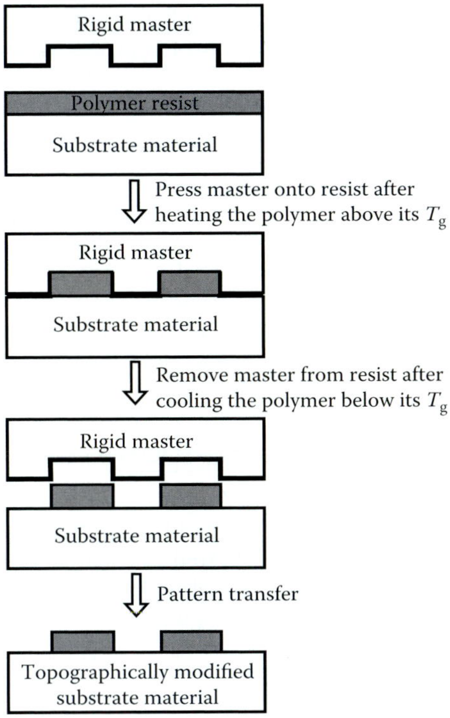

FIGURE 10.8 Schematic of the major steps involved in NIL.

for NIL, because it has a small thermal expansion coefficient of approximately 5×10^{-5} per °C and a small pressure shrinkage coefficient of approximately 3.8×10^{-7} psi. Note that the topography of the master can be designed with a desired geometry by photolithography and reactive ion etching (RIE). The second step is the pattern transfer, where an anisotropic etching process (e.g., RIE) is used to remove the residual resist in the compressed area. This step transfers the thickness of the contrast pattern onto the entire resist, leaving polymer patterns on the substrate material.

Recently, Yim et al. created a nanopatterned silicon substrate using a polymer thin film made of PMMA, which comprised gratings with a 350-nm line width, 700-nm pitch, and 350-nm depth, by NIL in order to study the efficacy of the nanopatterned system in regulating cell behaviors [94,95]. To evaluate the cellular responses to the patterned surfaces, smooth muscle cells were plated on these substrates, and their morphology and concomitant orientation were studied with respect to their elongation and alignment. The cell culture results showed that a patterned surface with nanotopographical features could effectively direct cell orientation and function.

10.5 TECHNIQUES FOR THREE-DIMENSIONAL PATTERNING

The most promising and versatile methods for constructing mimics of native tissues are techniques that enable the creation of 3-D substrates. Most tissue engineering approaches in current use involve random seeding of cells within porous polymer structures. Although this has yielded major advances in the field, the generation of complex tissue structures may require control over the localization of behavior of multiple cell types in 3-D. Moreover, cells cultured in 3-D environment often behave more physiologically than cells cultured on 2-D surfaces [96].

One of the techniques developed for the creation of 3-D cellular patterns is *microfluidics*. In this method, microchannels are used to deliver fluids or cells to selected areas of a substrate, resulting in patterning of the material. Unlike conventional *in vitro* cell culture methods, microfluidics can provide miniature and complex structures mimicking the *in vivo* cellular environment. Hydrogels are frequently used in 3-D tissue engineering applications because of their biocompatibility. Khademhosseini et al. engineered contractile cardiac organoids by using microfluidic patterning of hyaluronic acid [97]. Microfluidic patterning also allows for the formation of 3-D structures consisting of multiple cell types [98–100]. The ability to pattern multiple cell types will pave the way to studying the functional significance of tissue architecture at the resolution of individual cells and the molecular interactions between different cells.

Over the last few years, the field of *bioprinting* has gained significant momentum to demonstrate the potential of printing cells, gels, or a combination of both in a controlled manner on a substrate [101–103]. Xu et al. showed that neuronal cells survive the printing process with a commercial inkjet printer and show healthy morphology and physiology in a 3-D patterned construct. Now, special "cell printers" are available on the market, and direct patterning of cells on 3-D substrates became commercially feasible [101,102].

10.6 INTEGRATION OF CELLS INTO ENGINEERED CONSTRUCTS FOR COMMERCIAL APPLICATIONS

There are several emerging fields of tissue engineering where the integration of living cells with engineered materials is a crucial problem. Live cells are ideal biosensors, can be used for the identification of unknown toxins, or test possible side effects of promising drug candidates. Complex cellular constructs have been created for organ replacement or repair, or as an *in vitro* functional model of an organ system such as the lung or the immune system. Engineered neuronal networks could help in understanding the mechanism of information processing in the brain and led to the development of more efficient computers.

10.6.1 FUNCTIONAL *IN VITRO* DRUG SCREENING PLATFORMS

The development of high-throughput functional drug screening platforms using human cells would have high significance in drug development, personalized disease and drug testing, and personalized therapies. We have initiated several projects to develop such drug screening platforms. We have shown that neonatal rat and stem cell-derived human cardiomyocytes could be patterned with PEG background and fibronectin foreground and could be integrated with substrate-embedded electrodes for high-throughput interrogation (Figure 10.9) [104]. This system would enable the measurement of cardiac side effects of drugs on human cells in a high-throughput screen.

We have also successfully integrated skeletal muscle with cantilevers for high-throughput noninvasive interrogation of muscle functions (Figure 10.10) [105].

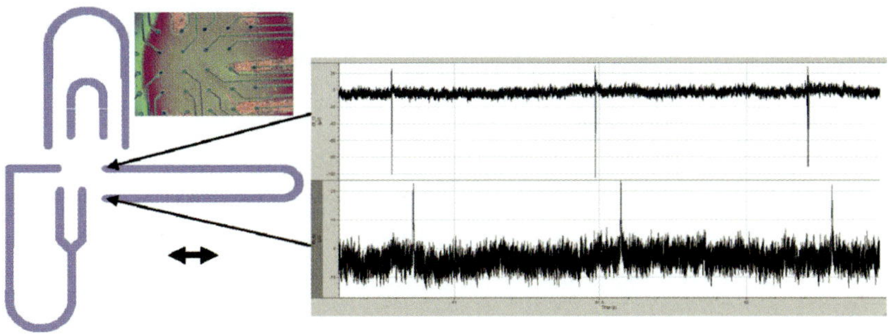

FIGURE 10.9 Extracellular electrophysiological measurements on patterned cardiac myocytes with substrate-embedded electrodes. Long continuous excitation propagation pathways were created from neonatal rat cardiac myocytes using surface patterning. Spontaneous or evoked activity was recorded using extracellular electrodes with high temporal resolution.

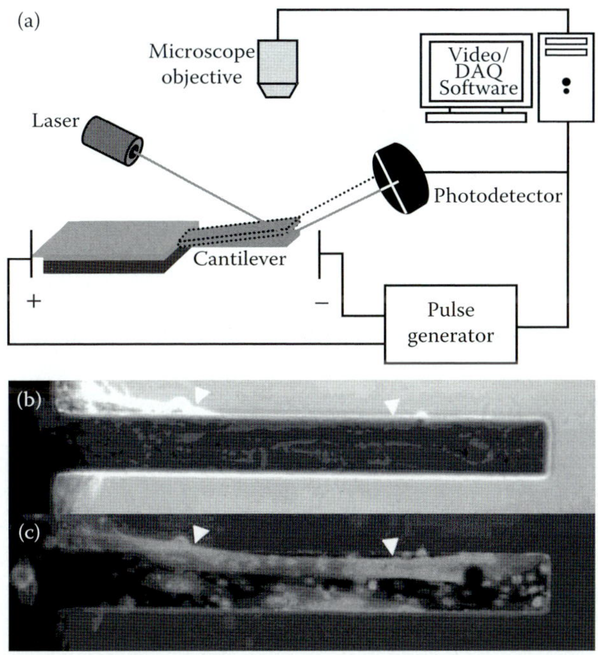

FIGURE 10.10 (a) Simplified schematic of AFM detection system; (b) phase contrast image of C2C12 myotube grown for eight days on silicon cantilevers; (c) fluorescence image of myotube stained for a-actin. Arrow heads indicate the location of the myotube. Skeletal muscle myotubes plated on atomic force microscopy cantilevers. (Wilson, K. et al., *Lab on a Chip* 7, 920–922, 2007. Reproduced with permission from the Royal Society of Chemistry.)

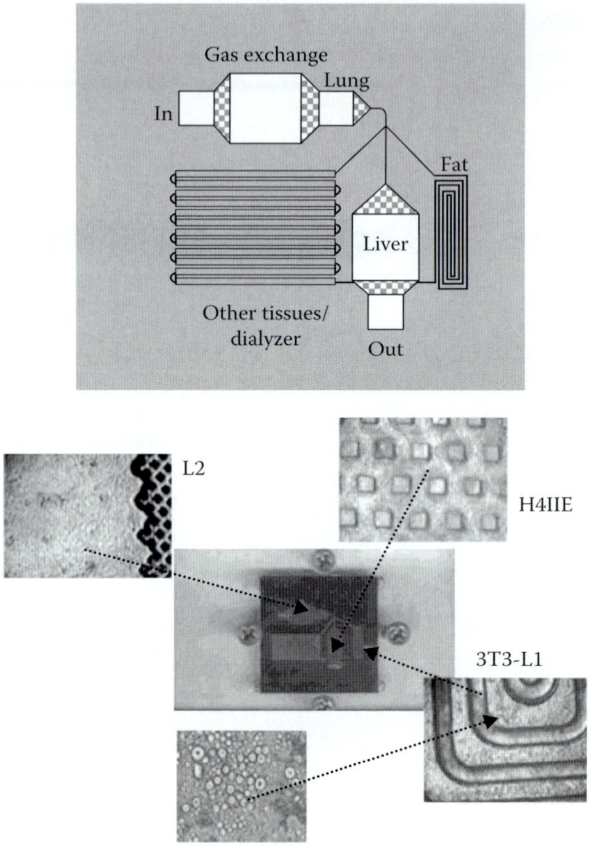

FIGURE 10.11 Body-on-a chip model: Microscopic cell culture analog system with four chambers. Cells are cultured as monolayers on silicon surfaces modified by the adsorption of polylysine and collagen. (Park, T.H. and Shuler, M.L.: *Biotechnology Progress*. 2003. 19: 243–253. Copyright Wiley-VCH Verlag GmBH & Co. KGaA. Reproduced with permission.)

Shuler and Esch of Cornell University are working on the creation of a whole-body metabolic model on a microchip [106]. This device consists of multiple compartments connected with engineered microfluidic channels containing different cell types modeling the body organ systems. The interconnecting microchannel system models the circulation (Figure 10.11). This realistic model could be used for *in vitro* high-content and high-throughput drug screening, replacing *in vivo* experiments. Currently, we are integrating our cardiac muscle, skeletal muscle, and nervous system models into this more complex system.

10.6.2 Engineered Functional Neuronal Networks

The development of a functional nervous system analog is very important for several tissue engineering and medical applications. They could be used for the understanding of information processing in the brain, the development of new antidementia

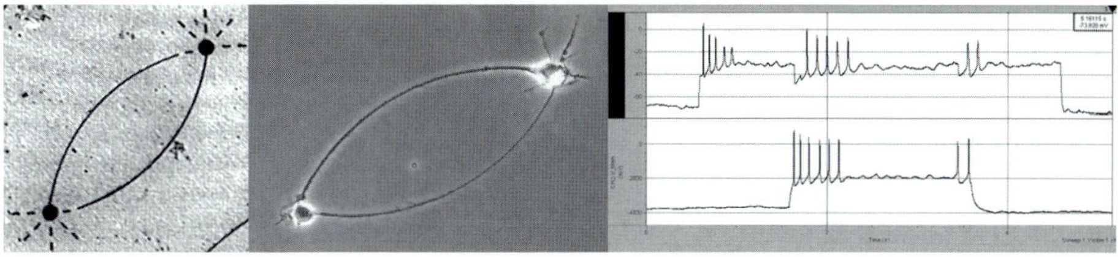

FIGURE 10.12 Engineered neuronal networks. The interconnection of two rat hippocampal neurons was engineered and implemented using surface chemistry and photolithography. Patch clamp electrophysiological recordings proved that the neurons had normal physiology and formed synaptic connections with each other.

drugs, or the engineering of future computers [107,108]. Several groups have already reported partial success in the realization of engineered functional neuronal networks [109,110]. We used photolithography to pattern single neurons and neuronal layers on the top of substrate-embedded microelectrode arrays for drug screening applications and for a functional model of information processing in the brain (Figure 10.12) [108].

10.6.3 Accelerated Biocompatibility Studies

The interface between any engineered material and a biological system is the surface of the biomaterial. Surface treatment could be used to engineer the interaction between the scaffold or implant and the host tissue. We do not have enough theoretical knowledge to predict this interaction, but we can use an experimental approach where we systematically create a series of test surfaces using SAMs and test their effects on cultured cells in a relatively high-throughput way [111]. The above-mentioned 3-D patterning methods could revolutionize tissue engineering by enabling the creation of complex tissue analogs. By engineering surface properties and cues, we can guide the development and integration of the different living components in the scaffold.

10.7 SUMMARY

The development of complex functional tissue analogs or manipulation of single cells in lab-on-a-chip applications became a crucial task in tissue engineering. In the last decade, several novel methods have been developed to engineer the extracellular environment of the cells and determine their adhesion, growth, and differentiation. Other techniques have also been invented to selectively place the cells or materials to engineered positions in 2-D or 3-D scaffolds. These enabling technologies were already used in attempts to create commercial devices for high-throughput drug screening, complex functional tissue analogs, or implants. As our knowledge extends concerning interactions of cells with their environment and more effective methods will be developed for the spatial and temporal patterning of extracellular signals, cells, and biomaterials, much more tissue engineering applications will find their way from the laboratory to the market, changing medical and clinical practice in a fundamental way. Tissue engineered hybrid devices could be cheap enough and have the appropriate throughput to satisfy the demanding requirements for individual testing and the development of personalized medicines. Tissue engineered organs and implants using human cells, personalized drug testing, and therapies might become a reality in the near future.

ACKNOWLEDGMENT

The authors thank the U.S. National Institutes of Health for funding this work through several K01, R21, and R01 research grants.

REFERENCES

1. Murugan, R., P. Molnar, K.P. Rao, and J.J. Hickman. Biomaterial surface patterning of self-assembled monolayers for controlling neuronal cell behavior. *International Journal of Biomedical Engineering and Technology*, 2. pp. 104–134, 2009.
2. Fink, J., M. Thery, A. Azioune, R. Dupont, F. Chatelain, M. Bornens, and M. Piel. Comparative study and improvement of current cell micropatterning techniques. *Lab on a Chip*, 7. pp. 672–680, 2007.
3. Molnar, P., W.S. Wang, A. Natarajan, J.W. Rumsey, and J.J. Hickman. Photolithographic patterning of C2C12 myotubes using vitronectin as growth substrate in serum-free medium. *Biotechnology Progress*, 23. pp. 265–268, 2007.
4. Shin, H. Fabrication methods of an engineered microenvironment for analysis of cell–biomaterial interactions. *Biomaterials*, 28. pp. 126–133, 2007.
5. Yap, F.L. and Y. Zhang. Protein and cell micropatterning and its integration with micro/nanoparticles assembly. *Biosensors and Bioelectronics*, 22. pp. 775–788, 2007.
6. Spatz, J.P. and B. Geiger. Molecular engineering of cellular environments: Cell adhesion to nanodigital surfaces. In *Cell Mechanics*. Elsevier Academic Press Inc., San Diego, CA, p. 89–111, 2007.
7. Altomare, L. and S. Fare. Cells response to topographic and chemical micropatterns. *Journal of Applied Biomaterials and Biomechanics*, 6. pp. 132–143, 2008.
8. Quist, A.P. and S. Oscarsson. Micropatterned surfaces: Techniques and applications in cell biology. *Expert Opinion on Drug Discovery*, 5. pp. 569–581, 2010.
9. Rosines, E., K. Johkura, X. Zhang, H.J. Schmidt, M. DeCambre, K.T. Bush, and S.K. Nigam. Constructing kidney-like tissues from cells based on programs for organ development: Toward a method of *in vitro* tissue engineering of the kidney. *Tissue Engineering—Part A*, 16A. pp. 2441–2455, 2010.
10. Sasaki, J.I., T. Matsumoto, H. Egusa, T. Nakano, T. Ishimoto, T. Sohmura, and H. Yatani. *In vitro* engineering of transitional tissue by patterning and functional control of cells in fibrin gel. *Soft Matter*, 6. pp. 1662–1667, 2010.
11. Khademhosseini, A., R. Langer, J. Borenstein, and J.P. Vacanti. Microscale technologies for tissue engineering and biology. *Proceedings of the National Academy of Sciences of the United States of America*, 103. pp. 2480–2487, 2006.
12. Lim, J.Y. and H.J. Donahue. Cell sensing and response to micro- and nano-structured surfaces produced by chemical and topographic patterning. *Tissue Engineering*, 13. pp. 1879–1891, 2007.
13. Pancrazio, J.J., F. Wang, and C.A. Kelley. Enabling tools for tissue engineering. *Biosensors and Bioelectronics*, 22. pp. 2803–2811, 2007.
14. Hannachi, I.E., M. Yamato, and T. Okano. Cell sheet technology and cell patterning for biofabrication. *Biofabrication*, 1. p. 13, 2009.

15. Kotov, N.A., J.O. Winter, I.P. Clements, E. Jan, B.P. Timko, S. Campidelli, S. Pathak et al. Nanomaterials for neural interfaces. *Advanced Materials*, 21. pp. 3970–4004, 2009.
16. Rao, S.S. and J.O. Winter. Adhesion molecule-modified biomaterials for neural tissue engineering. *Frontiers in Neuroengineering*, 2. p. 6, 2009.
17. Alves, N.M., I. Pashkuleva, R.L. Reis, and J.F. Mano. Controlling cell behavior through the design of polymer surfaces. *Small*, 6. pp. 2208–2220, 2010.
18. Winther, M , Berezin, V and Walmod. NCAM2/OCAM/RNCAM: cell adhesion molecule with a role in neuronal compartmentalization, The international journal of biochemistry & cell biology, 44. pp. 441–446, 2012.
19. Woodward, J. Crossing the endothelium: E-selectin regulates tumor cell migration under flow conditions. *Cell Adhesion and Migration*, 2. pp. 151–152, 2008.
20. Barczyk, M., S. Carracedo, and D. Gullberg. Integrins. *Cell and Tissue Research*, 339. pp. 269–280, 2010.
21. Murugan, R., S.S. Liao, S. Ramakrishna, P. Molnar, Z.M. Huang, M. Kotaki, K.P. Rao, and J.J. Hickman. Skeletal regenerative nanobiomaterials. *Biomaterials for Bone Regenerative Medicine*. Trans Tech Publications Inc., Durnten-Zurich, 2009.
22. Yamada, K.M. Cell surface interactions with extracellular materials. *Annual Review of Biochemistry*, 52. pp. 761–799, 1983.
23. Wilson, C.J., R.E. Clegg, D.I. Leavesley, and M.J. Pearcy. Mediation of biomaterial–cell interactions by adsorbed proteins: A review. *Tissue Engineering*, 11. pp. 1–18, 2005.
24. Charest, J.L., M.T. Eliason, A.J. García, and W.P. King. Combined microscale mechanical topography and chemical patterns on polymer cell culture substrates. *Biomaterials*, 27. pp. 2487–2494, 2006.
25. Papenburg, B.J., E.D. Rodrigues, M. Wessling, and D. Stamatialis. Insights into the role of material surface topography and wettability on cell–material interactions. *Soft Matter*, 6. pp. 4377–4388, 2010.
26. Pollard, J.W. *Animal Cell Culture. Methods in Molecular Biology*, Vol. 5. J.M. Walker, editor. Springer, Clifton, NJ, 1990.
27. Spargo, B.J., M.A. Testoff, T.B. Nielsen, D.A. Stenger, J.J. Hickman, and A.S. Rudolph. Spatially controlled adhesion, spreading, and differentiation of endothelial-cells on self-assembled molecular monolayers. *Proceedings of the National Academy of Sciences of the United States of America*, 91. pp. 11070–11074, 1994.
28. Durbeej, M. Laminins. *Cell and Tissue Research*, 339. pp. 259–268, 2010.
29. Goody, M.F. and C.A. Henry. Dynamic interactions between cells and their extracellular matrix mediate embryonic development. *Molecular Reproduction and Development*, 77. pp. 475–488, 2010.
30. Aplin, A.E., A. Howe, S.K. Alahari, and R.L. Juliano. Signal transduction and signal modulation by cell adhesion receptors: The role of integrins, cadherins, immunoglobulin-cell adhesion molecules, and selectins. *Pharmacological Reviews*, 50. pp. 197–264, 1998.
31. van Kooten, T.G. and A.F. von Recum. Cell adhesion to textured silicone surfaces: The influence of time of adhesion and texture on focal contact and fibronectin fibril formation. *Tissue Engineering*, 5. pp. 223–240, 1999.
32. Kato, M. and M. Mrksich. Using model substrates to study the dependence of focal adhesion formation on the affinity of integrin–ligand complexes. *Biochemistry*, 43. pp. 2699–2707, 2004.
33. Lee, M.H., D.A. Brass, R. Morris, R.J. Composto, and P. Ducheyne. The effect of nonspecific interactions on cellular adhesion using model surfaces. *Biomaterials*, 26. pp. 1721–1730, 2005.

34. Li, L.Y., S.F. Chen, and S.Y. Jiang. Protein adsorption on alkanethiolate self-assembled monolayers: Nanoscale surface structural and chemical effects. *Langmuir*, 19. pp. 2974–2982, 2003.
35. Martins, M.C.L., B.D. Ratner, and M.A. Barbosa. Protein adsorption on mixtures of hydroxyl- and methyl-terminated alkanethiols self-assembled monolayers. *Journal of Biomedical Materials Research, Part A*, 67A. pp. 158–171, 2003.
36. Faucheux, N., R. Schweiss, K. Lutzow, C. Werner, and T. Groth. Self-assembled monolayers with different terminating groups as model substrates for cell adhesion studies. *Biomaterials*, 25. pp. 2721–2730, 2004.
37. Qian, L.C. and W.M. Saltzman. Improving the expansion and neuronal differentiation of mesenchymal stem cells through culture surface modification. *Biomaterials*, 25. pp. 1331–1337, 2004.
38. Li, M., T. Cui, D.K. Mills, Y.M. Lvov, and M.J. McShane. Comparison of selective attachment and growth of smooth muscle cells on gelatin- and fibronectin-coated micropatterns. *Journal of Nanoscience and Nanotechnology*, 5. pp. 1809–1815, 2005.
39. Li, N. and A. Folch. Integration of topographical and biochemical cues by axons during growth on microfabricated 3-D substrates. *Experimental Cell Research*, 311. pp. 307–316, 2005.
40. Hayashi, Y., M.K. Furue, T. Okamoto, K. Ohnuma, Y. Myoishi, Y. Fukuhara, T. Abe, J.D. Sato, R.I. Hata, and M. Asashima. Integrins regulate mouse embryonic stem cell self-renewal. *Stem Cells*, 25. pp. 3005–3015, 2007.
41. Sodunke, T.R., K.K. Turner, S.A. Caldwell, K.W. McBride, M.J. Reginato, and H.M. Noh. Micropatterns of Matrigel for three-dimensional epithelial cultures. *Biomaterials*, 28. pp. 4006–4016, 2007.
42. Wittmer, C.R., J.A. Phelps, W.M. Saltzman, and P.R. Van Tassel. Fibronectin-terminated multilayer films: Protein adsorption and cell attachment studies. *Biomaterials*, 28. pp. 851–860, 2007.
43. Jose, A. and L.K. Krishnan. Effect of matrix composition on differentiation of nestin-positive neural progenitors from circulation into neurons. *Journal of Neural Engineering*, 7. pp. 10, 2010.
44. Lin, Y.M., A. Zhang, H.J. Rippon, A. Bismarck, and A.E. Bishop. Tissue engineering of lung: The effect of extracellular matrix on the differentiation of embryonic stem cells to pneumocytes. *Tissue Engineering, Part A*, 16A. pp. 1515–1526, 2010.
45. So, K.H., Y.J. Han, H.Y. Park, J.G. Kim, D.J. Sung, Y.M. Bae, B.C. Yang, S.B. Park, S.K. Chang, E.Y. Kim, and S.P. Park. Generation of functional cardiomyocytes from mouse induced pluripotent stem cells. *International Journal of Cardiology*, 2010. pp. 24, 2010.
46. Wink, T., S.J. van Zuilen, A. Bult, and W.P. van Bennekom. Self-assembled monolayers for biosensors. *Analyst*, 122. pp. R43–R50, 1997.
47. Silver, J.H., J.-C. Lin, F. Lim, V.A. Tegoulia, M.K. Chaudhury, and S.L. Cooper. Surface properties and hemocompatibility of alkyl-siloxane monolayers supported on silicone rubber: Effect of alkyl chain length and ionic functionality. *Biomaterials*, 20. pp. 1533–1543, 1999.
48. Bain, C.D. and G.M. Whitesides. Molecular-level control over surface order in self-assembled monolayer films of thiols on gold. *Science*, 240. pp. 62–63, 1988.
49. Stenger, D.A., J.H. Georger, C.S. Dulcey, J.J. Hickman, A.S. Rudolph, T.B. Nielsen, S.M. McCort, and J.M. Calvert. Coplanar molecular assemblies of aminoalkylsilane and perfluorinated alkylsilane—Characterization and geometric definition of mammalian-cell adhesion and growth. *Journal of the American Chemical Society*, 114. pp. 8435–8442, 1992.
50. Hickman, J.J., S.K. Bhatia, J.N. Quong, P. Shoen, D.A. Stenger, C.J. Pike, and C.W. Cotman. Rational pattern design for *in vitro* cellular networks using surface photochemistry. *Journal of Vacuum Science and Technology, A: Vacuum, Surfaces, and Films*, 12. pp. 607–616, 1994.

51. Hickman, J.J., K. Foster, D.A. Stenger, A. Schaffner, and J. Barker. The use of monolayers as templates for biocompatibility studies. *Abstracts of Papers of the American Chemical Society*, 205. p. 146-COLL, 1993.
52. Vargo, T.G., P.M. Thompson, L.J. Gerenser, R.F. Valentini, P. Aebischer, D.J. Hook, and J.A. Gardella. Monolayer chemical lithography and characterization of fluoropolymer films. *Langmuir*, 8. pp. 130–134, 1992.
53. Ferguson, G.S., M.K. Chaudhury, H.A. Biebuyck, and G.M. Whitesides. Monolayers on disordered substrates—Self-assembly of alkyltrichlorosilanes on surface-modified polyethylene and poly(dimethylsiloxane). *Macromolecules*, 26. pp. 5870–5875, 1993.
54. Ulman, A. *Ultrathin Organic Films: From Langmuir–Blodgett to Self-Assembly*. Academic Press, Boston, MA, 1991.
55. Fields, G.B., J.L. Lauer, Y. Dori, P. Forns, Y.C. Yu, and M. Tirrell. Protein-like molecular architecture: Biomaterial applications for inducing cellular receptor binding and signal transduction. *Biopolymers*, 47. pp. 143–151, 1998.
56. Fields, G.B. Induction of protein-like molecular architecture by self-assembly processes. *Bioorganic and Medicinal Chemistry*, 7. pp. 75–81, 1999.
57. Stenger, D.A., J.J. Hickman, K.E. Bateman, M.S. Ravenscroft, W. Ma, J.J. Pancrazio, K. Shaffer, A.E. Schaffner, D.H. Cribbs, and C.W. Cotman. Microlithographic determination of axonal/dendritic polarity in cultured hippocampal neurons. *Journal of Neuroscience Methods*, 82. pp. 167–173, 1998.
58. Kleinfeld, D., K.H. Kahler, and P.E. Hockberger, Controlled outgrowth of dissociated neurons on patterned substrates. *Journal of Neuroscience*, 8. pp. 4098–4120, 1988.
59. Cooper, E., L. Parker, C.A. Scotchford, S. Downes, G.J. Leggett, and T.L. Parker. The effect of alkyl chain length and terminal group chemistry on the attachment and growth of murine 3T3 fibroblasts and primary human osteoblasts on self-assembled monolayers of alkanethiols on gold. *Journal of Materials Chemistry*, 10. pp. 133–139, 2000.
60. Cheng, S.-S., K.K. Chittur, C.N. Sukenik, L.A. Culp, and K. Lewandowska. The conformation of fibronectin on self-assembled monolayers with different surface composition: An FTIR/ATR study. *Journal of Colloid and Interface Science*, 162. pp. 135, 1994.
61. Stenger, D.A., C.J. Pike, J.J. Hickman, and C.W. Cotman. Surface determinants of neuronal survival and growth on self-assembled monolayers in culture. *Brain Research*, 630. pp. 136–147, 1993.
62. Franco, M., P.F. Nealey, S. Campbell, A.I. Teixeira, and C.J. Murphy. Adhesion and proliferation of corneal epithelial cells on self-assembled monolayers. *Journal of Biomedical Materials Research*, 52. pp. 261–269, 2000.
63. Margel, S., E.A. Vogler, L. Firment, T. Watt, S. Haynie, and D.Y. Sogah. Peptide, protein and cellular interactions with self-assembled monolayer model surfaces. *Journal of Biomedical Materials Research*, 27. pp. 1463–1476, 1993.
64. Ostuni, E., R.G. Chapman, M.N. Liang, G. Meluleni, G. Pier, D.E. Ingber, and G.M. Whitesides. Self-assembled monolayers that resist the adsorption of proteins and the adhesion of bacterial and mammalian cells. *Langmuir*, 17. pp. 6336–6343, 2001.
65. Mrksich, M. Tailored substrates for studies of attached cell culture. *Cellular and Molecular Life Sciences*, 54. pp. 653–662, 1998.
66. Mrksich, M. A surface chemistry approach to studying cell adhesion. *Chemical Society Reviews*, 29. pp. 267–273, 2000.
67. Tidwell, C.D., S.I. Ertel, B.D. Ratner, B.J. Tarasevich, S. Atre, and D.L. Allara. Endothelial cell growth and protein adsorption on terminally functionalized, self-assembled monolayers of alkanethiolates on gold. *Langmuir*, 13. pp. 3404–3413, 1997.
68. Faucheux, N., R. Tzoneva, M.D. Nagel, and T. Groth. The dependence of fibrillar adhesions in human fibroblasts on substratum chemistry. *Biomaterials*, 27. pp. 234–245, 2006.

69. Hutt, D.A., E. Cooper, L. Parker, G.J. Leggett, and T.L. Parker. Fabrication of cellular wires on micropatterned monolayers of short-chain alkanethiols on gold. *Langmuir*, 12. pp. 5494–5497, 1996.
70. Zhu, B., T. Eurell, R. Gunawan, and D. Leckband. Chain-length dependence of the protein and cell resistance of oligo(ethylene glycol)-terminated self-assembled monolayers on gold. *Journal of Biomedical Materials Research*, 56. pp. 406–416, 2001.
71. Thomas, C.H., J.B. Lhoest, D.G. Castner, C.D. McFarland, and K.E. Healy. Surfaces designed to control the projected area and shape of individual cells. *Transactions of the ASME, Series K: Journal of Biomechanical Engineering*, 121. pp. 40–48, 1999.
72. Keselowsky, B.G., D.M. Collard, and A.J. García. Integrin binding specificity regulates biomaterial surface chemistry effects on cell differentiation. *Proceedings of the National Academy of Sciences of the United States of America*, 102. pp. 5953–5957, 2005.
73. Amirpour, M.L., P. Ghosh, W.M. Lackowski, R.M. Crooks, and M.V. Pishko. Mammalian cell cultures on micropatterned surfaces of weak-acid, polyelectrolyte hyperbranched thin films on gold. *Analytical Chemistry*, 73. pp. 1560–1566, 2001.
74. Das, M., P. Molnar, C. Gregory, L. Riedel, A. Jamshidi, and J.J. Hickman. Long-term culture of embryonic rat cardiomyocytes on an organosilane surface in a serum-free medium. *Biomaterials*, 25. pp. 5643–5647, 2004.
75. Acarturk, T.O., M.M. Peel, P. Petrosko, W. LaFramboise, P.C. Johnson, and P.A. DiMilla. Control of attachment, morphology, and proliferation of skeletal myoblasts on silanized glass. *Journal of Biomedical Materials Research*, 44. pp. 355–370, 1999.
76. Schaffner, A.E., J.L. Barker, D.A. Stenger, and J.J. Hickman. Investigation of the factors necessary for growth of hippocampal neurons in a defined system. *Journal of Neuroscience Methods*, 62. pp. 111–119, 1995.
77. Kingshott, P. and H.J. Griesser. Surfaces that resist bioadhesion. *Current Opinion in Solid State and Materials Science*, 4. pp. 403–412, 1999.
78. Das, M., P. Molnar, H. Devaraj, M. Poeta, and J.J. Hickman. Electrophysiological and morphological characterization of rat embryonic motoneurons in a defined system. *Biotechnology Progress*, 19. pp. 1756–1761, 2003.
79. Ista, L.K., S. Mendez, and G.P. Lopez. Attachment and detachment of bacteria on surfaces with tunable and switchable wettability. *Biofouling*, 26. pp. 111–118, 2010.
80. Liu, Y., L. Mu, B. Liu, and J. Kong. Controlled switchable surface. *Chemistry*, 11. pp. 2622–2631, 2005.
81. Mendes, P.M. Stimuli-responsive surfaces for bioapplications. *Chemical Society Reviews*, 37. pp. 2512–2529, 2008.
82. Hook, A.L., N.H. Voelcker, and H. Thissen. Patterned and switchable surfaces for biomolecular manipulation. *Acta Biomaterialia*, 5. pp. 2350–2370, 2009.
83. Shimizu, T., M. Yamato, Y. Isoi, T. Akutsu, T. Setomaru, K. Abe, A. Kikuchi, M. Umezu, and T. Okano. Fabrication of pulsatile cardiac tissue grafts using a novel 3-dimensional cell sheet manipulation technique and temperature-responsive cell culture surfaces. *Circulation Research*, 90. pp. E40–E48, 2002.
84. Quirk, R.A., B. Kellam, R.N. Bhandari, M.C. Davies, S.J.B. Tendler, and K.M. Shakesheff. Cell type–specific adhesion onto polymer surfaces from mixed cell populations. *Biotechnology and Bioengineering*, 81. pp. 625–628, 2003.
85. Liu, R.Y., R.S. Schmid, W.D. Snider, and P.F. Maness. NGF enhances sensory axon growth induced by laminin but not by the L1 cell adhesion molecule. *Molecular and Cellular Neuroscience*, 20. pp. 2–12, 2002.
86. Shen, H., J. Tan, and W.M. Saltzman. Surface-mediated gene transfer from nanocomposites of controlled texture. *Nature Materials*, 3. pp. 569–574, 2004.
87. Singhvi, R., A. Kumar, G.P. Lopez, G.N. Stephanopoulos, D.I. Wang, G.M. Whitesides, and D.E. Ingber. Engineering cell shape and function. *Science*, 264. pp. 696–698, 1994.

88. Liu, W.F. and C.S. Chen. Engineering biomaterials to control cell function. *Materials Today*, 8. pp. 28–35, 2005.
89. Dalby, M.J., M.O. Riehle, H.J.H. Johnstone, S. Affrossman, and A.S.G. Curtis. Polymer-demixed nanotopography: Control of fibroblast spreading and proliferation. *Tissue Engineering*, 8. pp. 1099–1108, 2002.
90. Harrison, R.G. On the stereotropism of embryonic cells. *Science*, 34. pp. 279, 1911.
91. Weiss, P. Experiments on cells and axon orientation *in vitro*: The role of colloidal exudates in tissue organization. *Journal of Experimental Zoology*, 100. pp. 353, 1945.
92. Curtis, A.S.G. and M. Varde. Control of cell behavior: Topographical factors. *Journal of the National Cancer Institute*, 33. p. 15, 1964.
93. Yang, F., R. Murugan, S. Wang, and S. Ramakrishna. Electrospinning of nano/micro-scale poly(L-lactic acid) aligned fibers and their potential in neural tissue engineering. *Biomaterials*, 26. pp. 2603–2610, 2005.
94. Yim, E.K.F. and K.W. Leong. Significance of synthetic nanostructures in dictating cellular response. *Nanomedicine: Nanotechnology, Biology and Medicine*, 1. pp. 10–21, 2005.
95. Yim, E.K.F., R.M. Reano, S.W. Pang, A.F. Yee, C.S. Chen, and K.W. Leong. Nanopattern-induced changes in morphology and motility of smooth muscle cells. *Biomaterials*, 26. pp. 5405–5413, 2005.
96. Sun, T., S. Jackson, J.W. Haycock, and S. MacNeil. Culture of skin cells in 3-D rather than 2-D improves their ability to survive exposure to cytotoxic agents. *Journal of Biotechnology*, 122. pp. 372–381, 2006.
97. Khademhosseini, A., G. Eng, J. Yeh, P.A. Kucharczyk, R. Langer, G. Vunjak-Novakovic, and M. Radisic. Microfluidic patterning for fabrication of contractile cardiac organoids. *Biomedical Microdevices*, 9. pp. 149–157, 2007.
98. Tan, W. and T.A. Desai. Microfluidic patterning of cellular biopolymer matrices for biomimetic 3-D structures. *Biomedical Microdevices*, 5. pp. 235–244, 2003.
99. Tan, W. and T.A. Desai. Layer-by-layer microfluidics for biomimetic three-dimensional structures. *Biomaterials*, 25. pp. 1355–1364, 2004.
100. Chiu, D.T., N.L. Jeon, S. Huang, R.S. Kane, C.J. Wargo, I.S. Choi, D.E. Ingber, and G.M. Whitesides. Patterned deposition of cells and proteins onto surfaces by using three-dimensional microfluidic systems. *Proceedings of the National Academy of Sciences of the United States of America*, 97. pp. 2408–2413, 2000.
101. Xu, T., C.A. Gregory, P. Molnar, X. Cui, S. Jalota, S.B. Bhaduri, and T. Boland. Viability and electrophysiology of neural cell structures generated by the inkjet printing method. *Biomaterials*, 27. pp. 3580–3588, 2006.
102. Xu, T., P. Molnar, C. Gregory, M. Das, T. Boland, and J.J. Hickman. Electrophysiological characterization of embryonic hippocampal neurons cultured in a 3-D collagen hydrogel. *Biomaterials*, 30. pp. 4377–4383, 2009.
103. Jakab, K., A. Neagu, V. Mironov, R.R. Markwald, and G. Forgacs. Engineering biological structures of prescribed shape using self-assembling multicellular systems. *Proceedings of the National Academy of Sciences of the United States of America*, 101. pp. 2864–2869, 2004.
104. Natarajan, A., M. Stancescu, V. Dhir, C. Armstrong, F. Sommerhage, J.J. Hickman, and P. Molnar. Patterned cardiomyocytes on microelectrode arrays as a functional, high information content drug screening platform. *Biomaterials*, 32. pp. 4267–4274, 2011.
105. Wilson, K., P. Molnar, and J. Hickman. Integration of functional myotubes with a Bio-MEMS device for non-invasive interrogation. *Lab on a Chip*, 7. pp. 920–922, 2007.
106. Shuler, M.L. and M.B. Esch. Body-on-a chip: Using microfluidic systems to predict human responses to drugs. *Pure and Applied Chemistry*, 82. pp. 1635–1645, 2010.
107. Pearce, T.M. and J.C. Williams. Microtechnology: Meet neurobiology. *Lab on a Chip*, 7. p. 30–40, 2007.

108. Molnar, P., J. Kang, N. Bhargava, M. Das, and J.J. Hickman. Synaptic connectivity in engineered neuronal networks. In *Patch-Clamp Methods and Protocols*, P. Molnar and J.J. Hickman, editors. Humana Press, New York, 2007.
109. Wheeler, B.C. and G.J. Brewer. Designing neural networks in culture: Experiments are described for controlled growth, of nerve cells taken from rats, in predesigned geometrical patterns on laboratory culture dishes. *Proceedings of the IEEE*, 98. pp. 398–406, 2010.
110. Rowe, L., M. Almasri, K. Lee, N. Fogleman, G.J. Brewer, Y. Nam, B.C. Wheeler, J. Vukasinovic, A. Glezer, and A.B. Frazier. Active 3-D microscaffold system with fluid perfusion for culturing *in vitro* neuronal networks. *Lab on a Chip*, 7. pp. 475–482, 2007.
111. Natarajan, A., C. Chun, J.J. Hickman, and P. Molnar. Growth and electrophysiological properties of rat embryonic cardiomyocytes on hydroxyl- and carboxyl-modified surfaces. *Journal of Biomaterials Science, Polymer Edition*, 19. pp. 1319–1331, 2008.
112. Park, T.H. and M.L. Shuler. *Biotechnology Progress*. 2003. 19. 243–253.

11 Electrospun Nanofibers as Next-Generation Bioactive Tissue Scaffolds

Victor Leung, Ryan Hartwell, Elodie Faure, Heejae Yang, Aziz Ghahary, and Frank Ko

CONTENTS

11.1 NEW FRONTIERS FOR TISSUE REGENERATION

Tissue repair has been an area of great interest in the medical field due to the massive demands stemming from patients worldwide suffering from injuries and diseases that necessitate repair or replacement of various tissues such as skin, bones, and internal organs. For example, in the United States, 500,000 burn patients each year require hospitalization (American Burn Association 2009), more than 40 million individuals are at high risk of bone fracture due to osteoporosis or low bone mass (NIH Osteoporosis and Related Bone Diseases National Resource Center 2011), and more than 110,000 patients wait for life-saving organ transplants, whereas only 24,000 transplantations were performed in 2010 (United Network for Organ Sharing 2010). Many of these conditions are dangerous for patients, as they lose vital body functions, and therefore, it is critical that any lost functions are restored fully and in a timely manner. Accordingly, the market for medical care is large, such as the pharmaceutical sector, which has an annual sales of $750 billion (Organization

2010), and the wound care sector, which is currently worth $13 billion (Research and Markets 2009).

Autografting is the current standard for tissue repair due to the reduction in allergic or immunogenic response, but donor site shortage and morbidity remain a challenge. As a result, many have examined synthetic analogs, typically known as scaffolds, as an alternative. The key advantage of using synthetic tissue scaffolds as opposed to autografting is that no donor site is required, which reduces the potential for further injuries and infection. Correspondingly, material choice and form becomes an important consideration. Typically, the scaffold must be biocompatible and allow cell attachment and migration. Biodegradable designs are also desirable, as they eliminate the need for removal surgeries. For cell activities, two scaffolding approaches are common: 1) porous scaffolds and 2) hydrogel scaffolds. Porous scaffolds, including pellets (Burg et al. 2000), sponges (Yannas and Burke 1980; Li and Zhang 2005), and fibers (Ko 1997; Flemming et al. 1999; Yang et al. 2005; Xie and Wang 2006; Leung et al. 2010), allow cells to attach, proliferate, and differentiate on surfaces, and the goal for applying porous scaffolds is to obtain sufficient coverage of the appropriate cells. Hydrogel scaffold, on the other hand, allows cells to migrate into the gel matrix, which is mainly composed of water, followed by proliferation and differentiation. Dimensions for porous scaffolds are often fixed, and therefore, structural integrity is a requirement, whereas hydrogel can be handled and applied as a liquid. In terms of applicability, hydrogels can be cast into specific shapes or injected into the repair site, whereas foams and pellets can be shaped and then inserted into the damaged area. Fiber, on the other hand, can be assembled into a range of textile forms for optimal application, including woven and nonwoven pads, and sutures (Ko 1997). In addition, fibrous structures can also be designed to withstand high loading, which is ideal for applications such as dressings under pressure garments and sutures. For example, fibrous scaffolds such as Resolut LT and BioFix have tensile strengths on the order of 10 MPa (Rho et al. 2006), more than two orders of magnitude higher than other forms such as hydrogel. Traditionally, fibers have played a mainly supportive role in tissue regeneration, such as structural backing. In the recent decade, however, a new generation of fibers with nanometric diameters have attracted attention in tissue repair. With a large aspect ratio, these nanofibers have a large surface area available for cell activities.

For nanofibers to adopt a more active role to effectively encourage tissue regeneration, unlike traditional fibrous scaffolds, recent research has focused on enhancing bioactivity via improving cell attachment on nanofibers and enabling cell signaling by the scaffolds. This chapter explores the latest advancement of nanofibers for tissue regeneration, with a focus on how nanofiber scaffolds can be tailored to better suit specific applications, including an introduction to nanofiber technology, a discussion on material choice, and the current trends in nanofiber technology and their potential impact on the tissue repair market.

11.2 ELECTROSPINNING TECHNOLOGY

Nanofiber technology combines the aforementioned desirable fibrous properties with a large surface area for biological and biochemical activities. In theory, several attributes of nanofibers are suitable for tissue repair, such as the structural resemblance to native extracellular matrix (ECM), which is composed of nanofibrous collagen, and the large surface area-to-volume ratio. From a processing point of view, a wider range of polymers can be fabricated into nanofibers, unlike other forms such as hydrogel and pellets. The flexibility in material choice, which will be further discussed in the next section, allows a wider range of material-specific properties to be manifested in the scaffolds, such as antimicrobial activity and absorbability. In addition, similar to other fibrous structures, nanofibers are connective, and therefore, it is possible to further fabricate nanofibers into higher order textile products, as mentioned previously. The ability to form higher order textile products with nanofibers opens the opportunities for fabricating hierarchical structures, which is the basis for many biological systems such as the skin, tendon, and bone.

There are several approaches for fabricating nanofibers, which can involve either top–down or bottom–up processes. Examples of top–down processes include drawing from a polymer solution with a micropipette tip, template synthesis using a nanoporous membrane, phase separation between a polymer solution and an incompatible solvent, and electrospinning that draws fibers from polymer solutions using static repulsion (Ramakrishna et al. 2005). Among the bottom–up approaches, drawing is the simplest but also has the lowest productivity. Template synthesis and phase separation are batch processes, and therefore, the throughput can be higher than drawing, but the fiber-forming processes are more complex while posing larger requirements on reagents. Electrospinning requires high voltage, but the process is simple and can be continuous. On the other hand, bottom–up approaches are based on the self-assembly of molecules such as proteins. Nanofibers from self-assembly have the greatest tailorability in material properties due to the possibility of customizing polymer constituents such as the self-assembly of peptide-amphiphiles or amino acid sequences for biochemically mimicking native ECM (Hartgerink et al. 2001; Niece et al. 2003). However, nanofiber fabrication from self-assembly is also the most complex due to stringent processing requirements, including temperature and pH, whereas productivity is often low.

Among the fabrication approaches, electrospinning has been the most explored due to its simplicity and feasibility for mass production. In addition, electrospinning enables control over a wide range of fiber properties, including diameter, porosity, orientation, and surface chemistry. The use of electrospinning to produce textile fibers was patented by Formhals in 1934 (Formhals 1934), although investigations on overcoming liquid surface tension with electrostatic charge had been conducted by Rayleigh in 1882 (Katti et al. 2005), and the resulting electrospraying technique had been patented by Cooley and Morton in 1902 (Hutmacher and Dalton 2011). The principle of electrospinning is the application of sufficient electrostatic charge onto a polymer solution to overcome its surface tension, thereby forming a cone and eventually a fiber jet that will travel toward a grounded collector (Ramakrishna et al. 2005). A schematic of a typical electrospinning setup is shown in Figure 11.1. Nanofiber fabrication by electrospinning, however, was not realized until the pioneering work of the Reneker group in the 1990s (Ko et al. 1998; Li and Xia 2004; Leung and Ko 2010), which revitalized attention on the technique as a new area of multidisciplinary studies involving, for example, chemistry, physics, and computational modeling.

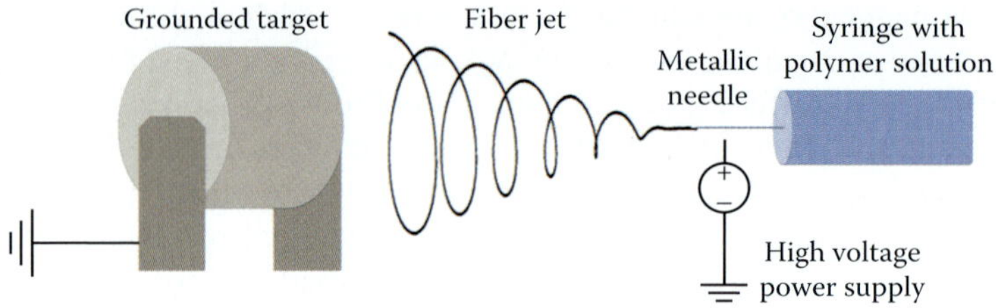

FIGURE 11.1 Typical setup for electrospinning, with the solution contained in a syringe, power supply, and a ground collector.

Since then, the number of studies on nanofibers, including those for biomedical applications, has exponentially increased. Much of the earlier studies aimed at verifying the desirable properties of nanofibers for tissue regeneration, including the ECM-mimicking structure, adjustable fiber diameter and porosity, and large surface area. Ko et al. (1998) presented one of the first studies on cell interaction with nanofibers, outlining the effect of the increased surface area on the nanofiber scaffold compared to a microfiber one. Building on these early results, Ko and Gandhi (2007) further reported that fibroblast cell proliferation on *Bombyx mori* silk nanofibers were superior to that on microfibers of the same silk, highlighting that cell activities on nanofibers are diameter sensitive. In addition, Ko and Gandhi (2007) showed that cells adapt differently on the microfiber and nanofiber scaffolds. Although fibroblast cells were more crimped on the microfibers, they were more spread on the nanofiber scaffold, resembling fibroblast morphology on native tissue. Besides silk, Ko et al. also demonstrated that nanofiber scaffolds can be fabricated from native tissue proteins such as collagen and elastin while maintaining the cell activities on these materials (Li et al. 2005a). Scaffold interactions with cells such as fibroblasts, chondrocytes, and mesenchymal stem cells (MSC) were also outlined in earlier studies by Li et al. (2005b), showing positive initial results, including comparable chondrogenesis between MSC in nanofiber scaffold and those in cell pellets and aggregates, as well as increased ECM production by chondrocytes on nanofiber scaffold compared to monolayer cultures on tissue culture plastics. In addition, Li et al. (2002) examined fibroblast interaction with a poly(lactic-*co*-glycolic acid) (PLGA) electrospun scaffold and suggested that the pore sizes, with the majority in the 25–100 μm range, are adequate for fibroblast migration. Besides cell interaction, Katti et al. (2005) studied the effect of different electrospinning parameters on nanofiber architecture and verified the feasibility of loading antibiotic drugs into the fibers without compromising the morphology. More recently, our study on alginate nanofibers also showed improved cell viability and proliferation compared to a commercial alginate-based microfibrous dressing (Leung et al. 2010).

Another benefit of electrospinning over other fabrication approaches is the freedom in nanofiber placement and orientation. By adjusting the position and rotation of the fiber collector, nanofiber assemblies with different morphologies, orientations, density, and patterns can be fabricated. Fiber morphology has been studied by Fridrikh et al. (2003) via modeling and experimental work, with early models (based on Newton's second law), focusing on the fiber jet formation; this has been summarized in the work of Wan et al. (2010). Wan et al. (2010) also presented a thermodynamic-based model that aimed at determining the final fiber diameter in electrospinning, as shown in Equation 11.1. Equation 11.1 shows the relationship between the fiber diameter r and spinning parameters such as the volumetric flow rate Q_v, current I, electric field density E, and axial coordinate z, as well as solution parameters, including density ρ and polymer weight concentration C. In a more recent work, D'Amore et al. (2010) demonstrated the use of an image-based algorithm for characterizing the topography of electrospun poly(ester urethane) fibers, including diameter, intersection density, angle distribution, and connectivity. In terms of experimental work, various groups have identified a range of process parameters that affect fiber morphology and porosity, including voltage, spinning distance, solution flow rate, and concentration.

$$r = \left(\frac{\rho Q_v^3 C^2}{4\pi^2 EIz} \right)^{1/4} \tag{11.1}$$

For tissue repair applications, several studies have also found that cell attachment and differentiation can be affected by controlling nanofiber orientation. For example, the earlier work of Li et al. (2007) suggested that inducing controlled anisotropy in electrospun scaffolds affected organization of actin filaments from the human MSC. Preliminary results from cell interaction with anisotropic nanofiber scaffolds prompted further investigations into their use on tissues in which directional properties are desired, such as cardiovascular and neural tissues, which are significant areas due to the lack of self-regeneration and challenges associated with the small-size scale. Lim et al. (2010) and Xie et al. (2009) seeded neural and embryonic stem cells onto aligned polycaprolactone (PCL) nanofibers, and both reported elongation of the neural stem cells along the fiber axis. In addition, when differentiation was initiated, the cells exhibited a higher extent of neuronal differentiation markers compared to randomly oriented fibers. Similar results were also reported in the work of Mukhatyar et al. (2011), who reported that Schwann cell migration and neurite outgrowth on polyacrylonitrile–methacrylate nanofibers were superior to those on a smooth film, which the authors attributed to increased adsorption of fibronectin. Using the knowledge gained on interactions between cells and aligned fibers, Orlova et al. (2011) demonstrated the ability to control the architecture of cardiac tissue cultures by manipulating the positions of electrospun polymethylglutarimide, with the cardiomyocytes seeded. In addition to controlling the architecture of the cardiac tissue culture, Kurpinski et al. (2010) also reported increased endothelial cell (EC) infiltration into aligned poly(L-lactic acid) nanofibers, which may be helpful in fabricating scaffolds for blood vessels, which are often difficult to repair, especially with small-diameter blood vessels.

Electrospinning can also be controlled such that individual fibers are assembled into specific patterns. In addition to aligned bundles of fibers for cell elongation, fibers can be assembled to form a layer with specific porosity and pore sizes to optimize cell attachment. For example, the Dalton group developed a patterning technique for polymer melt electrospinning that enabled formation of aligned lines of fiber into a gridlike structure (Detta et al. 2010; Hutmacher and Dalton 2011). Using their technique, Dalton and coworkers can also deposit lines of fibers above one another, forming a structure resembling an egg box, with variable spacing. The spacing, which can be microns in width, can be optimized to suit the attachment of different cells, which may be desirable for tissue scaffold fabrication.

Encouraging cell attachment and infiltration is only one of the potential uses for electrospun tissue repair scaffolds. In fact, there are also studies aimed at preventing cell and tissue adhesion, which is equally as important as tissue culturing, because in many postsurgical cases, bacteria or scars may cause tissue adhesion, which can lead to patient discomfort and reduced mobility. An example for preventing tissue adhesion is the demonstration of Yang et al. (2009), who implanted an electrospun layer of polylactic acid–polyethylene glycol copolymer between the cecum and peritoneal wall defects in rats and largely prevented cell and tissue adhesion on the fiber surface. Yang et al. (2009) attributed their observed phenomenon to the enhanced hydrophilicity from the introduction of polyethylene glycol (PEG), which is more mobile in body fluid and facilitates liquid diffusion. Zong et al. (2004) also reported reduction in abdominal adhesion by implanting an electrospun polylactic-glycolic acid scaffold.

The studies explored so far showed the potential of the nanofiber architecture in aiding cell regeneration. In addition to the structural property of the scaffold such as orientation and arrangement, the materials that are used to construct the scaffolds are equally important in determining the scaffold function.

11.3 MATERIALS

With numerous studies showing desirable cell activities, including adhesion, viability, proliferation, and differentiation, electrospun nanofibers have been established as a potential new standard for tissue repair. The focus on nanofibers for biomedical applications has shifted into several areas, with the ultimate goal being to bring the technology closer to the market. One of these areas is material, in which the aim is to transfer material-specific functional and mechanical properties to nanofiber scaffolds. As a result, material choice is an important consideration. Electrospinning predominately involves thermoplastic polymers, although electrospun thermosets have also been reported (Ramakrishna et al. 2005; Reddy et al. 2009). The main requirement is for the polymer to be biocompatible and not cause allergic or toxic response in the body. Polymers for electrospinning can be natural or synthetic, with some natural polymers showing resemblance to some features in native tissue or bioactive properties and synthetic polymer showing controllable structural properties. Figure 11.2 shows scanning electron microscope images, at 5000× magnification, of several polymers that have been spun into nanofibers.

Natural polymers can be derived from animals or plants, and therefore, they are mostly biocompatible, and many carry specific biological activity such as antimicrobial and cell binding. In addition, many natural polymers are renewable. Natural polymers that are common in electrospinning for tissue repair scaffolds include proteins such as collagen (Li et al. 2005a; Rho et al. 2006; Newton et al. 2009), keratin (Li et al. 2009a), silk (Sukigara et al. 2003; Sukigara et al. 2004; Gandhi et al.

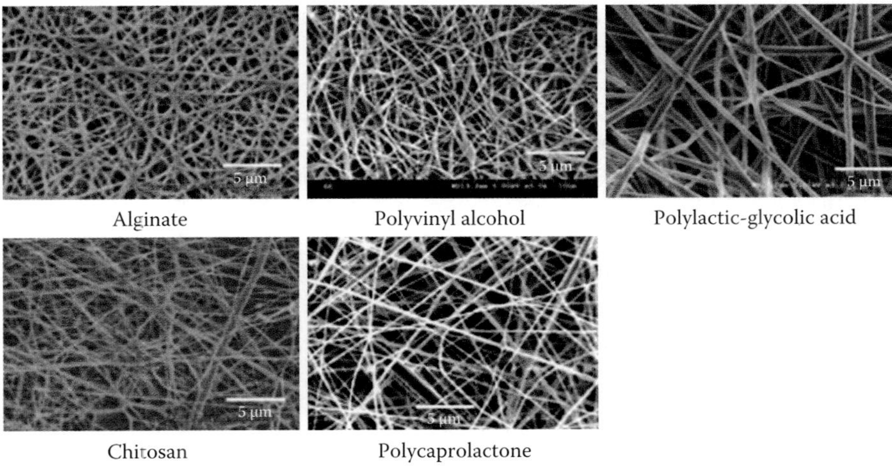

FIGURE 11.2 Examples of natural and synthetic polymers that can be electrospun into nanofibers.

2009b), and polysaccharides such as alginate (Leung and Ko 2010), hyaluronic acid (Ji et al. 2006), and chitosan (Desai et al. 2008; Chu et al. 2009). Several proteins that can be found in native tissues, such as collagen and keratin, have superior cell affinity due to the presence of arginine–glycine–aspartic sequences that can interact with integrin expressed by various cells. This enhanced cell affinity can be transferred to nanofibers via electrospinning, as verified by the studies of Li et al. (2009a) on keratin and Rho et al. (2006). In addition, Li et al. (2005a) compared human embryonic palatal mesenchymal cell proliferation on a range of substrates, including electrospun fibers of ECM proteins such as collagen and elastin, and reported superior proliferation on these electrospun proteins compared to the tissue culture plate controls. Silk is another commonly used protein that has been found to support cell adhesion and differentiation in several studies, such as the work of Mandal and Kundu (2010) and Sukigara and coworkers (Sukigara et al. 2003, 2004; Ayutsede et al. 2005; Gandhi et al. 2007). Besides native proteins, polysaccharides play a significant role in nanofiber scaffold fabrication. Alginate and chitosan, for example, can contribute to fluid absorbance and bacteria growth inhibition (Desai et al. 2008), respectively, enabling these polymers to be desirable candidates for various tissue scaffolds. In addition, the positive surface charge on chitosan nanofibers can be optimized for binding specific drugs or proteins. Hyaluronic acid is a polysaccharide that can be found in native ECM as part of the glycosaminoglycan and possess binding capability with cells and other ECM proteins, which has been confirmed in the early study of Ji et al. (2006). Recently, Hsu et al. (2010) have also reported that, by incorporating hyaluronic acid onto an electrospun collagen scaffold, the expression ratio of tissue inhibitors of metalloproteinases to matrix metalloproteinases is successfully reduced, which have implications in scar prevention for wound healing.

Although natural polymers often possess desirable bioactive properties, electrospun synthetic polymers for tissue repair applications have also been widely studied due to their often superior and customizable structural properties. Examples of synthetic polymers commonly used for tissue applications include PCL (Bölgen et al. 2007; Gandhi et al. 2009a), poly-L-lactic or poly-DL-lactic acid (PLLA or PDLLA) (Yang et al. 2005; Li et al. 2009a; Liu and Ma 2010), polyglycolide or polyglycolic acid (PGA), polyvinyl alcohol (PVA; Zeng et al. 2005; Asran et al. 2010), and polycaprolactone-ethyl ethylene phosphate (PCLEEP; Chua et al. 2005). When electrospun into nanofibers, many synthetic polymers are significantly more extensible compared to natural ones, as shown in Figure 11.3. The figure also shows that electrospun nanofibers have much higher tensile strength compared to artificial skins such as Integra developed by Yannas and Burke (1980), as well as injectable hydrogel scaffolds, whereas the values are in the same range as existing microfibrous scaffolds such as Resolut LT and BioFix (Rho et al. 2006). Natural polymer nanofibers remain less extensible compared to existing scaffolds. In addition, polymers such as poly(lactic acid) or polylactide (PLA) and PGA can be combined to form PLGA copolymer, with properties that can be controlled through the lactic acid to glycolic acid ratio (Li et al. 2002; Duan et al. 2006; Shin et al. 2006). Li et al. (2006) compared the structural properties of several electrospun synthetic polymers and reported that, electrospun PLLA and PCL formed compliant scaffolds, whereas PGA and PLGA formed the stiffest ones, although both PGA and PLGA were more degradable in aqueous environments. Moreover, Li et al. (2006) showed that PLLA and PCL scaffolds exhibited faster cell proliferation compared to PGA and PLGA.

Having known the properties of polymers commonly used in electrospinning, we can select an appropriate material or combinations for scaffold design such that the most suitable properties can be incorporated. In addition to biocompatibility and functional properties, biodegradability is an important aspect in scaffold design. For specific tissues, biodegradability should be controlled for optimal treatment. For example, for wound healing, the scaffold must remain intact as cells populate, but not excessively persistent such that the formation of new ECM is obstructed. The

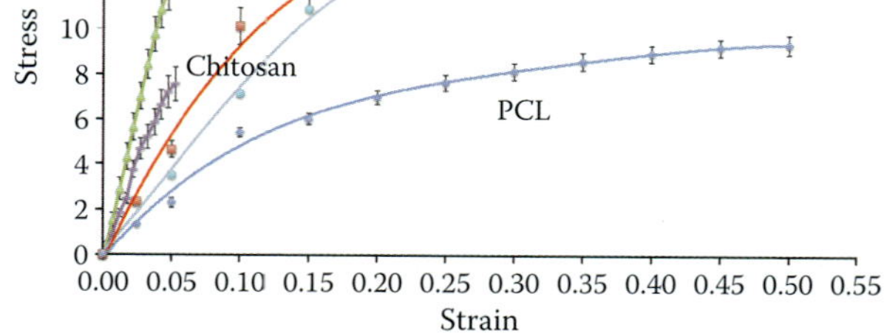

FIGURE 11.3 Tensile stress–strain behavior of natural and synthetic nanofiber scaffolds and comparison with existing scaffolds.

degradation profiles of several electrospun polymers in phosphate-buffered saline solutions are compared in Figure 11.4 below. It can be seen from the comparison that hydrophilic polymers degrade at a much faster rate, whereas PCL is marginally degradable. In addition to the comparison in Figure 11.4, the studies of Chew et al. (2005) also established PCLEEP as a marginally degradable polymer that is less degradable than PCL. In an attempt to more closely simulate degradation conditions in the body, Li et al. (2010) characterized the degradation of an electrospun PLGA scaffold under tensile load and reported that tensile strength and stiffness of scaffolds under load tend to initially increase in the degradation media, followed by a sharp decline that is faster than scaffolds not under load, thereby determining that nanofibers degrade faster under load. The findings of Li et al. (2010) is important, because in many situations, nanofiber scaffolds will be attached to surrounding tissues as a temporary replacement and sustain physiological loading, and, therefore, more accurate information on fiber degradation is required for scaffold design.

With material choice being a determinant factor in overall scaffold properties, it is possible to fine-tune scaffold properties through processing. For example, electrospun fibers can be annealed to induce crystallization, which modifies scaffold strength and extensibility. Gandhi et al. (2009b) have presented several examples of postspinning treatment and their effects on the properties of electrospun silk nanofibers. In addition to annealing, cross-linking is a common method of modifying scaffold structural properties. Our previous work on alginate has shown that cross-linking with divalent ions such as calcium could enhance alginate nanofiber integrity in aqueous environments, although extensibility of the fibers reduced significantly (Leung et al. 2010). Indeed, cross-linking has been suggested in electrospinning proteins such as collagen and gelatin, as well as other polysaccharides such as chitosan, in order for the scaffolds to withstand physiological loads (Luo et al. 2000; Newton et al. 2009).

In this section, the effects of material choice and postelectrospinning processing on the structural and functional properties of nanofiber scaffolds have been explored. The knowledge gained in electrospinning materials and related processing is an important aspect of nanofiber technology development. However, through the studies discussed so far, nanofibers have mainly played a supportive role as a platform

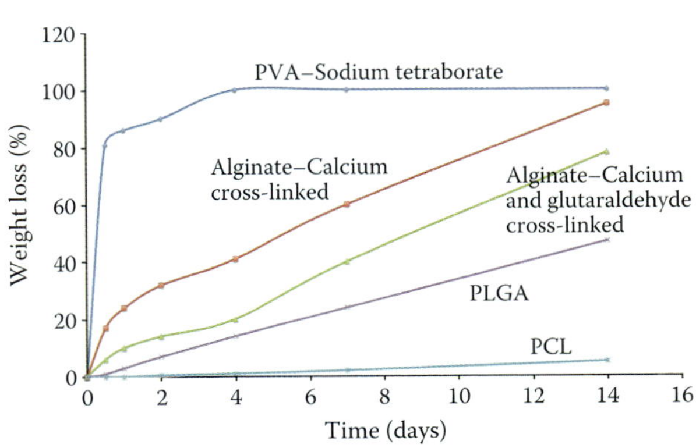

FIGURE 11.4 Degradation profiles of nanofibers spun from natural and synthetic nanofibers.

for cell adhesion and proliferation. To better demonstrate nanofiber technology for tissue regeneration, a more active role for scaffolds must be adopted. The next section discusses strategies toward designing active scaffolds.

11.4 TOWARD ACTIVE NANOFIBER SCAFFOLDS: SYSTEMS DESIGN APPROACH

Throughout the previous sections, nanofiber technology has been introduced as a versatile technique for processing a range of polymers into nanofibers with various morphologies, thereby combining large surface area and controllable porosity with desirable tailored material properties. However, in tissue regeneration applications, the objectives of the scaffolds are often multidimensional due to the intricate processes involved, and the versatility in nanofiber technology that has been discussed so far is insufficient in addressing all challenges. As a result, a different perspective in nanofiber scaffold development is a composite approach in which the advantages of multiple functional elements and matrices are combined into an effective scaffold. Indeed, current studies have shown the superior performance of composite scaffolds in terms of tissue regeneration and control over cell activities compared to traditional ones spun from single materials.

11.4.1 Methods for Fabricating Active Nanofiber Scaffolds

The first step in developing composite scaffolds is to determine the combination of materials for imparting the desired functions. A detailed discussion on matrix materials useful for tissue regeneration has been provided in the previous section. In addition to matrix materials, other functional elements can be added to the nanofiber scaffolds, including metallic or ceramic particles, drugs, proteins, and live cells. An important consideration is how multiple matrix materials and functional elements can be incorporated into a common scaffold. As a result, different methods of forming various nanofiber architectures have been explored. In the simplest form, multiple polymer matrices or functional elements can be directly blended and electrospun as a common phase, leading to homogeneous functional nanofibers. This direct blend method is feasible when the scaffold components are compatible with one another. For example, we have demonstrated the direct blending of a water-soluble model drug into PVA, as well as that of the model drug into a matrix composed of a mixture of sodium alginate and polyethylene oxide (PEO), with results shown later in the discussion.

For applications in which the desired combinations of materials are not compatible with one another, other methods can be employed to fabricate composite scaffolds. If scaffold functionality does not necessitate the scaffold components to be in a common phase, multinozzle electrospinning can be employed to spin each element separately into a scaffold. For example, nozzles carrying different polymer solutions can be placed side by side during electrospinning while ensuring that the charge is well distributed to all nozzles by using a linear charge source, or the nozzles can be placed on opposite ends of a rotating collector, with an individual power supply connected to each nozzle. Ladd et al. (2011) utilized such a multinozzle setup to fabricate a composite scaffold mimicking muscle tendon junctions. By spinning a

PCL/collagen blend on one side of a stationary mandrel and a PLLA/collagen blend on the other side, Ladd et al. (2011) created a dual-phase scaffold that has three ranges of mechanical properties due to PLLA, PCL, and an intermediate region containing PCL and PLLA. The establishment regional mechanical properties in the scaffold allow effective force transfer from the muscle to the tendon, thereby mimicking the muscle–tendon junction.

For applications in which scaffold components must be within the same phase, such as for drugs or a degradable polymer matrix that needs to be protected by a more robust matrix, emulsion and coaxial electrospinning can be employed. In emulsion electrospinning, a common phase can be obtained by suspending a functional phase in a the carrier phase, often with the aid of an emulsifier. The emulsified solution can then be electrospun into either fibers with heterogeneous phases within (Xu et al. 2005; Qi et al. 2006) or those with core–shell architecture. In the work of Wei et al. (2011), an aqueous solution containing a collagen-like protein was emulsified in a PLGA carrier using Polysorbate 80 as an emulsifier and obtained a core–shell architecture in which the core is composed of aqueous protein and the shell is composed of PLGA. In addition to emulsion electrospinning, the core–shell architecture can be achieved via coaxial electrospinning, which utilizes a primary nozzle containing a smaller secondary nozzle. The secondary nozzle delivers polymer solution that will be enveloped by the solution delivered by the primary nozzle, forming a core–shell fiber structure. In the coaxial setup, the core solution in the secondary does not need to be electrospinnable, as demonstrated in the work of Jiang et al. (2006), who electrospun PCL nanofibers containing dextran dispersed in bovine serum albumin (BSA) in the core using the coaxial setup. Figure 11.5 compares

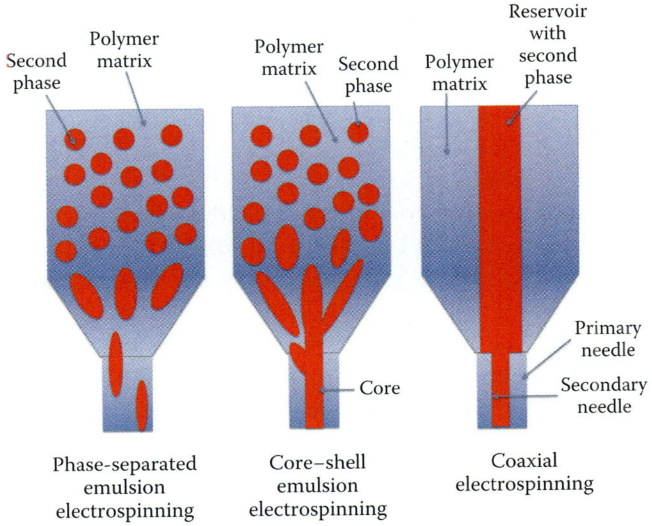

FIGURE 11.5 Methods of incorporating incompatible phases into a nanofiber scaffold, with the phase-separated and core–shell structure fabricated from emulsion electrospinning and core–shell nanofiber from coaxial electrospinning.

emulsion and coaxial electrospinning techniques for incorporating multiple phases into a nanofiber matrix.

11.4.2 Incorporation of Drugs and Functional Elements

Devising methods for combining materials into a scaffold is an important first step leading to the development of active scaffolds. In addition to the polymer matrices and proteins discussed above, inorganic functional elements can be incorporated into nanofiber scaffolds. Li et al. (2009b) demonstrated the blending of nanometric silver particles into a PLLA matrix. The evaluation of a bacteria inhibition zone size using *Escherichia coli* and *Staphylococcus* showed that the PLLA nanofibers containing silver had stronger antibacterial properties compared to plain PLLA nanofibers. In addition to silver nanoparticles, Chae et al. (2009) fabricated PLLA and alginate nanofiber scaffolds functionalized with hydroxyapatite, which was coated on the fibers via *in situ* mineralization. The hydroxyapatite–PLLA/alginate nanofibers showed superior osteoblast compatibility compared to PLLA or alginate scaffolds.

Another major area for active nanofiber scaffolds is the incorporation of drugs, which imparts a range of functions useful for different stages in the regeneration. Using polymers as drug carriers, nanofiber scaffolds provide a means for localized and controlled delivery, ensuring that drugs focus at the site of interest and their concentrations are within therapeutic window. Controlled delivery also eliminates the need for multiple drug applications. In terms of controlled delivery, it is important for the drug to be released according to specific schedules to ensure that the desired function is stimulated correctly and for an appropriate duration. For example, a scaffold for treating wounds at early stages may require quick release of antibacterial agents, whereas one for delivering cancer drugs may require a slow sustainable release of antitumor agents. In nanofibers, drug release occurs through three mechanisms: 1) drug detachment from the fiber surface; 2) diffusion from within the fibers to the surface; and 3) fiber degradation (Srikar et al. 2008). Matrix choice therefore plays a significant role in determining drug release behavior due to the importance of their degradability and hydrophilicity. Indeed, it has been found that nanofibers composed of polymers with varying degradation and swelling rate yielded vastly different release duration, ranging from 2 h using highly soluble polymer (PVA) (Yang et al. 2008a) to 10 days using PDLLA (Yang et al. 2008b), 60 days using PCL (Saraf et al. 2010), and up to 3 months using marginally swellable polymers (PCLEEP; Chew et al. 2005).

In addition to the drug release duration, it is equally important to control the release rate throughout the duration in order to ensure that drug concentration remains within the therapeutic window. As a result, several groups have explored models for quantifying drug release with respect to time to gain an understanding on the release mechanisms involved. Perhaps, the most famous example of a drug release model is that developed by Higuchi in the 1960s, which was derived based on Fick's first law of diffusion and intended for a film of ointment formulation (Siepmann and Peppas 2011). A typical Higuchi equation is shown in Equation 11.2 to relate the amount released M_t at time t, in which C_0 and C_s are the initial concentration and maximum soluble concentration, respectively, A is the contact area, and D is the diffusion coefficient. The Higuchi model was an important milestone in understanding physical phenomena involved in diffusional drug release. Following the Higuchi model, a more generalized equation, based on a more empirical approach, was later developed and was known as the power law, as shown in Equation 11.3 (Siepmann and Peppas 2001, 2011). In Equation 11.3, M_∞ is the amount of drugs released at infinite time, and k and n are fitting parameters, in which k depends on the matrix geometry, and n depends on the mode of release, with an n value of 0.5 indicating Fickian diffusion, that of unity indicating a zero-order release, and those values in between 0.5 and 1 indicating anomalous transport, possibly a combination of Fickian diffusion and zero-order release (Costa and Sousa Lobo 2001). More recently, Srikar et al. (2008) also derived a release model, shown in Equation 11.4, based on electrospun fibers, via physical analysis of small pores outside of a fiber, which was able to closely match experimental results obtained from a PCL scaffold loaded with dye. In Equation 11.4, M_{d0} is the amount of drug loaded into the nanofiber, α is the porosity factor based on the ratio between amount on the fiber surface and that in the entire fiber, and τ_r is the characteristic time, depending on the pore length and effective diffusion coefficient. These existing models have provided relatively accurate, easy-to-understand methods of quantifying drug release from polymer matrices that are hydrophobic and marginally degradable. For example, Luong-Van et al. (2006) demonstrated heparin release from electrospun PCL and showed that, for a hydrophobic matrix such as PCL, the release rate is proportional to the square root of time. To quantify drug release from polymers that can swell, a modified power law was derived, as shown in Equation 11.5, in which the first term addresses the diffusional release, and the second term addresses the zero-order release due to polymer chain relaxation upon swelling (Peppas and Sahlin 1989). However, Ferrero et al. (2010) also pointed out possible limitations in the modified power law when both swelling and polymer degradation occur in the matrix.

$$M_t = A\sqrt{(2C_0 - C_s)DC_s t} \tag{11.2}$$

$$\frac{M_t}{M_\infty} = kt^n \tag{11.3}$$

$$\frac{M_t}{M_{d0}} = \alpha\left[1 - \exp\left(\frac{-\pi^2 t}{8\tau_r}\right)\right] \tag{11.4}$$

$$\frac{M_t}{M_\infty} = k_1 t^n + k_2 t^{2n} \tag{11.5}$$

With the aid of release models, we can gain an understanding on the factors that affect the release behavior. The apparent next step is to use this knowledge to demonstrate release control on nanofiber scaffolds. Based on the earlier modeling work, diffusional release from the matrix and matrix swelling or degradation seem to have the most impact on the release rate. As a result, a common approach of controlling drug release is to add barrier layers to limit liquid diffusion into and out of a scaffold. This approach has been demonstrated in our work on a PVA scaffold containing a model drug, in which the burst release in the first 3 h was greatly reduced by adding a PCL barrier layer (Leung et al. 2011). Wei et al. (2011) also adopt a barrier layer design by encapsulating an aqueous solution of a collagen-like protein into a PLGA matrix, forming a core–shell nanofiber structure. The addition of a PLGA layer allowed a sustained protein release over a span of 36 days. Ji et al. (2010) presented a similar core–shell nanofiber design to demonstrate the release of BSA, which was encapsulated in a PCL shell. Through their design, Ji et al. (2010) also determined that, by adding PEG into the core phase, drug release was accelerated, highlighting the effect of hydrophilic polymer addition on enhancing the drug diffusion rate. In addition to barrier layers, the swelling rate can also be tailored for drug release control. Loh et al. (2010) showed that, by incorporating copolymers of PEG, polypropylene glycol (PPG), and PCL into a nanofibrous scaffold, the release rate of BSA can be controlled by changing the temperature. Due to the varying degrees of relaxation of the copolymer components at a specific temperature, the swelling behavior of the scaffold can thus change with the temperature, leading to different release profiles. Our previous work on the aforementioned model drug also demonstrated that, by reducing the swelling rate of the scaffolds via cross-linking, the burst release of the drug was greatly reduced (Leung et al. 2011). Figure 11.6 shows the effect of cross-linking a PVA matrix on burst release reduction, which led to a change from full release in 15 min for the as-electrospun PVA to full release in 8 h for the cross-linked PVA. Moreover, when the model drug was loaded into a cross-linked alginate scaffold containing water-soluble PEO as a sacrificial phase, the release was accelerated compared to the cross-linked PVA scaffold, which is likely due to the increase in void space in the scaffold as a result of PEO dissolution, indicating that sacrificial phases can also be an effective means of fine-tuning the drug release.

The incorporation of functional elements in nanofiber scaffolds enabled a new toolset for developers in fabricating scaffolds with specific functions. However, it is

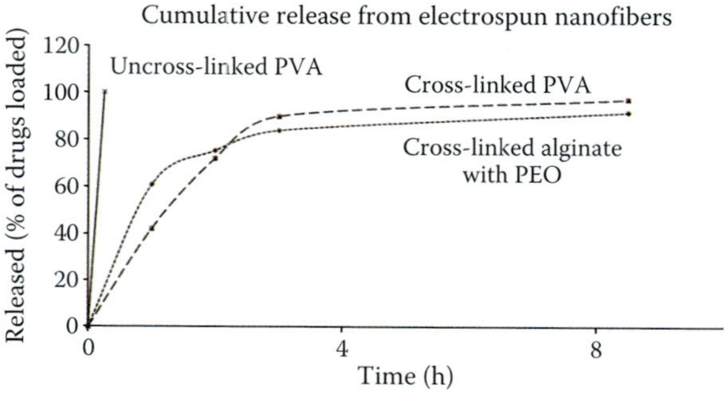

FIGURE 11.6 Release profiles of a model drug comparing the behavior of the PVA scaffold before and after cross-linking, as well as the effect of adding a PEO sacrificial phase in cross-linked alginate scaffolds.

also important to gain perspectives on how the advancement of nanofiber technology can be transferred to clinical applications. Indeed, many groups have demonstrated how nanofiber scaffolds with tailored properties and controlled functionalities can address challenges posed in current clinical challenges.

11.5 APPLICATION OF ACTIVE NANOFIBERS

Having an understanding on properties of relevant materials used in nanofiber fabrication and the selection of suitable functional elements and methods for incorporation into scaffolds, we can fabricate active nanofibers that can stimulate and control various activities involved in tissue regeneration. This section aims at exploring several recent examples in which nanofiber technology, as discussed so far, is applied to address current challenges.

11.5.1 Wound Healing

Acute and chronic wounds remain a challenge in the medical sector today, as patients can suffer from excessive fluid loss and wound infection, both of which can be fatal. The current standard for healing severe wounds is autografting but can lead to donor site morbidity. Accordingly, wound scaffolding is an important area for nanofiber studies. In addition to antimicrobial loaded scaffolds mentioned earlier, a popular approach for enhancing wound healing is to incorporate growth factors for cell signaling. Schneider et al. (2009) reported a timelier wound closure upon the application of an electrospun silk scaffold containing epithelial growth factor (EGF), which regulates cell attachment, proliferation, and differentiation. The scaffold was tested on a three-dimensional human skin wound model, produced by culturing fibroblast cells in collagen layers and inflicting a wound via punching and removing a 1.5-cm section from the culture. The silk/EGF scaffold showed feasibility in increasing the rate of wound closure by outperforming an as-spun silk scaffold by a factor of 4 in terms of the percentage of wound closure after 48 h. Yang et al. (2011) demonstrated the incorporation of basic fibroblast growth factor (bFGF) in the core of a nanofiber with a PEG–PLLA copolymer shell via emulsion electrospinning. The controlled release of bFGF from the core–shell nanofiber scaffold enabled significant reduction of wound area of diabetic ulcers on rats compared to the copolymer scaffold without bFGF.

11.5.2 Musculoskeletal Repair

In addition to the skin, the regeneration of tissues in the musculoskeletal system, such as the bone, cartilage, ligament, and tendon, have also been explored as part of active nanofiber feasibility studies. Tissue regeneration aids such as nanofiber scaffolds are important for the repair of musculoskeletal tissues, because those such as the ligament, cartilage, and tendon cannot self-regenerate (Leung and Ko 2010). For tendon repair, Sahoo et al. (2010) followed a similar approach to that in the wound healing work of Yang et al. (2011) by incorporating bFGF in a PLGA scaffold through direct blending. By releasing bFGF in a sustained manner over a 1-week period, the scaffold stimulated differentiation of MSC into tendon- and ligament-specific fibroblasts. Other than the tendon, Li et al. (2005b) fabricated a PCL nanofiber scaffold and showed that, in the presence of transforming growth factor-β, MSC can undergo chondrogenesis, which may lead to effective repairs for cartilages. The chondrogenesis observed on the PCL scaffold was comparable to the MSC maintained in cell pellets and aggregates (Li et al. 2005b). For bone repair, nanofibers coated with hydroxyapatite have been widely explored as a potential solution, as demonstrated in the work of Chae et al. (2009). Whited et al. (2011) also fabricated bone scaffold from PLLA coated with apatite and reported the upregulation of alkaline phosphatase activity and the increase in mineralization by osteoblast cells. Moreover, by incorporating water-soluble PEO fibers in the PLLA matrix, Whited et al. (2011) reported that the resulting increase in porosity allowed faster cell infiltration, suggesting that the PEO/PLLA matrix coated with apatite can be an effective system for osteoblast infiltration, proliferation, and differentiation, leading to bone regeneration. Cui et al. (2010) performed a study on surface effects on hydroxyapatite crystallization on electrospun nanofibers and determined that, for the PLA scaffold fabricated, an optimal ratio of carboxyl, hydroxyl, and amino groups exist to mineralize the finest and largest content of hydroxyapatite.

11.5.3 Cardiovascular Tissue Regeneration

Cardiovascular tissue repair has generally been carried out through grafting. However, autografts are limited in supply and may lead to donor site morbidity, whereas allografts may cause immunogenic response. In addition, grafting becomes difficult for small-diameter vessels, and the use of nanofiber scaffolds therefore enables dimensional and property control. The challenge in producing cardiovascular scaffolds with nanofibers is that blood tends to clot upon contact with biomaterial, prompting a biomimetic approach. Blood remains in fluid state in vessels as a result of their inner lining consisted of EC. Accordingly, it has been shown that coating vascular implants with EC can prevent thrombosis (Ma et al. 2005). To further mimic the properties of vascular tissues, however, several groups have examined multilayered approaches that support the growth of both EC and smooth muscle cells (SMC), thereby resembling native vessels composing of the intima, media, and adventitia layers, with EC being the lining inside the intima and SMC in the media. Stitzel et al. (2006) fabricated a tubular scaffold using a blend of PLGA, collagen, and elastin and cultured EC on the inner surface and SMC on the outer surface. The authors reported that both types of cells were viable on the scaffold and a monolayer of EC was also formed in the inner surface. The results of the work of Stitzel et al. (2006) were also confirmed by Ju et al. (2010), who fabricated a bilayer scaffold with a PCL/collagen blend by varying the fiber diameter and porosity during electrospinning. At the inner less porous layer with a finer fiber diameter, Ju et al. (2010) seeded EC, whereas at the more porous outer layer with larger fibers, they seeded SMC. The authors reported enhanced adhesion and orientation of EC in the inner layer, whereas SMC was able to infiltrate the scaffold from the outer layer. Although blood clotting on biomaterials can pose challenges for implant design, failure to clot can also be dangerous to patients when wounds are inflicted. Liao and Leong

(2011) developed a coaxial nanofiber scaffold for treating hemophilia by delivering angiogenic or lymphangiogenic growth factors intended to stimulate the transport of plasma factor VIII (FVIII). Under controlled delivery, the level of FVIII was enhanced in hemophilic mice for more than 2 months, and blood coagulation time was successfully reduced.

The short examples discussed in this section give a brief introduction to recent advancements in nanofiber technology for tissue repair based on the knowledge in material properties and tailoring strategies in electrospinning as previously discussed. Although the examples shown have demonstrated feasibility of nanofibers in a range of biomedical applications, challenges clearly remain. For example, scaffold implants, regardless of the tissue it is intending to replace, may face immunogenic responses. Although efforts have been taken to address any remaining challenges, such as the work of Holan et al. (2011) on controlled release of cyclosporine A, an immunosuppressive drug, from PLA electrospun scaffolds, it is important to continue investigation leading to the safe transfer in nanofiber technology to the clinical usage.

11.6 CONCLUSION

The recent effort in active nanofiber development has shown feasibility in tissue regeneration due to the desirable surface properties common to the nanofiber architecture and the ability to stimulate and manage biological processes through the use of functional elements. The major theme that we have explored so far is the ability to tailor nanofiber physical and therapeutic properties, which is a major factor for the success of nanofiber scaffolds in tissue repair. Through electrospinning, nanofibers with a range of diameters, porosities, orientations, and assemblies can be fabricated, enabling the optimization of surface properties for specific applications such as cell attachment and infiltration. Moreover, a wide range of materials can be incorporated into electrospinning, some of which have been previously discussed. The material choice allows further tailoring of physical scaffold properties, such as hydrophilicity, degradation, mechanical properties, and compatibility with cells and functional elements. In addition to matrix materials for constructing the scaffold, functional elements or multiple matrix phases can be incorporated into a scaffold via direct blending, emulsion, and coaxial electrospinning. Emulsion and coaxial electrospinning can also lead to core–shell nanofiber structure, which was shown to be effective for drug and protein delivery. Other than drugs, metallic and ceramic phases can be added to nanofiber scaffolds to achieve specific properties such as antimicrobial and encouragement of bone growth. With the degree of control over nanofiber properties currently achievable, several groups have explored the use of nanofiber technology to address challenges in areas such as wound healing and tissue repair including bone, tendon, cartilage, and vascular, as well as immunosuppression for implant applications.

Current examples of active nanofiber application, most of which show improved cell attachment, proliferation and differentiation, and tissue growth, have demonstrated the merit for tailored nanofiber scaffolds. However, it is clear that many clinical challenges remain, indicating that further development is needed in active nanofiber scaffolds. Although a relatively comprehensive idea on cell–scaffold interaction has been obtained through existing studies, it is desirable to explore scaffold–body interaction for the transition to clinical applications. From the viewpoint of scaffold design, more fundamental understanding on control mechanisms can help in the development of more effective scaffolds. For example, improved drug release models can inform developers on parameters that affect the release rate in more detail, allowing optimization for more precise release behavior, such as multiphase release. In terms of processing, it is desirable to develop scaffolds that are not only tailored for specific applications but can also respond to external stimulants, similar to the PEG/PPG/PCL scaffold developed by Loh et al. (2010), which can respond to temperature. In addition to temperature, scaffolds that can respond to conditions such as pH, plasma protein concentration, or blood glucose level will also be helpful in aiding tissue regeneration for chronic wounds such as diabetic ulcers. For application, on the other hand, it is important to explore the applicability of nanofibers as a scaffold for specific tissues, such as ensuring the correct mechanical response compared to the tissue that it is replacing, and methods of securing scaffolds in their intended areas under physiological loads. In this respect, it will be helpful to shift the focus from *in vitro* cell culturing to *in vivo* testing. Combining the several areas discussed above, we can better harvest the potential of nanofibers in addressing challenges in existing tissue repair technology.

REFERENCES

American Burn Association. (2009). Burn Incidence Fact Sheet. Retrieved January 20, 2010, from http://www.ameriburn.org/resources_factsheet.php.

Asran, A.S., S. Henning et al. (2010). Polyvinyl alcohol–collagen–hydroxyapatite biocomposite nanofibrous scaffold: Mimicking the key features of natural bone at the nanoscale level. *Polymer* 51(4): 868–876.

Ayutsede, J., M. Gandhi et al. (2005). Regeneration of *Bombyx mori* silk by electrospinning—Part 3: Characterization of electrospun nonwoven mat. *Polymer* 46(5): 1625–1634.

Bölgen, N., I. Vargel et al. (2007). *In vivo* performance of antibiotic embedded electrospun PCL membranes for prevention of abdominal adhesions. *Journal of Biomedical Materials Research—Part B: Applied Biomaterials* 81B(2): 530–543.

Burg, K.J.L., S. Porter et al. (2000). Biomaterial developments for bone tissue engineering. *Biomaterials* 21(23): 2347–2359.

Chae, T., H. Yang et al. (2009). Hydroxyapatite/alginate nanocomposite fibrous scaffolds for biomedical applications. *Proceedings of the 27th Conference of the Canadian Biomaterials Society*, Kingston, Ontario.

Chew, S.Y., J. Wen et al. (2005). Sustained release of proteins from electrospun biodegradable fibers. *Biomacromolecules* 6(4): 2017–2024.

Chu, X.H., X.L. Shi et al. (2009). Chitosan nanofiber scaffold enhances hepatocyte adhesion and function. *Biotechnology Letters* 31: 347–352.

Chua, K.N., W. Lim et al. (2005). Stable immobilization of rat hepatocyte spheroids on galactosylated nanofiber scaffold. *Biomaterials* 26: 2537–2547.

Costa, P. and J.M. Sousa Lobo (2001). Modeling and comparison of dissolution profiles. *European Journal of Pharmaceutical Sciences* 13: 123–133.

Cui, W., X. Li et al. (2010). Hydroxyapatite nucleation and growth mechanism on electrospun fibers functionalized with different chemical groups and their combinations. *Biomaterials* 31(17): 4620–4629.

D'Amore, A., J.A. Stella et al. (2010). Characterization of the complete fiber network topology of planar fibrous tissues and scaffolds. *Biomaterials* 31(20): 5345–5354.

Desai, K., K. Kit et al. (2008). Morphological and surface properties of electrospun chitosan nanofibers. *Biomacromolecules* 9(3): 1000–1006.

Detta, N., T.D. Brown et al. (2010). Melt electrospinning of polycaprolactone and its blends with poly(ethylene glycol). *Polymer International* 59(11): 1558–1562.

Duan, B., X. Yuan et al. (2006). A nanofibrous composite membrane of PLGA–chitosan/PVA prepared by electrospinning. *European Polymer Journal* 42(9): 2013–2022.

Ferrero, C., D. Massuelle et al. (2010). Towards elucidation of the drug release mechanism from compressed hydrophilic matrices made of cellulose ethers—Part II: Evaluation of a possible swelling-controlled drug release mechanism using dimensionless analysis. *Journal of Controlled Release* 141: 223–233.

Flemming, R.G., C.J. Murphy et al. (1999). Effects of synthetic micro- and nano-structured surfaces on cell behavior. *Biomaterials* 20(6): 573–588.

Formhals, A. (1934). Process and Apparatus for Preparing Artificial Threads. U.S.P. Office. U.S. 1,975,504.

Fridrikh, S.V., J.H. Yu et al. (2003). Controlling the fiber diameter during electrospinning. *Physical Reviews Letters* 90: 144502–144506.

Gandhi, M., R. Srikar et al. (2009a). Mechanistic examination of protein release from polymer nanofibers. *Molecular Pharmaceutics* 6(2): 641–647.

Gandhi, M., H. Yang et al. (2007). Regeneration of *Bombyx mori* silk by electrospinning: A comparative study of the biocompatibility of natural and synthetic polymers for tissue engineering applications. *Journal of Biobased Materials and Bioenergy* 1(2): 274–281.

Gandhi, M., H. Yang et al. (2009b). Postspinning modification of electrospun nanofiber nanocomposite from *Bombyx mori* silk and carbon nanotubes. *Polymer* 50: 1918–1924.

Hartgerink, J.D., E. Beniash et al. (2001). Self-assembly and mineralization of peptide-amphiphile nanofibers. *Science* 294(5547): 1684–1688.

Holan, V., M. Chudickova et al. (2011). Cyclosporine A-loaded and stem cell–seeded electrospun nanofibers for cell-based therapy and local immunosuppression. *Journal of Controlled Release* 156(3): 406–412.

Hsu, F.-Y., Y.-S. Hung et al. (2010). Electrospun hyaluronate–collagen nanofibrous matrix and the effects of varying the concentration of hyaluronate on the characteristics of foreskin fibroblast cells. *Acta Biomaterialia* 6(6): 2140–2147.

Hutmacher, D.W. and P.D. Dalton (2011). Melt electrospinning. *Chemistry—An Asian Journal* 6(1): 44–56.

Ji, W., F. Yang et al. (2010). Fibrous scaffolds loaded with protein prepared by blend or coaxial electrospinning. *Acta Biomaterialia* 6(11): 4199–4207.

Ji, Y., K. Ghosh et al. (2006). Electrospun three-dimensional hyaluronic acid nanofibrous scaffolds. *Biomaterials* 27(20): 3782–3792.

Jiang, H., Y. Hu et al. (2006). Modulation of protein release from biodegradable core–shell structured fibers prepared by coaxial electrospinning. *Journal of Biomedical Materials Research—Part B: Applied Biomaterials* 79B(1): 50–57.

Ju, Y.M., J.S. Choi et al. (2010). Bilayered scaffold for engineering cellularized blood vessels. *Biomaterials* 31(15): 4313–4321.

Katti, D.S., K.W. Robinsin et al. (2005). Bioresorbable nanofiber-based systems for wound healing and drug delivery: Optimization of fabrication parameters. *Journal of Biomedical Materials Research—Part B: Applied Biomaterials* 70B(2): 286–296.

Ko, F. and M. Gandhi (2007). Producing nanofibre structures by electrospinning for tissue engineering. *Nanofibers and Nanotechnology in Textiles*. P. Brown and K. Stevens, editors. Clemson, SC: Woodhead Publishing, p. 67.

Ko, F.K. (1997). *Medical Applications for Textile Structures*. Taiwan: Textile Asia.

Ko, F.K., C.T. Laurencin et al. (1998). The dynamics of cell–fiber architecture interaction. *Proceedings of the Annual Meeting of the Biomaterials Research Society*, San Diego, CA.

Kurpinski, K.T., J.T. Stephenson et al. (2010). The effect of fiber alignment and heparin coating on cell infiltration into nanofibrous PLLA scaffolds. *Biomaterials* 31(13): 3536–3542.

Ladd, M.R., S.J. Lee et al. (2011). Coelectrospun dual scaffolding system with potential for muscle–tendon junction tissue engineering. *Biomaterials* 32(6): 1549–1559.

Leung, V., R. Hartwell et al. (2011). Bioactive nanofibres for wound healing applications. *Journal of Fiber Bioengineering and Informatics* 4(1): 1–14.

Leung, V. and F.K. Ko (2010). Biomedical applications of nanofibers. *Polymers for Advanced Technologies* 22(3): 350–365.

Leung, V., H. Yang et al. (2010). Alginate nanofibre based tissue engineering scaffolds by electrospinning. *Proceedings of the Society for the Advancement of Materials and Process Engineering 2010 Conference*, Seattle, WA.

Li, D. and Y. Xia (2004). Electrospinning of nanofibers: Reinventing the wheel? *Advanced Materials* 16(14): 1151–1170.

Li, J., Y. Li et al. (2009a). Preparation and biodegradation of electrospun PLLA/keratin nonwoven fibrous membrane. *Polymer Degradation and Stability* 94(10): 1800–1807.

Li, L., Y. Li et al. (2009b). Antibacterial properties of nanosilver PLLA fibrous membranes. *Journal of Nanomaterials* 2009: 1–5.

Li, M., M.J. Mondrinos et al. (2005a). Electrospun protein fibers as matrices for tissue engineering. *Biomaterials* 26(30): 5999–6008.

Li, P., X. Feng et al. (2010). Influences of tensile load on *in vitro* degradation of an electrospun poly(L-lactide-co-glycolide) scaffold. *Acta Biomaterialia* 6(8): 2991–2996.

Li, W.J., R.L. Mauck et al. (2007). Engineering controllable anisotropy in electrospun biodegradable nanofibrous scaffolds for musculoskeletal tissue engineering. *Journal of Biomechanics* 40: 1686–1693.

Li, W.J., R. Tuli et al. (2005b). A three-dimensional nanofibrous scaffold for cartilage tissue engineering using human mesenchymal stem cells. *Biomaterials* 26: 599–609.

Li, W.-J., J.A. Cooper, Jr. et al. (2006). Fabrication and characterization of six electrospun poly(α-hydroxy ester)-based fibrous scaffolds for tissue engineering applications. *Acta Biomaterialia* 2(4): 377–385.

Li, W.-J., C.T. Laurencin et al. (2002). Electrospun nanofibrous structure: A novel scaffold for tissue engineering. *Journal of Biomedical Materials Research* 60(4): 613–621.

Li, Z. and M. Zhang (2005). Chitosan–alginate as scaffolding material for cartilage tissue engineering. *Journal of Biomedical Materials Research—Part B: Applied Biomaterials* 75B(2): 485–493.

Liao, I.C. and K.W. Leong (2011). Efficacy of engineered FVIII-producing skeletal muscle enhanced by growth factor-releasing coaxial electrospun fibers. *Biomaterials* 32(6): 1669–1677.

Lim, S.H., X.Y. Liu et al. (2010). The effect of nanofiber-guided cell alignment on the preferential differentiation of neural stem cells. *Biomaterials* 31(34): 9031–9039.

Liu, X. and P.X. Ma (2010). The nanofibrous architecture of poly(L-lactic acid)-based functional copolymers. *Biomaterials* 31(2): 259–269.

Loh, X.J., P. Peh et al. (2010). Controlled drug release from biodegradable thermoresponsive physical hydrogel nanofibers. *Journal of Controlled Release* 143(2): 175–182.

Luo, Y., K.R. Kirker et al. (2000). Cross-linked hyaluronic acid hydrogel films: New biomaterials for drug delivery. *Journal of Controlled Release* 69(1): 169–184.

Luong-Van, E., L. Grondahl et al. (2006). Controlled release of heparin from poly(e-caprolactone) electrospun fibers. *Biomaterials* 27: 2042–2050.

Ma, Z., M. Kotaki et al. (2005). Surface engineering of electrospun polyethylene terephthalate (PET) nanofibers towards development of a new material for blood vessel engineering. *Biomaterials* 26: 2527–2536.

Mandal, B.B. and S.C. Kundu (2010). Biospinning by silkworms: Silk fiber matrices for tissue engineering applications. *Acta Biomaterialia* 6(2): 360–371.

Mukhatyar, V.J., M. Salmerón-Sánchez et al. (2011). Role of fibronectin in topographical guidance of neurite extension on electrospun fibers. *Biomaterials* 32(16): 3958–3968.

Newton, D., R. Mahajan et al. (2009). Regulation of material properties in electrospun scaffolds: Role of cross-linking and fiber tertiary structure. *Acta Biomaterialia* 5(1): 518–529.

Niece, K.L., J.D. Hartgerink et al. (2003). Self-assembly combining two bioactive peptide-amphiphile molecules into nanofibers by electrostatic attraction. *Journal of the American Chemical Society* 125(24): 7146–7147.

NIH Osteoporosis and Related Bone Diseases National Resource Center. (2011). Osteoporosis. Retrieved October 10, 2011, from http://www.niams.nih.gov/Health_Info/Bone/osteoporosis/.

Organization, W.H. (2010). Good Governance for Medicines. Retrieved February 1, 2010, from http://www.who.int/medicines/ggm/en/index.html.

Orlova, Y., N. Magome et al. (2011). Electrospun nanofibers as a tool for architecture control in engineered cardiac tissue. *Biomaterials* 32(24): 5615–5624.

Peppas, N.A. and J.J. Sahlin (1989). A simple equation for the description of solute release—Part III: Coupling of diffusion and relaxation. *International Journal of Pharmaceutics* 57: 169–172.

Qi, H., P. Hu et al. (2006). Encapsulation of drug reservoirs in fibers by emulsion electrospinning: Morphology characterization and preliminary release assessment. *Biomacromolecules* 7(8): 2327–2330.

Ramakrishna, S., K. Fujihara et al. (2005). *An Introduction to Electrospinning and Nanofibers*. Singapore: World Scientific Publishing.

Reddy, C.S., A. Arinstein et al. (2009). Fabrication of thermoset polymer nanofibers by coelectrospinning of uniform core–shell structures. *Journal of Materials Chemistry* 19(39): 7198–7201.

Research and Markets (2009). The Future of Wound Care Management Market to 2015. http://www.marketresearch.com/Global-Markets-Direct-v3480/Future-Wound-Care-Management-2493651/.

Rho, K.S., L. Jeong et al. (2006). Electrospinning of collagen nanofibers: Effects on the behavior of normal human keratinocytes and early-stage wound healing. *Biomaterials* 27: 1452–1461.

Sahoo, S., L.T. Ang et al. (2010). Bioactive nanofibers for fibroblastic differentiation of mesenchymal precursor cells for ligament/tendon tissue engineering applications. *Differentiation* 79: 102–110.

Saraf, A., L.S. Baggett et al. (2010). Regulated nonviral gene delivery from coaxial electrospun fiber mesh scaffolds. *Journal of Controlled Release* 143(1): 95–103.

Schneider, A., X.Y. Wang et al. (2009). Biofunctionalized electrospun silk mats as a topical bioactive dressing for accelerated wound healing. *Acta Biomaterialia* 5(7): 2570–2578.

Shin, H.J., C.H. Lee et al. (2006). Electrospun PLGA nanofiber scaffolds for articular cartilage reconstruction: Mechanical stability, degradation and cellular responses under mechanical stimulation in vitro. *Journal of Biomaterials Science, Polymer Edition* 17: 103–119.

Siepmann, J. and N.A. Peppas (2001). Modeling of drug release from delivery systems based on hydroxypropyl methylcellulose. *Advanced Drug Delivery Reviews* 48: 139–157.

Siepmann, J. and N.A. Peppas (2011). Higuchi equation: Derivation, applications, use and misuse. *International Journal of Pharmaceutics* 418: 6–12.

Srikar, R., A.L. Yarin et al. (2008). Desorption-limited mechanism of release from polymer nanofibers. *Langmuir* 24: 965–974.

Stitzel, J., J. Liu et al. (2006). Controlled fabrication of a biological vascular substitute. *Biomaterials* 27: 1088–1094.

Sukigara, S., M. Gandhi et al. (2003). Regeneration of *Bombyx mori* silk by electrospinning—Part 1: Processing parameters and geometric properties. *Polymer* 44(19): 5721–5727.
Sukigara, S., M. Gandhi et al. (2004). Regeneration of *Bombyx mori* silk by electrospinning—Part 2: Process optimization and empirical modeling using response surface methodology. *Polymer* 45(11): 3701–3708.
United Network for Organ Sharing (2010). 2010 Annual Report. Richmond, VA: UNOS.
Wan, Y., W. Gao et al. (2010). Diameter prediction for electrospun nanofibers. *International Journal of Nonlinear Sciences and Numerical Simulation* 11: 151–153.
Wei, K., Y. Li et al. (2011). Emulsion electrospinning of a collagen-like protein/PLGA fibrous scaffold: Empirical modeling and preliminary release assessment of encapsulated protein. *Macromolecular Bioscience* 11(11): 1526–1536.
Whited, B.M., J.R. Whitney et al. (2011). Preosteoblast infiltration and differentiation in highly porous apatite-coated PLLA electrospun scaffolds. *Biomaterials* 32(9): 2294–2304.
Xie, J. and C.H. Wang (2006). Electrospun micro- and nano-fibers for sustained delivery of paclitaxel to reat C6 glioma *in vitro*. *Pharmaceutical Research* 23(8): 1817–1826.
Xie, J., S.E. Willerth et al. (2009). The differentiation of embryonic stem cells seeded on electrospun nanofibers into neural lineages. *Biomaterials* 30: 354–362.
Xu, X., L. Yang et al. (2005). Ultrafine medicated fibers electrospun from W/O emulsions. *Journal of Controlled Release* 108(1): 33–42.
Yang, D.-J., F. Chen et al. (2009). Tissue antiadhesion potential of biodegradable PELA electrospun membranes. *Acta Biomaterialia* 5(7): 2467–2474.
Yang, D.Z., Y.H. Long et al. (2008a). Release of lysozyme from electrospun PVA/lysozyme–gelatin scaffolds. *Frontiers of Materials Science in China* 2(3): 261–265.
Yang, F., R. Murugan et al. (2005). Electrospinning of nano/micro-scale poly (L-lactic acid) aligned fibers and their potential in neural tissue engineering. *Biomaterials* 26: 2603–2610.
Yang, Y., X. Li et al. (2008b). Release pattern and structural integrity of lysozyme encapsulated in core–sheath structured poly(D,L-lactide) ultrafine fibers prepared by emulsion electrospinning. *European Journal of Pharmaceutics and Biopharmaceutics* 69(1): 106–116.
Yang, Y., T. Xia et al. (2011). Promotion of skin regeneration in diabetic rats by electrospun core–sheath fibers loaded with basic fibroblast growth factor. *Biomaterials* 32(18): 4243–4254.
Yannas, I.V. and J.F. Burke (1980). Design of an artificial skin—Part I: Basic design principles. *Journal of Biomedical Materials Research* 14(1): 65–81.
Zeng, J., A. Aigner et al. (2005). Poly(vinyl alcohol) nanofibers by electrospinning as a protein delivery system and the retardation of enzyme release by additional polymer coatings. *Biomacromolecules* 6: 1484–1488.
Zong, X., S. Li et al. (2004). Prevention of postsurgery-induced abdominal adhesions by electrospun bioabsorbable nanofibrous poly(lactide-co-glycolide)-based membranes. *Annals of Surgery* 240(5): 910–915.

12 Mechanics of Fiber-Reinforced Scaffolds and Tissues Formed from Organized Electrospun Assemblies

Matthew B. Fisher and Robert L. Mauck

CONTENTS

12.1 INTRODUCTION

The composition and organization of the extracellular matrix (ECM) of biologic soft tissues of the musculoskeletal system are tailored to fulfill the specific function of the tissue *in vivo*. For many such tissues, the mechanical behavior, that is, the load–extension and stress–strain relationships, is of paramount importance. These characteristics derive directly from the ECM, which is composed primarily of fibrillar collagen molecules, arranged into hierarchical dense bundles, as well as other minor components, including elastin, small proteoglycans (PGs), and other matrix molecules. Within the musculoskeletal system, tendons, ligaments, the menisci of the knee, and the annulus fibrosus (AF) of the intervertebral disc (IVD) all transmit tensile loads through their aligned ECM. Moreover, cells interact with this ECM on the nanoscale, with the loading of the tissue at the macroscale translated to microscopic cellular deformations, which, in turn, govern the natural homeostatic response of the tissue.

Given the importance of the ECM in tissue function, a major focus of scaffold development for tissue engineering and regenerative medicine applications has been on the development of materials that can recapitulate certain unique features of the ECM. One of the many design criteria for these materials is to match the biomechanical properties of the native tissue. Another goal is to provide an appropriate template for tissue formation and maturation by modulating cellular activity, including cell adhesion, proliferation, differentiation, and organized matrix production. The end goal of these activities is to generate a "neotissue" with characteristics similar to that of the native tissue (Butler et al. 2000; Langer and Vacanti 1993). To meet these design criteria, a number of scaffold processing techniques have been employed, including phase separation, particle leaching, gas foaming, fabrication of woven and nonwoven felts, emulsification, self-assembly, and hydrogel formation [see the review of Stella et al. (2010)].

In addition to the above methods, one prominent current technique for the creation of nanofibrous materials is the process of electrospinning. Although electrospinning has existed since the 1930s, its potential as a simple and cost-effective approach for making scaffolds with broad applications in tissue engineering and regenerative medicine has led to a recent explosion in its use. In fact, a Scopus search for the topic of electrospinning reveals an exponential increase in publications over the last decade, with more than 1500 new studies published in 2010 (Figure 12.1). The general enthusiasm of the field for utilizing this technique for regenerative medicine applications lies in the ability to easily fabricate tunable mats consisting of fine fibers from as small as 50 nm up to several microns in diameter. These fibers are many times smaller than most mammalian cells (which are ~10–20 μm in diameter) and, in fact, are of similar size to collagen fibers normally present in the ECM (Provenzano and Vanderby 2006; Riesle et al. 1998; Woo et al. 2006b). This nanoscale and microscale topography modulates cell signaling pathways (Nur et al. 2005) and elicits superior metabolic and matrix forming activities from attached cells (Bashur et al. 2009; Li et al. 2006). Nanofibrous meshes are also porous structures with a continuous distribution of pore sizes in the range of 2–465 μm and void volumes of 80%–95% (Li et al. 2002). This implies that nutrient transport should be readily accomplished and sufficient void volumes exist for robust ECM formation.

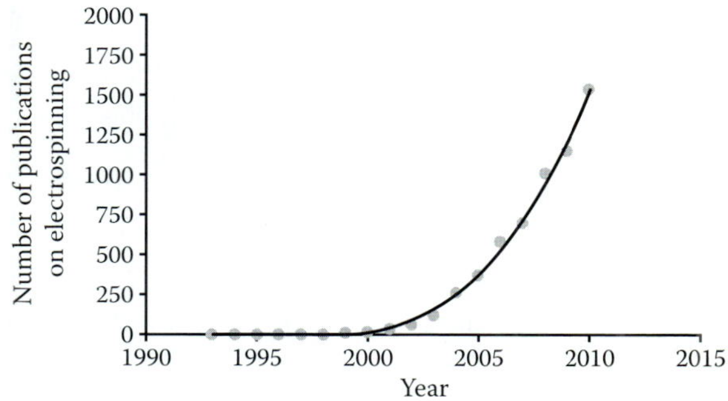

FIGURE 12.1 Exponential rise in the number of peer-reviewed publications on the topic of electrospinning, as provided through the Scopus publication search engine.

Due to its popularity and widespread use in tissue engineering and regenerative medicine, as well as the importance of mechanical properties for these applications, this chapter focuses on the method of electrospinning and the resulting mechanical behavior of electrospun scaffolds. First, we briefly discuss the electrospinning process and the mechanics of single electrospun fibers, as well as the bulk mechanics of scaffolds formed from a number of different polymers. We also discuss specific scaffold characteristics that influence the resulting mechanical properties of the scaffolds. Following this discussion, specific applications where electrospun scaffolds have been employed for orthopedic soft tissues are highlighted. For each, we show how electrospun scaffolds can be designed to fit the needs of each application. Finally, we conclude with a discussion of the challenges and future directions that remain for this exciting technology.

12.2 ELECTROSPINNING

12.2.1 Overview of the Process

The typical electrospinning apparatus consists of one or more polymers, a power supply to generate an electric field, a spinneret for polymer focusing, and a collector plate (Figure 12.2). The polymer of choice is dissolved in a compatible solvent to generate a viscous solution, with the delivery rate of the polymer generally controlled via a syringe pump in combination with a thin metal needle (the spinneret). The application of a high voltage (in the range of 10–30 kV) results in charge buildup and charge–charge repulsion among the individual polymer chains within the extruded solution. Eventually, these intermolecular electrostatic forces overcome the surface tension holding the droplet in place, and a fine charge-carrying polymer jet is formed. If the collector plate is electrically grounded (or oppositely charged), the established electric field will drive fibers from the spinneret to the collector (Brenner et al. 2001a,b; Doshi and Reneker 1995; Reneker and Chun 1996; Shin et al. 2001). Solvent evaporation and whipping instability as the fibers travel to the collector produce ultrafine fibers, as the hydrodynamic forces stretch the individual polymer fibers in transit (Shin et al. 2001; Tan et al. 2001). Over time, a mesh of randomly oriented fibers is deposited on the collector.

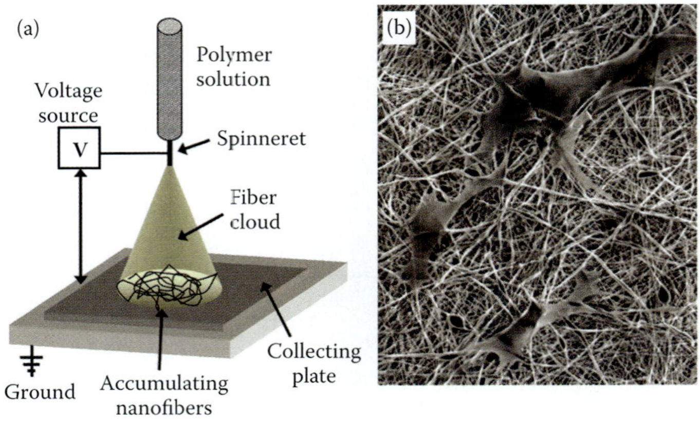

FIGURE 12.2 Electrospinning ultrafine fibrous networks. (a) Schematic of a common electrospinning setup. These simple systems consist of a polymer source flowing (actively or passively) through a highly charged spinneret, where charge repulsion in the solution generates a dispersion of fiber jets that travel rapidly to a grounded collecting plate or mandrel. (b) Electrospun fibers have diameters that range from 50 nm to several micrometers, providing a surface topography in which cells interact simultaneously with multiple fibers. Shown are human MSCs on a nonaligned PCL nanofibrous sheet. (Reprinted from Mauck R.L. et al., *Tissue Eng, Part B* 15B, 171–193, 2009. With permission from Mary Ann Liebert, Inc.)

Nanofiber features in the network depend on the polymer composition and several controllable processing variables (Tan et al. 2001). Common intrinsic parameters that can be varied include the solvent type and composition, the mass concentration and molecular weight of the polymer, solution viscosity, applied voltage, electric field strength, spinneret-to-collector distance, and polymer flow rate (Fridrikh et al. 2003; Kidoaki et al. 2006; Reneker and Chun 1996; Tan et al. 2001; Tuan et al. 2005). Additional extrinsic parameters include atmospheric conditions such as ambient temperature and humidity. These individual parameters are tuned to generate fibers that meet the criteria of the intended application, and many review articles on this subject already exist in the literature (Barnes et al. 2007b; Li and Xia 2004; Li et al. 2005c; Mauck et al. 2009; Pham et al. 2006a; Stella et al. 2010). Instead, we focus our discussion on the mechanical properties of individual electrospun polymer fibers, scaffolds that comprise these individual fibers, and important characteristics of the scaffold that influence the mechanical response of the scaffolds. However, it should be recognized that the experimental variables mentioned above help tune each of the characteristics discussed in this chapter.

12.2.2 Brief Discussion of the Evaluation of Mechanical Properties of Nanofibrous Scaffolds (NFS) and Biologic Tissues

In order to quantitatively assess the response of both electrospun scaffolds and biologic tissues to deformation, these materials are often isolated and subjected to

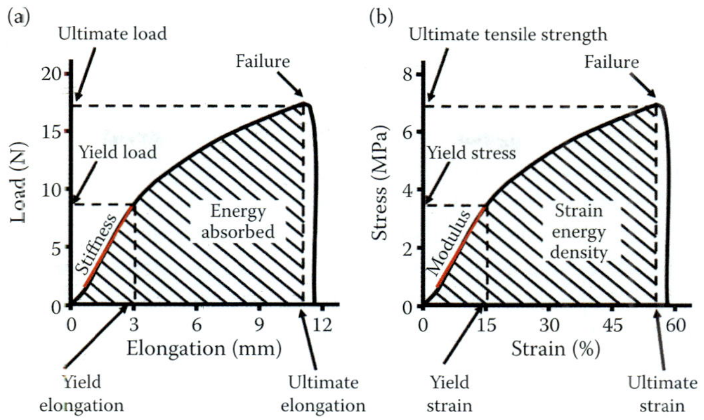

FIGURE 12.3 Typical parameters obtained from (a) load–elongation and (b) stress–strain curves obtained via tensile testing.

tensile loading using material testing machines. Many mechanical testing modalities are available; however, uniaxial tensile testing is often carried out as a primary test as a simple method to evaluate material properties. When loaded to failure (Figure 12.3a), the load–elongation behavior of the specimen is generally described by several parameters, including stiffness (the slope of the linear portion of the load–elongation curve), yield load and elongation (values at the point where the slope of the load–elongation curve deviates from the linear portion), ultimate load (the maximum load at which the material fails), ultimate elongation (elongation corresponding to deformation at the maximum load), and energy absorbed at failure (the area under the curve to the maximum load).

To determine the intrinsic quality of the specimen irrespective of size (i.e., its mechanical properties), it is common to normalize the load–elongation behavior of the tissues. This is accomplished with knowledge of the specimen's cross-sectional area and local tissue deformation. Given the deformable nature of NFS and soft tissues, noncontact approaches are generally preferred to measure the cross-sectional area of the specimen, usually accomplished via laser methods or digital image capture (Fung 1993; Lee and Woo 1988; Woo et al. 1983). The measurement of strain is also typically done using digital imaging techniques and reflective markers, ink patterns, or other markers placed on the tissue substance (Fung 1993; Scheffler et al. 2001; Soslowsky et al. 2010; Weiss et al. 2005; Yin et al. 1972). Stress is obtained by dividing the load by the cross-sectional area, and strain is obtained by computing the change in marker-to-marker distance during the test relative to the original configuration of the markers. Based on the resulting stress–strain curve (Figure 12.3b), parameters that are often reported include tangent modulus (the slope of the linear portion of the stress–strain curve), yield stress and strain (values at the point where the slope of the stress–strain curve deviates from the linear portion), tensile strength (stress at failure), ultimate strain (strain corresponding with the tensile strength), and strain energy density (the area under the stress–strain curve until failure).

12.2.3 Single-Fiber Mechanical Properties

The macroscale mechanics of NFS represent the aggregate contributions of many thousands of individual polymer fibers. Although these fibers can interact with one another via welds that form at interstices or by frictional interactions as fibers slide relative to one another, the mechanical properties of individual nanofibers are the dominant contributor to the bulk scaffold properties and are useful information for scaffold design. In addition, the isolation of a single nanofiber allows the direct study of factors such as fiber size on fiber mechanics. Despite the experimental challenges in the mechanical testing of a single fiber, these data have been collected for a number of polymers (Table 12.1). These properties are generally obtained through either microtensile testing (MT) or three-point bending using atomic force microscopy (AFM).

In MT, single fibers are collected on parallel surfaces separated by a small air gap. For example, Tan et al. (2005b) found that single poly(ε-caprolactone) (PCL) fibers (1.0–1.7 μm in diameter) that extended until failure had a modulus of ~120 MPa. Alternatively, AFM and optical microscopy have been used to manipulate and test single fibers. Tan et al. (2005a) fixed one end of a nanofiber and used the AFM cantilever to apply and measure a tensile force generated in poly(ethylene oxide) (PEO) nanofibers, measuring a modulus of ~45 MPa. Similarly, the AFM probe tip has been used to carry out three-point bending tests of individual poly(L-lactide) (PLLA) fibers placed over a groove microetched into a silicon wafer (Tan and Lim 2006). Carlisle et al. (2010) used an AFM tip to apply three-point bending to individual electrospun collagen fibers. They obtained a modulus of approximately 3 GPa up to 5.6% strain. In the case of fibrinogen, a modulus of 17.5 MPa was found (Carlisle et al. 2009), showing the wide range of mechanical properties observed in ECM constituents in the nanofibrous format. Specialized microelectromechanical devices that couple with AFM cantilevers have been constructed and used to show the importance of applied strain rate on the mechanical properties of polyacrylonitrile nanofibers (Naraghi et al. 2007).

A number of factors can influence the mechanical responses of individual nanofibers. First, the testing technique (tensile versus bending forces) may yield different results. Interestingly, isolated polyacrylonitrile fibers have been evaluated through both AFM and nanotensile testing, with similar mechanical properties observed (0.5–6.3 and 1–7 GPa, respectively; Liao et al. 2011; Naraghi et al. 2011). Whether this characteristic holds for other polymers or natural proteins is not yet clear. Fiber diameter appears to be a crucial variable in the measured properties, as many studies note a sharp increase in the modulus as the fiber diameter decreases past a critical threshold in the range of 200–700 nm, depending on the material tested (Carlisle et al. 2009; Chew et al. 2006; Wong et al. 2008). For PCL fibers subjected to tensile elongation, the modulus increased linearly as the fiber size decreased below 700 nm (Wong et al. 2008). More drastic changes were observed in electrospun collagen fibers, as the modulus increased more than 100-fold as the fiber radius decreased from 783 nm to 160 nm (Carlisle et al. 2010). The mechanism by which stiffening occurs may be that, as the diameter of the fiber decreases, the size of the fiber "shell" (or thin outer layer) possessing higher mechanical properties reaches a similar cross-sectional size as the fiber core. Thus, the relative contribution of the shell

TABLE 12.1
Tensile Properties of Individual Electrospun Nanofibers

	Polymer	Testing Method	Fiber Diameter (nm)	Modulus (MPa)	Yield Strain (%)	Ultimate Strain (%)	Yield Stress (MPa)	Ultimate Tensile Strength (MPa)	References
1	Poly(ε-caprolactone)	MT	1000–1700	120	20	200	13	40	Tan et al. 2005b
2	Poly(ethylene oxide)	AFM	700	~45	n/a	n/a	n/a	n/a	Tan et al. 2005a
3	Poly(propylene)	MT	670	478	81	52	n/a	n/a	Watanabe et al. 2011
4	Poly(acrylonitrile)	MT	900–150	1000–7000	n/a	n/a	n/a	n/a	Naraghi et al. 2011
		AFM	826–221	540–6290	n/a	n/a	n/a	n/a	Liao et al. 2011
5	Collagen (dry)	AFM	160–783 356–179	2800 1400–7500	12	33	n/a	25	Carlisle et al. 2010; Yang et al. 2007
	Collagen (dry, cross-linked)	AFM	450–191	1300–7800	n/a	n/a	n/a	n/a	Yang et al. 2007
	Collagen (wet, cross-linked)	AFM	425–210	70–260	n/a	n/a	n/a	n/a	Yang et al. 2007
6	Fibrinogen (wet)	AFM	208	17.5	n/a	130	n/a	22	Carlisle et al. 2009

Note: MT, microtensile testing; AFM, atomic force microscopy; n/a, not available.

TABLE 12.2
Tensile Properties of Common Electrospun Scaffolds

	Polymer	Alignment	Modulus (MPa)	Yield Strain (%)	Ultimate Strain (%)	Yield Stress (MPa)	Ultimate Tensile Strength (MPa)	References
1	Thermoplastic polyurethane	Both	n/a	n/a	n/a	n/a	n/a	Pedicini and Farris 2003
2	Polydioxanone–elastin blend	NA	5–10	n/a	60–200	n/a	3–5	Sell et al. 2006
3	Poly(ester-urethane)urea	NA	8	n/a	2.2	n/a	13	Stankus et al. 2004
	Poly(ester-urethane) urea–collagen blends	NA	1–7	n/a	160–280	n/a	2–11	Stankus et al. 2004
4	Biodegradable polyurethane	NA	5.7	61	79	61	1.3	Rockwood et al. 2007
5	Poly(D,L-lactide-*co*-glycolide) (85:15)	NA	323	n/a	96	n/a	23	Li et al. 2002
6	Poly(p-dioxanone-*co*-L-lactide)-block-poly(ethylene glycol)	NA	n/a	n/a	n/a	n/a	1.3–1.4	Bhattarai et al. 2004
7	Silk fibroin	NA	625	n/a	4	n/a	13.6	Jin et al. 2004
8	Poly(L-lactide-*co*-ε-caprolactone)	NA	156	n/a	127	n/a	5	Xu et al. 2004
9	Collagen–poly(ethylene oxide)	NA	12	n/a	n/a	n/a	0.37	Huang et al. 2001
10	Collagen	AL	52.3 26.1	n/a	n/a	n/a	1.5 0.7	Matthews et al. 2002
11	Poly(ethylene glycol)-g-chitosan	NA	n/a	n/a	78–112	n/a	8–17	Jiang et al. 2004b
12	Gelatin	NA	105	n/a	64	n/a	2.5	Zhang et al. 2009

	Gelatin–poly(ε-caprolactone)	NA	5	n/a	126	n/a	2.7	Zhang et al. 2009
	Poly(ε-caprolactone)	NA	31	n/a	138	n/a	1.3	Zhang et al. 2009
13	Poly(vinyl alcohol)–cellulose acetate blends	NA	44	n/a	420	10	11	Ding et al. 2004
	Poly(vinyl alcohol)	NA	18–34	n/a	214–334	5.9–7.2	7.0–9.4	Ding et al. 2004
	Cellulose acetate	NA	3	n/a	175	1.5	2	Ding et al. 2004
14	Poly(L-lactic acid)	NA	8.5	1.9	20–25	0.1	n/a	Li et al. 2006
	Poly(D,L-lactic acid)	NA	70	1.9	n/a	1.1	n/a	Li et al. 2006
	Poly(D,L-lactic-*co*-glycolic acid) (50:50)	NA	144	2.1	38	2.8	n/a	Li et al. 2006
	Poly(D,L-lactic-*co*-glycolic acid) (85:15)	NA	114	1.9	n/a	1.9	n/a	Li et al. 2006
	Poly(glycolic acid)	NA	138	2.1	n/a	2.7	n/a	Li et al. 2006
	Poly(ε-caprolactone)	NA	8.5	7.5	n/a	0.7	n/a	Li et al. 2006
15	Poly(ε-caprolactone)	AL	3.3–4.7 2.7–3.9	n/a	200–231 216–277	0.6–0.7 0.6–0.4	1.4–2.1 1.2–2.0	Lee et al. 2003
16	Poly(glycolic acid)	AL	55–105 20–55	n/a	60–80 105–130	n/a	n/a	Boland et al. 2001
	Poly(glycolic acid)	NA	20–95	n/a	80–100	n/a	n/a	Boland et al. 2001

Source: Reprinted from Mauck R.L. et al., *Tissue Eng, Part B* 15B, 171–193, 2009. With permission from Mary Ann Liebert, Inc.

Note: NA, nonaligned; AL, aligned; n/a, not available.

increases, resulting in improved mechanical properties (Arinstein and Zussman 2011). Hydration can be another important factor, as Yang et al. (2008) demonstrated that hydrating cross-linked collagen fibers led to a decrease in the modulus of two orders of magnitude. Chew et al. (2006) found that individual fiber properties change when additional components, such as model proteins and pharmacologics, are incorporated. The subsequent work of Lim et al. (2008) showed that higher crystallinity resulted in higher fiber stiffness, even when controlling for fiber diameter, and crystallinity may increase as fiber diameters decrease during formation.

12.2.4 Bulk NFS Mechanical Properties

Electrospun meshes have been successfully produced from numerous polymers, including polyurethanes (Khil et al. 2003), biodegradable polyesters [e.g., PCL (Li et al. 2002; Venugopal et al. 2005; Xu et al. 2004; Yoshimoto et al. 2003), polyglycolic acid (Boland et al. 2004), polylactic acid (PLA; Li et al. 2006; Yang et al. 2004, 2005), and polydioxanone (Boland et al. 2005)], and natural biopolymers, including collagen (Buttafoco et al. 2006; Li et al. 2005b; Matthews et al. 2002; Rho et al. 2006; Venugopal et al. 2005), elastin (Buttafoco et al. 2006; Li et al. 2005b), silk fibroin (Min et al. 2004a,b), chitosan (Bhattarai et al. 2005; Geng et al. 2005), dextran (Jiang et al. 2004a), and wheat gluten (Woerdeman et al. 2005), resulting in a wide range of mechanical properties (Table 12.2). For example, nonaligned polymers such as poly(D,L lactide-*co*-glycolide) (PLGA) produce meshes that have a high initial modulus but yield at low levels of strain. On the other hand, scaffolds composed of PCL have a linear modulus one order of magnitude lower but remain elastic over a wider range of strain (up to 8%–10%; Figure 12.4; Li et al. 2006).

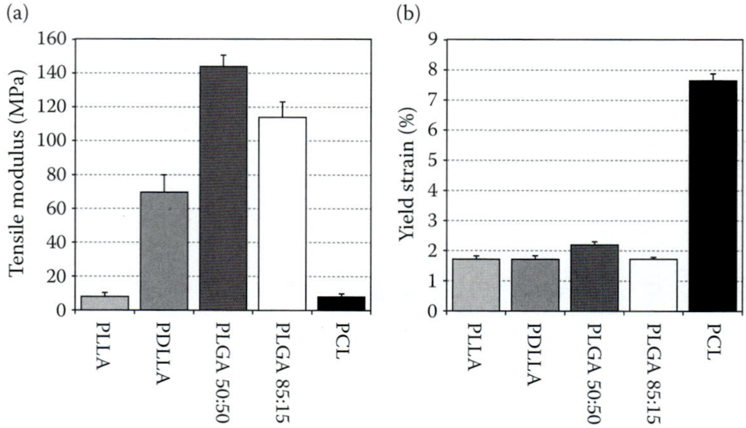

FIGURE 12.4 Tensile properties for a common set of biodegradable polymers. (a) Tensile modulus and (b) yield strain for nonaligned NFS provide a range of properties, but few polymers can be distended to greater than 2% elongation before yielding. (Adapted from Li W.J. et al., *Acta Biomater* 2, 377–385, 2006. With permission from Elsevier.)

Multijet electrospinning systems have also been developed to allow the creation of composite scaffolds (with a different polymer in each jet) to create a wider range of mechanical properties reflective of the properties and ratios of the individual components (Baker et al. 2008, 2009; Ding et al. 2004; Kidoaki et al. 2005; Kwon et al. 2005; Madhugiri et al. 2003). To increase the stiffness of scaffolds, cross-linking agents (genipin or methacrylate-based copolymers) have been utilized (Kim et al. 2005; Pornsopone et al. 2005; Tan et al. 2008). Blends of biosynthetic and natural components have been electrospun (with components mixed in every fiber) to create meshes with enhanced cell compatibility (Lee et al. 2007; Sell et al. 2006). Additional studies have modified fiber surfaces to enhance cell binding and/or growth factor retention (Casper et al. 2005, 2007; Ma et al. 2005). Finally, hydrolytically degradable elastomers have been electrospun, allowing more control over the degradation of these materials to match the matrix synthesis of cells seeded within these scaffolds (Ifkovits et al. 2008). In addition to properties imparted by the polymer composition itself, several other structural features regarding the organization and combination of individual fibers within a given scaffold factors may influence the measured mechanical properties. These are discussed in detail in the following sections.

12.2.4.1 Effects of Alignment

The most studied factor impacting the mechanics of electrospun scaffolds is their organization, specifically, the degree to which fibers are aligned. When deposited on a stationary and uniform collector, the direction of the fibers is random, leading to generally isotropic mechanical properties (independent of the direction of testing). Given that most biological tissues are anisotropic, several methods have been developed to introduce anisotropy into NFS. Fibers can be pulled into alignment simply by stretching the scaffolds (Johnson et al. 2007). Due to the nonrecoverable mechanical behavior of the polymer and the physical changes induced by this deformation, these fibers are then locked into place. Another set of methods operate by altering the geometry of the collector and, as a result, directing the orientation of fibers as they are collected. One of the earliest examples collected fibers on the thin edge of a rotating disc (Theron et al. 2001). Additional examples have used conducting materials separated by an air gap (Li et al. 2004, 2005a). These can be arranged as two parallel surfaces (Kakade et al. 2007), as a grid of electrodes (Li et al. 2004, 2005a), or as a wire drum (Katta et al. 2004). Kakade et al. (2007) found that, for PEO fibers, this technique aligns the molecular backbone of the polymer in addition to the fibers themselves, which could result in a multiplicative increase in scaffold mechanical properties. In a more advanced technique, the collectors can be polarized to create a magnetic field at the collector. With the inclusion of a small amount of magnetic nanoparticles in the polymer solution, the fibers will align along the direction of the magnetic field (Yang et al. 2007). For even greater control, near-field electrospinning has been developed (Sun et al. 2006). A point spinneret is dipped in polymer solution and can "write" ordered structures onto the grounded surface distance a few microns away from the tip.

However, the most common method of inducing fiber alignment has been by using a cylindrical collecting mandrel and moving it at a high rate of speed (Ayres et al.

2007; Baker and Mauck 2007; Boland et al. 2001; Courtney et al. 2006; Gaumer et al. 2009; Li et al. 2007; Nerurkar et al. 2007; Sundaray et al. 2004; Yeganegi et al. 2010). Numerous studies have found that, when the surface velocity of the mandrel passes a certain threshold (~2 m/s), it overcomes the speed of a single fiber from the source spinneret to the grounded surface; therefore, the fibers are pulled along the direction of mandrel rotation (Courtney et al. 2006; Li et al. 2007). The degree of alignment steadily increases after this point before reaching a plateau at ~10 m/s, at which time the vast majority of fibers are aligned. More recently, the rotating mandrel approach has been expanded to include an additional electric field near the mandrel, which causes a larger whipping area along the direction of mandrel rotation, allowing further elongation of the fibers and increased alignment (Lee et al. 2009).

Compared to nonaligned scaffolds, aligned scaffolds exhibit very different mechanical behavior, depending on the degree of alignment and other experimental factors (Table 12.2). For example, nonaligned scaffolds made from PCL exhibit a relatively linear stress–strain response in the preyield region and extend linearly after yield (Figure 12.5). These properties are independent of the direction of testing. Conversely, aligned PCL scaffolds tested in the fiber direction are more nonlinear at low levels of strain, followed by a sharper but linear increase in stress with increasing strain. In these aligned scaffolds, yield and failure occur at much lower levels of strain compared to nonaligned scaffolds. When these same scaffolds are tested in the transverse direction, a much lower and more linear stress–strain profile is observed. As such, the ratio of properties, such as linear modulus and ultimate tensile strength,

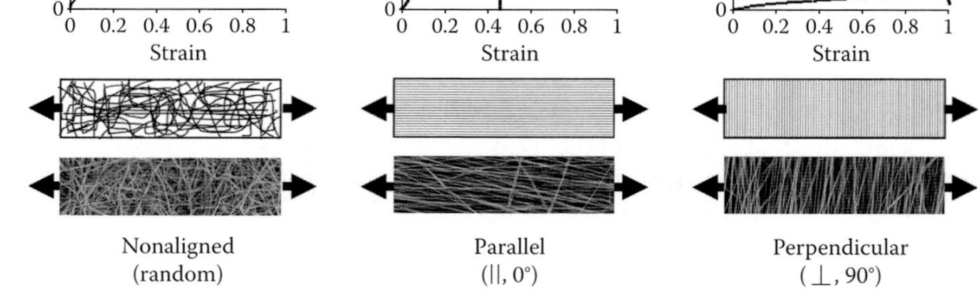

FIGURE 12.5 Fiber alignment influences mechanical properties of NFS. Stress–strain response of PCL NFS (a) produced in a nonaligned configuration, (b) produced in an aligned configuration and tested in the prevailing fiber direction, and (c) produced in an aligned configuration and tested in the transverse direction. Schematics and images illustrate fiber organization and testing direction for each scaffold. (Reprinted from Mauck R.L. et al., *Tissue Eng, Part B* 15B, 171–193, 2009. With permission from Mary Ann Liebert, Inc.)

in the fiber direction compared to transverse to the fiber direction can increase by 10- to 20-fold with increasing alignment (Ayres et al. 2007; Baker and Mauck 2007; Courtney et al. 2006; Gaumer et al. 2009; Li et al. 2007).

12.2.4.2 Effects of Porosity

Aside from fiber alignment, another major contributor to scaffold mechanics is the overall porosity and pore size or, in other words, the relative spacing between the fibers. Again, several methods have been devised to alter the number and size of pores within the scaffold. For example, increasing the individual fiber diameter to 5–10 μm (by changing the concentration of the polymer solution or altering other variables) can increase the average pore size to 50–100 μm, which is several folds higher than values typically observed (Balguid et al. 2009; Pham et al. 2006b; Rnjak-Kovacina et al. 2011). Although cells can enter these scaffolds more readily, this approach sacrifices the nanoscale cellular signals provided by smaller fibers. Given this limitation, some studies have proposed that it may be beneficial to include both nanoscale and microscale fibers intermixed throughout or in distributed in layers through the scaffold (Kidoaki et al. 2005; Pham et al. 2006b). In order to maintain the nanoscale fibers while producing larger pore sizes, Wright et al. (2011) added a salt solution within PLLA fibers. Upon incubation in an aqueous solution after spinning, the salt particles were leached out, leaving pores eight times larger. Although promising, this approach also resulted in a fivefold decrease in scaffold mechanical properties. Alternatively, our group has produced composite scaffolds containing a stable PCL fiber fraction along with a temporary water-soluble fiber fraction (PEO; Figure 12.6). The modulus of these composite scaffolds decreases linearly as the original PEO content increases following incubation in saline (Baker et al. 2008, 2011b). Other studies have proposed changes in the geometry of the collector to induce larger pores, but again, the mechanics can decrease by one-half or more (Blakeney et al. 2011; Vaquette and Cooper-White 2011). These results highlight the need to balance the design criteria of mechanical properties and pore size for each specific application.

12.2.4.3 Additional Factors

Beyond fiber alignment and porosity, additional factors can influence the overall mechanical function of electrospun scaffolds. Using scanning electron microscopy (SEM), Stella et al. (2009) found that, as the alignment of scaffolds composed of poly(ester-urethane) urea (PEUU) increased, the tortuosity of the fibers increased (i.e., the ratio of the total fiber length to the end-to-end fiber length). This is counterintuitive, as one might suspect that the fibers would be "pulled" into alignment and made straighter, and also shows the complexity of electrospinning scaffolds (Stella et al. 2009). More recently, D'Amore et al. (2010) developed an automated algorithm to detect and quantify scaffold features such as fiber orientation, connectivity, diameter, and intersection points from SEM images. A decrease in fiber–fiber connections strongly correlated with an increase in fiber alignment and scaffold mechanical properties (Figure 12.7). However, the nature of these connections may be important. Johnson et al. (2007) found that "point bonds" created in PCL scaffolds led to a decrease in mechanical properties. Conversely, Kidoaki et al. (2006) studied fiber–fiber welding as a function of the amount of slowly evaporating *N*,*N*-dimethylformamide (DMF)

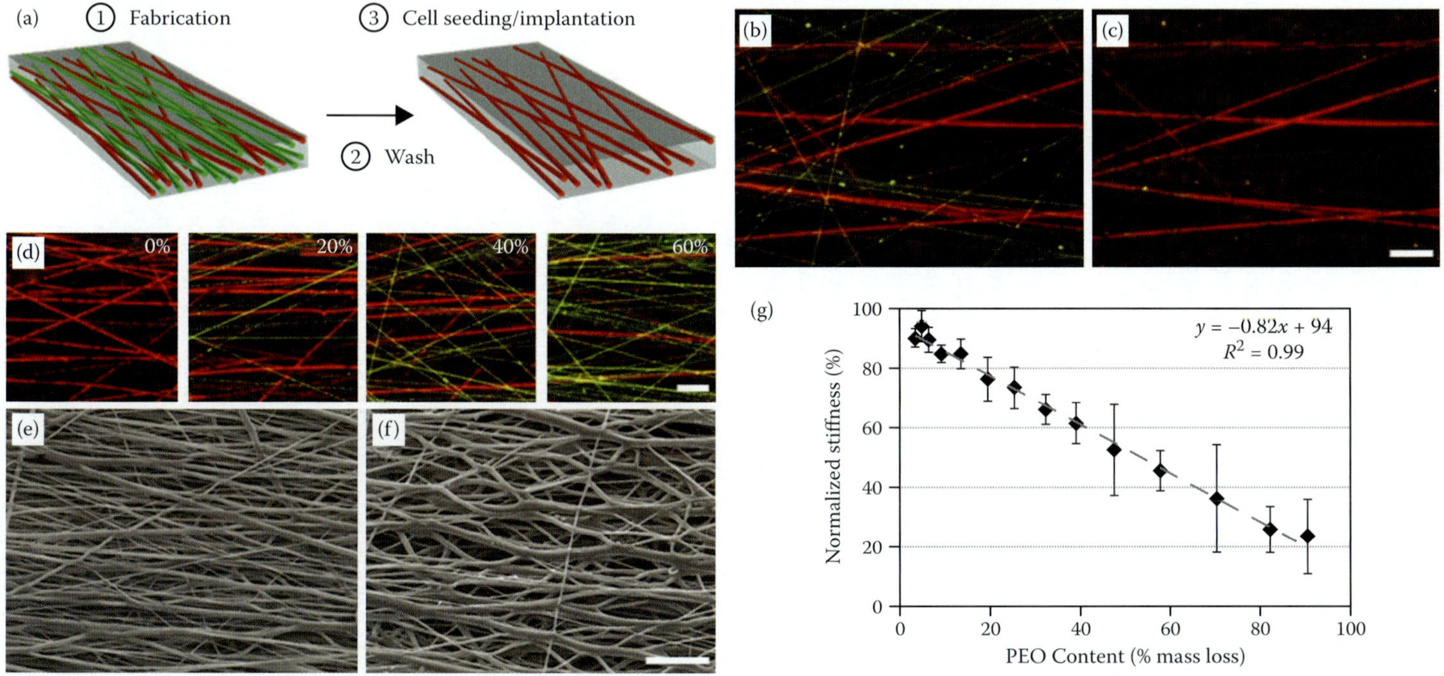

FIGURE 12.6 (a) Schematic of the fabrication of PCL (red) and PEO (green) fibers. (b) and (c) Presence of fluorescently labeled fibers before and after aqueous incubation showing the removal of the PEO fibers. (d) Prewash images showing the increased presence of PEO. (e) and (f) SEM of PCL and PCL/PEO scaffolds after aqueous incubation. Scale bar = 10 µm. (g) Scaffold PEO content correlated with changes in stiffness when samples were tested in the fiber direction. Correlation was significant with $p < 0.001$. (Adapted from Baker B.M. et al., *Biomaterials* 29, 2348–2358, 2008; Baker B.M. et al., Instruction without Impediment: Tunable Fibrous Scaffolds for Engineering Dense Connective Tissues, Annual Meeting of the Orthopaedic Research Society, Long Beach, CA, 2011. With permission from the Elsevier and the Orthopaedic Research Society.)

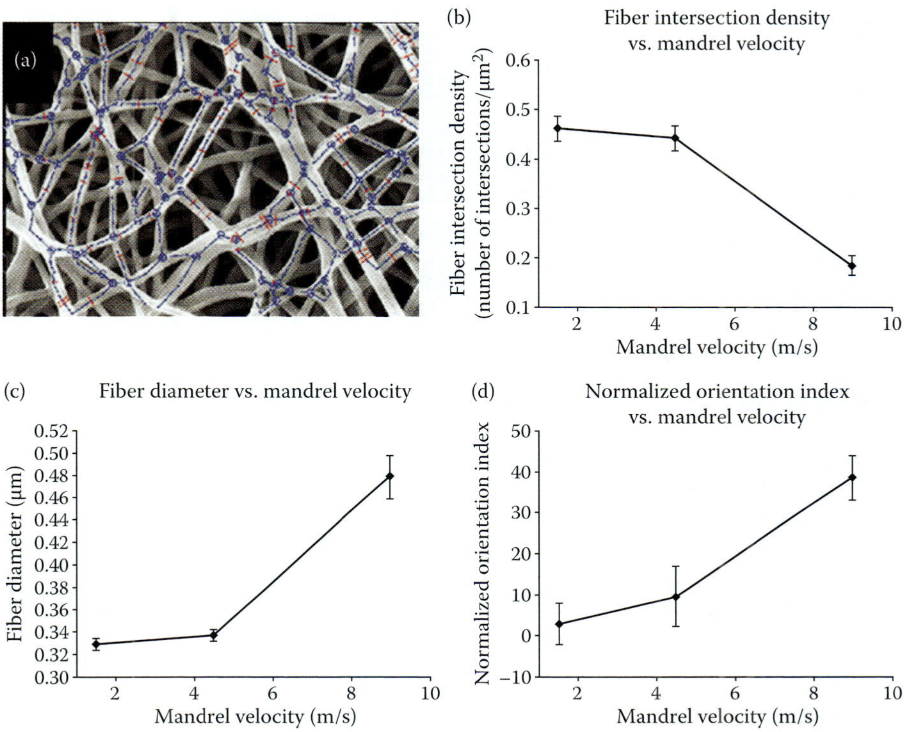

FIGURE 12.7 Illustration of complex relationships of ultrastructural properties of electrospun scaffolds. (a) Methodology for the automated determination of fiber characteristics. (b)–(d) Fiber intersection, fiber diameter, and fiber orientation as a function of mandrel velocity. (Adapted from D'Amore A. et al., *Biomaterials* 31, 5345–5354, 2010. With permission from Elsevier.)

in the solution and noted a fourfold increase in properties of electrospun scaffolds with greater welding (Figure 12.8). Using the automated approach of D'Amore et al. (2010), Amoroso et al. (2011) have recently investigated the impact of cospinning with a separate solution including cells. Interestingly, the addition of cells increased the mechanical anisotropy of the scaffolds. The authors hypothesized that the cells function as bonding points without creating stress concentrations in the fibers, which increases the overall mechanical function of the scaffold.

Environmental conditions in which the analysis is performed may also play a role in the apparent scaffold mechanics. For example, as mentioned above, some materials show profound changes in mechanical properties when they are tested in a dry versus hydrated state. For example, collagen nanofibers (which must be cross-linked before being hydrated) decrease in tensile properties by approximately 100-fold with hydration (Barnes et al. 2007a; Carlisle et al. 2010; Shields et al. 2004; Yang et al. 2008). Moreover, these scaffolds can degrade under physiologic conditions (temperature, pH, ionic concentrations, and the presence of enzymes) or with cell-mediated matrix deposition, and thus, it is also important to track the changes in mechanical

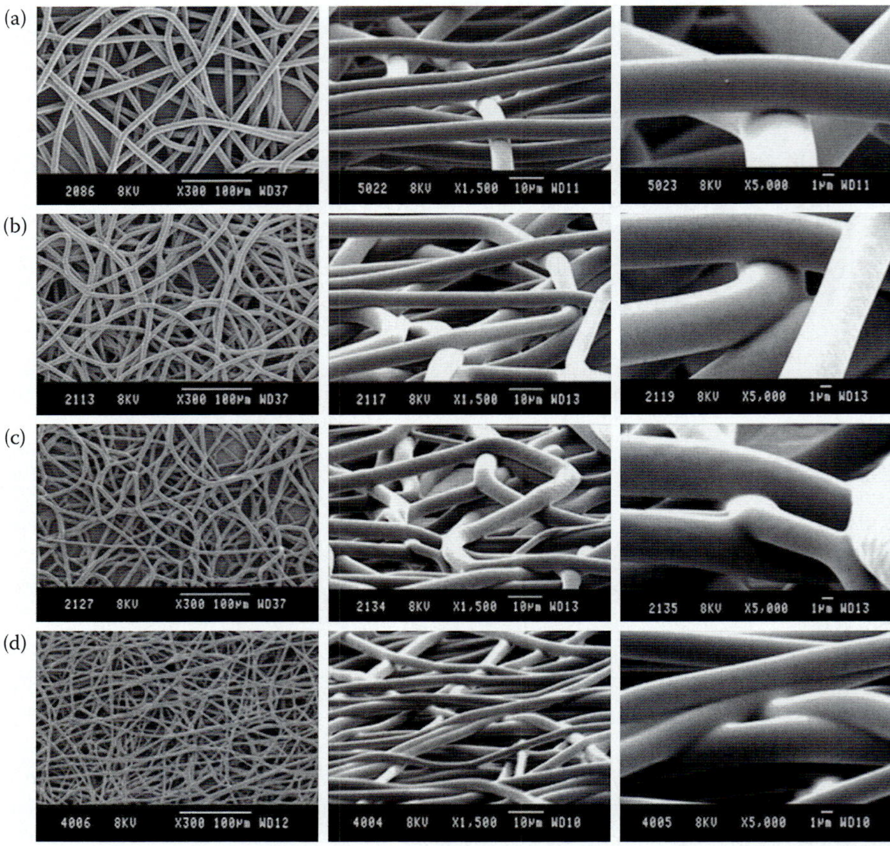

FIGURE 12.8 Effect of mixed solvent constituents on fiber–fiber welding. Relative DMF/THF concentrations: 0% (a), 5% (b), 10% (c), and 30% (d). (Reprinted from Kidoaki S. et al., *J Biomed Mater Res—Part B: Appl Biomater* 76B, 219–229, 2006. With permission from Wiley.)

properties over time. In all of these cases, it should be noted that experimental parameters such as humidity, temperature, electric field strength, solution mass concentration, and spinneret-to-collector distance can impact the final results and that these conditions must be optimized to obtain the desired scaffold characteristics.

12.3 STRUCTURE–FUNCTION OF FIBROUS SOFT TISSUES OF MUSCULOSKELETAL SYSTEM

12.3.1 Overview of the Structure and Function of Orthopedic Fibrous Soft Tissues

The functions of the soft tissues of the musculoskeletal system are varied. Tendons and ligaments connect the muscle to the bone and one bone to another bone, respectively, and guide normal joint motion. Generally, these tissues experience uniaxial tensile loads. Other tissues experience more complex loading patterns. The meniscus transmits loads between the tibia and femur by resolving the geometrical incongruence between the relatively flat tibial plateau and the convex femoral condyles. As such, it is exposed to compression, tension, and shear, as the femur moves relative to the tibia. The IVD lies between the bony vertebral bodies of the spine and is responsible for the transmission of complex multiaxial loads of the spine. To accommodate such a wide variety of loading patterns, the IVD is composed of the nucleus pulposus (NP), a hydrated gelatinous structure, and the AF, a multilamellar fibrocartilage that surrounds the NP between the vertebral bodies and experiences tension, compression, shear, torsion, and bending.

To support these varying functions, the ECM of these tissues is highly evolved to optimize their mechanical function (Figure 12.9a through f). Ligaments and tendons consist mainly of type I collagen (70%–85% of the dry weight), which is the major load-bearing structure and is generally well aligned along the direction of tensile

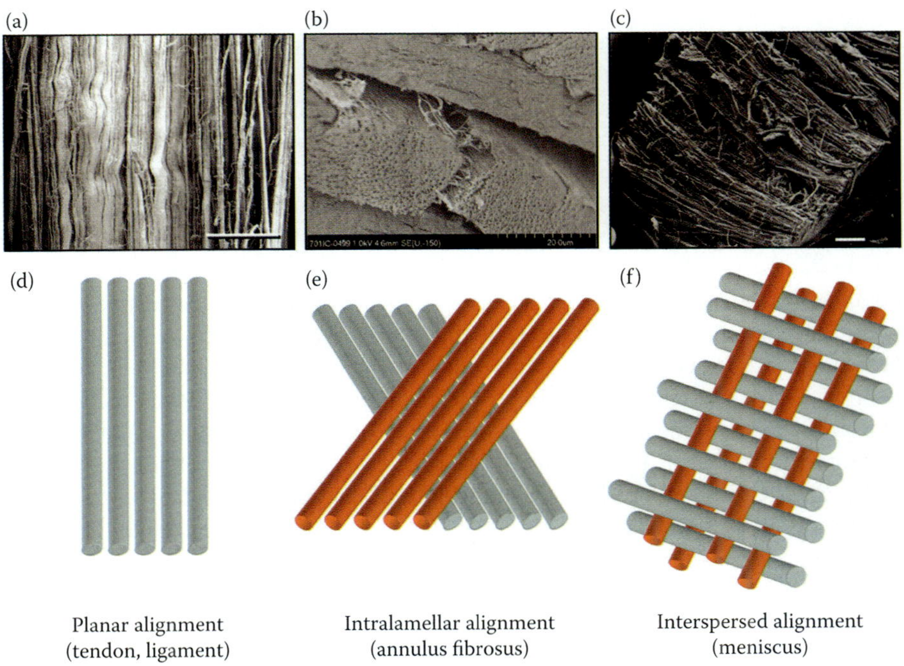

FIGURE 12.9 Fiber organization in dense connective tissues. Most tendons and simple ligaments show fiber alignment in a common prevailing direction over which load is transferred. AF and meniscus achieve fiber reinforcement in more complicated loading environments through multilamellar planar alignment of ECM or via interspersion of fibers perpendicular to the predominant fiber direction. Scale bars = 100 μm (a, d), 200 μm (b, e), and 500 μm (c, f). (Image on the left from Provenzano P.P. et al., *J Orthop Res* 20, 975–983, 2002. With permission from Wiley. Middle image from Iatridis J.C. et al., *J Biomech* 31, 535–544, 1998. With permission from Elsevier. Right image from Petersen W. and Tillmann B., *Anat Embryol* 197, 317–324, 1998. With permission from Springer.)

loading. Other collagen types, including III, V, IX, X, XI, and XII, elastin, and PGs exist only in minor quantities but play important roles with regard to regulated collagen fibril diameter, wound healing, and limiting stress concentrations (Birk and Mayne 1997; Fukuta et al. 1998; Goh et al. 2003; Linsenmayer et al. 1993; Liu et al. 1995; Niyibizi et al. 1995, 1996; Sagarriga Visconti et al. 1996). Interestingly, there is a bimodal distribution of collagen fibril diameters (40–75 nm and 100–150 nm; Dyer and Enna 1976; Eyden and Tzaphlidou 2001; Goh et al. 2003), which is thought to allow better functional properties via a high surface/volume ratio for better interfibrillar binding, fibril packing, and strength (Ottani et al. 2001).

The meniscus and IVD are fibrocartilaginous. In the meniscus, PGs make up <2%–3% of the dry weight (Adams and Hukins 1992; McDevitt and Webber 1990) and are concentrated in the inner regions, where compressive loading predominates. However, the major constituent is collagen (85%–95%; McDevitt and Webber 1990). A mixture of type I and type II collagen can be found in the inner region, whereas only type I exists in the outer region. Collagen fibers are mostly oriented in the circumferential direction, with the occasional radial "tie" fiber or sheet interspersed within this network (Petersen and Tillmann 1998). The architecture of this collagen component is paramount for the meniscus mechanical function. When compressive loads impinge on the wedge-shaped meniscus, the PG-rich inner zone projects this load outward to engage the circumferential fibers of the outer zone. These fibers resist these hoop stresses and enforce joint congruency and promote axial load transmission.

In terms of the IVD, the NP is structurally and mechanically isotropic and contains a network of type II collagen interspersed with PGs, resulting in a high water content within the tissue, whereas the fibrocartilaginous AF has a high degree of structural organization over multiple length scales—aligned bundles of collagen fibers reside within each lamella and the direction of alignment alternates from one lamella to the next by 30° above and below the transverse plane (Cassidy et al. 1989; Marchand and Ahmed 1990). The resulting angle-ply laminate possesses pronounced mechanical anisotropy and nonlinearity, allowing the AF to support tension, shear, compression, and torsion (Best et al. 1994; Ebara et al. 1996; Elliott and Setton 2001; Fujita et al. 1997, 2000; Guerin and Elliott 2007; Iatridis and Gwynn 2004; Iatridis et al. 1998; Klisch and Lotz 2000; Perie et al. 2005; Yao et al. 2002). For example, when the disc is compressed, the hydrostatic pressure in the NP increases, resulting in stresses along the radial and circumferential directions of the AF.

12.3.1.1 Tensile Properties of Orthopedic Fibrous Soft Tissues

Although many testing modalities exist, uniaxial tensile loading has been utilized as a simple method of quantifying the biomechanical properties of these tissues. For ligaments and tendons, testing is generally conducted on the entire bone–ligament–bone or muscle–tendon–bone complex, with tissue insertion sites left anatomically intact to determine the structural properties of the complex. For the meniscus and AF, sections of tissue are isolated; however, given that these tissues are no longer "tissue structures," these properties are not really indicative of their function. Instead, it is common to obtain the mechanical properties of the tissue, which are independent of the size of the specimen, as discussed in Section 12.2. From the resulting stress–strain curve, these tissues generally demonstrate nonlinear behavior, beginning with a concave toe region which is governed by the straightening and recruitment of collagen fibers. This is followed by a linear region in which the fibers are stretched. With continued deformation, yield and eventual failure occur.

The different functions of ligaments and tendons, the meniscus, and AF can be summarized by their uniaxial tensile properties (Table 12.3). The range of mechanical properties of various ligaments and tendons is very large, as they are designed to function in different joints. For the human medial collateral ligament (MCL), the tangent modulus and the ultimate tensile strength have been reported to be 332.2 ± 58.3 MPa and 38.6 ± 4.8 MPa, respectively (Quapp and Weiss 1998), whereas comparable values for the inferior glenohumeral ligament ranged from

TABLE 12.3
Mechanical Properties of Orthopedic Human Soft Tissues

	Linear Modulus (MPa)	Ultimate Tensile Strength (MPa)	Ultimate Strain (%)	Yield Strain (%)	References
		Ligaments and Tendons			
Medial collateral ligament	332 ± 58	39 ± 5	17 ± 2		Quapp and Weiss 1998
Anterior cruciate ligament	65–447	13–46	15–44		Butler et al. 1992, 1986; Noyes and Grood 1976
Glenohumeral ligament	5–42	1–6	8–33		Bigliani et al. 1992; Moore et al. 2004
Patellar tendon	504–680	54–65	12–15		Butler et al. 1984, 1986; Johnson et al. 1994
Achilles tendon	819 ± 208	79 ± 22	9 ± 2		Wren et al. 2001
		Meniscus			
Circumferential	43–294	15–18	21–31	3–11	Bursac et al. 2009;
Radial	3–60	2–4	33–55	10–20	Fithian et al. 1990; Lechner et al. 2000; Tissakht and Ahmed 1995
		Annulus Fibrosus			
Circumferential	17–50	1–3	10–20		Acaroglu et al. 1995;
Axial/Radial	0.45–0.96	n/a	n/a	20–30	Ebara et al. 1996; Elliott and Setton 2001; Fujita et al. 1997; Guerin and Elliott 2006

5 to 42 MPa and from 1 to 6 MPa, respectively, representing approximately 10% of those for the MCL (Bigliani et al. 1992). Moreover, the mechanical properties are strongly dependent on the prevailing fiber direction. Specimens along the fiber direction of the human MCL displayed a tangent modulus of 332.2 ± 58.3 MPa and a tensile strength of 38.6 ± 4.8 MPa; whereas the transverse specimens were 11.0 ± 0.9 MPa and 1.7 ± 0.5 MPa, respectively, a difference of about 30-fold (Quapp and Weiss 1998).

Ligaments and tendons are well aligned; however, the meniscus and AF have more varied alignment, which is captured by the mechanical properties. For the meniscus, the circumferential modulus ranges from 48 to 198 MPa and the radial modulus from 3 to 70 MPa (Bursac et al. 2009; Fithian et al. 1990; Lechner et al. 2000; Tissakht and Ahmed 1995). In terms of the AF, the uniaxial tensile modulus is one to two orders higher along the circumferential direction (18–45 MPa) compared to the radial direction (Acaroglu et al. 1995; Ebara et al. 1996; Elliott and Setton 2001; Fujita et al. 1997; Guerin and Elliott 2006). However, it should be noted that, although uniaxial extension is a simple and valuable testing modality, other loading conditions such as biaxial loading, shear, and torsion can provide a more complete picture of tissue function, as the tissue can be deformed in a more physiologic manner. Furthermore, it should be noted that the properties obtained from these testing methodologies are subject to many biological and experimental variables, including species, age, specimen orientation, strain rate, handling, storage, and hydration (Beckstein et al. 2008; Bursac et al. 2009; Guerin and Elliott 2006; Noyes and Grood 1976; Proctor et al. 1989; Setton et al. 1999; Woo et al. 1986, 1990, 1991).

12.3.1.2 Healing of Fibrous Soft Tissues and Considerations for Tissue Engineering and Regenerative Medicine

Unfortunately, the same demanding mechanical environment that helps define and maintain the structure of fibrous soft tissues also predisposes them to damage. Given the dense ECM and the hypocellularity and avascularity observed in adult musculoskeletal tissues (Bray et al. 1996; Hildebrand et al. 2004; Lo et al. 2002; Manske 1988; Woo et al. 2000), their healing response is suboptimal or nonexistent, particularly for intra-articular tissues. As such, the function of these tissues is compromised following injury, leading to negative secondary effects on the other tissue structures, for example, degeneration of the articular cartilage.

With the clinical need for a mechanically functional implantable tissue or tissue substitute following injury, many have focused efforts on utilizing electrospun NFS for these applications. It is essential to tailor the scaffolds to the tissue of interest and the specific tissue engineering or regenerative medicine approach taken. For example, the demands of stimulating better tissue healing are different from those for tissue replacement. In addition, some fibrous soft tissues have complex organization, which should be reflected in the scaffolds. Furthermore, a construct that was intended for immediate implantation will require different properties than one that is matured *in vitro* before implantation. In either case, by the time of implantation, the material should be functional, particularly within the physiological range of loading and movement, which are quite different and more important than failure characteristics. To illustrate these principles, we discuss how NFS have been applied to different fibrous soft tissues within the musculoskeletal system.

12.3.2 NFS for Ligament and Tendon Tissue Engineering

Injury to the ligaments and tendons in and around diarthrodial joints are prevalent, making their treatment one of the most common orthopedic procedures. In 2004, 1.3 million cases of either outpatient or inpatient hospitalization for dislocations, sprains, and strains were reported (data collected by the American Academy of Orthopedic Surgeons), and a significant subset included damage to ligaments and tendons that required surgery (American Academy of Orthopaedic Surgeons 2008). For example, rotator cuff tears are common, with 8%–26% of the U.S. population having a full-thickness tear (Cotton and Rideout 1964) and 13%–37% having a partial-thickness tear (Breazeale and Craig 1997; Fukuda 1994, 2000). In the case of the anterior cruciate ligament (ACL), there are more than 150,000 tears treated each year in the United States (Beaty 1999; Cole et al. 2005), and a majority of patients require surgical reconstruction using soft tissue autografts and allografts to restore initial knee stability (Aglietti et al. 1994; Armour et al. 2004; Arnoczky et al. 1982; Brand et al. 2000; Burks et al. 2005; Kurosaka et al. 1987; Nakamura et al. 2002; Ohkoshi et al. 1998; Papastergiou et al. 2006; Spindler et al. 2004; Viola et al. 2000; Yunes et al. 2001).

Most ligaments and tendons are able to mount some form of healing response, consisting of overlapping stages with different time scales—inflammation (days), proliferation of a provisional and disorganized scar tissue (days to weeks), and gradual remodeling with an increased deposition of aligned collagen (weeks to years; Frank et al. 1983; Woo et al. 2006a). Often, this response is capable of restoring the load–elongation behavior of the tissue within physiological ranges (Ohland et al. 1991; Weiss et al. 1991), but with significant proliferation, leading to reduced quality of the healing tissue and, thus, its mechanical properties. With a limited blood supply (e.g., within the intra-articular environment), very limited or no healing may occur. As such, tissue replacement of tissues, such as the ACL of the knee, is common. Thus, applications for NFS depend on the level of the natural healing response.

For ligaments and tendons that heal well, efforts are aimed at modulating the healing response to allow better quality healing. Numerous studies have noted that aligned NFS can positively influence cellular morphology, gene expression, and matrix production of both mesenchymal stem cells (MSCs) and native fibroblasts (Bashur et al. 2009; Lee et al. 2005; Moffat et al. 2009; Yin et al. 2010). For example, Bashur et al. (2009) seeded mesenchymal progenitor cells on aligned PEUU scaffolds. Interestingly, as the fiber diameter of the scaffolds decreased to 280 nm, the expression of coll I, decorin, and tenomodulin (an indicator of differentiation into a tenocyte-like phenotype) increased compared to scaffolds made from higher diameter fibers (820 nm or 2.3 μm). Others have utilized the ability for electrospun fibers to retain growth factors to use them as a carrier to promote better healing (Teh et al. 2011; Vaquette et al. 2010). However, conclusive results are still needed to determine whether an aligned matrix can result in better matrix organization within the healing tissue *in vivo*.

For ACL replacement, Sahoo et al. (2010) spun PLGA fibers loaded with basic fibroblast growth factor onto a knitted silk scaffold. The knitted silk provided mechanical strength, whereas the PLGA fibers instructed seeded cells and released growth factors to influence their activities. Others have utilized electrospinning in combination with other materials processing techniques to create even more complex scaffolds. For example, Sato et al. (2010) took electrospun fibers and used weaving techniques to create fiber bundles in an attempt to recreate the hierarchical nature of the ECM of ligaments and tendons. Despite these efforts, much work remains to match the complex mechanical properties of the native ACL.

Another potential use of scaffolds would be to serve as a mechanical support to protect the tissue from overloading during early ligament and tendon healing. Moffat et al. (2009) utilized aligned PLGA (85:15) fibers as a mechanical augmentation for rotator cuff repair. With aligned fibers, these scaffolds had a modulus of ~341 MPa, more than three times that of nonaligned scaffolds (107 MPa). However, these scaffolds still had limited extensibility before yield (<5%) and had a limited toe region, so it is unclear how well these scaffolds will provide mechanical support under repeated physiological loading. More recently, PLGA nanofibers with and without the addition of hydroxyapatite have been utilized to create biphasic scaffolds that simulate the mineralized and nonmineralized regions of the tendon-to-bone insertion. When used along with the suture repair of the supraspinatus tendon of the shoulder in the rat model, a more fibrocartilaginous transition from the tendon to the bone was observed compared to the normal healing response (Moffat et al. 2011).

Another area of interest is using electrospun NFS to recapitulate the complex insertion of tendons and ligaments into the muscle and/or the bone, which is not restored after injury. Furthermore, these junctions are slow to heal and contribute to reduced stability and reinjury. In an application designed for tissue engineering of the musculotendinous junction, Ladd et al., utilized two polymer solutions (PLA/collagen and PCL/collagen) at varying distances along the mandrel, creating scaffolds with higher mechanical properties on one end and lower mechanical properties on the other end, with a gradient in the middle (Ladd et al. 2011). This was intended to match the spatial variation of the natural junction, where the tendon has a modulus a few orders of magnitude higher than the muscle. Indeed, video strain analysis revealed a similar strain distribution under tensile loading for the composite scaffolds and the native tissue (Figure 12.10). Yet, much work remains to capture the exact mechanical properties of the junction, as the scaffold properties were many times higher than the native tissue. To simulate the ligament or tendon insertion into the bone, Li et al. (2009) devised a simple yet clever way of creating a gradient in mineral content over a physiologic scale. Electrospun PLGA or PCL fibers were placed at an angle within an aqueous environment (Figure 12.11). By adding drops on simulated body fluid (SBF) to the solution on one end of the scaffold, a linear gradient of SBF was deposited. The mechanical properties followed the same linear pattern (Figure 12.11), and preosteoblast (MC3T3) cells adhered and proliferated preferentially with increasing mineral content. These novel advances allowing gradients in mechanics may provide valuable solutions to what are significant problems in soft tissue repair of structures with varying mechanical properties and functions.

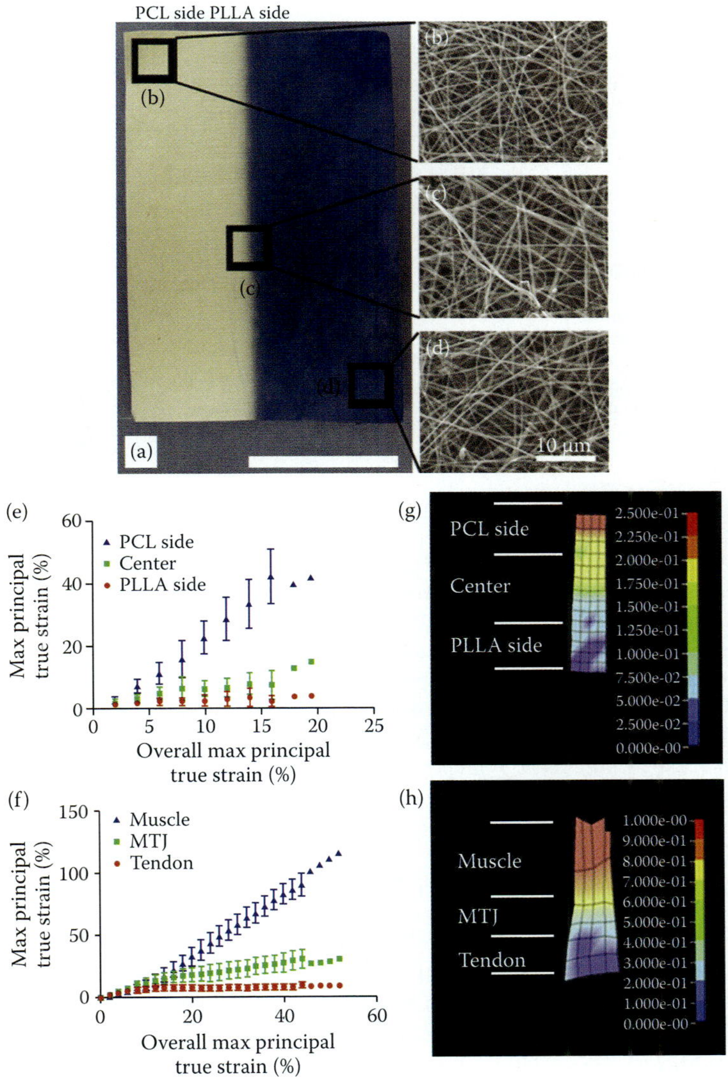

FIGURE 12.10 NFS with graded mechanical properties and composition. (a) Gross picture and (b)–(d) SEM images of a scaffold showing the three regions of interest: (b) PCL side, (c) center, and (d) PLLA side, with methylene blue dye added. Strain analysis of composite NFS and native musculotendinous junction. Maximum principal strain for each material plotted as a function of the strain of the composite for the (e) scaffolds and (f) native tissue. Color maps of the maximum principal strain of the (g) scaffolds and (h) native tissue during uniaxial tensile testing to failure. (Adapted from Ladd M.R. et al., *Biomaterials* 32, 1549–1559, 2011. With permission from Elsevier.)

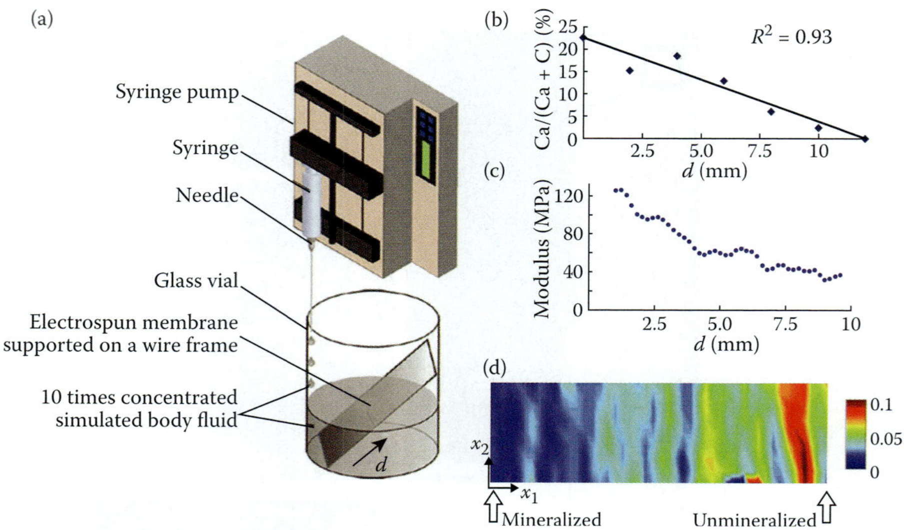

FIGURE 12.11 (a) Schematic of the procedure for generating a graded coating of calcium phosphate on a nonwoven mat of electrospun nanofibers. Ten times concentrated SBF was added at a constant rate to linearly increase the deposition time from the bottom to the top end of the substrate. The parameter *d* refers to the distance from the bottom edge of the substrate. (b) Linear deposition of calcium phosphate was achieved along the length of the scaffold. (c) Tensile modulus decreased with decreasing calcium phosphate content. (d) Localized strains were highest on the unmineralized side of the scaffold. (Adapted from Li X. et al., *Nano Lett* 9, 2763–2768, 2009. With permission from American Chemical Society.)

12.3.3 NFS for Meniscus Tissue Engineering

As introduced above, the medial and lateral menisci are wedge-shaped semilunar structures residing between the femur and tibia (Rath and Richmond 2000) and are crucial for proper load transmission through the knee joint. Surgery for the repair or replacement of the menisci is one of the most common procedures in orthopedic surgery, with more than 750,000 operations each year (Greis et al. 2002). Yet, the menisci have a poor intrinsic ability to heal due to the synovial environment and the acellular and avascular nature of the tissue, particularly the inner regions (Adams and Hukins 1992; Arnoczky 1992; Clark and Ogden 1983; McDevitt and Webber 1990; Petrosini and Sherman 1996). In cases of severe injury or degeneration, the injured portion of the meniscus is either removed (meniscectomy) or the entire meniscus is replaced using a meniscal allograft. However, meniscectomy does not restore the proper mechanics of the native tissue and can lead to early osteoarthritis (Ahmed 1992; Petrosini and Sherman 1996; Rath and Richmond 2000), whereas meniscus transplantation has incomplete cellular recolonization (Arnoczky et al. 1992) and is limited by the availability and proper size of donor tissues, concerns related to disease transmission, and the observed decline in tissue properties as remodeling events transpire postimplantation. Many have proposed tissue engineering as an alternative to overcome the limitations associated with these procedures. Although many acellular and cell-based tissue engineering therapies have emerged for tissue engineering of the meniscus (Aufderheide and Athanasiou 2007; Ballyns et al. 2008; Bruns et al. 1998; Cook et al. 2006; Heijkants et al. 2004; Ibarra et al. 2000; Klompmaker et al. 1996; Kobayashi et al. 2005; Rodkey et al. 2008; Sommerlath and Gillquist 1993; Stone et al. 1992; van Tienen et al. 2002; Veth et al. 1986), no single approach has shown functional *in vivo* repair and reconstitution of fiber architecture of the native tissue.

Noting the crucial role of the collagen architecture to the function of the meniscus, our group has focused on the application of NFS for tissue engineering of the meniscus. The stress–strain response of the meniscus (tested in the fiber direction) is shown in Figure 12.12. Also shown on this plot are the stress–strain profiles of several aligned nanofibrous constructs of differing polymer compositions (all tested in the aligned fiber direction). Some of these polymers result in scaffolds that are stiffer and more brittle than the native tissue and fail at much higher stress levels, but at lower strain levels. Other polymers produce scaffolds that are considerably less stiff, matching at best the toe region modulus of the native tissue. Of the many possible polymers, electrospun scaffolds made from PCL can match the native tissue in its toe region, show some nonlinearity in the stress–strain response, and extend to greater than 10% strain before undergoing permanent deformation, i.e., yielding (Baker and Mauck 2007; Li et al. 2007).

In one recent study, we have evaluated the role of scaffold alignment on the long-term *in vitro* maturation of constructs after seeding with either MSCs or meniscal fibrochondrocytes (MFCs; Baker and Mauck 2007). The aligned NFS

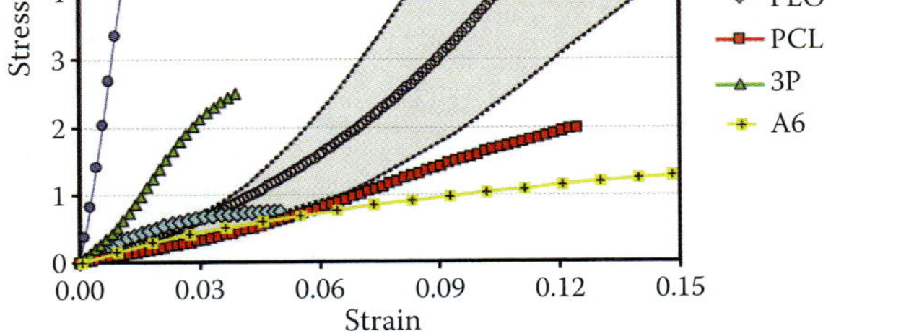

FIGURE 12.12 Benchmarks for the engineering of fibrous tissues. Healthy meniscus tissue tested in the fiber direction produces a nonlinear stress–strain response (circles indicate average response; boundaries indicate high and low range of responses from 10 separate samples). Superimposed on this physiologic range are the stress–strain responses of aligned NFS of varying composition tested in the fiber direction. (Reprinted from Mauck R.L. et al., *Tissue Eng, Part B* 15B, 171–193, 2009. With permission from Mary Ann Liebert, Inc.)

directed cell polarity in the scaffold for both MSCs and MFCs. At the outset of the study, aligned scaffolds had a higher linear region tensile modulus (~12 MPa) than nonaligned scaffolds (~4 MPa). When seeded with MSCs or MFCs, both aligned and nonaligned scaffolds increased in tensile properties following culture in defined prochondrogenic medium for 10 weeks (Figure 12.13). Aligned constructs increased by ~10 MPa, whereas nonaligned constructs increased by only ~1 MPa when seeded with either cell type. The most interesting finding in this study was

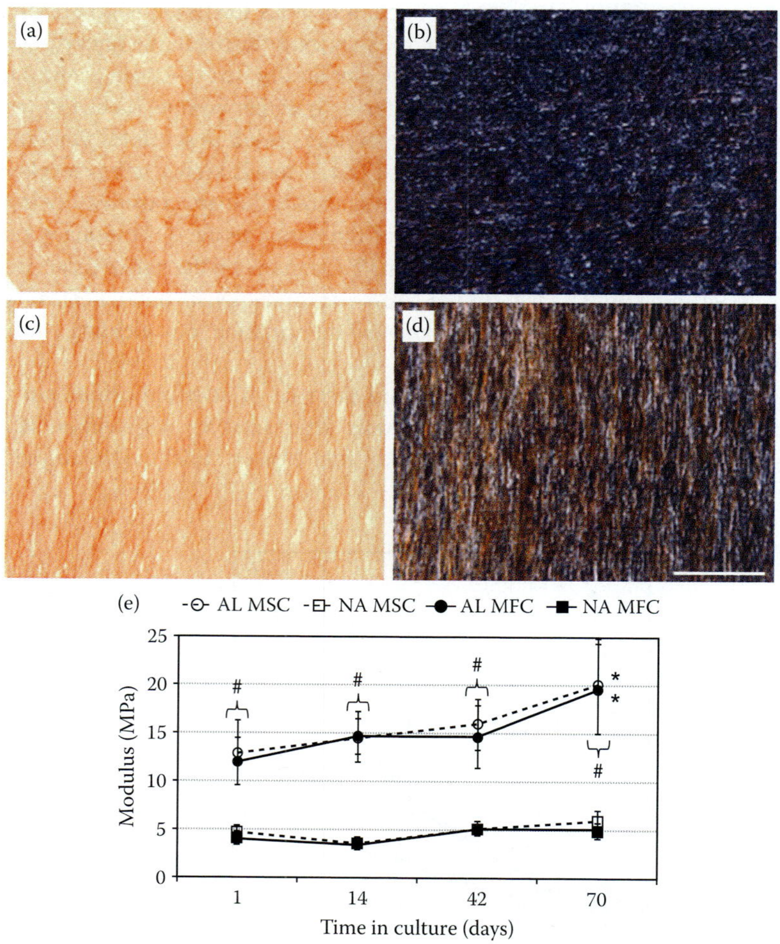

FIGURE 12.13 Scaffold architecture influences the organization of forming neotissue with long-term culture. (a) and (c) Bright-field and (b) and (d) polarized light microscopy images of *en face* sections of (a) and (b) nonaligned and (c) and (d) aligned scaffolds seeded with MSCs on day 70. Scale bar: 200 mm. (e) Tensile modulus of MFC- and MSC-laden aligned and nonaligned NFS with time in culture. *, $p < 0.001$ versus day 1 within group; # $p < 0.001$ versus NA within same cell type at same time point. (Adapted from Baker B.M. and Mauck R.L., *Biomaterials* 28, 1967–1977, 2007. With permission from Elsevier.)

that, although cells on aligned and nonaligned scaffolds produce similar amounts of ECM (PG and collagen content), marked increases in tensile properties were observed only on aligned scaffolds. This finding may be explained by the organization of newly deposited ECM. On aligned scaffolds, polarized light microscopy showed ordered collagen deposition, whereas deposited collagen was disorganized on nonaligned scaffolds (Figure 12.13). Follow-up studies using human meniscus cells showed similar potential for forming function meniscus-like scaffolds from a potentially autologous cell source (Baker et al. 2009a). Because this technology has the potential to be used with both MFCs and MSCs, it has broad application for tissue regeneration and/or replacement in a clinically relevant fashion. However, the long time required for cell colonization (42–70 days) and the need for space for robust matrix accumulation motivated additional development of these novel scaffolds and culture systems.

As a method of increasing pore sizes in electrospun scaffolds (see Section 12.2.4.2), we have recently fabricated dual-polymer constructs with two polymers, PCL as above, and PEO (Baker et al. 2008). PEO was selected, as it dissolves rapidly in an aqueous environment and thus serves as a sacrificial element to increase pore size. After seeding with human MFCs and culturing for 12 weeks in a chemically defined medium, quantification of cross sections stained to visualize cell nuclei revealed significantly higher fractions of cells within more central regions of the scaffolds with increased PEO content (Figure 12.14). Most interestingly, the ability to allow cellular infiltration resulted in larger increases in construct mechanics over the culture period. Although scaffolds containing 60% PEO have initial mechanical properties much lower than pure PCL, as the culture time increases, the tensile modulus of the 60% PEO scaffolds far surpasses that of the pure PCL scaffolds and reaches levels on the order of the native meniscus (Figure 12.14). Histological evaluation confirmed an increase in collagen deposition throughout the 60% PEO scaffolds, whereas the pure PCL scaffolds had deposition limited to the outer one-third.

One consideration moving toward clinical translation is the ability of our scaffolds to integrate with the native meniscus. As a first assessment, we utilized an *in vitro* explant model (Ionescu et al. 2011a). Circular explants from bovine meniscus were obtained using an 8-mm sterile biopsy punch. To assess scaffold integration to the meniscus, the 4-mm internal core was removed and replaced by scaffolds with varying PEO content. By 4 weeks, punch-out testing revealed that increased PEO content (and, thus, pore size) led to increased integration strength of the repair. Thus, increased pore size improves both the maturation of tissue-engineered constructs and the potential for integration to the native tissue. Ongoing studies evaluate how these scaffolds operate within the challenging environment within the knee.

A final consideration is how we can best scale up from a single nanofibrous mat (<1 mm thick) to a full construct with an appropriate geometry. Again, this depends strongly on the application at hand. If the goal is to interpose the scaffold between the injured sides of a bucket-handle tear to improve tissue integration, then a single layer of aligned fibers may be sufficient. However, for larger replacement applications, where a substantial portion of the meniscus is to be replaced,

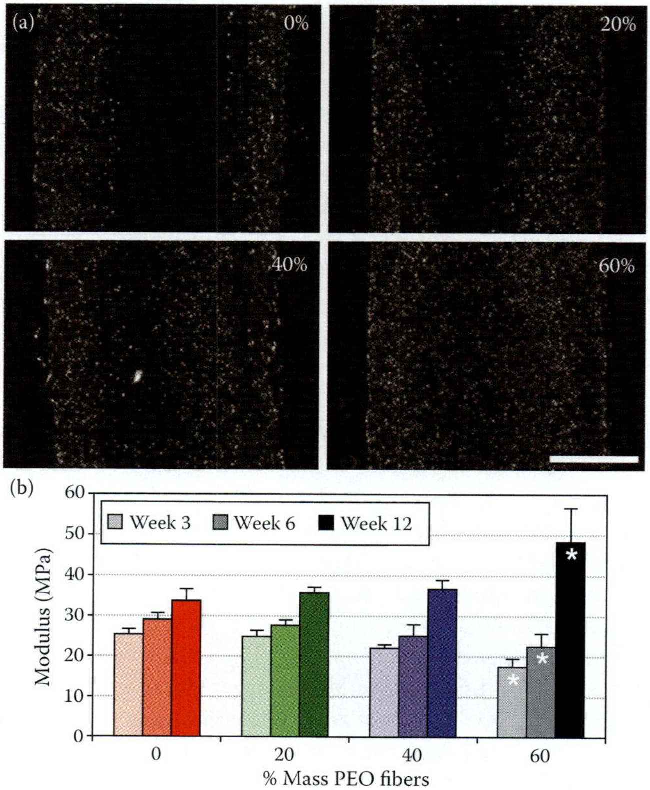

FIGURE 12.14 (a) Removal of sacrificial fibers promotes cell infiltration. Cross sections showing cell nuclei from 0%, 20%, 40%, and 60% PEO constructs. Scale bar: 500 μm. (b) Increasing sacrificial content improved construct tensile properties with time in culture. Tensile modulus of 0%, 20%, 40%, and 60% PEO constructs at weeks 3, 6, and 12. $^*p < 0.05$ compared with time point–matched 0% constructs. (Adapted from Baker B.M. et al., Instruction without Impediment: Tunable Fibrous Scaffolds for Engineering Dense Connective Tissues, Annual Meeting of the Orthopaedic Research Society, Long Beach, CA, 2011. With permission from the Orthopaedic Research Society.)

this would not be appropriate. As an alternative, multiple layers could be combined together to form a construct with appropriate geometry (Figure 12.15). Using a heated probe, these layers can be spot welded together to ensure mechanical stability between the layers. For even larger defects or full tissue replacement of the entire meniscus, it will also be necessary for the fibers to align circumferentially, similar to the native meniscus. Currently, most efforts to create aligned NFS result in uniform alignment on the macroscale. This is not the case for the native meniscus, in which the fibers are circumferentially aligned. As such, we have devised a novel method of creating scaffolds with microscale fiber alignment and macroscale circumferential alignment, fully restoring the macroscale structure of the native meniscus (Figure 12.15).

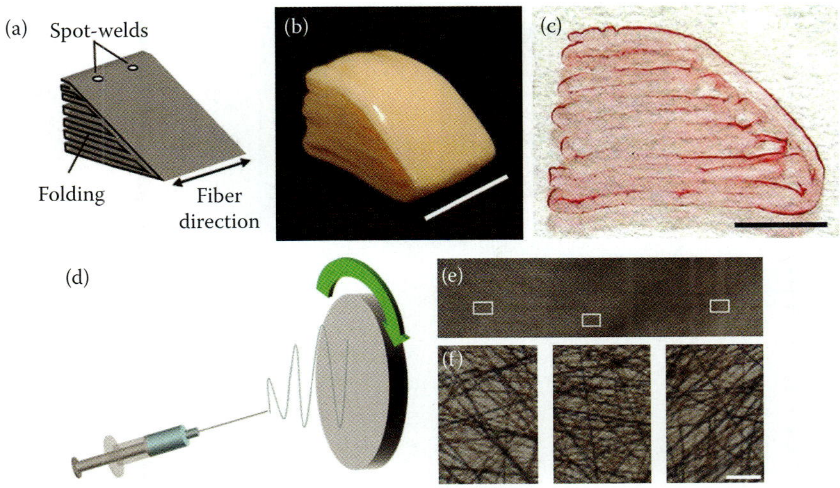

FIGURE 12.15 Engineering anatomic form of meniscus with NFS. Custom wrapping of aligned NFS can produce (a) cell-laden wedge-shaped structures (b) for meniscus applications that can be colonized by (c) seeded cells (scale bar = 1 cm). (d) Alternatively, NFS can be electrospun onto a rotating disc to create angled macrostructure. (e) Bright-field images of angled fibers with (f) magnified images showing local alignment (scale bar = 10 μm). (Adapted from Mauck R.L. et al., *Tissue Eng, Part B* 15B, 171–193, 2009. With permission from Mary Ann Liebert, Inc.)

12.3.4 NFS for AF Tissue Engineering

Disc degeneration is a ubiquitous problem with multiple causal factors, including age and genetics, mechanical fatigue, or injury through trauma, continued exposure to vibrational loading, or mechanical overload. The annual prevalence of low back pain in the United States has been estimated to be more than 20% of the general population and more than 50% of the population over the age of 55 (Deyo et al. 2006; Katz 2006). Specifically, lumbar IVD degeneration has been strongly implicated as a causative factor of low back pain (Albert et al. 2008). One mechanism of disc degeneration originates in tears within the AF region, which propagate from the central region and can result in a complete tear of the AF involving leakage of the NP (disc herniation). There is often a loss of annulus tissue as well (Ahlgren et al. 2000; Heuer et al. 2008). Importantly, reherniation rates following the most common surgical treatment for disc herniation, for example, microdiscectomy, can range from 5% to 15% (Fakouri et al. 2011; Thome et al. 2005; Veresciagina et al. 2010; Yeung and Tsou 2002). As such, tissue engineering and regenerative medicine approaches have been proposed as an alternative (Bowles et al. 2011; Chang et al. 2007, 2010; Gruber et al. 2009; Mizuno et al. 2006; Nerurkar et al. 2009, 2010, 2011b; Rong et al. 2002; Sato et al. 2003; Shao and Hunter 2007).

Using NFS, we have focused on engineering a single annulus layer. In an initial study, rectangular samples were removed from aligned fibrous meshes at angles that were not coincident to the fiber direction, replicating the fiber angles seen in individual AF lamellae (Figure 12.16a; Nerurkar et al. 2007). Tensile testing of

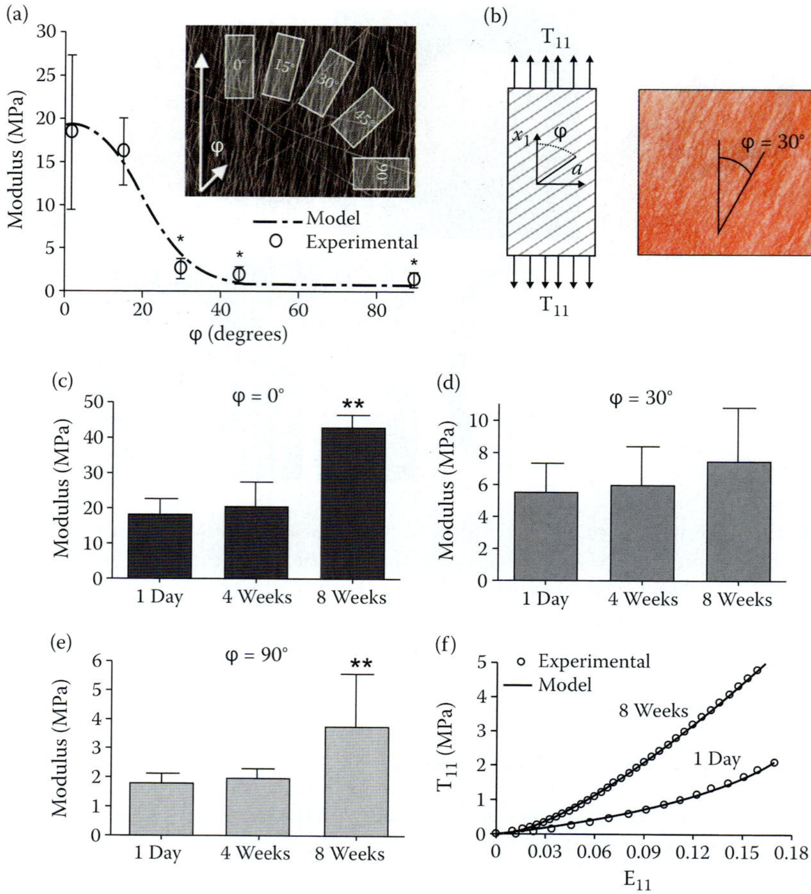

FIGURE 12.16 NFS for AF tissue engineering. (a) Mechanical response and properties of acellular aligned scaffolds tested in directions that are not coincident (see inset) with the prevailing fiber direction can be captured with simple composite models. (b) When seeded with AF cells, matrix deposition (collagen staining) is coincident with the direction of the underlying fibers. Mechanical properties of cell-seeded scaffolds of different orientations increase with time in culture: (c) 0°, in fiber direction; (d) 30°, off axis to the scaffold direction; and (e) 90°, perpendicular to the fiber direction. (f) Hyperelastic models capture the temporal evolution of functional matrix. Representative curves for engineered AF ($\varphi = 0°$) demonstrate a nonlinear response in uniaxial tension. With time in culture, ECM deposition increases the nonlinearity and modulus from 1 day to 8 weeks. The model (solid lines) successfully fit experimental tensile behavior (open circles) at each time point, as shown. Model parameters for (g) matrix and (h) fiber moduli increase with culture duration. (Adapted from Nerurkar N.L. et al., *J Orthop Res* 25, 1018–1028, 2007; Nerurkar N.L. et al., *Spine* 33, 2691–2701, 2008. With permission from Wiley and Wolters Kluwer Health.)

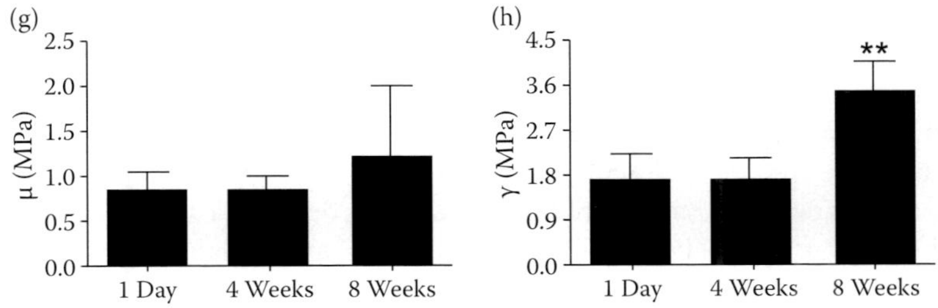

FIGURE 12.16 (Continued)

these scaffolds revealed that the modulus decreased drastically as the fiber direction changed from 15° to 30° relative to the direction of loading, with only limited changes as the angle increased further. The mechanics of these acellular meshes were further characterized using a homogenization model that we have previously applied to native AF tissue, and we found good agreement between modeled and measured linear region moduli (Nerurkar et al. 2007).

Having established a material framework that can mimic the lamellar orientation of the AF, cells were then isolated from the AF of bovine caudal discs and seeded onto these oriented NFS (Nerurkar et al. 2008). Fiber orientations of 0°, 30°, and 90° with respect to the longitudinal axis of the sample were excised from an aligned sheet (Figure 12.16b). These AF constructs increased in tensile moduli with time for each fiber orientation (Figure 12.16c–e). Notably, constructs tested in the fiber direction (0°) increased by ~25 MPa (140%), whereas oblique (30°) and transverse (90°) constructs increased by only ~2 MPa (33%–100%). More importantly, cell-seeded scaffolds with a fiber orientation similar to the AF (30°) matched the circumferential modulus of native inner AF (5.6 MPa) after just 8 weeks of culture (7.3 MPa). However, the 0° fiber direction modulus (43.3 MPa) remained below the single lamella of the native AF (80–120 MPa) (Skaggs et al. 1994; Holzapfel et al. 2005). It should be noted that the orientation of collagen coincided with the scaffold fiber direction (Figure 12.16b). Therefore, in the range of small deformations, constructs achieved comparable tensile properties to the native AF.

To describe the tensile stress–strain curves of these samples and, specifically, quantify the changes observed with time in culture, a hyperelastic constitutive model was developed (Nerurkar et al. 2008). The model considered both the nonfibrillar matrix (e.g., PGs and elasti) and fibrillar component (collagen) and included four scalar material parameters: 1) nonfibrillar (matrix) stiffness μ; 2) matrix compressibility ν; 3) fiber stiffness γ; and 4) fiber nonlinearity ξ. Both data for the 0° and 90° samples could be fit well ($R^2 > 0.98$). More importantly, the obtained values for the material parameters yielded excellent predictions for the 30° samples (Figure 12.16f; $R^2 > 0.98$), thus providing validation for the model. With time in culture, the material parameters that represent the matrix and fibers both increased, with the largest differences for the fiber stiffness γ (Figure 12.16g–h). Because the scaffolds alone do not change in mechanical properties during culture, these results suggest that the functional improvement of the constructs is based on the deposition of new ECM, particularly aligned collagen matrix. The use of such a model demonstrates the effectiveness of a combined experimental and computational approach to characterize the full nonlinear behavior of constructs, elucidate the physical and biochemical mechanisms of functional growth, and guide efforts for creating tissue-engineered AF tissue.

The studies described in the previous paragraphs dealt with single lamellar structures; however, it should be noted that the native AF and eventual tissue-engineered scaffolds for AF repair or replacement will require a multilamellar fabrication strategy to mimic native tissue hierarchical organization. As such, we next examined the maturation of bilayers made from NFS (Nerurkar et al. 2009). In this study, MSCs were seeded on scaffolds with a fiber orientation ±30° to the long edge of the scaffold. After 2 weeks of preculture in a chemically defined media, scaffolds were organized into bilamellar angle-ply structures (APS) with parallel (+30°/+30°) or opposing (±30°) fiber orientations (Figure 12.17a). By 10 weeks of culture, we observed fully infiltrated constructs, with deposition of collagen throughout the depth. Opposing bilayers achieved a modulus (14.2 ± 2.5 MPa) that was higher than parallel bilayers and very close to the modulus of native AF (~18–45 MPa) (Figure 12.17b). When sections across two fiber planes were viewed using polarized light microscopy, a multiscale collagen architecture that mirrors the cross-ply form of the

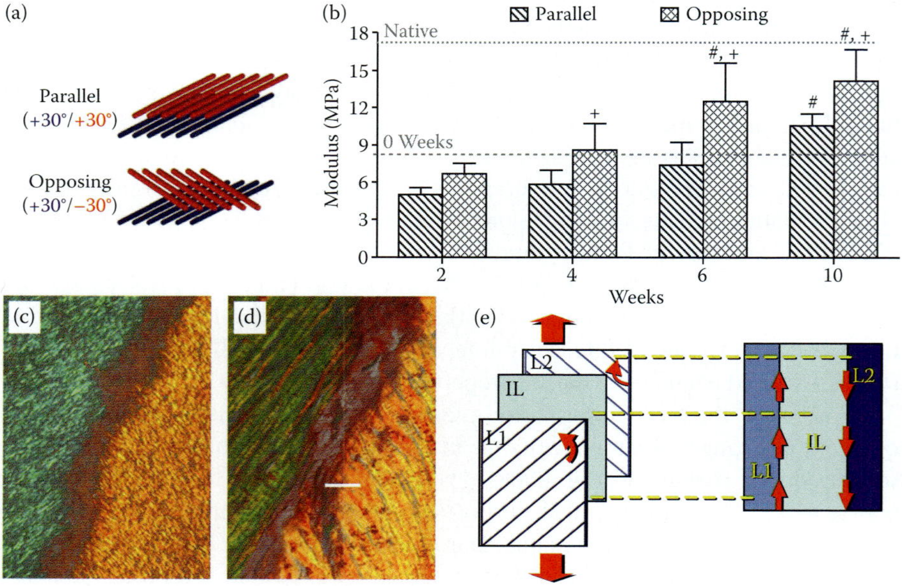

FIGURE 12.17 Hierarchy in nanofibrous laminates. (a) Construction methods for producing parallel and opposing bilayers. (b) Tensile modulus of opposing and parallel bilayers. # $p \leq 0.05$ vs. single lamellar modulus at 0 weeks; +, $p \leq 0.05$ compared to time-matched parallel modulus; Native circumferential tensile modulus of the human AF. Polarized light images of (c) engineered and (d) native tissues at the lamellar interface. (e) Proposed mechanism for the role of interlamellar material in tensile response. (Adapted from Nerurkar N.L. et al., *Nat Mater* 8, 986–992, 2009. With permission from Nature Publishing Group.)

native AF was observed—two layers of aligned, opposing collagen (Figure 12.17c,d). These data demonstrate the fundamental reinforcing effect of alternating directions in angle-ply laminates (Figure 12.17e) and suggest that APS are a promising tissue engineering strategy to achieve native mechanical properties and provide functional repair of the damaged AF.

Finally, to take the next step toward a fully functional tissue-engineered AF, we sought to consider the full anatomy of the AF, which consists of 15–25 lamellae. Thus, we engineered a disc-like angle-ply structure (DAPS) that replicates the multiscale architecture of the IVD using a novel construction tool (Nerurkar et al. 2011b). This allows the formation of a DAPS with lamellae of alternating fiber orientation ±30° with respect to the circumferential direction of the construct (Figure 12.18a through c). This was wrapped around an agarose hydrogel and seeded with bovine MSCs to form a tissue-engineered IVD construct. Early results show that mechanical characteristics of the disc under compression and torsion were qualitatively similar to native tissue, although lesser in magnitude, and increased with time in culture. Cells seeded into both AF and NP regions adopted morphologies

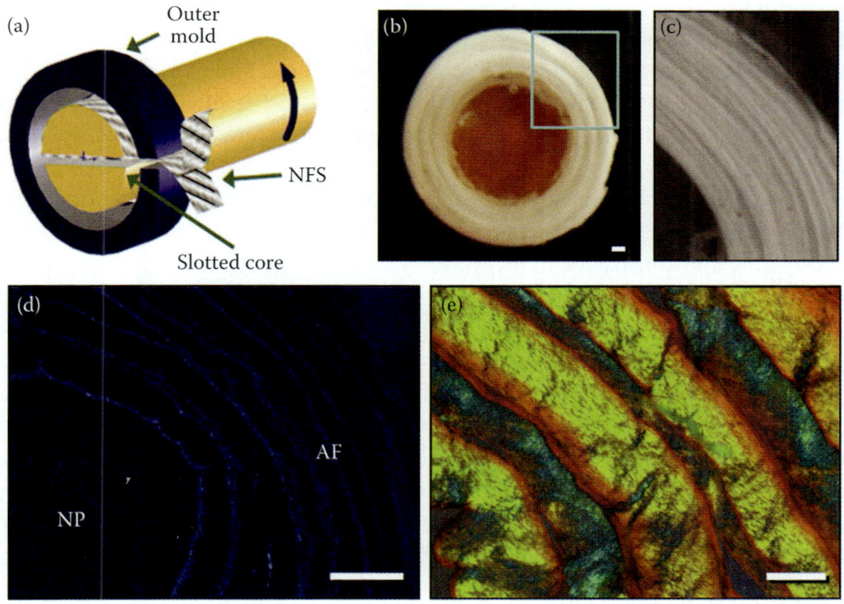

FIGURE 12.18 (a) Schematic showing the fabrication process for the formation of engineered DAPS (NFS). (b) Gross morphology of DAPS with nanofibrous AF region and agarose NP region. Scale bar: 1 mm. (c) Close-up view of the AF region. (d) DAPI staining of transverse section of DAPS at 1 week, showing homogeneous distribution of MSCs in the "NP" region and lamellar organization of MSCs in the "AF" region. Scale bar: 500 μm. (e) Polarized light image of Picrosirius Red–stained oblique section of "AF" region at 6 weeks showing birefringent material in opposing orientations with progression through adjacent lamellae. Scale bar: 250 μm. (Adapted from Nerurkar N.L. et al., *Spine* 35, 867–873, 2011. With permission from Wolters Kluwer Health.)

that mirror those seen in native tissue, with rounded cells in the NP and more elongated cells in the AF regions (Figure 12.18d). Moreover, in the AF region, the collagenous matrix deposited followed the angle-ply configuration of the scaffold (Figure 12.18e). Thus, although much work remains to fully replicate the multiscale form and function of the IVD, these studies provide a solid foundation toward eventual disc replacement.

12.4 DISCUSSION, LIMITATIONS, AND FUTURE FOCUS

In recent years, the large body of literature on the use of NFS for tissue engineering and regenerative medicine has shown their potential for providing novel treatment strategies to enhance tissue regeneration and replacement. Because the mechanical function of these tissues is an important design criterion, many efforts using NFS have directed attention to the important factors governing their mechanical properties, and the achievements to date have been numerous. In the last 10 years, much work has been done on quantifying the mechanics of a single nanofiber as well as scaffolds made from these nanofibers. Many polymers have been evaluated and provide a wide range of mechanical properties from which to base a specific tissue engineering or regenerative medicine approach. The individual factors underlying the mechanics of these scaffolds have been explored, and novel techniques have been developed to control them and create scaffolds with the desired characteristics. Most importantly, NFS have been broadly applied as promising alternatives to improve treatment following injury or disease in a number of systems, including for musculoskeletal soft tissues.

Despite the advances discussed in this chapter, there are several caveats to keep in mind when using mechanical properties of the native tissue as a benchmark for tissue-engineered constructs. First, typical values reported in the literature represent simplified loading patterns in many cases (e.g., uniaxial tensile testing) compared to the complex loading environment *in vivo*, particularly for the meniscus and AF. Second, due to the many experimental variables that contribute to the determination of these properties, there is much variability in values between studies. Thus, these data at best provide a range of standards against which the scaffolds can be compared. Third, it is likely impossible to replicate the entire stress–strain response of the native tissue under all conditions, and thus, physiologic levels of loading must be identified, and the properties of the constructs in these regions must be given additional weight (Butler et al. 2000, 2008). Furthermore, the response of these constructs in response to thousands of cycles of physiologic loading must be considered, as nonrecoverable stress–strain in response to loading would lead to impair function and make further damage probable. Fourth, it is also important to note that the properties of these tissues are spatially heterogeneous, which may also be important criteria that must be matched by tissue-engineered constructs.

Finally, this chapter focuses on achieving native mechanical properties at the time of implantation. However, ultimate success of the tissue-engineered construct is highly dependent on constructive growth and remodeling *in vivo* to achieve a functional tissue in the long run. Subcutaneous models can provide information regarding immunogenicity, degradation, and cellular infiltration. However, orthotopic models of tissue injury and repair are needed to fully evaluate these scaffolds. For orthopedic tissues, the demanding loading environment and challenging nutritional limitations of the intra-articular space provide additional challenges for *in vivo* maturation and maintenance of function.

Despite these complex remaining challenges, electrospinning technologies represent a novel material platform that will provide new opportunities to better stimulate healing or match the mechanics of the native tissue. For example, the combination of multiple polymers can allow the production of NFS with a wide array of mechanical properties, allowing even greater control of scaffold characteristics and matching a multitude of native tissues (Baker et al. 2009b). The inclusion of growth factors and other bioactive agents will create additional biological functionality (Hong et al. 2008; Ionescu et al. 2011b; Verreck et al. 2003). Bioreactor systems can be used to further promote tissue growth and maturation via nutrient provision (Li et al. 2008; Nerurkar et al. 2011a) and/or mechanical stimulation (Baker et al. 2011a; Lee et al. 2005). In sum, the significant advances made using NFS have been impressive, and their use has only accelerated with time. With continued development and innovation, guided by mechanical function, this approach provides an exciting avenue to improve clinical treatment of injuries to fiber-reinforced tissues and will have a profound impact on health care.

REFERENCES

Acaroglu E.R., J.C. Iatridis, L.A. Setton et al. 1995. Degeneration and aging affect the tensile behavior of human lumbar anulus fibrosus. *Spine* 20: 2690–2701.

Adams M.E. and D.W.L. Hukins. 1992. The extracellular matrix of the meniscus. In: *Knee Meniscus: Basic and Clinical Foundations*, Mow, V.C., S.P. Arnoczky, and D.W. Jackson (eds.), pp. 15–28. New York: Raven Press, Ltd.

Aglietti P., R. Buzzi, G. Zaccherotti, and P. De Biase. 1994. Patellar tendon versus doubled semitendinosus and gracilis tendons for anterior cruciate ligament reconstruction. *Am J Sports Med* 22: 211–217; discussion 217–218.

Ahlgren B.D., W. Lui, H.N. Herkowitz, M.M. Panjabi, and J.P. Guiboux. 2000. Effect of anular repair on the healing strength of the intervertebral disc: A sheep model. *Spine* 25: 2165–2170.

Ahmed A.M. 1992. The load-bearing role of the knee meniscus. In: *Knee Meniscus: Basic and Clinical Foundations*, Mow, V.C., S.P. Arnoczky, and D.W. Jackson (eds.), pp. 59–73. New York: Raven Press, Ltd.

Albert H.B., P. Kjaer, T.S. Jensen et al. 2008. Modic changes, possible causes and relation to low back pain. *Med Hypotheses* 70: 361–368.

American Academy of Orthopaedic Surgeons. 2008. *United States Bone and Joint Decade: The Burden of Musculoskeletal Diseases in the United States*. Rosemont, IL: American Academy of Orthopaedic Surgeons.

Amoroso N.J., A. D'Amore, Y. Hong, W.R. Wagner, and M.S. Sacks. 2011. Elastomeric electrospun polyurethane scaffolds: The interrelationship between fabrication conditions, fiber topology, and mechanical properties. *Adv Mater* 23: 106–111.

Arinstein A. and E. Zussman. 2011. Electrospun polymer nanofibers: Mechanical and thermodynamic perspectives. *J Polym Sci, Part B: Polym Phys* 49B: 691–707.

Armour T., L. Forwell, R. Litchfield et al. 2004. Isokinetic evaluation of internal/external tibial rotation strength after the use of hamstring tendons for anterior cruciate ligament reconstruction. *Am J Sports Med* 32: 1639–1643.

Arnoczky S.P. 1992. Gross and vascular anatomy of the meniscus and its role in meniscal healing, regeneration, and remodeling. In: *Knee Meniscus: Basic and Clinical Foundations*, Mow, V.C., S.P. Arnoczky, and D.W. Jackson (eds.), pp. 1–14. New York: Raven Press, Ltd.

Arnoczky S.P., E.F. DiCarlo, S.J. O'Brien, and R.F. Warren. 1992. Cellular repopulation of deep-frozen meniscal autografts: An experimental study in the dog. *Arthroscopy* 8: 428–436.

Arnoczky S.P., G.B. Tarvin, and J.L. Marshall. 1982. Anterior cruciate ligament replacement using patellar tendon: An evaluation of graft revascularization in the dog. *J Bone Jt Surg, Am Vol* 64: 217–224.

Aufderheide A.C. and K.A. Athanasiou. 2007. Assessment of a bovine coculture, scaffold-free method for growing meniscus-shaped constructs. *Tissue Eng* 13: 2195–2205.

Ayres C.E., G.L. Bowlin, R. Pizinger et al. 2007. Incremental changes in anisotropy induce incremental changes in the material properties of electrospun scaffolds. *Acta Biomater* 3: 651–661.

Baker B.M., A.O. Gee, R.B. Metter et al. 2008. The potential to improve cell infiltration in composite fiber-aligned electrospun scaffolds by the selective removal of sacrificial fibers. *Biomaterials* 29: 2348–2358.

Baker B.M. and R.L. Mauck. 2007. The effect of nanofiber alignment on the maturation of engineered meniscus constructs. *Biomaterials* 28: 1967–1977.

Baker B.M., A.S. Nathan, G.R. Huffman, and R.L. Mauck. 2009a. Tissue engineering with meniscus cells derived from surgical debris. *Osteoarthritis Cartilage* 17: 336–345.

Baker B.M., N.L. Nerurkar, J.A. Burdick, D.M. Elliott, and R.L. Mauck. 2009b. Fabrication and modeling of dynamic multipolymer nanofibrous scaffolds. *J Biomech Eng* 131: 101012.

Baker B.M., R.P. Shah, A.H. Huang, and R.L. Mauck. 2011a. Dynamic tensile loading improves the functional properties of mesenchymal stem cell-laden nanofiber-based fibrocartilage. *Tissue Eng, Part A* 17A: 1445–1455.

Baker B.M., R.P. Shah, A.M. Silverstein et al. 2011b. Instruction without impediment: Tunable fibrous scaffolds for engineering dense connective tissues. *Proceedings of the Annual Meeting of the Orthopaedic Research Society*, Long Beach, CA.

Balguid A., A. Mol, M.H. van Marion et al. 2009. Tailoring fiber diameter in electrospun poly(epsilon-caprolactone) scaffolds for optimal cellular infiltration in cardiovascular tissue engineering. *Tissue Eng, Part A* 15A: 437–444.

Ballyns J.J., J.P. Gleghorn, V. Niebrzydowski et al. 2008. Image-guided tissue engineering of anatomically shaped implants via MRI and micro-CT using injection molding. *Tissue Eng—Part A* 14A: 1195–1202.

Barnes C.P., C.W. Pemble, D.D. Brand, D.G. Simpson, and G.L. Bowlin. 2007a. Cross-linking electrospun type II collagen tissue engineering scaffolds with carbodiimide in ethanol. *Tissue Eng* 13: 1593–1605.

Barnes C.P., S.A. Sell, E.D. Boland, D.G. Simpson, and G.L. Bowlin. 2007b. Nanofiber technology: Designing the next generation of tissue engineering scaffolds. *Adv Drug Deliv Rev* 59: 1413–1433.

Bashur C.A., R.D. Shaffer, L.A. Dahlgren, S.A. Guelcher, and A.S. Goldstein. 2009. Effect of fiber diameter and alignment of electrospun polyurethane meshes on mesenchymal progenitor cells. *Tissue Eng, Part A* 15A: 2435–2445.

Beaty J. 1999. Knee and leg: Soft tissue trauma. In: *OKU Orthopaedic Knowledge Update*, First edition, Arendt, E.A. (ed.), pp. xix, 442. Rosemont, IL: American Academy of Orthopaedic Surgeons.

Beckstein J.C., S. Sen, T.P. Schaer, E.J. Vresilovic, and D.M. Elliott. 2008. Comparison of animal discs used in disc research to human lumbar disc: Axial compression mechanics and glycosaminoglycan content. *Spine* 33: E166–E173.

Best B.A., F. Guilak, and L.A. Setton et al. 1994. Compressive mechanical properties of the human anulus fibrosus and their relationship to biochemical composition. *Spine* 19: 212–221.

Bhattarai N., D. Edmondson, O. Veiseh, F.A. Matsen, and M. Zhang. 2005. Electrospun chitosan-based nanofibers and their cellular compatibility. *Biomaterials* 26: 6176–6184.

Bhattarai S.R., N. Bhattarai, H.K. Yi et al. 2004. Novel biodegradable electrospun membrane: Scaffold for tissue engineering. *Biomaterials* 25: 2595–2602.

Bigliani L.U., R.G. Pollock, L.J. Soslowsky et al. 1992. Tensile properties of the inferior glenohumeral ligament. *J Orthop Res* 10: 187–197.

Birk D.E. and R. Mayne. 1997. Localization of collagen types I, III, and V during tendon development: Changes in collagen types I and III are correlated with changes in fibril diameter. *Eur J Cell Biol* 72: 352–361.

Blakeney B.A., A. Tambralli, J.M. Anderson et al. 2011. Cell infiltration and growth in a low density, uncompressed three-dimensional electrospun nanofibrous scaffold. *Biomaterials* 32: 1583–1590.

Boland E.D., B.D. Coleman, C.P. Barnes et al. 2005. Electrospinning polydioxanone for biomedical applications. *Acta Biomater* 1: 115–123.

Boland E.D., T.A. Telemeco, D.G. Simpson, G.E. Wnek, and G.L. Bowlin. 2004. Utilizing acid pretreatment and electrospinning to improve biocompatibility of poly(glycolic acid) for tissue engineering. *J Biomed Mater Res, Part B: Appl Biomater* 71B: 144–152.

Boland E.D., G.E. Wnek, D.G. Simpson, K.J. Pawlowski, and G.L. Bowlin. 2001. Tailoring tissue engineering scaffolds using electrostatic processing techniques: A study of poly(glycolic acid) electrospinning. *J Macromol Sci, Part A: Pure Appl Chem* 38A: 1231–1243.

Bowles R.D., H.H. Gebhard, R. Härtl, and L.J. Bonassar. 2011. Tissue-engineered intervertebral discs produce new matrix, maintain disc height, and restore biomechanical function to the rodent spine. *Proc Natl Acad Sci USA* 108: 13106–13111.

Brand J., Jr., A. Weiler, D.N. Caborn, C.H. Brown, Jr., and D.L. Johnson. 2000. Graft fixation in cruciate ligament reconstruction. *Am J Sports Med* 28: 761–774.

Bray R.C., R.M. Rangayyan, and C.B. Frank. 1996. Normal and healing ligament vascularity: A quantitative histological assessment in the adult rabbit medial collateral ligament. *J Anat* 188: 87–95.

Breazeale N.M. and E.V. Craig. 1997. Partial-thickness rotator cuff tears. Pathogenesis and treatment. *Orthop Clin North Am* 28: 145–155.

Brenner M.P., M.M. Hohman, M. Shin, and G. Rutledge. 2001a. Electrospinning and electrically forced jets. I. Stability theory. *Phys Fluids* 13: 2201–2220.

Brenner M.P., M.M. Hohman, M. Shin, and G. Rutledge. 2001b. Electrospinning and electrically forced jets—Part II: Applications. *Phys Fluids* 13: 2221–2236.

Bruns J., J. Kahrs, J. Kampen, P. Behrens, and W. Plitz. 1998. Autologous perichondral tissue for meniscal replacement. *J Bone Jt Surg, Br Vol* 80: 918–923.

Burks R.T., J. Crim, B.P. Fink, D.N. Boylan, and P.E. Greis. 2005. The effects of semitendinosus and gracilis harvest in anterior cruciate ligament reconstruction. *Arthroscopy* 21: 1177–1185.

Bursac P., A. York, P. Kuznia, L.M. Brown, and S.P. Arnoczky. 2009. Influence of donor age on the biomechanical and biochemical properties of human meniscal allografts. *Am J Sports Med* 37: 884–889.

Butler D.L., S.A. Goldstein, and F. Guilak. 2000. Functional tissue engineering: The role of biomechanics. *J Biomech Eng* 122: 570–575.

Butler D.L., E.S. Grood, F.R. Noyes, R.F. Zernicke, and K. Brackett. 1984. Effects of structure and strain measurement technique on the material properties of young human tendons and fascia. *J Biomech* 17: 579–596.

Butler D.L., Y. Guan, M.D. Kay et al. 1992. Location-dependent variations in the material properties of the anterior cruciate ligament. *J Biomech* 25: 511–518.

Butler D.L., N. Juncosa-Melvin, G.P. Boivin et al. 2008. Functional tissue engineering for tendon repair: A multidisciplinary strategy using mesenchymal stem cells, bioscaffolds, and mechanical stimulation. *J Orthop Res* 26: 1–9.

Butler D.L., M.D. Kay, and D.C. Stouffer. 1986. Comparison of material properties in fascicle–bone units from human patellar tendon and knee ligaments. *J Biomech* 19: 425–432.

Buttafoco L., N.G. Kolkman, P. Engbers-Buijtenhuijs et al. 2006. Electrospinning of collagen and elastin for tissue engineering applications. *Biomaterials* 27: 724–734.

Carlisle C.R., C. Coulais, and M. Guthold. 2010. The mechanical stress–strain properties of single electrospun collagen type I nanofibers. *Acta Biomater* 6: 2997–3003.

Carlisle C.R., C. Coulais, M. Namboothiry et al. 2009. The mechanical properties of individual, electrospun fibrinogen fibers. *Biomaterials* 30: 1205–1213.

Casper C.L., N. Yamaguchi, K.L. Kiick, and J.F. Rabolt. 2005. Functionalizing electrospun fibers with biologically relevant macromolecules. *Biomacromolecules* 6: 1998–2007.

Casper C.L., W. Yang, M.C. Farach-Carson, and J.F. Rabolt. 2007. Coating electrospun collagen and gelatin fibers with perlecan domain I for increased growth factor binding. *Biomacromolecules* 8: 1116–1123.

Cassidy J.J., A. Hiltner, and E. Baer. 1989. Hierarchical structure of the intervertebral disc. *Connect Tissue Res* 23: 75–88.

Chang G., H.J. Kim, D. Kaplan, G. Vunjak-Novakovic, and R.A. Kandel. 2007. Porous silk scaffolds can be used for tissue engineering annulus fibrosus. *Eur Spine J* 16: 1848–1857.

Chang G., H.J. Kim, G. Vunjak-Novakovic, D.L. Kaplan, and R. Kandel. 2010. Enhancing annulus fibrosus tissue formation in porous silk scaffolds. *J Biomed Mater Res, Part A* 92A: 43–51.

Chew S.Y., T.C. Hufnagel, C.T. Lim, and K.W. Leong. 2006. Mechanical properties of single electrospun drug-encapsulated nanofibres. *Nanotechnology* 17: 3880–3891.

Clark C.R. and J.A. Ogden. 1983. Development of the menisci of the human knee joint: Morphological changes and their potential role in childhood meniscal injury. *J Bone Jt Surg, Am Vol* 65: 538–547.

Cole D.W., T.A. Ginn, G.J. Chen et al. 2005. Cost comparison of anterior cruciate ligament reconstruction: Autograft versus allograft. *Arthroscopy* 21: 786–790.

Cook J.L., D.B. Fox, P. Malaviya et al. 2006. Long-term outcome for large meniscal defects treated with small intestinal submucosa in a dog model. *Am J Sports Med* 34: 32–42.

Cotton R.E. and D.F. Rideout. 1964. Tears of the humeral rotator cuff: A radiological and pathological necropsy survey. *J Bone Jt Surg, Br Vol* 46: 314–328.

Courtney T., M.S. Sacks, J. Stankus, J. Guan, and W.R. Wagner. 2006. Design and analysis of tissue engineering scaffolds that mimic soft tissue mechanical anisotropy. *Biomaterials* 27: 3631–3638.

D'Amore A., J.A. Stella, W.R. Wagner, and M.S. Sacks. 2010. Characterization of the complete fiber network topology of planar fibrous tissues and scaffolds. *Biomaterials* 31: 5345–5354.

Deyo R.A., S.K. Mirza, and B.I. Martin. 2006. Back pain prevalence and visit rates: Estimates from U.S. national surveys, 2002. *Spine* 31: 2724–2727.

Ding B., E. Kimura, T. Sato, S. Fujita, and S. Shiratori. 2004. Fabrication of blend biodegradable nanofibrous nonwoven mats via multijet electrospinning. *Polymer* 45: 1895–1902.

Doshi J. and D.H. Reneker. 1995. Electrospinning process and applications of electrospun fibers. *J Electrostat* 35: 151–160.

Dyer R.F. and C.D. Enna. 1976. Ultrastructural features of adult human tendon. *Cell Tissue Res* 168: 247–259.

Ebara S., J.C. Iatridis, L.A. Setton et al. 1996. Tensile properties of nondegenerate human lumbar anulus fibrosus. *Spine* 21: 452–461.

Elliott D.M. and L.A. Setton. 2001. Anisotropic and inhomogeneous tensile behavior of the human anulus fibrosus: Experimental measurement and material model predictions. *J Biomech Eng* 123: 256–263.

Eyden B. and M. Tzaphlidou. 2001. Structural variations of collagen in normal and pathological tissues: Role of electron microscopy. *Micron* 32: 287–300.

Fakouri B., V. Patel, E. Bayley, and S. Srinivas. 2011. Lumbar microdiscectomy versus sequestrectomy/free fragmentectomy: A long-term (>2 y) retrospective study of the clinical outcome. *J Spinal Disord Tech* 24: 6–10.

Fithian D.C., M.A. Kelly, and V.C. Mow. 1990. Material properties and structure–function relationships in the menisci. *Clin Orthop Relat Res* 19–31.

Frank C., D. Amiel, and W.H. Akeson. 1983. Healing of the medial collateral ligament of the knee: A morphological and biochemical assessment in rabbits. *Acta Orthop Scand* 54: 917–923.

Fridrikh S.V., J.H. Yu, M.P. Brenner, and G.C. Rutledge. 2003. Controlling the fiber diameter during electrospinning. *Phys Rev Lett* 90: 144502.

Fujita Y., N.A. Duncan, and J.C. Lotz. 1997. Radial tensile properties of the lumbar annulus fibrosus are site and degeneration dependent. *J Orthop Res* 15: 814–819.

Fujita Y., D.R. Wagner, A.A. Biviji, N.A. Duncan, and J.C. Lotz. 2000. Anisotropic shear behavior of the annulus fibrosus: Effect of harvest site and tissue prestrain. *Med Eng Phys* 22: 349–357.

Fukuda H. 2000. Partial-thickness rotator cuff tears: A modern view on Codman's classic. *J Shoulder Elbow Surg* 9: 163–168.

Fukuda H., K. Hamada, T. Nakajima, and A. Tomonaga. 1994. Pathology and pathogenesis of the intratendinous tearing of the rotator cuff viewed from en bloc histologic sections. *Clin Orthop Relat Res* 60–67.

Fukuta S., M. Oyama, K. Kavalkovich, F.H. Fu, and C. Niyibizi. 1998. Identification of types II, IX, and X collagens at the insertion site of the bovine Achilles tendon. *Matrix Biol* 17: 65–73.

Fung Y.C. 1993. *Biomechanics: Mechanical Properties of Living Tissues*. New York: Springer.

Gaumer J., A. Prasad, D. Lee, and J. Lannutti. 2009. Structure–function relationships and source-to-ground distance in electrospun polycaprolactone. *Acta Biomater* 5: 1552–1561.

Geng X., O.H. Kwon, and J. Jang. 2005. Electrospinning of chitosan dissolved in concentrated acetic acid solution. *Biomaterials* 26: 5427–5432.

Goh J.C., H.W. Ouyang, S.H. Teoh, C.K. Chan, and E.H. Lee. 2003. Tissue-engineering approach to the repair and regeneration of tendons and ligaments. *Tissue Eng* 9 (Suppl 1): S31–S44.

Greis P.E., M.C. Holmstrom, D.D. Bardana, and R.T. Burks. 2002. Meniscal injury—Part II: Management. *J Am Acad Orthop Surg* 10: 177–187.

Gruber H.E., G. Hoelscher, J.A. Ingram, E. and N. Hanley, Jr. 2009. Culture of human anulus fibrosus cells on polyamide nanofibers: Extracellular matrix production. *Spine* 34: 4–9.

Guerin H.A. and D.M. Elliott. 2006. Degeneration affects the fiber reorientation of human annulus fibrosus under tensile load. *J Biomech* 39: 1410–1418.

Guerin H.L. and D.M. Elliott. 2007. Quantifying the contributions of structure to annulus fibrosus mechanical function using a nonlinear, anisotropic, hyperelastic model. *J Orthop Res* 25: 508–516.

Heijkants R.G., R.V. van Calck, J.H. De Groot et al. 2004. Design, synthesis and properties of a degradable polyurethane scaffold for meniscus regeneration. *J Mater Sci: Mater Med* 15: 423–427.

Heuer F., S. Ulrich, L. Claes, H. and J. Wilke. 2008. Biomechanical evaluation of conventional anulus fibrosus closure methods required for nucleus replacement: Laboratory investigation. *J Neurosurg Spine* 9: 307–313.

Hildebrand K.A., C.B. Frank, and D.A. Hart. 2004. Gene intervention in ligament and tendon: Current status, challenges, future directions. *Gene Ther* 11: 368–378.

Hong Y., K. Fujimoto, R. Hashizume et al. 2008. Generating elastic, biodegradable polyurethane/poly(lactide-*co*-glycolide) fibrous sheets with controlled antibiotic release via two-stream electrospinning. *Biomacromolecules* 9: 1200–1207.

Holzapfel G.A., C.A. Schulze-Bauer, G. Feigl, and P. Regitnig. 2005. Single lamellar mechanics of the human lumbar anulus fibrosus. *Biomech Model Mechanobiol* 3: 125–140.

Huang L., K. Nagapudi, R.P. Apkarian, and E.L. Chaikof. 2001. Engineered collagen–PEO nanofibers and fabrics. *J Biomater Sci, Polym Ed* 12: 979–993.
Iatridis J.C. and I. ap Gwynn. 2004. Mechanisms for mechanical damage in the intervertebral disc annulus fibrosus. *J Biomech* 37: 1165–1175.
Iatridis J.C., L.A. Setton, R.J. Foster et al. 1998. Degeneration affects the anisotropic and nonlinear behaviors of human anulus fibrosus in compression. *J Biomech* 31: 535–544.
Ibarra C., J.A. Koski, and R.F. Warren. 2000. Tissue engineering meniscus: Cells and matrix. *Orthop Clin North Am* 31: 411–418.
Ifkovits J.L., R.F. Padera, and J.A. Burdick. 2008. Biodegradable and radically polymerized elastomers with enhanced processing capabilities. *Biomed Mater* 3: 034104.
Ionescu L.C., G.C. Lee, G.H. Garcia et al. 2011a. Maturation state-dependent alterations in meniscus integration: Implications for scaffold design and tissue engineering. *Tissue Eng—Part A* 17A: 193–204.
Ionescu L.C., G.C. Lee, B.J. Sennett, J.A. Burdick, and R.L. Mauck. 2011b. An anisotropic nanofiber/microsphere composite with controlled release of biomolecules for fibrous tissue engineering. *Biomaterials* 31: 4113–4120.
Jiang H., D. Fang, B.S. Hsiao, B. Chu, and W. Chen. 2004a. Optimization and characterization of dextran membranes prepared by electrospinning. *Biomacromolecules* 5: 326–333.
Jiang H.L., D.F. Fang, B.J. Hsiao, B.J. Chu, and W.L. Chen. 2004b. Preparation and characterization of ibuprofen-loaded poly(lactide-*co*-glycolide)/poly(ethylene glycol)-g-chitosan electrospun membranes. *J Biomater Sci, Polym Ed* 15: 279–296.
Jin H.J., J. Chen, V. Karageorgiou, G.H. Altman, and D.L. Kaplan. 2004. Human bone marrow stromal cell responses on electrospun silk fibroin mats. *Biomaterials* 25: 1039–1047.
Johnson G.A., D.M. Tramaglini, R.E. Levine et al. 1994. Tensile and viscoelastic properties of human patellar tendon. *J Orthop Res* 12: 796–803.
Johnson J., A. Ghosh, and J. Lannutti. 2007. Microstructure–property relationships in a tissue-engineering scaffold. *J Appl Polym Sci* 104: 2919–2927.
Kakade M.V., S. Givens, K. Gardner et al. 2007. Electric field-induced orientation of polymer chains in macroscopically aligned electrospun polymer nanofibers. *J Am Chem Soc* 129: 2777–2782.
Katta P., M. Alessandro, R.D. Ramsier, and G.G. Chase. 2004. Continuous electrospinning of aligned polymer nanofibers onto a wire drum collector. *Nano Lett* 4: 2215–2218.
Katz R.T. 2006. Impairment and disability rating in low back pain. *Clin Occup Environ Med* 5: 719–740, viii.
Khil M.S., D.I. Cha, H.Y. Kim, I.S. Kim, and N. Bhattarai. 2003. Electrospun nanofibrous polyurethane membrane as wound dressing. *J Biomed Mater Res, Part B: Appl Biomater* 67B: 675–679.
Kidoaki S., I.K. Kwon, and T. Matsuda. 2005. Mesoscopic spatial designs of nano- and microfiber meshes for tissue-engineering matrix and scaffold based on newly devised multilayering and mixing electrospinning techniques. *Biomaterials* 26: 37–46.
Kidoaki S., I.K. Kwon, and T. Matsuda. 2006. Structural features and mechanical properties of *in situ* bonded meshes of segmented polyurethane electrospun from mixed solvents. *J Biomed Mater Res—Part B: Appl Biomater* 76B: 219–229.
Kim S.H., S. Nair, and E. Moore. 2005. Reactive electrospinning of cross-linked poly(2-hydroxyethyl methacrylate) nanofibers and elastic properties of individual hydrogel nanofibers in aqueous solutions. *Macromolecules* 38: 3719–3723.
Klisch S. and M.J.C. Lotz. 2000. A special theory of biphasic mixtures and experimental results for human annulus fibrosus tested in confined compression. *J Biomech Eng* 122: 180–188.
Klompmaker J., R.P. Veth, H.W. Jansen et al. 1996. Meniscal replacement using a porous polymer prosthesis: A preliminary study in the dog. *Biomaterials* 17: 1169–1175.

Kobayashi M., Y.S. Chang, and M. Oka. 2005. A two-year *in vivo* study of polyvinyl alcohol–hydrogel (PVA-H) artificial meniscus. *Biomaterials* 26: 3243–3248.

Kurosaka M., S. Yoshiya, and J.T. Andrish. 1987. A biomechanical comparison of different surgical techniques of graft fixation in anterior cruciate ligament reconstruction. *Am J Sports Med* 15: 225–229.

Kwon I.K., S. Kidoaki, and T. Matsuda. 2005. Electrospun nano- to micro-fiber fabrics made of biodegradable copolyesters: Structural characteristics, mechanical properties and cell adhesion potential. *Biomaterials* 26: 3929–3939.

Ladd M.R., S.J. Lee, J.D. Stitzel, A. Atala, and J.J. Yoo. 2011. Coelectrospun dual scaffolding system with potential for muscle–tendon junction tissue engineering. *Biomaterials* 32: 1549–1559.

Langer R. and J.P. Vacanti. 1993. Tissue engineering. *Science* 260: 920–926.

Lechner K., M.L. Hull, and S.M. Howell. 2000. Is the circumferential tensile modulus within a human medial meniscus affected by the test sample location and cross-sectional area? *J Orthop Res* 18: 945–951.

Lee C.H., H.J. Shin, I.H. Cho et al. 2005. Nanofiber alignment and direction of mechanical strain affect the ECM production of human ACL fibroblast. *Biomaterials* 26: 1261–1270.

Lee H., H. Yoon, and G. Kim. 2009. Highly oriented electrospun polycaprolactone micro/nanofibers prepared by a field-controllable electrode and rotating collector. *Appl Phys A: Mater Sci Process* 97: 559–565.

Lee K.H., H.Y. Kim, M.S. Khil, Y.M. Ra, and D.R. Lee. 2003. Characterization of nano-structured poly(epsilon-caprolactone) nonwoven mats via electrospinning. *Polymer* 44: 1287–1294.

Lee S.J., J.J. Yoo, G.J. Lim, A. Atala, and J. Stitzel. 2007. *In vitro* evaluation of electrospun nanofiber scaffolds for vascular graft application. *J Biomed Mater Res—Part A* 83A: 999–1008.

Lee T.Q. and S.L. Woo. 1988. A new method for determining cross-sectional shape and area of soft tissues. *J Biomech Eng* 110: 110–114.

Li D., G. Ouyang, J.T. McCann, and Y.N. Xia. 2005a. Collecting electrospun nanofibers with patterned electrodes. *Nano Lett* 5: 913–916.

Li D., Y.L. Wang, and Y.N. Xia. 2004. Electrospinning nanofibers as uniaxially aligned arrays and layer-by-layer stacked films. *Adv Mater* 16: 361–366.

Li D. and Y.N. Xia. 2004. Electrospinning of nanofibers: Reinventing the wheel? *Adv Mater* 16: 1151–1170.

Li M., M.J. Mondrinos, M.R. Gandhi et al. 2005b. Electrospun protein fibers as matrices for tissue engineering. *Biomaterials* 26: 5999–6008.

Li W.J., J.A. Cooper, Jr., R.L. Mauck, and R.S. Tuan. 2006. Fabrication and characterization of six electrospun poly(alpha-hydroxy ester)-based fibrous scaffolds for tissue engineering applications. *Acta Biomater* 2: 377–385.

Li W.J., Y.J. Jiang, and R.S. Tuan. 2008. Cell–nanofiber-based cartilage tissue engineering using improved cell seeding, growth factor, and bioreactor technologies. *Tissue Eng—Part A* 14A: 639–648.

Li W.J., C.T. Laurencin, E.J. Caterson, R.S. Tuan, and F.K. Ko. 2002. Electrospun nanofibrous structure: A novel scaffold for tissue engineering. *J Biomed Mater Res* 60: 613–621.

Li W.J., R.L. Mauck, J.A. Cooper, X. Yuan, and R.S. Tuan. 2007. Engineering controllable anisotropy in electrospun biodegradable nanofibrous scaffolds for musculoskeletal tissue engineering. *J Biomech* 40: 1686–1693.

Li W.J., R.L. Mauck, and R.S. Tuan. 2005c. Electrospun nanofibrous scaffolds: Production, characterization, and applications for tissue engineering and drug delivery. *J Biomed Nanotechnol* 1: 259–275.

Li X., J. Xie, J. Lipner et al. 2009. Nanofiber scaffolds with gradations in mineral content for mimicking the tendon-to-bone insertion site. *Nano Lett* 9: 2763–2768.

Liao C.C., C.C. Wang, C.Y. Chen, and W.J. Lai. 2011. Stretching-induced orientation of polyacrylonitrile nanofibers by an electrically rotating viscoelastic jet for improving the mechanical properties. *Polymer* 52: 2263–2275.

Lim C.T., E.P.S. Tan, and S.Y. Ng. 2008. Effects of crystalline morphology on the tensile properties of electrospun polymer nanofibers. *Appl Phys Lett* 92: 141908.

Linsenmayer T.F., E. Gibney, F. Igoe et al. 1993. Type V collagen: Molecular structure and fibrillar organization of the chicken alpha 1(V) NH2-terminal domain, a putative regulator of corneal fibrillogenesis. *J Cell Biol* 121: 1181–1189.

Liu S.H., R.S. Yang, R. al-Shaikh, and J.M. Lane. 1995. Collagen in tendon, ligament, and bone healing: A current review. *Clin Orthop Relat Res* 265–278.

Lo I.K., Y. Ou, J.P. Rattner et al. 2002. The cellular networks of normal ovine medial collateral and anterior cruciate ligaments are not accurately recapitulated in scar tissue. *J Anat* 200: 283–296.

Ma Z., W. He, T. Yong, and S. Ramakrishna. 2005. Grafting of gelatin on electrospun poly(caprolactone) nanofibers to improve endothelial cell spreading and proliferation and to control cell Orientation. *Tissue Eng* 11: 1149–1158.

Madhugiri S., A. Dalton, J. Gutierrez, J.P. Ferraris, and K.J. Balkus, Jr. 2003. Electrospun MEH-PPV/SBA-15 composite nanofibers using a dual syringe method. *J Am Chem Soc* 125: 14531–14538.

Manske P.R. 1988. Flexor tendon healing. *J Hand Surg: Br Eur Vol* 13: 237–245.

Marchand F. and A.M. Ahmed. 1990. Investigation of the laminate structure of lumbar disc anulus fibrosus. *Spine* 15: 402–410.

Matthews J.A., G.E. Wnek, D.G. Simpson, and G.L. Bowlin. 2002. Electrospinning of collagen nanofibers. *Biomacromolecules* 3: 232–238.

Mauck R.L., B.M. Baker, N.L. Nerurkar et al. 2009. Engineering on the straight and narrow: The mechanics of nanofibrous assemblies for fiber-reinforced tissue regeneration. *Tissue Eng—Part B* 15B: 171–193.

McDevitt C.A. and R.J. Webber. 1990. The ultrastructure and biochemistry of meniscal cartilage. *Clin Orthop Relat Res* 252: 8–18.

Min B.M., L. Jeong, Y.S. Nam et al. 2004a. Formation of silk fibroin matrices with different texture and its cellular response to normal human keratinocytes. *Int J Biol Macromol* 34: 281–288.

Min B.M., G. Lee, S.H. Kim et al. 2004b. Electrospinning of silk fibroin nanofibers and its effect on the adhesion and spreading of normal human keratinocytes and fibroblasts in vitro. *Biomaterials* 25: 1289–1297.

Mizuno H., A.K. Roy, V. Zaporojan et al. 2006. Biomechanical and biochemical characterization of composite tissue-engineered intervertebral discs. *Biomaterials* 27: 362–370.

Moffat K.L., A.S. Kwei, J.P. Spalazzi et al. 2009. Novel nanofiber-based scaffold for rotator cuff repair and augmentation. *Tissue Eng—Part A* 15A: 115–126.

Moffat K.L., X. Zhang, S. Greco et al. 2011. *In vitro* and *in vivo* evaluation of a biphasic nanofiber scaffold for integrative rotator cuff repair. *Proceedings of the Orthopaedic Research Society Annual Meeting*, Long Beach, CA.

Moore S.M., P.J. McMahon, and R.E. Debski. 2004. Bidirectional mechanical properties of the axillary pouch of the glenohumeral capsule: Implications for modeling and surgical repair. *J Biomech Eng* 126: 284–288.

Nakamura N., S. Horibe, S. Sasaki et al. 2002. Evaluation of active knee flexion and hamstring strength after anterior cruciate ligament reconstruction using hamstring tendons. *Arthroscopy* 18: 598–602.

Naraghi M., S.N. Arshad, and I. Chasiotis. 2011. Molecular orientation and mechanical property size effects in electrospun polyacrylonitrile nanofibers. *Polymer* 52: 1612–1618.

Naraghi M., I. Chasiotis, H. Kahn, Y.K. Wen, and Y. Dzenis. 2007. Mechanical deformation and failure of electrospun polyacrylonitrile nanofibers as a function of strain rate. *Appl Phys Lett* 91: 151901.

Nerurkar N.L., B.M. Baker, S. Sen et al. 2009. Nanofibrous biologic laminates replicate the form and function of the annulus fibrosus. *Nat Mater* 8: 986–992.

Nerurkar N.L., D.M. Elliott, and R.L. Mauck. 2007. Mechanics of oriented electrospun nanofibrous scaffolds for annulus fibrosus tissue engineering. *J Orthop Res* 25: 1018–1028.

Nerurkar N.L., D.M. Elliott, and R.L. Mauck. 2010. Mechanical design criteria for intervertebral disc tissue engineering. *J Biomech* 43: 1017–1030.

Nerurkar N.L., R.L. Mauck, and D.M. Elliott. 2008. ISSLS prize winner: Integrating theoretical and experimental methods for functional tissue engineering of the annulus fibrosus. *Spine* 33: 2691–2701.

Nerurkar N.L., S. Sen, B.M. Baker, D.M. Elliott, and R.L. Mauck. 2011a. Dynamic culture enhances stem cell infiltration and modulates extracellular matrix production on aligned electrospun nanofibrous scaffolds. *Acta Biomater* 7: 485–491.

Nerurkar N.L., S. Sen, A.H. Huang, D.M. Elliott, and R.L. Mauck. 2011b. Engineered disc-like angle-ply structures for intervertebral disc replacement. *Spine* 35: 867–873.

Niyibizi C., C. Sagarriga Visconti, G. Gibson, and K. Kavalkovich. 1996. Identification and immunolocalization of type X collagen at the ligament–bone interface. *Biochem Biophys Res Commun* 222: 584–589.

Niyibizi C., C.S. Visconti, K. Kavalkovich, and S.L. Woo. 1995. Collagens in an adult bovine medial collateral ligament: Immunofluorescence localization by confocal microscopy reveals that type XIV collagen predominates at the ligament–bone junction. *Matrix Biol* 14: 743–751.

Noyes F.R. and E.S. Grood. 1976. The strength of the anterior cruciate ligament in humans and Rhesus monkeys. *J Bone Jt Surg, Am Vol* 58: 1074–1082.

Nur E.K.A., I. Ahmed, J. Kamal, M. Schindler, and S. Meiners. 2005. Three-dimensional nanofibrillar surfaces induce activation of Rac. *Biochem Biophys Res Commun* 331: 428.

Ohkoshi Y., C. Inoue, S. Yamane, T. Hashimoto, and R. Ishida. 1998. Changes in muscle strength properties caused by harvesting of autogenous semitendinosus tendon for reconstruction of contralateral anterior cruciate ligament. *Arthroscopy* 14: 580–584.

Ohland K.J., S.L.-Y. Woo, J.A. Weiss, S. Takai, and F.J. Shelley. 1991. Healing of combined injuries of the rabbit medial collateral ligament and its insertions: A long-term study on the effects of conservative versus surgical treatment. *Proceedings of the Winter Annual Meeting of the American Society of Mechanical Engineers*, Atlanta, GA.

Ottani V., M. Raspanti, and A. Ruggeri. 2001. Collagen structure and functional implications. *Micron* 32: 251–260.

Papastergiou S.G., H. Voulgaropoulos, P. Mikalef et al. 2006. Injuries to the infrapatellar branch(es) of the saphenous nerve in anterior cruciate ligament reconstruction with four-strand hamstring tendon autograft: Vertical versus horizontal incision for harvest. *Knee Surg Sports Traumatol Arthrosc* 14: 789–793.

Pedicini A. and R.J. Farris. 2003. Mechanical behavior of electrospun polyurethane. *Polymer* 44: 6857–6862.

Perie D., D. Korda, and J.C. Iatridis. 2005. Confined compression experiments on bovine nucleus pulposus and annulus fibrosus: Sensitivity of the experiment in the determination of compressive modulus and hydraulic permeability. *J Biomech* 38: 2164–2171.

Petersen W. and B. Tillmann. 1998. Collagenous fibril texture of the human knee joint menisci. *Anat Embryol* 197: 317–324.

Petrosini A.V. and O.H. Sherman. 1996. A historical perspective on meniscal repair. *Clin Sports Med* 15: 445–453.

Pham Q.P., U. Sharma, and A.G. Mikos. 2006a. Electrospinning of polymeric nanofibers for tissue engineering applications: A review. *Tissue Eng* 12: 1197–1211.

Pham Q.P., U. Sharma, and A.G. Mikos. 2006b. Electrospun poly(epsilon-caprolactone) microfiber and multilayer nanofiber/microfiber scaffolds: Characterization of scaffolds and measurement of cellular infiltration. *Biomacromolecules* 7: 2796–2805.

Pornsopone V., P. Supaphol, R. Rangkupan, and S. Tantayanon. 2005. Electrospinning of methacrylate-based copolymers: Effects of solution concentration and applied electrical potential on morphological appearance of as-spun fibers. *Polym Eng Sci* 45: 1073–1080.

Proctor C.S., M.B. Schmidt, R.R. Whipple, M.A. Kelly, and V.C. Mow. 1989. Material properties of the normal medial bovine meniscus. *J Orthop Res* 7: 771–782.

Provenzano P.P., K. Hayashi, D.N. Kunz, M.D. Markel, and R. Vanderby Jr. 2002. Healing of subfailure ligament injury: Comparison between immature and mature ligaments in a rat model. *J Orthop Res* 20: 975–983.

Provenzano P.P. and R. Vanderby Jr. 2006. Collagen fibril morphology and organization: Implications for force transmission in ligament and tendon. *Matrix Biol* 25: 71–84.

Quapp K.M. and J.A. Weiss. 1998. Material characterization of human medial collateral ligament. *J Biomech Eng* 120: 757–763.

Rath E. and J.C. Richmond. 2000. The menisci: Basic science and advances in treatment. *Br J Sports Med* 34: 252–257.

Reneker D.H. and I. Chun. 1996. Nanometer diameter fibres of polymer produced by electrospinning. *Nanotechnology* 7: 216–223.

Rho K.S., L. Jeong, G. Lee et al. 2006. Electrospinning of collagen nanofibers: Effects on the behavior of normal human keratinocytes and early-stage wound healing. *Biomaterials* 27: 1452–1461.

Riesle J., A.P. Hollander, R. Langer, L.E. Freed, and G. Vunjak-Novakovic. 1998. Collagen in tissue-engineered cartilage: Types, structure, and crosslinks. *J Cell Biochem* 71: 313–327.

Rnjak-Kovacina J., S.G. Wise, Z. Li et al. 2011. Tailoring the porosity and pore size of electrospun synthetic human elastin scaffolds for dermal tissue engineering. *Biomaterials* 32: 6729–6736.

Rockwood D.N., K.A. Woodhouse, J.D. Fromstein, D.B. Chase, and J.F. Rabolt. 2007. Characterization of biodegradable polyurethane microfibers for tissue engineering. *J Biomater Sci, Polym Ed* 18: 743–758.

Rodkey W.G., K.E. DeHaven, and W.H. Montgomery, 3rd et al. 2008. Comparison of the collagen meniscus implant with partial meniscectomy: A prospective randomized trial. *J Bone Jt Surg, Am Vol* 90: 1413–1426.

Rong Y., G. Sugumaran, J.E. Silbert, and M. Spector. 2002. Proteoglycans synthesized by canine intervertebral disc cells grown in a type I collagen–glycosaminoglycan matrix. *Tissue Eng* 8: 1037–1047.

Sagarriga Visconti C., K. Kavalkovich, J. Wu, and C. Niyibizi. 1996. Biochemical analysis of collagens at the ligament–bone interface reveals presence of cartilage-specific collagens. *Arch Biochem Biophys* 328: 135–142.

Sahoo S., S.L. Toh, and J.C. Goh. 2010. A bFGF-releasing silk/PLGA-based biohybrid scaffold for ligament/tendon tissue engineering using mesenchymal progenitor cells. *Biomaterials* 31: 2990–2998.

Sato M., T. Asazuma, M. Ishihara et al. 2003. An atelocollagen honeycomb-shaped scaffold with a membrane seal (ACHMS-scaffold) for the culture of annulus fibrosus cells from an intervertebral disc. *J Biomed Mater Res, Part A* 64A: 248–256.

Sato M., M. Maeda, H. Kurosawa et al. 2000. Reconstruction of rabbit Achilles tendon with three bioabsorbable materials: Histological and biomechanical studies. *J Orthop Sci* 5: 256–267.

Scheffler S.U., T.D. Clineff, C.D. Papageorgiou et al. 2001. Structure and function of the healing medial collateral ligament in a goat model. *Ann Biomed Eng* 29: 173–180.

Sell S.A., M.J. McClure, C.P. Barnes et al. 2006. Electrospun polydioxanone–elastin blends: Potential for bioresorbable vascular grafts. *Biomed Mater* 1: 72–80.

Setton L.A., F. Guilak, E.W. Hsu, and T.P. Vail. 1999. Biomechanical factors in tissue-engineered meniscal repair. *Clin Orthop Relat Res* (367 Suppl): S254–272.

Shao X. and C.J. Hunter. 2007. Developing an alginate/chitosan hybrid fiber scaffold for annulus fibrosus cells. *J Biomed Mater Res, Part A* 82A: 701–710.

Shields K.J., M.J. Beckman, G.L. Bowlin, and J.S. Wayne. 2004. Mechanical properties and cellular proliferation of electrospun collagen type II. *Tissue Eng* 10: 1510–1517.

Shin Y.M., M.M. Hohman, M.P. Brenner, and G.C. Rutledge. 2001. Electrospinning: A whipping fluid jet generates submicron polymer fibers. *Appl Phys Lett* 78: 1149–1151.

Skaggs D.L., M. Weidenbaum, J.C. Iatridis, A. Ratcliffe, and V.C. Mow. 1994. Regional variation in tensile properties and biochemical composition of the human lumbar anulus fibrosus. *Spine (Phila Pa 1976)* 19: 1310–1319.

Sommerlath K. and J. Gillquist. 1993. The effects of an artificial meniscus substitute in a knee joint with a resected anterior cruciate ligament: An experimental study in rabbits. *Clin Orthop Relat Res* 276–284.

Soslowsky L.J., S.P. Lake, K.S. Miller, and D.M. Elliott. 2010. Tensile properties and fiber alignment of human supraspinatus tendon in the transverse direction demonstrate inhomogeneity, nonlinearity, and regional isotropy. *J Biomech* 43: 727–732.

Spindler K.P., J.E. Kuhn, K.B. Freedman et al. 2004. Anterior cruciate ligament reconstruction autograft choice: Bone–tendon–bone versus hamstring: Does it really matter? A systematic review. *Am J Sports Med* 32: 1986–1995.

Stankus J.J., J. Guan, and W.R. Wagner. 2004. Fabrication of biodegradable elastomeric scaffolds with sub-micron morphologies. *J Biomed Mater Res—Part A* 70A: 603–614.

Stella J.A., A. D'Amore, W.R. Wagner, and M.S. Sacks. 2010. On the biomechanical function of scaffolds for engineering load-bearing soft tissues. *Acta Biomater* 6: 2365–2381.

Stella J.A., W.R. Wagner, and M.S. Sacks. 2009. Scale-dependent fiber kinematics of elastomeric electrospun scaffolds for soft tissue engineering. *J Biomed Mater Res—Part A* 93A: 1032–1042.

Stone K.R., W.G. Rodkey, R.J. Webber, L.A. McKinney, and J.R. Steadman. 1992. Development of a prosthetic meniscal replacement. In: *Knee Meniscus: Basic and Clinical Foundations*, Mow, V.C., S.P. Arnoczky, and D.W. Jackson (eds.), pp. 165–173. New York: Raven Press, Ltd.

Sun D.H., C. Chang, S. Li, and L.W. Lin. 2006. Near-field electrospinning. *Nano Letters* 6: 839–842.

Sundaray B., V. Subramanian, T.S. Natarajan et al. 2004. Electrospinning of continuous aligned polymer fibers. *Appl Phys Lett* 84: 1222–1224.

Tan A.R., J.L. Ifkovits, B.M. Baker et al. 2008. Electrospinning of photocrosslinked and degradable fibrous scaffolds. *J Biomed Mater Res—Part A* 87A: 1034–1043.

Tan E.P.S., C.N. Goh, C.H. Sow, and C.T. Lim. 2005a. Tensile test of a single nanofiber using an atomic force microscope tip. *Appl Phys Lett* 86: 073115.

Tan E.P.S. and C.T. Lim. 2006. Characterization of bulk properties of nanofibrous scaffolds from nanomechanical properties of single nanofibers. *J Biomed Mater Res—Part A* 77A: 526–533.

Tan E.P.S., S.Y. Ng, and C.T. Lim. 2005b. Tensile testing of a single ultrafine polymeric fiber. *Biomaterials* 26: 1453–1456.

Tan N.C.B., J.M. Deitzel, J. Kleinmeyer, and D. Harris. 2001. The effect of processing variables on the morphology of electrospun nanofibers and textiles. *Polymer* 42: 261–272.

Teh T.K., S.L. Toh, and J.C. Goh. 2011. Aligned hybrid silk scaffold for enhanced differentiation of mesenchymal stem cells into ligament fibroblasts. *Tissue Eng—Part C* 17C: 687–703.

Theron A., E. Zussman, and A.L. Yarin. 2001. Electrostatic field-assisted alignment of electrospun nanofibres. *Nanotechnology* 12: 384–390.

Thome C., M. Barth, J. Scharf, and P. Schmiedek. 2005. Outcome after lumbar sequestrectomy compared with microdiscectomy: A prospective randomized study. *J Neurosurg Spine* 2: 271–278.

Tissakht M. and A.M. Ahmed. 1995. Tensile stress–strain characteristics of the human meniscal material. *J Biomech* 28: 411–422.

Tuan R.S., W.J. Li, and R.L. Mauck. 2005. Electrospun nanofibrous scaffolds: Production, characterization, and applications for tissue engineering and drug delivery. *J Biomed Nanotechnol* 1: 259–275.

van Tienen T.G., R.G. Heijkants, P. Buma et al. 2002. Tissue ingrowth and degradation of two biodegradable porous polymers with different porosities and pore sizes. *Biomaterials* 23: 1731–1738.

Vaquette C. and J.J. Cooper-White. 2011. Increasing electrospun scaffold pore size with tailored collectors for improved cell penetration. *Acta Biomater* 7: 2544–2557.

Vaquette C., C. Kahn, C. Frochot et al. 2010. Aligned poly(L-lactic-*co*-e-caprolactone) electrospun microfibers and knitted structure: A novel composite scaffold for ligament tissue engineering. *J Biomed Mater Res, Part A* 94A: 1270–1282.

Venugopal J., L.L. Ma, T. Yong, and S. Ramakrishna. 2005. *In vitro* study of smooth muscle cells on polycaprolactone and collagen nanofibrous matrices. *Cell Biol Int* 29: 861–867.

Veresciagina K., B. Spakauskas, and K.V. Ambrozaitis. 2010. Clinical outcomes of patients with lumbar disc herniation, selected for one-level open-discectomy and microdiscectomy. *Eur Spine J* 19: 1450–1458.

Verreck G., I. Chun, J. Peeters, J. Rosenblatt, and M.E. Brewster. 2003. Preparation and characterization of nanofibers containing amorphous drug dispersions generated by electrostatic spinning. *Pharm Res* 20: 810–817.

Veth R.P., H.W. Jansen, J.W. Leenslag et al. 1986. Experimental meniscal lesions reconstructed with a carbon fiber–polyurethane-poly(L-lactide) graft. *Clin Orthop Relat Res* 286–293.

Viola R.W., W.I. Sterett, D. Newfield, J.R. Steadman, and M.R. Torry. 2000. Internal and external tibial rotation strength after anterior cruciate ligament reconstruction using ipsilateral semitendinosus and gracilis tendon autografts. *Am J Sports Med* 28: 552–555.

Watanabe K., Y. Enomoto, I.S. Kim, and B.S. Kim. 2011. Fabrication of uniaxially aligned poly(propylene) nanofibers via handspinning. *Macromol Mater Eng* 296: 568–573.

Weiss J.A., T.J. Lujan, S.P. Lake, T.A. Plaizier, and B.J. Ellis. 2005. Simultaneous measurement of three-dimensional joint kinematics and ligament strains with optical methods. *J. Biomech Eng* 127: 193–197.

Weiss J.A., S.L. Woo, K.J. Ohland, S. Horibe, and P.O. Newton. 1991. Evaluation of a new injury model to study medial collateral ligament healing: Primary repair versus nonoperative treatment. *J Orthop Res* 9: 516–528.

Woerdeman D.L., P. Ye, S. Shenoy et al. 2005. Electrospun fibers from wheat protein: Investigation of the interplay between molecular structure and the fluid dynamics of the electrospinning process. *Biomacromolecules* 6: 707–712.

Wong S.C., A. Baji, and S.W. Leng. 2008. Effect of fiber diameter on tensile properties of electrospun poly(epsilon-caprolactone). *Polymer* 49: 4713–4722.

Woo S., K.-N. An, C. Frank et al. 2000. Anatomy, biology and biomechanics of tendon and ligaments. In: *Orthopaedic Basic Science: Biology and Biomechanics of the Musculoskeletal System*, 2nd edition, Einhorn, T.A., S.R. Simon, and The American Academy of Orthopaedic Surgeons (eds.), pp. 581–616. Rosemont, IL: American Academy of Orthopaedic Surgeons.

Woo S.L., S.D. Abramowitch, R. Kilger, and R. Liang. 2006a. Biomechanics of knee ligaments: Injury, healing, and repair. *J Biomech* 39: 1–20.

Woo S.L., M.A. Gomez, Y. Seguchi, C.M. Endo, and W.H. Akeson. 1983. Measurement of mechanical properties of ligament substance from a bone–ligament–bone preparation. *J Orthop Res* 1: 22–29.

Woo S.L., J.M. Hollis, D.J. Adams, R.M. Lyon, and S. Takai. 1991. Tensile properties of the human femur–anterior cruciate ligament–tibia complex: The effects of specimen age and orientation. *Am J Sports Med* 19: 217–225.

Woo S.L., C.A. Orlando, J.F. Camp, and W.H. Akeson. 1986. Effects of postmortem storage by freezing on ligament tensile behavior. *J Biomech* 19: 399–404.

Woo S.L., R.H. Peterson, K.J. Ohland, T.J. Sites, and M.I. Danto. 1990. The effects of strain rate on the properties of the medial collateral ligament in skeletally immature and mature rabbits: A biomechanical and histological study. *J Orthop Res* 8: 712–721.

Woo S.L., Y. Takakura, R. Liang, F. Jia, and D.K. Moon. 2006b. Treatment with bioscaffold enhances the the fibril morphology and the collagen composition of healing medial collateral ligament in rabbits. *Tissue Eng* 12: 159–166.

Wren T.A., S.A. Yerby, G.S. Beaupre, and D.R. Carter. 2001. Mechanical properties of the human Achilles tendon. *Clin Biomech (Bristol, Avon)* 16: 245–251.

Wright L.D., T. Andric, and J.W. Freeman. 2011. Utilizing NaCl to increase the porosity of electrospun materials. *Mater Sci Eng, C* 31: 30–36.

Xu C., R. Inai, M. Kotaki, and S. Ramakrishna. 2004. Electrospun nanofiber fabrication as synthetic extracellular matrix and its potential for vascular tissue engineering. *Tissue Eng* 10: 1160–1168.

Yang D.Y., B. Lu, Y. Zhao, and X.Y. Jiang. 2007. Fabrication of aligned fibirous arrays by magnetic electrospinning. *Adv Mater* 19: 3702–3706.

Yang F., R. Murugan, S. Wang, and S. Ramakrishna. 2005. Electrospinning of nano/micro scale poly(L-lactic acid) aligned fibers and their potential in neural tissue engineering. *Biomaterials* 26: 2603–2610.

Yang F., C.Y. Xu, M. Kotaki, S. Wang, and S. Ramakrishna. 2004. Characterization of neural stem cells on electrospun poly(L-lactic acid) nanofibrous scaffold. *J Biomater Sci, Polym Ed* 15: 1483–1497.

Yang L., C.F.C. Fitie, K.O. van der Werf et al. 2008. Mechanical properties of single electrospun collagen type I fibers. *Biomaterials* 29: 955–962.

Yao H., M.A. Justiz, D. Flagler, and W.Y. Gu. 2002. Effects of swelling pressure and hydraulic permeability on dynamic compressive behavior of lumbar annulus fibrosus. *Ann Biomed Eng* 30: 1234–1241.

Yeganegi M., R.A. Kandel, J.P. Santerre. 2010. Characterization of a biodegradable electrospun polyurethane nanofiber scaffold: Mechanical properties and cytotoxicity. *Acta Biomater* 6: 3847–3855.

Yeung A.T. and P.M. Tsou. 2002. Posterolateral endoscopic excision for lumbar disc herniation: Surgical technique, outcome, and complications in 307 consecutive cases. *Spine* 27: 722–731.

Yin, Frank C.P., W.R. Tompkins, M. Intaglietta, and K.L. Peterson. 1972. A video-dimension analyzer. *IEEE Trans Biomed Eng* BME-19: 376–381.

Yin Z., X. Chen, J.L. Chen et al. 2010. The regulation of tendon stem cell differentiation by the alignment of nanofibers. *Biomaterials* 31: 2163–2175.

Yoshimoto H., Y.M. Shin, H. Terai, and J.P. Vacanti. 2003. A biodegradable nanofiber scaffold by electrospinning and its potential for bone tissue engineering. *Biomaterials* 24: 2077–2082.

Yunes M., J.C. Richmond, E.A. Engels, and L.A. Pinczewski. 2001. Patellar versus hamstring tendons in anterior cruciate ligament reconstruction: A meta-analysis. *Arthroscopy* 17: 248–257.

Zhang X., V. Thomas, and Y.K. Vohra. 2009. In vitro biodegradation of designed tubular scaffolds of electrospun protein/polyglyconate blend fibers. *J Biomed Mater Res—Part B: Appl Biomater* 89B: 135–147.

13 Regulation of Stem Cell Behavior Using Nanobiomaterials

Andrew M. Handorf, Patrick T. Brown, John G. Barber, and Wan-Ju Li

CONTENTS

13.1 INTRODUCTION

Regenerative medicine and tissue engineering are rapidly evolving areas of medicine and research. Often used interchangeably, these closely related research fields work cohesively, with the goal of altering the course and treatment of major injuries and chronic disease, ranging from orthopedic defects to diabetes (Furth and Atala 2008; Chiu et al. 2011; Wang and Guan 2010; Levenberg and Langer 2004). According to the McGowan Institute for Regenerative Medicine, there are four main concentrations in regenerative medicine: 1) medical devices and artificial organs; 2) tissue engineering and biomaterials; 3) cellular therapies; and 4) clinical translation (McGowan Institute for Regenerative Medicine n.d.). The use and combination of research practices in these four areas provides the backbone for future medical research in tissue repair and replacement. Currently, large tissue defects are treated by the surgical removal of the defective tissue and replacement with nonbiological prostheses that only need to match the structural or mechanical criteria of the native tissue (Williams 2001). As clinical practice trends shift toward minimally invasive procedures, the ability to address lesions earlier in development will provide regenerative medicine applications with many opportunities to replace defunct tissue with a cellular-based therapy. In the long term, cell-based regenerative medical therapy has the potential for greatly affecting clinical medicine by addressing the shortage in donor organs through the use of tissue-specific stem cells or stem cells derived from adult or embryonic sources for organ regeneration. Although 50,000 individuals need transplants, many will never receive them. However, this can be alleviated through the regeneration of organs *de novo* to overcome the donor shortage, the need to immunophenotype donor and patient, and the immunosuppressive therapy resulting from organ transplantation (Williams 2001). Tissue engineering offers the potential for using cell therapies and biomaterials to regenerate tissues and restore organ function, with the hopes of one day eliminating the need for organ transplants.

Although regeneration of organs *in vivo* may still be years away, current research has targeted smaller defects, including portions of organs, by using stem cells and biomaterials to restore organ function. Stem cells have become an attractive choice for cellular therapy in regenerative medicine research because of their ability for self-renewal and differentiation into various cell types (van der Sanden et al. 2010). A variety of stem cells, including embryonic stem cells (ESCs), induced pluripotent stem cells (iPSCs), mesenchymal stem cells (MSCs) derived from bone marrow and adipose tissue, and a variety of other tissue-specific stem cells have been used in regenerative medicine and tissue engineering applications *in vitro* and *in vivo* (Beltrami et al. 2003; Bruder et al. 1997; Cao et al. 2005; Wang and Guan 2010; Zuk et al. 2002). ESCs and iPSCs are able to undergo seemingly unlimited cell division while still maintaining their ability to differentiate (Thomson et al. 1998; Keller 2005). Both ESCs and iPSCs are pluripotent and thus can be directed toward ectodermal, endodermal, and mesodermal tissues, whereas hMSCs, being multipotent, are more restricted in their differentiation capabilities (Furth and Atala 2008). Regardless, the use of stem cells in tissue regeneration applications allows for the development of complex tissues composed of several cell types to be derived from a single stem cell population.

However, even with the numerous advantages of using stem cells for tissue engineering research and clinical applications, they are not without limitations. The use of ESCs in regenerative medicine applications has been delayed not only because of ethical issues but also because of inefficiencies in controlling differentiation. Implantation of undifferentiated human ESCs (hESCs) into immunocompetent and immunocompromised mice results in uncontrolled teratoma growth (Thomson et al. 1998; Jukes et al. 2008; Wakitani et al. 2003). Adult stem cells, although not tumorigenic, are more restricted in differentiation potential and have limitations during *in vitro* culture. *In vivo*, adult stem cells occupy a specific niche in which they receive signals to remain in an undifferentiated, quiescent state. Once taken out of this native environment and expanded *in vitro*, they often undergo senescence, which severely limits their proliferative and differentiation capabilities (Brown et al. 2009; de Peppo et al. 2010; van der Sanden et al. 2010). Current *in vitro* differentiation protocols require a large number of cells to be expanded to obtain the cell numbers needed for tissue regeneration applications.

Nanobiomaterials have emerged as a promising material for use as scaffolds during cell seeding and implantation due to their unique structure, which is composed of nanometer-sized building components similar to natural extracellular matrix (ECM).

Nanostructures, such as three-dimensional (3-D) nanofibers or two-dimensional (2-D) nanotopography, induce a cellular response during *in vitro* culture that more closely mimics conditions seen in the cell's native environment (Li et al. 2006). *In vivo*, a cell resides in a complex microenvironment containing both chemical and physical cues. The physical cues that a cell may encounter range from the macroscale (like the overall shape of a tissue; Carrel and Burrows 1911), the microscale (like the overall arrangement of cells in the surrounding microenvironment; Chen et al. 1997), to the nanoscale (like the dimensionality of surrounding ECM proteins; Katz et al. 2000). Recent evidence indicates that interaction with nanostructures in a stem cell's native microenvironment can have profound effects on several aspects of cell behavior, including adhesion (Dalby et al. 2004), motility (Brammer et al. 2008), proliferation (Dalby et al. 2002), and differentiation (Kolambkar et al. 2010). In the following sections, we will provide examples of how various nanostructures have been used to control stem cell behavior *in vitro*. This discussion will include examples from both 2-D and 3-D cultures, with a particular emphasis on how 3-D nanofibrous scaffolds and various modifications to 2-D substrates; namely, the introduction of specific nanotopographies, affect cell behavior. The regulation of many biological processes holds true for both stem cells and fully differentiated somatic cells and, as such, cell response to nanostructures will include examples from differentiated cell types in addition to stem cells. However, examples given for the regulation of self-renewal, maintenance of pluripotency, and differentiation will invariably be restricted to stem cells. The regulation of cell behavior using nanostructures is ultimately driven by the control of cell morphology, cytoskeletal organization, and focal adhesion (FA) formation. This section aims to shed light on our current understanding of the regulation of these cellular features, with a focus on harnessing such knowledge to controllably drive tissue regeneration *ex vivo*.

13.2 CURRENT RESEARCH FINDINGS

13.2.1 Regulation of Cell Adhesion by Nanostructures

Before a stem cell can initiate biological processes, it must first adhere to the biomaterial. It is this initial adhesion event that ultimately determines the long-term viability of the stem cell (Ravichandran et al. 2009). Cells adhere readily to natural biomaterials, such as fibronectin, vitronectin, and collagen, through cell surface receptors called integrins. Integrins are obligate heterodimers containing one α and one β subunit (Humphries 2000). In mammals, eighteen α and eight β subunits have been characterized, and it is the combination of α and β subunits that determines which ECM protein the integrin will bind. For instance, integrin $\alpha_5\beta_1$ mediates adhesion to fibronectin (Hemler 1990), whereas integrin $\alpha_v\beta_3$ mediates adhesion to vitronectin (Hermann et al. 1999). Conversely, cell adhesion cannot occur spontaneously on most synthetic biomaterials. However, adhesion can be mediated on synthetic biomaterials in a few ways. First, if cells are cultured in the presence of serum, the serum proteins will adsorb onto the surface of many synthetic biomaterials and serve as binding sites for cells. Second, synthetic biomaterials can be functionalized with short peptides, such as arginine–glycine–aspartic acid (RGD) sequences, that are found in several native ECM proteins to promote cell adhesion. Third, the surface of biomaterials can be modified to yield certain surface properties that are more amenable to cell adhesion. Physicochemical parameters of a biomaterial surface, such as surface free energy, charge, roughness, composition, and topography, can be introduced to promote cell adhesion. Once the cell is securely anchored to the biomaterial, it can then initiate other cell activities, such as migration, proliferation, and differentiation (Grinnell 1978).

Importantly, extracellular structures presented in a cell's native microenvironment have dimensionality on the nanoscale. Not surprisingly, a variety of nanostructures have been shown to enhance cell adhesion compared with microstructured or unstructured substrates of identical composition *in vitro*. For instance, Thapa et al. (2003) found that smooth muscle cells in the bladder showed enhanced cell adhesion on nanostructured poly(lactic-*co*-glycolic acid) and polyurethane surfaces with dimensions ranging from 50 nm to 100 nm compared with microstructured controls. Similarly, Tian et al. (2008) investigated the adhesion of NIH3T3 fibroblasts on composite poly(glycolic acid)/collagen nanofibers with the poly(glycolic acid) composition ranging from 7% to 86% and fiber diameters ranging from 500 nm to 10 μm. They found that regardless of the composition, there were significantly more cells attached on the 500-nm fibers compared with the 3-μm to 5-μm and 10-μm fibers.

Generally, it is believed that nanostructures more effectively promote cell adhesion through the enhanced and selective adsorption of adhesion molecules (Woo et al. 2003). For example, Webster et al. (2000) cultured osteoblasts on nanophase alumina, titania and hydroxyapatite, or untextured controls to compare cell adhesion. They found that osteoblast adhesion was significantly greater on all nanophase substrates, and the enhanced adhesion was likely driven by the selective adsorption of vitronectin. Similarly, Woo et al. (2003) fabricated poly(L-lactic acid) (PLLA) nanofibrous scaffolds with fiber diameters of 50 nm to 500 nm, or with micron-scaled fiber diameters, to compare the effect of fiber diameter on protein adsorption. They found that nanofibrous scaffolds adsorbed four times more human serum proteins than microfibrous scaffolds, and that fibronectin and vitronectin were selectively adsorbed. Cell adhesion was likewise increased almost twofold on these nanofibrous scaffolds. However, inconsistencies in cell adhesion have been reported. For example, Gallagher et al. (2002) found that fibroblasts showed reduced adhesion to ordered nanopits, whereas Wan et al. (2005) found that nanopits (450 nm) enhanced osteoblast adhesion compared with both micropits (2.2 μm) and smooth surface controls. This may suggest greater mechanistic complexity than simply the selective adsorption of specific adhesion molecules. As discussed in the next section, the interplay between nanostructure dimensionality and spacing likely plays a role in promoting cell adhesion by mediating integrin clustering and FA formation.

13.2.2 Regulation of Cell Morphology by Nanostructures through the Regulation of FA Formation and Actin Cytoskeletal Organization

Once a cell is securely adhered to a biomaterial, it responds first through dramatic morphological changes, like cell spreading. The degree and dynamics of cell spreading are largely dictated by the dimensionality and distribution of constituent structural features within a biomaterial. Specifically, nanostructures modulate cell spreading, and ultimately a cell's overall behavior, by regulating the formation of FA complexes and organization of the actin cytoskeleton (Chen et al. 2003). FAs are large receptor complexes that transmit mechanical forces and regulatory cues from the extracellular environment intracellularly. Importantly, FAs serve as potent regulators of cell behavior largely through their ability to serve as dynamic signaling hubs (Giancotti and Ruoslahti 1999). Factors like FA density and length are both linked to changes in cell behavior and can be controlled by nanofeatures *in vitro* (Dalby et al. 2003; Cavalcanti-Adam et al. 2006). FAs can contain more than 100 different proteins based on their maturity (Zaidel-Bar et al. 2004), thus suggesting great functional diversity depending on the identity of the associated proteins. Many of the proteins that associate with FAs, such as talin, α-actinin, and vinculin, are adaptor proteins linking cell surface integrin receptors to the actin cytoskeleton, whereas others, most notably focal adhesion kinase (FAK), serve as intracellular signaling mediators relaying external cues to the cytoplasm of a cell (Schaller et al. 1992). FAK, which is constitutively associated with the β-integrin subunit (Hildebrand et al. 1993), relays integrin-dependent signaling to downstream intracellular signaling proteins in an adhesion-dependent manner (Schaller et al. 1996). In particular, the mitogen-activated protein kinase (MAPK) signaling pathway is activated by FA elongation through phosphorylation of extracellular signal-regulated kinase (ERK) by FAK (Klees et al. 2005). Upon activation, ERK translocates into the nucleus to modulate gene expression in a context-dependent manner (Kokubu et al. 2009). Depending on the maturity of the FA formed, and thus the identity of proteins associated with the FA, distinct morphologies and cytoskeletal organizations can be promoted.

For instance, well-developed FAs are typically associated with the formation of actin stress fibers, high-order actin structures consisting of cross-linked actin microfilaments and myosin II motor proteins that span the length of the cell (Byers et al. 1984). Stress fibers allow cells to generate the contractile forces required for filament growth and reorganization to move and reshape the cell (Small et al. 1998). In fact, it seems that cytoskeletal remodeling is critical for the morphological changes observed in response to nanotopographies. Gerecht et al. (2007) cultured hESCs on fibronectin-coated poly(dimethyl siloxane) substrates with ridges that are 600 nm wide and 600 nm tall and spaced 600 nm apart to study the effect of cytoskeletal remodeling on cell alignment and indeed culturing hESCs on the nanotopographic surfaces induced cell alignment and modulated the organization of the actin cytoskeleton. However, cytochalasin D, a cell-permeable compound that disrupts actin filaments by binding actin monomers and preventing their polymerization, attenuated the cell alignment induced by the nanotopographical features, demonstrating a functional interplay between a cell's cytoskeleton and the observed morphological changes occurring in response to nanotopographical substrates. However, actin stress fibers are rarely observed in cells *in vivo* and are thought to be an artifact of 2-D culture *in vitro* (Byers et al. 1984). Thus, it will be important for *in vitro* culture systems to deter stress fiber formation and more closely recapitulate the key morphological features of cells found *in vivo*.

Massia and Hubbell (1991) showed that there exist different thresholds for adhesive ligand spacing that regulate FA formation and cell spreading. By covalently grafting RGD adhesive peptides at well-defined spacings onto glass substrates that are otherwise poorly adhesive, it was found that when RGD peptides were spaced 440 nm apart or less, fibroblasts become fully spread within 4 h of plating. However, FAs did not form unless peptides were spaced within 140 nm of each other, suggesting that the clustering of integrin receptors drives FA formation. Furthermore, they showed that cells can become fully spread without the formation of well-developed FAs, demonstrating that cells can adhere to *in vitro* substrates without showing the aforementioned artifacts of 2-D culture.

Recently, elegant studies have more conclusively shown that integrin clustering indeed drives FA formation. By spacing adhesive RGD nanodots of less than 8 nm diameter per dot (allowing the binding of one integrin per dot) with high precision at 28, 58, 73, and 85 nm apart, Arnold and colleagues demonstrated that a separation of RGD peptides of greater than 73 nm greatly limited cell spreading and FA formation (Figure 13.1d and e), whereas elongated FAs formed readily when RGD peptides were spaced less than 73 nm apart (Figure 13.1a through c; Arnold et al. 2004). They concluded that restricted integrin clustering, rather than an insufficient number of adhesive ligands, affects cell adhesion because patterned substrates consisting of alternating fields with dense or no nanodots are capable of supporting adhesion. In fact, Arnold and colleagues suggested that spacing adhesive ligands 58 nm to 73 nm apart may represent a universal length scale for integrin clustering and activation, because these adhesive properties were shared by a variety of cell types (Figure 13.1f).

A variety of nanotopographical features, including pillars, pits, grooves, and roughness, have been shown to control the maturity of FAs formed by providing limited surface area for cell attachment. For example, Biggs et al. (2008) cultured human osteoblasts on ordered nanoscale groove/ridge arrays with grooves of varying width and 330 nm depth. As shown in Figure 13.2, they found that both planar control substrates and 100-μm-wide grooves induced fully spread cells with mature FAs (Figure 13.2a through d). Conversely, 25-μm-wide grooves were associated with a reduction in mature FA formation (Figure 13.2e and f), whereas 10-μm-wide grooves significantly reduced osteoblast adhesion (Figure 13.2g and h). Similarly, Teixeira et al. (2003) cultured epithelial cells on nanogrooves of varying widths and depths to investigate trends in cell morphology, FA formation, and cytoskeletal organization. They found that epithelial cells aligned and elongated in the direction of the nanogrooves, with a greater percentage of epithelial cells aligning on deeper groove patterns. Furthermore, although all cells adhered through FAs, the sizes of the FAs were dependent on ridge width, with wider ridges allowing the formation of larger FAs. Clark et al. (1990) further demonstrated that groove width affects the degree of cell alignment for grooves of more than 2 μm. In addition, cells will not align if groove widths are much greater than a cell's diameter, because the cell simply will not recognize the structure, further illustrating the importance of nanoscaled features.

In another study, human osteoblasts were cultured on nanopits, nanocraters, and nanoislands, and it was found that both nanocraters and nanoislands reduced FA formation, whereas nanopits directly inhibited the formation of mature FAs, suggesting

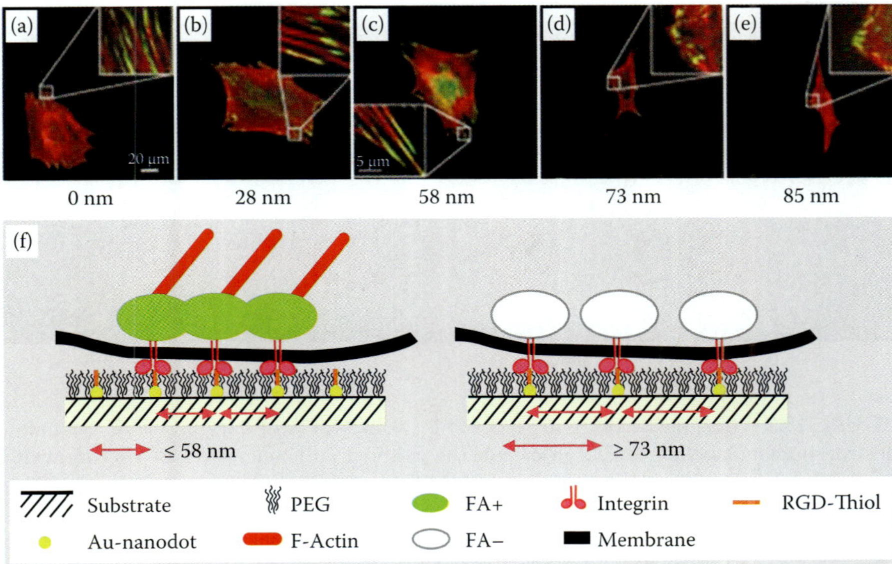

FIGURE 13.1 Characterization of FA formation, actin cytoskeletal organization, and cell spreading in osteoblasts attached to substrates with different spacings of RGD adhesive nanodots. Confocal micrographs of MC3T3-osteoblasts stained for actin (red) and vinculin (green) 24 h after plating. Cells were plated on glass coverslips homogeneously coated with (a) RGD peptides or RGD nanodot patterns of different spacings: (b) 28, (c) 58, (d) 73, and (e) 85 nm. Mature FAs are formed when RGD nanodots are spaced less than 73 nm apart, whereas greater spacings do not induce FA formation. (f) Scheme of biofunctionalized nanopattern to control integrin clustering in cell membranes. Surfaces are functionalized with RGD nanodots, and the areas between cell-adhesive RGD nanodots are passivated by PEG against cell adhesion. Therefore, cell adhesion is mediated entirely via RGD nanodots. An RGD nanodot separation of ≥73 nm causes limited FA formation, actin stress fiber formation, and cell spreading, demonstrating that nanostructures that control integrin clustering can be used to control key features of cell morphology. This is indicated by no FA activation (FA−), whereas distances between dots of ≤58 nm caused FA activation (FA+). (Adapted with permission from Arnold M. et al. *ChemPhysChem* 5(3): 383–388, 2004.)

that FA formation varies depending on the nanofeature investigated (Biggs et al. 2009). Lim et al. (2007) similarly looked at osteoblast adhesion to 14-, 29-, and 45-nm-deep nanopits and found that 14- and 29-nm-deep nanopits increased osteoblast adhesion and spreading compared with 45-nm-deep nanopits or flat control surfaces. Accordingly, the 14- and 29-nm-deep nanopits showed selective integrin expression of the α_v subunit, increased paxillin protein synthesis, colocalization with actin stress fibers, and increased FAK expression and phosphorylation, suggesting that the nanodimensionality of specific nanofeatures must be tightly controlled to precisely regulate cell behavior. Lastly, Pennisi et al. (2011) introduced nanoroughness into an otherwise 2-D substrate and found that nanorough substrates reduced the formation of FA complexes and impaired stress fiber formation.

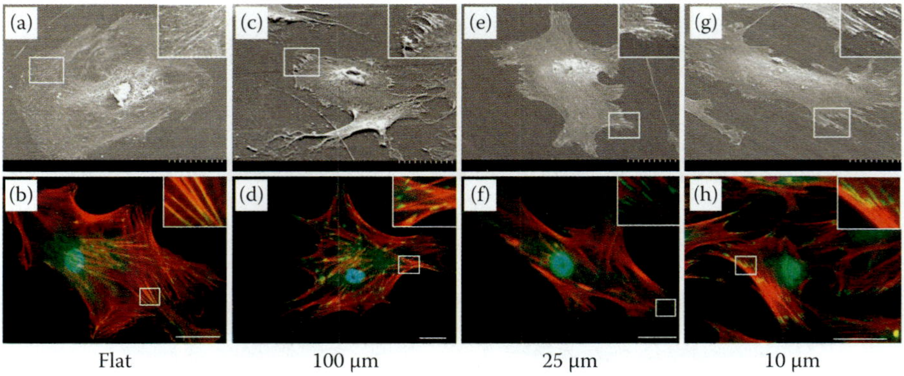

FIGURE 13.2 Cell morphology in response to nanogrooves of varying width. Scanning electron immunolabeling of FAs in osteoblasts cultured on nanogrooves of varying width showed that (a) osteoblasts cultured on flat PMMA control substrates formed numerous FAs. (c) Diffuse 100-μm grooves did not affect FA formation, whereas HOBs on (e) 25-μm grooves formed adhesions predominantly on the raised ridge areas (indicated in inset). Osteoblasts cultured on (g) 10-μm grooves influenced cellular morphology and FA orientation, aligning both osteoblasts and adhesion complexes with groove direction. (b) Fluorescent labeling of actin (red), FAs (green), and nuclei (blue) in osteoblasts on experimental substrates showed that HOBs cultured on planar controls formed well-organized stress fibers and numerous well-developed FAs. (d) Osteoblasts on 100-μm grooves possessed a well-developed actin cytoskeleton. FAs were large and numerous. (f) Osteoblasts cultured on 25-μm grooves developed an organized actin cytoskeleton with reduced FA formation relative to planar controls. (h) Osteoblasts cultured on 10 μm groove arrays disrupted cellular spreading. Adhesion formation was influenced by groove orientation. Bar, 50 μm. (Adapted with permission from Biggs M. J. et al. *J. R. Soc. Interface* 5(27):1231–1242, 2008.)

The dimensionality of nanofibers has likewise been shown to affect cell adhesion, morphology, FA formation, and cytoskeletal organization. For example, Shih et al. (2006) compared FA formation on type I collagen nanofibers and 2-D plastic substrates and found significantly reduced FA formation on nanofibers that were 200–500 nm in diameter. Similarly, Hsia et al. (2011) compared FA and stress fiber formation on nanofibers and microfibers and found smaller FA complexes and fewer actin stress fibers in fibroblasts cultured on nanofibers. Accordingly, Li et al. (2006) observed dramatic differences in morphology between primary chondrocytes seeded on PLLA nanofibers and chemically identical microfibers (Figure 13.3). Chondrocytes seeded on the PLLA microfibers displayed a well-spread morphology that spanned several fibers (Figure 13.3a), whereas chondrocytes on nanofibers were found to have a rounded morphology (Figure 13.3b). Interestingly, the distinct cell morphologies observed on microfibers and nanofibers occurred concomitantly with changes in actin cytoskeletal organization. The chondrocytes seeded on microfibers displayed prominent actin stress fibers that spanned the length of the cell (Figure 13.3c), whereas chondrocytes on nanofibers displayed a less well-developed cortical actin cytoskeletal organization (Figure 13.3d), reminiscent of that found *in vivo*.

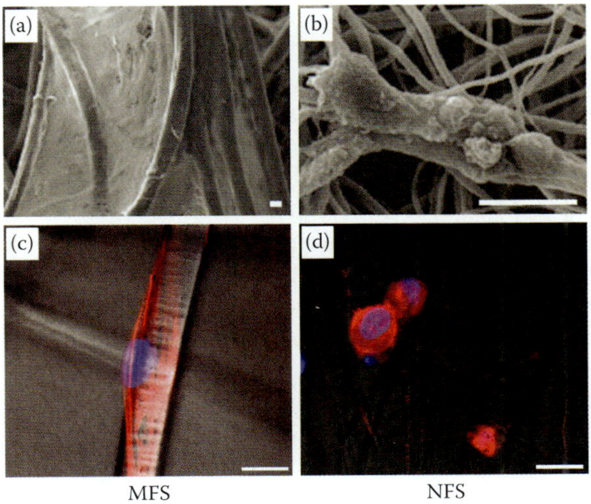

FIGURE 13.3 Chondrocyte morphology and actin cytoskeletal organization seeded on microfibrous (MFS) and nanofibrous scaffolds (NFS). SEM image of chondrocytes seeded in fibrous scaffolds showed (a) well-spread fibroblast-like cells spanning across multiple microfibers after 28 days of culture, whereas (b) globular chondrocyte-like cells grew on nanofibers. Bar: 10 μm. Confocal micrographs of chondrocytes stained for actin (red) and nuclei (blue) seeded in fibrous scaffolds show (c) large well-organized actin stress fibers on microfibers and (d) short disorganized actin filaments in chondrocytes seeded in nanofibers. Bar: 25 μm. (Adapted with permission from Li W. et al. *Tissue Eng.* 12(7): 1775–1785, 2006.)

These studies thus demonstrate that nanofibers promote the acquisition of a more *in vivo*-like cell morphology and cytoskeletal organization. Taken together, both nanotopographical features on 2-D substrates and 3-D nanofibrous environments can be used to promote the formation of more *in vivo*-like adhesion complexes and cytoskeletal organization, likely through mimicry of the physical structure of a cell's native microenvironment.

13.2.3 Regulation of Cell Migration by Nanostructures

Because FA formation and cytoskeletal organization have both been implicated in modulating intracellular signaling, understanding how to use nanostructures to control the formation and dynamics of these features will ultimately allow researchers to control complex cell fate processes. For instance, FAs and the actin cytoskeleton play a critical role in cell migration. Cell migration is a highly coordinated process involving the dynamic assembly and disassembly of FAs, as well as continual membrane protrusion and retraction (Lauffenburger and Horwitz 1996). Actin polarization and polymerization drive membrane protrusion and retraction through regulation of the Rho GTPases RhoA, Rac, and Cdc42 (Nobes and Hall 1995). At the leading edge, Rac mediates the formation of lamellipodia—actin projections that drive membrane protrusions outward (Mitra et al. 2005). Lamellipodia formation requires the activation of FAK and is coordinated with the continuous formation of transient focal contacts (Hall 1998). Importantly, the formation of stable FAs slows migration (Gupton and Waterman-Storer 2006). Thus, nanostructures can play a pivotal role in modulating cell migration through the control of FA dynamics.

Cavalcanti-Adam et al. (2007) characterized the effect of nanoscale variation in the spacing between adhesive RGD ligands on FA dynamics and cell migration, and found that the formation of FAs is less persistent and undergoes rapid turnover in cells plated on the 108-nm-spaced pattern than on the 58-nm-spaced pattern. The apparent instability of FAs in cells on the 108-nm-spaced pattern, in turn, resulted in increased cell polarization and enhanced cell migration, suggesting that nanostructures that control FA dynamics may promote cell migration. Furthermore, a dynamic actin cytoskeleton is required to extend lamellipodia at the leading edge. Thus, the formation of stress fibers would hinder migration. Indeed, Cavalcanti-Adam et al. (2007) did not observe stress fibers on the 108-nm-spaced pattern, whereas stress fibers were abundant in cells cultured on the 58-nm-spaced pattern.

Aligned nanostructures have also been shown to enhance cell migration. For instance, Patel et al. (2007) observed enhanced dermal fibroblast migration, as well as neurite outgrowth, on aligned nanofibers compared with randomly oriented fibers during wound healing. Furthermore, the rate of stem cell migration was shown to be higher on nanofibrous scaffolds compared with 2-D cultures. Rahjouei et al. (2011) showed that neural progenitors derived from hESCs displayed an increased rate of migration and neurite extension on nanofibers compared with 2-D plastic culture. Overall, these studies demonstrate the ability of nanostructures to control cell migration through the regulation of FA and actin cytoskeleton dynamics by varying the spacing and alignment of nanostructures.

13.2.4 Regulation of the Stem Cell Self-Renewal and Maintenance of Potency by Nanostructures

The response of various mature cell types to nanostructures in terms of cell morphology, FA formation, and cytoskeletal organization is observed in stem cells, which have the additional capabilities of extensive self-renewal and multilineage differentiation. However, as stem cells proliferate *in vitro*, their self-renewal may be slowed or halted by two phenomena: spontaneous differentiation and senescence. Spontaneous differentiation refers to the uncontrolled differentiation of stem cells into undesired mature cell types and can happen to both embryonic and adult stem cells. Senescence refers to the loss of a cell's ability to proliferate and is primarily a problem encountered by adult stem cells as they are progressively cultured in 2-D culture systems *in vitro*. Senescence also leads to the loss of potency in adult stem cells, essentially exhausting a stem cell of its "stemness." Interestingly, as stem cells undergo senescence, they begin to exhibit actin stress fibers, suggesting a connection between the actin cytoskeleton and stem cell self-renewal. Furthermore, given the ability of nanostructures to regulate actin cytoskeletal organization, this observation suggests that nanostructures may provide a promising avenue to prevent senescence. Indeed, nanostructures have recently been shown to promote stem cell self-renewal and preserve their potency, potentially providing state-of-the-art culture systems for the *ex vivo* expansion of stem cells. The following section will explore the regulation of stem cell self-renewal and potency in nanobiomaterials.

Topographical features have been shown to modulate stem cell self-renewal and, not surprisingly, the effect on self-renewal seems to be largely dependent on nanotopographical alterations in actin cytoskeletal organization. For instance, the proliferation of hESCs and murine ESCs (mESCs) was reduced in response to nanogrooves of 600 nm width and depth (Gerecht et al. 2007) and in micropillars with diameters of 5 μm (Biehl et al. 2009), respectively. Furthermore, proliferation of mESC-derived neural precursors was inversely correlated with surface roughness in the range of 0 nm to 30 nm (Bakeine et al. 2009). Conversely, nanoscale grooves on 2-D substrates of varying pitch from 400 nm up to 4 μm were shown to promote hESC self-renewal in the presence of a feeder layer (McFarlin et al. 2009). Importantly, cells showing reduced proliferation in response to nanogrooves and micropillars displayed highly organized actin stress fibers, whereas cells showing enhanced proliferation displayed a disorganized actin cytoskeleton. Furthermore, the reduction in proliferation caused by nanogrooves/ridges was attenuated by the addition of cytochalasin D (Gerecht et al. 2007), whereas the inhibitory effect of microscale topography was reversed by blocking Rho activity (Biehl et al. 2009), indicating that actin cytoskeletal organization and tension play an important role in mediating the effect of topographical cues on the proliferation of ESCs. Taken together, these data suggest that nanotopographical features that prevent stress fiber formation promote stem cell self-renewal.

Nanofibrous scaffolds have consistently been shown to enhance stem cell self-renewal, as well, and Nur-E-Kamal et al. (2006) found that the enhanced proliferation and self-renewal of mESCs on nanofibrillar surfaces were correlated with the activation of the small GTPase Rac and the phosphoinositide 3-kinase (PI3K) pathway compared with 2-D surfaces without nanofibers. Indeed, sustained activation of Rac led to cytoskeletal reorganization and increased cell proliferation (Nur-E-Kamal et al. 2005), suggesting that Rac is an important signaling molecule in directing stem cell self-renewal in 3-D nanofibrous culture. Interestingly, the 3-D nanofibrillar scaffolds also enhanced the expression of *Nanog*, a transcription factor required for the maintenance of pluripotency, and inhibition of the PI3K pathway reduced the expression of *Nanog*, suggesting that nanofibers may enhance stem cell self-renewal and potency through activation of the PI3K pathway. Nanofibrous scaffolds have also been shown to upregulate the expression of *Oct3/4* and *Sox2*, two other transcription factors necessary for maintaining pluripotency in hMSCs (Barber et al. 2011), demonstrating the ability of nanobiomaterials to promote stem cell potency in a variety of stem cell types.

Similarly, neural progenitor cells have been shown to increase their rate of proliferation when cultured on collagen nanofibers compared with collagen-coated 2-D surfaces, and neural progenitor cells proliferated more quickly on aligned nanofibers than on randomly oriented nanofibers (Wang et al. 2011). Interestingly, blocking the β1 integrin with an anti-β1 integrin antibody, or incubation of neural progenitor cells with a pharmacological inhibitor of the MAPK pathway, eliminated the enhanced rate of proliferation observed on the nanofibers. Furthermore, cyclin D1 and cyclin-dependent kinase 2, regulators of cell cycle progression downstream of β1 integrin/MAPK signaling, were correspondingly affected, suggesting that nanobiomaterials regulate stem cell proliferation by promoting cell cycle progression through β1 integrin/MAPK signaling. Taken together with the work by Nur-E-Kamal et al., it is likely that multiple signaling pathways are concomitantly regulated in stem cells and function cooperatively to enhance self-renewal in response to nanobiomaterials.

In addition to promoting self-renewal, nanofibrous scaffolds have been shown to maintain the potency of *ex vivo*–expanded stem cells. Mauney et al. (2004) showed that collagen type I nanofibrillar matrices can preserve the osteogenic (Mauney et al. 2004) and adipogenic (Mauney et al. 2005) potentials of hMSCs compared with conventional 2-D tissue culture plastic substrates. Although the mechanism underlying the retention of differentiation potential remains to be fully elucidated, they found that expansion on nanofibrillar matrices significantly reduced a prominent manifestation of cellular aging, the expression of the protective stress response protein HSP70 (Mauney et al. 2005). Interestingly, the reduction in HSP70 expression was accompanied by an increase in proliferation and reduction in morphological changes that signify cellular senescence, such as extensive cellular spreading and formation of actin stress fibers. Given the potent ability of nanostructures to prevent stress fiber formation, it would be interesting to examine whether the enhanced proliferation and maintenance of potency in response to nanostructures is a direct effect of modulated actin cytoskeletal organization.

13.2.5 Regulation of Stem Cell Differentiation in Nanobiomaterials

The ability of nanobiomaterials to provide a biomimetic microenvironment capable of promoting stem cell self-renewal while maintaining their differentiation potential demonstrates their promising role in attaining a large quantity of quality stem cells for tissue engineering and regenerative medicine applications. Although invaluable, it would be ideal for a tissue engineering construct to also promote stem cell differentiation into the desired mature cell type(s) while being chemically induced. Together, a single nanobiomaterial would thus drive the process of tissue regeneration *ex vivo*. The following section will explore the regulation of stem cell differentiation in nanobiomaterials.

Differentiation is a complex process whereby stem cells gradually obtain a mature cellular phenotype. Many interactions ultimately dictate the success of differentiation, and indeed, FA formation and actin cytoskeletal organization play critical roles in the process (Hart et al. 2007). Thus, nanostructures have been shown to regulate stem cell differentiation into specific cell types depending on the degree of FA formation and cytoskeletal organization. In general, FA maturation and subsequent activation of FAK act to regulate various signaling pathways, most notably the MAPK pathway, and cytoskeletal organization to influence differentiation in response to nanostructures. For example, the actin cytoskeleton has long since been known to have an instrumental role in driving hMSC lineage commitment. One of the earliest reports of the actin cytoskeleton controlling differentiation was back in 1966 when Abbott and Holtzer found that chondrocytes lost their differentiated phenotype when isolated and expanded *in vitro* in 2-D tissue culture plastics (Abbott and Holtzer 1966). Chondrocyte “dedifferentiation” was marked by cellular spreading, the appearance of actin stress fibers, and a concomitant loss in chondrocyte-related gene expression. Since then, the actin cytoskeleton has been shown more directly to regulate stem cell differentiation. For instance, disrupting the actin cytoskeleton with the actin depolymerizing drug, cytochalasin D, markedly enhanced the differentiation of chondroprogenitor cells cultured in 2-D culture (Zanetti and Solursh 1984).

Perhaps the most studied example of nanobiomaterials promoting differentiation is the induction of osteogenesis by using nanotopography. A variety of nanotopographical features have been used to modulate osteogenic differentiation, and the effectiveness of differentiation is strongly correlated to the resulting cellular morphology. In general, any nanofeature that increases cellular spreading, FA and stress fiber formation, and subsequent activation of FAK promotes osteogenic differentiation, whereas a reduction in cellular adhesion and subsequent deactivation of FAK induces adipogenic differentiation (Rodriguez Fernandez 1989). Oh et al. (2009) demonstrated the ability to regulate hMSC differentiation simply by altering the areal size of the nanosurface presented to the cell. Human MSCs were seeded onto highly ordered titania nanotubes with diameters ranging from 30 nm to 100 nm. hMSCs seeded on larger nanotubes displayed higher expression of osteogenic markers compared with cells seeded on smaller nanotubes. The upregulation in expression of osteogenic markers likely resulted from the almost tenfold increase in cell elongation observed on the 100 nm nanotubes (Oh et al. 2009). The elongated morphology resulted in higher actin cytoskeletal tension, which, in turn, promoted osteogenic differentiation without any requirement for chemical induction cues.

In another study, Sjöström et al. (2009) cultured hMSCs on 15 nm titania nanopillar structures and found that they displayed a highly organized cytoskeleton, large FA complexes, and significantly increased cell areas (Figure 13.4b and d) compared

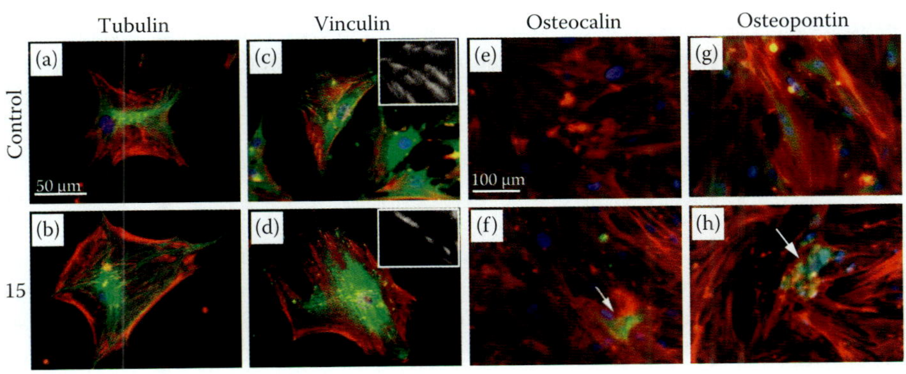

FIGURE 13.4 Cytoskeleton and FA staining in hMSCs cultured on a flat control surface and 15 nm nanopillar surface. (a)–(d) Actin microfilaments (red), tubulin microtubules, or vinculin focal contacts (green), and nucleus (blue). The 15-nm nanopillar surface supports a well-spread morphology with a highly organized cytoskeleton and larger FAs compared with controls. Inset shows FA sites. (e) and (f) Osteocalcin and (g) and (h) osteopontin immunohistochemical staining on control and 15 nm titanium surfaces. Actin microfilaments (red), osteocalcin or osteopontin (green), and nucleus (blue). The arrows show sites of bone matrix nodule formation on the 15 nm nanopillar surface, not seen on the control surface. Bar: 50 μm (a)–(d); 100 μm (e)–(h). (Adapted from Sjöström T. et al. *Acta Biomater* 5(5): 1433–1441, 2009. With permission.)

with cells seeded on smooth surface controls (Figure 13.4a and c). Indeed, only those cells displaying a well-spread morphology with large FAs stained positive for the osteoblast markers osteocalcin and osteopontin (Figure 13.4f and h). It was concluded that the formation of larger and more numerous FAs is critical for osteogenic differentiation, because larger adhesions are linked to increased mechanotransductive signaling and cytoskeletal tension (Biggs et al. 2008; Dalby et al. 2007a,b). Lipski et al. (2008) modified biomaterial surfaces with nanoparticles to produce nanoroughness for subsequent examination of osteoblast progenitor differentiation. They found that cell spreading and FA formation increased with nanoparticle size due to the increased surface area in contact with the cell membrane. The same research group demonstrated that the addition of nanoparticles on the surface of titanium significantly increased the expression of bone sialoprotein, a critical protein for bone mineralization, in hMSCs compared with nonmodified titanium surfaces, demonstrating the ability to regulate differentiation through actin cytoskeletal arrangement based on surface roughness alone (Lipski et al. 2007). Additionally, canonical Wnt signaling may be playing a role in promoting osteogenic differentiation in response to nanoroughness (Galli et al. 2010). Taken together, these studies highlight the ability of nanotopographical cues to regulate differentiation through modulation of cell spreading, FA formation, and actin cytoskeletal organization. Thus, a nanosurface engineered to maximize cell spreading would be optimal for bone tissue engineering applications (Dalby et al. 2006a,b).

More recent evidence supports the notion that nanotopography can direct stem cells to differentiate into specific lineages. For example, synthetic nanostructures have been used to direct neurogenesis of hMSCs without induction medium (Yim et al. 2007). In one such study, hMSCs were cultured on 350-nm-, 1-μm-, and 10-μm-wide gratings and compared for cell morphology and neurogenic differentiation. Cells cultured on the nanograting aligned along the grating axis and had a greater number of extended cell protrusions. These nanosurface-dependent cell behaviors facilitated neurogenic lineage commitment, as mature neuronal markers, such as MAP2 and β-tubulin III, and the synapse marker synaptophysin, were highly expressed in stem cells cultured on nanosurfaced substrates. Moreover, upregulation of adhesion molecule signaling was observed, suggesting that the induction of neuronal differentiation may be associated with changes in FA development.

Nanofibrous scaffolds have likewise been shown to support hMSC differentiation into the chondrogenic, adipogenic, and osteogenic lineages (Li et al. 2005). However, directing the differentiation of hMSCs into specific lineages using nanofibrous scaffolds is not well understood. Differentiation can be driven into other lineages, though, simply by adding another degree of structural complexity. Instead of fabricating scaffolds composed of randomly oriented nanofibers, constructs that have nanofibers aligned along one axis can be produced. In doing so, seeded stem cells preferentially align along the axis of the nanofibers, and this is accompanied by an alignment of their actin cytoskeletons. Interestingly, aligned nanofibers have been shown to induce a significant upregulation in the expression of tendon-specific genes compared with that on randomly oriented nanofibers (Yin et al. 2010). Furthermore, aligned nanofibrous scaffolds seem capable of discouraging osteogenic differentiation. Although randomly oriented nanofibrous scaffolds induced alkaline phosphatase activity and calcified matrix accumulation (positive markers of bone formation) in the presence of osteogenic supplements, the aligned scaffolds showed no signs of osteogenic differentiation.

Aligned nanofibers have also been shown to promote neuronal differentiation from a variety of stem cell types. For example, Mahairaki et al. (2011) cultured hESC-derived neural precursors on polycaprolactone fiber matrices of different diameters and orientations. They found that aligned fibrous scaffolds yielded a higher percentage of cells committed to neuronal differentiation than randomly oriented scaffolds by facilitating the acquisition of a polarized cell morphology with processes extending along the axis of the fiber. Furthermore, nanofibers were more effective at promoting neuronal differentiation than microfibers in the presence of retinoic acid, an inducer of neurogenic differentiation. Additionally, Lim et al. (2007) found that aligned nanofibers induced the alignment of adult neural stem cells, and upon induction with retinoic acid, promoted the neuronal differentiation of a higher fraction of cells compared with cells cultured in randomly oriented nanofibrous meshes. Interestingly, the aligned nanofiber-induced changes in cell morphology led to an upregulation in canonical Wnt signaling, and treatment with retinoic acid further potentiated Wnt signaling, thus providing a potential mechanism by which changes in stem cell morphology work cooperatively with biochemical cues to drive differentiation into specific lineages. Although further work is needed to elucidate the underlying mechanisms translating changes in cell morphology with directed differentiation, one possible theory is that changes in cell morphology lead to changes in cytoskeletal organization, which is, in turn, transmitted to the nucleus through specific cytoskeletal-based signaling pathways or changes in chromatin packaging.

13.3 CONCLUSION AND FUTURE DIRECTIONS

Stem cells are an indispensable tool for tissue engineers, given their extensive proliferative capacity and ability to differentiate into multiple mature cell types. This is especially important as researchers strive to regenerate more complex tissues consisting of multiple cell types. Many factors, including cell–cell, cell–ECM, and cell–soluble factor signaling, ultimately dictate stem cell fate decision. However, future advances must be made to more precisely control stem cell behavior *in vitro*. Nanobiomaterials hold great promise for tissue engineering and regenerative medicine applications because of their ability to mimic key features of native ECM proteins. As a result, cells recognize these nanofeatures and respond in an *in vivo*-like manner. That is, cells are able to acquire a morphology, FA maturity and stability, and cytoskeletal organization reminiscent of that found *in vivo*. Cell morphology and, more specifically, the degree of cellular spreading seems to be dictated by the overall surface area, whereas FA formation and actin cytoskeletal organization tend to be regulated by the dimensionality and spacing of underlying nanofeatures. Through the control of these parameters, researchers therefore have the ability to regulate cell migration, proliferation, and differentiation in a reproducible fashion.

As the field advances, the next step will be to design nanobiomaterials capable of controlling the spatiotemporal presentation of important regulatory signals. For instance, a single nanobiomaterial could instruct a stem cell population to first proliferate, and then differentiate into the desired cell type(s) through temporal changes in ligand presentation, thus mimicking normal tissue morphogenesis during development. Furthermore, a nanobiomaterial could drive different cell fate decisions based on location within the construct. For example, a biphasic construct could be fabricated in which the bottom portion drives osteogenesis whereas the top portion induces chondrogenesis, thus producing an osteochondral plug resembling that found at the ends of long bones. However, our understanding of how stem cells will respond to specific nanofeatures must advance for these "smart" biomaterials to become a reality. Only then will researchers be able to predictably control cell behavior and thus successfully orchestrate tissue regeneration *ex vivo.*

REFERENCES

Abbott, J., and H. Holtzer. 1966. The loss of phenotypic traits by differentiated cells—Part 3: The reversible behavior of chondrocytes in primary cultures. *Journal of Cell Biology* 28: 473.

Arnold, M., E. A. Cavalcanti-Adam, R. Glass, J. Blummel, W. Eck, M. Kantlehner, H. Kessler, and J. P. Spatz. 2004. Activation of integrin function by nanopatterned adhesive interfaces. *ChemPhysChem* 5(3): 383–388.

Bakeine, G. J., J. Ban, G. Grenci, A. Pozzato, S. Dal Zilio, M. Prasciolu, L. Businaro, M. Tormen, and M. E. Ruaro. 2009. Design, fabrication and evaluation of nanoscale surface topography as a tool in directing differentiation and organisation of embryonic stem cell-derived neural precursors. *Microelectronic Engineering* 86: 1435–1438.

Barber, J. G., A. M. Handorf, T. J. Allee, and W. J. Li. 2011. Braided nanofibrous scaffold for tendon and ligament tissue engineering. *Tissue Engineering Part A.* (in press).

Beltrami, A. P., L. Barlucchi, D. Torella, M. Baker, F. Limana, S. Chimenti, H. Kasahara et al. 2003. Adult cardiac stem cells are multipotent and support myocardial regeneration. *Cell* 114 (6): 763–776.

Biehl, J. K., S. Yamanaka, T. A. Desai, K. R. Boheler, and B. Russell. 2009. Proliferation of mouse embryonic stem cell progeny and the spontaneous contractile activity of cardiomyocytes are affected by microtopography. *Developmental Dynamics* 238 (8): 1964–1973.

Biggs, M. J., R. G. Richards, N. Gadegaard, R. J. McMurray, S. Affrossman, C. D. Wilkinson, R. O. Oreffo, and M. J. Dalby. 2009. Interactions with nanoscale topography: Adhesion quantification and signal transduction in cells of osteogenic and multipotent lineage. *Journal of Biomedical Materials Research Part A* 91(1): 195–208.

Biggs, M. J., R. G. Richards, S. McFarlane, C. D. Wilkinson, R. O. Oreffo, and M. J. Dalby. 2008. Adhesion formation of primary human osteoblasts and the functional response of mesenchymal stem cells to 330-nm-deep microgrooves. *Journal of the Royal Society Interface* 5(27): 1231–1242.

Brammer, K. S., S. Oh, J. O. Gallagher, and S. Jin. 2008. Enhanced cellular mobility guided by TiO_2 nanotube surfaces. *Nano Letters* 8(3): 786–793.

Brown, S. E., W. Tong, and P. H. Krebsbach. 2009. The derivation of mesenchymal stem cells from human embryonic stem cells. *Cells Tissues Organs* 189(1–4): 256–260.

Bruder, S. P., N. Jaiswal, S. E. Haynesworth. 1997. Growth kinetics, self-renewal, and the osteogenic potential of purified human mesenchymal stem cells during extensive subcultivation and following cryopreservation. *Journal of Cellular Biochemistry* 64: 278.

Byers, H. R., G. E. White, and K. Fujiwara. 1984. Organization and function of stress fibers in cells *in vitro* and *in situ*: A review. *Cell and Muscle Motility* 5: 83–137.

Cao, B., B. M. Deasy, J. Pollett, and J. Huard. 2005. Cell therapy for muscle regeneration and repair. *Physical Medicine and Rehabilitation Clinics of North America* 16(4): 889–907, viii.

Carrel, A., and M. Burrows. 1911. Culture *in vitro* of malignant tumors. *Journal of Experimental Medicine* 12: 571–575.

Cavalcanti-Adam, E. A., A. Micoulet, J. Blummel, J. Auernheimer, H. Kessler, and J. P. Spatz. 2006. Lateral spacing of integrin ligands influences cell spreading and focal adhesion assembly. *European Journal of Cell Biology* 85(3–4): 219–224.

Cavalcanti-Adam, E. A., T. Volberg, A. Micoulet, H. Kessler, B. Geiger, and J. P. Spatz. 2007. Cell spreading and focal adhesion dynamics are regulated by spacing of integrin ligands. *Biophysical Journal* 92(8): 2964–2974.

Chen, C. S., J. L. Alonso, E. Ostuni, G. M. Whitesides, and D. E. Ingber. 2003. Cell shape provides global control of focal adhesion assembly. *Biochemical and Biophysical Research Communications* 307(2): 355–361.

Chen, C. S., M. Mrksich, S. Huang, G. M. Whitesides, and D. E. Ingber. 1997. Geometric control of cell life and death. *Science* 276(5317): 1425–1428.

Chiu, L. L. Y., Z. Chu, and M. Radisic. 2011. Tissue engineering. In: *Comprehensive Nanoscience and Technology*, edited by L. A. David, D. S. Gregory, and P. W. Gary. Amsterdam: Academic Press.

Clark, P., P. Connolly, A. S. Curtis, J. A. Dow, and C. D. Wilkinson. 1990. Topographical control of cell behavior—Part II: Multiple grooved substrata. *Development* 108(4): 635–644.

Dalby, M. J., M. J. Biggs, N. Gadegaard, G. Kalna, C. D. Wilkinson, and A. S. Curtis. 2007a. Nanotopographical stimulation of mechanotransduction and changes in interphase centromere positioning. *Journal of Cellular Biochemistry* 100(2): 326–338.

Dalby, M. J., N. Gadegaard, P. Herzyk, D. Sutherland, H. Agheli, C. D. Wilkinson, and A. S. Curtis. 2007b. Nanomechanotransduction and interphase nuclear organization influence on genomic control. *Journal of Cellular Biochemistry* 102(5): 1234–1244.

Dalby, M. J., D. Giannaras, M. O. Riehle, N. Gadegaard, S. Affrossman, and A. S. Curtis. 2004. Rapid fibroblast adhesion to 27-nm-high polymer demixed nanotopography. *Biomaterials* 25(1): 77–83.

Dalby, M. J., D. McCloy, M. Robertson, H. Agheli, D. Sutherland, S. Affrossman, and R. O. Oreffo. 2006a. Osteoprogenitor response to semiordered and random nanotopographies. *Biomaterials* 27(15): 2980–2987.

Dalby, M. J., D. McCloy, M. Robertson, C. D. Wilkinson, and R. O. Oreffo. 2006b. Osteoprogenitor response to defined topographies with nanoscale depths. *Biomaterials* 27(8): 1306–1315.

Dalby, M. J., M. O. Riehle, H. J. H. Johnstone, S. Affrossman, and A. S. G. Curtis. 2002. Polymer-demixed nanotopography: control of fibroblast spreading and proliferation. *Tissue Engineering* 8(6): 1099–1108.

Dalby, M. J., M. O. Riehle, H. J. Johnstone, S. Affrossman, and A. S. Curtis. 2003. Nonadhesive nanotopography: fibroblast response to poly(n-butyl methacrylate)-poly(styrene) demixed surface features. *Journal of Biomedical Materials Research—Part A* 67(3): 1025–1032.

de Peppo, G. M., S. Svensson, M. Lenneras, J. Synnergren, J. Stenberg, R. Strehl, J. Hyllner, P. Thomsen, and C. Karlsson. 2010. Human embryonic mesodermal progenitors highly resemble human mesenchymal stem cells and display high potential for tissue engineering applications. *Tissue Engineering Part A* 16(7): 2161–2182.

Furth, M. E., and A. Atala. 2008. Current and future perspectives of regenerative medicine. In *Principles of Regenerative Medicine*. San Diego: Academic Press.

Gallagher, J. O., K. F. McGhee, C. D. Wilkinson, and M. O. Riehle. 2002. Interaction of animal cells with ordered nanotopography. *IEEE Transactions on Nanobioscience* 1(1): 24–28.

Galli, C., G. Passeri, F. Ravanetti, E. Elezi, M. Pedrazzoni, and G. M. Macaluso. 2010. Rough surface topography enhances the activation of Wnt/beta-catenin signaling in mesenchymal cells. *Journal of Biomedical Materials Research A* 95A(3): 682–690.

Gerecht, S., C. J. Bettinger, Z. Zhang, J. T. Borenstein, G. Vunjak-Novakovic, and R. Langer. 2007. The effect of actin disrupting agents on contact guidance of human embryonic stem cells. *Biomaterials* 28(28): 4068–4077.

Giancotti, F. G., and E. Ruoslahti. 1999. Integrin signaling. *Science* 285(5430): 1028–1032.

Grinnell, F. 1978. Cellular adhesiveness and extracellular substrata. *International Review of Cytology* 53: 65–144.

Gupton, S. L., and C. M. Waterman-Storer. 2006. Spatiotemporal feedback between actomyosin and focal-adhesion systems optimizes rapid cell migration. *Cell* 125(7): 1361–1374.

Hall, A. 1998. Rho GTPases and the actin cytoskeleton. *Science* 279(5350): 509–514.

Hart, A., N. Gadegaard, C. D. Wilkinson, R. O. Oreffo, and M. J. Dalby. 2007. Osteoprogenitor response to low-adhesion nanotopographies originally fabricated by electron beam lithography. *Journal of Materials Science. Materials in Medicine* 18(6): 1211–1218.

Hemler, M. E. 1990. VLA proteins in the integrin family: Structures, functions, and their role on leukocytes. *Annual Review of Immunology* 8: 365–400.

Hermann, P., M. Armant, E. Brown, M. Rubio, H. Ishihara, D. Ulrich, R. G. Caspary et al. 1999. The vitronectin receptor and its associated CD47 molecule mediates proinflammatory cytokine synthesis in human monocytes by interaction with soluble CD23. *Journal of Cell Biology* 144(4): 767–775.

Hildebrand, J. D., M. D. Schaller, and J. T. Parsons. 1993. Identification of sequences required for the efficient localization of the focal adhesion kinase, pp125(FAK), to cellular focal adhesions. *Journal of Cell Biology* 123(4): 993–1005.

Hsia, H. C., M. R. Nair, R. C. Mintz, and S. A. Corbett. 2011. The fiber diameter of synthetic bioresorbable extracellular matrix influences human fibroblast morphology and fibronectin matrix assembly. *Plastic and Reconstructive Surgery* 127(6): 2312–2320.

Humphries, M. J. 2000. Integrin structure. *Biochemical Society Transactions* 28(4): 311–339.

Jukes, J. M., L. Moroni, C. A. van Blitterswijk, and J. de Boer. 2008. Critical steps toward a tissue-engineered cartilage implant using embryonic stem cells. *Tissue Engineering Part A* 14(1): 135–147.

Katz, B. Z., E. Zamir, A. Bershadsky, Z. Kam, K. M. Yamada, and B. Geiger. 2000. Physical state of the extracellular matrix regulates the structure and molecular composition of cell–matrix adhesions. *Molecular Biology of the Cell* 11(3): 1047–1060.

Keller, G. 2005. Embryonic stem cell differentiation: Emergence of a new era in biology and medicine. *Genes and Development* 19(10): 1129–1155.

Klees, R. F., R. M. Salasznyk, K. Kingsley, W. A. Williams, A. Boskey, and G. E. Plopper. 2005. Laminin-5 induces osteogenic gene expression in human mesenchymal stem cells through an ERK-dependent pathway. *Molecular Biology of the Cell* 16(2): 881–890.

Kokubu, E., D. W. Hamilton, T. Inoue, and D. M. Brunette. 2009. Modulation of human gingival fibroblast adhesion, morphology, tyrosine phosphorylation, and ERK 1/2 localization on polished, grooved and SLA substratum topographies. *Journal of Biomedical Materials Research Part A* 91(3): 663–670.

Kolambkar, Y. M., A. Peister, A. K. Ekaputra, D. W. Hutmacher, and R. E. Guldberg. 2010. Colonization and osteogenic differentiation of different stem cell sources on electrospun nanofiber meshes. *Tissue Engineering Part A* 16(10): 3219–3230.

Lauffenburger, D. A., and A. F. Horwitz. 1996. Cell migration: A physically integrated molecular process. *Cell* 84(3): 359–369.

Levenberg, S., and R. Langer. 2004. Advances in tissue engineering. *Current Topics in Developmental Biology* 61: 113–134.

Li, W. J., Y. J. Jiang, and R. S. Tuan. 2006. Chondrocyte phenotype in engineered fibrous matrix is regulated by fiber size. *Tissue Engineering* 12(7): 1775–1785.

Li, W. J., R. Tuli, X. Huang, P. Laquerriere, and R. S. Tuan. 2005. Multilineage differentiation of human mesenchymal stem cells in a three-dimensional nanofibrous scaffold. *Biomaterials* 26(25): 5158–5166.

Lim, J. Y., A. D. Dreiss, Z. Zhou, J. C. Hansen, C. A. Siedlecki, R. W. Hengstebeck, J. Cheng, N. Winograd, and H. J. Donahue. 2007. The regulation of integrin-mediated osteoblast focal adhesion and focal adhesion kinase expression by nanoscale topography. *Biomaterials* 28(10): 1787–1797.

Lipski, A. M., C. Jaquiery, H. Choi, D. Eberli, M. Stevens, I. Martin, I. W. Chen, and V. P. Shastri. 2007. Nanoscale engineering of biomaterial surfaces. *Advanced Materials* 19: 553–557.

Lipski, A. M., C. Pino, F. Haselton, I. W. Chen, and V. P. Shastri. 2008. The effect of silica nanoparticle-modified surfaces on cell morphology, cytoskeletal organization and function. *Biomaterials* 29: 3836–3846.

Mahairaki, V., S. H. Lim, G. T. Christopherson, L. Xu, I. Nasonkin, C. Yu, H. Q. Mao, and V. E. Koliatsos. 2011. Nanofiber matrices promote the neuronal differentiation of human embryonic stem cell-derived neural precursors *in vitro*. *Tissue Engineering—Part A* 17(5–6): 855–863.

Massia, S. P., and J. A. Hubbell. 1991. An RGD spacing of 440 nm is sufficient for integrin alpha V beta 3-mediated fibroblast spreading and 140 nm for focal contact and stress fiber formation. *Journal of Cell Biology* 114(5): 1089–1100.

Mauney, J. R., D. L. Kaplan, and V. Volloch. 2004. Matrix-mediated retention of osteogenic differentiation potential by human adult bone marrow stromal cells during *ex vivo* expansion. *Biomaterials* 25(16): 3233–3243.

Mauney, J. R., V. Volloch, and D. L. Kaplan. 2005. Matrix-mediated retention of adipogenic differentiation potential by human adult bone marrow-derived mesenchymal stem cells during *ex vivo* expansion. *Biomaterials* 26(31): 6167–6175.

McFarlin, D. R., K. J. Finn, P. F. Nealey, and C. J. Murphy. 2009. Nanoscale topography modulates human embryonic stem cell self-renewal. *Journal of Biomimetics, Biomaterials, and Tissue Engineering* 2:15–26.

McGowan Institute for Regenerative Medicine. n.d. Available from http://www.mirm.pitt.edu/.

Mitra, S. K., D. A. Hanson, and D. D. Schlaepfer. 2005. Focal adhesion kinase: In command and control of cell motility. *Nature Reviews. Molecular Cell Biology* 6(1): 56–68.

Nobes, C. D., and A. Hall. 1995. Rho, rac, and cdc42 GTPases regulate the assembly of multimolecular focal complexes associated with actin stress fibers, lamellipodia, and filopodia. *Cell* 81(1): 53–62.

Nur-E-Kamal, A., I. Ahmed, J. Kamal, M. Schindler, and S. Meiners. 2005. Three-dimensional nanofibrillar surfaces induce activation of Rac. *Biochemical and Biophysical Research Communications* 331(2): 428–434.

Nur-E-Kamal, A., I. Ahmed, J. Kamal, M. Schindler, and S. Meiners. 2006. Three-dimensional nanofibrillar surfaces promote self-renewal in mouse embryonic stem cells. *Stem Cells* 24(2): 426–433.

Oh, S., K. S. Brammer, Y. S. J. Li, D. Teng, A. J. Engler, S. Chien, and S. Jin. 2009. Stem cell fate dictated solely by altered nanotube dimension. *Proceedings of the National Academy of Sciences of the United States of America* 106(7): 2130–2135.

Patel, S., K. Kurpinski, R. Quigley, H. Gao, B. S. Hsiao, M. M. Poo, and S. Li. 2007. Bioactive nanofibers: Synergistic effects of nanotopography and chemical signaling on cell guidance. *Nano Letters* 7(7): 2122–2128.

Pennisi, C. P., A. Dolatshahi-Pirouz, M. Foss, J. Chevallier, T. Fink, V. Zachar, F. Besenbacher, and K. Yoshida. 2011. Nanoscale topography reduces fibroblast growth, focal adhesion size and migration-related gene expression on platinum surfaces. *Colloids and Surfaces. B, Biointerfaces* 85(2): 189–197.

Rahjouei, A., S. Kiani, A. Zahabi, N. Z. Mehrjardi, M. Hashemi, and H. Baharvand. 2011. Interactions of human embryonic stem cell-derived neural progenitors with an electrospun nanofibrillar surface *in vitro*. *The International Journal of Artificial Organs* 34(7): 559–570.

Ravichandran, R., S. Liao, C. Ch Ng, C. K. Chan, M. Raghunath, and S. Ramakrishna. 2009. Effects of nanotopography on stem cell phenotypes. *World Journal of Stem Cells* 1(1): 55–66.

Rodriguez Fernandez, J. L., and A. Ben-Ze'ev. 1989. Regulation of fibronectin, integrin and cytoskeleton expression in differentiating adipocytes: Inhibition by extracellular matrix and polylysine. *Differentiation* 42(2): 65–74.

Schaller, M. D., C. A. Borgman, B. S. Cobb, R. R. Vines, A. B. Reynolds, and J. T. Parsons. 1992. pp125FAK a structurally distinctive protein-tyrosine kinase associated with focal adhesions. *Proceedings of the National Academy of Sciences of the United States of America* 89(11): 5192–5196.

Schaller, M. D., Y. Shen, J. M. Broome, T. Sasaki, and M. A. Cooley. 1996. pp125(FAK) signaling: Role of the C-terminus. *Molecular Biology of the Cell* 7: 3422–3422.

Shih, Y. R., C. N. Chen, S. W. Tsai, Y. J. Wang, and O. K. Lee. 2006. Growth of mesenchymal stem cells on electrospun type I collagen nanofibers. *Stem Cells* 24(11): 2391–2397.

Sjöström, T., M. J. Dalby, A. Hart, R. Tare, R. O. Oreffo, and B. Su. 2009. Fabrication of pillar-like titania nanostructures on titanium and their interactions with human skeletal stem cells. *Acta Biomaterialia* 5(5): 1433–1441.

Small, J. V., K. Rottner, I. Kaverina, and K. I. Anderson. 1998. Assembling an actin cytoskeleton for cell attachment and movement. *Biochimica et Biophysica Acta* 1404(3): 271–281.

Teixeira, A. I., G. A. Abrams, P. J. Bertics, C. J. Murphy, and P. F. Nealey. 2003. Epithelial contact guidance on well-defined micro- and nano-structured substrates. *Journal of Cell Science* 116(Pt 10): 1881–1892.

Thapa, A., D. C. Miller, T. J. Webster, and K. M. Haberstroh. 2003. Nanostructured polymers enhance bladder smooth muscle cell function. *Biomaterials* 24(17): 2915–2926.

Thomson, J. A., J. Itskovitz-Eldor, S. S. Shapiro, M. A. Waknitz, J. J. Swiergiel, V. S. Marshall, and J. M. Jones. 1998. Embryonic stem cell lines derived from human blastocysts. *Science* 282(5391): 1145–1147.

Tian, F., H. Hosseinkhani, M. Hosseinkhani, A. Khademhosseini, Y. Yokoyama, G. G. Estrada, and H. Kobayashi. 2008. Quantitative analysis of cell adhesion on aligned micro- and nano-fibers. *Journal of Biomedical Materials Research Part A* 84(2): 291–299.

van der Sanden, B., M. Dhobb, F. Berger, and D. Wion. 2010. Optimizing stem cell culture. *Journal of Cellular Biochemistry* 111(4): 801–807.

Wakitani, S., K. Takaoka, T. Hattori, N. Miyazawa, T. Iwanaga, S. Takeda, T. K. Watanabe, and A. Tanigami. 2003. Embryonic stem cells injected into the mouse knee joint form teratomas and subsequently destroy the joint. *Rheumatology (Oxford)* 42(1): 162–165.

Wan, Y., Y. Wang, Z. Liu, X. Qu, B. Han, J. Bei, and S. Wang. 2005. Adhesion and proliferation of OCT-1 osteoblast-like cells on micro- and nano-scale topography structured poly(L-lactide). *Biomaterials* 26(21): 4453–4459.

Wang, F., and J. Guan. 2010. Cellular cardiomyoplasty and cardiac tissue engineering for myocardial therapy. *Advanced Drug Delivery Reviews* 62(7–8): 784–797.

Wang, Y., M. Yao, J. Zhou, W. Zheng, C. Zhou, D. Dong, Y. Liu et al. 2011. The promotion of neural progenitor cells proliferation by aligned and randomly oriented collagen nanofibers through beta1 integrin/MAPK signaling pathway. *Biomaterials* 32(28): 6737–6744.

Webster, T. J., C. Ergun, R. H. Doremus, R. W. Siegel, and R. Bizios. 2000. Specific proteins mediate enhanced osteoblast adhesion on nanophase ceramics. *Journal of Biomedical Materials Research* 51(3): 475–483.

Williams, D. F. 2001. Tissue engineering concepts. In *Encyclopedia of Materials: Science and Technology*, Buschow, K.H.J., Cahn R.W., Flemings, M.C., Ilschner, B., Kramer, E.J., Mahajan, S., Veyssiere, P., editors. Oxford: Elsevier.

Woo, K. M., V. J. Chen, and P. X. Ma. 2003. Nano-fibrous scaffolding architecture selectively enhances protein adsorption contributing to cell attachment. *Journal of Biomedical Materials Research Part A* 67(2): 531–537.

Yim, E. K., S. W. Pang, and K. W. Leong. 2007. Synthetic nanostructures inducing differentiation of human mesenchymal stem cells into neuronal lineage. *Experimental Cell Research* 313(9): 1820–1829.

Yin, Z., X. Chen, J. L. Chen, W. L. Shen, T. M. Hieu Nguyen, L. Gao, and H. W. Ouyang. 2010. The regulation of tendon stem cell differentiation by the alignment of nanofibers. *Biomaterials* 31(8): 2163–2175.

Zaidel-Bar, R., M. Cohen, L. Addadi, and B. Geiger. 2004. Hierarchical assembly of cell–matrix adhesion complexes. *Biochemical Society Transactions* 32(Pt3): 416–420.

Zanetti, N. C., and M. Solursh. 1984. Induction of chondrogenesis in limb mesenchymal cultures by disruption of the actin cytoskeleton. *Journal of Cell Biology* 99(1 Pt 1): 115–123.

Zuk, P. A., M. Zhu, P. Ashijian, D. A. De Ugarte, J. I. Huang, H. Mizuno, Z. C. Alfonso, J. K. Fraser, P. Benhaim, and M. H. Hedrick. 2002. Human adipose tissue is a source of multipotent stem cells. *Molecular and Cellular Biology* 13: 4279.

14 Biomaterials and Their Use in Craniofacial Surgery

Kurt Laedrach

CONTENTS

14.1 INTRODUCTION

The objective of craniofacial surgery is to treat patients presenting with congenital and acquired deformities (resulting from birth defects, trauma, or tumor surgery) affecting the hard and soft tissues of the head and face. The best possible functional and aesthetic outcome is essential for the successful reintegration of these individuals. In cases of extensive soft tissue and bone loss, surgery to repair missing structures using the patients' own local tissue does not always obtain the desired results. The increasing knowledge and safety of microsurgical procedures has opened new frontiers in craniofacial reconstruction. However, although autologous tissues are generally preferred, advances in medical technology nowadays allow the application of reliable synthetic materials as an attractive alternative.

Biocompatibility is the ability of a material to perform with an appropriate host response in a specific application (Williams' definition). Implemented in craniofacial surgery, natural or synthetic materials are used to treat, enhance, augment, or replace living tissue or function in the facial region. The ideal biomaterial should provide the following characteristics: it should be mechanically stable, chemically inert, easily contourable, able to maintain shape and volume, incapable of producing any foreign body reaction or tissue hypersensitivity, noncarcinogenic, and replaced or ingrown by the surrounding living tissue or completely resorbed over time (if designed to do so). An optimal biomaterial that meets all these requirements probably does not exist.

Two major groups of biomaterials can be differentiated for craniofacial applications: osteosynthetic materials applied in traumatic fractures or osteotomies and materials as a substitute or augmentation of bony defects.

This chapter reviews our experience in the clinical use of the most important biomaterials, such as metal implants, biodegradable plates, bone substitutes, and augmentation materials, including titanium, polylactides, hydroxyapatite cements, bioactive glass, and polymer products—all with particular emphasis on craniofacial reconstruction.

14.2 INTERNAL FIXATION

14.2.1 Metallic Implants

The first metal alloy specifically developed for the stabilization of bone fractures in humans was vanadium steel in the early 1900s. However, complications such as osteomyelitis, early implant loss, and mechanical failures due to corrosion and poor biocompatibility were encountered. As a result of advances in technology, selection, and design of new biomedical materials, stainless steel implants are rarely used for craniofacial indications today.

14.2.1.1 Titanium and Titanium Alloys

The literature on the biological reaction to titanium is extensive. Because of a thin surface oxide layer, it is able to interact with proteins and cells. The most favorable characteristics of titanium include its corrosion resistance, nontoxic state, and biocompatibility, paired with its high strength-to-weight ratio and low elasticity modulus that is close to that of human bone.[1] The capacity for osseointegration—the formation of a direct contact with the bone without interposed soft tissue—makes it incredibly durable and long-lasting when used *in vivo*.[2] Pure titanium contains only trace amounts of other elements whereas, for most applications, titanium is alloyed with a small percentage (4%–6%) of other materials (aluminum, vanadium, molybdenum, zirconium, or iron), achieving favorable mechanical properties and superior biocompatibility.

In 1984, Leibinger (Howmedica Leibinger GmbH, Freiburg, Germany) was the first company worldwide to introduce a new implant system (made of commercially pure titanium) for maxillofacial surgery. The Würzburg System was revolutionary, with more than 20 plate variations, self-tapping bone screws of 2.0 mm, and specially designed instruments. Nowadays, newer systems still share the same principle of a customizable modular concept, in which the surgeon can choose from a broad range of plates, meshes, and screws in different profiles and flexibilities, covering every known major indication for bone fixation for all cranio-maxillofacial fractures, orthognathic, and reconstructive needs (Figure 14.1). Surface-anodized titanium is available in a wide spectrum of colors, allowing coding for implant and instrument identification purposes (Figure 14.2).

Orbital fractures represent some of the most common fractures of the midface. Depending on the severity, in particular, the presence of globe motility distraints, we recommend early surgical treatment. The aim is to restore the anatomical orbital cone according to its original volume and internal shape to achieve normal eye globe position. Trapped soft tissue must be lifted from the fracture line. Incorrect reduction

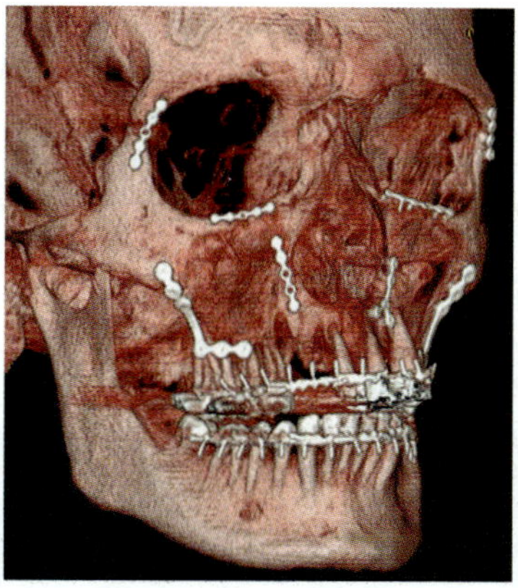

FIGURE 14.1 Postoperative 3-D CT scan with titanium miniplates and microplates for midface Le Fort III fracture fixation: dental arch bars for maxillo-mandibular *in situ*.

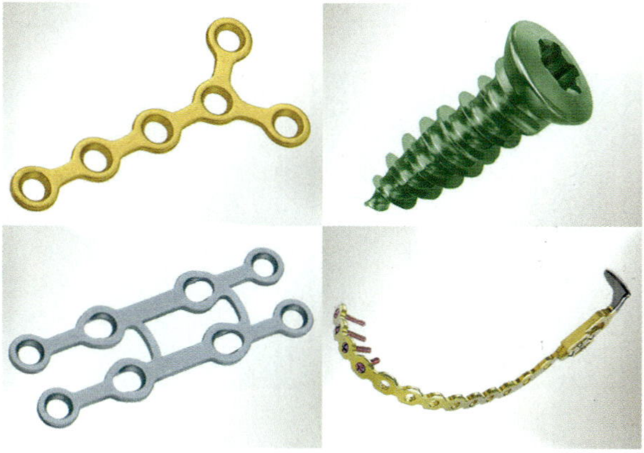

FIGURE 14.2 Various Medartis® (Basel, Switzerland) titanium implants for different indications in a wide spectrum of colors, allowing coding for implant identification.

of the orbit may lead to severe complications such as persistent diplopia, hypoglobus, or enophthalmos. Specially designed three-dimensional (3-D) titanium meshes are easily malleable, maintaining adequate rigidity for bridging of defects bigger than 1.5 cm^2 in nonload-bearing areas (Figure 14.3). However, reconstruction of the anteroposterior inclination of the medial aspect of the orbital floor is a challenging task. Sterilizable anatomical midface models to preform the 3-D meshes or computed tomography (CT)–based computer plannings with mirroring of the unaffected side onto the affected fracture side, including intraoperative control of the virtual contours, allow better adjustment of the implants. Because these plates are very flexible, undesired shaping or disalignment with consecutive globe motility constraints is possible. Therefore, postoperative CT control is mandatory in all cases (Figure 14.4).

Mandibular reconstruction in tumor and reconstructive surgery demands stable, long-term and rigid internal stabilization, well tolerated even in highly irradiated tissue. Further innovations in the development of titanium mandibular reconstruction plating systems included the principle of the titanium locking screw technology, first described by Joram Raveh, Bern, Switzerland.[3] In most current systems, the plate–screw combination is similar and mostly produced by a cold-welding process between two different titanium alloys. Because of the rigid fixation of the screw head to the plate, stability is independent from the plate–bone contact and avoids undue compression of the bone under the plate, which can induce resorption and the incidence of plate fractures (Figure 14.5a through c). Moreover, the frictional connection also allows implants to be of smaller dimensions with greater loading capacity. For

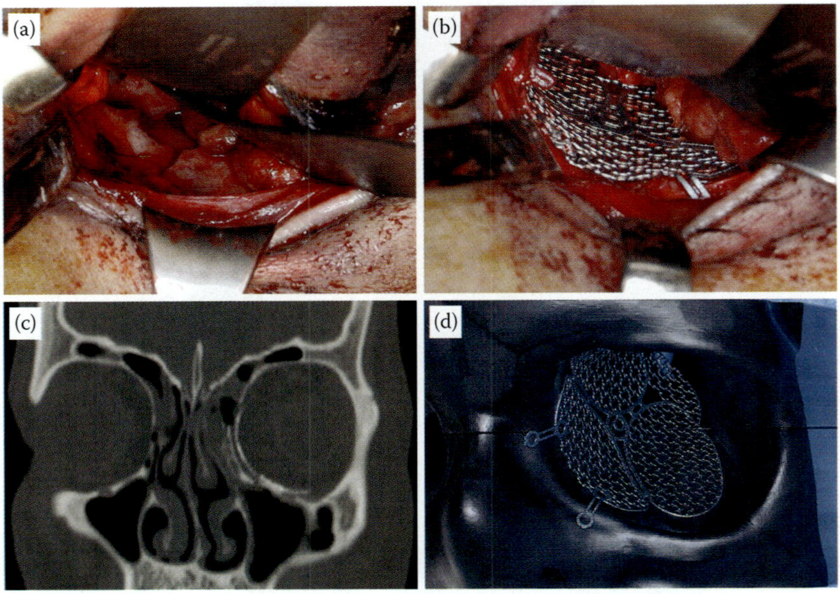

FIGURE 14.3 Multifragment displaced medial orbital wall and floor fracture. (a) Transconjunctival exposure. (b) Preformed 3-D mesh Medartis® (Basel, Switzerland) MODUS OPS 1.5 on anatomical midface model. (c) Mesh in place and fixed. (d) Coronal CT scan demonstrating correct alignment.

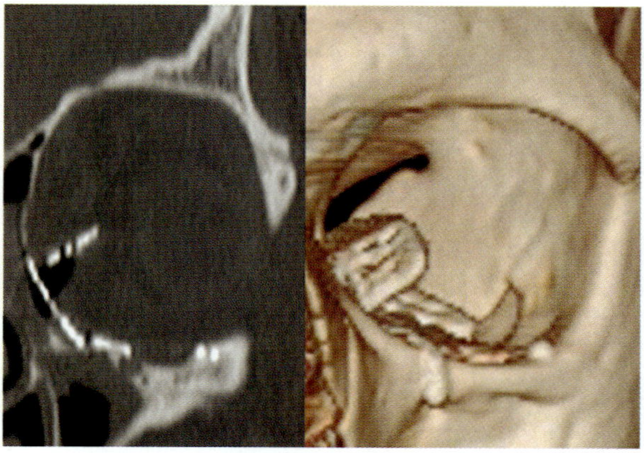

FIGURE 14.4 Postoperative CT showing disalignment of a 3-D mesh Medartis® (Basel, Switzerland) MODUS OPS 1.5 due to its flexibility.

this reason, low-profile plates are suitable in primary microvascular fibula osteocutaneous free flap reconstruction of the mandible (Figure 14.6).

Certain temporomandibular joint pathologies, such as bony ankylosis, absence of anatomical structures due to trauma, congenital deformity, or tumor with loss of the condylar head, require reconstruction with a temporary total condylar head prosthesis. However, only a few TMJ systems have been established to this day.

Consider the huge benefit of titanium in the world of dentistry. Titanium dental implants have become the most widely accepted and successfully used type of implant around the world because of its ability to osseointegrate (Figure 14.6c).

Because of its nonferromagnetic property, patients with titanium implants can be safely examined with MRI. The need for removing asymptomatic titanium implants

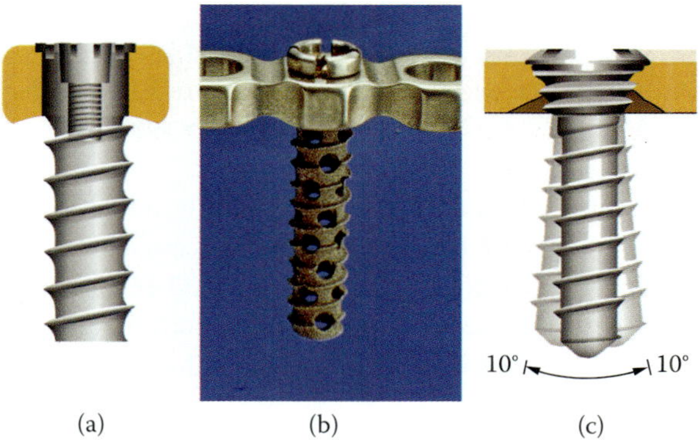

FIGURE 14.5 Locking screw technology. Historical THORP Titanium Hollow Screw Reconstruction Plating system, first described by Raveh, Sutter, and Schroeder in 1979 (Straumann AG, Waldenburg, Switzerland, a, b). Expansion screws were inserted into the head of the titanium bone anchoring hollow screw to lock the plate to the screw and the bone. Modern concepts allowing multidirectional and angular fixation (Stryker® Leibinger GmbH, Freiburg, Germany, c). Point wedge-locking creates a connection between the head and the screw.

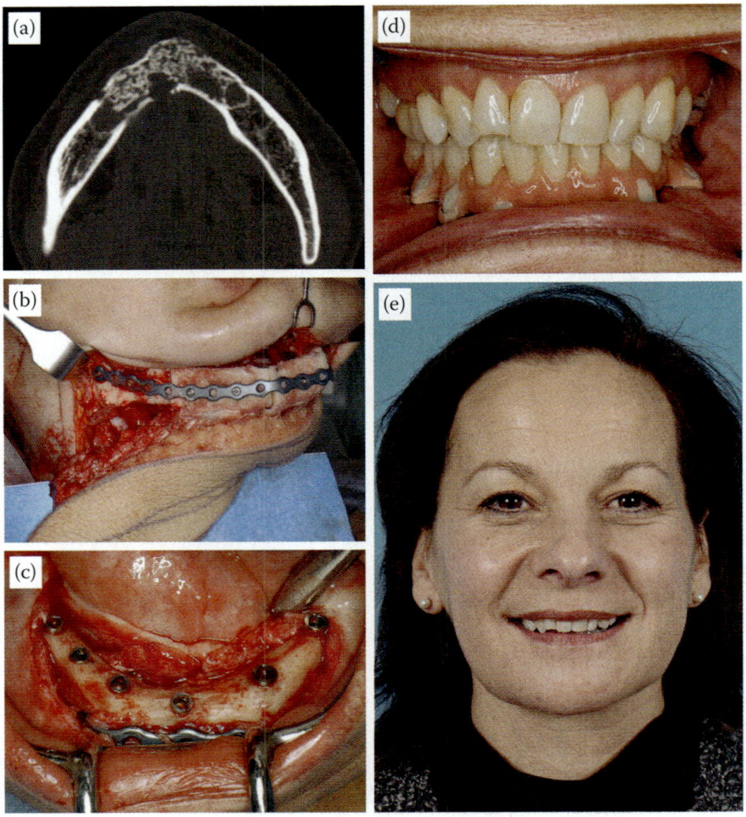

FIGURE 14.6 Ameloblastoma of the mandible. Axial CT demonstrating bone involvement (a). Mandibular reconstruction plate (b) Medartis® (Basel, Switzerland) MODUS TriLock 2.3 with osteocutaneous fibula free flap (Microsurgery: A. Banic, Department of Plastic and Reconstructive Surgery). (c) Second step of the operation: insertion of dental implants (Straumann, reconstruction plate *in situ*). (d) Dental suprastructure (R. Mericse, Department of Prosthodontics, Dental School). (e) Patient's postoperative appearance.

after the healing process is still controversial. Generally, a low complication rate can be observed when the material remains *in situ*. The indications for plate removal include infections, plate extrusion or fracture, thermal sensitivity, palpability and visibility, or growth disturbances in children. Considering all these facts, titanium has become the metal of choice in the field of cranio-maxillofacial surgery for trauma and osteotomies.

14.2.2 Resorbable Plating Systems

Pediatric facial fractures and osteotomies have the ability to heal in a very short time through the assistance of growth. The risk of possible growth restriction or long-term metal migration in pediatric patients and the potential elimination of the need for a secondary surgery demanded resorbable osteosynthesis.

The first clinical trials with resorbable implants started in the 1970s[4]; however, it was not until the 1990s that biodegradable plates and screws were routinely inserted for the fixation of fractures or osteotomies.[5] The most widely used synthetic materials today are high molecular weight α-hydroxy acid polymers, particularly polyglycolide acid and polylactid acid, providing an optimal balance of mechanical strength and resorption profile by varying their ratio. The resorption process occurs by hydrolytic fragmentation into monomers, metabolized by macrophages. Approximately 70% of the initial strength is retained during 8 weeks, allowing for complete bone healing process. Complete resorption occurs within 12–24 months. Because of the self-reinforcing technique, using polymeric fibers, resorbable plating systems show high initial strength and mechanical stability similar to titanium, and are therefore suitable for fracture repair and reconstructive procedures in nonload-bearing areas[6–8] (Figure 14.7). For contouring, the

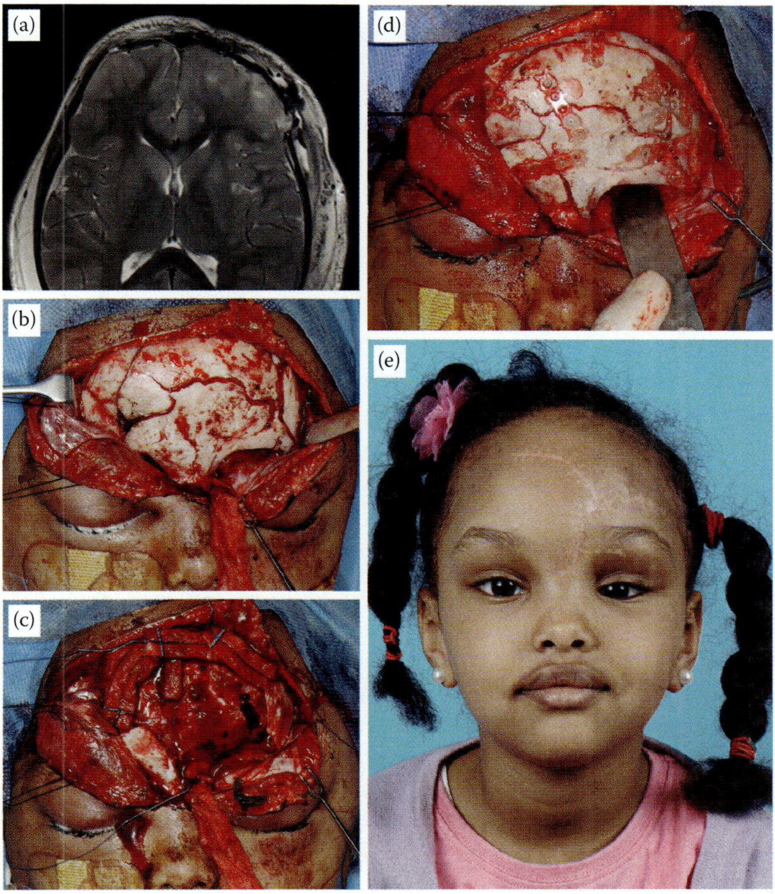

FIGURE 14.7 Extensive fronto-orbital injury in a 5 1/2-year-old girl. (a) Axial MR demonstrating intracranial haemorrhage. (b) Frontal bone disruption. (c) Dural lacerations after craniectomy and evacuation of intracranial haematoma. (d) Cranial reconstruction using resorbable plates (KLS Martin SonicWeld). (e) Girl's postoperative appearance 6 months later.

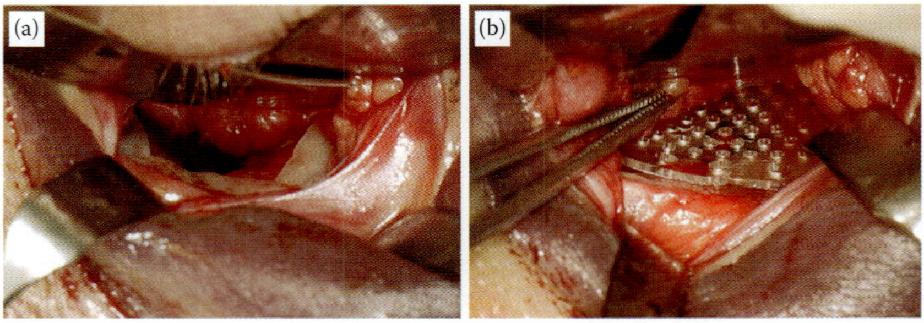

FIGURE 14.8 Orbital floor fracture. (a) Transconjunctival exposure with trapped soft tissue is lifted from the defect. (b) Resorbable orbital floor plate (Synthes®) PolyMax in place.

devices are heated in a hot water bath until a malleable state is achieved, before again becoming rigid at room temperature.

A number of different plates and meshes are available, providing easy adaptation through both hot and cold bending, and various screw and tack insertion techniques. Triangular polylactide plates are used to repair relatively large orbital floor and wall defects, leaving a stable bridge of healed bone or soft tissue after complete degradation[9] (Figure 14.8). Although these implants are generally well tolerated and reliable, based on our experience, they are afflicted with a certain incidence of local inflammatory foreign body reactions, such as soft tissue swelling, osteolysis, or infections after a follow-up period of several weeks. This mainly occurs when the implants have been placed directly under the skin with little subcutaneous tissue. In a young patient with orbital floor and wall repair using a P(L/DL)LA 70/30 Polymax plate, intermittent periorbital swelling appeared. At reoperation, crystalline particles of the polymer could be removed.

14.3 BONE SUBSTITUTES

14.3.1 Metallic Mesh and Implant

Loss of fronto-orbital segments or full-thickness craniectomy defects require immediate reconstruction to prevent major aesthetic deformities and to protect the underlying brain.[10] Titanium meshes in different thicknesses are ideal bridging materials for these nonload-bearing areas. The meshes can be cut with scissors to the correct size and bent to the desired contour with special bulged bending pliers. The space between the underlying tissue and the mesh must be filled with a suitable vital or avital material. A temporary option is given with autologous, abdominal fat, although it has a tendency toward infection or necrosis (Figure 14.9). For secondary reconstruction of contour irregularities or major defects, individual 3-D skull models can be used to correctly adapt the mesh preoperatively to the desired anatomical shape (Figure 14.10), or full titanium custom-made implants can be fabricated. The methods described are also useful in combination with microvascular free flap application (Figure 14.11).

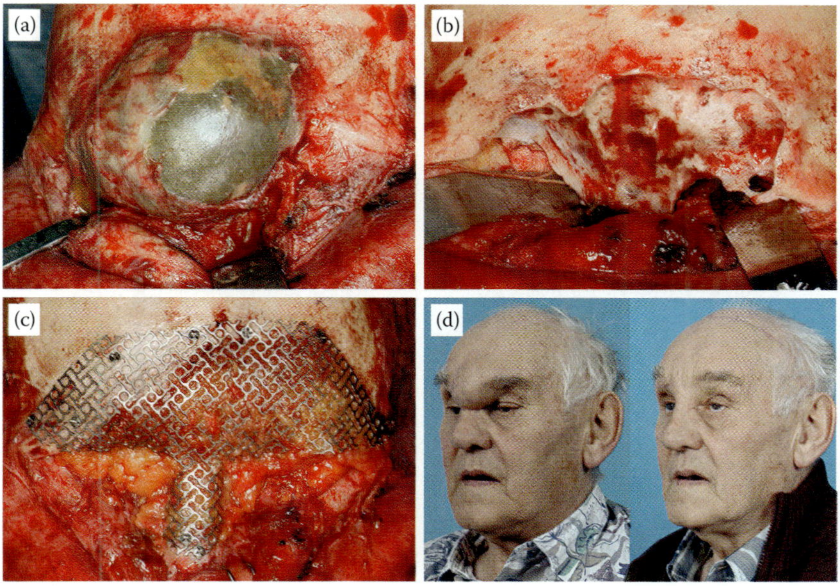

FIGURE 14.9 Huge frontal sinus mucocele. (a) Intraoperative view of the mucocele. (b) After meticulous mucocele removal. (c) Sinus obliteration with abdominal fat and titanium mesh in place. (d) Preoperative and postoperative appearance.

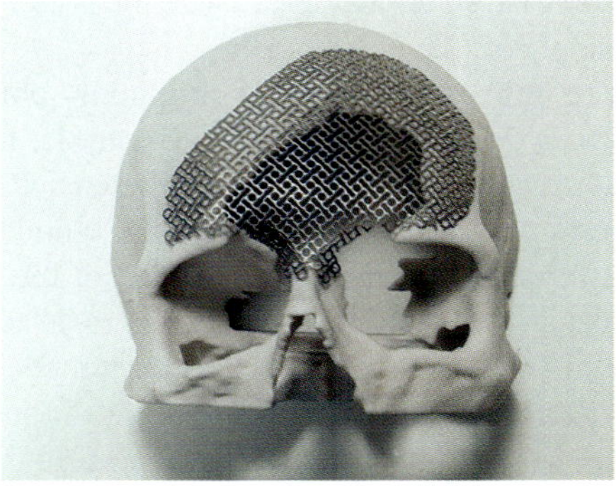

FIGURE 14.10 Secondary reconstruction of fronto-orbital defect with prebent titanium mesh on a 3-D skull model.

14.3.2 Autologous Bone Grafts

Autologous bone transplantation is still considered the gold standard for bone repair throughout the craniofacial skeleton. Because of its natural osteoinductive, osteogenic, and osteoconductive character, it demonstrates the ability to regenerate completely with minimal risk of graft rejection and disease transmission. The decision

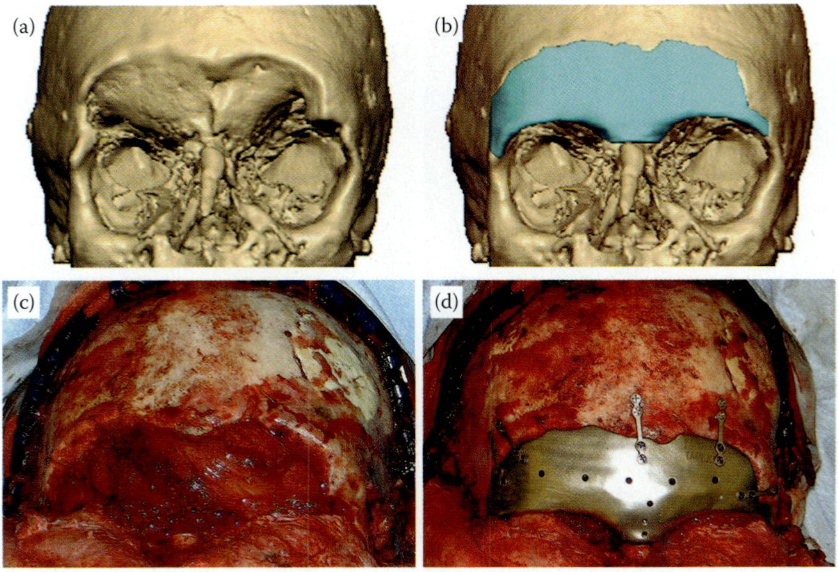

FIGURE 14.11 Frontal defect after multiple surgeries and infections in a syndromic patient. (a) and (b) Virtual 3-D models with patient-specific implant proposal. (c) Intraoperative condition with anterolateral free flap ALT adapted. (d) Patient-specific implant in full titanium (Synthes®).

about which graft to use is based on the defect size, the type and amount of missing tissue, and the blood supply of the recipient site.

Free nonvascularized bone transplants are commonly obtained as block grafts or bone marrow aspirates from the mandible (symphysis, ascending ramus, and coronoid process), iliac crest, tibia, or calvarium, whereas vital bone grafts, including part of the periosteum and accompanying nutrient blood vessels, are harvested from the fibula, radial forearm, scapula, or iliac crest of the same individual. The most important negative aspects of autologous bone grafts are the prolonged operation time, the potential complication risks, and the related donor site morbidity along with increased postoperative pain and scarring—and uncommonly—even fractures. In addition, the amount and availability of autogenous bone may be insufficient for many clinical cases, especially in children. The limited initial possibility of shaping and contouring of the grafts, as well as the unpredictable resorption and remodeling estimates, may also jeopardize the long-term functional or aesthetic outcome.

In oral and maxillofacial surgical procedures, autologous bone grafts are frequently used such as in alveolar ridge augmentation (Figure 14.12), dental implant and sinus lifting surgery, reconstruction of various osseous defects resulting from trauma, inflammatory periodontal diseases, cystic and neoplastic deformities, and congenital anomalies. Autologous block bone grafts also qualify for the skeletal reconstruction of the nose, anterior maxillary or frontal sinus walls, orbit, and

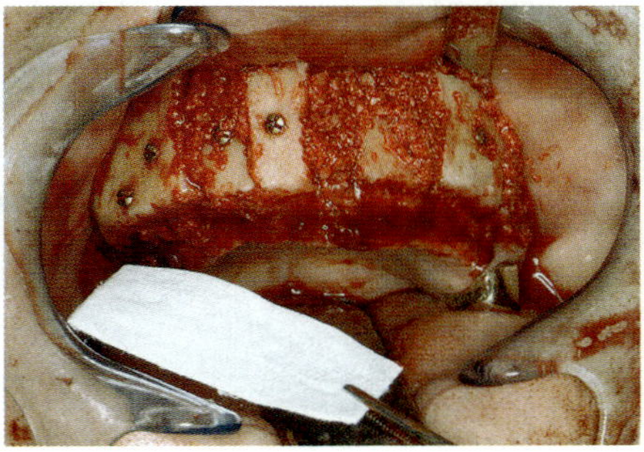

FIGURE 14.12 Alveolar ridge augmentation with autologous iliac crest cortico-spongious bone, membrane ready for adaptation.

cranial vault. Free bone grafts exposed to the sinonasal or pharyngeal cavity are vulnerable to infection and necrosis; therefore, covering the grafts with vascularized tissue should reduce the occurrence of these complications.

14.3.3 Bone Allografts

Allograft bone is obtained from human living or cadaveric donors—inactivated, denatured, and sterilized to an organic bone matrix with normal bone structure and preservation of the osteoinductive proteins. Three types of bone allografts are clinically available: fresh-frozen, freeze-dried, and demineralized freeze-dried, mostly stored in bone banks. Bone allografts are available as cancellous and cortical granules or chips, with the same convenience in handling as fresh bone. Therefore, they are considered a suitable alternative to autologous bone, avoiding donor site complications and reducing operation time; providing an almost unlimited quantity. However, freeze-dried bone undergoes significant resorption and remodeling, takes longer to heal, and has a higher infection rate. There are also reasonable safety risk concerns of disease transmission; for example, hepatitis and HIV. Although fresh-frozen bone was described for the reconstruction of maxillary alveolar ridges[11] and other indications, this kind of implant is mainly used in orthopedic surgery.

14.3.4 Cartilage Allograft

Commercially available lyophilized allogenic sterile costal grafts have undergone an inactivating multistep process against viruses, prions, and microbes without affecting structure and biomechanics.[12] For many years—in the absence of suitable synthetic materials—we frequently used lyophilized cartilage for orbital floor repair and as disc replacement in open temporomandibular joint arthroplasty. However, total resorption, destruction, and fragmentation were noted in several of these cases. Today, lyophilized costal cartilage still has great value in augmentation rhinoplasty procedures (Figure 14.13a). The use of finely carved or crushed cartilage, wrapped in cellulose membrane and moistened with an antibiotic mixed with the patient's

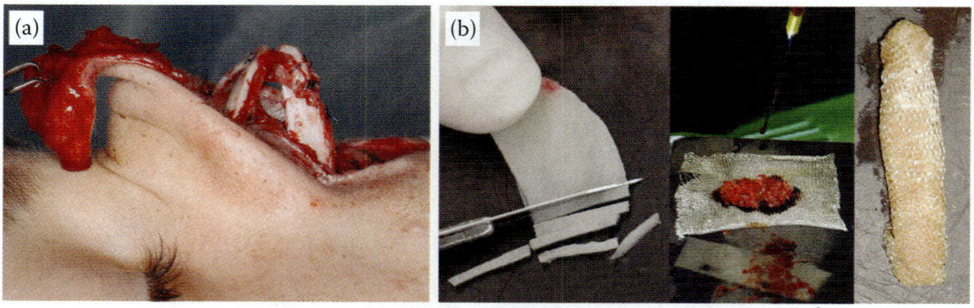

FIGURE 14.13 Open rhinoplasty. (a) Tip refinement with lyophilized costal cartilage grafts. (b) "Turkish delight" (see text).

blood, was first suggested by Erol in 2000 and is nowadays popularly known as a "Turkish delight."[13] This smooth pliable graft is then molded into a cylindrical form and inserted under the dorsal nasal skin (Figure 14.13b).

14.3.5 Bone Xenograft

Xenograft bone originates from nonhuman, mostly bovine, or coral sources, and is usually distributed as a calcified matrix. Due to biofunctionality, including a level of porosity close to human bone, supporting optimal ingrowth of osteoblasts and blood vessels, some commercially available products have been tested in a wide range of applications in dental implantology, periodontology, and oral surgery worldwide (Figure 14.14). Despite the high safety standards and certification, some concerns are present and have been mentioned previously.

14.3.6 Demineralized Bone Matrix

Demineralized bone is produced by the removal of minerals from cortical human bone, consisting of collagen and growth factors. Bone morphogenic proteins are

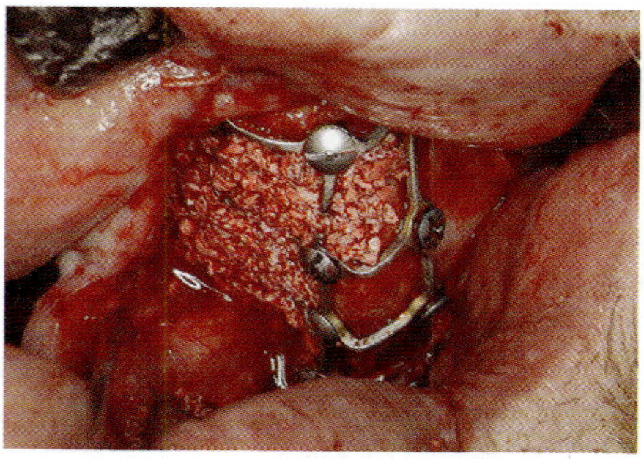

FIGURE 14.14 Alveolar ridge augmentation with bovine Bio-Oss® (Geistlich, Switzerland) bone xenograft.

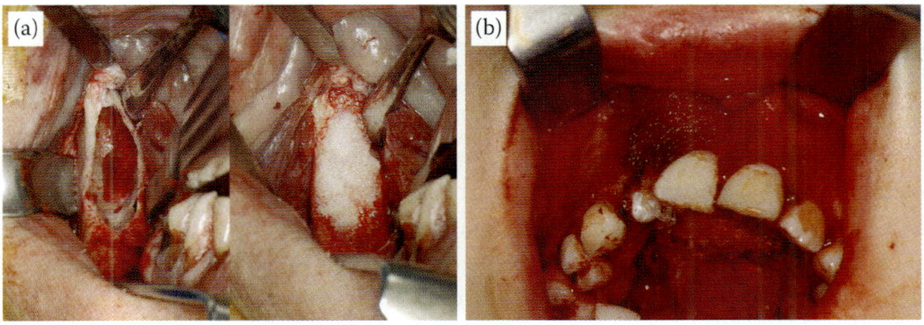

FIGURE 14.15 Demineralized bone matrix as filler for bone voids. (a) Cystic lesion in the retromolar region of the mandible. (b) Alveolar bone grafting in the patient with a cleft lip palate (Grafton®, Osteotech).

responsible for the osteoinductive stimulation and degeneration of new bone, formed via stem cells into bone-forming cells, serving as an osteoconductive matrix for new bone ingrowth within 4–6 months. The lyophilized material is available as a moldable paste or putty consistency and mixes easily with autogenous bone marrow or blood. Thanks to its excellent handling characteristics, demineralized bone matrix is being used with greater frequency in a wide variety of clinical craniofacial applications, mainly as a filler for bone voids or gaps of any size and shape. Classical indications include extraction socket grafting, bone cysts (Figure 14.15a), local alveolar ridge augmentation, sinus lift, or aesthetic augmentations. Alveolar bone grafting in patients with cleft lip palate for stabilization of the maxilla and new bone formation for developing teeth has been carried out by the author, but is still controversial (Figure 14.15b). The bone matrix can either be placed in the defect alone or in combination with a covering membrane underneath the soft tissue flap. Many publications attest to the safety of demineralized bone regarding immunogenity and transmission of infectious diseases.

14.3.7 Synthetic Bone Graft Substitutes

14.3.7.1 Acrylates

Polymethylmethacrylate (PMMA), generally called acrylic glass—although chemically unrelated to glass—is a synthetic transparent thermoplastic polymer, whose development began in the late nineteenth century. In 1933, it was popularized under the trademark Plexiglas. Due to numerous new applications in electronics, architecture or automotive industry, the worldwide demand today is booming.

There are many uses for acrylates in the field of medicine: hard contact lenses, ocular prostheses or full denture bases, all color-matched to patients' tissue, are often made using PMMA. Methacrylates are also the main organic component of white dental filling composite materials. In cosmetic surgery, acrylic microspheres have been injected under the skin for the permanent reduction of wrinkles. However, with more than 35 years of proven clinical success in orthopedic surgery and neurosurgery, PMMA is frequently used as a cement for screw fixation in bone and filler for bone cavities and cranial defects. For this purpose, powdered granules of methylmethacrylate polymer are mixed with methylmethacrylate liquid, as the monomer polymerizes to a moldable dough-like mass and hardens in approximately 10 min. PMMA is a radiolucent, extremely durable material with excellent fatigue strength. It is also available as a homogenous cement/gentamycin antibiotic mixture. PMMA cements may be used for primary reconstruction of craniotomy defects, even if the resection is of unknown dimension in advance, or as secondary, planned reconstruction using patient-specific implants derived from CT data. Different from autogenic bone, methacrylic implants remain at a constant volume and are not absorbed—on the contrary—due to stress shielding, resorption of the surrounding bone may occur. Although PMMA is biologically compatible, it does not bond to either the bone or the implant with subsequent fibrous encapsulation with implant displacement (Figure 14.16). Furthermore, it is considered to be an irritant and possible

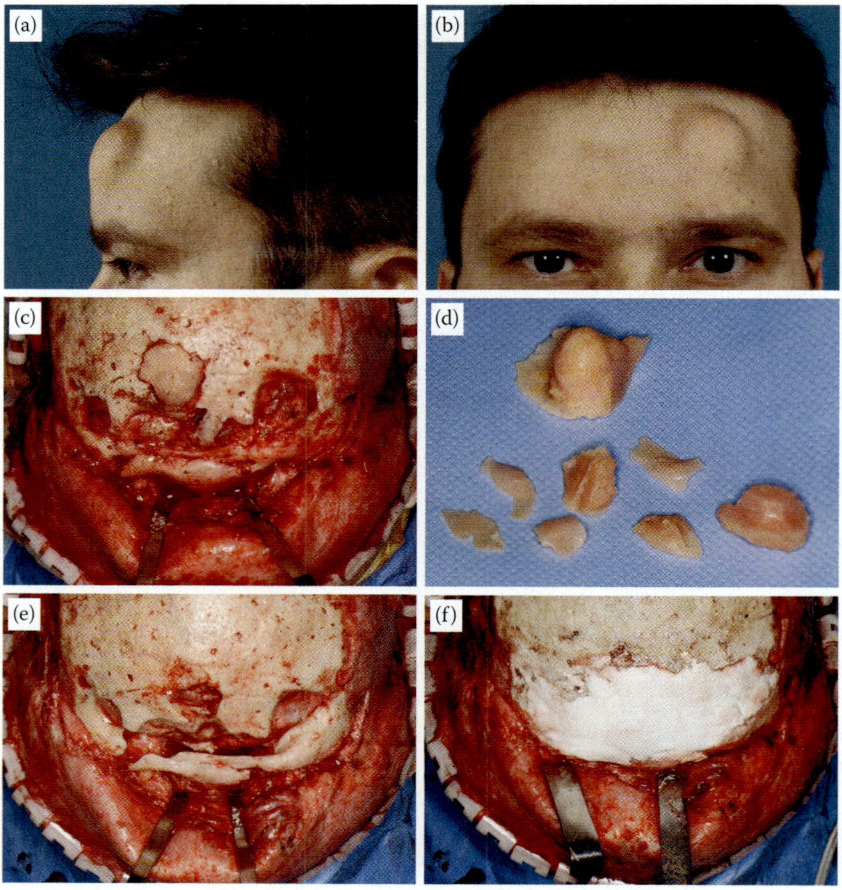

FIGURE 14.16 Years after defect reconstruction with PMMA. (a) and (b) Sudden frontal bossing. (c) Intraoperative foundings with (d) multiple PMMA fragments removed. (e) Meticulous debridement of the defect area. (f) Defect reconstruction and contouring with hydroxyapatite cement (BoneSource, Stryker Leibinger).

carcinogen. A major disadvantage is the exothermic reaction occurring near the end of the polymerization process, with potential necrosis of the surrounding tissues. Rare but serious complications, such as transient hypotension, complete heart block, cardiac arrest, and other arrhythmias after the insertion of methacrylate cements have also been reported in the literature. Despite all these negative aspects, PMMA remains a popular and preferred material because of its moderate properties, easy handling, and low cost. However, PMMA is clearly contraindicated in a manifested history of infection.

14.3.7.2 Ceramics

Ceramics are nonorganic, nonmetallic materials (typically crystalline in nature), and are compounds formed between metallic and nonmetallic elements.

14.3.7.2.1 Hydroxyapatite—Calcium Phosphate Ceramics

Calcium phosphate, in particular, hydroxyapatite, is the principal naturally occurring anorganic mineral component of teeth and bones. It can be produced synthetically by a number of methods, including conversion of natural bone or sea corals. Calcium phosphate implants have been used clinically for many years in porous granular form as filling material in both soft and osseous tissues,[14] or implanted as blocks into osteotomy gaps. However, their inherent fragility, difficulty in contouring, and lack of compressive strength were significant limitations in the use of earlier forms of hydroxyapatite for skeletal reconstruction.

Since 1992, hydroxyapatite has been available in nonceramic form as hydroxyapatite cement, whose properties have overcome the inherent previous limitations. Consisting of tetracalcium and dicalcium phosphate powder, mixed with water or a sodium phosphate buffer, this self-setting cement forms an easily sculptable paste that converts to hydroxyapatite within 20–30 min. During the hardening process, no thermal emissions occur. Once set, the cement can be further drilled and shaped. Other cement types consist of modified components, presenting slightly different physical properties. Generally, hydroxyapatite cement exhibits excellent biocompatibility, remodeling partially into new vital bone by osteoclastic resorption, osteoinduction, and osteoconduction, yet maintaining its shape and volume over time. According to experimental and clinical evidence, there are fewer long-term bone ingrowths into the micropores of the cement than in implants with macroporous architecture[15] (Gosain). The latest advances in bone substitution technology have made the calcium phosphate cement injectable, fast-setting, and sculptable even in a wet field environment, eliminating the need to meticulously dry the operative site before implantation.

The wide variety of clinical applications mainly includes the repair of full-thickness cranial skull defects in nonload-bearing areas, allowing improved implant osseointegration. Hydroxyappatite cement is safe and effective in aesthetic contour augmentation in the craniofacial skeleton[16] (Figure 14.16a). For cranial defects with a surface larger than approximately 25 cm^2, more mechanical stability and protection of the cement against repetitive trauma from dural pulsations are provided by the application of an additional titanium mesh (Figure 14.17).

According to our considerable experience, hydroxyapatite cement represents a valuable option for frontal sinus obliteration after mucocele resection or in trauma,

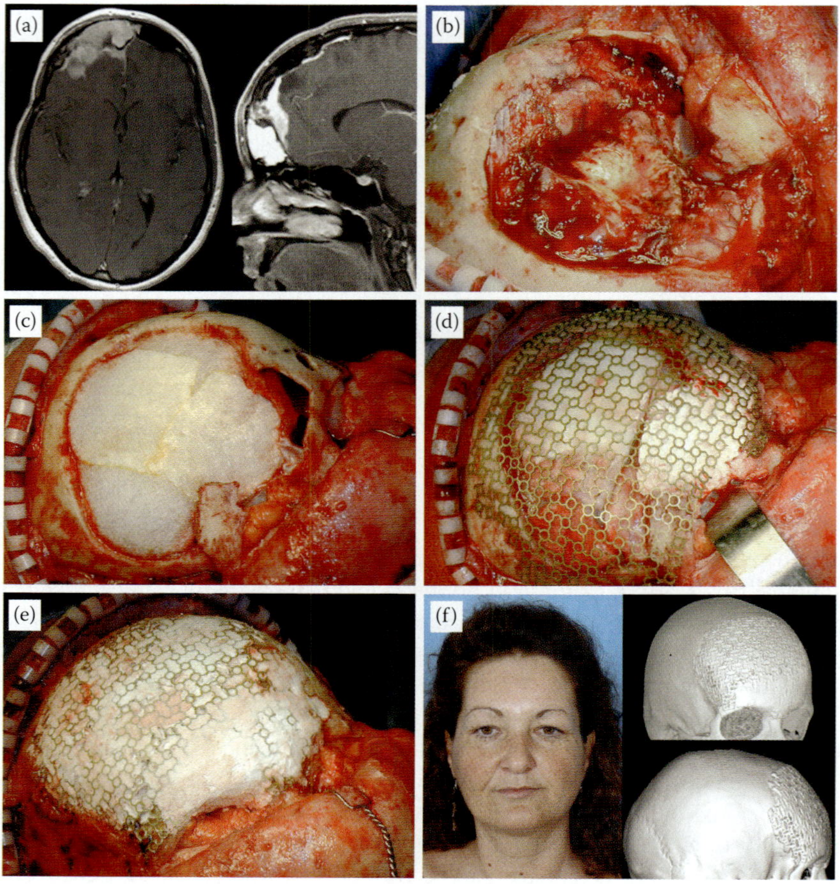

FIGURE 14.17 Recurrent meningioma of the transitorial type. (a) Axial and sagittal MRI depict tumor involvement of the frontal sinus and frontal bone. (b) Intraoperative view of the craniectomy defect and resected dura. (c) Dural repair with fleece-bound tissue sealing (TachoSil® Nycomed). The frontonasal duct closed with the muscle and fascia. (d) After obliteration of the frontal sinus with hydroxyapatite cement, a titanium mesh is adapted. (e) Contouring of the supraorbital region and orbital roof. (f) Frontal and lateral 3-D CT scans show the exact shape of the reconstructed fronto-orbital region and patient's postoperative appearance.

allowing for the option of not replacing the anterior table if destroyed. For the prevention of failures, careful attention must be paid to the meticulous removal of all mucosa and shaping of the posterior wall with a high-speed diamond drill, and the tight closure of the frontonasal ducts with muscle or fascia. Even some rare recurrent cerbrospinal fluid-fistulae in the sphenoid sinus have been successfully sealed. In contrast to several reports in the literature, hydroxyapatite cement has proved to be a well-tolerated material in our series, with essentially minimal inflammatory response and a very low infection rate.[17,18] The cement had to be removed and replaced with fat in only one case because of recurrent unknown inflammation.[19]

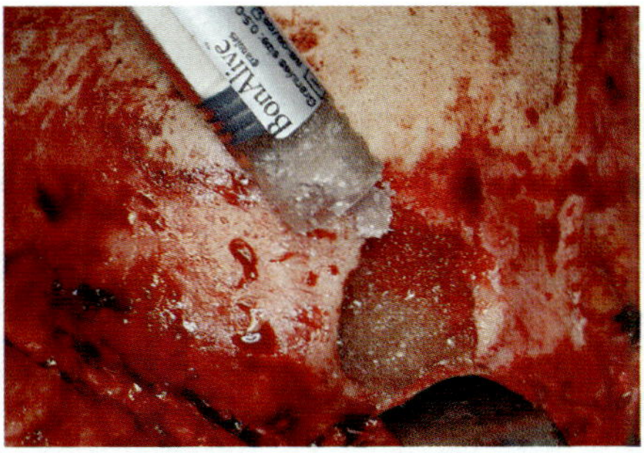

FIGURE 14.18 Frontal sinus obliteration with BonAlive® (Turku, Finland) bioactive glass granules. Ethisorb® (Johnson & Johnson Inc.) membrane allows proper contouring of the orbital roof and prevents adhesion of the granules against the eye.

14.3.7.2.2 Bioactive Glass

Bioactive glasses are a group of surface-reactive glass–ceramics, first described by Larry Hench and colleagues in the late 1960s during the Vietnam War. Further developments were carried out worldwide—particularly recently in Finland.[20] The base components usually include silicates (SiO_2), sodium (Na_2O), calcium (CaO), and phosphates (P_2O_5). The physical and biological properties can be varied by alteration of the corresponding fraction ratio. Bioglass binds to bone by the fast formation of a surface-layered structure of silica and hydroxyapatite, reacting with collagen, mucopolysaccharides, and glycoproteins. The relatively fragile, amorphous but brittle, glass meshwork has proven to be a promising bone-bonding and osteoconductive antimicrobial implant material with minimal fibrous encapsulation. Compared with hydroxyapatite cement, increased new bone formation with fewer substitutions of the implanted material was found. Most important medical applications for bioactive glass granules include alveolar ridge augmentation, osteoplastic frontal sinus obliteration (Figure 14.18), and craniofacial defect reconstructions in non-load-bearing areas. Additionally, rigid bioactive glass plates have been successfully used as a reliable well-tolerated material for orbital floor repair.

Because bioactive glass cannot be bent, shaped, or drilled during surgery, novel composite implants containing PMMA coated with bioactive glass fiber reinforcement to enhance bone ingrowth have been introduced recently for the reconstruction of large calvarial defects. Based on patients' CT data and using rapid prototyping technologies, custom-made implants are characterized by a very high accuracy and may even be used for very large calvarial defects (Figure 14.19).

14.3.7.3 Polymers

14.3.7.3.1 Silicone

Silicones are inert, synthetic, nonporous rubber-like compounds with no tissue ingrowth. Therefore, infections and implant migration often occur due to a fibrous capsule formation around the implant, making their removal necessary. Because of the advantage of being affordable and easy to reshape, solid silicone blocks were

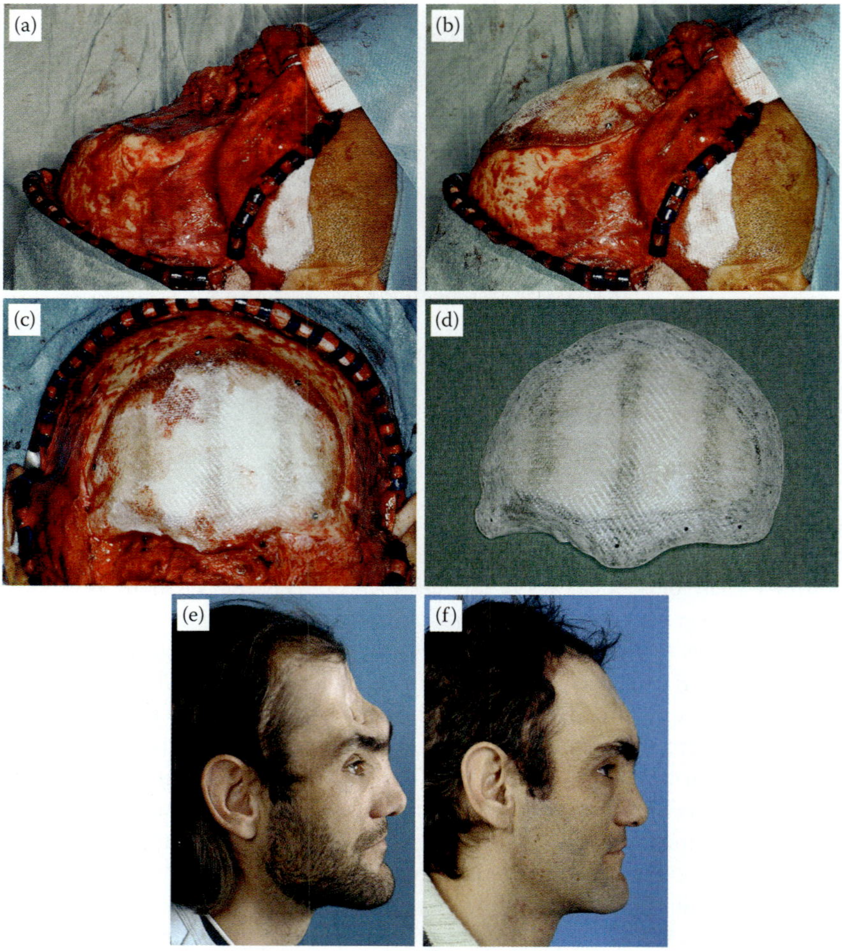

FIGURE 14.19 Forehead reconstruction with custom-made bioactive glass implant. (a) Intraoperative condition demonstrating frontal defect, side view. (b) Perfect fit of the implant, side view. (c) Perfect fit of the implant, front view. (d) Custom-made specimen. (e) Patient preoperation with posttraumatic forehead defect. (f) Postoperative appearance.

frequently used in nasal, chin, and malar augmentation, as well as for joint replacements in temporomandibular joint ankylosis. With time, silicone came into disfavor and many displaced or extruded implants were seen as a result of trauma. In our experience, silastic sheets are helpful in frontal sinus obliteration procedures, preventing the adhesion of the hydroxyapatite cement against the eye and allowing proper contouring of the orbital roof. They must be removed after hardening of the cement (Figure 14.17e). For scar revision or soft tissue defect procedures, tissue expanders can be used for skin flap enlargement. Inflatable devices in different configurations and shapes consist of a polyester mesh–reinforced silicone elastomer and are for temporary subcutaneous or submuscular implantation that can be used for no longer than 6 months (Figure 14.20).

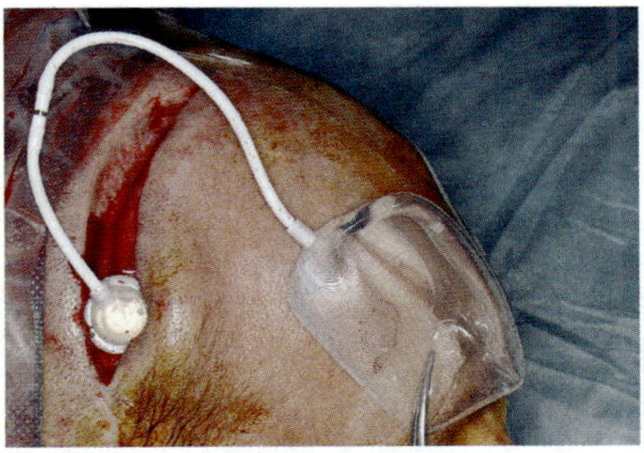

FIGURE 14.20 Silicone-based tissue expander in the forehead area. The device has a remote injection dome that is attached to the expander with a length of silicone elastomer tubing.

14.3.7.3.2 Porous Polyethylene

Polyethylene is a thermoplastic polymer with a long history of use in surgical implants, best known today under the brand name of Medpor® (Stryker). The porosity permits tissue ingrowth, and the firm nature allows carving with a knife. A wide variety of blocks, sheets, and anatomically preformed shapes are commercially available for augmentation or aesthetic contour restoration in the facial skeleton, particularly for the nasal dorsum, chin, zygoma, or mandibular angle. For orbital floor and wall repair, thin sheets of polyethylene are strengthened with a titanium mesh embedded in it, allowing easy bending and contouring whereas minimizing sharp edges, and thus facilitating insertion. Orbital wedges provide volume to restore the orbital cone to its normal shape in cases of enophthalmos, and special spheres are suitable as biocompatible filler after enucleation. Finally, preformed Medpor ear implants in two-piece designs are suitable for primary or secondary repair.

14.3.7.3.3 Polyester

Polyesters are polymers with an ester functional group in their main chain. Synthetic polyglactine 910 and polydioxanone are commonly used for the preparation of absorbable suturing materials and surgical patches. Vicryl® (Johnson & Johnson Inc.) is the most used braided synthetic absorbable suture currently available. Ethisorb absorbable flexible patches are suitable for the repair of small-to-moderate orbital floor fracture defects,[21] and for bridging dural leaks. The fleece-like, porous structure allows for progressive integration of fibrous, connective tissue with minimal risk of inflammation or foreign body reaction. Absorption is completed approximately 90 days after implantation.

14.4 CUSTOM-MADE IMPLANTS

Reconstruction of large cranial defects still represents a challenge in craniofacial surgery and neurosurgery. Surgeons are constantly in search of the perfect anatomically fitting implant that does not cause infection and resorption, provides good aesthetical results, and reduces operating times. For this purpose, custom-made implants are derived from CT data to create an equivalent 3-D model of the bone defect involved, and from which an individual implant design is manufactured (Figure 14.21).

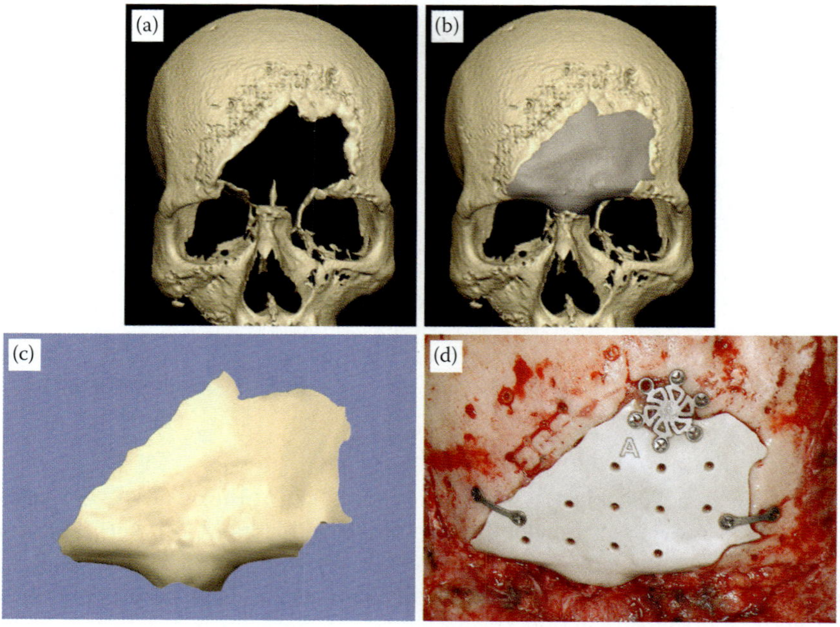

FIGURE 14.21 Custom-made implants. (a) 3-D computer images of the bone defect involved, derived from CT data. (b) Three-dimensional computer image of the implant fitted into the defect. (c) Final implant (Synthes® PEEK Optima-LT Polyetherketone). (d) Anatomical fit and fixation of the implant with conventional plates and self-tapping screws.

Different biomaterials, most of which are discussed previously, are used for the final fabrication; for example, pure titanium, polyetheretherketone, hydroxyapatite, bioglass, and others (Figures 14.11 and 14.19).

14.5 BLOOD-DERIVED BIOMATERIALS

14.5.1 Tissue Sealants

Fibrin sealants are allogen blood-derived biomaterials derived from plasma donations, which are physiologically compatible with human tissue. Once a solution of concentrated fibrinogen and coagulation factor XIII is activated with thrombin, a strong fibrin clot is formed that is physiologically compatible with human tissues. Tissue sealants may be used in many surgical fields for hemostasis,[22] tissue sealing, and promotion of wound healing, it also serves as a matrix for the bone marrow in the repair of bone defects. In skull base surgery, fibrin glue frequently serves as dural sealing for duraplasty using autologous fascia lata layers.

Under the commercial name of TachoSil® (Nycomed), a fixed combination of an equine-derived collagen patch sponge coated with a dry layer of the human coagulation factors fibrinogen and thrombin is available, ready for quick and easy tissue sealing without reconditioning or thawing before application. In contact with blood or other fluids, it provides quick and strong adhesive properties, creating an air- and liquid-tight seal. The resistant collagen carrier offers additional mechanical support to the tissue surface. TachoSil degrades and absorbs within 12 weeks. Fleece-bound tissue sealing has gained increasing popularity for hemorrhage control and duraplasty within the neurosurgical and craniofacial disciplines (Figure 14.17c).

14.6 PEARLS AND PITFALLS

The wide range of craniofacial procedures and the size, location, and vascularity of the defect to be reconstructed often determine the type of implant material. The increasing variety of medical biomaterials available in craniofacial surgery today can make this decision difficult. Numerous potential complications may occur with any implant-related procedure, wherein the most frequent is the risk of infection. Mild early postoperative infections—often caused by wound dehiscence—are managed by intravenous antibiotics and secondary wound closure, whereas more serious late infections with implant exposure or extrusion mostly require immediate implant removal. Decreased vascularity of the recipient site secondary to preexisting scar formation or high doses of radiation compromise the normal healing process and acceptance of any biomaterial (Figure 14.22). For a reduction of these risks, incisions should not be placed directly over the implants, enabling a tension-free multilayer closure and sufficient soft tissue coverage. Sharp edges must be smoothed to prevent soft tissue irritation, palpability, or skin erosion. Adequate implant fixation to the underlying tissue should avoid motion and migration. Finally, progress in microvascular techniques have allowed the insertion of large volumes of well-vascularized tissue into the defect area, avoiding dead space formation around implants.

Despite all enthusiastic clinical successes with the use of biomaterials, there is often a lack of long-term knowledge concerning the tissue reaction and healing process in the human body. The surgeon must be aware of this and take appropriate responsibility. In the event of implant failure, all inserted alloplastic materials should be registered in the patient's chart. Informed consent, including the nature, potential risks, and possible alternatives of the biomaterial to be used, takes a central position in this field of surgery and is of utmost importance.

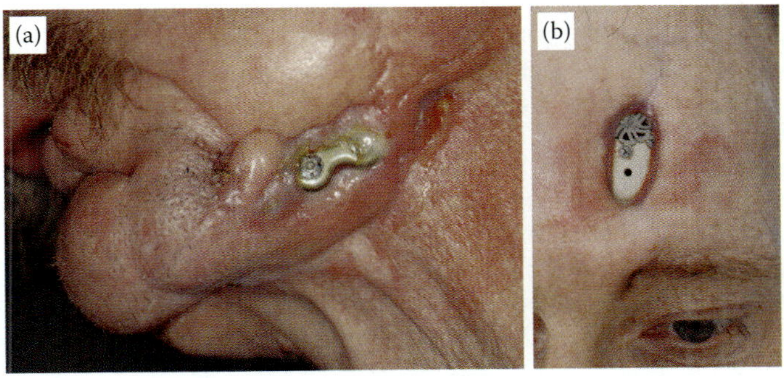

FIGURE 14.22 Implant failures. (a) Plate extrusion in highly irradiated tissue. (b) Wound breakdown and dehiscency over patient-specific implant and titanium plate due to decreased vascularity of the recipient site secondary to preexisting scar formation.

Although alloplastic materials are very popular in craniofacial reconstruction, their cost-effectiveness must be kept in mind. The extensive use of these expensive materials is justified only if the donor site morbidity and the operation times are significantly reduced and relevant complications with related revision surgeries can be avoided. The increased use of medical technologies contributes significantly to annual cost increases, and controlling this technology is the most important factor in reducing them.[23]

REFERENCES

1. Brunette, D. M., P. Tengvall, M. Textor, and P. Thomsen. 2001. Titanium in Medicine: Material Science, Surface Science, Engineering, Biological Responses and Medical Applications. New York: Springer.
2. Brånemark, P. I. 1983. Osseointegration and its experimental background. *Journal of Prosthetic Dentistry* 50, 399–410.
3. Raveh, J., H. Stich, and F. Sutter. 1984. Use of titanium-coated hollow-screw and reconstruction system in bridging of lower jaw defects. *Journal of Oral and Maxillofacial Surgery* 42, 28.
4. Suuronen, R., and C. Lindqvist. 2002. Bioresorbable materials for bone fixation: Review of biological concepts and mechanical aspects. In *Craniomaxillofacial Reconstructive and Corrective Bone Surgery: Principles of Internal Fixation Using the AO/ASIF Technique*. Chapter 13, edited by Greenberg, A. M. and Prein, J., New York: Springer, section II, 113–123.
5. Eppley, B. L. 2005. Use of resorbable plates and screws in pediatric facial fractures: Clinical controversies in oral and maxillofacial surgery: part two. *Journal of Oral and Maxillofacial Surgery* 63, 385–391.
6. Cheung, L. K., L. K. Chow, and W. K. Chiu. 2004. A randomized controlled trial of resorbable versus titanium fixation for orthognathic surgery. *Oral Surgery, Oral Medicine, Oral Pathology, Oral Radiology and Endodontics* 98(4), 386–397.
7. Uckan, S., F. Veziroglu, S. Soydan, and E. Uckan. 2009. Comparison of stability of resorbable and titanium fixation systems by finite element analysis after maxillary advancement surgery. *Journal of Craniofacial Surgery* 20(3), 775–779.
8. Kulkarni, R. K., E. G. Moore, A. F. Hegyeli, and F. Leonard. 1971. Biodegradable poly(lactic acid) polymers. *Journal of Biomedical Materials Research* 5, 169–181.
9. Lieger, O., B. Schaller, J. Zix, F. Kellner, and T. Iizuka. 2010. Repair of orbital floor fractures using bioresorbable poly-L/DL-lactide plates. *Archives of Facial Plastic Surgery* 12(6), 399–404.
10. Laedrach, K., A. Lukes, and J. Raveh. 2007. Reconstruction of skull base and fronto-orbital defects following tumor resection. *Skull Base: An Interdisciplinary Approach* 17(1), 59–72.
11. Contar, C. M., J. R. Sarot, J. Bordini, Jr., G. H. Galvão, V. N. Gastão, G. V. Nicolau, and M. A. Machado. 2009. Maxillary ridge augmentation with fresh-frozen bone allografts. *Journal of Oral and Maxillofacial Surgery* 67, 1280–1285.
12. Sailer, H. F. 1983. Transplantation of lyophilized cartilage in maxillo-facial surgery. *Experimental Foundations and Clinical Success*. Basel, Switzerland: S. Karger Verlag.
13. Erol, C. O. 2000. The Turkish delight: A pliable graft for rhinoplasty. *Plastic and Reconstructive Surgery* 105(6), 2229–2241.
14. Burstein, F. D., S. R. Cohen, R. Hudgins, and W. Boydston. 1997. The use of porous granular hydroxyapatite in secondary orbitocranial reconstruction. *Plastic and Reconstructive Surgery* 100(4), 869–874.

15. Gosain, A. K. 2003. Biomaterial in facial reconstruction. *Plastic and Reconstructive Surgery* 9(1), 23–30.
16. Friedman, C. D., P. D. Costantino, C. D. Snyderman, L. C. Chow, and S. Takagi. 2000. Reconstruction of the frontal sinus and frontofacial skeleton with hydroxyapatite cement. *Archives of Facial Plastic Surgery* 2, 124–129.
17. Turk, J. B., and A. Parhiscar. 2000. Bone source for craniomaxillofacial reconstruction. *Facial Plastic Surgery* 16(1), 7–14.
18. Agarwala, S., and A. Bhagwat. 2005. Hydroxyapatite as a bone substitute: Use in cortical and cancellous bone. *Indian Journal of Orthopedics* 39(4), 254–256.
19. Taghizadeh, F., A. Kroemer, and K. Laedrach. 2006. Evaluation of hydroxyapatite cement for frontal sinus obliteration after mucocele resection. *Archives of Facial Plastic Surgery* 8, 1–8.
20. Peltola, M. J., P. K. Vallitu, V. Vuorinen, A. A. J. Aho, A. Puntala, and K. M. Aitasalo. 2012. Novel composite implant in craniofacial bone reconstruction. *European Archives of Oto-Rhino-Laryngology* 269(2), 623–628.
21. Büchel, P., A. Rahal, I. Seto, and T. Iizuka. 2005. Reconstruction of orbital floor fracture with polyglactin 910/polydioxanon patch (Ethisorb): A retrospective study. *Journal of Oral and Maxillofacial Surgery* 63(5), 646–650.
22. Peng, T. 2010. Biomaterials for hemorrhage control. *Trends in Biomaterials and Artificial Organs* 24(1), 27–68.
23. Callahan, D. 2008. Health care costs and medical technology. In *Birth to Death and Bench to Clinic: The Hastings Center Bioethics Briefing Book for Journalists, Policymakers, and Campaigns*. Edited by Crowley, M., Garrison, NY: The Hastings Center, 79–82.

15 Nanobiomaterials for Bone Tissue Engineering

Meng Deng, Roshan James, Cato T. Laurencin, and Sangamesh G. Kumbar

CONTENTS

15.1 INTRODUCTION

The bone constitutes an internal support system for all higher vertebrates. It is a highly specialized connective tissue that consists of a structural framework of mineralized matrix and a heterogeneous cell population. It is characterized by its marked rigidity and hardness and is a complex living tissue that continuously undergoes rebuilding, which is critical to maintaining its function. The main functions of the bone are as follows: 1) to provide mechanical support for the body and serve as levers for muscle action; 2) to provide protection of internal vital organs; 3) to provide an environment for marrow (blood cell production); and 4) to maintain calcium homeostasis and act as an ion reservoir for phosphate, magnesium, potassium, and bicarbonate (Bilezikian et al. 1996). In addition, the bone matrix maintains the storage of growth factors and fatty acids, buffers against excessive pH changes in the blood, and is involved in endocrine signaling. The bone is a very dynamic organ that constantly undergoes self-remodeling and has within itself the unique ability to repair/regenerate to a certain extent after injury. Each year, more than 6.5 million fractures occur in the United States (Braddock et al. 2001). Sales of bone graft and bone graft substitutes in the United States alone amounted to $1.5 billion in 2009 (American Academy of Orthopaedic Surgeons 2010). Thus, there is a great need for the development of functional bone grafts for fracture repair, and there are significant opportunities to improve the existing graft materials.

15.2 BONE GRAFTS

As illustrated previously, the repair and replacement of damaged or lost bone due to trauma, pathological degeneration, or congenital deformity of the tissue is a major clinical problem in the United States and around the world. These figures will continue to grow as the life expectancy and the total population increase in the United States (Tomford 2003). The healing of a bone fracture involves four sequential stages: hematoma (inflammation), soft callus formation, hard callus formation, and bone remodeling. It is a unique repair process through which cells, growth factors, and extracellular matrix (ECM) closely interact with each other and result in scarless tissue regeneration (Einhorn 1998; McKibbin 1978; Carano and Filvaroff 2003; Schindeler et al. 2008). In cases of fracture nonunions (up to 10% of the fractures) and large mass bone loss associated with osteoporosis, osteosarcoma, and revision total joint replacement, this natural process does not occur, thereby requiring surgical intervention to assist healing (Khan and Laurencin 2008).

15.2.1 Autografts

Reconstructive surgery is often performed to replace defective tissues with viable alternatives. Autografts, allografts, and a variety of bone graft substitutes are used for surgical treatments (Laurencin et al. 2005; Ilan and Ladd 2003; Laurencin and Khan 2012). Autografts, which constitute approximately 58% of the bone substitutes, are typically tissues harvested from the patient's own iliac crest. They are considered the gold standard for bone repair, because they possess all the properties necessary for new bone growth. Upon implantation, the grafts are able to support the attachment and migration of new osteoblasts and osteoprogenitor cells (osteoconductivity), *in situ* mineralization of the collagen matrix produced by osteoblasts to form new bone (osteogenicity), the recruitment and differentiation of stem cells or osteoprogenitor cells into osteoblasts (osteoinductivity), and formation of intimate bonding between the newly formed mineralized tissues and surrounding bone tissues (osteointegrativity). However, they are limited in availability and often associated with donor site morbidity and increased operative blood loss, particularly when a large graft is required (Goulet et al. 1997).

15.2.2 Allografts

Allografts are tissues obtained from banked freeze-dried bones of human cadavers and represent approximately 34% of the bone substitutes. They are osteoconductive and have fewer limitations on supply (Laurencin et al. 2005). However, allografts are usually not osteoinductive or osteogenic and are associated with risks of immunological reaction or disease transmission. Furthermore, they possess insufficient mechanical properties for load-bearing bone applications. Studies have shown that the failure rate of allografts is between 25% and 35% (Berrey et al. 1990; Ito et al. 2005).

15.2.3 Bone Graft Substitutes

As alternatives to the autografts and allografts, a variety of bone graft substitutes have been developed and account for approximately 8% of bone grafts (Laurencin et al. 2005). On the basis of material composition, they can be classified as allograft-based, factor-based, cell-based, ceramic-based, and polymer-based bone graft substitutes (Laurencin et al. 2005). They have been developed for use in bone repair because of its unlimited supply, ease of sterilization, and ease of storage. However, each suffers from a number of disadvantages. Human-derived allograft-based bone graft substitutes can be potentially associated with immunogenicity and disease transmission. Factor- and cell-based bone graft substitutes often need additional structural support. Ceramic-based bone graft substitutes are brittle and possess inappropriate mechanical properties for use in load-bearing sites. In addition, some of the ceramic-based bone graft substitutes do not remodel over time.

To overcome the limitations associated with the current bone graft options, bone tissue engineering has emerged as a promising strategy that provides *de novo* regenerated tissue for bone repair. Tissue engineering has been defined as "the application of biological, chemical, and engineering principles toward the repair, restoration or regeneration of tissues using biomaterials, cells, and factors alone or in combination" (Laurencin et al. 1999). All approaches in bone tissue engineering involve one or more of three key components: 1) three-dimensional (3-D) porous matrices with optimum properties to support and promote tissue regeneration; 2) a healthy population of harvested cells; and 3) biological growth factors that can induce cellular migration, differentiation, and tissue growth *in vivo* (Langer and Vacanti 1993). In a scaffold-based bone tissue engineering approach, a 3-D biodegradable scaffold is seeded with cells (osteoblasts or osteoprogenitor cells) with or without growth factors. After *in vitro* culture, the cell-seeded scaffolds are implanted *in vivo* to induce and direct the growth of new bone bridging the defects. Thus, the biodegradable scaffold plays a crucial role in bone tissue engineering. During the repair/regeneration process, the biodegradable scaffold provides for structural and mechanical restoration of the damaged tissues, gradually degrades into biocompatible products, presents an interconnected porous structure to accommodate cell infiltration and vascularization, and promotes matrix synthesis (Laurencin et al. 1999; Langer 2000; Khan et al. 2008). Furthermore, it can be designed to deliver growth factors or antibiotics (or both) to treat bone defects and promote wound healing.

15.3 MATERIAL AND BIOLOGY CONSIDERATIONS OF BONE

15.3.1 Overview of Bone and Bone Mechanics

At the macroscopic level, a mature bone is differentiated into compact (cortical) and trabecular (cancellous or spongy) bones that significantly vary in density (Bilezikian et al. 1996). Cortical bone constitutes 80% of the total human skeleton by mass. It is a very dense material with 5% to 10% porosity (Bilezikian et al. 1996). In cortical bone, three to eight lamellae of continuous collagen fibrils (50–500 nm) arrange in a concentric manner around a central harversian canal to form osteons or Harversian systems (200–250 μm in diameter; Giraud-Guille 1988; Ross et al. 2003; Weiner and Traub 1992). They are cylindrical in longitudinal section and contain gaps between the lamellae (3–7 μm thickness) where mature bone cells are housed. The arrangement of the collagen fibers in a lamella of an osteon has been proposed as a twisted plywood or helicoidal structure by Giraud-Guille (1988). Continuous collagen fibrils composed of parallel collagen molecules form the lamellae structured osteon (Weiner and Wagner 2003). These hierarchical features largely determine the structural properties of bone. Volkman canals connect Harversian canals to one another. These canals are occupied by blood vessels to supply nutrients. Cortical bone is important in maintaining the structural and mechanical functions of the skeleton (Pearson and Lieberman 2004). On the other hand, trabecular bone is in a sponge-like form with a loosely organized trabeculae network. The trabeculae can be configured in a number of combinations, which all involve one or more of the basic cellular structures such as rod–rod, rod–plate, or plate–plate. The dimension of a trabecular rod is approximately 50–300 μm in diameter. It is highly porous with 50% to 90% porosity, and plays an important role in maintaining metabolic activity (Bilezikian et al. 1996). In the epiphysis, where complex forces are encountered, the trabecular network provides reinforcement to the outer cortical bone (Bilezikian et al. 1996). The outer surface of bone is covered by a periosteum. Periosteum consists of an inner layer that contains committed osteogenic cells and an outer fibrous layer. Periosteal bone formation remains active during our lifespan. In the case of bone fractures, the periosteum promotes the formation of cartilaginous callus, followed by ossification. The blood vessels and nerves connect to the bone through the periosteum, and it serves as attachment sites for surrounding tendons and muscles. The inner vascular thin layer lining the marrow cavity is called the endosteum and it also harbors bone precursor cells. The cortical bone functions mechanically in tension, compression, and torsion, whereas trabecular bone functions mainly in compression. The mechanical properties of human cancellous and cortical bone are listed in Table 15.1 (Murugan and Ramakrishna 2005).

TABLE 15.1
Bone Biomechanical Properties

Properties	Cortical Bone	Trabecular Bone
Elastic modulus (GPa)	14–20	0.05–0.5
Compressive strength (MPa)	170–193	7–10
Tensile strength (MPa)	50–150	10–20
Fracture toughness (MPa m $^{1/2}$)	2–12	0.1
Strain to failure	1–3	5–7

Source: Murugan, R., and S. Ramakrishna. *Compos Sci Technol* 65(15–16): 2385–2406, 2005. With permission.

By examining the organization and pattern of collagen fibrils of bone matrix, bone can also be classified into woven or lamellar bone (Buckwalter et al. 1995). Woven bone, composed of randomly organized collagen fibers, is typically mechanically weak. On the other hand, mechanically strong lamellar bone is characterized by a regular arrangement of parallel collagen fibers organized into lamellae sheets. Woven bone is often produced during the fetal bone development and is then resorbed and replaced by remodeling and the deposition of lamellar bone. Woven bone is rarely found in the mature human skeleton. However, in the repair of a bone fracture, woven bone appears during the rapid formation of new bone.

15.3.2 Bone Matrix and Nanoscale Components

The successful design of a bone graft substitute that will modulate repair and regeneration after an injury requires an acute understanding of the composition of bone, its hierarchical architecture, and its organization into the bone matrix. Collagens, such as type I collagen, are the major structural proteins of the bone ECM (Bilezikian et al. 1996; Wiesmann et al. 2004). Osteoblasts secrete collagen molecules that self-assemble into collagen fibrils having a specific tertiary structure. The tertiary structure includes a 67 nm periodicity and 40 nm gaps between the collagen molecules (Rho et al. 1998). Collagen chains are characterized by highly organized fibrils forming a continuous triple helix, which consists of three intertwining helical polypeptides (van der Rest and Garrone 1991). The organized fibrils possess high tensile strength. In body tissues such as bone, tendons, and ligaments, the structured collagen fibrils are responsible for providing mechanical support and function. For example, type I collagen fibrils form the fiber bundles in tendons, whereas they compose a concentric network of fibrils encapsulating hydroxyapatite crystals in the gaps between collagen molecules, as in bone. The inorganic phase accounts for 60% of bone ECM and is primarily composed of hydroxyapatite ($Ca_{10}(PO_4)_6(OH)_2$). At the nanoscale, plate-like apatite crystals form discretely and discontinuously with a specific crystalline orientation within the collagen fibrils (Weiner and Traub 1992). It is found that the *c* axis of the crystals (2–3 nm thickness) is in parallel to the long axis of the collagen fibrils (Landis 1995; Ziv and Weiner 2009). The crystal plates formed has a dimension of 50 × 25 nm (length × width). Several other components, such as HPO_4^{2-}, Na^+, Mg^{2+}, citrate, carbonate, and K^+ are also present in the bone apatite, which is characterized to be nanocrystalline hydroxyapatite without hydroxyl groups (Rey et al. 1995). The presence of noncollagenous proteins such as osteocalcin, osteopontin, sialoprotein, and osteonectin may contribute to the regulation of crystal size, orientation, and mineral deposition by binding calcium or helping release phosphate. Thus, collagen fibers are important constituents and forms structures that contribute to the outstanding properties of bone. They are further reinforced by the impregnated mineral hydroxyapatite.

The formation and maintenance of bone tissue is the result of highly organized and coordinated multicellular actions. Three distinctly different cell types are found within the bone tissue, namely, osteoblasts, osteoclasts, and osteocytes, which account for about 90% of all cells in the adult skeleton. In addition, two other distinct cells types, osteoprogenitor cells and bone-lining cells, are associated with bone function.

During the development of the vertebrate embryo, cells derived from the neural crest are responsible for the development of the craniofacial skeleton (Beddington and Robertson 1999). Sclerotome cells, which are of mesodermal origin, give rise to the axial skeleton, and the lateral mesoderm is the point of origin for the formation of the limb bud and, eventually, the development of the long bone. During morphogenesis, the mesenchymal cells undergo either direct differentiation into osteoblasts that proceed to form bone, or differentiation proceeds via chondrocytes, which form a cartilaginous template that later ossify in a process known as endochondral ossification. These progenitors are termed osteoprogenitor cells or bone precursor cells. Functionally, osteoblasts are the cells within bone that are responsible for deposition of the new ECM. These cells are derived from the progenitor population, appear cuboidal in shape, and are located at the bone surface in a monolayer (Aubin 1998; Kierszenbaum 2002; Jee and Cowin 2001). Osteoblasts are highly anchorage dependent, and start by secreting collagen and then coat them with noncollagenous proteins that have the ability to bind minerals such as calcium and phosphate, and thus regulate mineralization. The osteoblasts maintain cellular function, and respond to stimuli facilitated by cell–matrix and cell–cell communications conducted through a variety of transmembranous proteins and specific receptors (Ferrari et al. 2000; Lecanda et al. 1998). As the newly formed bone mineralizes, some osteoblasts become enclosed in their own calcified matrix, and will change their phenotype, developing into osteocytes or mature osteoblasts. Osteocytes are the most abundant cells in bone, and remain connected with other osteocytes and with the bone-lining cells present at the bone's surface, allowing for intercellular communication that may be responsible for the spatial and temporal recruitment of cells that form and resorb bone, and will also allow the transport of minerals between bone and blood (Jee and Cowin 2001; Liu and Ma 2004). Osteoclasts are found at the surface of the bone and their responsibility is to resorb fully mineralized bone (Jee and Cowin 2001; Li et al. 2006; Blair et al. 1989). They originate from hematopoietic stem cells, and are highly specialized as they release acids and enzymes to dissolve the minerals and collagens present in mature bone. The dissolved minerals recycle back into the blood, and their transportation in and out of the bone is regulated by the bone-lining cells. All these cellular processes regulated by the various distinct cell types must be in equilibrium to coordinate the processes of bone formation and resorption. This will ensure the formation of healthy bone, its maintenance and renewal as necessary. This process, termed "bone modeling," is necessary for the development of normal bone architecture, to adequately maintain bones biomechanically and metabolically, and to repair microdamages. These processes occur naturally throughout the life of the tissue and quite simply, are described as shaping and repairing processes.

The hierarchical structures of bone at different length scales are of great importance in maintaining the chemical, mechanical, and biological properties of bone (Rho et al. 1998; Weiner and Wagner 2003). As illustrated in Figure 15.1, the structural hierarchy of bone architecture includes five levels on the basis of the length scales discussed in the previous section: 1) the macrostructures of cortical bone and trabecular bone; 2) the microstructures (10–500 μm) of the osteons and trabeculae; 3) the submicrostructures (1–10 μm) of bone lamella (Marotti 1993); 4) the nanostructures (from a few hundred nanometers to 1 μm) of collagen fibrils; and 5) the

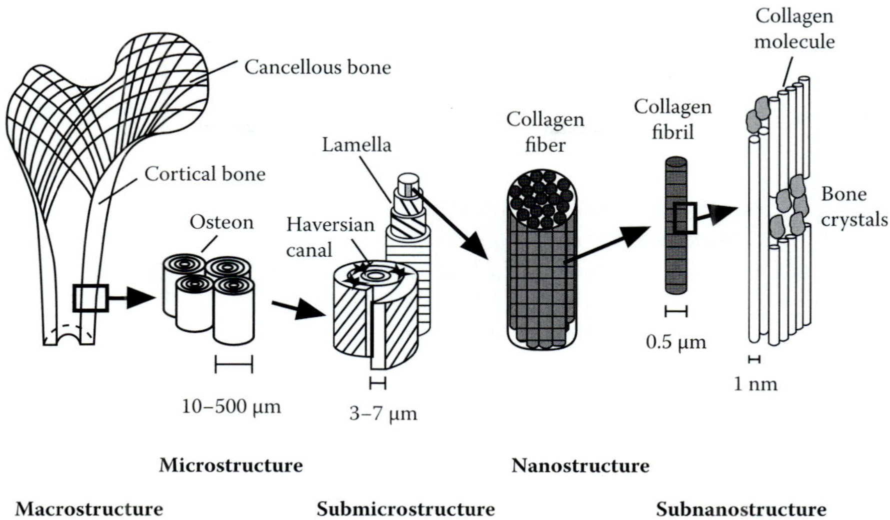

FIGURE 15.1 Hierarchical structural organization of the bone where the macrostructure is composed of cortical and cancellous bone, which is made up of osteons with haversian systems and lamellae. Fibers, laminae, and pores are present at different size scales resulting in various macrostructures to subnanostructures. This microstructure is composed of collagen fibers made up of nanodiameter collagen fibrils. The smallest structural units in the bone are bone mineral crystals, collagen molecules, and noncollagenous proteins. Such material hierarchical arrangement is essential for the mechanical functions of bone. (Reproduced from Rho, J.-Y. et al. *Med Eng Phys* 20(2): 92–102, 1998. With permission.)

subnanostructures (less than a few hundred nanometers) of collagen molecules, bone crystals, and noncollagenous organic proteins. The basic nanostructural components are arranged in a coordinated fashion to enable the physicochemical and biological functions of bone.

15.4 DESIGN CONSIDERATIONS FOR IDEAL BONE TISSUE ENGINEERING SCAFFOLDS

In the repair of a bone defect, the scaffold is designed to maintain the structure of the defect and restore the bone function. Ideally, the scaffold should satisfy a number of design criteria to achieve comparable properties present in autologous grafts: 1) the scaffold should be biocompatible so that it is integrated with host tissues without any immune responses; 2) it should be mechanically competent to tolerate the local forces (Athanasiou et al. 2000). This is most critical to protect the tissues and transmit the compressive and tensile force and mechanical cues across the defect to the regenerative cells; 3) it should be biodegradable with nontoxic degradation products that can be metabolized and excreted by the body, and with controllable degradation kinetics to match the rate of bone healing process so that the newly formed tissue compensates the mechanical and mass loss of the degraded matrices (Einhorn 1998;

Schindeler et al. 2008); 4) it should be osteoconductive with a porous structure to allow cell infiltration and proliferation, neovascularization, and nutrient transport. Porosity is the percentage of void space within a solid object, and macroporosity (pore size >50 μm) in bone grafts support the ingrowth of new bone and vasculature (Kuboki et al. 1998; D'Lima et al. 1998; Sul et al. 2002). The interconnected pore spaces enable the transport of oxygen and nutrients, and surface roughness promotes cellular adhesion, proliferation, and differentiation of anchorage-dependent cells; and 5) It should be able to integrate with the surrounding osseous tissue through the formation of bone bonding. Additionally, seeding cells on the scaffold contributes to the osteogenicity of the scaffold while adding growth factors such as bone morphogenetic proteins (BMPs) is beneficial for the osteoinductivity of the scaffold. The success of a 3-D scaffold in regenerating tissues *de novo* is largely dependent on its nature, composition, and associated properties. Biomaterials, including synthetic biodegradable polymers and composites, have shown great promise in a tissue engineering approach (Nair and Laurencin 2007; Deng et al. 2008, 2010a,b,c,d, 2011a). Tremendous efforts have been focused on the development of ideal biomaterials and their fabrication into appropriate 3-D matrices that mimic the composition and architecture of native bone (Deng et al. 2011a). In that direction, the development of composite materials consisting of polymer/polymer and polymer/ceramics exhibiting mimicry of the bone ECM at the nanoscale may potentially enhance cellular responses and consequent regenerative efficacy.

15.5 NANOBIOMATERIALS FOR BONE TISSUE ENGINEERING

Recent studies have demonstrated the importance of nanotopographical features in regulating cell behavior, including preferential adhesion, migration, proliferation, and expression of cell-specific phenotypes (Kumbar et al. 2008a). With the advances of nanofabrication techniques, biomaterials can be fabricated into nanostructures that mimic the native tissue hierarchical structures. As illustrated in Figure 15.2, nanobiomaterials play an important role in enhancing the cellular processes and biological factor delivery through various nanotopographical and surface cues, and hold the promise for the development of an ideal bone graft substitute with similar properties as autografts (Grauer et al. 2003; Chan et al. 2006).

15.5.1 Nanocomposites

The optimal bone graft material and its structural form should encompass the properties discussed in the previous section. Native bone is a natural nanocomposite of hydroxyapatite and organic components composed mainly of collagen and proteoglycans. Composites are defined as being made up of two or more materials that possess significantly different properties that remain distinct within the resulting finished structure. It may be difficult for single phase materials to satisfy the requirements needed for a successful bone graft substitute. The design of a nanocomposite material to mimic the composition and structure of native bone in combination with or without stimulatory biological factors potentially allows for the development of a composite having appropriate mechanical strength, osteoconductive, osteoinductive,

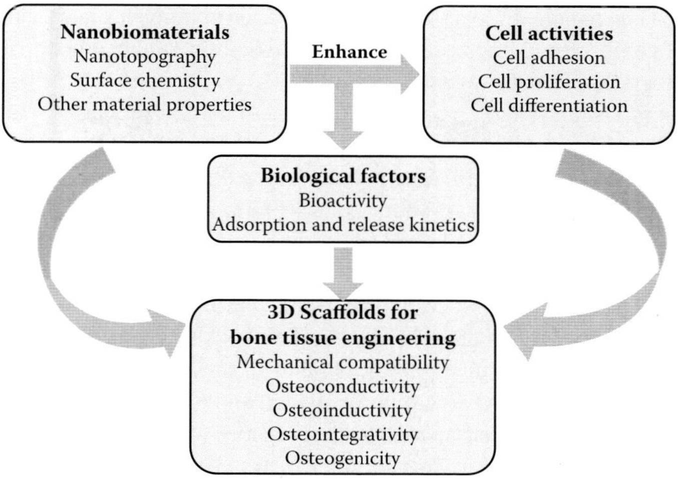

FIGURE 15.2 Schematics illustrating the important role of nanobiomaterials in a scaffold-based bone tissue engineering approach to create ideal bone graft substitutes with comparable properties to native tissue. Engineered 3-D scaffolds composed of novel nanobiomaterials and various ECM cues are combined with appropriate cells or biological factors, alone or in combination to regenerate *de novo* functional bone. A critical need exists for the development of advanced nanobiomaterials with different nanotopographical and surface cues to enhance cellular processes such as cell adhesion, proliferation, and differentiation.

osteointegrative, and osteogenic properties. Hydroxyapatite, the inorganic component of bone is thus the logical biomaterial choice for developing bone grafts. Combining hydroxyapatite with collagen could give a bioactive composite that will be very similar to native bone composition. Studies have demonstrated the effectiveness of a hydroxyapatite/collagen composite in treating long bone fractures in combination with BMA. The mechanical properties of the bone graft substitute are determined by the spatial orientation of hydroxyapatite and collagen molecules. In native bone, the components of the ECM provide informational cues to the various cell types present in bone. At the structural level, the inorganic phase is composed of nanocrystalline calcium phosphate (Christenson et al. 2007; Webster and Ahn 2007). In hydroxyapatite/collagen microcomposites, fabricated by mixing hydroxyapatite particulates into a collagen matrix, the hydroxyapatite are large-sized crystallites and thus it is difficult to prepare a homogenous hydroxyapatite/collagen composite. The microcomposite of hydroxyapatite and collagen may take far longer to remodel into bone postimplantation and, in some scenarios, may exhibit poor mechanical properties when used for load-bearing applications. The weaker mechanical property of the bone graft substitute might be caused by the lack of strong interfacial bonding between the organic and inorganic phases and because of the bulky nature of the constituents (Webster et al. 1999). Improved bone cell function has been correlated with reduced grain size of the hydroxyapatite constituent (Webster et al. 2000, 2001). Compared with micro grain-sized hydroxyapatite components, nanosized particle dimensions further improve osteointegration within the host bone tissue, and may enhance compressive strength, thus being more suitable for load-bearing applications. The nanohydroxyapatite (nHA)/collagen composites more closely mimic native bone composition and structure, and significantly enhance bone cell function and host integration leading to faster recovery.

A nanocomposite system composed of nHA/collagen/osteoblasts was developed in conjunction with poly(lactic acid) (PLA; Liao and Cui 2004; Liao et al. 2004). The novel graft substitute (nHAC/PLA) supported cell adhesion, proliferation and migration into the matrix to a depth of about 200–400 μm in a period of 12 days. *In vivo*, these nanocomposites were evaluated in a rabbit segmental defect model, in which they reported integration of the defect and evidence of new bone formation. Pek et al. (2008) prepared a porous nanocomposite bone scaffold containing collagen fibers and apatite nanocrystals to match the chemical composition and crystalline structure of natural bone. As shown in Figure 15.3a and b, the freeze-dried nanocomposite scaffold composed of 32.5 wt% of type I collagen and 67.5 wt% of a nanocrystalline apatite mixture of carbonated apatite/hydroxyapatite (4:1, w/w) displayed a porous microstructure with a pore size distribution that was similar to trabecular bone. The

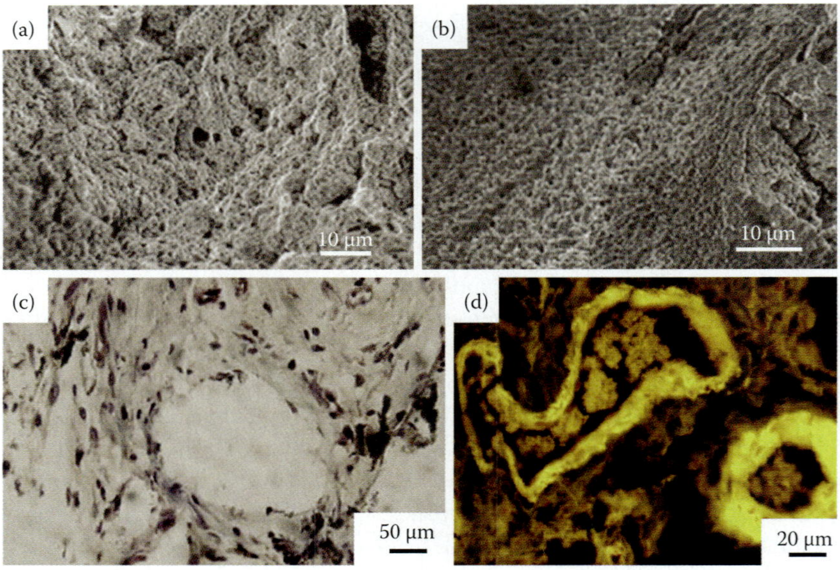

FIGURE 15.3 Nanocomposite bone scaffold composed of 32.5 wt% of type I collagen and 67.5 wt% of a nanocrystalline apatite mixture of carbonated apatite/hydroxyapatite (4:1, w/w). SEM images of (a) nanocomposite scaffold and (b) trabecular bone illustrating microstructure similarity of the nanocomposite scaffold to trabecular bone. (c) Masson trichrome staining of nanocomposite scaffold showing a new bone canal resulting from osteoblastic activity after 4 weeks of ectopic implantation in Swiss albino mice. (d) Fluorescence microscopy image of nanocomposite scaffold after 4 weeks of ectopic implantation in mice injected with tetracycline, which binds specifically to calcium and exhibits a bright yellow fluorescence. The nanocomposite scaffold was able to promote the attachment and proliferation of MC3T3 cells and resulted in the formation of new bone structures and scaffold mineralization. (Reproduced from Pek, Y. S. et al. *Biomaterials* 29(32): 4300–4305, 2008. With permission.)

compressive modulus and yield strength of the nanocomposite scaffold were found to be 37.3 ± 2.2 and 2.7 ± 0.1 MPa, respectively. The nanocomposite scaffold was able to promote attachment and proliferation of MC3T3 cells (Figure 15.3c) and resulted in the formation of new bone structures and scaffold mineralization 4 weeks after ectopic implantation in Swiss albino mice (Figure 15.3d). The potential of using the composite scaffold was also demonstrated to heal a nonunion fracture in rat femur and a critical-sized defect in pig tibia.

In another study, nHA particles were surface grafted to PLA (g-HA) and fabricated into g-HA/poly(lactide-*co*-glycolide) (PLAGA) scaffolds using PLAGA (80:20) by solvent casting/particulate leaching method (Zhang et al. 2009). BMP-2/g-HA/PLAGA, HA/PLAGA, and PLAGA scaffolds were similarly fabricated for comparative studies. These novel biocomposites were evaluated in critical-sized defects created in the radius of a rabbit forelimb. At 2 weeks postsurgery, the radiographs at the repaired areas of BMP-2/g-HA/PLAGA, g-HA/PLAGA, and HA/PLAGA groups seem to be similar and the implants exhibited distinct mineralization. The implants in all the experimental groups, except the PLAGA group, exhibited bridging between the ends of neighboring bones. Comparatively, there was more new bone formation in the BMP-2 protein supplemented group. In contrast, in the PLAGA group, only the surface and the end areas of the implant closest to the host bone tissues showed gentle mineralization. At 4 and 8 weeks postsurgery, the densities of the repaired areas in all groups increased gradually. There was still a large area of bone defect (half of the original defect) remaining in the PLAGA group at 8 weeks postsurgery, although the bone formation was guided on the surface of the implant from the periosteum and resulted in a slender bridging. The X-ray results indicated that the composite of PLAGA with g-HA or HA allowed for more rapid mineralization, integration with host bone, and osteogenesis compared with the pure PLAGA scaffold. The incorporation of BMP-2 could improve the mineralization and osteogenesis of g-HA/PLAGA, and speed the bone healing process. Without the nHA component, the polymer scaffold elicits bone formation mainly from the periosteum of the nearby radius.

In other reports, nanometer-sized carbon fibers have been noted to promote mineral deposition in osteoblast cells as compared with conventional and larger sized fibers (Elias et al. 2002; Price et al. 2003). Carbon nanotubes and alumoxane nanoparticles have been used as fillers for biodegradable materials (Mistry et al. 2007; Horch et al. 2004; Shi et al. 2006). When these small particles are uniformly dispersed within the polymer matrix, the mechanical properties are significantly reinforced and can be modulated to match the native tissue. The increased interaction between the polymeric chains and the uniformly dispersed nanoparticles result in increased flexural modulus without any loss of compressive or flexural strength. Titania nanoparticles have been dispersed by sonication within PLAGA to fabricate nanocomposites that enhanced osteoblast adhesion and long-term functions (Huinan et al. 2005; Webster and Smith 2005). The greatest osteoblast adhesion was observed on nanocomposites that have surface roughness closest to the natural bone tissue. Up to three times more osteoblasts adhered to titania/PLAGA nanocomposites than on conventional titania/PLAGA composites having similar weight ratio and porosity. Composites made of PLA reinforced with 40 wt% and 50 wt% nanophase (i.e., alumina, titania, and hydroxyapatite) had bending moduli values closer to bone and greater than that of composites with conventional coarser-grained ceramics.

In summary, nanocomposites are successful in inducing better cellular responses as compared with standard composites because of their similarity with the natural bone structure and, additionally, can be processed to have mechanical properties that are more closely related to native bone.

15.5.2 Nanofiber Scaffolds

Nanostructures such as nanofibers and nanoparticles represent an interesting class of tissue engineering scaffolds with great potential for bone regeneration. The widely used nanocomposite constituents are hydroxyapatite crystals that are 20–80 nm long and 2–5 nm thick, and collagen fibrils with diameters of less than 500 nm. Nanofibers are ECM-mimicking scaffolds that are characterized by high porosity with a wide range of pore sizes, high surface area, unusual surface properties, and morphological similarity to native bone ECM (Li et al. 2002; Nair and Laurencin 2008; Nair et al. 2004a). Nanofiber scaffolds with interconnecting porous structures provide high surface area for cell attachment, growth, and differentiation, as well as nutrient transport. In addition, nanoparticle-based materials have received significant attention for bone tissue engineering applications (Christenson et al. 2007; Nair and Laurencin 2008). A number of studies have documented the favorable responses of bone cells toward synthetic nanofibers or nanophase particles. Li et al. have demonstrated enhanced osteoblast adhesion and proliferation on polymeric nanofiber scaffolds that exhibit mimicry of the ECM. The extracellular mimetic structure increased protein adsorption and to the selective adsorption of cell adhesion proteins such as fibronectin and vitronectin on the nanostructures compared with structures with solid pore walls (Woo et al. 2003). Furthermore, the nanofiber scaffold environment has been shown to support mesenchymal stem cells to differentiate along the osteogenic lineage (Pelled et al. 2007). Nanophase materials composed of alumina, titania, and hydroxyapatite with grain sizes of <100 nm have been shown to elicit favorable osteoblast and osteoclast activity compared with micrometer-sized particles of the same materials (Webster et al. 2000, 2001).

Electrospinning holds great promise to engineer nanofiber scaffolds for bone tissue engineering due to the ease of fabrication, efficient control over the process, and easy scale-up (Huang et al. 2003). Zeleny first studied the electrospinning process, which was initially known as electrostatic spinning in 1914 (Zeleny 1914). Formhals et al. (1934) reported the first patent describing the process of developing polymeric nanofibers via electrospinning. Fibers with diameters ranging from a few nanometers to several micrometers can be obtained via the electrospinning process (Kumbar et al. 2008b). Unlike the other two processes, electrospinning allows for the development of complex structures such as aligned nanofibers. In addition to providing a high surface area for cell adhesion and proliferation, aligned nanofibers offer the ability to guide the orientation of cytoskeletal proteins of cells (Li et al. 2003; Patel et al. 2007; Ma et al. 2005).

Thus far, electrospun nanofiber matrices have shown great promise for scaffold-based tissue engineering applications (Deng et al. 2011b). More than 100 different polymers (natural and synthetic) and their polymer composites have been electrospun thus far (Huang et al. 2003; Kumbar et al. 2008c). Because of their unique potential of being replaced by the newly synthesized ECM, biodegradable polymeric scaffolds are highly preferable in tissue engineering. For example, polyesters such as PLAGA and polycaprolactone (PCL) have been widely investigated for the fabrication of electrospun tissue engineered scaffolds (Li et al. 2002). *In vitro* studies have shown favorable cellular responses with these materials. Nanofiber composites consisting of PLAGA and amorphous TCP nanoparticles (PLAGA/TCP, 60:40) were prepared by low-temperature electrospinning (Schneider et al. 2009). The potential of bone regeneration was investigated in a 6-mm critical-sized cranial defect model in New Zealand white rabbits. Through histological and microcomputed topographical analysis, TCP-containing PLAGA fiber scaffolds showed a significant increase in the area fraction of newly formed bone 4 weeks postimplantation when compared with pure PLAGA fiber scaffolds. The newly formed bone in the nanocomposite-treated defects exhibited a spongiosa-like structure, whereas solid cortical bone formed around defects treated with Bio-Oss, a porous bovine-derived mineral. Yeo et al. (2010) fabricated a new hierarchical scaffold composed of a melt-plotted β-TCP-PCL composite and embedded electrospun collagen nanofibers. *In vitro* evaluation with osteoblast-like cells (MG63) demonstrated a synergistic effect of collagen nanofibers and β-TCP resulting in improved cell adhesion and proliferation on the hierarchical composite scaffold. Natural polymers such as chitosan and their blends/composites have also been electrospun into nanofiber scaffolds as ECM-mimicking analogues for bone regeneration due to their osteocompatibility, biocompatibility, and the ability to incorporate functional groups through amine and hydroxyl groups within the polymer chain (Bhattarai et al. 2005; Zhang et al. 2008; Yang et al. 2008). Bhattarai et al. (2005) fabricated the electrospun chitosan/PEO nanofibers that promoted the adhesion of osteoblast (MG-63) cells and maintained characteristic cell morphology and cell phenotype during *in vitro* culture. Zhang et al. (2008) reported the synthesis of produced chitosan/hydroxyapatite nanocomposite nanofibers. Through *in vitro* cell culture with human fetal osteoblast cells, the incorporation of hydroxyapatite has significantly stimulated the bone-forming ability as evidenced by the observation of cell proliferation, mineral deposition, and morphology, due to the excellent osteoconductivity of hydroxyapatite.

In other studies, silica-based sol–gel glass ($70SiO_2{\cdot}25CaP{\cdot}5P_2O_5$) mixed with a polymer binder was electrospun into a nanofibrous mesh and heat-treated to produce fibers with diameter ranging from 84 nm to 640 nm by varying the sol concentration (Kim et al. 2006). This glass nanofiber induced the formation of a bone mineral-like apatite phase on the fiber surface in a simulated body fluid. This fibrous nanocomposite supported rat bone marrow stromal cells and promoted osteogenic differentiation better than on sintered bioactive glass or PCL polymer nanofiber alone. To mimic the bone ECM, a gelatin-hydroxyapatite nanofiber was designed; the gelatin and hydroxyapatite precipitates were dissolved in an organic solvent and composite nanofibers were fabricated by an electrospinning technique (Kim et al. 2005). In this composite, hydroxyapatite nanocrystallites were evenly distributed within the gelatin matrix throughout the nanofiber. The gelatin amino acid sequences directed the precipitation of hydroxyapatite crystals within, and the composite scaffold enhanced osteoblastic differentiation.

Our laboratory pioneered the use of biodegradable polyphosphazenes as unique candidate materials for electrospun nanofiber scaffolds due to synthetic flexibility, buffering and neutral degradation products, and excellent biocompatibility (Nair et al. 2004b; Bhattacharyya et al. 2006, 2009). Nair et al. (2004b) demonstrated that the electrospun nanofiber matrices of poly[bis(*p*-methylphenoxy)phosphazene] supported the adhesion and proliferation of osteoblasts such as MC3T3-E1 cells. Bhattacharyya et al. (2006, 2009) fabricated poly[bis(ethyl alanato)phosphazene] (PNEA) and PNEA–hydroxyapatite composite nanofiber matrices as scaffolds for bone tissue engineering applications. Such polyphosphazene nanofiber structures closely mimic the ECM architecture, and have shown excellent osteoconductivity and osteointegration (Bhattacharyya et al. 2006, 2009; Nair et al. 2004b; Conconi et al. 2006).

Although nanofiber matrices have already shown some great promise for regeneration applications, significant challenges exist in creating complex 3-D structures with appropriate compressive properties to support load-bearing defects. With a more in-depth investigation into the constituents of bone and their properties, several biomimetic scaffold platforms that closely mimic the physicochemical and mechanical characteristics of native bone have been developed in our laboratories (Laurencin et al. 2010; Deng et al. 2011b). Inspired by the hierarchical structures that enable bone function, Deng et al. (2011c) recently fabricated a mechanically competent 3-D scaffold mimicking the bone marrow cavity, as well as the lamellar structure of bone, by orienting electrospun dipeptide-based polyphosphazene-polyester blend nanofibers in a concentric manner with an open central cavity. The 3-D biomimetic scaffold exhibited a similar characteristic mechanical behavior to that of native bone. Compressive modulus of the scaffold was found to be within the range of human trabecular bone. As shown in Figure 15.4, *in vitro* studies using primary cell cultures demonstrated the ability of the biomimetic scaffold to support osteoblast proliferation and accelerated differentiation throughout the scaffold architecture, which resulted in a similar cell–matrix organization to that of native bone and maintenance of structure integrity. It was thus suggested that the concentric open macrostructures of nanofibers that structurally and mechanically mimic the native bone can be a potential scaffold design for accelerated bone healing.

In addition to electrospinning, self-assembly (Huang et al. 2003; Whitesides and Grzybowski 2002; Hartgerink et al. 2001, 2002) or phase separation (Huang et al. 2003; Ma and Zhang 1999) have also been actively explored to fabricate nanofiber scaffolds that emulate natural ECM both structurally and functionally. Stupp et al. have developed a class of self-assembling peptide amphiphiles (PAs) composed of a hydrophobic alkyl tail and a hydrophilic oligopeptide head (Hartgerink et al. 2001, 2002). The self-assembly of PAs in water induces a liquid-to-gel transformation due to the hydrophobic collapse of alkyl tail domains and hydrogen bonding between β-sheet regions of neighboring PA molecules, and result in nanofibers that are 6–12 nm in diameter and up to several microns in length. Furthermore, it was demonstrated that PAs having a phosphoserine residue and RGD cell-adhesion epitope directed the mineralization of hydroxyapatite to resemble the structural orientation between collagen and hydroxyapatite observed in bone. In a recent report, Lee et al. (2009) prepared a PA gel containing BMP receptor–binding peptides

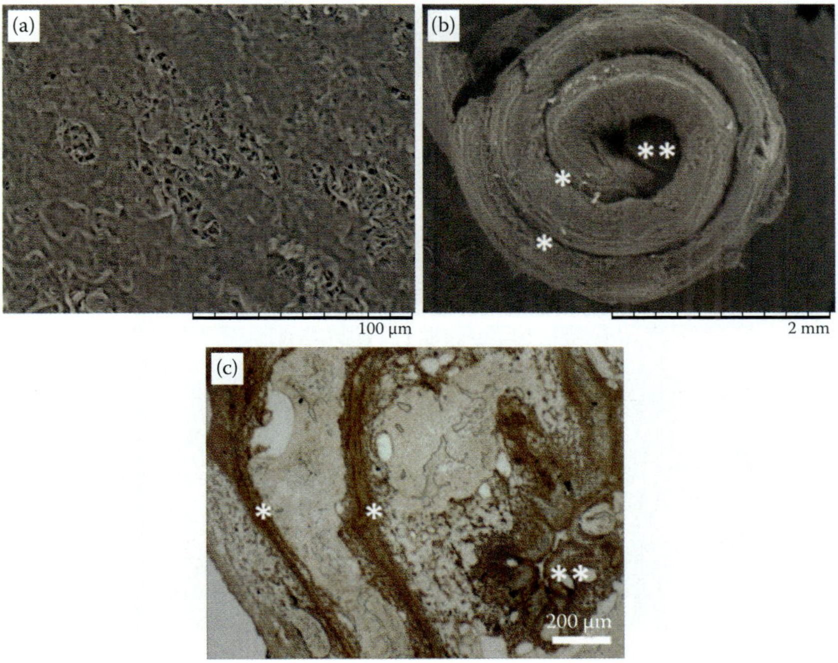

FIGURE 15.4 Biomimetic scaffolds composed of electrospun nanofibers of dipeptide-based polyphosphazene-polyester blends mimicking the collagen fibrils present in native bone ECM. (a) SEM image showing osteoblasts covering the blend nanofiber surface with a well-spread morphology by day 3. (b) SEM image illustrating the morphologies of cell-seeded 3-D biomimetic scaffolds after 28 days of culture. (c) Immunohistochemical staining for osteopontin (OPN), a prominent component of the mineralized ECM, illustrating a homogenous ECM distribution throughout the scaffold architecture at day 28. Interlamellar space (*); central cavity (**). (Reproduced from Deng, M. et al. *Adv Funct Mater* 21(14): 2641–2651, 2011. With permission.)

(osteopromotive domains; OPDs) identified from BMP-2, and hydrophobic alkyl chains to induce differentiation of human bone marrow stromal cells (hBMSCs) into osteoblastic cells. Mixing the OPD peptide-amphiphiles (OPDAs) solution with calcium ions led to the formation of nanofibrous structures (Figure 15.5a). The self-assembled OPDA gels showed enhanced proliferation and osteoblastic differentiation of hBMSCs compared with control matrices without OPD peptide. A significant amount of green fluorescent spots, indicating calcein uptake into newly mineralized ECM, suggested greater osteogenic mineralization when cells were cultured on OPDA gels as compared with those cultured in control PA gels and cells without gels (Figure 15.5b and c). However, these self-assembling gels often exhibit inferior mechanical properties to those of the native bone, which limits their applications at the load-bearing sites. Brown et al. (2010) developed a novel structure that combines the robust mechanical aspects of the sintered microsphere scaffold with a highly bioactive nanofiber structure to produce a composite scaffold that

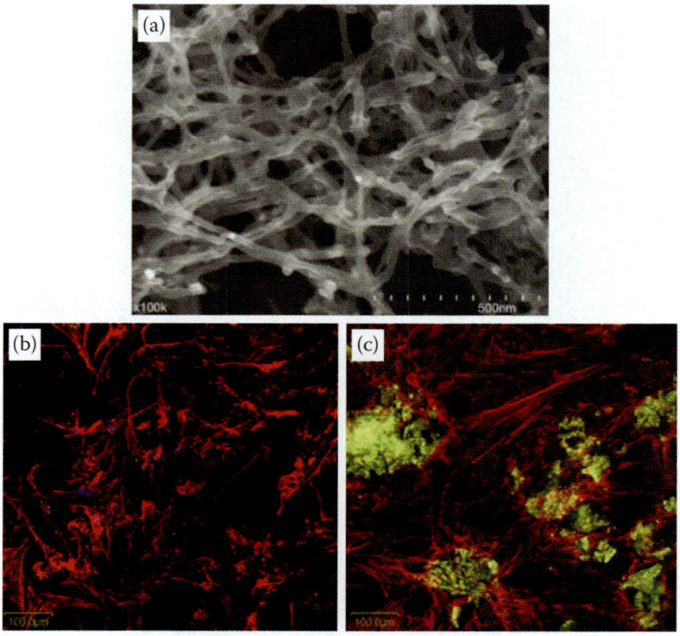

FIGURE 15.5 A self-assembly PA gel containing a BMP receptor–binding peptide OPD and hydrophobic alkyl chains to induce differentiation of hBMSCs into osteoblastic cells. (a) SEM image of a self-assembled OPDA nanofiber network. (b) and (c) Confocal micrographs of a specific calcium-binding fluorescent dye (calcein) incorporated into hBMSCs cultured in a (b) self-assembled control PA gel and (c) OPDA gel to examine newly formed minerals. A significant amount of green fluorescent spots indicating calcein uptake into newly mineralized ECM suggested greater osteogenic mineralization when cells were cultured on OPDA gels as compared with those cultured in control PA gels and cells without gels. (Reproduced from Lee, J.-Y. et al. *Biomaterials* 30(21): 3532–3541, 2009. With permission.)

demonstrates an ability to mimic the mechanical environment of trabecular bone while also promoting the osteoinduction of osteoblast progenitor cells. Exploiting the chemistry of two biodegradable polymers, a 3-D PLLA nanofiber mesh was successfully incorporated within the void spaces between sintered PNEPhA microspheres via thermally induced phase separation (Figure 15.6a and b). The nonload-bearing fiber portion of these scaffolds is sufficiently porous to allow cell migration and ECM matrix production throughout the fibrous portion of the scaffold (Figure 15.6c and d). These composite nanofiber/microsphere scaffolds promote osteoinduction through focal adhesion kinase activity. Ultimately, the focal adhesion kinase activity on the composite nanofiber/microsphere scaffolds demonstrated causality over the production of the mature osteoblast marker, osteocalcin, and the development of a calcified matrix. The phenotype progression of osteoblast progenitor cells on the composite nanofiber/microsphere scaffolds illustrated a stronger and more rapid progression, leading to fully matured osteoblasts by 21 days.

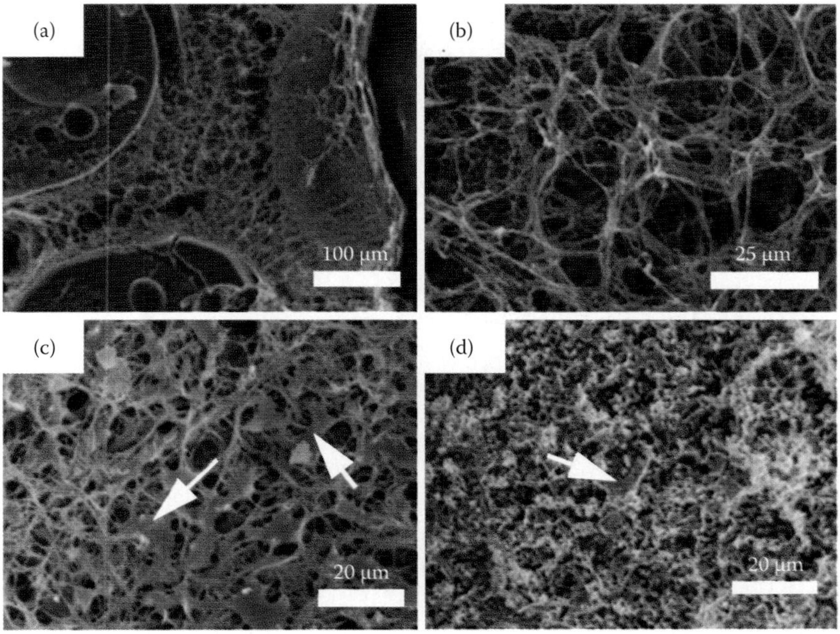

FIGURE 15.6 Composite nanofiber/microsphere scaffolds bridging nanoscale and microscale architectures to improve the bioactivity of mechanically competent constructs for bone regeneration. SEM images of the cross sections of the composite nanofiber/microsphere scaffold demonstrating the incorporation of nanofibers of less than 1 μm within the void space generated by sintering microspheres: (a) low-magnification image and (b) high-magnification image. (c) and (d) SEM image of preosteoblasts in the interior of composite scaffolds after (c) 3 and (d) 14 days of culture, demonstrating the presence of preosteoblasts and the accumulation of matrix proteins in the interior of the composite scaffold during cell culture. The arrows point to representative cells. Notably, the fibrous portion of the scaffold has been extensively modified through the accumulation of ECM proteins after 14 days, to the extent that identifying the cells is difficult. (Reproduced from Brown, J. L. et al. *J Biomed Mater Res A* 95(4): 1150–1158, 2010. With permission.)

Controlled delivery of biological factors to direct stem cell behavior using nanostructured materials is also one of the promising strategies for accelerated bone healing. Nie et al. (2008) prepared BMP-2–loaded PLAGA/nHA composite fiber scaffolds through either encapsulation into fibers during electrospinning or coating on fiber surface postelectrospinning. The composite scaffolds showed sustained release of BMP-2 in the time frame of 2–8 weeks. With the increase of nHA content, a greater release rate was observed due to the increased hydrophilicity of the scaffolds (Figure 15.7). It was confirmed that BMP-2 encapsulated into fibers retained its integrity and structural conformation. Furthermore, cell culture evaluation with hMSCs demonstrated the effect of nHA in enhancing cell attachment and viability. In an *in vivo* evaluation using a tibia nude defect model in mouse, the PLAGA/nHA composite scaffolds resulted in enhanced new bone formation and healing of

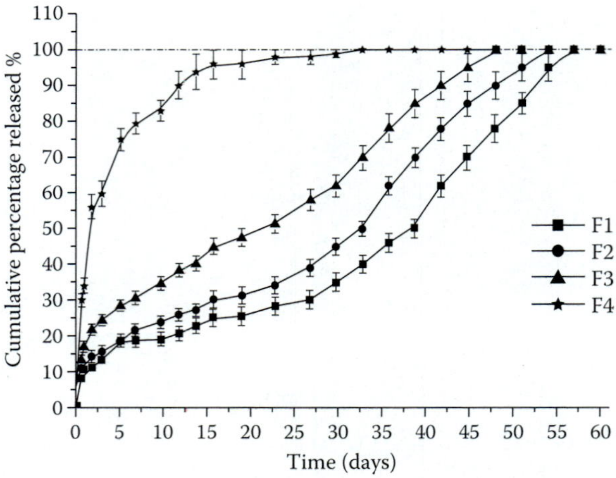

FIGURE 15.7 The cumulative *in vitro* release kinetics of BMP-2 in terms of the percentage mass released over the original mass of protein present over a course of 60 days from the electrospun fiber scaffolds including F1 (PLAGA), F2 (PLAGA/nHA, 20:1, w/w), F3 (PLAGA/nHA, 10:1, w/w), and F4 (PLAGA/nHA, 20:1, w/w). For F1–F3, the BMP-2 solution was encapsulated in the fibers by adding directly into the electrospinning solution, whereas the BMP-2 was adsorbed to electrospun fiber scaffolds for F4. In general, the composite scaffolds showed sustained release of BMP-2 in the time frame of 2–8 weeks. With the increase of nHA content, a greater release rate was observed due to the increased hydrophilicity of the scaffolds. (Reproduced from Nie, H. et al. *Biotechnol Bioeng* 99(1): 223–234, 2008. With permission.)

segmental defects, indicating the bioactivity of released BMP-2 (Fu et al. 2008). Liao et al. (2004) also showed the enhanced healing effect of BMP-2 released from a PLA/collagen nanofiber composite matrix with nHA. Such enhanced healing effect could be due to the ability of nanofibrous structures to better retain growth factors due to high surface area and nanotextured features.

These studies have collectively demonstrated the feasibility of achieving controlled cellular responses and effective bone regeneration with the use of nanostructured scaffolds alone or combined with cells or biological factors. In summary, nanofiber and nanosphere composites fabricated from polymer blends or polymer composites constitute a versatile biomaterial platform for developing nanostructured scaffolds for bone tissue engineering. The advancements in nanotechnology, in combination with the versatility of material chemistry, further allows for the development of biomimetic and bioactive scaffolds for enhanced fracture healing applications.

15.6 CONCLUSION AND FUTURE DIRECTIONS

The nanobiomaterial-based repair/regeneration strategies discussed previously provide an insight into the development of ideal bone graft substitutes. Nanotopographical and chemical cues play an important role in the scaffold regenerative efficacy. In particular, various polymeric composite/blend scaffolds with nanofeatures have been created and have shown great promise for bone tissue engineering applications.

However, regeneration of complex tissues, organs, or organ systems such as human limbs remains a significant challenge.

The need for reconstructing complex tissues and a whole organ structure, such as in the repair of severely damaged or diseased limbs, has necessitated the combination of classical top–down tissue engineering approach with bottom–up strategies used in regenerative biology, which represents a new multidisciplinary paradigm to engineer complex tissues, termed as "regenerative engineering" (Laurencin et al. 2010; Reichert et al. 2011). Regenerative biology provides an acute understanding on the bottom–up approach through which native tissues are self-assembled from highly proliferative and differentiated cell layers. For example, the bottom–up approach to limb regeneration is seeking to determine why humans lack the ability to regenerate lost organs and whether epimorphic regeneration, seen in urodele amphibians, could be translated to human regenerative medicine. In a traditional top–down approach, the limb structure as a whole is considered, developed tissue by tissue, and then integrated into one contiguous structure. Regenerative engineering is "the integration of materials science and tissue engineering with stem and developmental cell biology and regenerative medicine toward the regeneration of complex tissues, organs, or organ systems" (Laurencin et al. 2010; Reichert et al. 2011). It has elements of tissue engineering, regenerative medicine, and morphogenesis but is distinct from these individual disciplines in that regenerative engineering focuses specifically on the integrated response of stem cells to biomaterials and utilization of biomaterials to guide tissue development for regenerating multiple tissue types simultaneously into one intact and functional organ. Thus, regenerative engineering seeks to develop tissue-inducing biomaterials built on the knowledge learned from native tissue development to direct the same cell population toward different cell types through location-specific topographies and physicochemical cues within a continuous tissue phase.

The developmental phases of an embryo provide us with significant insights into the formation of various tissues and organs in the body. Specifically, morphogenesis provides a fundamental understanding of biological processes during the formation, maintenance, degeneration, and regeneration of tissues or organs. The emphasis of morphogenesis in a regenerative engineering approach is focused on studying the implications of regulated cellular activity in the formation of multicellular arrays and consequent complex tissue systems. It is critical to understand the molecular mechanisms governing cells and cell–material interaction to generate better scaffold and implant performance. The ability of stem cells to differentiate down different lineages enables the possibility of investigating the effect of various physicochemical and mechanical cues on cellular activity and capitalizing on the plasticity of the cells for complex tissue regeneration. The complexity within the native tissues or organs, such as human limbs, necessitates the development of biomimetic integrated graft systems to elicit precise control over stem cells into distinct lineages while maintaining a continuous tissue phase. The integration of physical, chemical, and mechanical cues built from the nanoscale will help design synthetic structures to regenerate complex biological tissues possessing tissue type heterogeneity, anisotropic mechanical properties, and well-defined tissue-implant interactions. The next step involves systematic evaluation of stem cell responses to various topographical features and chemical functionalities to achieve graft controls and focus on the development of integrated graft systems for the regeneration of multicomponent tissues. The emergence of induced pluripotent stem cell technology will strengthen the convergence of advanced nanotechnology, developmental biology and regenerative engineering for translational tissue repair.

ACKNOWLEDGMENTS

The authors acknowledge funding from NIH RO1 EB004051, RO1 AR052536, and NSF EFRI-0736002. Dr. Laurencin was the recipient of a Presidential Faculty Fellow Award from the National Science Foundation.

REFERENCES

American Academy of Orthopaedic Surgeons (AAOS). 2010. The evolving role of bone-graft substitutes. http://www.aatb.org/aatb/files/ccLibraryFiles/Filename/000000000322/BoneGraft Substitutes2010.pdf. Accessed February 2012.

Athanasiou, K. A., C-F. Zhu, D. R. Lanctot, C. M. Agrawal, and X. Wang. 2000. Fundamentals of biomechanics in tissue engineering of bone. *Tissue Engineering* 6(4): 361–381.

Aubin, J. E. 1998. Bone stem cells. Journal of Cellular Biochemistry. *Supplement* 30–31:73–82.

Beddington, R. S., and E. J. Robertson. 1999. Axis development and early asymmetry in mammals. *Cell* 96(2): 195–209.

Berrey, B. H., C. F. Lord, M. C. Gebhardt, and H. J. Mankin. 1990. Fractures of allografts. Frequency, treatment, and end-results. *Journal of Bone and Joint Surgery. American Volume* 72(6): 825–833.

Bhattacharyya, S., S. G. Kumbar, Y. M. Khan, L. S. Nair, A. Singh, N. R. Krogman, P. W. Brown, H. R. Allcock, and C. T. Laurencin. 2009. Biodegradable polyphosphazene-nanohydroxyapatite composite nanofibers: Scaffolds for bone tissue engineering. *Journal of Biomedical Nanotechnology* 5(1): 69–75.

Bhattacharyya, S., L. S. Nair, A. Singh, N. R. Krogman, Y. E. Greish, P. W. Brown, H. R. Allcock, and C. T. Laurencin. 2006. Electrospinning of poly[bis(ethyl alanato) phosphazene] nanofibers. *Journal of Biomedical Nanotechnology* 2(1): 36–45.

Bhattarai, N., D. Edmondson, O. Veiseh, F. A. Matsen, and M. Zhang. 2005. Electrospun chitosan-based nanofibers and their cellular compatibility. *Biomaterials* 26(31): 6176–6184.

Bilezikian, J. P., L. G. Raisz, and G. A. Rodan. 1996. Principles of Bone Biology. San Diego, CA: Academic Press.

Blair, H. C., S. L. Teitelbaum, R. Ghiselli, and S. Gluck. 1989. Osteoclastic bone resorption by a polarized vacuolar proton pump. *Science* 245(4920): 855–857.

Braddock, M., P. Houston, C. Campbell, and P. Ashcroft. 2001. Born again bone: Tissue engineering for bone repair. *News in Physiological Sciences* 16(5): 208–213.

Brown, J. L., M. S. Peach, L. S. Nair, S. G. Kumbar, and C. T. Laurencin. 2010. Composite scaffolds: Bridging nanofiber and microsphere architectures to improve bioactivity of mechanically competent constructs. *Journal of Biomedical Materials Research—Part A* 95(4): 1150–1158.

Buckwalter, J. A., M. J. Glimcher, R. R. Cooper, and R. Recker. 1995. Bone biology—Part I: Structure, blood supply, cells, matrix, and mineralization. *Journal of Bone and Joint Surgery. American Volume* 77-A(8): 1256–1275.

Carano, R. A. D., and E. H. Filvaroff. 2003. Angiogenesis and bone repair. *Drug Discovery Today* 8(21): 980–989.

Chan, C. K., T. S. Kumar, S. Liao, R. Murugan, M. Ngiam, and S. Ramakrishnan. 2006. Biomimetic nanocomposites for bone graft applications. *Nanomedicine* 1(2): 177–188.

Christenson, E. M., K. S. Anseth, J. J. van den Beucken, C. K. Chan, B. Ercan, J. A. Jansen, C. T. Laurencin, et al. 2007. Nanobiomaterial applications in orthopedics. *Journal of Orthopaedic Research* 25(1): 11–22.

Conconi, M. T., S. Lora, A. M. Menti, P. Carampin, and P. P. Parnigotto. 2006. *In vitro* evaluation of poly[bis(ethyl alanato)phosphazene] as a scaffold for bone tissue engineering. *Tissue Engineering* 12(4): 811–819.

Deng, M., R. James, C. T. Laurencin, and S. G. Kumbar. 2011b. Nanostructured polymeric scaffolds for orthopaedic regenerative engineering. *IEEE Transactions on Nanobioscience* 11(1): 3–14.

Deng, M., S. G. Kumbar, Kevin W. H. Lo, B. D. Ulery, and C. T. Laurencin. 2011a. Novel polymer-ceramics for bone repair and regeneration. *Recent Patents on Biomedical Engineering* 4(3): 168–184.

Deng, M., S. G. Kumbar, L. S. Nair, A. L. Weikel, H. R. Allcock, and C. T. Laurencin. 2011c. Biomimetic structures: Biological implications of dipeptide-substituted polyphosphazene-polyester blend nanofiber matrices for load-bearing bone regeneration. *Advanced Functional Materials* 21(14): 2641–2651.

Deng, M., S. G. Kumbar, Y. Wan, U. S. Toti, H. R. Allcock, and C. T. Laurencin. 2010a. Polyphosphazene polymers for tissue engineering: an analysis of material synthesis, characterization and applications. *Soft Matter* 6: 3119–3132.

Deng, M., L. S. Nair, S. P. Nukavarapu, S. G. Kumbar, J. L. Brown, N. R. Krogman, A. L. Weikel, H. R. Allcock, and C. T. Laurencin. 2010c. Biomimetic, bioactive etheric polyphosphazene-poly(lactide-co-glycolide) blends for bone tissue engineering. *Journal of Biomedical Materials Research—Part A* 92A(1): 114–125.

Deng, M., L. S. Nair, S. P. Nukavarapu, S. G. Kumbar, T. Jiang, N. R. Krogman, A. Singh, H. R. Allcock, and C. T. Laurencin. 2008. Miscibility and in vitro osteocompatibility of biodegradable blends of poly[(ethyl alanato) (p-phenyl phenoxy) phosphazene] and poly(lactic acid-glycolic acid). *Biomaterials* 29(3): 337–349.

Deng, M., L. S. Nair, S. P. Nukavarapu, S. G. Kumbar, T. Jiang, A. L. Weikel, N. R. Krogman, H. R. Allcock, and C. T. Laurencin. 2010d. In situ porous structures: a unique polymer erosion mechanism in biodegradable dipeptide-based polyphosphazene and polyester blends producing matrices for regenerative engineering. *Advanced Functional Materials* 20(17): 2794–2806.

Deng, M., L. S. Nair, S. P. Nukavarapu, T. Jiang, W. A. Kanner, X. Li, S. G. Kumbar, A. L. Weikel, N. R. Krogman, H. R. Allcock, and C. T. Laurencin. 2010b. Dipeptide-based polyphosphazene and polyester blends for bone tissue engineering. *Biomaterials* 31(18): 4898–4908.

D'Lima, D. D., S. M. Lemperle, P. C. Chen, R. E. Holmes, and C. W. Colwell, Jr. 1998. Bone response to implant surface morphology. *Journal of Arthroplasty* 13(8): 928–934.

Einhorn, T. A. 1998. The cell and molecular biology of fracture healing. Clinical Orthopaedics and Related Research 355: S7–21.

Elias, K. L., R. L. Price, and T. J. Webster. 2002. Enhanced functions of osteoblasts on nanometer diameter carbon fibers. *Biomaterials* 23(15): 3279–3287.

Ferrari, S. L., K. Traianedes, M. Thorne, M. H. Lafage-Proust, P. Genever, M. G. Cecchini, V. Behar et al. 2000. A role for N-cadherin in the development of the differentiated osteoblastic phenotype. *Journal of Bone and Mineral Research* 15(2): 198–208.

Formhals, A. 1934. Process and apparatus for preparing artificial threads. U.S. patent 1,975,504.

Fu, Y.-C., H. Nie, M.-L. Ho, C.-K. Wang, and C.-H. Wang. 2008. Optimized bone regeneration based on sustained release from three-dimensional fibrous PLGA/HAp composite scaffolds loaded with BMP-2. *Biotechnology and Bioengineering* 99(4): 996–1006.

Giraud-Guille, M. 1988. Twisted plywood architecture of collagen fibrils in human compact bone osteons. *Calcified Tissue International* 42(3): 167–180.

Goulet, J. A., L. E. Senunas, G. L. DeSilva, and M. L. Greenfield. 1997. Autogenous iliac crest bone graft: Complications and functional assessment. *Clinical Orthopaedics and Related Research* 339: 76–81.

Grauer, J. N., J. M. Beiner, B. K. Kwon, and A. R. Vaccaro. 2003. Bone graft alternatives for spinal fusion. *BioDrugs* 17(6): 391–394.

Hartgerink, J. D., E. Beniash, and S. I. Stupp. 2001. Self-assembly and mineralization of peptide-amphiphile nanofibers. *Science* 294(5547): 1684–1688.

Hartgerink, J. D., E. Beniash, and S. I. Stupp. 2002. Peptide-amphiphile nanofibers: A versatile scaffold for the preparation of self-assembling materials. *Proceedings of the National Academy of Sciences* 99(8): 5133–5138.

Horch, R. A., N. Shahid, A. S. Mistry, M. D. Timmer, A. G. Mikos, and A. R. Barron. 2004. Nanoreinforcement of poly(propylene fumarate)-based networks with surface modified alumoxane nanoparticles for bone tissue engineering. *Biomacromolecules* 5(5): 1990–1998.

Huang, Z. M., Y. Z. Zhang, M. Kotaki, and S. Ramakrishna. 2003. A review on polymer nanofibers by electrospinning and their applications in nanocomposites. *Composites Science and Technology* 63: 2223–2253.

Huinan, L. et al. 2005. Increased osteoblast functions on nanophase titania dispersed in polylactic-co-glycolic acid composites. *Nanotechnology* 16(7): S601.

Ilan, D. I., and A. L. Ladd. 2003. Bone graft substitutes. *Operative Techniques in Plastic Reconstrive Surgery* 9(4): 151–160.

Ito, H., M. Koefoed, P. Tiyapatanaputi, K. Gromov, J. J. Goater, J. Carmouche, X. Zhang et al. 2005. Remodeling of cortical bone allografts mediated by adherent rAAV-RANKL and VEGF gene therapy. *Nature Medicine* 11(3): 291–297.

Jee, W. S. S., and S. C. Cowin. 2001. Integrated bone tissue physiology: Anatomy and physiology. In Bone Mechanics Handbook: CRC Press LLC.

Khan, Y., and C. T. Laurencin. 2008. Fracture repair with ultrasound: Clinical and cell-based evaluation. *Journal of Bone and Joint Surgery. American Volume* 90(Suppl 1): 138–144.

Khan, Y., M. J. Yaszemski, A. G. Mikos, and C. T. Laurencin. 2008. Tissue engineering of bone: material and matrix considerations. *Journal of Bone and Joint Surgery. American Volume* 90(Suppl 1): 36–42.

Kierszenbaum, A. L. 2002. Connective tissue. In Histology and Cell Biology: An Introduction to Pathology. St. Louis: Mosby, Inc.

Kim, H. W., H. E. Kim, and J. C. Knowles. 2006. Production and potential of bioactive glass nanofibers as a next-generation biomaterial. *Advanced Functional Materials* 16(12): 1529–1535.

Kim, H. W., J. H. Song, and H. E. Kim. 2005. Nanofiber generation of gelatin–hydroxyapatite biomimetics for guided tissue regeneration. *Advanced Functional Materials* 15(12): 1988–1994.

Kuboki, Y., H. Takita, D. Kobayashi, E. Tsuruga, M. Inoue, M. Murata, N. Nagai et al. 1998. BMP-induced osteogenesis on the surface of hydroxyapatite with geometrically feasible and nonfeasible structures: Topology of osteogenesis. *Journal of Biomedical Materials Research* 39(2): 190–199.

Kumbar, S. G., M. Kofron, L. S. Nair, and C. T. Laurencin. 2008a. Cell behavior toward nanostructured surfaces. In Biomedical Nanostructures, edited by K. Gonsalves, C. Halberstadt, C. T. Laurencin, and L. S. Nair. New York: John Wiley & Sons.

Kumbar, S. G., S. P. Nukavarapu, R. James, M. V. Hogan, and C. T. Laurencin. 2008c. Recent patents on electrospun biomedical nanostructures: An overview. *Recent Patents on Biomedical Engineering* 1:68–78.

Kumbar, S. G., Roshan J., S. P. Nukavarapu, and C. T. Laurencin. 2008b. Electrospun nanofiber scaffolds: Engineering soft tissues. *Biomedical Materials* 3(3): 034002.

Landis, W. J. 1995. The strength of a calcified tissue depends in part on the molecular structure and organization of its constituent mineral crystals in their organic matrix. *Bone* 16(5): 533–544.

Langer, R. 2000. Tissue engineering. *Molecular Therapy* 1(1): 12–15.

Langer, R., and J. P. Vacanti. 1993. Tissue engineering. *Science* 260(5110): 920–926.

Laurencin, C. T., A. M. A. Ambrosio, M. D. Borden, and J. A. Cooper. 1999. Tissue engineering: Orthopedic applications. *Annual Review of Biomedical Engineering* 1(1): 19–46.

Laurencin, C. T., and Y. Khan. 2012. Bone graft substitute materials. http://emedicine.medscape.com/article/1230616-overview.

Laurencin, C., Y. Khan, and S. F. El-Amin. 2005. Bone graft substitutes. *Expert Review of Medical Devices* 3(1): 49–57.

Laurencin, C. T., S. G. Kumbar, M. Deng, and R. James. 2010. Nanostructured scaffolds for regenerative engineering. In Honorary Series in Translational Research in Biomaterials, 2010 AICHE Annual Meeting. Salt Lake City, UT.

Lecanda, F., D. A. Towler, K. Ziambaras, S. L. Cheng, M. Koval, T. H. Steinberg, and R. Civitelli. 1998. Gap junctional communication modulates gene expression in osteoblastic cells. *Molecular Biology of the Cell* 9(8): 2249–2258.

Lee, J.-Y., J.-E. Choo, Y.-S. Choi, J.-S. Suh, S.-J. Lee, C.-P. Chung, and Y.-J. Park. 2009. Osteoblastic differentiation of human bone marrow stromal cells in self-assembled BMP-2 receptor-binding peptide-amphiphiles. *Biomaterials* 30(21): 3532–3541.

Li, W. J., C. T. Laurencin, E. J. Caterson, R. S. Tuan, and F. K. Ko. 2002. Electrospun nanofibrous structure: A novel scaffold for tissue engineering. *Journal of Biomedical Materials Research* 60(4): 613–621.

Li, D., Y. Wang, and Y. Xia. 2003. Electrospinning of polymeric and ceramic nanofibers as uniaxially aligned arrays. *Nano Letters* 3(8): 1167–1171.

Li, Z., K. Kong, and W. Qi. 2006. Osteoclast and its roles in calcium metabolism and bone development and remodeling. *Biochemical and Biophysical Research Communications* 343(2): 345–350.

Liao, S. S., and F. Z. Cui. 2004. *In vitro* and *in vivo* degradation of mineralized collagen-based composite scaffold: Nanohydroxyapatite/collagen/poly(L-lactide). *Tissue Engineering* 10(1–2): 73–80.

Liao, S. S., F. Z. Cui, W. Zhang, and Q. L. Feng. 2004. Hierarchically biomimetic bone scaffold materials: Nano-HA/collagen/PLA composite. Journal of Biomedical Materials Research—Part B: *Applied Biomaterials*. 69B(2): 158–165.

Liu, X., and P. X. Ma. 2004. Polymeric scaffolds for bone tissue engineering. *Annals of Biomedical Engineering* 32(3): 477–486.

Ma, P. X., and R. Zhang. 1999. Synthetic nanoscale fibrous extracellular matrix. *Journal of Biomedical Materials Research* 46(1): 60–72.

Ma, Z., M. Kotaki, R. Inai, and S. Ramakrishna. 2005. Potential of nanofiber matrix as tissue-engineering scaffolds. *Tissue Engineering* 11(1–2): 101–109.

Marotti, G. 1993. A new theory of bone lamellation. *Calcified Tissue International* 53(0): S47–S56.

McKibbin, B. 1978. The biology of fracture healing in long bones. *Journal of Bone and Joint Surgery. British Volume* 60-B(2): 150–162.

Mistry, A. S., A. G. Mikos, and J. A. Jansen. 2007. Degradation and biocompatibility of a poly(propylene fumarate)-based/alumoxane nanocomposite for bone tissue engineering. *Journal of Biomedical Materials Research—Part A* 83(4): 940–953.

Murugan, R., and S. Ramakrishna. 2005. Development of nanocomposites for bone grafting. *Composites Science and Technology* 65(15–16): 2385–2406.

Nair, L. S., S. Bhattacharyya, J. D. Bender, Y. E. Greish, P. W. Brown, H. R. Allcock, and C. T. Laurencin. 2004b. Fabrication and optimization of methylphenoxy substituted polyphosphazene nanofibers for biomedical applications. *Biomacromolecules* 5(6): 2212–2220.

Nair, L. S., S. Bhattacharyya, and C. T. Laurencin. 2004a. Development of novel tissue engineering scaffolds via electrospinning. *Expert Opinion on Biological Therapy* 4(5): 659–668.

Nair, L. S., and C. T. Laurencin. 2007. Biodegradable polymers as biomaterials. *Progress in Polymer Science* 32(8–9): 762–798.

Nair, L. S., and C. T. Laurencin. 2008. Nanofibers and nanoparticles for orthopaedic surgery applications. *Journal of Bone and Joint Surgery. American Volume* 90(Suppl 1): 128–131.

Nie, H., B. W. Soh, Y.-C. Fu, and C.-H. Wang. 2008. Three-dimensional fibrous PLGA/HAp composite scaffold for BMP-2 delivery. *Biotechnology and Bioengineering* 99(1): 223–234.

Patel, S., K. Kurpinski, R. Quigley, H. Gao, B. S. Hsiao, M.-M. Poo, and S. Li. 2007. Bioactive nanofibers: Synergistic effects of nanotopography and chemical signaling on cell guidance. *Nano Letters* 7(7): 2122–2128.

Pearson, O. M., and D. E. Lieberman. 2004. The aging of Wolff's "law": Ontogeny and responses to mechanical loading in cortical bone. *American Journal of Physical Anthropology* 125(S39): 63–99.

Pek, Y. S., Shujun Gao, M. S. Mohamed Arshad, Kwong-Joo Leck, and Jackie Y. Ying. 2008. Porous collagen–apatite nanocomposite foams as bone regeneration scaffolds. *Biomaterials* 29(32): 4300–4305.

Pelled, G., K. Tai, D. Sheyn, Y. Zilberman, S. Kumbar, L. S. Nair, C. T. Laurencin, D. Gazit, and C. Ortiz. 2007. Structural and nanoindentation studies of stem cell-based tissue-engineered bone. *Journal of Biomechanics* 40(2): 399–411.

Price, R. L., M. C. Waid, K. M. Haberstroh, and T. J. Webster. 2003. Selective bone cell adhesion on formulations containing carbon nanofibers. *Biomaterials* 24(11): 1877–1887.

Reichert, W. M., B. D. Ratner, J. Anderson, A. Coury, A. S. Hoffman, C. T. Laurencin, and D. Tirrell. 2011. 2010 Panel on the biomaterials grand challenges. *Journal of Biomedical Materials Research—Part A* 96(2): 275–287.

Rey, C., J. L. Miquel, L. Facchini, A. P. Legrand, and M. J. Glimcher. 1995. Hydroxyl groups in bone mineral. *Bone* 16(5): 583–586.

Rho, J.-Y., L. Kuhn-Spearing, and P. Zioupos. 1998. Mechanical properties and the hierarchical structure of bone. *Medical Engineering and Physics* 20(2): 92–102.

Ross, M. H., G. I. Kaye, and W. Pawlina. 2003. Connective Tissue in Histology: A Text and Atlas. Philadelphia: Lippincott Williams and Wilkins.

Schindeler, A., M. M. McDonald, P. Bokko, and D. G. Little. 2008. Bone remodeling during fracture repair: The cellular picture. *Seminars in Cell and Developmental Biology* 19(5): 459–466.

Schneider, O. D., F. Weber, T. J. Brunner, S. Loher, M. Ehrbar, P. R. Schmidlin, and W. J. Stark. 2009. In vivo and in vitro evaluation of flexible, cottonwool-like nanocomposites as bone substitute material for complex defects. *Acta Biomaterialia* 5(5): 1775–1784.

Shi, X., J. L. Hudson, P. P. Spicer, J. M. Tour, R. Krishnamoorti, and A. G. Mikos. 2006. Injectable manocomposites of single-walled carbon nanotubes and biodegradable polymers for bone tissue engineering. *Biomacromolecules* 7(7): 2237–2242.

Sul, Y. T., C. B. Johansson, S. Petronis, A. Krozer, Y. Jeong, A. Wennerberg, and T. Albrektsson. 2002. Characteristics of the surface oxides on turned and electrochemically oxidized pure titanium implants up to dielectric breakdown: The oxide thickness, micropore configurations, surface roughness, crystal structure and chemical composition. *Biomaterials* 23(2): 491–501.

Tomford, W. W. 2003. Overview. In *Bone Graft Substitutes*, edited by C. T. Laurencin. West Conshohocken, PA: ASTM.

van der Rest, M., and R. Garrone. 1991. Collagen family of proteins. *FASEB Journal* 5(13): 2814–2823.

Webster, T. J., and E. S. Ahn. 2007. Nanostructured biomaterials for tissue engineering bone. *Advances in Biochemical Engineering/Biotechnology* 103: 275–308.

Webster, T. J., R. W. Siegel, and R. Bizios. 1999. Osteoblast adhesion on nanophase ceramics. *Biomaterials* 20(13): 1221–1227.

Webster, T. J., and T. A. Smith. 2005. Increased osteoblast function on PLGA composites containing nanophase titania. *Journal of Biomedical Materials Research—Part A* 74A(4): 677–686.

Webster, T. J., C. Ergun, R. H. Doremus, R. W. Siegel, and R. Bizios. 2000. Enhanced functions of osteoblasts on nanophase ceramics. *Biomaterials* 21(17): 1803–1810.

Webster, T. J., C. Ergun, R. H. Doremus, R. W. Siegel, and R. Bizios. 2001. Enhanced osteoclast-like cell functions on nanophase ceramics. *Biomaterials* 22(11): 1327–1333.

Weiner, S., and W. Traub. 1992. Bone structure: From angstroms to microns. *FASEB Journal* 6(3): 879–885.

Weiner, S., and H. D. Wagner. 2003. The material bone: Structure–mechanical function relations. *Annual Review of Materials Science* 28(1): 271–298.

Whitesides, G. M., and B. Grzybowski. 2002. Self-assembly at all scales. *Science* 295(5564): 2418–2421.

Wiesmann, H. P., U. Meyer, U. Plate, and H. J. Höhling. 2004. Aspects of collagen mineralization in hard tissue formation. *International Review of Cytology* 242: 121–156.

Woo, K. M., V. J. Chen, and P. X. Ma. 2003. Nano-fibrous scaffolding architecture selectively enhances protein adsorption contributing to cell attachment. *Journal of Biomedical Materials Research—Part A* 67A(2): 531–537.

Yang, D., Y. Jin, Y. Zhou, G. Ma, X. Chen, F. Lu, and J. Nie. 2008. *In situ* mineralization of hydroxyapatite on electrospun chitosan-based nanofibrous scaffolds. *Macromolecular Bioscience* 8(3): 239–246.

Yeo, M. G., H. Lee, and G. H. Kim. 2010. Three-dimensional hierarchical composite scaffolds consisting of polycaprolactone, β-tricalcium phosphate, and collagen nanofibers: Fabrication, physical properties, and *in vitro* cell activity for bone tissue regeneration. *Biomacromolecules* 12(2): 502–510.

Zeleny, J. 1914. The electrical discharge from liquid points, and a hydrostatic method of measuring the electric intensity at their surfaces. *Physical Review* 3(2): 69–91.

Zhang, P., Z. Hong, T. Yu, X. Chen, and X. Jing. 2009. *In vivo* mineralization and osteogenesis of nanocomposite scaffold of poly(lactide-co-glycolide) and hydroxyapatite surface-grafted with poly(l-lactide). *Biomaterials* 30(1): 58–70.

Zhang, Y., J. R. Venugopal, A. El-Turki, S. Ramakrishna, B. Su, and C. T. Lim. 2008. Electrospun biomimetic nanocomposite nanofibers of hydroxyapatite/chitosan for bone tissue engineering. *Biomaterials* 29(32): 4314–4322.

Ziv, V., and S. Weiner. 2009. Bone crystal sizes: A comparison of transmission electron microscopic and X-ray diffraction line width broadening techniques. *Connective Tissue Research* 30(3): 165–175.

16 Stem Cells, Growth Factors, and Biomaterials in Bone Surgery

George K. B. Sándor

CONTENTS

16.1 INTRODUCTION

Bone reconstruction is required when there are defects in bone, or when bone is absent with missing adjacent soft tissue components. There are numerous causes of bone defects. Congenital bony defects in the craniofacial and maxillofacial skeleton can arise as a result of failed development such as in patients with a cleft lip and palate (Carmichael and Sándor 2008; Figure 16.1). Ablative surgery, in which

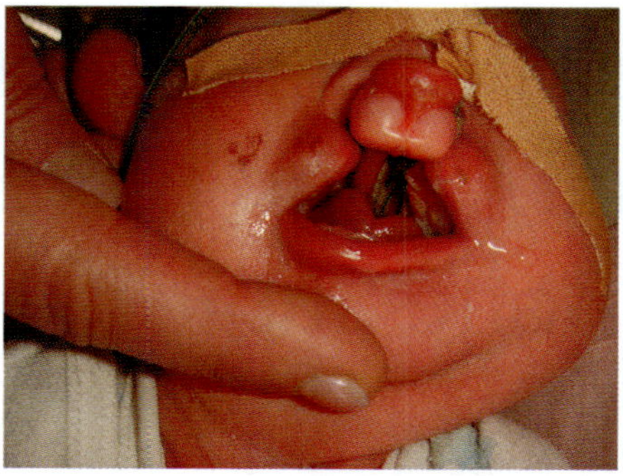

FIGURE 16.1 Classic complete bilateral cleft upper lip and palate deformity with missing bone, cartilage, mucosa, and skin.

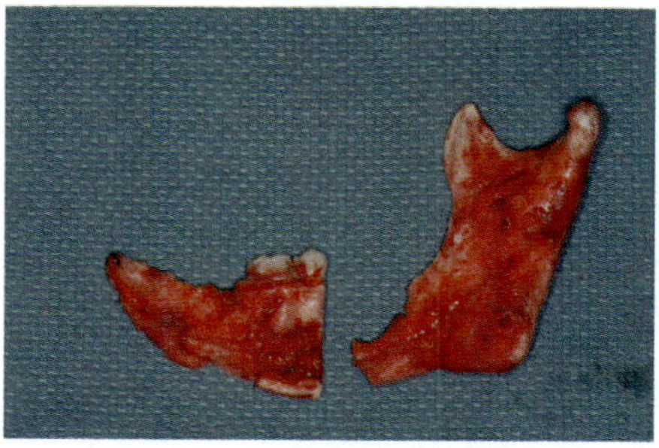

FIGURE 16.2 Resection specimen of the left mandibular body, ramus, coronoid process, and condyle due to refractory osteomyelitis leaving a large mandibular bone and soft tissue defect.

segments of bones are removed or resected to treat certain cysts, tumors, or infections, may produce large bony defects (Sándor et al. 2008; Figure 16.2). Large bony defects may result from accidental injury when tissue may have been traumatically avulsed (Sándor and Carmichael 2008). Such combined mucosal, osseous, and even cartilaginous defects can be reconstructed using flaps and bone grafts, or hopefully, in the future, with alloplastic biomaterials such as bone graft substitutes discussed previously in this book or even using cell therapy with tissue-engineered constructs (Sándor and Suuronen 2008). The healing of such complicated wounds always relies on the vascularity of the surrounding tissues (Fok et al. 2008).

16.2 BONE AS A REGENERATIVE TISSUE

16.2.1 Intramembranous versus Endochondral Bones

The head and neck are gifted areas for bony regeneration, because the blood supply is far more generous than the blood supply to the limbs and axial skeleton. Bones in the head and neck are formed by intramembranous ossification, without the need for cartilaginous precursors. This is unlike the bones of the axial skeleton and long bones, which require an additional step in endochondral ossification. In general, all bones of the head and neck (with the exception of the mandible) do not require a cartilaginous template, whereas all long bones and bones of the axial skeleton (with the exception of the scapula) require an endochondral process.

Great strides have been made at the clinical level in tissue engineering using cell therapy (Sándor and Suuronen 2008; Mesimäki et al. 2009); mainly, in those bones formed by the process of intramembranous ossification. This is largely due to the enriched blood supply found in the head and neck, and the lack of complex loading with large magnitudes of forces. The difference between the maxillofacial milieu and the orthopedic context lies in the differences between endochondral versus intramembranous origins, blood supply, and the order of magnitude of forces sustained in axial loading. This chapter will focus on the reconstruction of craniofacial bones formed by intramembranous ossification and will serve as a future stepping stone in the understanding of other complex problems of long bones in the axial skeleton.

16.2.2 Bone Reconstruction Principles

The reconstruction of a specific bone defect must follow logical and sound surgical principles. These authors use the concept of a reconstructive surgical ladder (Sándor et al. 2010a,b), in which techniques having a step-wise increasing complexity are used with a strong preference for the simplest possible procedure at the outset. The missing tissues are identified, including covering mucosa or skin, muscle, bone, cartilage, nerve, and blood vessels. Then, efforts ensue to replace missing tissue elements with similar or comparable tissue. For example, for the hard palate in the mouth, surgeons would always aim at replacing missing palatal mucosa with keratinized attached mucosa rather than totally dissimilar nonkeratinized unattached alveolar mucosa.

Then, flaps containing the missing tissue elements are planned using the following hierarchy, with the simplest reconstructive ladder technique being used first:

1. Local soft tissue grafts (free gingival grafts)
2. Local flaps (buccal advancement flaps or local pedicled flaps)
3. Regional flaps (tongue flaps)
4. Bone grafts harvested from intraoral sources (ramus, chin, or zygoma)
5. Bone grafts harvested from extraoral sources (iliac crest or tibia)
6. Distant pedicled flaps (delto-pectoral, sternocleidomastoid, or pectoralis flaps)
7. Distant microvascular flaps (fibula, iliac crest, latisimus dorsi, rectus abdominis, or scapula)
8. Tissue-engineered constructs consisting of alloplastic materials combined with growth factors or cell signaling molecules and mesenchymal stem cells (MSCs)

Reconstruction of membranous osseous defects in the oral and maxillofacial skeleton represent one of the most challenging tasks to the reconstructive surgeon and encompasses the paradigm of tissue engineering (Figure 16.3). Although significant advances have been made with bone substitutes, autogenous bone still remains the gold standard in the reconstruction of bone defects (Sándor et al. 2010b). To understand bone reconstruction, the surgeon must first fully understand bone healing and the healing of the surrounding soft tissues.

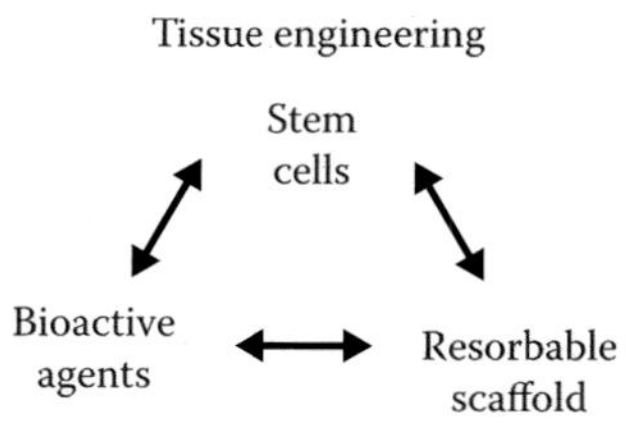

FIGURE 16.3 Classic tissue engineering triad representing the interaction between scaffolds or biomaterials, stem cells, and signaling molecules or growth factors.

16.2.3 Bone Healing

Inflammation, wound healing, vasculogenesis, and bone healing are all delicately intertwined. Three types of bone healing have been described including primary bone healing, secondary bone healing, and gap healing. The differences among the three are actually dependent on the size of the osseous defect and the rigidity of fixation. All three bone healing models provide information about the differences between bone regeneration within nongrafted as well as grafted defects.

Primary bone healing without callus formation occurs when bone ends are in direct contact, rigidly fixated, or anatomically reduced and are compressed by bone plates, for example. This process was initially described in 1949 after studying long bone fractures that were treated using rigid fixation with plating. Studying these fractures failed to show callus formation (Danis 1949). Osteoclasts were seen to begin cutting away cores on either side in the area of compression at a rate of 50–80 μm per day, progressing toward the fracture. Cutting cores were 200 μm wide, which provided space for vessel ingrowth, osteoblastic proliferation, and new bone formation (Simmons 1980).

Fractures treated without rigid fixation heal with secondary bone healing. A good example of nonrigid fixation is the wiring together of fractured jaws in maxillomandibular or intermaxillary fixation, or holding the ends of a fractured bone in near-apposition using a plaster cast in the distal radius. The initial injury elicits an inflammatory response and the activation of the complement cascade. Damage to surrounding blood vessels initiates cellular extravasation and cell signaling. Chemotactic factors (C5a, leukotriene B4) attract monocytes and macrophages (Marx et al. 1998). Activated macrophages then release fibroblast growth factor (FGF), which stimulate endothelial cells to release plasminogen activator and procollagenase. Growth factors such as platelet-derived growth factor (PDGF), transforming growth factor-β1 (TGF-β1), and transforming growth factor-β2 (TGF-β2) are released from the α granules of degranulating platelets. Early in wound healing, TGF-β1 and TGF-β2 are chemotactic stimulants for polymorphoneutrophils (PMNs), lymphocytes, monocytes, and macrophages. The blood clot acts as a hemostatic plug and contains the growth factors at the injured site and provides an environment for cell signaling.

Injured tissue is normally hypoxic, with low oxygen partial pressures of 5–10 mmHg, as well as being acidotic with pH between 4 and 6. Acidosis and hypoxia are required for PMN and macrophage stimulation (Marx et al. 1998). A proliferation phase begins with healing by day 3 and can last up to 40 days after fracture occurrence. A reparative phase follows with the appearance of new blood vessels, collagen, and cells. Osteoprogenitor cells are then stimulated to proliferate and differentiate into active chondroblasts and osteoblasts, laying down large amounts of extracellular matrices, and thereby forming a bridging callus.

In secondary bone healing, chondroblasts lay down amorphous chondroitin sulfate matrices which then become chondrocytes that eventually hypertrophy and die, leaving empty lacunae in a calcified matrix. These empty spaces (lacunae) allow for vascular ingrowth, resulting in a higher oxygen tension and normalized pH, which, in turn, favor the differentiation of osteoblasts. The cartilage is then removed by osteoclasts whereas osteoblasts lay down immature woven bone. Mobility at the regeneration site will disrupt the blood supply and result in cartilage and fibrous tissue predominance.

Defects that are larger and have no bony contact heal by gap healing and may require bone grafts to allow the bony wound to regenerate bone rather than produce a fibrous union or scar. Three essential processes are involved in the healing of bone grafts: osteogenesis, osteoinduction, and osteoconduction (Burchadt 1983):

1. Osteogenesis is defined as the formation of new bone using osteocompetent cells contained within bone tissue surrounding a defect or within a bone graft.
2. Osteoinduction is defined as the bone formation that results from primitive mesenchymal cells in the recipient bed that has been stimulated to differentiate into bone-forming cells by inductive proteins within the graft.
3. Osteoconduction is represented by the ingrowth of capillaries and osteoprogenitor cells from the recipient bed into and around the grafted material.

16.2.4 Cell Signaling Molecules

Cell signaling molecules or growth factors important in bone healing include the bone morphogenetic proteins (BMPs; Wozney et al. 1999), which are found in many species (Hu et al. 2011). BMPs are small, low molecular weight proteins of 19–30 kDa with a pH of 4.9–5.1 (Moghadam et al. 2001). Some BMPs stimulate MSCs to differentiate into osteoblasts during development and during bone healing (Ducy et al. 1997; Schmitt et al. 1999). BMPs are present in most tissues and play an important role in the remodeling of the adult skeleton. When osteoclasts resorb bone matrix, BMPs are released (Hu et al. 2009a,b), initiating further recruitment and differentiation of stem cell precursors to form osteoblasts and ultimately new bone (Dragoo et al. 2003).

BMP-7 or osteogenic protein-1 (OP-1) was first used in 2001 to treat a mandibular segmental defect in the craniofacial skeleton (Moghadam et al. 2001). BMP-7 was later used to aid in the healing of a maxillary Lefort I osteotomy (Warnke and Coren 2003). BMP-2 and BMP-7 are now approved for human use in maxillary sinus augmentation procedures. The recombinant human form, labeled rhBMP, can be produced using a Chinese hamster ovary cell line. Both laboratory and human studies using BMP show data with superior results in regenerating critical-sized defects (Moghadam et al. 2001, 2004). Clokie and Sándor's group were the first to report the clinical use of recombinant human BMP-7 in the human maxillofacial skeleton to manage mandibular postresection defects (Figure 16.4a). They documented ten cases of postresection mandibular defects successfully reconstructed with BMP-7 (OP-1) in

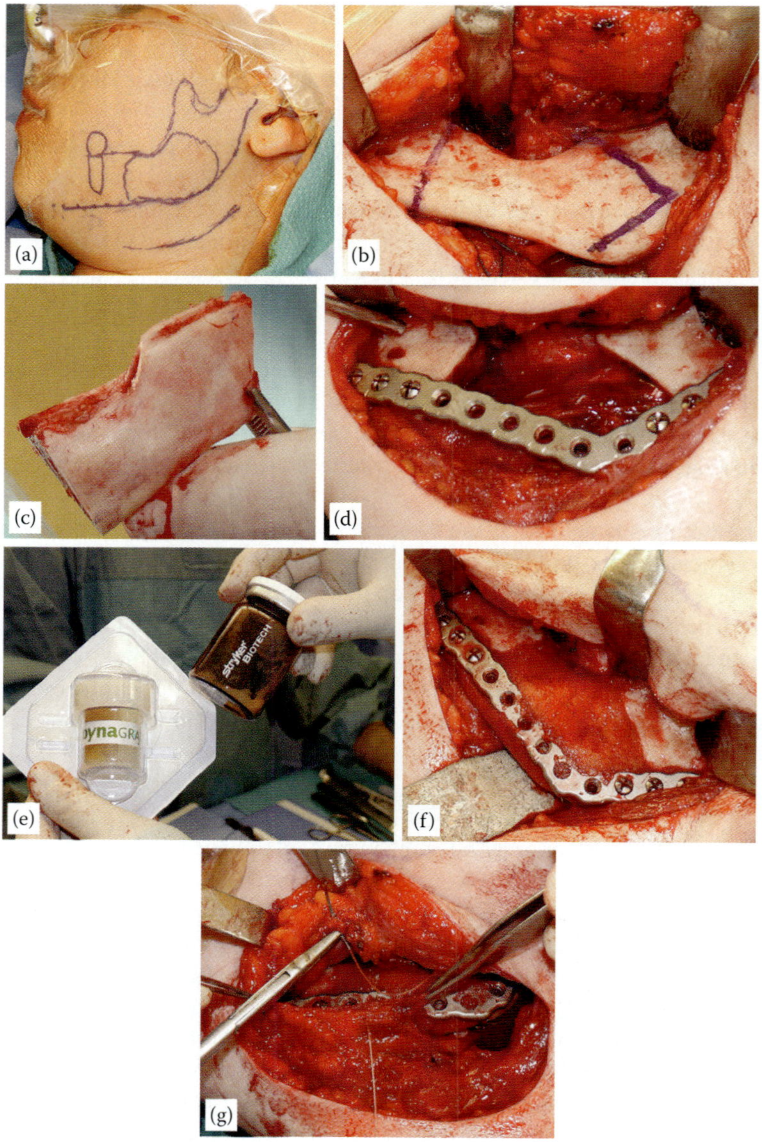

FIGURE 16.4 (a) Outline of mandible and ameloblastoma tumor drawn on the skin overlying the mandible. (b) Planned resection lines drawn on exposed mandible. (c) Resected fragment of mandible containing ameloblastoma tumor. (d) Defect caused by tumor resection is larger than a critical-sized defect and will not heal without some osteogenic intervention. (e) Components of the bioimplant include DBM mixed with BMP-7. (f) Bioimplant placed into the wound behind the reconstruction plate. (g) Muscle of pterygo-masseteric sling surrounds the BMP-DBM construct to provide maximal vascularity to the construct.

demineralized bone matrix (DBM) suspended in a reverse phase copolymer medium using poloxamer 407 (Figure 16.4a through g). Uneventful postoperative courses were reported, with dental rehabilitation completed 1 year after reconstruction (Moghadam et al. 2001; Clokie and Sándor 2008).

16.2.5 Blood Supply and the Soft Tissue Envelope

The establishment of a new blood supply is critical to wound healing (Bauer et al. 2005). This is demonstrated most clearly in pathologic states in which healing is compromised by diabetes, poor vascular supply, radiation, bisphosphonate exposure, or in which the normal healing process becomes pathologic, such as in instances of heterotopic bone formation in response to surgery or trauma (Du Val et al. 2007). In pathologic processes, the need for the reestablishment of nutrient and oxygen supply to support regeneration and repair of the wound exceeds the maintenance needs of the tissue that will ultimately form. Vascular endothelial growth factor (VEGF) is one growth factor involved in the recruitment, survival, and activity of endothelial cells, osteoblasts, and osteoclasts in wound healing (Bauer at al. 2005; Midy and Plouet 1994; Ferrara et al. 2003; Deckers and Karperien 2000; Engsig et al. 2000; Niida et al. 1999; Street et al. 2000).

Soft tissue wound healing is required to support bone healing and is characterized by four phases:

1. Coagulation (immediate–1 h)
2. Inflammatory (1 h–4 days)
3. Proliferative (3 days–21 days)
4. Remodeling (21 days onward).

Coagulation is characterized by endothelial damage and platelet activation resulting in a fibrin clot. Platelets release angiogenic cytokines including PDGF and TGF-β.

16.2.6 VEGF Effects

All tissue-engineered constructs are dependent on the vascular bed that surrounds them at the time of implantation. VEGF is one growth factor that has attracted a great deal of research attention because of its vasculogenic effects. Understanding VEGF is very important for future tissue engineering efforts. This is not only so that exogenous VEGF can be added to a wound, but more importantly, so that endogenous VEGF can be further harnessed in tissue engineering efforts.

Like other growth factors, VEGF is also released by monocytes, damaged keratinocytes, endothelial cells, and especially, active macrophages in the wound (Byrne et al. 2005; Sheikh et al. 2000). VEGF acts on nearby capillaries to induce vascular leakage, supporting inflammation in physiologic and pathologic conditions by enhancing vascular permeability (Midy and Plouet 1994). VEGF also attracts neutrophils and monocytes required for wound healing, and likely also facilitates the formation of granulation tissue. The inflammatory phase is characterized by leukocyte migration, which is necessary for bacterial killing. VEGF release also increases local leukocyte rolling by a threefold increase and adhesion by a fourfold increase in addition to increasing vascular permeability—further facilitating inflammatory processes. Monocytes differentiate into macrophages, begin phagocytosis, and become important in releasing the cytokines necessary for subsequent healing and granulation tissue formation. VEGF is known to play a role in the regulation of initiation and growth of new vascular structures (Midy and Plouet 1994) as well being necessary for endochondral ossification. Endothelial cells activated by VEGF initiate angiogenesis locally. Endothelial progenitors in bone marrow are also recruited to the wound for vasculogenesis. Together, both processes supply the neovascularization required for granulation tissue formation (Byrne et al. 2005), although angiogenesis dominates the process (Bauer at al. 2005).

The absence of VEGF inhibits angiogenesis in wound healing, impairs the recruitment of cells necessary for normal bone development, and interferes with the necessary cell-to-cell cytokine communication due to the absence of a developing vasculature. VEGF is important in bone repair and regeneration for similar reasons (Street et al. 2000). The potent angiogenic activity of the blood clot that forms immediately after the fracture has been attributed to VEGF. The importance of VEGF in fracture healing has been demonstrated in mice and rabbits. When VEGF was inhibited, the resulting reduction in blood flow resulted in the disruption of fracture repair and nonunion. Although the role of VEGF remains to be completely understood, VEGF is clearly necessary for healing and bone repair through the modulation of angiogenesis and new vessel formation (Street et al. 2000).

In addition to stimulating and regulating angiogenesis, VEGF influences osteoblast differentiation and is involved in the mineralization of the fracture callus (Schliephake 2002; Barr et al. 2008). Different growth factors influence wound and bone healing in a manner that is both distinct and synergistically interconnected (Barr et al. 2010). VEGF is required for other factors to function, as is the case with BMP-2 or FGF-2–induced angiogenesis (Yeh and Lee 1999). Inhibition of VEGF has been shown to block BMP-7–induced osteoblast differentiation and BMP-4–induced bone formation (Yeh and Lee 1999).

16.2.7 Angiogenesis

Angiogenesis is certainly important in soft tissue healing but it is also important for bone tissue differentiation, bone formation, maintenance, and remodeling. Bone repair follows four sequential steps:

1. Establishment of a thrombus
2. Formation of the soft callus
3. Formation of the hard callus
4. Bony remodeling

The first step in angiogenesis involves the formation of a platelet- and leukocyte-rich blood clot around the site of the fracture or bony defect. Activated platelets promote the chemotaxis of leukocytes to assist in clearing debris. Both platelets and leukocytes release several growth factors that will ultimately favor the formation of granulation tissue rich in immature vessels forming in and around the wound. These growth factors include tumor necrosis factors, BMPs, FGF, insulin-like growth factor (IGF), PDGF, TGF, and VEGF (Tatsuyama et al. 2000; Hu et al. 2009a and 2009b). Neovessels (Street et al. 2000) and fibroblasts proliferate, forming immature fibrous tissue that is responsible for the early soft callus.

The callus undergoes further maturation to form a so-called hard callus, which is eventually replaced by lamellar bone. Immature woven bone formed by osteoprogenitor cells will eventually undergo remodeling. Bone remodeling follows the maturation of the newly formed bone, characterized by maturation and reorganization, as the fracture or defect gains tensile strength. Each step in bone healing involves the migration, proliferation, differentiation, and activation of local and distant cells; this process is closely regulated (Tatsuyama et al. 2000; Mandracchia et al. 2001). Osteoprogenitor cells respond to the presence of growth factors including BMPs, FGF, IGF, PDGF, TGF, and VEGF (Tatsuyama et al. 2000; Mandracchia et al. 2001). Bone is an excellent source of most of these growth factors, which are released after remodeling or trauma, and this is responsible in part for the success of autogenous grafting compared with alternative sources of graft material such as the alloplastic bone substitutes currently used (Canalis et al. 1988).

Growth factors modulate the healing response and mediate the remodeling process. VEGF increases angiogenesis at the graft or fracture site, providing greater oxygen and nutrient supply, thus enhancing the differentiation of perivascular stem cells and MSCs into osteoblasts (Barr et al. 2010). BMPs, potent inducers of osteogenesis, are expressed during the bone deposition stage, although this is dependent on the formation of neovessels (Canalis et al. 1988).

A process similar to embryonic vasculogenesis may play a role in the coalescence of vessel progenitors (Sipola et al. 2007). VEGF-D and VEGF receptor-3 (VEGFR-3) are all important in osteoblast maturation, with VEGF-D and VEGFR-3 acting downstream from VEGF in osteogenesis (Sipola et al. 2007; Tarkka et al. 2003). Rat studies have revealed VEGF splice variants, VEGF121 and VEGF165, corresponding to human VEGF splice variants VEGF121 and VEGF165. These are expressed during fracture healing. VEGF is most strongly expressed in bone fracture callus during the early phase of healing, decreasing significantly at day 5 (Tarkka et al. 2003). The role of VEGF beyond 5 days in healing of the bony wound is not known. Angiogenic growth factors may mediate the interaction between endothelial cells, osteoblasts, and chondrocytes, all of which secrete endogenous VEGF. As healing progresses, the fibroblast dominates wound healing through the proliferative phase, forming collagen and scar tissue. Several factors assist in favoring the balance from repair to regeneration (Tarkka et al. 2003), supporting the formation of immature bone rather than fibrous tissue. For example, in a rabbit calvarial defect model, the addition of VEGF to DBM favors the regeneration of critical-sized bony wounds compared with the repair seen mainly in DBM or void controls (Emad 2006). Moreover, treatment of rabbit mandibular defects with supraphysiologic doses of exogenous human recombinant VEGF165 results in increased angiogenesis and bone regeneration (Kleinheinz 2005).

Bony wound healing, be it a fracture, a graft, or a defect, requires stability and adequate blood supply. Bony wound healing relies on osteogenic progenitors from adjacent and distant bone marrow. Bone possesses limited potential for regeneration, yet this potential can be enhanced (Fok et al. 2008; Jan et al. 2006, 2009, 2010). The periosteum is a key source of blood supply and when it remains intact, demonstrates extensive microvasculature needed for bone healing and regeneration (Glowacki 1998; Bourque et al. 1992). Bone adjacent to a defect has limited capability for neovascularization and osteogenesis. The periosteum and, in the case of cranial defects, the pericranium and even the dura stimulates bone formation and neovascularization (Bourque et al. 1992).

Augmented angiogenesis in bony wounds and bone grafting applications may improve graft survival, minimize resorption, and hopefully reduce reliance on autogenous grafts, reduce the amount of autogenous graft required for a given defect, or ultimately enable entirely bioengineered grafts. In addition, the ability to promote neoangiogenesis could potentially prove very useful in the treatment and prevention of osteoradionecrosis and bisphosphonate-related necrosis.

Efforts to use angiogenic growth factors for the treatment of chronic wounds have met with limited success. To date, only PDGF has been FDA-approved for use as a topical cytokine wound ointment. PDGF is active in all phases of the healing process and has been highly effective as an adjunct measure in the treatment of chronic wounds of the foot (Bauer et al. 2005).

Attempts to improve bony and chronic wound healing have been the focus of much attention. Hyperbaric oxygen (HBO) improves the ability of macrophages to phagocytose bacteria especially, because hypoxic tissues impair the oxidative function of macrophages. Moreover, macrophages release higher levels of angiogenic cytokines in response to HBO (Bourque et al. 1992). Another approach under investigation is VEGF gene therapy to locally increase neovascularization (Tarkka et al. 2003). However, results have been unpredictable, and disorganized vascular tissue and hematomas resulted rather than normal angiogenesis or vasculogenesis due in part to overexpression of VEGF (Tarkka et al. 2003).

VEGF has been shown to increase blood vessel formation, mineral density, and bone fill in rat calvarial defects (Fok et al. 2008) and rabbit mandibular defects. VEGF has also been shown to increase bone formation and improve demineralized bone allograft consolidation (Kleinheinz 2005). Most studies using VEGF in bony wounds and fractures involve local infusions of VEGF (Street et al. 2000; Kleinheinz 2005), a technique that has proven effective but is not practical or convenient as a long-term mode of delivery. The delivery of VEGF incorporated into an implantable membrane or allogeneic material, eventually along with other important growth factors, is a preferable and more viable mode of delivery in terms of applicability to patients (Schliephake et al. 2012a,b).

Because VEGF does not seem to drive osteoprogenitor cell differentiation, a combination therapy with BMPs may be ultimately required (Sándor and Suuronen 2008). Once combined with grafting materials, VEGF would be released initially to promote vascularization to the site followed by the release of BMP to promote osteoprogenitor cell differentiation and bone fill. This exploits the synergy between bone formation and angiogenesis by manufacturing bone grafting materials that promote angiogenesis in the early stages of injury, followed by treatment with bone growth factors, such as BMP, to stimulate osteoprogenitor cells to differentiate in the later stages. This type of combined treatment might lead to optimized wound healing.

It is clear that endogenous VEGF is required for normal bone fracture healing, remodeling, and osteoblast and osteoclast activity (Street et al. 2000). In addition, the autocrine activity of VEGF in osteoblasts adjacent to cortical defects has been demonstrated. Exogenous VEGF provided by a pump during a period of 48 h to 7 days increases early vascularization and bone healing in bone fracture healing and bone defect healing (Street et al. 2000; Kleinheinz 2005). The angiogenic role of VEGF has been studied separately in soft tissue porcine bladder models, and VEGF was shown to improve neovascularization in porcine bladder constructs at VEGF doses of up to 1000 times lower than in studies demonstrating effects on bone fracture and bone defect healing. This may be largely explained by the fact that most studies on bone have used recombinant human VEGF165 (rhVEGF165), whereas recombinant human VEGF121 (rhVEGF121) has been used in many of the soft tissue studies. It is difficult to compare studies as the dose and mode of delivery of VEGF varies in the reports. Furthermore, it is not known whether bone healing and angiogenesis have differing VEGF requirements. However, VEGF seems to have an important role in the paradigm of bone healing and potentially in tissue engineering scenarios (Broussard 2003; Fok et al. 2008; Beaumont et al. 2010). VEGF is not currently used clinically.

16.3 SURGICAL PROCEDURES TO INDUCE AND PROMOTE HEALING OF LARGE BONE DEFECTS

Some cystic cavities, such as those formed by simple bone cysts, will heal with curettage alone. Other larger bone defects, which are larger than critical size, will not heal. A critical-sized defect of bone is defined as a defect that is so large that it cannot regenerate and heal with bone spontaneously during the lifetime of the individual without bone augmentation of the defect (Schmitz and Hollinger 1986). Critical-sized defects are both species-specific and site-specific, being different in different animals and between their various bones. For example, the critical-sized defect of a skeletally mature male rabbit calvarium is 15 mm (Figure 16.5). The critical-sized defect model provides a good platform to study bone healing.

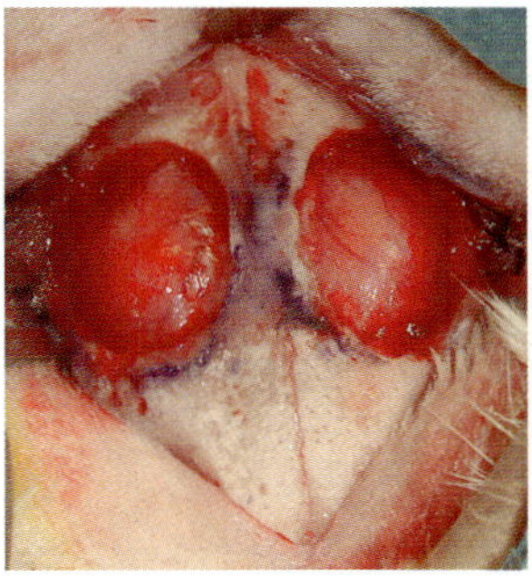

FIGURE 16.5 Bilateral critical-sized defects in the calvarium of a rabbit (15 mm in diameter). These defects will not heal with bone spontaneously without a bone graft or some other measure to induce bone healing.

16.3.1 Periosteal Bone Regeneration

Manipulating the periosteum is one method that surgeons have learned that induces bone regeneration. Periosteal stretching occurs naturally in periosteal new bone formation under pathologically stretched periosteum, in osteomyelitis for example. The periosteum is a key source of blood supply and when it remains intact, it demonstrates an extensive microvasculature that is necessary for bone healing and regeneration (Glowacki 1998; Bourque et al. 1992). Bone adjacent to a defect has a limited capability for neovascularization and osteogenesis. The periosteum and, in the case of cranial defects, even the dura stimulates bone formation and neovascularization (Bourque et al. 1992).

There are sites where periosteal bone regeneration can occur predictably in young children. The fibula or ribs are examples of sites that may totally regenerate when a large graft has been harvested in a child and the periosteum is maintained. The mandible has also been known to regenerate in a child when it has been resected and the periosteal sleeve was left intact.

These same principles of periosteal stretching and defect maintenance are important in sinus lifting, guided bone regeneration with membranes (Humber et al. 2010), and distraction osteogenesis. Defects that are not maintained result in periosteum that sags or collapses and, as a result, their alveolar ridges become nonideal in form.

16.3.2 Gradual Periosteal Stretching with Distraction Osteogenesis

Distraction osteogenesis involves osteotomy of a bone with minimal periosteal disruption and slow continuous traction on the ends of the osteotomized bone to produce a lengthening that is unsurpassed by standard osteotomy techniques and does not require harvesting of a bone graft (Sándor et al. 2008) (Figure 16.6). There is evidence that during this process, bone and other tissues are formed to compensate for this lengthening. Some refer to this process as distraction histiogenesis. Distraction osteogenesis actually recapitulates all the processes of embryonic bone formation to produce new bone without requiring a bone graft (Ylikontiola et al. 2012). The levels

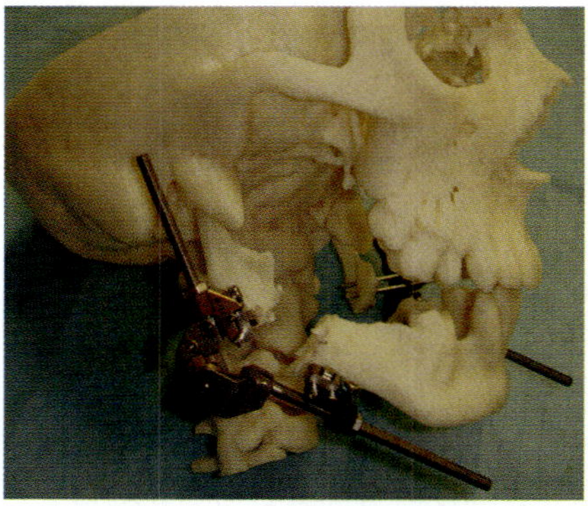

FIGURE 16.6 Distraction osteogenesis device placed on the mandible of the stereolythic model. Distraction osteogenesis can produce bone by gradual periosteal stretching.

of growth factors such as BMPs, PDGF, and VEGF are markedly elevated in the distraction regenerate (Sándor and Suuronen 2008).

16.3.3 Ancillary Therapy with HBO

One maneuver surgeons have learned is to amplify the oxygen tension in a poorly vascularized wound to help promote healing (Figure 16.7). Some wounds are poorly vascularized due to excessive scarring or from repeated failed surgical attempts (Sándor et al. 2010a). Radiation therapy also causes tissues in the irradiated field to become hypovascular. HBO can help in the healing of problem wounds in the dento-facial skeleton (Brown et al. 1998; Fok et al. 2008).

HBO therapy is defined as intermittent exposure to 100% oxygen under pressures greater than 1 atmosphere absolute (ATA). The concentration of oxygen in the atmosphere is 21%. At 1 ATA, the oxygen in the blood is almost entirely carried by hemoglobin. In arterial blood, 97% of the oxygen is carried chemically bound to hemoglobin whereas only 3% is dissolved in plasma. One gram of hemoglobin carries a maximum of 1.34 mL of oxygen. Fully saturated hemoglobin (100%) in 100 mL of blood carries approximately 20 mL of oxygen. Saturated hemoglobin (97%) in 100 mL of blood carries 19.5 mL of oxygen. This amount is reduced to 5 mL of oxygen while passing through capillaries. Increasing the oxygen-carrying capacity of blood by increasing hemoglobin saturation is not possible (Jan et al. 2006).

At the sea level, gas pressure is 760 mmHg or 1 ATA. Arterial hemoglobin saturation is 97%, whereas venous hemoglobin saturation is 70%. Inhalation of HBO increases the quantity of oxygen dissolved in plasma. At 1 ATA, the amount of dissolved oxygen in 100 mL of plasma is 0.449 mL. When inhaling 100% oxygen at 1 ATA, O2 concentration increases to 1.5 mL/100 mL of plasma. When inhaling 100%

FIGURE 16.7 Animal HBO chamber.

oxygen at 3 ATA, the amount of dissolved oxygen in 100 mL of plasma increases to 6.422 mL/100 mL of plasma, which is enough to meet the basic metabolic needs of healing tissues in the human body (Fok et al. 2008; Jan et al. 2006, 2009, 2010).

The driving force for oxygen diffusion from the capillaries to tissues can be estimated by the difference between the partial pressure of oxygen on the arterial side and the venous side of the capillaries. The difference in the partial pressure of oxygen from the arterial side to the venous side of the capillary system is approximately 37 times greater when breathing 100% oxygen at 3 ATA than air at 1 ATA (Jan et al. 2006, 2009, 2010; Fok et al. 2008).

HBO has been shown to promote the total healing of cranial defects in the rabbit critical-sized defect model of both critical-sized and supracritical-sized defects even without bone grafting (Figure 16.5; Jan et al. 2006). Increased VEGF expression has also been noted during bone healing with HBO at 6 weeks in this same model (Fok et al. 2008). The healing of bone grafts is also ameliorated by HBO in this same animal model.

Although HBO has been shown to increase VEGF levels in healing bony wounds in the rabbit calvarial model (Fok et al. 2008), its immunogenic effects have not been so promising in mice (Gassas et al. 2011). The clinical use of HBO has been established in the treatment of compromised wounds with poor tissue perfusion, and here, HBO had positive effects (Brown et al. 1998). From a tissue engineering point of view, HBO has been beneficial in studying bone healing and in understanding the relative efficacies of certain bone substitute materials (Jan et al. 2006, 2009, 2010; Fok et al. 2008; Humber et al. 2010).

16.3.4 Tissue Engineering with Growth Factors (BMPs)

The constraints of both autogenous bone harvesting morbidity and allogeneic bone quality have stimulated researchers to focus on the fabrication of initially synthetic bioimplants, which included new biomaterials, growth factors, and later incorporated stem cells. By the late 1980s, one active factor responsible for the induction of bone was identified by Urist et al. (1983) and was named BMPs. BMP is regarded as a morphogen; a protein that replicates the embryonic induction of bone formation. BMP can induce pluripotent MSCs to differentiate into bone-forming osteoblasts. BMP-2, BMP-4, and BMP-7 have been shown to stimulate *de novo*, *in vitro*, and *in vivo* bone formation in various animal models (Khanna-Jain et al. 2010a and 2010b). Many BMPs have been isolated, and with the exception of BMP-1, they are all members of the TGF-β superfamily (Clokie et al. 2002). In the early 1990s, the synthetic production of these proteins using recombinant technology was developed. In 2006, this lead to the development of products such as OP-1 (BMP-7; Stryker Biotech, Allendale, NJ) and Infuse (BMP-2; Medtronic, Fridley, MN), both of which are now available for clinical use. OP-1 is supplied as a simple powder and Infuse has a collagen sponge carrier (Moghadam et al. 2004; Barr et al. 2010).

One of the greatest challenges to the clinical application of BMP has been the development of an acceptable carrier (Barr et al. 2010). Over the past decade, Clokie and Urist (2000) have explored the use of a reverse phase block poloxamer medium as a carrier for BMP, with the first BMP bioimplant being successfully implanted in

1999 using poloxamer 407. Since that time, Clokie and Sándor (2008) have successfully reconstructed ten large human mandibular defects using bioimplants consisting of OP-1 (BMP-7; Stryker Biotech) and DynaGraft Putty (DBM in a reversed phase poloxamer medium; IsoTis, Irvine, CA).

The BMP bioimplant was created by manually mixing BMP-7 (OP-1; Stryker Biotech.) with 10 mL of DBM in a reversed phase medium (DynaGraft Putty; IsoTis) and then molding it to form the shape of the resected segment of mandible (Figure 16.4a through g). This was then inserted into the defect in the mandible and the pterygo-masseteric muscular sling surrounding the mandible was reapproximated to ensure complete muscular coverage of the bioimplant (Figure 16.4e). Patients were followed closely both clinically and radiographically, and mandibular regeneration allowing successful dental implant restoration was achieved (Figure 16.4f; Clokie and Sándor 2008). All of these patients' defects were reconstructed without the need to harvest any of the patients' own bone tissue. In the future, biogels that allow sustained release of growth factors and other alloplastic implants that may allow the sequential release of combinations of growth factors will be of interest.

Any discussion of using BMP clinically must include the disadvantage of its cost. Recombinant BMP is extremely expensive. Investigators are searching for new technologies to provide less expensive rhBMP using transgenic sources. BMP is used in supraphysiologic concentrations. In nature, BMP is found in nanogram quantities in bone, whereas BMP is used in milligram quantities in clinical application. When used clinically, patients exposed to BMP have a remarkable amount of postoperative edema—beyond the normal expected amount of swelling (Figure 16.8).

In cell cultures, BMP has failed to produce meaningful differences with osteoblastic differentiation (Khanna-Jain et al. 2010a and 2010b). In fact, osteogenic medium containing dexamethasone, glycerophosphate, and ascorbic acid has been more effective with regard to osteogenic differentiation than any of the tested BMPs. Future biomaterials with the capability of sustained release of the components of the osteogenic medium will be of great interest.

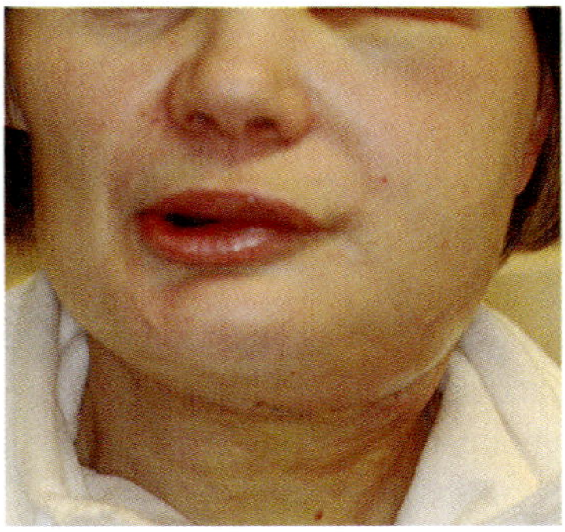

FIGURE 16.8 Sustained postoperative edema noted near the site of BMP treatment. This significant edema is a characteristic of patients treated with BMP.

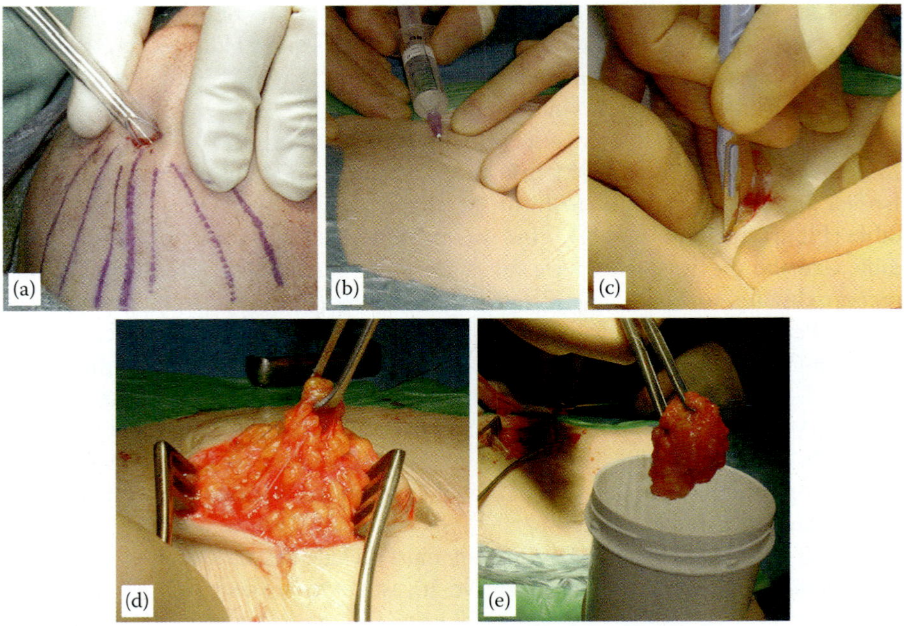

FIGURE 16.9 (a) Liposuction may be one convenient and minimally morbid way of future harvesting ASCs for cell culturing and tissue engineering purposes however at present the cell survival is insufficient. (b) Presently the harvesting of subcutaneous fat can easily be done through an existing scar such as the laparoscopy scar at the umbilicus of this patient. Here, the harvest is being performed using local anesthesia. (c) Incision is made using the preexisting scar. (d) Fat is easily removed from the wound. (e) Fat is placed directly into the storage vessel without contact with the skin to prevent any possible contamination.

16.3.5 Tissue Engineering with Stem Cells and Growth Factors

This chapter has discussed the use of allogeneic, alloplastic materials and growth factors. The addition of pluripotent MSCs has the promise of reconstructing larger bone defects with more predictable results. Cell therapy in this context has the goal of minimizing donor site morbidity by using easily harvested MSCs from subcutaneous fat.

Interdependency exists between adipogenesis, the formation of fat tissue from stem cells and osteogenesis, and the formation of bone from its stem cell precursors (Gimble et al. 2006). Adipose-derived adult MSCs may be useful in future bone regeneration and tissue engineering efforts using autogenous adipose-derived stem cells (ASC), growth factors, and resorbable biomaterials.

Tissue engineering efforts include the vital collaboration between cell biologists, biochemists, material scientists, engineers, and clinicians. To understand the complex role of the various components of tissue engineering, one can consider an equilateral triangle in which stem cells, resorbable scaffolds, and bioactive molecules such as growth factors continuously interact with each other (Figure 16.3). It is the understanding of the nature of the interactions between these three key components on which tissue engineering is built (Sándor and Suuronen 2008).

16.3.6 Sources of Stem Cells

The source of cells for tissue engineering depends on the structure that is to be replaced. Human embryonic stem cells (hESC) are pluripotent stem cells isolated from the inner mass of human blastocysts. Although these cells have great potential because of their differentiation capacity, there are problems that must be solved prior to their clinical use in tissue engineering. The problems with hESCs include its culture without exposure to animal proteins, avoidance of teratoma development, and immune rejection by the recipient host (Grinnemo et al. 2008). Currently, it is adult stem cells that are used clinically. Cells with the lowest morbidity in harvesting and cells that still retain a degree of pluripotentiality would be the most advantageous in the tissue engineering of bone for example.

One source of MSCs for bone regeneration is from adipose tissue that will provide ASCs. This should not be surprising as there is an interrelationship between adipogenesis and osteogenesis (Gimble et al. 2006). Certainly, the harvesting of adipose tissue is not morbid and may even be advantageous for some if liposuction were used as the harvesting method (Figure 16.9a through e). ASCs may be a hybrid source of cells that may provide material for both osteogenesis and vasculogenesis. Bone marrow aspirates are another source of MSCs. Although many of these cells are already growing in a bone lineage, their harvesting by bone marrow aspiration may be considerably more morbid compared with the simple harvesting of subcutaneous fat.

16.3.7 Physical Stimulation of Stem Cells

Using the tissue engineering model, autogenous ASCs could be harvested from a patient having a liposuction procedure and used to seed a resorbable scaffold (Suuronen and Asikainen 2004) that was made using CAD/CAM technology, to the precise dimensions of a missing segment of bone, for example. The seeded cells could be stimulated by physical means using vibration (Tirkkonen et al. 2011), magnetic, galvanic, ultrasound, or laser stimulation, along with hypoxic or hyperoxic gradients (Sándor et al. 2010a,b) or growth factors such as TGF-β1 (Clokie and Bell 2003) or the BMPs (Hu et al. 2005) to guide the differentiation and growth of the cells.

16.3.8 Biomaterials

In some cases, the surface configuration of a particular scaffold may stimulate stem cells to differentiate in a desired direction. Bioactive glasses, for example, are known to cause MSCs to differentiate into osteoblasts (Haimi et al. 2009). Future tissue engineering efforts will use such surface interactions of biomaterials and stem cells to guide cellular differentiation toward the desired tissue to be regenerated. Titanium or other metals will give way to resorbable biomaterials, even though titanium is so well tolerated by bone. One challenge for resorbable polymers is to fabricate them to the exact shapes required in a particular reconstruction, using CAD/CAM techniques, without changing the properties of the resorbable biomaterial.

Simple configurations of polymers already in widespread use as sutures such as polylactate, polyglycolate, and caprolactones will be modified to match their physical characteristics with the loading requirements of the reconstructed bones and desired degradation profiles. Hybrid biomaterials using combinations of resorbable polymers and bioactive glass, for example, may be used to guide cellular differentiation, with the goal of more immediate functional loading. Biomaterials also need to have optimal porosity to allow both cellular seeding that penetrates well beyond just the surface of the scaffold. The pores should also interconnect to allow the establishment of vascular and lymphatic channels.

Surgeons must carefully plan surgical procedures using resorbable biomaterial scaffolds. The biomaterial must match with the loading requirements of the bony wound. Granular materials provide no load-bearing possibilities, for example, even though they maximize the surface area to allow tissue ingrowth and resorption of the biomaterial. Surgeons must also be reminded not to exceed the capacity of a wound to degrade and resorb such biomaterials. The concept of biomaterial wound load becomes important when planning tissue-engineered constructs. Wounds that are superficial, without thick soft tissue cover, may become dehiscent over such biomaterials as they degrade.

16.3.9 Manipulating the Construct

Once the cells have populated the scaffold, the resulting bioimplant or construct could be transplanted into the patient to restore the defect. This *ex vivo*-derived reconstruction has one major obstacle: the vitality of the bioimplant is entirely dependent on the vascularity of the recipient bed. To this end, growth factors such as VEGF might be used to stimulate angiogenesis (Marx et al. 1998) to help vascularize the construct more rapidly after construct implantation. Currently, microsurgical vessel anastomosis provides such immediate profusion (Mesimäki et al. 2009). In the future, tissue engineering paradigms might eliminate the need for complex microsurgical repair, greatly shortening the operative procedure for the patient with the goal of reducing the morbidity of reconstruction.

16.3.10 Current Therapy

Cell therapy for large bone defects has become a clinical reality (Mesimäki et al. 2009). Autogenous ASCs are harvested from the subcutaneous fat of a patient with a bony defect, due to an ameloblastoma of the mandible, for example. Ameloblastoma provides a good model for mandibular reconstruction as it requires a mandibular resection for its treatment. Moreover, the resulting wound is larger than critical size. These tumors do not require radiation treatment, so that the vascular bed is unaltered, lowering the risk of wound healing complications.

The ASCs are transported to a cell culturing laboratory and separated from the surrounding tissue. The ASCs are cultured and expanded (Figure 16.10a and b), with appropriate quality assurance protocols acceptable to the responsible regulatory agencies. Granules of tricalcium phosphate (TCP) are seeded with the expanded ASCs, and BMP-2 is added to the culture solution. The seeded alloplast and BMP-2

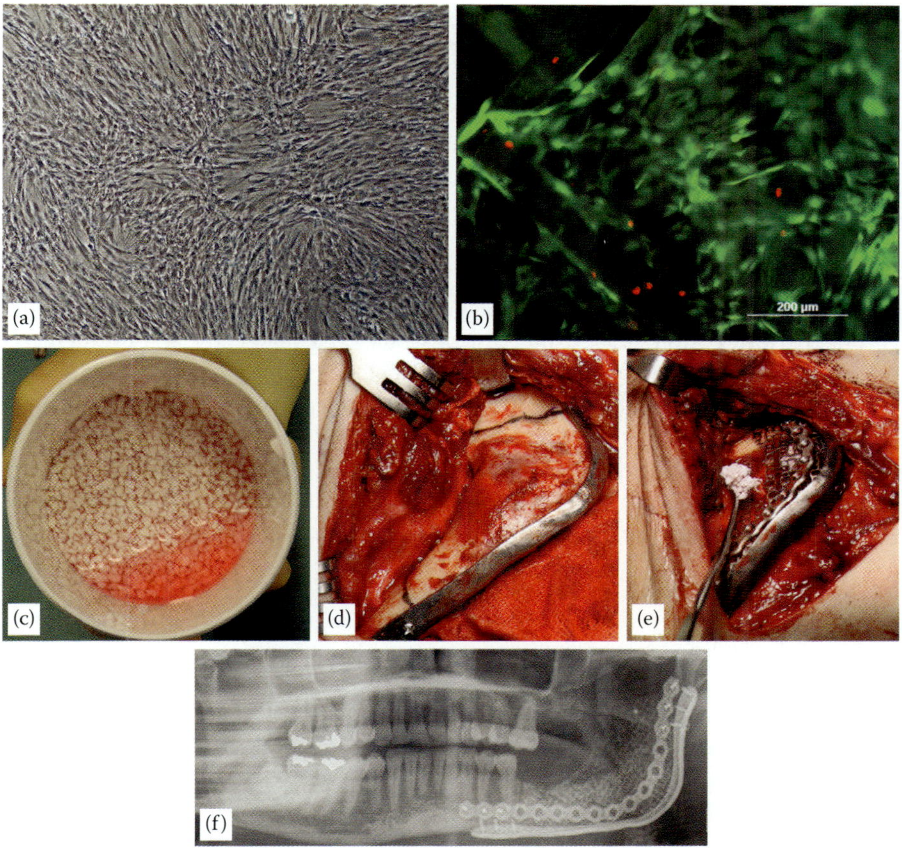

FIGURE 16.10 (a) ASCs grown in cell culture. Live/dead staining of adipose stem cells are attached onto a scaffold of resorbable biomaterial fibers (3-h time point). Scale: 200 μM. (b) Green cells are viable. (c) Granules of β-TCP seeded with ASCs and BMP-2 added to the culture solution. (d) Ameloblastoma resection outlined on mandible. (e) ASC seeded granules placed beneath titanium mesh to provide granules with immobility and stability. (f) Orthopantomogram of bioimplant *in situ*.

combination is transported from the laboratory approximately 3 weeks after the fat harvesting was performed (Figure 16.10c). The TCP is placed into a titanium mesh cage and once the resection is completed, the construct is inserted into the wound (Figure 16.10d through f). Nine months after placement, and with radiographic control, dental implants can be placed into the remodeling construct and the mandibular defect can be rehabilitated with an implant-borne prosthesis.

16.3.11 Future Directions

Another source of stem cells could be from suction trap aspirates of bone during the removal of the mandibular third molar (Lindholm et al. 2003; Mesimäki et al. 2009). This technique would allow stem cell harvesting during one of the most common oral surgical procedures performed today. Third molar removal may also present some other new opportunities. The removed developing third molar follicle may yield follicular cells, cementoblast-like cells, and dental pulp stem cells that can also be cultured and studied. The receptors of these cells can be characterized and this is an important first step in the understanding of these cells and their possible future use (Khanna-Jain et al. 2010a and 2010b).

16.4 CONCLUSION

Surgeons can augment the healing potential of large bone defects by acute and chronic periosteal stretching, using growth factors and stem cells, and by augmenting angiogenesis indirectly with the use of HBO therapy.

Although these modalities are of interest, costs, access to the various treatment modalities, and practicality remain as major issues. Future surgeons will likely combine the useful features of stem cell therapy, new biomaterials, and cell signaling molecules to simplify reconstructive measures with the goal of making the reconstruction of bone defects more predictable and less morbid.

REFERENCES

Barr T., N. M. Carmichael, and G. K. Sándor. 2008. Juvenile idiopathic arthritis: A chronic pediatric musculoskeletal condition with significant dental manifestations. *Journal—Canadian Dental Association*. 74(9): 813–821.

Barr T., A. J. McNamara, G. K. Sándor, S. A. Peel, and C. M. Clokie. 2010. Comparison of the osteoinductivity of bioimplants containing rhBMP-2 (Infuse®) and rhBMP-7 (OP-1®). *Oral Surgery, Oral Medicine, Oral Pathology, Oral Radiology, and Endodontics* 109(4): 531–540.

Bauer S. M., R. J. Bauer, and O. C. Velazquez. 2005. Angiogenesis, vasculogenesis, and induction of healing in chronic wounds. *Vascular and Endovascular Surgery* 39(4): 293–306.

Beaumont M., M. G. Du Val, Y. Loai, W. A. Farhat, G.K. Sándor, H. L. Cheng. 2010. Monitoring angiogenesis in soft-tissue engineered constructs for calvarium bone regeneration: an in-vivo longitudinal DCE-MRI study. *Nuclear Medicine Resonance in Biomedicine*. 23(1): 48–55.

Bourque W. T., M. Gross, and B. K. Hall. 1992. A reproducible method for producing and quantifying the stages of fracture repair. *Laboratory Animal Science* 42(4): 369–374.

Broussard C. L. 2003. Hyperbaric oxygenation and wound healing. *J Wound Ostomy Continence Nursing*. 30(4):210–216.

Brown D. A., A. W. Evans, and G. K. B. Sándor. 1998. Hyperbaric oxygen therapy in the management of osteoradionecrosis of the mandible. *Advances in Otorhinolaryngology*, 54: 14–32.

Burchadt H. 1983. The biology of bone graft repair. *Clinical Orthopaedics* 174(1): 28–36.

Byrne A. M., D. J. Bouchier-Hayes, and J. H. Harmey. 2005. Angiogenic and cell survival functions of vascular endothelial growth factor (VEGF). *Journal of Cellular and Molecular Medicine* 9(4): 777–794.

Canalis E., T. McCarthy, and M. Centrella. 1988. Growth factors and the regulation of bone remodeling. *Journal of Clinical Investigation* 81(2): 277–281.

Carmichael R. P., and G. K. Sándor. 2008. Use of dental implants in the management of cleft lip and palate. *Atlas of the Oral and Maxillofacial Clinics of North America* 16(1): 61–82.

Clokie C. M., and R. C. Bell. 2003. Human transforming growth factor B-1 and its effects on osseointegration. *Journal of Craniofacial Surgery* 14(3): 268–277.

Clokie C. M., H. Moghadam, M. T. Jackson, and G. K. Sándor. 2002. Closure of critical sized defects with allogenic and alloplastic bone substitutes. *Journal of Craniofacial Surgery* 2002; 13(1): 111–121.

Clokie C. M., and G. K. Sándor. 2008. Reconstruction of 10 major mandibular defects using bioimplants containing BMP-7. *Journal—Canadian Dental Association* 74(1): 67–72.

Clokie C. M., and M. R. Urist. 2000. Bone morphogenetic protein excipients: comparative observations on poloxamer. *Plastic and Reconstructive Surgery* 105(2): 628–637.

Danis R. 1949. Theorie et pratique de l'osteosynthese. *Libraries de L'Academie de Medicine* 76, 409–429.

Deckers M., and M. Karperien. 2000. Expression of vascular endothelial growth factors and their receptors during osteoblast differentiation. *Endocrinology* 141(5): 1667–1674.

Dragoo J. L., J. Y. Choi, J. R. Lieberman, J. Huang, P. A. Zuk, J. Zhang, M. H. Hedrick, and P. Benhaim. 2003. Bone induction by BMP-2 transduced stem cells derived from human fat. *Journal of Orthopaedic Research* 21(4): 622–629.

Ducy P., R. Zhang, V. Geoffroy, A. L. Ridall, and G. Karsenty. 1997. Osf2/Cbfa1: A transcriptional activator of osteoblast differentiation. *Cell* 89(4): 747–754.

Du Val M. G., S. Davidson, A. Ho, R. Cohen, M. Park, S. Nourian, G. Baker, and G. K. Sándor. 2007. Albright's hereditary osteodystrophy with extensive heterotrophic ossification of the oral and maxillofacial region: How fetuin research may help a seemingly impossible condition. *Journal—Canadian Dental Association* 73(9): 845–850.

Emad B. 2006. Vascular endothelial growth factor augments the healing of demineralized bone matrix grafts. *International Journal of Surgery* 4(3): 160–166.

Engsig M. T., Q. J. Chen, T. H. Vu, A. C. Pedersen, B. Therkidsen, L. R. Lund, K. Henriksen, T. Lenhard, N. T. Foged, Z. Werb, and J. M. Delaissé. 2000. Matrix metalloproteinase 9 and vascular endothelial growth factor are essential for osteoclast recruitment into developing long bones. *Journal of Cell Biology* 151(4): 879–889.

Ferrara N., H. P. Gerber, and J. LeCouter 2003. The biology of VEGF and its receptors. *Nature Medicine* 9(6): 669–676.

Fok, T. C. O., A. Jan, S. A. Peel, C. M. L. Clokie, and G. K. Sándor. 2008. Hyperbaric oxygen results in an increase in vascular endothelial growth factor (VEGF) protein expression in rabbit calvarial critical sized defects. *Oral Surgery, Oral Medicine, Oral Pathology, Oral Radiology, and Endodontics* 105(4): 417–422.

Gassas A., W. Min, A. W. Evans, S. Carter, G. K. Sándor, and E. Grunebaum. 2011. Hyperbaric oxygen therapy as a sole agent is not immunosuppressant in a highly immunogenic mouse model. *Bone Marrow Research* 579268. As per pubmed.

Gimble J. M., S. Zvonic, Z. E. Floyd, M. Kassem, and M. Nuttall. 2006. Playing with bone and fat. *Journal of Cellular Biochemistry* 98(2): 251–266.

Glowacki J. 1998. Angiogenesis in fracture repair. *Clinical Orthopaedics and Related Research* 355 (Suppl 1998): S82–89.

Grinnemo K. H., C. Syven, O. Hovatta, G. Dellgren, and M. Corbascio. 2008. Immunogenicity of human embryonic stem cells. *Cell and Tissue Research* 331(1): 67–68.

Haimi S., G. Gorianc, L. Moimas, B. Lindroos, H. Huhtala, S. Räty, H. Kuokkanen et al. 2009. Characterization of zinc-releasing three dimensional bioactive glass scaffolds and their effect on adipose stem cell proliferation and osteogenic differentiation. *Acta Biomaterialia* 5(8): 3122–3131.

Hu Z. M., S. A. Peel, S. K. Ho, G. K. B. Sándor, and C. M. Clokie. 2005. Role of bovine bone morphogenetic proteins in bone matrix protein and osteoblast-related gene expression during rat bone marrow stromal cell differentiation. *Journal of Craniofacial Surgery* 16(6): 1006–1014.

Hu Z., S. Peel, S. K. Ho, G. K. Sándor, and C. M. Clokie. 2009a. Comparison of platelet-rich plasma, bovine BMP, and rhBMP-4 on bone matrix protein expression *in vitro*. *Growth Factors* 27(2): 280–289.

Hu Z., S. A. Peel, S. K. Ho, G. K., Sándor, and C. M. Clokie. 2009b. Platelet-rich plasma induces Mrna expression of VEGF and PDGF in rat bone marrow stromal cell differentiation. *Oral Surgery, Oral Medicine, Oral Pathology, Oral Radiology, and Endodontics* 107(1): 43–48.

Hu Z., S. A. Peel, T. C. Lindholm, G. K. B. Sándor, C. M. Clokie, and Y. Su. 2011. Osteoinductivity of partially purified bovine, ostrich and emu bone morphogenetic proteins *in vitro*. *Journal of Biomedical Material Research A* 98(3): 473–477.

Humber C., G. K. Sándor, J. M. Davis, S. A. F. Peel, B. M. B. Brkovic, Y. D. Kim, H. I. Holmes, and C. M. L. Clokie. 2010. Bone healing with an *in situ* formed bioresorbable PEG membrane in rabbit calvarial defects. *Oral Surgery, Oral Medicine, Oral Pathology, Oral Radiology, and Endodontics* 109(3): 372–384.

Jan A., G. K. B. Sándor, D. Iera, A. Mhawi, S. A. Peel, and C. M. L. Clokie. 2006. Hyperbaric oxygen results in an increase in rabbit calvarial critical-sized defects. *Oral Surgery, Oral Medicine, Oral Pathology, Oral Radiology, and Endodontics* 101(2): 144–149.

Jan A., G. K. B. Sándor, B. M. B. Brkovic, S. A. Peel, A. W. Evans, and C. M. L. Clokie. 2009. Effects of hyperbaric oxygen on grafted and nongrafted on calvarial critical-sized defects. *Oral Surgery, Oral Medicine, Oral Pathology, Oral Radiology, and Endodontics* 107(2): 157–163.

Jan A., G. K. B. Sándor, B. M. B. Brkovic, S. A. Peel, Y. D. Kim, W. Z. Xiao, A. W. Evans, and C. M. L. Clokie. 2010. Effects of hyperbaric oxygen on demineralized bone matrix and biphasic calcium phosphate bone substitutes. *Oral Surgery, Oral Medicine, Oral Pathology, Oral Radiology, and Endodontics* 109(1): 60–67.

Khanna-Jain R., H. Agata, A. Vuorinen, G. K. Sándor, R. Suuronen, and S. Miettinen. 2010a. Osteogenic differentiation of human periodontal ligament cells induced by osteogenic supplements with or without bone morphogenetic proteins 2 or 6. *Growth Factors* 28(6): 437–446.

Khanna-Jain R., A. Vuorinen, G. K. Sándor, R. Suuronen, and S. Miettinen. 2010b. Vitamin D_3 metabolites induce osteogenic differentiation in human dental pulp and human dental follicle cells. *Journal of Steroid Biochemistry and Molecular Biology* 122(4): 133–141.

Kleinheinz J. 2005. VEGF-activated angiogenesis during bone regeneration. *Journal of Oral and Maxillofacial Surgery* 63(9): 1310–1316.

Lindholm T. C., S. A. F. Peel, C. M. L. Clokie, and G. K. Sándor. 2003. Cortical bone grafts used to culture bone cells to be used for increasing efficacy of bone morphogenetic proteins in tissue engineered bone substitutes. *Journal of Oral and Maxillofacial Surgery* 61(8) Suppl 1: 74.

Mandracchia V. J., S. C. Nelson, and E. A. Barp. 2001. Current concepts of bone healing. *Clinics in Podiatric Medicine and Surgery* 18(1): 55–77.

Marx R., E. Carlson, R. Eichstaedt, S. Schimmele, J. Strauss, and K. Georgeff. 1998. Platelet rich plasma growth factor enhancement for bone grafts. *Oral Surgery, Oral Medicine, Oral Pathology, Oral Radiology, and Endodontics* 85(6): 638–646.

Mesimäki K., B. Lindroos, J. Törnwall, J. Mauno, C. Lindqvist, R. Kontio, S. Miettinen, and R. Suuronen. 2009. Novel maxillary reconstruction with ectopic bone formation by GMP adipose stem cells. *Int Journal of Oral and Maxillofacial Surgery* 38(3): 201–209.

Midy V., and J. Plouet. 1994. Vasculotropin/vascular endothelial growth factor induces differentiation in cultured osteoblasts. *Biochemical and Biophysical Research Communications* 199(1): 380–386.

Moghadam H. G., M. R. Urist, G. K. Sándor, and C. M. Clokie. 2001. Successful mandibular reconstruction using a BMP bioimplant. *Journal of Craniofacial Surgery* 12(2): 119–127.

Moghadam H. G., G. K. Sándor, H. I. Holmes, and C. M. Clokie. 2004. Histomorphometric evaluation of bone regeneration using allogeneic and alloplastic bone substitutes. *Journal of Oral and Maxillofacial Surgery* 62(2): 202–213.

Niida S., M. Kaku, H. Amano, H. Yoshida, H. Kataoka, S. Nishikawa, K. Tanne, N. Maeda, S. Nishikawa, and H. Kodama. 1999.Vascular endothelial growth factor can substitute for macrophage colony-stimulating factor in the support of osteoclastic bone resorption. *Journal of Experimental Medicine* 190(2): 293–298.

Sándor G. K., and R. P. Carmichael. 2008. Rehabilitation of trauma using dental implants. *Atlas of the Oral and Maxillofacial Clinics of North America* 16(1): 83–105.

Sándor G. K., L. P. Ylikontiola, W. Serlo, P. M. Pirttiniemi, and R. P. Carmichael. 2008. Midfacial Distraction Osteogenesis. *Atlas of the Oral and Maxillofacial Surgical Clinics of North America* 16(2): 249–272.

Sándor G. K., and R. Suuronen. 2008. Combining adipose-derived stem cells, resorbable scaffolds and growth factors: an overview of tissue engineering. *Journal - Canadian Dental Association,* 74(2): 167–170.

Sándor G. K, R. P. Carmichael, and A. Binahmed. 2008. Reconstruction of ablative defects using dental implants. *Atlas of the Oral and Maxillofacial Clinics of North America* 16(1): 107–123.

Sándor G. K., R. P. Carmichael, and B. M. Brkovic. 2010a. Dental implants placed into alveolar clefts reconstructed with bone grafts and tongue flaps. *Oral Surgery, Oral Medicine, Oral Pathology, Oral Radiology, and Endodontics* 109(2): e1–e7.

Sándor G. K , D. K. Lam, L. P. Ylikontiola, V. Kainulainen, K. Oikarinen, and C. M. Clokie. 2010b. *Autogenous Bone Harvesting Techniques in Oral and Maxillofacial Surgery* (eds. Kahnberg K. E., L. Anderson, and A. Pogrel). Oxford, United Kingdom: Blackwell Munksgaard, 383–403.

Schliephake H. 2002. Bone growth factors in maxillofacial skeletal reconstruction. *International Journal of Oral and Maxillofacial Surgery* 31(5): 469–484.

Schliephake H., N. Strecker, A. Förster, B. Schwenzer, J. Reichert, D. Scharnweber. 2012a. Angiogenic functionalisation of titanium surfaces using nano-anchored VEGF – an in vitro study. *European Cell Materials* 13(23): 161–169.

Schliephake H., C. Bötel, A. Förster, B. Schwenzer, J. Reichert, D. Scharnweber. 2012b. Effect of oligonucleotide mediated immobilization of bone morphogenic proteins on titanium surfaces. *Biomaterials* 33(5): 1315–1322.

Schmitt J. M., K. Hwang, S. R. Winn, and J. O. Hollinger. 1999. Bone morphogenetic proteins: an update on basic biology and clinical relevance. *Journal of Orthopaedic Research* 17(3): 269–278.

Schmitz J. P. and J. O. Hollinger. 1986. The critical size defect as an experimental model for craniomandibulofacial nonunions. *Clinical Orthopaedic Research* 205: 299–308.

Sheikh A. Y., J. J. Gibson, M. D. Rollins, H. W. Hopf, Z. Hussain, and T. K. Hunt. 2000. Effect of hyperoxia on vascular endothelial growth factor levels in a wound model. *Archives of Surgery* 135(11): 1293–1297.

Simmons, D. 1980. Fracture healing, In *Fundamental and Clinical Bone Physiology* (ed. Urist M. R.). Philadelphia: JB Lippincott, 283–330.

Sipola A., J. Ilvesaro, E. Birr, P. Jalovaara, F. Stenbäck, T. Hautala, and J. Tuukkanen. 2007. Endostatin inhibits endochondral ossification. *Journal of Gene Medicine* 9(2): 1057–1064.

Suuronen R., and A. Asikainen. 2004. Biodegradable materials as tools to fasten bone in face and jaw surgery. *Duodecim* 120(16): 2002–2007.

Tarkka T., A. Sipola, T. Jämsä, Y. Soini, S. Ylä-Herttuala, J. Tuukkanen, and T. Hautala. 2003. Adenoviral VEGF-A gene transfer induces angiogenesis and promotes bone formation in healing osseous tissue. *Journal of Gene Medicine* 5(4): 560–566.

Tatsuyama K., Y. Maezawa, H. Baba, Y. Imamura, and M. Fukuda. 2000. Expression of various growth factors for cell proliferation and cytodifferentiation during fracture repair of bone. *European Journal of Histochemistry* 44(3): 269–278.

Tirkkonen L., H. Halonen, J. Hyttinen, H. Kuokkanen, H. Sievänen, A-M. Koivisto, B. Mannerström et al. 2011. The effects of vibration loading on adipose stem cell number, viability and differentiation towards bone-forming cells. *Journal of the Royal Society Interface* 8(65): 1736–174.

Urist M. R., R. J. DeLange, and G. A. Finerman. 1983. Bone cell differentiation and growth factors. *Science* 220(4598): 680–686.

Warnke P. H., and A. J. Coren. 2003. First experiences with recombinant human bone morphogenetic protein 7 (osteogenic protein 1) in a human case in maxillofacial surgery. *Plastic and Reconstructive Surgery* 111(10): 2471–2472.

Wozney J. M., S. E. Lynch, R. J. Genco, and R. E. Marx. 1999. *Biology and Clinical Application of rhBMP-2: Tissue Engineering: Applications for Maxillofacial Surgery and Periodontics.* Chicago: Quintessence, 103–123.

Yeh L. C., and J. C. Lee. 1999. Osteogenic protein-1 increases gene expression of vascular endothelial growth factor in primary cultures of fetal rat calvaria cells. *Molecular and Cellular Endocrinology* 153(1–2): 113–124.

Ylikontiola L. P., G. K. Sándor, N. Salokorpi, W. Serlo. 2012. Experience with craniosynostosis treatment using posterior cranial vault distraction osteogenesis. *Annals of Maxillofacial Surgery* 2(1): 4–7.

17 Nanomaterials for Cartilage Regeneration

Shirley Keolebogile Motaung and A. Hari Reddi

CONTENTS

17.1 INTRODUCTION

Articular cartilage is an avascular, aneural tissue in the joints, providing a smooth friction-free bearing surface for movement. It consists of surface, middle, and deep layers with a characteristic extracellular matrix (ECM) composition. Articular cartilage has a limited innate potential for repair and regeneration. There are three key requirements for the regeneration of articular cartilage:[1–8]

1. *Inductive morphogens as signals.* These are growth and differentiation factors and inductive signaling molecules.
2. *Responding stem/progenitor cells.* Acquisition of the appropriate source of stem/progenitor cells is a major challenge in the field of regeneration. Methodologies enabling proliferation and differentiation of cells are important in tissue engineering.

3. *ECM scaffolding*. These include natural and synthetic biomaterials that permit the growth and differentiation of stem/progenitor cells in the formation of functional tissues.

17.2 CARTILAGE REGENERATION BY TISSUE ENGINEERING

Although the repair of cartilage defects by transplantation of autologous chondrocytes is well known,[9] this procedure is associated with increased donor site morbidity and it is often difficult to harvest chondrocytes in sufficient amounts. In addition, in older osteoarthritis patients, the source of autologous chondrocytes is very limited. This has led several investigators to explore tissue engineering and regenerative medicine for cartilage regeneration. Tissue engineering has been defined as the science of design and manufacturing of new tissues for the functional restoration of impaired organs and replacement of lost parts due to disease and trauma.[1,10] The most common concept of tissue engineering is to seed cells on a three-dimensional polymer scaffold for culture *in vitro* or implantation *in vivo* to form a bioconstruct at the target tissue site. Tissue engineering brings together the different disciplines of biology, materials science, and biomedical engineering toward long-term repair and regeneration of damaged tissues and organs. These multidisciplinary approaches focus on tissue engineering and include the design, fabrication, and characterization of scaffolds; the isolation of stem/progenitor cells; and the manipulation of cell proliferation and differentiation. Several studies have been published in which proof of concept has been adequately demonstrated to engineer different tissues such as bone,[11] cartilage,[12] nerves,[13] and skin.[14] This article is focused on nanomaterials and scaffolds for articular cartilage regeneration.

17.2.1 Dedifferentiation and Redifferentiation of Chondrocytes

Repair of cartilage defects by autologous chondrocyte implantation (ACI) is currently used in many countries. In clinical practice, articular cartilage chondrocytes are isolated from the patient, expanded *in vitro* by a monolayer culture, assembled in a collagen scaffold, and then implanted autologously into the same patient.[9] The main advantage of ACI is that it can form new cartilaginous tissue without causing any immune response in the patients. However, when chondrocytes are cultured in a monolayer, it leads to the dedifferentiation of chondrocytes.[15,16] During dedifferentiation, chondrocytes change their normal polyhedral morphology into spindle-shaped fibroblast-like cells (Figure 17.1) due to phenotypic changes. They lose their ability to express articular cartilage-specific ECM, such as collagen type II and proteoglycan aggrecan, and produce fibroblast-specific ECM collagen type I, which is a marker of fibrous cartilage.[15–17] Various methods have been used to inhibit dedifferentiation and enable the redifferentiation of dedifferentiated chondrocytes. It is well documented that transforming growth factor-β (TGF-β) isoforms can stimulate the change of dedifferentiated cells to redifferentiated chondrocytes.[18,19] In addition to the use of TGF-β isoforms, three-dimensional scaffolds of agarose,[17] alginate,[20] fibrin glue,[21] collagen type II sponge,[22] and PLGA-collagen hybrid mesh[23] have aided

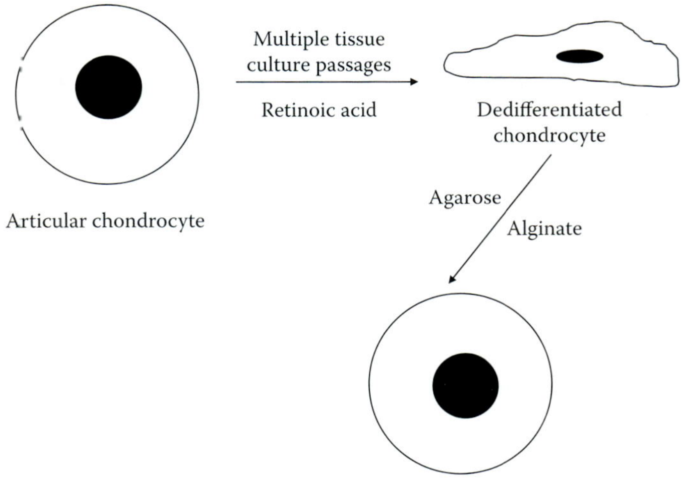

FIGURE 17.1 Schematic showing the dedifferentiation and redifferentiation of articular chondrocytes.

cartilage tissue engineering. Recently, in a study conducted by Yao and colleagues,[24] a TGF-β3 gene was delivered into dedifferentiated chondrocytes using recombinant lentiviral vectors. Chondrocytes dedifferentiate in the presence or absence of transgenic TGF-β3 embedded in two-dimensional and three-dimensional hydrogel scaffolds. Expression of TGF-β3 (LV-T) and Col I-targeting shRNA (LV-D) promoted chondrocyte redifferentiation in a three-dimensional culture within alginate hydrogen, but not in a two-dimensional culture. Thus, this study has demonstrated the utility of lentiviral vector-mediated expression of a TGF-β3 gene to stimulate the redifferentiation of chondrocytes.

17.3 SCAFFOLDS

Scaffolds are made of different materials such as polymers, ceramics, and composite biomaterials. Many scaffolds support three-dimensional tissue formation *in vitro*. Three-dimensional scaffolds are key components in articular cartilage tissue engineering and regeneration strategies. They serve as an optimal microenvironment for chondrocyte differentiation and biosynthesis of the ECM. To achieve the optimal goal of cartilage regeneration, scaffolds should meet the criteria listed in Table 17.1.[25–28]

17.4 BIOMIMETIC BIOMATERIALS FOR SCAFFOLDS

These are biomaterials that mimic extracellular matrices. Various materials, both natural and synthetic, have been investigated for the fabrication of porous scaffolds for articular cartilage regeneration. However, there are disadvantages that are associated with the use of synthetic polymers as materials for scaffolds, because they produce high local concentrations of acid by-products, leading to an inflammatory

TABLE 17.1
Biological Characteristics of Scaffolds

Characteristics	Properties
Biocompatible, nontoxic	
Nonimmunogenic	
Biodegradable	This will facilitate absorption by the surrounding tissues without the need for surgical removal.
High porosity and adequate pore size	This will facilitate cell seeding and diffusion throughout the structure of both cells and nutrients, anchorage for cell adhesion, and provide necessary space for neovascularization.
Delivery of morphogens and stem cell	Growth factor is important in guiding and controlling cell behavior in tissue regeneration. The combination of growth factors with the designed scaffold may improve cartilage tissue engineering. For stem cells, the use of relevant sources is important to determine the effectiveness of tissue engineering strategies.
Mechanical strength	This will maintain the spaces required for cells in growth and matrix formation. It must also supply sufficient temporary mechanical support. It also plays an important role in homeostasis, remodeling, and repair of load-bearing tissues.

response.[29] Several biodegradable polymers have been used for tissue engineering of articular cartilage such as hyaluronic acid (HA), hydrogels, and nanomaterials, which include alginate, agarose, collagens, and chitosan.

17.4.1 Hyaluronic Acid

HA is a major component of the ECM of the normal human cartilage and plays an important role in cartilage homeostasis.[30,31] Various studies have shown that HA is recognized by specific cell receptors regulating cell behavior, inflammation, and the wound healing process and it acts as a selective and protective coat around the cell membrane.[30] HA is prepared from different animal tissues and also by microbial fermentation.[32] HA promotes cell proliferation[33] and cell differentiation and migration, and also provides an optimal cellular environment.[34–37] HA, however, also has some limitations as a biomaterial scaffold.[38] Human primary chondrocytes were embedded with or without HA-treated gelatin scaffolds, and it was found that in HA-treated gelatin scaffolds, there was a significantly increased amount of chondrocyte proliferation and adhesive activity compared with the untreated scaffolds.

17.4.2 Hydrogels

Hydrogels are prepared from naturally derived materials such as agarose,[17] alginate,[39] gelatin,[41] chitosan, fibrin[21] and collagen type 1 gels[42] (Table 17.2), or from synthetic polymers such as polyethylene glycol (PEG) and polyvinyl alcohol (PVA). Hydrogels swell excessively in water[43,44] and are optimal for the delivery of cells

TABLE 17.2
Various Natural Molecules as Hydrogels for Chondrocytes

Polymer	Chemical Compositions	Origin	Properties
Alginate	1-4-linked β-D-mannuronic acid and α-L-guluronic acid	Marine algae	Used for drug delivery, cell encapsulation, and antiadhesion material. Injectable Mechanical weakness, does not resorb well
Agarose	D-Galactose and 3,6-anhydro-L-galactopyranose	Derived from seaweed	Easy preparation, maintains chondrocyte phenotype. Cells are uniformly distributed within implant. Injectable Mechanical weakness, difficult to handle and sterilize, impurities affect material properties
Collagen	Glycine, proline, and hydroxyproline	Animal cartilaginous tissues	Low antigenicity and good cell-binding properties. Low biomechanical stiffness and rapid biodegradation. Toxicity of some cross-linking agents
Chitosan	Polysaccharide β-1,4-linked N-glycosamine	Arthropod exoskeleton	Hydrophilic surface promotes cell adhesion, proliferation, and differentiation. Good biocompatibility, acceptable host response, and antibacterial activity Mechanical weakness and instability, incapacity to maintain a predefined shape, impurities affecting material properties
Hyaluronan	α-1,4-D-glucronic acid and β-1,3-N-acetyl-D-glucosamine	Animal tissues, microbial fermentation	Promotes cartilage formation by encapsulated cells, no immunogenic properties, interacts with cell surface receptors, good integration with host tissues May induce chondrolysis Repair cartilage is thinner than native cartilage

and bioactive factors. Naturally derived hydrogels lack the mechanical properties to withstand physiological loads.[46–50] Hydrogels play an important role in promoting chondrocyte attachment similar to that of the cartilage ECM.[17,49,51–53]

Elisseeff and colleagues[54] embedded chondrocytes in polyethylene oxide-based hydrogels and these chondrocytes survived for seven weeks and produced neocartilage. The hydrogel constructs at low cell density were incubated in a medium containing 10% fetal bovine serum, and they induced the redifferentiation of chondrocytes and effectively enhanced matrix synthesis.[55–58]

17.4.3 Nanomaterials

Nanomaterials can be defined as materials with a dimension of 1–100 nm. There are various types of material that have been used to produce nanomaterials and these include ceramics, metals, nanofibers, nanoparticles, nanofilms, nanoclusters, and polymers in the form of nanocrystals. Different methods have been used to generate these nanomaterials. These methods include self-assembling, electrospinning, nanoimprinting, phase separation, and photolithography. The use of self-assembling peptide hydrogels in the regeneration of articular cartilage is increasing in this ever-expanding field of cartilage tissue engineering.[59–61] The properties of a self-assembled hydrogel should include being noninvasive and injectable, as well as having the ability to mix with other bioactive signals such as morphogens and growth factors.[60,62] Many hydrogels stimulate proliferation and extracellular matrices (ECM) that lead to the formation of tissue.[63] Self-assembly of cartilage has been developed by Hu and Athanasiou,[4] in which they suspended primary chondrocytes in a nonadherent well at high cell densities. When growth factors and hydrostatic pressure are applied, the self-assembled cartilages achieve mechanical properties that are similar to native tissues.[64] The most common types of material used for nanomaterials are peptide amphiphiles and self-assembled peptides.[59,61] Peptide amphiphiles are a class of molecules consisting of hydrophobic nonpeptide tails covalently bound to peptide sequences. Self-assembling biomaterials were developed previously based on this broad class of molecules. The peptide amphiphiles are then self-assembled from aqueous medium into supramolecular nanofibers. These molecules serve as ECM, consisting of a peptide segment covalently bound to a hydrophobic segment.[62,65] Recently, in a study conducted by Stupp,[60] a unique self-assembled peptide amphiphile molecule binding to TGF-β1 was designed and evaluated for use in an articular cartilage animal model. Initially, the study was performed *in vitro* to establish the ability of peptide amphiphile molecules to support mesenchymal stem cell (MSC) viability and chondrogenic differentiation. It was discovered that self-assembled peptide amphiphile scaffolds can support MSC viability and chondrogenic differentiation. The presence of TGF-β1 can significantly stimulate the gene expression of cartilage-specific markers in a slower release manner. Full thickness chondral defects were created *in vivo* to evaluate the potential of peptide amphiphiles to induce cartilage regeneration. The defects were treated using a combination of peptide amphiphiles and TGF-β1. They found that peptide amphiphiles synthesized with a peptide-binding sequence of TGF-β1 significantly stimulated the microfractured chondral defects. Various types of peptides have been investigated (Table 17.3).[59, 61, 66–71]

17.4.4 Alginates

Alginates are generally prepared from marine algae. Alginates maintain chondrocytic phenotype.[45–50] Although alginates are degradable, these processes cannot be easily controlled.[72] They are used as a biomaterial for drug delivery and cell-encapsulating scaffolds.[45,72–77] When alginates are used alone, they do not interact with cells, and this consequently leads to limited bioactivity. Several studies have been done to overcome these limitations and to use its benefit as a biocompatible material. In a study performed by Sultzbaugh and Speaker[75] the surface of alginates

TABLE 17.3
Functional Peptide Sequences for Coupling to Hydrogels for Functional Cartilage Tissue Engineering and Regeneration

Peptides	Properties
RGDs (arginine–glycine–aspartic acid)	Found in fibronectin, ECM proteins, plays a role in cell adhesion
IKVAV	Cell attachment, migration, and neural attachment
YIGSR	Laminin and epithelial cell attachment
KLD4	Cartilage regeneration and cell proliferation
KLD12	Repair of articular cartilage defect and cell proliferation
VEVE	Forms cylindrical nanofibers
RAD16-I	Cell proliferation and cell attachment
$EAKA_4$	Cell attachment and proliferation
$RADA_4$	Cell attachment and proliferation

was coated with lectin possessing carbohydrate-specific binding properties. Lectin has the ability to retain ligand-specific properties and this was validated by the uptake of radiolabeled ligands when incubated with alginate gels. Mosahebi and colleagues[78] have demonstrated that alginate gels promoted the viability and function of Schwann cells for neural applications. In a study conducted by Cohen and colleagues,[79] it was demonstrated that alginate gel can be used as a substrate in the *in situ* biofabrication of a geometrically matched hydrogel *ex vivo* into a cartilage defect using additive printing methods.

17.4.5 Agarose

Agarose is one of the hydrogels studied in the field of tissue engineering because of its gelation and cell encapsulation[45] properties. It is well documented that chondrocytes cultured in agarose gel promote the production of an ECM and maintain the phenotype of the chondrocytes.[45,80] A mechanical environment can also modify the metabolic activity and the biomechanical properties of articular cartilage. According to the literature review, it was shown that when chondrocytes were seeded in agarose gel, there was an increased accumulation of proteoglycan that is subjected to dynamic loading.[46,81] There are other factors that increase the accumulation of ECM components in chondrocytes seeded in agarose gel, such as the extracellular pH, the addition of collagen hydrolysate to the culture medium, and the seeding density of the cells.[82–84]

17.4.6 Collagens

Collagens are prepared from animal cartilaginous tissue, have high swelling ratios, and promote the formation of cartilage by cell encapsulation. The fact that collagens are prepared from animal products can however cause immune responses.

Chondrocytes embedded with collagen gels via integrins promote cell proliferation and production of ECM components.[49] It was found that when MSCs were embedded in type II collagen gel, it promoted chondrogenic differentiation compared with type I collagen gel.[85–87] Autologous MSCs were embedded in type I collagen gel and implanted into cartilage defects from a sheep model for 6 months. It demonstrated that MSCs differentiated into chondrocytes and repaired the defects with zonally organized cartilage, but there were still areas of incomplete integration as well as high variability within the groups.[87] Cysts formed in the subchondral bone, which could have been caused by a lack of mechanical support from the collagen gel. According to the literature review, type I collagen gels have been used as a delivery for bone marrow transplantation into human cartilage,[88] which demonstrated good results in healthy knees[89–91] but not in osteoarthritic knees.[92] Collagen hydrogels have been used to study the behavior of chondrocytes and stem cells *in vitro* under various conditions such as during compression and in the presence of growth factors.[95,117,137–139]

17.4.7 Chitosan

Chitosan is derived from the exoskeleton of arthropods and has the same structure as the glycosaminoglycans found in cartilage.[36,93] Chemical cross-linking can stimulate the gelation of chitosan.[94] Certain properties such as acetylation can control the breakdown of chitosan by cell-secreted lysozyme.[96,97] The biochemical properties of chitosan cartilage can be modified to match various methods.[98–101] Chitosan is a biodegradable polymer that can be combined with other materials. When chondrocytes are seeded on chitosan-coated coverslips, it maintained the morphology of the chondrocytes and promoted the expression of collagen type II and aggrecan[94, 98,102,103] nanofibers,[104,107] gels,[105–108] and scaffolds.[109–111] Hsu and colleagues[112] found that combining chitosan with both alginate and hyaluronate, and coating with an arginine–glycine–aspartic acid (RGD) sequence of fibronectin, led to increased production of glycosaminoglycans (GAGs), cell proliferation, and collagen type II. In addition, mixing chitosan with hyaluronate improves cell adhesion of chondrocytes to the scaffolds, increases the rate of proliferation, stimulates accumulation of aggrecan and collagen type II, and maintains cell morphology compared with chitosan alone.[103] Hybridizing chitosan with collagen type I and chondroitin sulfate led to stable chondrocyte morphology, proper proliferation, and good adhesion properties, as well as stimulating the production of cartilage markers.[113] In a study conducted by Hao et al.,[101] chondrocytes encapsulated in injectable chitosan hydrogel were implanted in a defect in a sheep, and demonstrated good integration with the surrounding tissues.

17.4.8 Polyethylene Glycol

PEG has properties that make it an attractive polymer for biomedical applications. These properties include hydrophilicity and biocompatibility,[114,115] limited immunogenicity, antigenicity, and minimal cell adhesion.[116] PEG was introduced as an injectable and transdermally photopolymerizable hydrogel for cartilage tissue engineering.[11,54,116,118–122] When MSCs, chondrocytes, and embryonic stem cells are embedded into PEG gel with the addition of growth factors, cartilage tissue was formed.[123,131] PEG hydrogel can easily be modified to change cell behavior.[115,121,124,125] Currently, PEG is being investigated in clinical trials in combination with a bioadhesive based on chondroitin sulfate.[140]

17.4.9 Polyvinyl Alcohol

PVA can be formed by either chemical cross-linking or physical cross-linking. PVA can be chemically cross-linked into hydrogels through the formation of acetaldehyde or photopolymerization.[114] During the annealing, dehydration, and freeze–thaw cycling, PVA can be physically cross-linked through the formation of crystallites.[44,126,127] Different properties have been used to compare PVA hydrogels with articular cartilage. The compressive modulus, shear modulus, tensile modulus, and permeability values were similar to those of articular cartilage[128–132] (Table 17.4). Hydrogel values are obtained in the presence of water so that chemical structure is maintained. PVA has been used as a model system to study the properties of articular cartilage, such as swelling and lubricating,[43,133] because of their frictional and lubricating properties, which are similar to articular cartilage. Chemical cross-linking can cause harsh environments that can be toxic or damaging to the cells. To culture or encapsulate the cells using PVA, it must first be cross-linkable under gentle conditions or must have pores after preparation.[124,134–136] Bryant and colleagues[124] have used methacrylate groups to modify PVA hydrogels to allow photoencapsulation of cells. The compressive moduli ranged from 5 kPa to 2600 kPa depending on the concentration of the PVA. The spherical morphology of encapsulated chondrocytes was maintained for 3 days *in vitro*. Chondroitin sulfate was used to modify the PVA hydrogel, providing control over the properties of swelling, biochemical similarities to cartilage, and breaking down by cell-secreted chondroitinase.[124] In another example, Schmedlen et al.[141] incorporated PVA hydrogels into adhesive peptide arginine–glycine–aspartic acids to support cell–matrix interactions.

TABLE 17.4
Comparison between PVA Hydrogels and Native Articular Cartilage

Property	Test	Concentration of Polymer	Hydrogel Value	Articular Cartilage Value
Compressive modulus	Unconfined compression	20% w/w	0.25 MPa	0.3–0.8 MPa (bovine)
Aggregate modulus	Unconfined compression	10% w/w	0.11 MPa	0.33 MPa (bovine)
Shear modulus permeability	Torsion, unconfined compression	25% w/w 10% w/w	0.17–0.43 3.2×10^{-14} m4/Ns (Newton seconds)	0.13–0.22 (canine) 0.546×10^{-14} m4/Ns (bovine)
Coefficient of friction	Continuous sliding versus cartilage	15% w/w	0.152–0.171	0.029 (human)
Coefficient of friction	Cyclic load versus cartilage	15% w/w	0.046–0.058	0.029 (human)

17.5 SUMMARY

The articular cartilage consists of chondrocytes surrounded by vast expanses of ECM. The cell membrane ECM continuum functions at the organ, tissue, and cellular levels. ACI is used for the regeneration of articular cartilage. This requires large numbers of chondrocytes to be expanded to obtain the optimal amounts. However, during this process, chondrocytes dedifferentiate. Currently, it is a challenge to redifferentiate the dedifferentiated chondrocytes to their original morphology and phenotype. A key method for redifferentiation of dedifferentiated chondrocytes is by using agarose and alginate scaffolds. Different types of hydrogels have been investigated for the maintenance and regeneration of articular cartilage. The synthetic hydrogels have more advantages than natural ones because they allow control over the properties of the material whereas natural hydrogels have more benefits in terms of their biocompatibility, biodegradability, and control over cell matrix properties. It is noteworthy that nanomaterials can provide excellent properties, including biochemical and mechanical cues to elicit the differentiation of cells, for clinical applications. These nanomaterials can also be used to regenerate bone and cartilage. One of the studies demonstrated the potential of self-assembling peptide scaffolds to maintain differentiated chondrocytes and to stimulate the synthesis of ECM. Finally, it is exciting to recognize the potential use of self-assembling peptide hydrogels for cartilage regeneration.[60,61]

ACKNOWLEDGMENTS

This work was supported by the Lawrence Ellison Endowed Chair. We thank Gordon Peng for his excellent assistance with the figures, and the Tshwane University of Technology for encouragement and support, including the study leave.

REFERENCES

1. Reddi, A. H. 1998. Role of morphogenetic proteins in skeletal tissue engineering and regeneration. *Nature Biotechnology* 16: 247–252.
2. Ahsan, T., and R. L. Sah. 1999. Biomechanics of integrative cartilage repair. *Osteoarthritis and Cartilage* 7: 29–40
3. Hunziker, E., M. Spector, J. Libera et al. 2006. Translation from research to applications. *Tissue Engineering* 12: 3341–3364.
4. Hu, J. C., and K. A. Athanasiou. 2006. A self-assembling process in articular cartilage tissue engineering. *Tissue Engineering* 12: 969–976.
5. Reddi, A. H. 2008. *Morphogenesis and Morphogenetic Proteins. Principles of Regenerative Medicine*. San Diego: Academic Press, pp. 498–511.
6. Guilak, F., D. M. Cohen, B. T. Estes et al. 2009. Control of stem cell fate by physical interactions with the extracellular matrix. *Cell Stem Cell* 5: 17–26.
7. Becerra, J., L. Santos-Ruiz, J. A. Andrades, and M. Mari-Beffa. 2010. The stem cell niche should be a key issue for cell therapy in regenerative medicine. *Stem Cell Reviews* 7: 2, 248–255.
8. George, A., and S. Ravindran. 2010 Protein templates in hard tissue engineering. *Nano Today* 5: 254–266.
9. Brittberg, M., A. Lindahl, A. Nilsson et al. 1994. Treatment of deep cartilage defects in the knee with autologous chondrocyte transplantation. *New England Journal of Medicine* 331: 889–895.

10. Reddi, A. H. 2000. Morphogenesis and tissue engineering of bone and cartilage: Inductive signals, stem cells, and biomimetic biomaterials. *Tissue Engineering* 6: 351–359.
11. Williams, J. M., A. Adewunmi, R. M. Schek et al. 2005. Bone tissue engineering using polycaprolactone scaffolds fabricated via selective laser sintering. *Biomaterials* 26: 4817–4827.
12. Mercier, N. R., H. R. Costantino, M. A. Tracy, and L. J. Bonassar. 2005. Poly(lactide-co-glycolide) microspheres as a moldable scaffold for cartilage tissue engineering. *Biomaterials* 26: 1945–1952.
13. Lietz, M., L. Dreesmann, M. Hoss et al. 2006. Neuro tissue engineering of glial nerve guides and the impact of different cell types. *Biomaterials* 27: 1425–1436.
14. Dai, N. T., M. R. Williamson, N. Khammo et al. 2004. Composite cell support membranes based on collagen and polycaprolactone for tissue engineering of skin. *Biomaterials* 25: 4263–4271.
15. von der Mark, K., V. Gaus, H. von der Mark et al. 1977. Relationship between cell shape and type of collagen synthesized as chondrocytes lose their cartilage phenotype in culture. *Nature* 267: 531–532.
16. Elima, K., and E. Vuorio. 1989. Expression of mRNAs for collagens and other matrix components in dedifferentiating and redifferentiating human chondrocytes in culture. *FEBS Letters* 258: 195–198.
17. Benya, P. D., and J. D. Shaffer. 1978. Dedifferentiated chondrocytes reexpress the differentiated collagen phenotype when cultured in agarose gels. *Cell* 30: 215–224.
18. Na, K., S. Kim, D. G. Woo et al. 2007. Synergistic effect of TGFbeta-3 on chondrogenic differentiation of rabbit chondrocytes in thermoreversible hydrogel constructs blended with hyaluronic acid by *in vivo* test. *Journal of Biotechnology* 128: 412–422.
19. Yun, K., and H. T. Moon, 2008. Inducing chondrogenic differentiation in injectable hydrogels embedded with rabbit chondrocytes and growth factor for neocartilage formation. *Journal of Bioscience and Bioengineering* 105: 122–126.
20. Bonaventure, J., N. Kadhom, L. Cohen-Solal et al. 1994. Reexpression of cartilage-specific genes by dedifferentiated human articular chondrocytes cultured in alginate beads. *Experimental Cell Research* 212: 97–104.
21. Ting, V., C. D. Sims, L. E. Brecht et al. 1998. *In vitro* prefabrication of human cartilage shapes using fibrin glue and human chondrocytes. *Annals of Plastic Surgery* 40: 413–420; discussion 420–411.
22. Mukaida, T., K. Urabe, K. Naruse et al. 2005. Influence of three-dimensional culture in a type II collagen sponge on primary cultured and dedifferentiated chondrocytes. *Journal of Orthopaedic Science* 10: 521–528.
23. Chen, G., T. Sato, T. Ushida et al. 2003. Redifferentiation of dedifferentiated bovine chondrocytes when cultured *in vitro* in a PLGA–collagen hybrid mesh. *FEBS Letters* 542: 95–99.
24. Yao, Y., F. Zhang, P. X. Pang et al. 2011. *In vitro* study of chondrocyte redifferentiation with lentiviral vector-mediated transgenic TGF-beta3 and shRNA suppressing type I collagen in three-dimensional culture. *Journal of Tissue Engineering and Regenerative Medicine* 5: e219–227.
25. Hutmacher, D. W. 2000. Scaffolds in tissue engineering bone and cartilage. *Biomaterials* 21: 2529–2543.
26. Coutts, R. D., R. M. Healey, R. Ostrander et al. 2001. Matrices for cartilage repair. *Clinical Orthopaedics and Related Research*: S271–279.
27. Leong, K. F., C. M. Cheah, and C. K. Chua. 2003. Solid freeform fabrication of three-dimensional scaffolds for engineering replacement tissues and organs. *Biomaterials* 24: 2363–2378.

28. Karageorgiou, V., and D. Kaplan. 2005. Porosity of 3D biomaterial scaffolds and osteogenesis. *Biomaterials* 26: 5474–5491.
29. Grande, D. A., C. Halberstadt, G. Naughton et al. 1997. Evaluation of matrix scaffolds for tissue engineering of articular cartilage grafts. *Journal of Biomedical Materials Research* 34: 211–220.
30. Fraser, J. R., T. C. Laurent, and U. B. Laurent. 1997. Hyaluronan: Its nature, distribution, functions and turnover. *Journal of Internal Medicine* 242: 27–33.
31. Allison, D. D., and K. J. Grande-Allen. 2006. Review. Hyaluronan: A powerful tissue engineering tool. *Tissue Engineering* 12: 2131–2140.
32. Chong, B. F., L. M. Blank, R. McLaughlin, and L. K. Nielsen. 2005. Microbial hyaluronic acid production. *Applied Microbiology and Biotechnology* 66: 341–351.
33. Ehlers, E. M., P. Behrens, L. Wunsch et al. 2001. Effects of hyaluronic acid on the morphology and proliferation of human chondrocytes in primary cell culture. *Annals of Anatomy* 183: 13–17.
34. Chow, G., C. B. Knudson, G. Homandberg, and W. Knudson. 1995. Increased expression of CD44 in bovine articular chondrocytes by catabolic cellular mediators. *Journal of Biological Chemistry* 270: 27734–27741.
35. Akmal, M., A. Singh, A. Anand et al. 2005. The effects of hyaluronic acid on articular chondrocytes. *Journal of Bone and Joint Surgery* 87: 1143–1149.
36. Chen, Y. L., H. P. Lee, H. Y. Chan et al. 2007. Composite chondroitin-6-sulfate/dermatan sulfate/chitosan scaffolds for cartilage tissue engineering. *Biomaterials* 28: 2294–2305.
37. Puppi, D., F. Chiellini, A. M. Piras, and E. Chiellini. 2010. Polymeric materials for bone and cartilage. *Progress in Polymer Science* 35: 403–440.
38. Campoccia, D., P. Doherty, M. Radice et al. 1998. Semisynthetic resorbable materials from hyaluronan esterification. *Biomaterials* 19: 2101–2127.
39. Stevens, M. M., H. F. Qanadilo, R. Langer, and V. Prasad Shastri. 2004. A rapid-curing alginate gel system: Utility in periosteum-derived cartilage tissue engineering. *Biomaterials* 25: 887–894.
40. Gerard, C., C. Catuogno, C. Amargier-Huin et al. 2005. The effect of alginate, hyaluronate and hyaluronate derivatives biomaterials on synthesis of nonarticular chondrocyte extracellular matrix. *Journal of Materials Science. Materials in Medicine* 16: 541–551.
41. Ibusuki, S., Y. Fujii, Y. Iwamoto, and T. Matsuda. 2003. Tissue-engineered cartilage using an injectable and *in situ* gelable thermoresponsive gelatin: Fabrication and *in vitro* performance. *Tissue Engineering* 9: 371–384.
42. Chaipinyo K., B. W. Oakes, and M. P. Van Damme. 2004. The use of debrided human articular cartilage for autologous chondrocyte implantation: Maintenance of chondrocyte differentiation and proliferation in type I collagen gels. *Journal of Orthopaedic Research* 22: 446–455.
43. Broom, N. D., and A. Oloyede. 1998. The importance of physicochemical swelling in cartilage illustrated with a model hydrogel system. *Biomaterials* 19: 1179–1188.
44. Lowman, A. M., and N. A. Peppas. 1999. Solute transport analysis in pH-responsive, complexing hydrogels of poly(methacrylic acid-g-ethylene glycol). *Journal of Biomaterials Science. Polymer Edition* 10: 999–1009.
45. Guo, J. F., G. W. Jourdian, and D. K. MacCallum. 1989. Culture and growth characteristics of chondrocytes encapsulated in alginate beads. *Connective Tissue Research* 19: 277–297.
46. Mauck, R. L., M. A. Soltz, C. C. Wang et al. 2000. Functional tissue engineering of articular cartilage through dynamic loading of chondrocyte-seeded agarose gels. *Journal of Biomechanical Engineering* 122: 252–260.
47. Wong, M., M. Siegrist, X. Wang, and E. Hunziker. 2001. Development of mechanically stable alginate/chondrocyte constructs: Effects of guluronic acid content and matrix synthesis. *Journal of Orthopaedic Research* 19: 493–499.

48. Mauck, R. L., S. B. Nicoll, S. L. Seyhan et al. 2003. Synergistic action of growth factors and dynamic loading for articular cartilage tissue engineering. *Tissue Engineering* 9: 597–611.
49. Yamaoka, H., H. Asato, T. Ogasawara et al. 2006. Cartilage tissue engineering using human auricular chondrocytes embedded in different hydrogel materials. *Journal of Biomedical Materials Research Part A* 78: 1–11.
50. Selmi, T. A., P. Verdonk, P. Chambat et al. 2008. Autologous chondrocyte implantation in a novel alginate-agarose hydrogel: outcome at two years. *Journal of Bone and Joint Surgery. British Volume* 90: 597–604.
51. Passaretti, D., R. P. Silverman, W. Huang et al. 2001. Cultured chondrocytes produce injectable tissue-engineered cartilage in hydrogel polymer. *Tissue Engineering* 7: 805–815.
52. Cushing, M. C., and K. S. Anseth. 2007. Materials science. Hydrogel cell cultures. *Science* 316: 1133–1134.
53. Appelman, T. P., J. Mizrahi, J. H. Elisseeff, and D. Seliktar. 2010. The influence of biological motifs and dynamic mechanical stimulation in hydrogel scaffold systems on the phenotype of chondrocytes. *Biomaterials* 32: 1508–1516.
54. Elisseeff, J., K. Anseth, D. Sims et al. 1999. Transdermal photopolymerization for minimally invasive implantation. *Proceedings of the National Academy of Sciences of the United States of America* 96: 3104–3107.
55. Kato, Y., and D. Gospodarowicz. 1984. Growth requirements of low-density rabbit costal chondrocyte cultures maintained in serum-free medium. *Journal of Cellular Physiology* 120: 354–363.
56. Grunder, T., C. Gaissmaier, J. Fritz et al. 2004. Bone morphogenetic protein (BMP)-2 enhances the expression of type II collagen and aggrecan in chondrocytes embedded in alginate beads. *Osteoarthritis and Cartilage* 12: 559–567.
57. Woodward, W. A., and R. S. Tuan. 1999. N-Cadherin expression and signaling in limb mesenchymal chondrogenesis: Stimulation by poly-L-lysine. *Developmental Genetics* 24: 178–187.
58. Aszodi, A., E. B. Hunziker, C. Brakebusch, and R. Fassler. 2003. Beta1 integrins regulate chondrocyte rotation, G1 progression, and cytokinesis. *Genes and Development* 17: 2465–2479.
59. Kisiday, J., M. Jin, B. Kurz et al. 2002. Self-assembling peptide hydrogel fosters chondrocyte extracellular matrix production and cell division: implications for cartilage tissue repair. *Proceedings of the National Academy of Sciences of the United States of America* 99: 9996–10001.
60. Stupp, I. 2010. Self-assembly and biomaterials. *Nano Letters* 10: 4783–4786.
61. Shah, R. N., N. A., Shah, M. M. Del Rosario Lim et al. 2010. Supramolecular design of self-assembling nanofibers for cartilage regeneration. *Proceedings of the National Academy of Sciences of the United States of America* 107: 3293–3298.
62. Behanna, L., K. Rajangam, and I. Stupp. 2006. Modulation of fluorescence through coassembly of molecules in organic nanostructures. *Nano Letters* 6: 2086–2090.
63. Reddi, A. H., J. Becerra, and J. A. Andrades. 2011. Nanomaterials and hydrogel scaffolds for articular cartilage regeneration. *Tissue Engineering. Part B, Reviews* 17: 301–305.
64. Elder, B. D., and K. A. Athanasiou. 2008. Synergistic and additive effects of hydrostatic pressure and growth factors on tissue formation. *PLoS One* 3: e2341.
65. Hartgerink, J. D., E. Beniash, and S. I. Stupp. 2002. Peptide-amphiphile nanofibers: A versatile scaffold for the preparation of self-assembling materials. *Proceedings of the National Academy of Sciences of the United States of America* 99: 5133–5138.
66. Niece, K. L., J. D. Hartgerink, J. J. Donners, and S. I. Stupp. 2003. Self-assembly combining two bioactive peptide-amphiphile molecules into nanofibers by electrostatic attraction. *Journal of the American Chemical Society* 125: 7146–7147.

67. Cui, H., T. Muraoka, G. Cheetman et al. 2008. Self-assembly of giant peptide nanobelts. *Nano Letters* 9: 945–951.
68. Pashuck, E. T., H. Cui, and I. Stupp. 2009. Tuning supramolecular rigidity of peptide fibers through molecular structure. *Journal of the American Chemical Society* 132: 6041–6046.
69. Cui, H., M. J. Webber, and I. Stupp. 2009. Self-assembly of peptide amphiphiles: from molecules to nanostructures to biomaterials. *Peptide Science* 94: 1–18.
70. Sun, J., and Q. Zheng. 2009. Experimental study on self-assembly of KLD-12 peptide hydrogel and 3-D culture of MSC encapsulated within hydrogel *in vitro*. *Journal of Huazhong University of Science and Technology (Medical Sciences)* 29: 512–516.
71. Kopesky, P. W., E. J. Vanderploeg, J. S. Sandy et al. 2010. Self-assembling peptide hydrogels modulate *in vitro* chondrogenesis of bovine bone marrow stromal cells. *Tissue Engineering—Part A* 16: 465–477.
72. Lee, K. Y., and D. J. Mooney. 2001. Hydrogels for tissue engineering. *Chemical Reviews* 101: 1869–1879.
73. Hauselmann, H. J., M. B. Aydelotte, B. L. Schumacher et al. 1992. Synthesis and turnover of proteoglycans by human and bovine adult articular chondrocytes cultured in alginate beads. *Matrix* 12: 116–129.
74. Van Susante, J. L., P. Buma, G. J. van Osch, D. Versleyen et al. 1995. Culture of chondrocytes in alginate and collagen carrier gels. *Acta Orthopaedica Scandinavica* 66: 549–556.
75. Sultzbaugh, K. J., and T. J. Speaker. 1996. A method to attach lectins to the surface of spermine alginate microcapsules based on the avidin biotin interaction. *Journal of Microencapsulation* 13: 363–376.
76. Masuda, K., R. L. Sah, M. J. Hejna, and E. J. Thonar. 2003. A novel two-step method for the formation of tissue-engineered cartilage by mature bovine chondrocytes: the alginate-recovered-chondrocyte (ARC) method. *Journal of Orthopaedic Research* 21: 139–148.
77. Homicz, M. R., S. H. Chia, B. L. Schumacher et al. 2003. Human septal chondrocyte redifferentiation in alginate, polyglycolic acid scaffold, and monolayer culture. *Laryngoscope* 113: 25–32.
78. Mosahebi, A., M. Simon, M. Wiberg, and G. Terenghi. 2001. A novel use of alginate hydrogel as Schwann cell matrix. *Tissue Engineering* 7: 525–534.
79. Cohen, D. L., J. I. Lipton, L. J. Bonassar, and H. Lipson, 2010 Additive manufacturing for *in situ* repair of osteochondral defects. *Biofabrication* 2: 035004.
80. Ng, K. W., C. C. Wang, R. L. Mauck et al. 2005. A layered agarose approach to fabricate depth-dependent inhomogeneity in chondrocyte-seeded constructs. *Journal of Orthopaedic Research* 23: 134–141.
81. Chowdhury, T. T., D. L. Bader, J. C. Shelton, and D. A. Lee, 2003. Temporal regulation of chondrocyte metabolism in agarose constructs subjected to dynamic compression. *Archives of Biochemistry and Biophysics* 417: 105–111.
82. Wu, G., Y. Liu, T. Iizuka, and E. B. Hunziker. 2010. The effect of a slow mode of BMP-2 delivery on the inflammatory response provoked by bone-defect-filling polymeric scaffolds. *Biomaterials* 31: 7485–7493.
83. Ng, K. W., J. D. Saliman, E. Y. Lin et al. 2007. Culture duration modulates collagen hydrolysate-induced tissue remodeling in chondrocyte-seeded agarose hydrogels. *Annals of Biomedical Engineering* 35: 1914–1923.
84. Mauck, R. L., S. L. Seyhan, G. A. Ateshian, and C. T. Hung. 2002. Influence of seeding density and dynamic deformational loading on the developing structure/function relationships of chondrocyte-seeded agarose hydrogels. *Annals of Biomedical Engineering* 30: 1046–1056.

85. Bosnakovski, D., M. Mizuno, G. Kim et al. 2004. Chondrogenic differentiation of bovine bone marrow mesenchymal stem cells in pellet cultural system. *Experimental Hematology* 32: 502–509.
86. Lu, Z., B. Z. Doulabi, C. Huang et al. 2010. Collagen type II enhances chondrogenesis in adipose tissue-derived stem cells by affecting cell shape. *Tissue Engineering—Part A* 16: 81–90.
87. Zscharnack, M., P. Hepp, R. Richter et al. 2010. Repair of chronic osteochondral defects using predifferentiated mesenchymal stem cells in an ovine model. *American Journal of Sports Medicine* 38: 1857–1869.
88. Wakitani, S., T. Okabe, S. Horibe et al. 2011. Safety of autologous bone marrow-derived mesenchymal stem cell transplantation for cartilage repair in 41 patients with 45 joints followed for up to 11 years and 5 months. *Journal of Tissue Engineering and Regenerative Medicine* 5: 146–150.
89. Wakitani, S., T. Mitsuoka, N. Nakamura et al. 2004. Autologous bone marrow stromal cell transplantation for repair of full-thickness articular cartilage defects in human patellae: two case reports. *Cell Transplantation* 13: 595–600.
90. Kuroda, R., K. Ishida, T. Matsumoto et al. 2007. Treatment of a full-thickness articular cartilage defect in the femoral condyle of an athlete with autologous bone-marrow stromal cells. *Osteoarthritis and Cartilage* 15: 226–231.
91. Wakitani, S., A. Kawaguchi, Y. Tokuhara, and K. Takaoka. 2008. Present status of and future direction for articular cartilage repair. *Journal of Bone and Mineral Metabolism* 26: 115–122.
92. Wakitani, S., K. Imoto, T. Yamamoto et al. 2002. Human autologous culture expanded bone marrow mesenchymal cell transplantation for repair of cartilage defects in osteoarthritic knees. *Osteoarthritis and Cartilage* 10: 199–206.
93. Berger, J., M. Reist, J. M. Mayer et al. 2004. Structure and interactions in chitosan hydrogels formed by complexation or aggregation for biomedical applications. *European Journal of Pharmaceutics and Biopharmaceutics* 57: 35–52.
94. Chenite, A., C. Chaput, D. Wang et al. 2000. Novel injectable neutral solutions of chitosan form biodegradable gels *in situ*. *Biomaterials* 21: 2155–2161.
95. Koga, H., T. Muneta, T. Nagase et al. 2008. Comparison of mesenchymal tissues-derived stem cells for *in vivo* chondrogenesis: Suitable conditions for cell therapy of cartilage defects in rabbit. *Cell and Tissue Research* 333: 207–215.
96. Lee, K. Y., W. S. Ha, and W. H. Park. 1995. Blood compatibility and biodegradability of partially N-acylated chitosan derivatives. *Biomaterials* 16: 1211–1240.
97. Ganji, F., M. J. Abdekhodaie, and A. Ramzani. 2007. Gelation time and degradation rate of chitosan-based injectable hydrogel. *Journal of Sol-Gel Science and Technology* 42: 47–50.
98. Hoemann, C. D., J. Sun, A. Legare et al. 2005. Tissue engineering of cartilage using an injectable and adhesive chitosan-based cell-delivery vehicle. *Osteoarthritis and Cartilage* 13: 318–329.
99. Jin, R., L. S. Moreira Teixeira, P. J. Dijkstra et al. 2009. Injectable chitosan-based hydrogels for cartilage tissue engineering. *Biomaterials* 30: 2544–2551.
100. Park, K. M., S. Y. Lee, Y. K. Joung et al. 2009. Thermosensitive chitosan-pluronic hydrogel as an injectable cell delivery carrier for cartilage regeneration. *Acta Biomaterialia* 5: 1956–1965.
101. Hao, T., N. Wen, J. K. Cao et al. 2010. The support of matrix accumulation and the promotion of sheep articular cartilage defects repair *in vivo* by chitosan hydrogels. *Osteoarthritis and Cartilage* 18: 257–265.
102. Khor, E., and L. Y. Lim. 2003. Implantable applications of chitin and chitosan. *Biomaterials* 24: 2339–2349.

103. Yamane, S., N. Iwasaki, T. Majima et al. 2005. Feasibility of chitosan-based hyaluronic acid hybrid biomaterial for a novel scaffold in cartilage tissue engineering. *Biomaterials* 26: 611–619.
104. Jayakumar, R., M. Prabaharan, S. V. Nair, and H. Tamura. 2010. Novel chitin and chitosan nanofibers in biomedical applications. *Biotechnology Advances* 28: 142–150.
105. Tamura, H., M. Sawada, H. Nagagama et al. 2006. Influence of amide content on the crystal structure of chitin. *Holzforschung* 60: 480–484.
106. Jayakumar, R., H. Nagahama, T. Furuike, and H. Tamura. 2008. Synthesis of phosphorylated chitosan by novel method and its characterization. *International Journal of Biological Macromolecules* 42: 335–339.
107. Jayakumar, R., V. V. Divya Rani, K. T. Shalumon et al. 2009. Bioactive and osteoblast cell attachment studies of novel alpha- and beta-chitin membranes for tissue-engineering applications. *International Journal of Biological Macromolecules* 45: 260–264.
108. Jayakumar, R., K. P. Chennazhi, S. Srinivasan et al. 2011. Chitin scaffolds in tissue engineering. *International Journal of Molecular Sciences* 12: 1876–1887.
109. Maeda, Y., R. Jayakumar, H. Nagahama et al. 2008. Synthesis, characterization and bioactivity studies of novel beta-chitin scaffolds for tissue-engineering applications. *International Journal of Biological Macromolecules* 42: 463–467.
110. Madhumathi, K., P. T. Sudheesh Kumar, K. C. Kavya et al. 2009. Novel chitin/nanosilica composite scaffolds for bone tissue engineering applications. *International Journal of Biological Macromolecules* 45: 289–292.
111. Peter, M., P. T. Sudheesh Kumar, N. S. Binulal et al. 2009. Development of novel chitin/nanobioactive glass ceramic nanocomposite scaffolds for tissue engineering applications. *Carbohydrate Polymers* 78: 926–931.
112. Hsu, S. H., S. W. Whu, S. C. Hsieh et al. 2004. Evaluation of chitosan–alginate–hyaluronate complexes modified by an RGD-containing protein as tissue-engineering scaffolds for cartilage regeneration. *Artificial Organs* 28: 693–703.
113. Yan, J., N. Qi, and Q. Zhang. 2007. Rabbit articular chondrocytes seeded on collagen-chitosan-GAG scaffold for cartilage tissue engineering *in vivo*. *Artificial Cells, Blood Substitutes, and Immobilization Biotechnology* 35: 333–344.
114. Peppas, N., J. Z. Hilt, A. Khademhosseini, and R. Langer. 2006. Hydrogels in biology and medicine: From molecular principles to bionanotechnology. *Advanced Materials* 18: 1345–1350.
115. Martens, P. J., S. J. Bryant, and K. S. Anseth. 2003. Tailoring the degradation of hydrogels formed from multivinyl poly(ethylene glycol) and poly(vinyl alcohol) macromers for cartilage tissue engineering. *Biomacromolecules* 4: 283–292.
116. Elisseeff, J., W. McIntosh, K. Fu et al. 2001. Controlled-release of IGF-I and TGF-beta1 in a photopolymerizing hydrogel for cartilage tissue engineering. *Journal of Orthopaedic Research* 19: 1098–1104.
117. Li, Y., J. Qin, B. Lin, and W. Zhang. 2010. The effects of insulin-like growth factor-1 and basic fibroblast growth factor on the proliferation of chondrocytes embedded in the collagen gel using an integrated microfluidic device. *Tissue Engineering. Part C, Methods* 16: 1267–1275.
118. Elisseeff, J., W. McIntosh, K. Anseth et al. 2000. Photoencapsulation of chondrocytes in poly(ethylene oxide)-based semi-interpenetrating networks. *Journal of Biomedical Materials Research* 51: 164–171.
119. Elisseeff, J., W. McIntosh, K. Fu et al. 2001. Controlled-release of IGF-I and TGF-beta1 in a photopolymerizing hydrogel for cartilage tissue engineering. *Journal of Orthopaedic Research* 19: 1098–1104.
120. Burdick, J. A., and K. S. Anseth. 2002. Photoencapsulation of osteoblasts in injectable RGD-modified PEG hydrogels for bone tissue engineering. *Biomaterials* 23: 4315–4323.

121. Bryant, S. J., and K. S. Anseth. 2003. Controlling the spatial distribution of ECM components in degradable PEG hydrogels for tissue engineering cartilage. *Journal of Biomedical Materials Research—Part A* 64: 70–79.
122. Appelman, T. P., J. Mizrahi, J. H. Elisseeff, and D. Seliktar. 2009. The differential effect of scaffold composition and architecture on chondrocyte response to mechanical stimulation. *Biomaterials* 30: 518–522.
123. Hwang, N. S., S. Varghese, P. Theprungsirikul et al. 2006. Enhanced chondrogenic differentiation of murine embryonic stem cells in hydrogels with glucosamine. *Biomaterials* 27: 6015–6023.
124. Bryant, S. J., R. J. Bender, K. L. Durand, and K. S. Anseth. 2004. Encapsulating chondrocytes in degrading PEG hydrogels with high modulus: Engineering gel structural changes to facilitate cartilaginous tissue production. *Biotechnology and Bioengineering* 86: 747–755.
125. Kloxin, A. M., A. M. Kasko, C. N. Salinas, and K. S. Anseth. 2009. Photodegradable hydrogels for dynamic tuning of physical and chemical properties. *Science* 324: 59–63.
126. Peppas, N. A., K. B. Keys, M. Torres-Lugo, and A. M. Lowman. 1999. Poly(ethylene glycol)-containing hydrogels in drug delivery. *Journal of Controlled Release* 62: 81–87.
127. Bodugoz-Senturk, H., C. E. Macias, J. H. Kung, and O. K. Muratoglu. 2009. Poly(vinyl alcohol)-acrylamide hydrogels as load-bearing cartilage substitute. *Biomaterials* 30: 589–596.
128. Stammen, J. A., S. Williams, D. N. Ku, and R. E. Guldberg. 2001. Mechanical properties of a novel PVA hydrogel in shear and unconfined compression. *Biomaterials* 22: 799–806.
129. Ibusuki, S., Y. Fujii, Y. Iwamoto, and T. Matsuda. 2003. Tissue-engineered cartilage using an injectable and *in situ* gelable thermoresponsive gelatin: Fabrication and *in vitro* performance. *Tissue Engineering* 9: 371–384.
130. Kobayashi, M., and M. Oka. 2004. Characterization of a polyvinyl alcohol-hydrogel artificial articular cartilage prepared by injection molding. *Journal of Biomaterials Science. Polymer Edition* 15: 741–751.
131. Grant, C., P. Twigg, A. Egan et al. 2006. Poly(vinyl alcohol) hydrogel as a biocompatible viscoelastic mimetic for articular cartilage. *Biotechnology Progress* 22: 1400–1406.
132. Spiller, K. L., S. J. Laurencin, D. Charlton et al. 2008. Superporous hydrogels for cartilage repair: Evaluation of the morphological and mechanical properties. *Acta Biomaterialia* 4: 17–25.
133. Holloway, J. L., K. L. Spiller, A. M. Lowman, and G. R. Palmese. 2010. Analysis of the *in vitro* swelling behavior of poly(vinyl alcohol) hydrogels in osmotic pressure solution for soft tissue replacement. *Acta Biomaterialia* 7: 2477–2482.
134. Spiller, K. L., S. A., Maher, and A. M. Lowman. 2010. Hydrogels for the repair of articular cartilage defects. *Tissue Engineering. Part B, Reviews* 17: 281–299.
135. Spiller, K. L., J. L. Holloway, M. E. Gribb, and A. M. Lowman. 2011. Design of semi-degradable hydrogels based on poly(vinyl alcohol) and poly(lactic-co-glycolic acid) for cartilage tissue engineering. *Journal of Tissue Engineering and Regenerative Medicine* 5: 636–647.
136. Bichara, D. A., X. Zhao, H. Bodugoz-Senturk et al. 2011. Porous poly(vinyl alcohol)-hydrogel matrix-engineered biosynthetic cartilage. *Tissue Engineering—Part A* 17: 301–309.
137. Galois, L., S. Hutasse, D. Cortial et al. 2006. Bovine chondrocyte behaviour in three-dimensional type I collagen gel in terms of gel contraction, proliferation and gene expression. *Biomaterials* 27: 79–90.
138. Hirano, Y., N. Ishiguro, M. Sokabe et al. 2008. Effects of tensile and compressive strains on response of a chondrocytic cell line embedded in type I collagen gel. *Journal of Biotechnology* 133: 245–252.

139. Hui, T. Y., K. M. Cheung, W. L. Cheung et al. 2008. *In vitro* chondrogenic differentiation of human mesenchymal stem cells in collagen microspheres: Influence of cell seeding density and collagen concentration. *Biomaterials* 29: 3201–3212.
140. Wand, D. A., S. Varghese, B. Sharma et al. 2007. Multifunctional chondroitinsulphate for cartilage tissue–biomaterial integration. *Nature Materials* 6: 385–392.
141. Schmedlen, R. H., K. S. Masters, and J. L. West. 2002. Photocrosslinkable polyvinyl alcohol hydrogels that can be modified with cell adhesion peptides for use in tissue engineering. *Biomaterials* 23: 4325–4332.

18 Cartilage Tissue Engineering

Advances in Stem Cell-Based Approaches

Sibylle Grad, Martin J. Stoddart, David Eglin, and Mauro Alini

CONTENTS

18.1 INTRODUCTION

Effective treatment of articular cartilage defects remains one of the most demanding clinical problems in orthopedic surgery. Articular cartilage is a highly organized structure comprised largely of an extracellular matrix (ECM) that confers the complex biomechanical properties of the tissue. In synovial joints, layers of articular cartilage face the synovial cavity on one side, whereas they present the connection to the subchondral bone by a narrow layer of calcified cartilage on the other side. The cartilaginous layer functions to provide a low-friction, wear-resistant surface to support and transmit load and motion between bones. Although the tissue is characterized by extensive durability, its poor intrinsic capability for natural healing requires intervention after traumatic or degenerative damage. Left untreated, articular lesions may result in debilitating pain and dysfunction of the joint and degenerating disorders such as osteoarthritis.

During development, articular cartilage represents a comparably thick layer of nonstratified tissue and isotropically distributed cells, whereas distinct zones can be distinguished in mature cartilage, along with improved resistance to compressive and shear forces, and increased stiffness. The superficial tangential zone, the middle transitional zone, the deep radial zone, and the underlying calcified cartilage differ in their cell densities and phenotypes, biochemical compositions, organization of matrix molecules, and biomechanical properties.[1] Maintenance of the appropriate spatial ECM is the function of the resident cells, the articular chondrocytes, which are present at low density. Water accounts for approximately 80% and 65% of the tissue wet weight at the surface and in the deep zone, respectively. The superficial zone also has the highest collagen content (overall, 15%–22%), whereas the amount of proteoglycans (overall, 4%–7%) is lower in the superficial compared with the middle and deep zones. The main macromolecules include type II collagen and the large proteoglycan aggrecan, with smaller amounts of other collagen types (e.g., I, III, V, VI, IX, XI, and XII), small and large proteoglycans and glycosaminoglycans (e.g., hyaluronan), and other noncollagenous proteins such as cartilage oligomeric matrix protein, fibronectin, or elastin. Among large proteoglycans, versican, perlecan, and proteoglycan-4 [homologous to lubricin or superficial zone protein (SZP)] can be found in articular cartilage. Proteoglycan-4 is uniquely expressed at the superficial zone and has an essential role in maintaining the low-friction surface and in protecting the underlying zones. Functional cell and matrix properties thus vary with tissue depth. This has also been shown in studies of strain patterns generated during physiologically relevant compression loading in cartilage explants.[2] The ultimate aim of successful cartilage repair is to restore this organized cell–matrix structure and regain full function.

Current cartilage repair strategies include debridement, bone marrow stimulation techniques, and cell-based, cell scaffold-based, and whole tissue transplantation procedures. Bone marrow stimulation is the most frequently used method for the treatment of small symptomatic lesions of the knee cartilage. Microfracture leads to the recruitment of mesenchymal stem cells (MSCs) from the bone marrow into the defect. These cells are able to differentiate into fibrochondrocytes and contribute to fibrocartilage repair of the lesion. Successful outcome relies on the formation of a stable blood clot that optimally fills the chondral defect[3] and potentially on the number and quality of the MSCs, which may decline with age. Reparative fibrocartilage does not resemble the native surrounding hyaline cartilage. It consists of varying amounts of type I, type II, and type III collagens; in particular, less type II collagen is formed,[4] which may impair the mechanical competence of the repair tissue. This may be the reason why even with refined surgical techniques and careful patient selection, the outcome of marrow stimulation often weakens over time. Deterioration over time has been observed in athletes in different studies[5–7] and should not be neglected when considering microfracture as the first-line treatment for small chondral lesions. In addition, osseous overgrowth has been described as another relatively frequent adverse effect of marrow stimulation treatment.[8]

Autologous osteochondral transplantation or mosaicplasty is regularly used for effective resurfacing of osteochondral defects. The advantages of whole osteochondral tissue transplantation include the superior quality of the repair tissue, that is, viable hyaline cartilage and the relatively short rehabilitation time. However, major limitations exist, including donor site morbidity and the restricted availability of grafts that can be harvested from the lower weight-bearing locations of the articular surface. Furthermore, differences in thickness, orientation, and structural arrangement may lead to failure of integration of the repair tissue. Such impaired chondral integration, despite a satisfying osseous integration of the subchondral bone, has been described in the clinical setting.[9] On the other hand, hypertrophy of the chondral graft has been observed as an adverse effect. Concerns about the viability of chondrocytes at the surface of osteochondral grafts have also been raised, and a recent study describes findings of cell necrosis, apoptosis, and matrix degradation after press-fit implantation of osteochondral grafts in a rabbit model.[10]

A more innovative technique for cartilage restoration is autologous chondrocyte implantation, which was first described in 1994.[11] It was suggested that the implantation of mature chondrocytes would lead to the development of a repair tissue composed of hyaline cartilage, rather than fibrocartilage, with improved long-term outcome. However, dedifferentiation of chondrocytes in monolayer culture implies that the ultimately implanted cell population has an altered phenotype and is, in fact, likely to generate fibrocartilage. This loss of phenotype is an essential shortcoming of the procedure, aside from other difficulties, such as leakage of the cell suspension after injection and uneven cell distribution.[12] Additional limitations include the requirement for two operations, the potential donor site morbidity, and the long recovery time that is necessary to allow tissue maturation. Although the overall complication rate is relatively low, serious adverse events may occur that often require recurrent surgical interventions. These adverse events include graft failure, delamination, and hypertrophy of the periost flap used to retain the cells.[13] Although the traditional autologous chondrocyte transplantation technique is demanding for the surgeon, and results have not been highly predictable and reproducible, new techniques have been developed in which the chondrocytes are delivered on absorbable scaffolds. These scaffolds provide support for the cells during both the preimplantation and the postoperative healing phases. Membranes of type I/III collagen have been used in the "matrix-associated chondrocyte implantation" method for seeding of the autologous chondrocytes to be delivered to the articular defect. Furthermore, scaffolds of hyaluronan derivatives have been used for the culture of chondrocytes in a three-dimensional system before implantation. Procedures that use carrier systems for cell delivery have important advantages such as a more even cell distribution, avoidance of periosteal harvest to cover the lesion, and improved technical applicability.

The most advanced techniques aim to engineer cartilaginous tissues *in vitro/ex vivo* before implanting them into the defect site.[14] Isolated autologous chondrocytes are seeded onto a matrix or scaffold and the cell–scaffold construct may also be subjected to mechanical stimulation using a bioreactor system that applies hydrostatic, compressive, or shear load for at least 1 week. It has been widely accepted that mechanical stimuli promote the chondrocytic phenotype and improve the functional properties of engineered cartilage.[15,16] Although these recently developed technologies offer great promise for the restoration of high-quality cartilage, some considerable limitations remain. First, the need for two operations is a major burden for the patient and is cost-intensive. Second, the complex culture of the grafts before implantation, which requires the highest quality standard laboratories and lengthy culture times, is also associated with extensive costs. Finally, the difficulties of authentic implant structure and of proper implant integration have still not been sufficiently resolved.

Research therefore has recently focused on alternative methods to not only facilitate and shorten the technical procedures but also make the outcome more predictable and reproducible. Strategies include: 1) the application of MSCs as an alternative cell source, thus avoiding the chondrocyte harvesting procedure; 2) supporting the formation of a structured tissue by novel materials and selective mechanics; and 3) activation of endogenous cellular repair mechanisms.

18.2 NEW STRATEGIES FOR CARTILAGE REPAIR

18.2.1 Adult MSCs and Chondrogenesis

Although the occurrence of stem cells with chondrogenic potential have been described in various tissues, including synovial membranes, periosteum, skeletal muscle, and umbilical cord blood, the most common and best studied sources of MSCs are adipose tissue and bone marrow. MSCs represent only a minor fraction of the total nucleated cell population in these tissues; the proportion is approximately 2×10^2 and 2×10^4 per 10^6 mononucleated cells for bone marrow and adipose tissue, respectively.[17] Bone marrow-derived MSCs are therefore usually expanded *in vitro* to obtain sufficient amounts of cells for experimental or clinical use. To induce chondrogenic differentiation, the monolayer cultured cells are then maintained either in micromass culture or in a three-dimensional scaffold such as a polymer, sponge, hydrogel, or in microspheres.[18–20] In addition, defined culture conditions are required for chondrogenesis of MSCs. It has been shown that growth factors in the transforming growth factor-β (TGF-β) superfamily are necessary for successful chondrogenesis, whereby TGF-β1, TGF-β2, TGF-β3, bone morphogenetic proteins (BMP-2, -4, -6, and -7), and insulin-like growth factor I (IGF-1) in various combinations have been used.[21] Furthermore, serum-free conditions and specific culture supplements including dexamethasone and ascorbic acid are commonly used. When different methods are compared, it seems evident that, depending on the cell source, the optimal chondrogenic growth factors can vary, which is likely due to tissue-specific stem cell populations.[21] Other factors including hypoxia and mechanical stimulation have been shown to promote chondrogenesis of MSCs. Given that *in vivo* chondrocytes reside in a low oxygen and mechanical environment, these factors have been recognized as important chondrogenic triggers and regulators (for a review, see Egli et al.[16]).

The most striking difficulty with *in vitro*-induced MSC chondrogenesis is that differentiation generally does not stop at the prehypertrophic stage, that is, the cells do not express a stable chondrocytic phenotype but tend to further develop toward a hypertrophic profile, expressing hypertrophy and finally osteogenic markers. Basically, the mechanisms of chondrogenic differentiation of MSCs, in particular of bone marrow-derived MSCs, can be compared with embryonic limb development. This implies that the cells condensate, express chondrogenic markers such as SRY (sex determining region Y)-box 9 (Sox-9), and subsequently collagen type II but eventually become terminally differentiated. The latter stage is characterized by the expression of matrix metalloproteinase 13, collagen type X, and alkaline phosphatase.[22,23] When ectopically transplanted *in vivo*, these constructs start to mineralize, which is highly undesirable for repair of articular cartilage.[24] Different protocols have been developed with the aim of constraining terminal differentiation, primarily focusing on the application of molecules with a known inhibitory effect on growth plate development. In particular, the addition of parathyroid hormone-related peptide (PTHrP) during the *in vitro* chondrogenesis of MSCs has been shown to suppress the expression of the hypertrophic marker collagen type X.[25–27] Appropriate timing of PTHrP application seems to be of crucial importance, as the expression of the chondrocytic gene collagen type II also decreases through PTHrP. By comparing a number of inducing, promoting, and inhibiting factors, Weiss et al.[26] found that within a developmental window of collagen type II positive/collagen type X negative cells, basic fibroblast growth factor and PTHrP may allow the inhibition of further differentiation toward hypertrophy to obtain a stable chondrocytic phenotype. Furthermore, the addition of growth and differentiation factor 5 as a replacement for TGF-β as a chondrogenic factor may prevent the induction of type X collagen, although the chondrogenic potential is also generally lower.[28,29] Cotransfection of the Sox trio (Sox-5, Sox-6, and Sox-9) was recently shown to not only enhance chondrogenesis but also suppress hypertrophic marker genes in human MSC in pellet culture,[30] suggesting potential to inhibit terminal differentiation. Another approach consists of targeting the Smad signaling pathways at distinct stages of differentiation. Although the Smad 2/3 route has been found to obstruct terminal differentiation,[31] the Smad 1/5/8 route is activated by BMPs, which seem to be required for terminal differentiation and endochondral ossification.[32,33] Blocking Smad 1/5/8 phosphorylation after 2 weeks of chondrogenic induction of human MSCs was able to prevent expression of hypertrophic markers and inhibit mineralization while sustaining further cartilage matrix production.[34] Hence, blocking this pathway after chondrogenic differentiation may improve the outcome of MSC-based cartilage repair. Given that the exact timing of Smad inhibitor administration is crucial, and considerable interdonor variations in the rate of chondrogenesis are apparent, the application of such agents for tissue engineering *in vitro* or *in vivo* is likely to be challenging.

Results from a cartilage defect model in minipigs indicate that unlike *in vitro* chondrogenic induction of MSCs, the signaling molecules and the biomechanical environment *in vivo* may provide the proper conditions for MSCs to differentiate into chondrocytic cells with a stable phenotype.[35] The authors thus suggest that until better *in vitro* induction methods are identified for stable chondrogenesis of MSCs, MSC predifferentiation in a tissue engineering setting before implantation may be critical. As inappropriate terminal differentiation seems to be less likely in the accurate *in vivo* environment, coculture systems have been developed in an attempt to expose the MSCs to chondrocyte-derived signaling molecules that should prevent them from progressing down the hypertrophic pathway. In fact, it was shown that both soluble factors secreted by human articular chondrocytes and direct coculture are capable of improving chondrogenesis and, at the same time, suppressing hypertrophic phenotypes in MSCs. In particular, PTHrP is secreted by the chondrocytes, contributing to this stabilizing effect.[36] Similarly, coculture of human MSC with primary, but not with expanded osteoarthritic chondrocytes, promoted MSC chondrogenesis while downregulating hypertrophic markers. It was suggested that soluble chondrocyte-derived morphogens are responsible for this effect.[37] Moreover, a recently described bilaminar cell pellet system consisting of human MSCs encased in juvenile chondrocytes was reported to facilitate chondrogenic differentiation without hypertrophy as well as being relatively resistant to hypoxic and inflammatory conditions.[38] On the other hand, using a micromass coculture system, it has been found that human chondrocytes with retained intrinsic chondrogenic potential not only failed to induce chondrogenic differentiation of MSCs but also failed to prevent MSC hypertrophy induced by chondrogenic factors.[39] Apparently, the ratio of cocultured MSCs and chondrocytes plays a significant role in the regulation of the expression of the cartilaginous versus osseous phenotype. Coculture of human MSCs with xenogenic chondrocytes in alginate hydrogel constructs in a 2:1 ratio not only accelerated the expression of a cartilaginous phenotype during the culture period, but also initiated osteogenesis of the MSCs.[40] These conflicting results accentuate the importance of the cell source, cell status, donor age, and culture environment for *in vitro* model systems and translation into the *in vivo* situation.

The coculture of MSCs and articular chondrocytes in hyaluronic acid hydrogels has also been suggested for the generation of tissue-engineered grafts with improved matrix content, mechanical quality, and reduced hypertrophic potential.[41] However, these coculture studies are more of interest for improving our understanding of cell–cell interactions and soluble factors that may act as regulators of cell differentiation than for ultimate tissue engineering approaches; as the main advantage of using MSCs is to finally avoid the procedure of harvesting cartilage biopsies and isolating, and potentially expanding, autologous chondrocytes. Results from an *in vivo* ectopic chondrogenesis model, in correlation with corresponding *in vitro* coculture experiments, indicate that paracrine signaling of soluble chondrogenic factors provided by chondrocytes is a key mechanism for MSC chondrogenesis.[42] The authors suggest that a tissue-specific niche is created by chondrocytes in the subcutaneous environment similar to the articular milieu, although the expression of hypertrophic markers was not assessed.

Because it has been recognized that the cartilaginous phenotype acquired by bone marrow-derived MSCs is, in general, endochondral rather than permanent cartilage and will terminally differentiate, the potential of using progenitor cells from the cartilage itself has been investigated. It is hypothesized that these cells hold the developmental catalogue of the native tissue and would generate the authentic repair tissue. Previously, the identification of a chondroprogenitor population in the surface zone of bovine articular cartilage has been documented.[43,44] Additional studies have reported the presence of progenitor cell populations in normal and osteoarthritic human cartilage.[45,46] Using a differential adhesion assay to fibronectin, it was found that human articular cartilage contains a cartilage progenitor population that retains a stem cell-like phenotype until appropriate induction of chondrogenic differentiation is initiated.[47] Importantly, these progenitor cells residing in human cartilage demonstrated the ability to maintain chondrogenic potential even after extensive expansion; moreover, it was shown that they neither undergo terminal chondrogenic nor osteogenic differentiation.[47] Another study used the efflux of the Hoechst 33342 dye by stem/progenitor cells, which permits the isolation of a so-called "side population" of cells with stem cell properties.[48] These side population cells were present only in the superficial but not in the middle and deep zones of bovine articular cartilage.[49] Investigations revealed that these side population cells possessed progenitor characteristics for both superficial and middle zone chondrocytes, as assessed by the expression of collagen type II and SZP after stimulation with TGF-β or BMP-7.[49]

Such observations are of crucial importance, considering the differences in cellular phenotypes between the chondrocytes of the superficial, middle, and deep zones of articular cartilage. Specifically, some studies have investigated sources of SZP-producing progenitor cells for superficial zone cartilage tissue regeneration.[50–52] SZP has been suggested as a zonal molecular marker for the surface zone of articular cartilage, because little or no accumulation of SZP has been detected in middle and deep zone chondrocytes.[53,54] Controlling the localization of SZP in engineered cartilage may be critical for creating functional cartilage with effective lubrication properties. The fabrication of a functional tissue-engineered cartilage could depend on SZP-secreting cells at the surface to acquire the desired low-friction surface characteristics. Recent results show that mesenchymal progenitor cells from the synovium and infrapatellar fat pad not only exhibited high chondrogenic potential but also maintained or increased SZP production even after multiple passages.[55] Compared with cells from the infrapatellar fat pad, synovium-derived progenitors demonstrated higher colony-forming, proliferative, and chondrogenic potential, and greater SZP secretion after chondrogenic induction. Interestingly, TGF-β or BMP-7 treatment (or both) increased the release of SZP from cell pellets, whereas both the accumulation of SZP within the pellets and the mRNA expression of SZP were reduced upon growth factor stimulation. The same downregulation of SZP mRNA expression in the presence of exogenous TGF-β has also been seen in bone marrow-derived MSCs.[56] This suggests that additional signals may play a role for the regulation of SZP expression in progenitor cells.

18.2.2 Biomaterial and Mechanical Regulation of Progenitor Cells

In articular cartilage, similar to most other tissues, the arrangement of the cells and the ECM is not random, but reveals a distinct pattern and defined spatial orientation. Oriented nanofibrous scaffolds may thus support the differentiation of progenitor cells toward the desired phenotype. The superficial zone of natural articular cartilage is characterized by ellipsoid-like cells and a very dense organization of collagen type II fibers (approximately 20 nm in diameter) that are oriented parallel to the plane of the articular surface. The superficial zone has a higher tensile strength compared with the underlying zones, which is vital for the resistance of the articular surface to shear and tensile forces.[57,58] With the aim of generating a tissue-engineered construct that will mimic the superficial zone of articular cartilage, electrospun biocompatible polycaprolactone (PCL) scaffolds were seeded with human MSCs and cultured under chondrogenic conditions.[59] Cell orientation, proliferation, and chondrogenic marker expression were assessed, whereby oriented nanofibrous and microfibrous electrospun PCL scaffolds were compared with a random porous PCL film. Fiber-guided orientation was independent of the fiber size and was maintained for up to 5 weeks, whereas collagen type II expression seemed to be highest in cells that were seeded on the nanofibrous scaffold (approximately

500 nm diameter). It is postulated that the differential formation of focal adhesion that is cell type dependent may regulate stem cell behavior in the microenvironment and nanoenvironment.

Efforts have been made by a number of investigators to engineer tissues that recapitulate the different layers of native cartilage (for a review, see Klein et al.[60] and Chung and Burdick[61]). For example, articular chondrocytes from the superficial and deep zones were photoencapsulated in separate hydrogels as well as in adjacent layers of a bilayered hydrogel.[62] The bilayered constructs demonstrated greater shear and compressive strength than homogeneous cell constructs, indicating the importance and need for a zonal structure in articular cartilage, although, in this case, the organization was driven by the cells rather than the material used. Another biomaterial-based approach was undertaken with the aim of directing the differentiation of a single stem cell population into chondrocytic cells with the different phenotypic characteristics that are evident in the various zones of articular cartilage.[63] It was hypothesized that by tailoring the culture conditions and environmental signals through biomaterial compositions, directed differentiation toward the superficial, transitional, and deep zones could be achieved. The investigators could demonstrate that distinct combinations of synthetic and natural polymeric materials create unique milieus that can direct mouse bone marrow-derived MSC to acquire a zone-specific phenotype. The incorporation of chondroitin sulfate- and matrix metalloproteinase-sensitive peptides into a polyethylene glycol hydrogel induced a high expression of type II collagen and low proteoglycan expression, resulting in a low compressive modulus, which is seen in the superficial zone. Intermediate levels of collagen type II and proteoglycan were produced in the polyethylene glycol hydrogel functionalized by chondroitin sulfate; whereas the combination of polyethylene glycol with hyaluronic acid hydrogel resulted in low collagen type II and high proteoglycan levels and thus a high compressive modulus, similar to the characteristics of the deep zone. Moreover, layer-by-layer arrangement of these specific biomaterial compositions could create a three-dimensional environment that allowed single MSC populations to differentiate in a zone-specific manner and organize into a defined tissue structure.[64] However, an increase in type X collagen was observed in all the composite materials, indicating that hypertrophic differentiation could not be prevented and additional cues may be required to achieve a stable phenotype. Nevertheless, it can be concluded that spatially varying biomaterial compositions within single three-dimensional scaffolds may be able to stimulate the regeneration of multilayered complex tissues from a single stem cell population. Importantly, this approach is not only of great interest for *in vitro* cartilage tissue engineering, but also for the *in vivo* migration of MSCs from the bone marrow to the site of cartilage defects during microfracture procedures.

Generally, biomaterials that are optimized for MSC differentiation into zonal chondrocytes are likely to be particularly favorable for effective cartilage repair. Recent work from Gong et al.[65] showed a novel sandwich-type model for engineering multilayer cartilage constructs by seeding chondrocytes on thin (10 μm) decellularized cartilage sheets and stacking them layer by layer. The engineered cartilage was stable and maintained its phenotype after implantation in nude mice. The favorable mechanical properties and the possibility of defining the shape of the sheets is an added advantage. Although zonal organization was not addressed in this study, the approach seems promising, as MSCs might be used as a cell source and the properties of the sheets may be modulated by combining spatially varying matrices that represent the different layers.

In vivo cartilage is exposed to complex loading patterns, and hence the mechanical environment may provide the key stimuli for successful chondrogenesis of MSCs. A variety of bioreactor systems have been developed and used to condition stem/progenitor cell populations in different biomaterial scaffolds or as scaffold-free cultures.[15,16,66] Constructs have mainly been exposed to cyclic axial compression or cyclic hydrostatic pressure in short-term and long-term studies; however, because of the different culture conditions, species, loading regimes, and compositions of the constructs, it is difficult to compare the outcomes. The shortcoming of a number of studies is that experiments were performed in chondrogenic differentiation medium, that is, with the addition of growth factors, mainly TGF-β. Thus, the moderate, potentially physiologic, chondrogenic effects of the mechanical stimulus may be overwhelmed by the extensive chondrogenic induction caused by growth factors. It is generally accepted that although dynamic compressive and hydrostatic loading may promote chondrogenic differentiation, the results are not always consistent and are sometimes conflicting. More complex loading patterns or a combination with supporting biomaterials may be required for successful chondrogenic induction and stabilization of the phenotype.

Using a custom-built multiaxial loading device (Figure 18.1),[67] we have investigated the role of multiaxial load on the differentiation of human bone marrow-derived MSCs. The cells are embedded in a clinically relevant fibrin hydrogel before seeding into a mechanically compliant biodegradable polyurethane scaffold. Within this system, multiaxial load alone has the capacity to induce chondrogenic differentiation

FIGURE 18.1 Two stations of a four-station loading device that allows the application of multiaxial load and motion to cell-seeded scaffolds. A commercially available ceramic ball is used to apply axial compression and simultaneous surface motion to cell–scaffold constructs.

in the presence of dexamethasone and does so by inducing the endogenous synthesis of TGF-β.[56] At higher concentrations of growth factor, the response was diminished because of a saturation effect. As expected, the response could be modulated by varying the amplitude and frequency of the load applied.[68] A combined approach using mechanical load with adenoviral transduction of Sox-9 (in the absence of both TGF-β and dexamethasone) demonstrated that classical chondrogenic genes were differentially regulated.[69] Aggrecan was predominantly a Sox-9-induced gene, whereas SZP and collagen X only responded to multiaxial load. Collagen II was maximally expressed by the combination of the two stimuli. This suggests that multiple stimuli may be required for a more complete chondrogenesis. Also relevant for future applications, the same study demonstrated that transduction of Sox-9 alone was able to increase the expression of its two cofactors, L-Sox-5 and Sox-6,[69] suggesting that the transfer of a single gene may be sufficient. Groups using various uniaxial load bioreactors have demonstrated that uniaxial load alone is not sufficient to induce a chondrogenic phenotype but is capable of enhancing TGF-β-induced chondrogenesis.[70,71] Using the multiaxial device described previously, we have performed a study directly comparing the individual effects of axial compression, surface shear, and the combination of the two stimuli. The results confirmed those of the uniaxial bioreactors in that compression alone was not sufficient to induce chondrogenesis. It was demonstrated that shear is a critical component for chondrogenic induction in the absence of exogenous factors.[72] This offers the tantalizing possibility that the application of a defined rehabilitation protocol may be useful in enhancing MSC-based repair of cartilage defects.

18.2.3 *In Situ* Regeneration by MSCs

As outlined previously, cells with the features of (chondrogenic) progenitor cells have been discovered not only in the cartilaginous tissue itself but also in other tissues of the articular joint such as the synovium or the infrapatellar fat pad.[55] In fact, it has been found that synovium-derived progenitor cells may have a particularly high potential for chondrogenic differentiation.[73] In addition, either osteochondral lesion or microfracture treatment results in the release of bone marrow including its stem/progenitor cell fraction. It is therefore tempting to postulate that these endogenous cells may be stimulated to initiate and support the regeneration of a functional tissue. The activation of progenitor cells *in situ* by applying appropriate biochemical, biomaterial, and biomechanical methods will have important advantages over complex and lengthy *in vitro* tissue engineering approaches; in addition, regeneration will directly take place in the true physiological joint environment.

The occurrence of functional MSC niches in adult mouse knee joint synovium *in vivo* was recently described.[74] Specifically, slow cycling cells that were discovered in the synovium started to proliferate after the creation of a cartilage defect in the mouse knee. The cells that were identified stained positive for MSC markers, were phenotypically heterogeneous, and located in topographically distinct niches in the lining layer and the subsynovial tissue. After injury, chondrogenic differentiation of these cells was observed as assessed by staining for Sox-9 and type II collagen and surrounding metachromatic ECM. Furthermore, recent results from a rabbit cartilage defect model indicate that endogenous cell homing, initiated by the administration of exogenous TGF-β, could prompt the regeneration of articular cartilage with superior structural and mechanical properties.[75] The articular surface was surgically excised and replaced with scaffolds spatially infused with TGF-β3 adsorbed collagen hydrogel. TGF-β delivery resulted in a repair tissue with not only improved thickness, density, compressive, and shear properties, but also good integration into the subchondral bone. Importantly, approximately 130% more cells were recruited by the addition of TGF-β compared with spontaneous migration into the regenerative cartilage. Endogenous cell homing was associated with the regeneration of both avascular cartilage and vascularized bone, demonstrating their ability to restore complex tissues.[75]

As several stem/progenitor cell populations exist in tissues adjacent to the cartilage defect, different sources of endogenous stem-like cells were investigated with respect to their potential to migrate and differentiate in response to chemotactic cytokines.[76] Specifically, TGF-β3 and stromal-derived factor-1β (SDF-1β) were control released from gelatin microspheres and embedded in an alginate gel into a collagen scaffold, with underlying adipose, bone marrow, or synovium-derived MSCs in monolayer culture. Delivery of TGF-β or SDF-1β (or both) led to a significant recruitment of all three cell types into the collagen scaffold. Moreover, codelivery of both factors induced significant upregulation of aggrecan and type II collagen expression in all cell types, indicating chondrogenic differentiation. These results suggest that recruitment of MSCs from adjacent tissues into a cell-free bioactive scaffold may serve as an alternative approach to cell transplantation for cartilage repair.[76] Previously, the capability of a variety of growth factors and cytokines to induce the migration of human articular chondrocytes and bone marrow-derived MSCs had been evaluated using Boyden chamber assays.[77] Although only platelet-derived growth factor (PDGF) induced a significant increase in articular chondrocyte cell migration, human bone marrow-derived MSCs demonstrated a significant chemotactic response to several factors including PDGF, vascular endothelial growth factor, IGF-1, interleukin-8, and BMPs. These distinct differences between mature chondrocytes and MSCs indicate that progenitor cells are more responsive to a broader spectrum of chemotactic factors.

A migratory progenitor cell population has also been identified and isolated from late-stage human osteoarthritic tissue.[78] These cells were shown to adhere to, and to be influenced by, ECM components and to migrate toward diseased cartilage tissue explants. In addition, their chondrogenic potential could be enhanced by downregulation of Runx-2. The exact origin of this progenitor cell population remains to be elucidated; however, recruitment and manipulation of these cells may induce sustained chondrogenic potential in diseased cartilage. Taken together, if the regeneration of damaged cartilage can be achieved by chemotaxis and homing of endogenous cells rather than by cell transplantation (Figure 18.2), the development of regenerative therapies may be simplified, avoiding lengthy and costly cell isolation and *ex vivo* manipulation procedures.

Although this article essentially addresses advances in chondral repair, the restoration of full osteochondral tissue is of equal importance (Figure 18.3).[79] Brady et al.[80] reported on the creation of a biphasic cartilage/bone construct using preconditioned

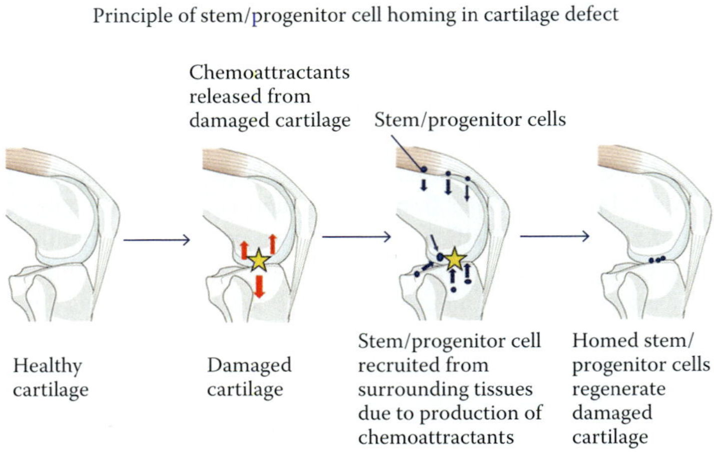

FIGURE 18.2 Basic principle of endogenous cell homing into a cartilage defect.

FIGURE 18.3 Microtomographic reconstruction of an osteochondral biphasic scaffold with a cartilage region, a bone region, and a membrane interface (scale bar = 1 mm).

(expanded and osteogenically or chondrogenically differentiated) MSCs seeded at different locations and densities in a continuous collagen gel. After 7 days *in vitro* in the collagen scaffold, zonal mineralization and cartilage-like matrix production could be observed in the construct, including a good interface simulating the tidemark layer. This relatively simple approach could be further emulated to create not only osteochondral tissue but also further zonal organization within the cartilage phase using a single source of human MSCs by advancing the reported protocol. The development of vehicles for the delivery of growth factors for stem cell differentiation has increasingly focused on nano-sized materials. A recent study investigated the potential of TGF-β-loaded nanoparticles coencapsulated with MSCs in fibrin scaffolds for the stimulation of chondrogenesis *in vitro* and *in vivo*.[81] Heparin-bound TGF-β was used for the preparation of matrices to increase the stability and bioactivity of growth factors. The results show that the combined delivery of cells and factor could help increase transplanted MSC differentiation, both *in vitro* and *in vivo*, in nude mouse and rabbit cartilage defect models. Finally, an effective strategy for the simultaneous regeneration of osseous and cartilaginous tissues has been presented using a bilayered gene-activated scaffold, consisting of a chitosan–gelatin layer for chondrogenesis and a hydroxyapatite/chitosan–gelatin layer for osteogenesis.[82] MSCs cultured in the bilayer scaffold, and induced by plasmids encoding for TGF-β and BMP-2 in different layers, generated osteochondral tissue *in vitro*, which could achieve good restoration of the osteochondral architecture in a rabbit defect model *in vivo*. These studies provide evidence that the combination of biomimetic and multiphasic scaffolds, controlled gene or protein delivery systems, and multilineage differentiation of a single stem cell population, may represent a future strategy for facilitating the development of complex tissues.

18.3 CONCLUSION AND OUTLOOK

Bearing in mind the advantages of minimally invasive harvesting and chondrogenic differentiation potential, MSCs are an attractive cell source for cartilage tissue engineering and regeneration. A major concern of MSC-based tissue engineering is cell hypertrophy and hence calcification. Fibrogenesis and heterotopic tissue formation have also been documented *in vivo*. Whether chondrogenic preconditioning of MSCs before implantation is beneficial, reducing the formation of fibrocartilage, or disadvantageous, increasing the risk for hypertrophy, is still ambiguous. In a rabbit full-thickness cartilage defect model, the regenerative potential of allogeneic chondrogenic predifferentiated MSCs and undifferentiated MSCs were compared.[83] The study showed that either chondrogenic differentiated MSCs or undifferentiated MSCs led to superior healing when compared with cartilage defects that were untreated. However, there was no difference in the outcome between the cell types. In an ovine osteochondral defect model, the transplantation of chondrogenic *in vitro* predifferentiated autologous MSCs embedded in collagen I hydrogel was compared with undifferentiated MSCs and with matrix-associated autologous chondrocyte transplantation (MACT).[84] After 12 months *in vivo*, predifferentiated MSC gels showed a better histological outcome compared with undifferentiated MSC and MACT gels. No degradation of the repair tissue was observed and no signs of hypertrophy or ectopic calcification were evident. However, cyst formation and new bone formation were occasionally observed in each group, leading to high variability in outcomes.

Because of the current uncertainty of the optimal differentiation stage of MSCs before implantation and the high costs implicated in their isolation and culture, the *in situ* recruitment of MSCs seems to be more attractive than extended *in vitro* manipulation procedures. A new generation of biomaterial-supported marrow stimulation techniques has been introduced to improve the outcome of this relatively simple and minimally invasive single-step method. It combines the microfracture procedure with the use of a specific biological matrix; generally, a collagen I/III membrane, fixed with fibrin glue or suture.[85–87] Although the method can be considered safe and effective for the treatment of symptomatic full-thickness chondral lesions of the knee in appropriately selected cases, additional studies with long-term follow-up are needed to determine whether structural and functional integrity can be maintained over time.[85] It has also been documented that the surgical technique remains essential and affects the final results. For example, when autologous concentrated bone marrow was used to potentially enhance the outcome, differences in the amount and chondrogenic potential of MSCs were observed between iliac crest and knee subchondral bone-derived bone marrow.[86] In fact, bone marrow from the created microfracture showed inferior quality. Novel biomaterial-assisted gene/molecule delivery systems to improve endogenous MSC recruitment, activation, and differentiation bear the ultimate goal of a one-step regeneration system.

REFERENCES

1. Mow, V. C., A. Ratcliffe, and A. R. Poole. 1992. Cartilage and diarthrodial joints as paradigms for hierarchical materials and structures. *Biomaterials* 13: 67–97.
2. Neu, C. P., M. L. Hull, J. H. Walton. 2005. Heterogeneous three-dimensional strain fields during unconfined cyclic compression in bovine articular cartilage explants. *Journal of Orthopaedic Research* 23: 1390–1398.
3. Frisbie, D. D., J. T. Oxford, L. Southwood, G. W. Trotter, W. G. Rodkey, J. R. Steadman, J. L. Goodnight, and C. W. McIlwraith. 2003. Early events in cartilage repair after subchondral bone microfracture. *Clinical Orthopaedics and Related Research* 215–227.
4. Bae, D. K., K. H. Yoon, and S. J. Song. 2006. Cartilage healing after microfracture in osteoarthritic knees. *Arthroscopy* 22: 367–374.
5. Mithoefer, K., R. J. Williams III, R. F. Warren, T. L. Wickiewicz, and R. G. Marx. 2006. High-impact athletics after knee articular cartilage repair: A prospective evaluation of the microfracture technique. *American Journal of Sports Medicine* 34: 1413–1418.
6. Mithoefer, K., and J. M. Scopp. 2007. Mandelbaum BR articular cartilage repair in athletes. *Instructional Course Lectures* 56: 457–468.
7. Gobbi, A., P. Nunag, and K. Malinowski. 2005. Treatment of full thickness chondral lesions of the knee with microfracture in a group of athletes. *Knee Surgery, Sports Traumatology, Arthroscopy* 13: 213–221.
8. Mithoefer, K., R. J. Williams III, R. F. Warren, H. G. Potter, C. R. Spock, E. C. Jones, T. L. Wickiewicz, and R. G. Marx. 2005. The microfracture technique for the treatment of articular cartilage lesions in the knee. A prospective cohort study. *Journal of Bone and Joint Surgery. American Volume* 87: 1911–1920.
9. Rose, T., S. Craatz, P. Hepp, C. Raczynski, J. Weiss, C. Josten, and H. Lill. 2005. The autologous osteochondral transplantation of the knee: Clinical results, radiographic findings and histological aspects. *Archives of Orthopaedic and Trauma Surgery* 125: 628–637.
10. Gulotta, L. V., J. R. Rudzki, D. Kovacevic, C. C. Chen, D. Milentijevic, and R. J. Williams III. 2009. Chondrocyte death and cartilage degradation after autologous osteochondral transplantation surgery in a rabbit model. *American Journal of Sports Medicine* 37: 1324–1333.
11. Brittberg, M., A. Lindahl, A. Nilsson, C. Ohlsson, O. Isaksson, and L. Peterson. 1994. Treatment of deep cartilage defects in the knee with autologous chondrocyte transplantation. *New England Journal of Medicine* 331: 889–895.

12. Marlovits, S., P. Zeller, P. Singer, C. Resinger, and V. Vecsei. 2006. Cartilage repair: Generations of autologous chondrocyte transplantation. *European Journal of Radiology* 57: 24–31.
13. Wood, J. J., M. A. Malek, F. J. Frassica, J. A. Polder, A. K. Mohan, E. T. Bloom, M. M. Braun, and T. R. Cote. 2006. Autologous cultured chondrocytes: Adverse events reported to the United States Food and Drug Administration. *Journal of Bone and Joint Surgery. American Volume* 88: 503–507.
14. Crawford, D. C., C. M. Heveran, W. D. Cannon, Jr., L. F. Foo, and H. G. Potter. 2009. An autologous cartilage tissue implant NeoCart for treatment of grade III chondral injury to the distal femur: Prospective clinical safety trial at 2 years. *American Journal of Sports Medicine* 37: 1334–1343.
15. Grad, S., D. Eglin M. Alini, and M. J. Stoddart. 2011. Physical stimulation of chondrogenic cells *in vitro*: A review. *Clinical Orthopaedics and Related Research* 469: 2764–2772.
16. Egli, R. J., E. Wernike, S. Grad, and R. Luginbuhl. 2011. Physiological cartilage tissue engineering effect of oxygen and biomechanics. *International Review of Cell and Molecular Biology* 289: 37–87.
17. Kastrinaki, M. C., I. Andreakou, P. Charbord, and H. A. Papadaki. 2008. Isolation of human bone marrow mesenchymal stem cells using different membrane markers: Comparison of colony/cloning efficiency, differentiation potential, and molecular profile. *Tissue Engineering. Part C, Methods* 14: 333–339.
18. Pittenger, M. F., A. M. Mackay, S. C. Beck, R. K. Jaiswal, R. Douglas, J. D. Mosca, M. A. Moorman, D. W. Simonetti, S. Craig, and D. R. Marshak. 1999. Multilineage potential of adult human mesenchymal stem cells. *Science* 284: 143–147.
19. Hui, T. Y., K. M. Cheung, W. L. Cheung, D. Chan, and B. P. Chan. 2008. *In vitro* chondrogenic differentiation of human mesenchymal stem cells in collagen microspheres: Influence of cell seeding density and collagen concentration. *Biomaterials* 29: 3201–3212.
20. Li, Z., L. Kupcsik, S. J. Yao, M. Alini, and M. J. Stoddart. 2009. Chondrogenesis of human bone marrow mesenchymal stem cells in fibrin–polyurethane composites. *Tissue Engineering Part A* 15: 1729–1737.
21. Freyria, A. M., and F. Mallein-Gerin. 2012. Chondrocytes or adult stem cells for cartilage repair: The indisputable role of growth factors. *Injury* 43: 259–265.
22. Cancedda, R., P. Castagnola, F. D. Cancedda, B. Dozin, and R. Quarto. 2000. Developmental control of chondrogenesis and osteogenesis. *International Journal of Developmental Biology* 44: 707–714.
23. Mueller, M. B., and R. S. Tuan. 2008. Functional characterization of hypertrophy in chondrogenesis of human mesenchymal stem cells. *Arthritis and Rheumatism* 58: 1377–1388.
24. Pelttari, K., A. Winter, E. Steck, K. Goetzke, T. Hennig, B. G. Ochs, T. Aigner, and W. Richter. 2006. Premature induction of hypertrophy during *in vitro* chondrogenesis of human mesenchymal stem cells correlates with calcification and vascular invasion after ectopic transplantation in SCID mice. *Arthritis and Rheumatism* 54: 3254–3266.
25. Mwale, F., G. Yao, J. A. Ouellet, A. Petit, and J. Antoniou. 2010. Effect of parathyroid hormone on type X and type II collagen expression in mesenchymal stem cells from osteoarthritic patients. *Tissue Engineering—Part A* 16: 3449–3455.
26. Weiss, S., T. Hennig, R. Bock, E. Steck, and W. Richter. 2010. Impact of growth factors and PTHrP on early and late chondrogenic differentiation of human mesenchymal stem cells. *Journal of Cellular Physiology* 223: 84–93.
27. Kim, Y. J., H. J. Kim, and G. I. Im. 2008. PTHrP promotes chondrogenesis and suppresses hypertrophy from both bone marrow-derived and adipose tissue-derived MSCs. *Biochemical and Biophysical Research Communications* 373: 104–108.

28. Gantenbein-Ritter, B., L. M. Benneker, M. Alini, and S. Grad. 2011. Differential response of human bone marrow stromal cells to either TGF-beta(1) or rhGDF-5. *European Spine Journal* 20: 962–971.
29. Stoyanov, J. V., B. Gantenbein-Ritter, A. Bertolo, N. Aebli, M. Baur, M. Alini, and S. Grad. 2011. Role of hypoxia and growth and differentiation factor-5 on differentiation of human mesenchymal stem cells towards intervertebral nucleus pulposus-like cells. *European Cells and Materials* 21: 533–547.
30. Kim, H. J., and G. I. Im. 2011. Electroporation-mediated transfer of SOX trio genes (SOX-5, SOX-6, SOX-9) to enhance the chondrogenesis of mesenchymal stem cells. *Stem Cells and Development* 20: 2103–2114.
31. Ferguson, C. M., E. M. Schwarz, P. R. Reynolds, J. E. Puzas, R. N. Rosier, and R. J. O'Keefe. 2000. Smad2 and 3 mediate transforming growth factor-beta1-induced inhibition of chondrocyte maturation. *Endocrinology* 141: 4728–4735.
32. Cao, X., and D. Chen. 2005. The BMP signaling and *in vivo* bone formation. *Gene* 357: 1–8.
33. Retting, K. N., B. Song, B. S. Yoon, and K. M. Lyons. 2009. BMP canonical Smad signaling through Smad1 and Smad5 is required for endochondral bone formation. *Development* 136: 1093–1104.
34. Hellingman, C. A., E. N. Davidson, W. Koevoet, E. L. Vitters, W. B. van den Berg, G. J. van Osch, and P. M. van der Kraan. 2011. Smad signaling determines chondrogenic differentiation of bone-marrow-derived mesenchymal stem cells: Inhibition of Smad1/5/8P prevents terminal differentiation and calcification. *Tissue Engineering—Part A* 17: 1157–1167.
35. Steck, E., J. Fischer, H. Lorenz, T. Gotterbarm, M. Jung, and W. Richter. 2009. Mesenchymal stem cell differentiation in an experimental cartilage defect: Restriction of hypertrophy to bone-close neocartilage. *Stem Cells and Development* 18: 969–978.
36. Fischer, J., A. Dickhut, M. Rickert, and W. Richter. 2010. Human articular chondrocytes secrete parathyroid hormone-related protein and inhibit hypertrophy of mesenchymal stem cells in coculture during chondrogenesis. *Arthritis and Rheumatism* 62: 2696–2706.
37. Aung, A., G. Gupta, G. Majid, and S. Varghese. 2011. Osteoarthritic chondrocyte-secreted morphogens induce chondrogenic differentiation of human mesenchymal stem cells. *Arthritis and Rheumatism* 63: 148–158.
38. Cooke, M. E., A. A. Allon, T. Cheng, A. C. Kuo, H. T. Kim, T. P. Vail, R. S. Marcucio, R. A. Schneider, J. C. Lotz, and T. Alliston. 2011. Structured three-dimensional co-culture of mesenchymal stem cells with chondrocytes promotes chondrogenic differentiation without hypertrophy. *Osteoarthritis Cartilage* 19: 1210–1218.
39. Giovannini, S., J. Diaz-Romero, T. Aigner, P. Heini, P. Mainil-Varlet, and D. Nesic. 2010. Micromass co-culture of human articular chondrocytes and human bone marrow mesenchymal stem cells to investigate stable neocartilage tissue formation *in vitro*. *European Cells and Materials* 20: 245–259.
40. Mo, X. T., S. C. Guo, H. Q. Xie, L. Deng, W. Zhi, Z. Xiang, X. Q. Li, and Z. M. Yang. 2009. Variations in the ratios of co-cultured mesenchymal stem cells and chondrocytes regulate the expression of cartilaginous and osseous phenotype in alginate constructs. *Bone* 45: 42–51.
41. Bian, L., D. Y. Zhai, R. L. Mauck, and J. A. Burdick. 2011. Coculture of human mesenchymal stem cells and articular chondrocytes reduces hypertrophy and enhances functional properties of engineered cartilage. *Tissue Engineering Part A* 17: 1137–1145.
42. Liu, X., H. Sun, D. Yan, L. Zhang, X. Lv, T. Liu, W. Zhang, W. Liu, Y. Cao, and G. Zhou. 2010. *In vivo* ectopic chondrogenesis of BMSCs directed by mature chondrocytes. *Biomaterials* 31: 9406–9414.

43. Dowthwaite, G. P., J. C. Bishop, S. N. Redman, I. M. Khan, P. Rooney, D. J. Evans, L. Haughton et al. 2004. The surface of articular cartilage contains a progenitor cell population. *Journal of Cell Science* 117: 889–897.
44. Khan, I. M., J. C. Bishop, S. Gilbert, and C. W. Archer. 2009. Clonal chondroprogenitors maintain telomerase activity and Sox9 expression during extended monolayer culture and retain chondrogenic potential. *Osteoarthritis Cartilage* 17: 518–528.
45. Alsalameh, S., R. Amin, T. Gemba, and M. Lotz. 2004. Identification of mesenchymal progenitor cells in normal and osteoarthritic human articular cartilage. *Arthritis and Rheumatism* 50: 1522–1532.
46. de la Fuente, R., J. L. Abad, J. Garcia-Castro, G. Fernandez-Miguel, J. Petriz, D. Rubio, C. Vicario-Abejon, P. Guillen, M. A. Gonzalez, and A. Bernad. 2004. Dedifferentiated adult articular chondrocytes: A population of human multipotent primitive cells. *Experimental Cell Research* 297: 313–328.
47. Williams, R., I. M. Khan, K. Richardson, L. Nelson, H. E. McCarthy, T. Analbelsi, S. K. Singhrao et al. 2010. Identification and clonal characterisation of a progenitor cell sub-population in normal human articular cartilage. *PLoS One* 5: e13246.
48. Goodell, M. A., K. Brose, G. Paradis, A. S. Conner, and R. C. Mulligan. 1996. Isolation and functional properties of murine hematopoietic stem cells that are replicating *in vivo*. *Journal of Experimental Medicine* 183: 1797–1806.
49. Hattori, S., Oxford C., and Reddi A. H. 2007. Identification of superficial zone articular chondrocyte stem/progenitor cells. *Biochemical and Biophysical Research Communications* 358: 99–103.
50. Gleghorn, J. P., A. R. Jones, C. R. Flannery, and L. J. Bonassar. 2007. Boundary mode frictional properties of engineered cartilaginous tissues. *European Cells and Materials* 14: 20–28.
51. Yamane, S., and A. H. Reddi. 2008. Induction of chondrogenesis and superficial zone protein accumulation in synovial side population cells by BMP-7 and TGF-beta1. *Journal of Orthopaedic Research* 26: 485–492.
52. Lee, S. Y., T. Nakagawa, and A. H. Reddi. 2008. Induction of chondrogenesis and expression of superficial zone protein (SZP)/lubricin by mesenchymal progenitors in the infrapatellar fat pad of the knee joint treated with TGF-beta1 and BMP-7. *Biochemical and Biophysical Research Communications* 376: 148–153.
53. Schumacher, B. L., J. A. Block, T. M. Schmid, M. B. Aydelotte, and K. E. Kuettner. 1994. A novel proteoglycan synthesized and secreted by chondrocytes of the superficial zone of articular cartilage. *Archives of Biochemistry and Biophysics* 311: 144–152.
54. Li, Z., S. Yao, M. Alini, and S. Grad. 2007. Different response of articular chondrocyte subpopulations to surface motion. *Osteoarthritis Cartilage* 15: 1034–1041.
55. Lee, S. Y., T. Nakagawa, and A. H. Reddi. 2010. Mesenchymal progenitor cells derived from synovium and infrapatellar fat pad as a source for superficial zone cartilage tissue engineering: Analysis of superficial zone protein/lubricin expression. *Tissue Engineering Part A* 16: 317–325.
56. Li, Z., L. Kupcsik, S. J. Yao, M. Alini, and M. J. Stoddart. 2010. Mechanical load modulates chondrogenesis of human mesenchymal stem cells through the TGF-beta pathway. *Journal of Cellular and Molecular Medicine* 14: 1338–1346.
57. Temenoff, J. S., and A. G. Mikos. 2000. Review: Tissue engineering for regeneration of articular cartilage. *Biomaterials* 21: 431–440.
58. Poole, A. R., T. Kojima, T. Yasuda, F. Mwale, M. Kobayashi, and S. Laverty. 2001. Composition and structure of articular cartilage: A template for tissue repair. *Clinical Orthopaedics and Related Research* S26–S33.
59. Wise, J. K., A. L. Yarin, C. M. Megaridis, and M. Cho. 2009. Chondrogenic differentiation of human mesenchymal stem cells on oriented nanofibrous scaffolds: Engineering the superficial zone of articular cartilage. *Tissue Engineering—Part A* 15: 913–921.

60. Klein, T. J., J. Malda, R. L. Sah, and D. W. Hutmacher. 2009. Tissue engineering of articular cartilage with biomimetic zones. *Tissue Engineering Part B, Reviews* 15: 143–157.
61. Chung, C., and J. A. Burdick. 2008. Engineering cartilage tissue. In: *Emerging Trends in Cell-Based Therapies*. vol. 60, 2nd ed., 243–262.
62. Sharma, B., C. G. Williams, T. K. Kim, D. Sun, A. Malik, M. Khan, K. Leong, and J. H. Elisseeff. 2007. Designing zonal organization into tissue-engineered cartilage. *Tissue Engineering* 13: 405–414.
63. Nguyen, L. H., A. K. Kudva, N. L. Guckert, K. D. Linse, and K. Roy. 2011. Unique biomaterial compositions direct bone marrow stem cells into specific chondrocytic phenotypes corresponding to the various zones of articular cartilage. *Biomaterials* 32: 1327–1338.
64. Nguyen, L. H., A. K. Kudva, N. S. Saxena, and K. Roy. 2011. Engineering articular cartilage with spatially-varying matrix composition and mechanical properties from a single stem cell population using a multi-layered hydrogel. *Biomaterials* 32: 6946–6952.
65. Gong, Y. Y., J. X. Xue, W. J. Zhang, G. D. Zhou, W. Liu, and Y. Cao. 2011. A sandwich model for engineering cartilage with acellular cartilage sheets and chondrocytes. *Biomaterials* 32: 2265–2273.
66. Schumann, D., R. Kujat, M. Nerlich, and P. Angele. 2006. Mechanobiological conditioning of stem cells for cartilage tissue engineering. *Biomedical Materials Engineering* 16: S37–S52.
67. Wimmer, M. A., S. Grad, T. Kaup, M. Hanni, E. Schneider, S. Gogolewski, and M. Alini. 2004. Tribology approach to the engineering and study of articular cartilage. *Tissue Engineering* 10: 1436–1445.
68. Li, Z., S. J. Yao, M. Alini, and M. J. Stoddart. 2010. Chondrogenesis of human bone marrow mesenchymal stem cells in fibrin–polyurethane composites is modulated by frequency and amplitude of dynamic compression and shear stress. *Tissue Engineering—Part A* 16: 575–584.
69. Kupcsik, L., M. J. Stoddart, Z. Li, L. M. Benneker, and M. Alini. 2010. Improving chondrogenesis: Potential and limitations of SOX9 gene transfer and mechanical stimulation for cartilage tissue engineering. *Tissue Engineering—Part A* 16: 1845–1855.
70. Thorpe, S. D., C. T. Buckley, T. Vinardell, F. J. O'Brien, V. A. Campbell, and D. J. Kelly. 2008. Dynamic compression can inhibit chondrogenesis of mesenchymal stem cells. *Biochemical and Biophysical Research Communications* 377: 458–462.
71. Thorpe, S. D., C. T. Buckley, T. Vinardell, F. J. O'Brien, V. A. Campbell, and D. J. Kelly. 2010. The response of bone marrow-derived mesenchymal stem cells to dynamic compression following TGF-beta3 induced chondrogenic differentiation. *Annals of Biomedical Engineering* 38: 2896–2909.
72. Schätti, O., S. Grad, J. Goldhahn, G. Salzmann, Z. Li, M. Alini, and M. J. Stoddart. 2011. A combination of shear and dynamic compression leads to mechanically induced chondrogenesis of human mesenchymal stem cells. *European Cells and Materials* 22: 214–225.
73. Sakaguchi, Y., I. Sekiya, K. Yagishita, and T. Muneta. 2005. Comparison of human stem cells derived from various mesenchymal tissues: Superiority of synovium as a cell source. *Arthritis and Rheumatism* 52: 2521–2529.
74. Kurth, T. B., F. Dell'Accio, V. Crouch, A. Augello, P. T. Sharpe, B. C. De. 2011. Functional mesenchymal stem cell niches in adult mouse knee joint synovium *in vivo*. *Arthritis and Rheumatism* 63: 1289–1300.
75. Lee, C. H., J. L. Cook, A. Mendelson, E. K. Moioli, H. Yao, and J. J. Mao. 2010. Regeneration of the articular surface of the rabbit synovial joint by cell homing: A proof of concept study. *Lancet* 376: 440–448.
76. Mendelson, A., E. Frank, C. Allred, E. Jones, M. Chen, W. Zhao, and J. J. Mao. 2011. Chondrogenesis by chemotactic homing of synovium, bone marrow, and adipose stem cells *in vitro*. *FASEB Journal* 25: 3496–3504.

77. Mishima, Y., and M. Lotz. 2008. Chemotaxis of human articular chondrocytes and mesenchymal stem cells. *Journal of Orthopaedic Research* 26: 1407–1412.
78. Koelling, S., J. Kruegel, M. Irmer, J. R. Path, B. Sadowski, X. Miro, and N. Miosge. 2009. Migratory chondrogenic progenitor cells from repair tissue during the later stages of human osteoarthritis. *Cell Stem Cell* 4: 324–335.
79. Martin, I., S. Miot, A. Barbero, M. Jakob, and D. Wendt. 2007. Osteochondral tissue engineering. *Journal of Biomechanics* 40: 750–765.
80. Brady, M. A., S. Sivananthan, V. Mudera, Q. Liu, J. Wiltfang, and P. H. Warnke. 2011. The primordium of a biological joint replacement: Coupling of two stem cell pathways in biphasic ultrarapid compressed gel niches. *Journal of Cranio-Maxillofacial Surgery* 39: 380–386.
81. Park, J. S., H. N. Yang, D. G. Woo, S. Y. Jeon, and K. H. Park. 2011. Chondrogenesis of human mesenchymal stem cells in fibrin constructs evaluated *in vitro* and in nude mouse and rabbit defects models. *Biomaterials* 32: 1495–1507.
82. Chen, J., H. Chen, P. Li, H. Diao, S. Zhu, L. Dong, R. Wang, T. Guo, J. Zhao, and J. Zhang. 2011. Simultaneous regeneration of articular cartilage and subchondral bone *in vivo* using MSCs induced by a spatially controlled gene delivery system in bilayered integrated scaffolds. *Biomaterials* 32: 4793–4805.
83. Dashtdar, H., H. A. Rothan, T. Tay, R. E. Ahmad, R. Ali, L. X. Tay, P. P. Chong et al. 2011. A preliminary study comparing the use of allogenic chondrogenic pre-differentiated and undifferentiated mesenchymal stem cells for the repair of full thickness articular cartilage defects in rabbits. *Journal of Orthopaedic Research* 29: 1336–1342.
84. Marquass, B., R. Schulz, P. Hepp, M. Zscharnack, T. Aigner, S. Schmidt, F. Stein, R. Richter, G. Osterhoff, G. Aust, C. Josten, and A. Bader. 2011. Matrix-associated implantation of predifferentiated mesenchymal stem cells versus articular chondrocytes: *In vivo* results of cartilage repair after 1 year. *American Journal of Sports Medicine* 39: 1401–1412.
85. Gille, J., E. Schuseil, J. Wimmer, J. Gellissen, A. P. Schulz, and P. Behrens. 2010. Mid-term results of autologous matrix-induced chondrogenesis for treatment of focal cartilage defects in the knee. *Knee Surgery, Sports Traumatology, Arthroscopy* 18: 1456–1464.
86. DeGirolamo, L., G. Bertolini, M. Cervellin, G. Sozzi, and P. Volpi. 2010. Treatment of chondral defects of the knee with one step matrix-assisted technique enhanced by autologous concentrated bone marrow: *In vitro* characterisation of mesenchymal stem cells from iliac crest and subchondral bone. *Injury* 41: 1172–1177.
87. Benthien, J. P., and P. Behrens. 2011. The treatment of chondral and osteochondral defects of the knee with autologous matrix-induced chondrogenesis (AMIC): Method description and recent developments. *Knee Surgery, Sports Traumatology, Arthroscopy* 19: 1316–1319.

19 Stem Cells and Nanobiomaterials for Cardiac Tissue Regeneration

Debasish Mondal

CONTENTS

19.1 INTRODUCTION

Cardiovascular disease is the leading cause of death worldwide (Rosamond et al. 2007; Taylor and Zenovich 2008). Myocardial infarction (MI) and congestive heart failure are major cardiovascular diseases. MI is a principal cause of congestive heart failure. MI, more commonly known as a heart attack, is caused by the abrupt occlusion of one or more of the blood vessels (coronary arteries) supplying blood to the heart. This reduces the supply of nutrients and oxygen to the heart muscle (myocardium). If blood flow is not restored rapidly, irreversible cell death occurs within the blood-deprived myocardium, eventually impairing cardiac performance. Patients who survive the acute event may eventually develop heart failure, which is defined as the clinical state resulting from the inability of the heart to pump enough blood to meet the body's metabolic requirements. The adult human heart is unable to self-regenerate to a significant degree, and a fibrous noncontractile scar tissue is formed in the infarcted myocardial region.

Heart transplantation is the ultimate solution to end-stage heart failure. Despite advances in surgical techniques, scientists and surgeons are continuously looking for alternative approaches to regenerate the injured heart because of the lack of organ donors and the complications associated with immunosuppressive treatments. Although heart muscle has the ability to regenerate through the activation of resident cardiac stem cells (CSCs) or through recruitment of a stem cell population from other tissues (Beltrami et al. 2003; Germani et al. 2007), the regenerative capability of the heart itself after MI cannot compensate for the large-scale tissue loss (Taylor and Zenovich 2008; McMullen and Pasumarthi 2007). Therefore, it is essential to transplant cells to the infarcted region to help in cardiac repair. Cardiac cell therapy has become a potential new therapeutic option for heart disease. A number of cell types, particularly stem/progenitor cells, have been delivered to the cardiac injury site to restore blood flow and contractility, aiming at repairing the infarcted, scarred, and dysfunctional heart. Currently, two main strategies have been taken for cellular transplantation: *in situ* cellular cardiomyoplasty and cardiac tissue engineering. *In situ* cellular cardiomyoplasty involves the injection of cells either directly into the infarct or intravenously. A variety of cell sources have been explored, including bone marrow-derived cells, skeletal myocytes, cardiac stem/progenitor cells, and embryonic stem cells (ESCs; Orlic et al. 2001; Taylor et al. 1998; Klug et al. 1996). However, this strategy is limited in terms of cell type, cell dose, cell retention, cell survival (engrafted cells), and cell differentiation, as well as the integration of transplanted cells within the host tissue. Cardiac tissue engineering, an alternate strategy to *in situ* cellular cardiomyoplasty, is designed to repair infarcted myocardium using cells, biomaterials, and regulative factors (for example, growth factors). Therefore, efforts have been made to incorporate cells into biomaterials that might provide a refined, three-dimensional environment to facilitating cell survival, proliferation, and differentiation. Nanobiomaterials have been fabricated in nanometer-scale dimensions by combining the chemical, mechanical, biological, and electrical aspects of materials for tissue engineering applications. Synthetic polymers have the advantage of having adjustable mechanical properties, whereas natural polymers offer cell recognition sites necessary for cell adhesion and proliferation. Multiple biomaterials have been evaluated as scaffolds for myocardial tissue engineering, including naturally occurring materials such as fibrin, alginate, gelatin, collagen, and chitosan (Tomita et al. 2002; Murry et al. 2005; Xiang et al. 2006; Li et al. 1999; Dvir et al. 2009); and synthetic and biodegradable polymers such as polycaprolactone (PCL), poly(L-lactic acid), poly(lactic-co-glycolic acid), poly(glycerol-sebacate), and polyurethane (Bursac et al. 1999; Gerecht-Nir et al. 2006; Eschenhagen et al. 2002; Radisic et al. 2005, 2008; Pego et al. 2003; Fujimoto et al. 2007a,b).

However, despite good results in myocardial tissue engineering, with the current methodologies and intense research activities, the gap between *in vitro* studies and clinical applications is still far from being overcome. There are significant limitations: 1) the stiffness of the tissue constructs is often mismatched with that of the myocardium. Such myocardium-incompatible biomechanical microenvironments may not allow cells to engraft and function properly in the infarcted heart (Kofidis et al. 2004; Davis et al. 2005). The mismatched stiffness also might not effectively decrease myocardial wall stress and reconstruct a myocardial-like biomechanical microenvironment for attenuating cardiac dilation; 2) the tissue constructs do not possess similar anisotropic and nanofibrous structural properties as that of the extracellular matrix (ECM) in the myocardium; and 3) cells within the tissue constructs have poor engraftment when implanted *in vivo* due to inadequate angiogenesis and low cell survival in an ischemic environment. Successful cardiac tissue engineering is largely dependent on the appropriate combination of scaffolds and cells (Radisic et al. 2004; Zimmermann et al. 2006). Therefore, nanobiomaterials should be able to crosstalk, on the molecular level (nanoscale), with cells in a precise and controlled manner, similar to the natural interactions existing between cells and the native ECM. At the same time, the basic requirements of a biomaterial should be kept; that is, the materials and their degradation products must be nontoxic and nonimmunogenic, and their degradation rate should match the rate of new tissue formation.

Thus, the recent trend has been to design bioactive nanomaterials that, on one hand, will have the appropriate physical strength as well as the degradation kinetics of synthetic polymers and, on the other hand, will have the biological specificity of collagen, fibronectin, and laminin—the major ECM components of myocardium.

19.2 MI AND REGENERATION

The heart wall is composed of the endocardial, myocardial, and epicardial layers (Kierszenbaum 2002). The major layer, the myocardium, provides the strength for the systole and diastole. The myocardium has a unique construction in which multiple layers made of ECM and extremely oriented cells are arranged together. Therefore, the ECM is considered to be an important factor directing the cardiomyocytes' alignment (Kim et al. 2006). MI, most commonly known as a heart attack, is caused by the abrupt occlusion of one or more of the blood vessels (coronary arteries) supplying blood to the heart. After injury, the human heart cannot regenerate in a significant manner because adult cardiac myocytes are terminally differentiated and cannot replicate (Pasumarthi and Field 2002). After MI, heart failure mainly occurs due to the substantial loss of cardiomyocytes in the infarct zone but more often is precipitated by the delayed and progressive pathological remodeling of the left ventricle. After MI, if the blood flow is restored within 30 min, the injured myocardium can revert to the normal state, although a depression in contractility may remain for hours. After 30 min, an irreversible cell death occurs within the blood-deprived myocardium, resulting in perturbed cardiac performance. The adult human heart is unable to self-regenerate to a significant manner, and a fibrous noncontractile scar tissue is formed in the infarcted domain (Figure 19.1).

Although angioplasty and thrombolytic agents can relieve the cause of the infarction, timely reperfusion is essential for acute MI (Schwarz et al. 1982). Bypass surgery can be performed to restore blood flow to the blocked arteries by grafting a short vein. Nevertheless, if the myocardium has already undergone permanent damage, reperfusion cannot retrieve the lost cardiac muscle due to the nonregenerative nature of adult cardiomyocytes. Currently, no medication or clinically used procedure has shown efficacy in replacing myocardial scar tissue with functioning contractile tissue. Thus, there is a demand for new therapeutics to regenerate normal cardiomyocytes. The major goal of post-MI therapy is to regenerate the new cardiomyocyte-based muscle. This can be accomplished by a cell delivery system consisting of a supporting matrix

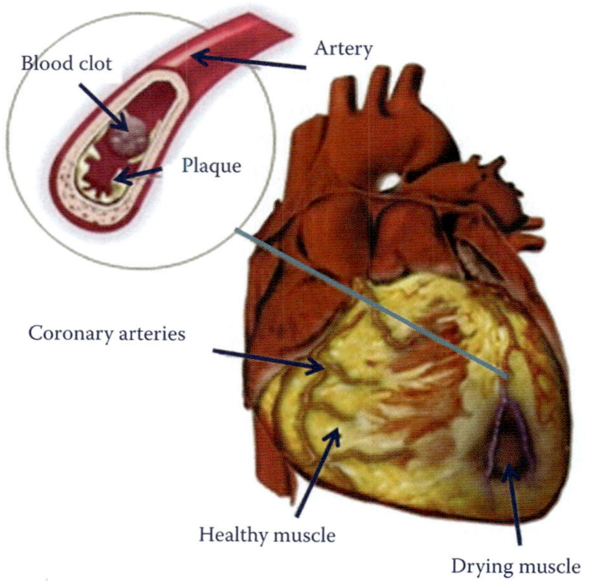

FIGURE 19.1 Schematic of the damage caused by a MI in human heart. (Adapted from MedicineNet.com; http://www.medicinenet.com/heart_attack/page2.htm.)

TABLE 19.1
Advantages and Disadvantages of Different Strategies for Cardiac Tissue Engineering

Strategies	Advantages	Disadvantages
	Biomaterials	
Injectable system	Nonimmunogenic	Need for chemical factors
	Support placed where required	Poor repair without cells
External restraints	Nonimmunogenic	Passive device
	Simple design	Only prevents spread of damage
	Biomaterials and Cells	
Scaffolds/meshes	Consistent support for cells	Surgical risks for implantation
	Degrade safely as cells incorporate	Limited vascularization
Cell sheets and patches	High cell density and organization	Surgical difficulty
		Limited vascularization
ECM-like gels	Similar composition to body	Natural components can be inflammatory
	Cells readily attach	Slow vascularization
Cells	No structure to design	High amount of cell exit or loss
	Less invasive delivery	Poor cell integration

and suitable cells. The main function of the supporting matrix is to stop or slow down the remodeling and scar tissue formation process by mechanically preventing the cardiac muscle from dilating (Etzion et al. 2001). Moreover, desirable cells are delivered to replace the dead cardiomyocytes and integrate with the neighboring cardiac tissue. The supporting matrix should also have a cell-favorable environment to support the survival of the delivered cells and accelerate the regeneration process. Several strategies for cardiac tissue engineering are summarized in Table 19.1.

19.3 STEM CELLS FOR MYOCARDIAL REGENERATION

The high plasticity of the stem cells enables them to be potential candidates for myocardial tissue engineering. Stem cells have been used to fabricate tissue constructs for the improvement of cardiac function or myocardial regeneration. After implantation, the differentiation of stem cells into cardiomyocytes and endothelial cells under a myocardial environment is beneficial for the improvement of cardiac function. The ideal stem cell types are those that are capable of differentiation into cardiomyocytes. Stem cells capable of inducing angiogenesis can also be used to reestablish the vasculature within the infarcted myocardium. The success of cardiomyocyte generation *in vitro* has been reported with adult bone marrow-derived mesenchymal stem cells (MSCs), adult adipose-derived MSCs, ESCs, and recently, with induced pluripotent stem (iPS) cells (Zhang et al. 2006, 2008).

19.3.1 Mesenchymal Stem Cells

MSCs are adult stem cells derived from the bone marrow. MSCs are multipotent and have multilineage differentiation capacity (Pittenger et al. 1999). Bone marrow-derived MSCs were quickly advanced into clinical trials for cardiac therapy due to its easy accessibility and processing, its autologous origin, and its putative ability to transdifferentiate into either myocardial or vascular cells. However, conflicting therapeutic benefits have been reported (Murry et al. 2005). In culture, MSCs can maintain an undifferentiated, stable phenotype over many generations. However, controversy still exists regarding their precise phenotype, and there are no adequate markers to allow the selection of purified cell populations. Regeneration of an ischemic heart can be accomplished by delivering culture-expanded MSCs into the coronary arteries or directly into the myocardium to amplify the endogenous regenerative pool. Janssens et al. (2006) described the first randomized controlled trial of autologous bone marrow MSC implantation in patients with MI. Stem cell therapy provided significant reductions in myocardial infarct size and improved the recovery rates of regional systolic function after 4 months of follow-up. Nevertheless, there was no significant benefit in terms of left ventricular ejection fraction, myocardial perfusion, and cardiac metabolism.

Studies have reported that bone marrow-derived MSCs demonstrated increased expression of cardiomyocyte markers and differentiation into spontaneously beating cells with a reagent (5-azacytidine, which is known to be a DNA demethylating agent; Tomita et al. 1999; Fukuda 2001; Makino et al. 1999). Some investigators believe that the efficacy of cardiogenesis is low in this cell type (Shiota et al. 2007). Moreover, cell therapy studies in animals suggested that MSCs' contribution to improved myocardial function after infarction is most likely due to their vasculogenic potential rather than their contribution to a cardiomyocyte population (Mangi et al. 2003).

Adipose tissue-derived MSCs (AD-MSCs) can be obtained easily from liposuction aspirates and appear like bone marrow-MSCs with regard to their differentiating features. They can differentiate into beating cardiomyocytes either spontaneously (Planat-Benard et al. 2004) or when treated with 5-aza-deoxycytidine (Rangappa et al. 2003; van Dijk et al. 2008). It was recently demonstrated that AD-MSCs can differentiate into contractile cardiomyocytes when cocultured with rat cardiomyocytes *in vitro* and *in vivo* (Choi et al. 2010a,b). Nevertheless, currently, no human or animal tissue-engineered cardiac tissue construct has been generated using either BM-MSC or AD-MSC, indicating their low differentiating capability into cardiomyocytes for clinical use.

19.3.2 Embryonic Stem Cells

ESCs are the most primitive of all stem cell populations that develop as the inner cell mass at day 5 after fertilization in the human blastocyst. At this early stage, ES cells have immense developmental potential and they give rise to the cells of the three embryonic germ layers. Mouse and human ES cells can undergo an undetermined number of cell doublings and the capacity to differentiate into specific cell types, including cardiomyocytes, when isolated and transferred to appropriate culture media (Thomson et al. 1998; Maltsev et al. 1993). ESCs seem to be the most efficient cell source for myocardial tissue engineering (Kehat et al. 2001, 2004; Mummery et al. 2002). ES cell-derived cardiomyocytes have a high proliferative potentiality, which is a great advantage considering that a large number of cells are needed when repairing a damaged heart (Snir et al. 2003; McDevitt et al. 2005). ESC pluripotency permits their differentiation into all components of the myocardium: cardiomyocytes, fibroblasts, and endothelial cells (Yang et al. 2008). ESC-derived cardiomyocytes can be used to produce spontaneous contractile cardiac tissues, and *in vivo* studies in rodent models have shown promising results (Caspi et al. 2007). Nevertheless, the potential risk of uncontrolled ESC differentiation into teratoma, immunogenicity, and the ethical issues associated with their derivation are major restrictions to their clinical application (Nussbaum et al. 2007).

19.3.3 Induced Pluripotent Stem Cells

Recently, iPS cells produced by reprogramming somatic cells using retroviral overexpression of Oct4, Sox2, c-Myc, and Klf4 have emerged. iPS cells share some of the properties of ES cells and are very promising for cardiac cells regeneration (Takahashi and Yamanaka 2006; Takahashi et al. 2007; Yu et al. 2007; Nakagawa et al. 2008). These cells avoid the legal and ethical controversies associated with ES cells and also enable the production of patient-specific cell therapeutics by using autologous cells. iPS cell-derived cardiomyocytes were noticed in mouse models using methods similar to those demonstrated for ES cells (Narazaki et al. 2008; Mauritz et al. 2008; Schenke-Layland et al. 2008). Myocardial regeneration via intramyocardial injection of human iPS cells differentiated into cardiac, smooth muscle, and endothelial tissue *in situ* was demonstrated in a mouse model of acute MI (Nelson et al. 2009).

19.3.4 Cardiac Stem Cells

Most preferably, CSCs would be the best candidate for cardiac tissue regeneration. CSCs, progenitor cells in the heart, could be described as undifferentiated cells that are devoted to the cardiac lineage and can create differentiated progenies, including cardiomyocytes, smooth muscle cells, and endothelial cells. Surface markers such as c-Kit and Sca-1, used in establishing stem cell populations in other tissues, were subsequently applied to identify CSCs. A c-Kit-expressing population negative for hematopoietic (CD34−) and endothelial (CD31−) markers was observed in adult rat hearts (Beltrami et al. 2003). CSCs are clonogenic, with self-renewal capacities and the ability to differentiate into cardiomyocytes, smooth muscle cells, and endothelial cells both *in vitro* and *in vivo*. When injected intracoronarily or directly into the infarcted rat myocardium, these cells led to myocardial regeneration, attenuated myocardial remodeling, and improved cardiac function and gave rise to large blood vessels (Beltrami et al. 2003; Tillmanns et al. 2008; Dawn et al. 2005).

19.4 NANOBIOMATERIALS FOR CARDIAC TISSUE REGENERATION

Nanostructured biomaterials are currently being investigated for use as a new generation of cardiac tissue engineering matrices. The main advantage of using nanotechnology in tissue engineering is that some specific characteristics of nanomaterials provide the cell–materials interaction and other cellular functions much more efficiently than conventional materials. Nanotechnology strategies can be used to fabricate nanofibers, nanopatterns, and self-assembled peptide hydrogels that are especially useful for cardiac tissue regeneration. Bionanotechnology approaches are particularly useful for mimicking native tissues because many biological cues, such as ECM fibers, are in the range of a few nanometers (Pham et al. 2006). For example, polymeric nanofibers that mimic collagen fibers can be fabricated by electrospinning and self-assembly (Zhang 2003; Thierry et al. 2003). Generally, the synthesis of nanobiomaterials can be produced by using one of these two approaches. In one approach, nanomaterials are synthesized by designing existing materials with nanoscale resolution. These techniques include electrospinning and nanopatterning. In another approach, molecular build-up, such as layer-by-layer deposition (Zhang 2003) and self-assembly (Thierry et al. 2003), can be used to create nanomaterials.

19.4.1 Electrospun Nanofibrous Scaffolds

In recent years, electrospinning, which was invented a long time ago, has emerged as a popular technique due to its potential use in the biomedical arena (Boundriot et al. 2006). Electrospinning is the most popular, simple, and cost-effective method, and is able to produce continuous nanofibers of various materials from polymers to ceramics. Furthermore, electrospinning seems to be the only method that can be further formulated for the large-scale production of continuous nanofibers for industrial applications. Pure polymers as well as composites and blends of both natural and synthetic polymers have been successfully electrospun into nanofiber scaffolds. Electrospun scaffolds have shown great potential in tissue engineering because of its high surface area to volume ratio, high porosity, and variable pore size distribution and structures mimicking natural ECM (Boundriot et al. 2006; Kim et al. 2005; Edror et al. 2003; Jun et al. 2005; Demir et al. 2004).

Electrospinning is one of the most popular methods of fabricating nanofibers and was first patented by Cooley (1902) and Morton (1902). An electrospinning device fundamentally consists of a high-voltage power supply, which creates an electric field between a collector and a syringe, with a metal spinneret of small diameters filled with polymer solution. Concisely, polymer jet is ejected from the surface of a charged polymer solution when a sufficiently high voltage is applied to a liquid droplet and the applied electric potential overcomes the surface tension of the polymer. Then, the ejected jet under the influence of the applied electrical field travels rapidly to the grounded collector. The solution jet, which is applied with electrostatic charges, undergoes stretching before reaching the collector. Finally, the jet is eventually solidified on the collector in the form of nanofibers with evaporation of solvents. Ishii et al. (2005) fabricated PCL nanofibrous scaffolds with fiber diameters of 250 nm, and neonatal rat cardiomyocytes were seeded in five layers. Their study demonstrated that the adherent cells established an electrical communication between the cell layers and created synchronized cell beating. Shin et al. (2004) developed PCL nanofibers with ECM-like morphology and the fiber was suspended across a wire ring that acted as a passive load to contracting cardiomyocytes. They noticed that neonatal Lewis rat cardiomyocytes that were cultured on these scaffolds expressed cardiac-specific proteins such as troponin I and connexin-43. Their results indicated contractile cardiac graft formation *in vitro* during 14 days of cell culture (Shin et al. 2004). Jin et al. (2009) demonstrated the epicardial implantation of poly(lactide-co-ε-caprolactone) patch seeded with MSCs to the infarct heart of rat models, and their results indicated the enhanced expression of cardiac markers such as troponin I and myosin heavy chain compared with infarct heart injected with MSCs alone. In another study, Zong et al. (2005) developed electrospun poly-L-lactide/polyglycolide/polyethyleneglycol (PLLA-PLGA-PEG)-based scaffolds in various compositions [100% PLLA, PLLA/PLGA (25:75), and PEG-PLA/PLGA (15:85)] and these researchers observed that the pure PLLA scaffold supported the cell growth and proliferation of rat myocytes better than the composite scaffolds (Zong et al. 2005). They also observed better cell attachment and actin expression of cardiomyocytes on random nanofibers of the PLLA compared with the PLGA blend scaffold. Cardiomyocytes were also found to assist in the development of the contractile behavior of cardiac tissues on PLLA scaffolds (Zong et al. 2005). Forte et al. (2008) developed various polymeric scaffolds such as PLLA, PLGA, and PCL by pressure-assisted microsyringe methods, and demonstrated comparative studies of MSCs and cardiac progenitor cells cultured on polymeric scaffolds. They noticed that the cells expressed functional adhesion proteins, thereby preserving their multipotential capability. Kai et al. (2011) developed random and aligned electrospun poly(e-caprolactone)/gelatin (PG) composite nanofibrous scaffolds to mimic the oriented ECM microenvironment. Their results indicated that PG nanofibers showed anisotropic wetting characteristics and mechanical properties, which closely match the anisotropy of the native cardiac tissue. Rabbit cardiomyocytes were seeded on random and aligned electrospun nanofibers to evaluate the biocompatibility of scaffolds, together with its potential for cell guidance.

Cardiomyocytes seeded on aligned PG nanofibrous scaffolds depicted higher cell attachment and greater cell alignment compared with PCL scaffolds (Kai et al. 2011).

Recently, Stout et al. (2011) developed PLGA (50:50 wt.%) carbon nanofiber (CNF) composite scaffolds for myocardial tissue engineering applications. CNF were added to biodegradable PLGA to increase the conductivity and cytocompatibility of pure PLGA. Various PLGA/CNF ratios (100:0, 75:25, 50:50, 25:75, and 0:100 wt.%) were used to evaluate the conductivity as well as cytocompatibility of cardiomyocytes and neurons. Their demonstration indicated that cardiomyocyte density increases with greater amounts of CNF in PLGA (up to 25:75 wt.% PLGA/CNF) for up to 5 days. They also observed similar trends for neuronal cells. Their results suggested that these conductive materials promoted the adhesion and proliferation of two cell types important for myocardial tissue engineering applications (Stout et al. 2011). Recently, in another study, Mukherjee et al. (2011) fabricated poly(L-lactic acid)-*co*-poly(-caprolactone)/collagen nanofibrous composite scaffolds to mimic the native myocardial environment of freshly isolated cardiomyocytes from rabbit heart. This bioactive nanofibrous composite scaffolds made of PLACL/collagen blend provide superior attachment and growth of adult cardiac cells preferring native myocardium-like alignment of newly seeded cardiac cells compared with purely synthetic poly(l-lactic acid)-*co*-poly(-caprolactone) (PLACL) scaffolds. Furthermore, PLACL/collagen permits cell–cell interaction without attenuating the functional activity of the cells. Their results indicated that the strongest expressions of troponin T, cardiac myosin heavy chain, α-actinin, and connexin 43 were obtained in cells grown on PLACL/collagen (Figure 19.2). These nanofibrous scaffolds have an elastic modulus of a magnitude nearing that of native heart tissue. Their demonstration suggested that the PLACL/collagen nanobiomaterials could potentially be applied for myocardial tissue engineering (Mukherjee et al. 2011).

19.4.2 Nanopattern Substrates

The myocardium is an anisotropic tissue in terms of its biophysical and anatomic characteristics. The parallel and cylindrical structure of the myocardium, together with fast ion transfer, enables the synchronous contraction of cardiomyocytes and produces enough force to pump blood (Henderson and Anderson 2009; Severs 2000). It is vital to maintain an appropriate structural organization of cardiomyocytes and cardiac fibroblasts to develop functional cardiac patches via a tissue engineering approach. Therefore, it is crucial to investigate how the behavior of cardiomyocytes, such as their elongation, orientation, and differentiation, is affected by various physical cues of substrates such as rigidity, stiffness, and surface topography (Deutsch et al. 2000; Fujita et al. 2006; Motlagh et al. 2003; Yin et al. 2004; Au et al. 2007). Recently, it was reported that nanotopography had greater influence on cellular functions and affects activities such as cell adhesion, proliferation, and differentiation.

Au et al. (2007) investigated the effect of topographic cue on rat cardiomyocytes cultured on polyvinyl carbonate surfaces with a range of micropatterns (peak distance, 3–13 μm). They noticed that the topographic cue influenced the orientation of cardiomyocytes significantly more than electrical stimulation. In another study, Kim et al. (2010) indicated that the size of the nanogrooves was a crucial factor in regulating rat cardiomyocyte behavior, such as morphology, conduction velocity, and the

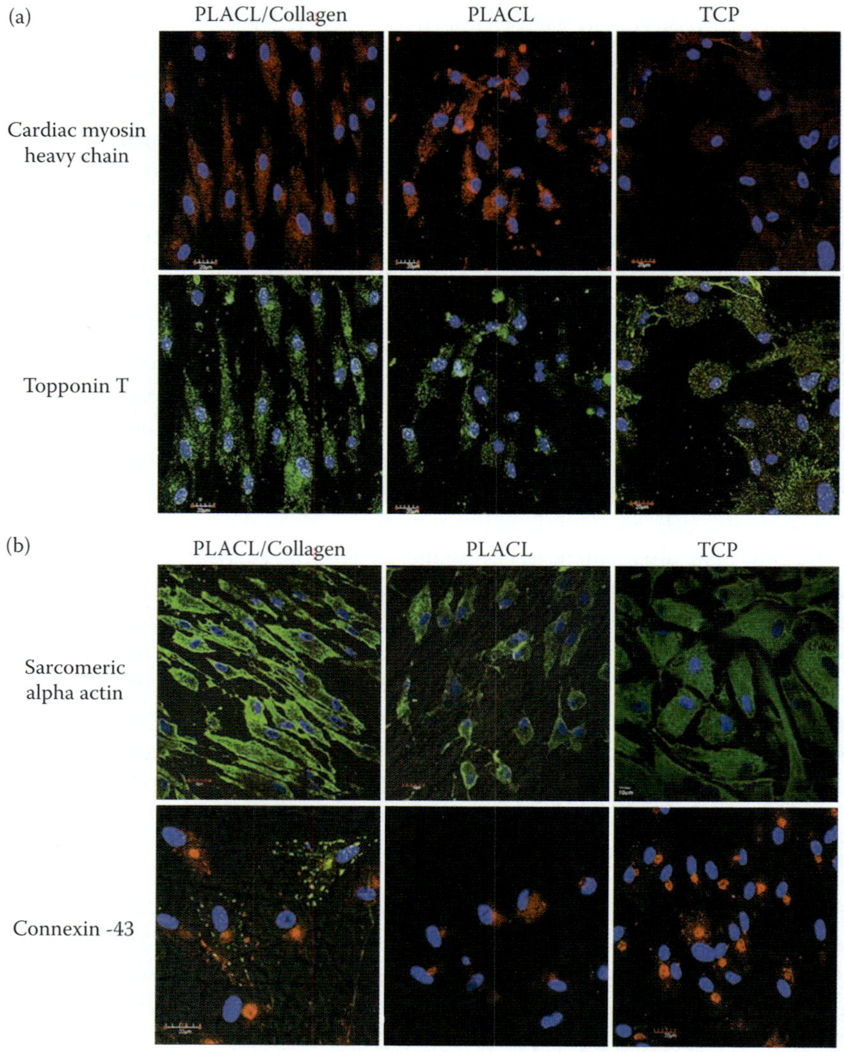

FIGURE 19.2 Immunofluorescence of cardiomyocytes on random nanofibers of PLACL/collagen, PLACL, and TCP stained for cardiac-specific protein at day 15. (a) Cardiac myosin heavy chain–Alexa Fluor 594 and troponin T–Alexa Fluor 488 expression (PLACL/collagen shows the strongest expression). (b) Sarcomeric α-actinin–Alexa Fluor 488 and connexin 43–Alexa Fluor 594. (Adapted from Mukherjee S. et al. *Adv. Funct. Mater.* 21: 2291–2300, 2011. With permission.)

expression of cell coupling proteins. The ECM microenvironment serves as a scaffold for cell adhesion and also provides signal transduction to anchorage-dependent cells via transmembrane proteins such as integrins and syndecan-4 through cellular feedback to local matrix rigidity (Boublik et al. 2005; Engler et al. 2008; Jawad et al. 2007; Jacot et al. 2008; McDevitt et al. 2003). Earlier studies depicted that gap junction formation in systole of cardiomyocytes was enhanced on substrates with suitable elasticity, such as polyurethane (McDevitt et al. 2003; Alperin et al. 2005). Moreover, cardiomyocytes lost their contractile ability when cultured on substrates without enough stiffness (Hidalgo-Bastida et al. 2007).

19.4.3 Self-Assembled Nanobiomaterials

Self-assembled nanostructured biomaterials can be generated from different materials, such as hyaluronan, chitosan, and peptide amphiphile. Various methods, such as layer-by-layer deposition, pH induction, and biomimetic coating, can be used to induce self-assembly. Molecular self-assembly of peptides and proteins can be used to produce hydrogels for tissue engineering applications (Zhang 2003). Self-assembled peptides usually contain hydrophobic and hydrophilic regions that assemble into fibers or sheets, which can be further assembled into hydrogels by charge shielding. Peptide amphiphile molecules were designed to self-assemble into nanofibers, which led to the generation of aqueous gels from pH changes (Hartgerink et al. 2002).

Self-assembling peptide nanofiber scaffolds with more than 99% water content have been shown to be good bioactive nanomaterials for cell culture and tissue regeneration. The self-assembling peptide nanofiber microenvironment has been shown to promote the arrangement and survival of many different cell types, including cardiomyocytes (Davis et al. 2006) and endothelial cells (Narmoneva et al. 2005). LaNasa and Bryant (2009) demonstrated the proteins (collagen 1, laminin, etc.) and peptide (RGD)-modified hydrogel for cardiac muscle tissue engineering, and they cultured skeletal myoblast cells on these hydrogels. Their findings indicated that skeletal myoblasts showed good bioactivity with the modified hydrogels, resulting in myoblast attachment and the development of an intracellular contractile network after 1 day (Figure 19.3). They also noticed that neonatal cardiomyocytes exhibited abnormal myocytic morphologies on RGD-modified hydrogel surfaces, resulting in

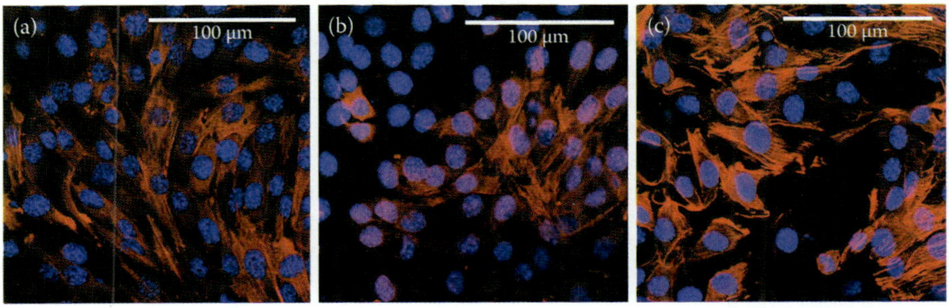

FIGURE 19.3 Morphology and cytoskeletal architecture of skeletal myoblasts cultured on hydrogels. (a) Modified with collagen I. (b) Modified with laminin. (c) Modified with RGD after 24 h of culture. The cytoskeletal architecture was assessed by immunostaining for tropomyosin, an intracellular contractile protein (red), and the nuclei were counterstained with DAPI (blue). (Adapted from LaNasa S. M. and Bryant S. J. *Acta Biomater.* 5: 2929–2938, 2009. With permission.)

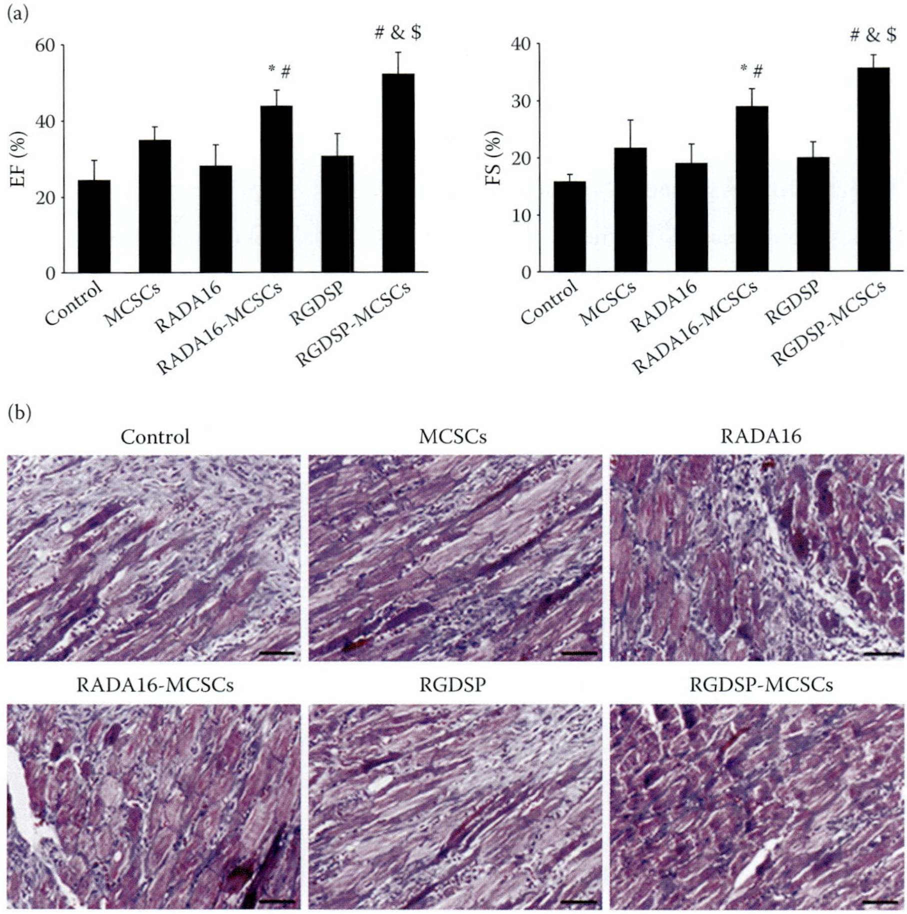

FIGURE 19.4 Improvement of the left ventricle contractile function and collagen deposition at 4 weeks after transplantation. (a) EF (%) and FS (%) were compared among the different treatments. *, $p < .01$ versus the RADA16 group. #, $p < .01$ versus the MCSC group. &, $p < .01$ versus the RGDSP group. $, $p < .01$ versus the MCSC + RADA16 group; $n = 10$. (b) Photomicrographs of representative myocardial sections stained with Masson's trichrome. Myocardium (red) was replaced significantly by fibrous tissue (blue) in the control group. Scale bar: 50 μm. (c) Quantitative analysis demonstrating that the collagen volume fraction in the MCSC + RGDSP group was significantly smaller than in the MCSC alone, RGDSP alone, or MCSC + RADA16 groups. Compared with the MCSC- and RADA16-alone groups, the collagen content in the MCSC + RADA16 group was lower. *, $p < .01$ versus the RADA16 group. #, $p < .01$ versus the MCSC group. &, $p < .01$ versus the RGDSP group. $, $p < .01$ versus the MCSC + RADA16 group; $n = 10$. (Adapted from Guo H. et al. *Biochem. Biophys. Res. Commun.* 399: 42–48, 2010. With permission.)

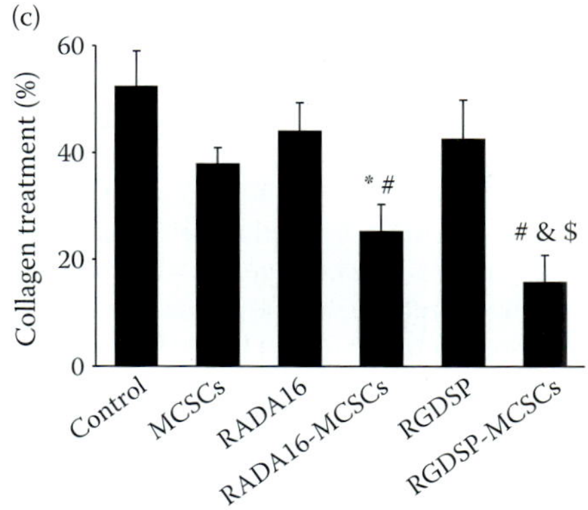

FIGURE 19.4 (Continued)

cell aggregation, but readily spread on protein-modified hydrogel surfaces. Overall, their results indicated that modification of hydrogels with collagen I or laminin resulted in superior cardiomyocyte attachment, proliferation, and development of a contractile cytoarchitecture with increased SERCA2a expression.

Recently, Guo et al. (2010) modified the self-assembling peptide AcN-RADARADARADARADA-CONH2 (RADA16) with a functional bioactive peptide RGDSP to facilitate the adhesion and survival of stem cells, promoting the differentiation of stem cells toward cardiomyocytes. They noticed that the biomimetic designer self-assembling peptide scaffold enhanced the adhesion, survival, and differentiation of marrow-derived CSCs (MCSCs). They established that MCSC transplantation in RGDSP scaffolds decreased collagen deposition and improved heart function compared with the transplantation of MCSCs alone or MCSCs in RADA16 (Figure 19.4c). MCSC transplantation in RGDSP scaffolds was more conducive to the survival and differentiation of the transplanted cells and thereby improved the cardiac function as presented in Figure 19.4a (Guo et al. 2010).

19.5 CONCLUSION

Cardiac tissue engineering aims at regenerating functional myocardium to repair the diseased and injured heart. Cardiomyocytes and various stem cells are used to repair and regenerate the cardiac tissue after MI. However, the injection of cells directly into the infarcted area involves the problems of low cell retention and engraftment rate. A proper supporting matrix can hold cells at the infarcted area initially, and further provide support for cell survival and functioning. Myocardial tissue engineering has encountered a variety of disputes, which are grouped into three categories associated with materials, cells, and vascularization. Nanobiomaterials and stem cells strategy could be promising to produce an advanced nanoengineered tissue construct with the required mechanical, chemical, electrical, biological, and contractile properties that could replace the damaged myocardium. Therefore, a combined prototype produced by stem cells, nanobiomaterials, and signaling molecules will be needed for the ischemic myocardial tissue.

REFERENCES

Alperin, C., P. W. Zandstra, and K. A. Woodhouse. 2005. Polyurethane films seeded with embryonic stem cell-derived cardiomyocytes for use in cardiac tissue engineering applications. *Biomaterials* 26: 7377–7386.

Au, H. T., I. Cheng, M. F. Chowdhury, and M. Radisic. 2007. Interactive effects of surface topography and pulsatile electrical field stimulation on orientation and elongation of fibroblasts and cardiomyocytes. *Biomaterials* 28: 4277–4293.

Beltrami, A. P., L. Barlucchi, D. Torella et al. 2003. Adult cardiac stem cells are multipotent and support myocardial regeneration. *Cell* 114: 763–776.

Boublik, J., H. Park, M. Radisic et al. 2005. Mechanical properties and remodeling of hybrid cardiac constructs made from heart cells, fibrin, and biodegradable, elastomeric knitted fabric. *Tissue Engineering* 11: 1122–1132.

Boundriot, U., R. Dersch, A. Greiner et al. 2006. Electrospinning approaches toward scaffold engineering—a brief overview. *Artificial Organs* 30: 785–792.

Bursac, N., M. Papadaki, R. J. Cohen et al. 1999. Cardiac muscle tissue engineering: Toward an *in vitro* model for electrophysiological studies. *American Journal of Physiology* 277: H433–H444.

Caspi, O., A. Lesman, Y. Basevitch et al. 2007. Tissue engineering of vascularised cardiac muscle from human embryonic stem cells. *Circulation Research* 100: 263–272.

Choi, Y. S., G. J. Dusting, S. Stubbs et al. 2010a. Differentiation of human adipose derived stem cells into beating cardiomyocytes. *Journal of Cellular and Molecular Medicine* 14: 878–889.

Choi, Y. S., K. Matsuda, G. J. Dusting, W. A. Morrison, and R. J. Dilley. 2010b. Engineering cardiac tissue *in vivo* from human adipose-derived stem cells. *Biomaterials* 31: 2236–2242.

Cooley, J. 1902. US Patent No. 692631.

Davis, M. E., J. P. Motion, and D. A. Narmoneva. 2005. Injectable self-assembling peptide nanofibers create intramyocardial microenvironments for endothelial cells. *Circulation* 111: 442–450.

Davis, M. E., P. C. Hsieh, T. Takahashi et al. 2006. Local myocardial insulin-like growth factor 1 (IGF-1) delivery with biotinylated peptide nanofibers improves cell therapy for myocardial infarction. *Proceedings of the National Academy of Sciences of the United States of America* 103: 8155–8160.

Dawn, B., A. B. Stein, K. Urbanek et al. 2005. Cardiac stem cells delivered intravascularly traverse the vessel barrier, regenerate infarcted myocardium, and improve cardiac function. *Proceedings of the National Academy of Sciences of the United States of America* 102: 3766–3771.

Demir, M. M., M. A. Gulgun, Y. Z. Menceloglu et al. 2004. Palladium nanoparticles by electrospinning from poly(acrylonitrile-co-acrylic acid)–PdC12 solutions. Relation between preparation conditions, particle size, and catalytic activity. *Macromolecules* 37: 1787–1792.

Deutsch, J., D. Motlagh, B. Russell, and T. A. Desai. 2000. Fabrication of microtextured membranes for cardiac myocyte attachment and orientation. *Journal of Biomedical Materials Research* 53: 267–275.

Dvir, T., A. Kedem, E. Ruvinov et al. 2009. Prevascularization of cardiac patch on the omentum improves its therapeutic outcome. *Proceedings of the National Academy of Sciences of the United States of America* 106: 14990–14995.

Edror, Y., W. Salalha, R. L. Khalfin et al. 2003. Carbon nanotubes embedded in oriented polymer nanofibers by electrospinning. *Langmuir* 19: 7012–7020.

Engler, A. J., C. Carag-Krieger, C. P. Johnson et al. 2008. Embryonic cardiomyocytes beat best on a matrix with heart-like elasticity: Scar-like rigidity inhibits beating. *Journal of Cell Science* 121: 3794–3802.

Eschenhagen, T., M. Didié, F. Münzel et al. 2002. 3D engineered heart tissue for replacement therapy. *Basic Research in Cardiology* 97: I146–I152.

Etzion, S., L. H. Kedes, R. A. Kloner, J. Leor. 2001. Myocardial regeneration: Present and future trends. *American Journal of Cardiovascular Drugs* 1: 233–244.

Forte, G., F. Carotenuto, F. Pagliari et al. 2008. Stem cell-derived cardiac patches: A tissue engineering approach to cardiac healing. *Journal of Molecular and Cellular Cardiology* 44: 711–806.

Fujimoto, K. L., J. Guan, H. Oshima et al. 2007a. *In vivo* evaluation of a porous, elastic, biodegradable patch for reconstructive cardiac procedures. *Annals of Thoracic Surgery* 83: 648–654.

Fujimoto, K. L., K. Tobita, W. D. Merryman et al. 2007b. An elastic, biodegradable cardiac patch induces contractile smooth muscle and improves cardiac remodeling and function in subacute myocardial infarction. *Journal of the American College of Cardiology* 49: 2292–2300.

Fujita, A., K. Fujita, O. Nakamura, T. Matsuda, and S. Kawata. 2006. Control of cardiomyocytes orientation on a microscaffold fabricated by photopolymerization with laser beam interference. *Journal of Biomedical Optics* 11: 021015.

Fukuda, K. 2001. Development of regenerative cardiomyocytes from mesenchymal stem cells for cardiovascular tissue engineering. *Artificial Organs* 25: 187–193.

Gerecht-Nir, S., M. Radisic, H. Park et al. 2006. Biophysical regulation during cardiac development and application to tissue engineering. *International Journal of Developmental Biology* 50: 233–243.

Germani, A., F., Limana, and M. C. Capogrossi. 2007. Pivotal advances: High-mobility group box 1 protein—a cytokine with a role in cardiac repair. *Journal of Leukocyte Biology* 81: 41–45.

Guo, H., G. Cui, H. Wang, and Y. Tan. 2010. Transplantation of marrow-derived cardiac stem cells carried in designer self-assembling peptide nanofibers improves cardiac function after myocardial infarction. *Biochemical and Biophysical Research Communications* 399: 42–48.

Hartgerink, J. D., E. Beniash, and S. I. Stupp. 2002. Peptide-amphiphile nanofibers: A versatile scaffold for the preparation of self-assembling materials. *Proceedings of the National Academy of Sciences of the United States of America* 99: 5133–5138.

Henderson, D. J., and R. H. Anderson. 2009. The development and structure of the ventricles in the human heart. *Pediatric Cardiology* 30: 588–596.

Hidalgo-Bastida, L. A., J. J. Barry, N. M. Everitt et al. 2007. Cell adhesion and mechanical properties of a flexible scaffold for cardiac tissue engineering. *Acta Biomaterialia* 3: 457–462.

Jacot, J. G., A. D. McCulloch, J. H. Omens. 2008. Substrate stiffness affects the functional maturation of neonatal rat ventricular myocytes. *Biophysical Journal* 95: 3479–3487.

Janssens, S., C. Dubois, J. Bogaert et al. 2006. Autologous bone marrow-derived stem-cell transfer in patients with ST-segment elevation myocardial infarction: Double-blind, randomised controlled trial. *Lancet* 367: 113–121.

Jawad, H., N. N. Ali, A. R. Lyon, Q. Z. Chen, S. E. Harding, and A. R. Boccaccini. 2007. Myocardial tissue engineering: A review. *Journal of Tissue Engineering and Regenerative Medicine* 1: 327–342.

Jin, J., S. I. Jeong, Y. M. Shin et al. 2009. Transplantation of mesenchymal stem cells within a poly(lactide-*co*-e-caprolactone) scaffold improves cardiac function in a rat myocardial infarction model. *European Journal of Heart Failure* 11: 147–153.

Jun, Z., A. Aigner, F. Czubayko, T. Kissel, J. H. Wendorff, and A. Greiner. 2005. Poly(vinyl alcohol) nanofibers by electrospinning as a protein delivery system and retardation of enzyme release by additional polymer coatings. *Biomacromolecules* 6: 1484–1488.

Ishii, O., M. Shin, T. Sueda, and J. P. Vacanti. 2005. *In vitro* tissue engineering of a cardiac graft using a degradable scaffold with an extracellular matrix-like topography. *Journal of Thoracic and Cardiovascular Surgery* 130: 1358–1363.

Kai, D., M. P. Prabhakaran, G. Jin, and S. Ramakrishna. 2011. Guided orientation of cardiomyocytes on electrospun aligned nanofibers for cardiac tissue engineering. *Journal of Biomedical Materials Research B: Applied Biomaterials* 98B: 379–386.

Kehat, I., D. Kenyagin-Karsenti, M. Snir et al. 2001. Human embryonic stem cells can differentiate into myocytes with structural and functional properties of cardiomyocytes. *Journal of Clinical Investigation* 108: 407–414.

Kehat, I., L. Khimovich, O. Caspi et al. 2004. Electromechanical integration of cardiomyocytes derived from human embryonic stem cells. *Nature Biotechnology* 22: 1282–1289.

Kierszenbaum, A. L. 2002. *Histology and Cell Biology*; New York: Elsevier Health Sciences, p. 648.

Kim, G. M., A. Wutzler, and H. J. Radusch. 2005. One-dimensional arrangement of gold nanoparticles by electrospinning. *Chemistry of Materials* 17: 4949–4957.

Kim, D., P. Kim, I. Song et al. 2006. Guided three-dimensional growth of functional cardiomyocytes on polyethylene glycol nanostructures. *Langmuir* 22: 5419–5426.

Kim, D. H., E. A. Lipke, P. Kim et al. 2010. Nanoscale cues regulate the structure and function of macroscopic cardiac tissue constructs. *Proceedings of the National Academy of Sciences of the United States of America* 107: 565–570.

Klug, M. G., M. H. Soonpaa, G. Y. Koh, and L. J. Field. 1996. Genetically selected cardiomyocytes from differentiating embryonic stem cells form stable intracardiac grafts. *Journal of Clinical Investigation* 98: 216–224.

Kofidis, T., J. L. de Bruin, and G. Hoyt. 2004. Injectable bioartificial myocardial tissue for large-scale intramural cell transfer and functional recovery of injured heart muscle. *Journal of Thoracic and Cardiovascular Surgery* 128: 571–578.

LaNasa, S. M., and S. J. Bryant. 2009. Influence of ECM proteins and their analogs on cells cultured on 2-D hydrogels for cardiac muscle tissue engineering. *Acta Biomaterialia* 5: 2929–2938.

Li, R. K., Z. Q. Jia, R. D. Weisel et al. 1999. Survival and function of bioengineered cardiac grafts. *Circulation* 100: 1163–1169.

Makino, S., K. Fukuda, S. Miyoshi et al. 1999. Cardiomyocytes can be generated from marrow stromal cells *in vitro*. *Journal of Clinical Investigation* 103: 697–705.

Maltsev, V. A., J. Rohwedel, J. Hescheler, and A. M. Wobus. 1993. Embryonic stem cells differentiate *in vitro* into cardiomyocytes representing sinusnodal, atrial and ventricular cell types. *Mechanisms of Development* 44: 41–50.

Mangi, A. A., N. Noiseux, D. Kong et al. 2003. Mesenchymal stem cells modified with Akt prevent remodeling and restore performance of infarcted hearts. *Nature Medicine* 9: 1195–1201.

Mauritz, C., K. Schwanke, M. Reppel et al. 2008. Generation of functional murine cardiac myocytes from induced pluripotent stem cells. *Circulation* 118: 507–517.

McDevitt, T. C., K. A. Woodhouse, S. D. Hauschka, C. E. Murry, and P. S. Stayton. 2003. Spatially organized layers of cardiomyocytes on biodegradable polyurethane films for myocardial repair. *Journal of Biomedical Materials Research Part A* 66: 586–595.

McDevitt, T. C., M. A. Laflamme, and C. E. Murry. 2005. Proliferation of cardiomyocytes derived from human embryonic stem cells is mediated via the IGF/PI 3-kinase/Akt signaling pathway. *Journal of Molecular and Cellular Cardiology* 39: 865–873.

McMullen, N. M., and K. B. Pasumarthi. 2007. Donor cell transplantation for myocardial disease: Does it complement current pharmacological therapies? *Canadian Journal of Physiology and Pharmacology* 85: 1–15.

Morton, W. US Patent No. 705,691, 1902.

Motlagh, D., T. J. Hartman, T. A. Desai, and B. Russell. 2003. Microfabricated grooves recapitulate neonatal myocyte connexin 43 and N-cadherin expression and localization. *Journal of Biomedical Materials Research—Part A* 67: 148–157.

Mukherjee, S., J. R. Venugopal, and R. Ravichandran. 2011. Evaluation of the biocompatibility of PLACL/collagen nanostructured matrices with cardiomyocytes as a model for the regeneration of infarcted myocardium. *Advanced Functional Materials* 21: 2291–2300.

Mummery, C., D. Ward, C. E. van den Brink et al. 2002. Cardiomyocyte differentiation of mouse and human embryonic stem cells. *Journal of Anatomy* 200: 233–242.

Murry, C. E., L. J. Field, and P. Menasche. 2005. Cell-based cardiac repair: Reflections at the 10-year point. *Circulation* 112: 3174–3183.

Nakagawa, M., M. Koyanagi, K. Tanabe et al. 2008. Generation of induced pluripotent stem cells without Myc from mouse and human fibroblasts. *Nature Biotechnology* 26: 101–106.

Narazaki, G., H. Uosaki, M. Teranishi et al. 2008. Directed and systematic differentiation of cardiovascular cells from mouse induced pluripotent stem cells. *Circulation* 118: 498–506.

Narmoneva, D. A., O., Oni, A. L. Sieminski et al. 2005. Self-assembling short oligopeptides and the promotion of angiogenesis. *Biomaterials* 26: 4837–4846.

Nelson, T. J., A. Martinez-Fernandez, S. Yamada et al. 2009. Repair of acute myocardial infarction by human stemness factors induced pluripotent stem cells. *Circulation* 120: 408–416.

Nussbaum, J., E. Minami, M. A. Laflamme et al. 2007. Transplantation of undifferentiated murine embryonic stem cells in the heart: Teratoma formation and immune response. *FASEB Journal* 21: 1345–1357.

Orlic, D., J. Kajstura, S. Chimenti et al. 2001. Bone marrow cells regenerate infarcted myocardium. *Nature* 410: 701–705.

Pasumarthi, K. B., and L. J. Field. 2002. Cardiomyocyte cell cycle regulation. *Circulation Research* 90: 1044–1054.

Pego, A. P., B. Siebum, M. J. Van Luyn et al. 2003. Preparation of degradable porous structures based on 1,3-trimethylene carbonate and D,L-lactide (co)polymers for heart tissue engineering. *Tissue Engineering* 9: 981–994.

Pham, Q. P., U. Sharma, and A. G. Mikos. 2006. Electrospinning of polymeric nanofibers for tissue engineering applications: A review. *Tissue Engineering* 12: 1197–1211.

Pittenger, M. F., A. M. Mackay, S. C. Beck et al. 1999. Multilineage potential of adult human mesenchymal stem cells. *Science* 284: 143–147.

Planat-Benard, V., C. Menard, M. Andre et al. 2004. Spontaneous cardiomyocytes differentiation from adipose tissue stroma cells. *Circulation Research* 94: 223–229.

Radisic, M., H. Park, H. Shing et al. 2004. Functional assembly of engineered myocardium by electrical stimulation of cardiacmyocytes cultured on scaffolds. *Proceedings of the National Academy of Sciences of the United States of America* 101: 18129–18134.

Radisic, M., W. Deen, R. Langer et al. 2005. Mathematical model of oxygen distribution in engineered cardiac tissue with parallel channel array perfused with culture medium containing oxygen carriers. *American Journal of Physiology. Heart and Circulatory Physiology* 288: H1278–H1289.

Radisic, M., H. Park, T. P. Martens et al. 2008. Pretreatment of synthetic elastomeric scaffolds by cardiac fibroblasts improves engineered heart tissue. *Journal of Biomedical Materials Research. Part A* 86: 713–724.

Rangappa, S., C. Fen, E. H. Lee et al. 2003. Transformation of adult mesenchymal stem cells isolated from the fatty tissue into cardiomyocytes. *Annals of Thoracic Surgery* 75: 775–779.

Rosamond, W., K. Flegal, G. Friday et al. 2007. Heart disease and stroke statistics—2007 update: A report from the American Heart Association Statistics Committee and Stroke Statistics Subcommittee. *Circulation* 115: e69–e171.

Schenke-Layland, K., K. E. Rhodes, E. Angelis et al. 2008. Reprogrammed mouse fibroblasts differentiate into cells of the cardiovascular and hematopoietic lineages. *Stem Cells* 26: 1537–1546.

Schwarz, F., G. Schuler, H. Katus et al. 1982. Intracoronary thrombolysis in acute myocardial infarction: Duration of ischemia as a major determinant of late results after recanalization. *American Journal of Cardiology* 50: 933–937.

Severs, N. J. 2000. The cardiac muscle cell. *BioEssays* 22: 188–199.

Shin, M., O. Ishii, T. Sueda et al. 2004. Contractile cardiac grafts using a novel nanofibrous mesh. *Biomaterials* 25: 3717–3723.

Shiota, M., T. Heike, M. Haruyama et al. 2007. Isolation and characterization of bone marrow-derived mesenchymal progenitor cells with myogenic and neuronal properties. *Experimental Cell Research* 313: 1008–23.

Snir, M., I. Kehat, A. Gepstein et al. 2003. Assessment of the ultrastructural and proliferative properties of human embryonic stem cell-derived cardiomyocytes. *American Journal of Physiology. Heart and Circulatory Physiology* 285: H2355–H2363.

Stout, D. A., B. Basu, and T. J. Webster. 2011. Poly(lactic-co-glycolic acid): Carbon nanofiber composites for myocardial tissue engineering applications. *Acta Biomaterialia* 7: 3101–3112.

Takahashi, K., and S. Yamanaka. 2006. Induction of pluripotent stem cells from mouse embryonic and adult fibroblast cultures by defined factors. *Cell* 126: 663–676.

Takahashi, K., K. Tanabe, M. Ohnuki et al. 2007. Induction of pluripotent stem cells from adult human fibroblasts by defined factors. *Cell* 131: 861–872.

Taylor, D. A., and A. G. Zenovich. 2008. Cardiovascular cell therapy and endogenous repair. *Diabetes, Obesity and Metabolism* 10: 5–15.

Taylor, D. A., B. Z. Atkins, P. Hungspreugs et al. 1998. Regenerating functional myocardium: Improved performance after skeletal myoblast transplantation. *Nature Medicine* 4: 929–933.

Thierry, B., F. M. Winnik, Y. Merhi et al. 2003. Nanocoatings onto arteries via layer-by-layer deposition: Toward the *in vivo* repair of damaged blood vessels. *Journal of the American Chemical Society* 125: 7494–7495.

Thomson, J. A., J. Itskovitz-Eldor, S. S. Shapiro et al. 1998. Embryonic stem cell lines derived from human blastocysts. *Science* 282: 1145–1147.

Tillmanns, J., M. Rota, T. Hosoda et al. 2008. Formation of large coronary arteries by cardiac progenitor cells. *Proceedings of the National Academy of Sciences of the United States of America* 105: 1668–1673.

Tomita, S., R. K. Li, R. D. Weisel et al. 1999. Autologous transplantation of bone marrow cells improves damaged heart function. *Circulation* 100: 247–256.

Tomita, S., D. A. G. Mickle, R. D. Weisel et al. 2002. Improved heart function with myogenesis and angiogenesis after autologous porcine bone marrow stromal cell transplantation. *Journal of Thoracic and Cardiovascular Surgery* 123: 1132–1140.

van Dijk, A., HW, Niessen B. Zandieh-Doulabi, F. C. Visser, and F. J. van Milligen. 2008. Differentiation of human adipose-derived stem cells towards cardiomyocytes is facilitated by laminin. *Cell and Tissue Research* 334: 457–467.

Xiang, Z., R. Liao, M. S. Kelly, and M. Spector. 2006. Collagen-GAG scaffolds grafted onto myocardial infarcts in a rat model: A delivery vehicle for mesenchymal stem cells. *Tissue Engineering* 12: 2467–2478.

Yang, L., M. H. Soonpaa, E. D. Adler et al. 2008. Human cardiovascular progenitor cells develop from a KDR+ embryonic-stem-cell-derived population. *Nature* 453: 524–528.

Yin, L., H. Bien, E. Entcheva et al. 2004. Scaffold topography alters intracellular calcium dynamics in cultured cardiomyocyte networks. *American Journal of Physiology. Heart and Circulatory Physiology* 287: H1276–H1285.

Yu, J., M. A. Vodyanik, K. Smuga-Otto et al. 2007. Induced pluripotent stem cell lines derived from human somatic cells. *Science* 318: 1917–1920.

Zhang, S. 2003. Fabrication of novel biomaterials through molecular self-assembly. *Nature Biotechnology* 21: 1171–1178.

Zhang, G., X. Wang, Z. Wang et al. 2006. A PEGylated fibrin patch for mesenchymal stem cell delivery. *Tissue Engineering* 12: 9–19.

Zhang, G., Q. Hu, E. A. Braunlin et al. 2008. Enhancing efficacy of stem cell transplantation to the heart with a PEGylated fibrin biomatrix. *Tissue Engineering—Part A* 14: 1025–1036.

Zimmermann, W. H., I. Melnychenko, G. Wasmeier et al. 2006. Engineered heart tissue grafts improve systolic and diastolic function in infarct rat hearts. *Nature Medicine* 12: 452–458.

Zong, X., H. Bien, C. Y. Chung et al. 2005. Electrospun fine-textured scaffolds for heart tissue constructs. *Biomaterials* 26: 5330–5338.

20 Nanotechnology for Engineering Cellular Microenvironment and Gene Delivery

Janice H. Lai, Anusuya Ramasubramanian, Shaheen Jeeawoody, and Fan Yang

CONTENTS

20.1 INTRODUCTION

Tissue engineering aims at repairing and restoring lost tissue structure and functions caused by disease processes or traumatic events (Langer and Vacanti 1993). Driven by clinical demands, extensive research efforts have been dedicated toward regenerating a broad range of tissues, from the skin, cartilage, bone, and blood vessels to more complex organ-level systems such as the bladder. Most tissue engineering strategies involve using one or a combination of three key components: (1) cells; (2) scaffolds; and (3) biological signals. Cells are the building blocks of biological tissues and cell-based therapy involves delivering cells directly into the body to help regenerate the lost tissue. For instance, autologous chondrocytes transplantation is a cell-based therapy for cartilage repair that involves harvesting healthy cartilage biopsy and then transplanting expanded cartilage cells back to a defected cartilage site in the same patient (Brittberg et al. 1994). Bone marrow transplantation is another example of cell-based therapy and has been widely used for a few decades to treat diseases such as leukemia. In addition to fully differentiated cell types, stem cells have gained tremendous attention as promising cell sources for tissue regeneration because of their ability to self-renew and differentiate into multiple types of cells in our body.

The success of tissue engineering is dependent on the ability to promote the desired cellular processes. Before stem cells can be broadly used for clinical applications, methods must be developed to control their lineage-specific differentiation with functional stability. Scaffolds and biological signals can be used collectively with cells to promote tissue repair and regeneration. Given the intricate processes involved in tissue development and regeneration, it is crucial to understand how microenvironmental cues regulate cell behavior. Such knowledge can then guide the rational design of tissue engineering strategies to promote the desired cellular processes and tissue formation.

20.1.1 The Multifactorial Cellular Microenvironment

In the natural extracellular matrix (ECM), cell behavior is regulated by a complex network of microenvironmental cues including biochemical and biophysical signals. Biochemical signals can arise from cell–cell interactions, soluble signaling, or insoluble ECM. Such biochemical signals play an important role in influencing cellular processes such as adhesion, migration, proliferation, and differentiation. Tissue development and remodeling are also tightly regulated by these signals. For example, in wound healing, the ECM orchestrates a cascade of biological events such as cell migration and proliferation, matrix synthesis, and angiogenesis. The ECM also serves as a reservoir for many growth factors, such as vascular endothelial growth factor (VEGF), basic fibroblast growth factors, and transforming growth factor β (TGF-β), sequestering and protecting these growth factors from degradation (Schultz and Wysocki 2009). After tissue injury, the ECM modulates the release of these growth factors through proteolytic mechanisms and facilitates the wound healing process (Ferrara 2010).

Biophysical signals can regulate cell behavior via either extrinsic mechanical forces or intrinsic matrix stiffness. Fluid shear stresses in blood vessels can directly influence endothelial cell gene expression and biosynthetic activities, which, in turn, regulate blood vessel remodeling processes. Fluid shear and hydrostatic compression in bone tissue facilitate bone cell mechanical adaptation and tissue remodeling (Papachroni et al. 2009). In addition to extrinsic mechanical forces, intrinsic matrix stiffness has also been shown to play a critical role in regulating cell fate. Mesenchymal stem cells (MSC) grown on two-dimensional substrates with various elasticities demonstrated preferential differentiation toward tissue lineages with similar elasticities (Engler et al. 2006). Similar cellular behavior in response to matrix stiffness has been recently reported in three-dimensions as well (Banerjee et al. 2009; Huebsch et al. 2010; Pek et al. 2010).

Aside from macroscopic biophysical cues, microstructures and nanostructures of the ECM, such as surface topography and organization of different ECM components, also regulate cell behavior. Microscale and nanoscale topological features, such as grooves, ridges, pores, and pits, are present in many tissues. For example, collagen fibers, an abundant component in connective tissue, possess hierarchical features on the microscale and nanoscale level (Buehler 2006). In the tendon, 20–280 nm diameter collagen fibrils are organized into fiber bundles of 1–300 μm diameter. These collagen fiber bundles are further aligned and organized into larger structural units, fassicles, to provide maximum tensile strength (Silver et al. 2003). Bone is a composite material that possesses hierarchical structures at the macroscale (cortical and cancellous bone), microscale (osteons and lamallae), and nanoscale (collagen fibrils, collagen molecules, and bone crystals) level (Rho et al. 1998). Not only do these multiscale hierarchical features confer mechanical and structural integrity to connective tissues, but they also interact with cells and bind to other ECM components (Kolácná et al. 2007).

Advances in nanotechnology facilitate better understanding of cell–matrix interactions at the nanoscale and inspire new strategies to control cell behavior. Numerous studies have demonstrated the effects of nanotopographical cues on cellular processes. When cultured on nanopatterned surfaces, both the cytoskeleton and nuclei of smooth muscle cells aligned to nanoscale gratings; proliferation was reduced and polarization pattern was altered in wound healing assays (Yim et al. 2005). Lamers and coworkers showed that the adhesion, morphology, and motility of osteoblasts are regulated by nanotopographical cues. Specifically, surface anisotropy regulated osteoblast morphology whereas spacing (ridge-to-groove ratio) of nanotopographical cues influenced osteoblast motility (Lamers et al. 2010). In addition to cell adhesion, morphology, motility, and proliferation, nanotopographical cues may also regulate the differentiation of stem cells. By controlling scaffold nanotopographical features, MSCs have been shown to undergo osteogenesis without osteogenic supplements (Dalby et al. 2007; Oh et al. 2009). Dalby et al. (2007) showed that a semidisorder array of nanopits of 120 nm diameter and 100 nm deep induced MSC osteogenesis, with enhanced osteoblastic gene expression and mineral deposition. In another study, human MSCs cultured on vertically oriented nanotubes of 70–100 nm diameter, but not less than 30 nm, differentiated into osteoblasts even in the absence of osteogenic supplement (Oh et al. 2009). All these findings highlighted the importance of nanostructures on influencing cell behavior. As such, designing scaffolds with optimized nanoscale features may provide a powerful tool to direct cell fate for tissue engineering applications.

20.1.2 Designing Scaffolds with Nanoscale Features

Scaffold design is an important tissue engineering strategy to control cellular microenvironment and promote tissue regeneration. Important criteria to consider for designing scaffolds include mechanical and physical properties, biocompatibility, and biodegradability. In contrast to delivering cells alone, scaffolds provide a three-dimensional environment in which the cells can proliferate, migrate, and differentiate. Scaffolds also allow the incorporation of biochemical or physical cues that may promote the desired cellular processes and tissue formation (Lutolf and Hubbell 2005; Shin et al. 2003). Cell adhesion peptides conjugated to scaffolds can enhance cell attachment in three-dimensions, and incorporating protease-sensitive peptide into hydrogel network allows cell-mediated matrix remodeling. Biological signals, such as growth factors, can also be conjugated to the scaffold and released through cell-mediated mechanisms.

Recapitulating microenvironmental cues at the nanometer scale has gained increasing attention as a way to better mimic cell–matrix interactions. Designing scaffolds with more refined control of scaffold functionalities in both a spatial and temporal fashion would provide a powerful tool to guide the desired tissue development processes. Different types of nanomaterials have been reported for a wide range of tissue engineering-related applications (Figure 20.1). Natural biomaterials and synthetic polymers can form nanofibers through electrospinning and can be used as scaffolds for tissue engineering applications for bone, skin, and blood vessels (Kumbar et al. 2008; Ma et al. 2005; Matthews et al. 2002). For example, electrospun poly(ε-caprolactone) nanofiber scaffolds have been shown to support the osteogenesis of MSCs with increased mineralization and type I collagen production (Yoshimoto et al. 2003). Nanofibrous scaffolds based on collagen I with a coating of ECM molecules facilitated cell adhesion of human keratinocytes and accelerated early-stage wound healing in a rat model (Rho et al. 2006). Aside from forming nanofibers via electrospinning, amphiphilic peptides and polymeric dendrimers can self-assemble into nanofibers (Figure 20.2; Ma et al. 2005). These nanofibers have

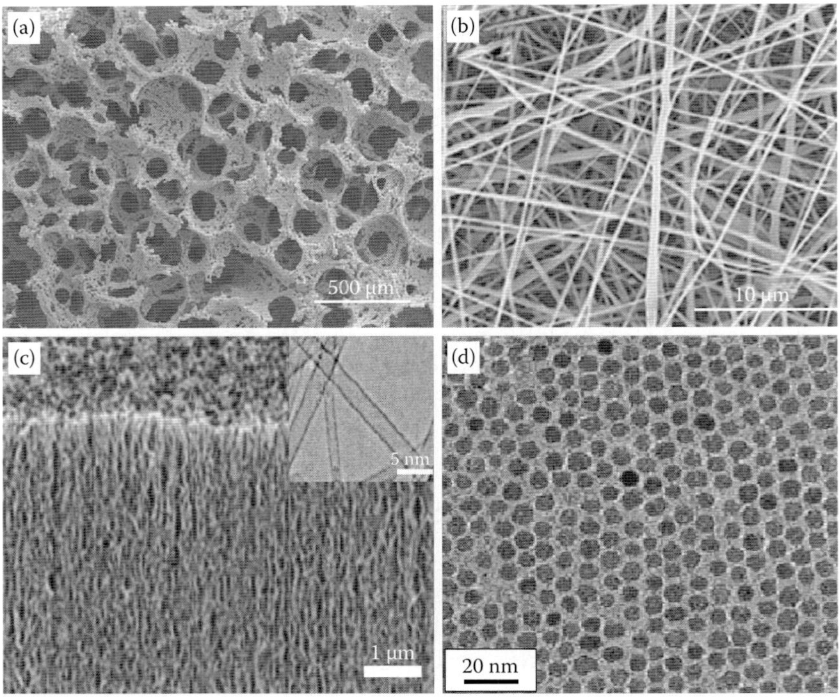

FIGURE 20.1 Examples of nanomaterials used in tissue engineering. (a) Poly(L-lactic acid) nanofibrous scaffold with interconnected spherical macropores. (b) Electrospun composite nanofibrous scaffold consists of synthetic biodegradable poly(ε-caprolactone), hydroxyapatite, and natural polymer gelatin (c) Vertically aligned, single-walled carbon nanotube forest. (d) Fe_3O_4 nanoparticles. (Reprinted from Zhang L. and Webster T. J. *Nano Today* 4(1), 2009. With permission.)

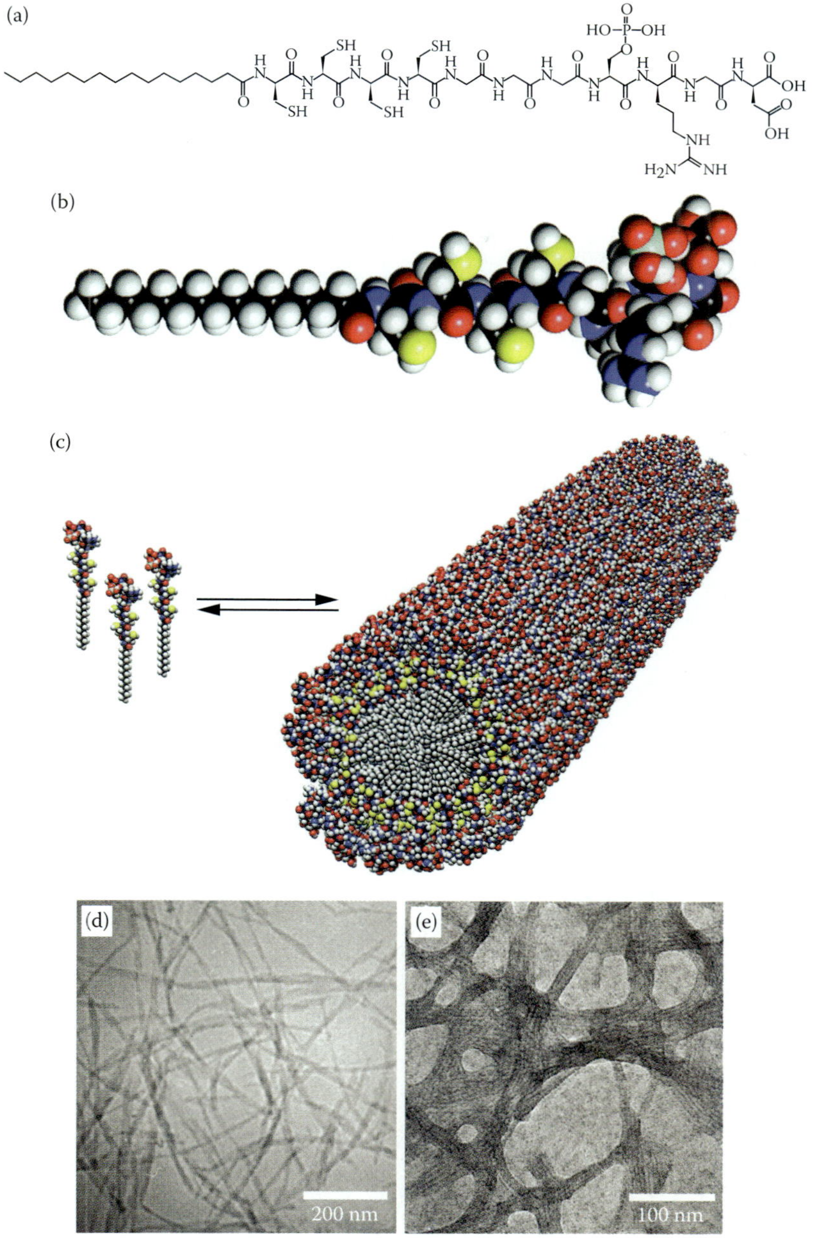

FIGURE 20.2 Self-assembled nanofiber. (a) Chemical structure and (b) molecular model of peptide amphiphile highlighting the hydrophobic tail and peptide region functionalized with cell-adhesive ligand RGD. (c) Schematic of the self-assembly process of peptide amphiphile molecules into a nanofiber. (d) Transmission electron microscopic image of self-assembled nanofibers with 7.6 ± 1 nm diameter. (e) Self-assembled nanofibers after oxidative cross-linking. (Adapted from Hartgerink J. D. et al. *Science* 294(5547), 1684–1688, 2001. With permission.)

been used as scaffolds to support the proliferation and differentiation of many cell types such as neurons, chondrocytes, and MSCs (Holmes et al. 2000; Kisiday et al. 2002; Hosseinkhani et al. 2006). Biomimetic signals such as adhesive ligands and growth factor binding domains can be incorporated into the supramolecular and molecular design of these self-assembled polymers. For instance, amphiphilic

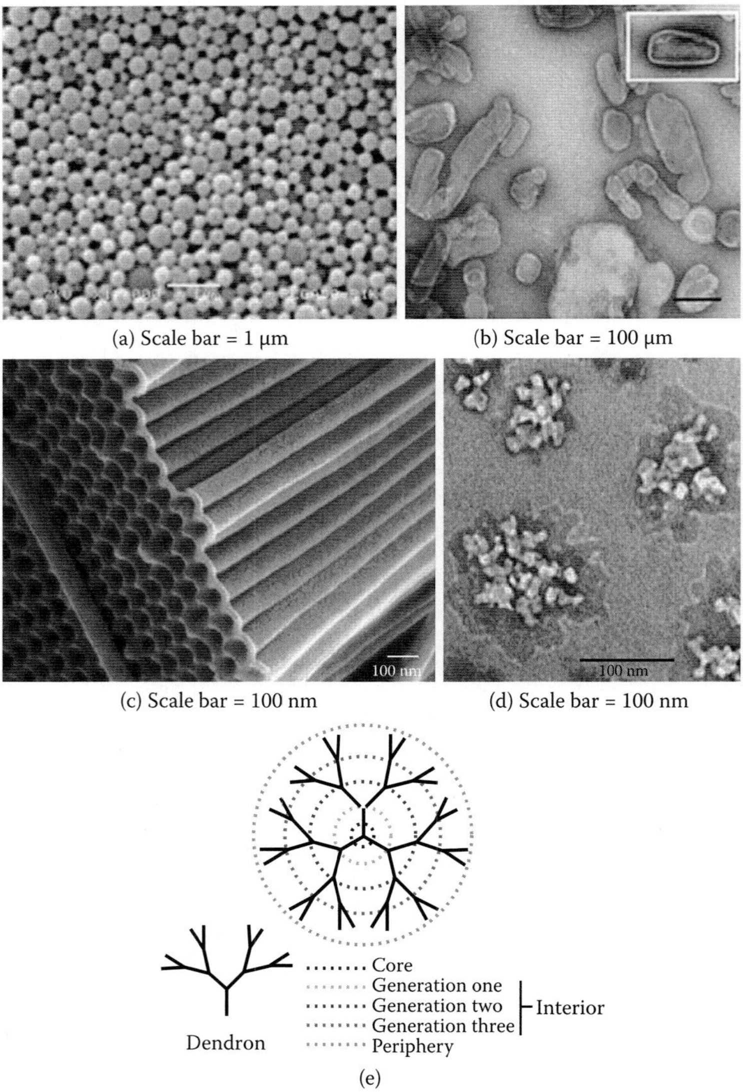

FIGURE 20.3 Examples of nanomaterials used for drug-delivery. (a) Nanoparticles. (b) Nanocapsules. (c) Nanotubes. (d) Nanogels. (e) Dendrimers. (Reprinted from Goldberg M. et al. *J. Biomater. Sci. Polym. Ed.* 18(3), 241–268, 2007. With permission.)

peptides with TGF-binding domains were shown to promote chondrogenic differentiation of human MSCs and enhance cartilage tissue regeneration in a full-thickness chondral defect rabbit model (Shah et al. 2010).

20.1.3 Biological Signals to Enhance Tissue Regeneration

Biological signals, such as growth factors and nucleic acids, are often used to promote specific cellular processes and tissue regeneration. Scaffolds can serve as a

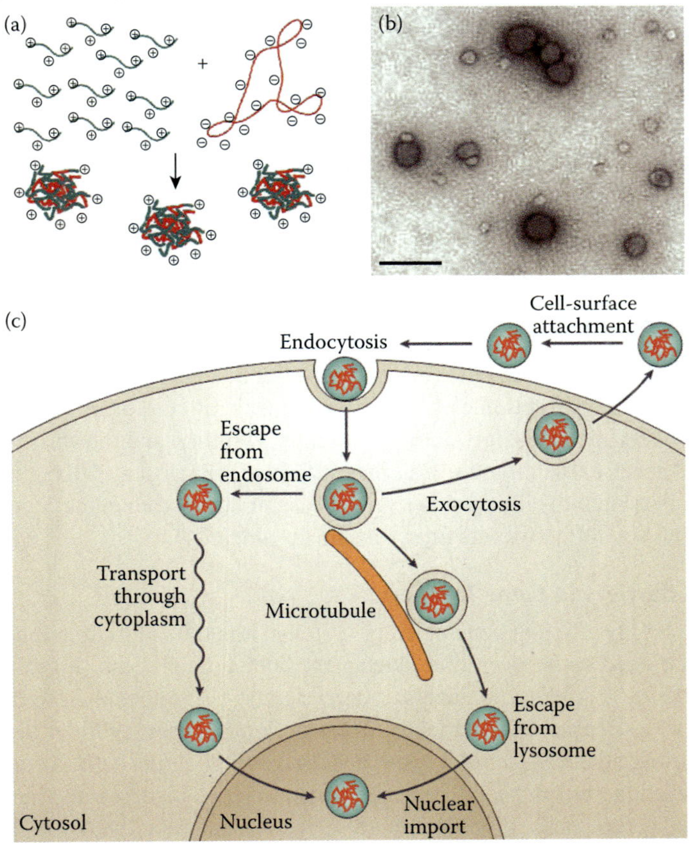

FIGURE 20.4 Gene delivery with cationic material. (a) Schematic of spontaneous polyplex formation by the electrostatic interaction between polycations (e.g., cationic lipids and polymers) and DNA. (b) Transmission electron microscopy (TEM) image of nanoparticles formed by plasmid DNA and PEI (scale bar = 200 nm). (c) Intracellular barriers to gene delivery. After attaching to the targeted cell surface, the polyplex must be internalized (e.g., by endocytosis), escape from the endosome, transport through the cytoplasm, move toward the nucleus and across the nuclear membrane, and unpackage. (Adapted from Pack D. W. et al. *Nat. Rev. Drug Discov.* 4(7), 581–593, 2005. With permission.)

delivery depot for bioactive molecules via various methods such as surface adsorption or covalent modification (Babensee et al. 2000; Tessmar and Göpferich 2007). Nanomaterials such as nanoparticles, nanocapsules, nanotubes, nanogels, and dendrimers have also been explored as delivery vehicles to enhance the controlled release and cellular uptake for various types of biomolecules (Figure 20.3; Goldberg et al. 2007). Advances in gene therapy provide a powerful tool to promote lineage-specific differentiation via directly regulating the intrinsic signals of stem cells. Today, technology is being developed with the potential to either "turn on" a target gene, through DNA delivery, or "turn off" a gene by small interfering RNA (siRNA) delivery. Current challenges in gene therapy include the need to identify optimal therapeutic gene targets as well as the lack of safe and efficient delivery methods. Cationic polymers can self-assemble with DNA to form nanoparticles via electrostatic interactions (Figure 20.4a and b), and different polymer designs have been used to overcome the hurdles involved in different steps during gene delivery. In this chapter, we will discuss recent advances in applying gene delivery as a tool to direct stem cell fate, with special focus on polymer-based gene delivery strategies using a nanomaterials approach.

20.2 GENE DELIVERY

20.2.1 Considerations for Effective Gene Delivery

Successful gene therapy requires a safe and effective delivery system to transport the DNA to the target cell nuclei. The basic considerations for developing carriers for gene delivery include safety, biocompatibility, stability, and the ability to transfer therapeutic genes to the target cells at high efficiency (Pack et al. 2005). To facilitate smooth translation into clinical practice, the ease of large-scale manufacture, administration, and cost-effectiveness must also be considered.

20.2.1.1 Barriers to Gene Delivery

To develop highly efficient gene delivery systems, it is crucial to first understand the barriers that need to be overcome during the transport process (Figure 20.4c). In transit to the target cells, the genetic material must be protected from degradation and preserve its properties for effective transfection (Papisov 1998). It also needs to navigate through the extracellular space and arrive at the target cells. To achieve cell-specific targeting, receptor-mediated mechanisms can be incorporated into the delivery vehicle (Ferkol et al. 1996; Wu et al. 1991; Wu and Wu 1988). Cellular uptake usually occurs via endocytosis, which leads the carrier/gene complexes into an acidic intracellular vesicle called an endosome. The genetic material needs to escape from the endosome, continue to transport through the cytoplasm, and finally transport into the nucleus and get expressed. During this multistep transport process, the genetic material continues to encounter barriers such as the possibility of exocytosis, degradation in the endosome, or the presence of cytosolic nucleases (Pollard et al. 2001). Finally, in the cases in which a delivery system, such as a synthetic polymer, is used, the delivery vehicle must release the genetic material and permit transcription once the genetic material–carrier complex is delivered to the nucleus.

20.2.2 Methods for Gene Delivery

Gene delivery methods can be broadly divided into two categories, viral and nonviral approaches. The viral approach uses viruses as the delivery vehicle and takes advantage of the ability of viruses to efficiently transfer genetic information into cells. Despite its high gene transfer efficiency, clinical translation of the viral-based approach is limited by safety concerns such as pathogenesis and potential immunogenicity (Pack et al. 2005). Nonviral methods include the direct injection of naked genetic material, physical methods, and delivery with a gene transfer carrier such as synthetic polymers and lipids. Nonviral methods are potentially much safer, but often suffer from significantly lower transfection efficiency.

20.2.2.1 Viral Methods

In viral methods, a replication-deficient viral vector can be formed by replacing the coding region of the viral genome with a therapeutic gene, turning the virus into a gene delivery vehicle. Viral vectors can generally be classified as integrating or nonintegrating vectors. Retrovirus and adeno-associated viruses (AAV) can integrate their genome into the DNA of the host cell. They can achieve stable expressions in dividing cells, and are suitable for the treatment of chronic diseases, in which long-term gene expression is needed (Pack et al. 2005). Nonintegrating viruses, such as adenovirus and herpes simplex virus type I, transfer their genomes into the nucleus of the target cell as episomes without integration.

Retroviruses are enveloped RNA viruses that are present in all vertebrates. The three main classes of retroviruses are oncoretrovirus, lentivirus, and spumavirus. Oncoretrovirus was the first viral vector developed and has been widely used in clinical trials to treat diseases such as severe combined immunodeficiency (Heilbronn and Weger 2010; Thomas et al. 2003). A major limitation of using oncoretrovirus for gene delivery is that the virus can only be used to transduce dividing cells. Lentivirus, on the other hand, can enter the nucleus of both dividing and nondividing cells. Lentivirus has been shown to successfully deliver short hairpin RNAs into a wide range of dividing and nondividing cell types, including primary T cells, stem cells, and single-cell embryos, and achieve stable gene silencing (Rubinson et al. 2003). Lentiviral vectors have also been used to reprogram human somatic cells into pluripotent stem cells (Yu et al. 2007). The AAV is another commonly used vector for gene delivery due to their nonpathogenic nature and inability to self-replicate. The major limitation of AAV vector is that it can only accommodate small gene products of up to 5 kb (Kootstra and Verma 2003).

Unlike integrating viral vectors, the genome of nonintegrating vectors remains episomal, eliminating the possibility of insertional mutagenesis. The capacity to carry large genetic materials, the ability to transfect nondividing cells, and high gene transfer efficiency in a variety of tissues make adenovirus vectors and herpes simplex virus type I promising gene delivery vehicles (Kay et al. 2001; Thomas et al. 2003). However, high immunogenicity and transient transgene expression limit the clinical potential of these nonintegrating viral vectors as gene delivery vehicles.

Naturally evolved as effective delivery vehicles, viruses have been shown to deliver therapeutic genes at high efficiency in many *in vitro* and *in vivo* studies (Goldberg et al. 2008; Pack et al. 2005; Thomas et al. 2003). Although more than

70% of gene therapy-based clinical trials (as of July 2007) used virus-mediated gene delivery, there is still no FDA-approved viral vector-based gene therapy on the market (Check 2005; Green et al. 2008; Hollon 2000). Acute toxicity, oncogenesis, mutagenesis, and carcinogenicity are among the safety risks associated with viral gene therapy (Goldberg et al. 2007; Green et al. 2008; Pack et al. 2005; Thomas et al. 2003). High manufacturing costs, low cargo capacity, low quality control, and resistance to repeated infection are some of the additional characteristics of viral gene therapy that makes it an undesirable gene delivery method.

20.2.2.2 Nonviral Approaches

20.2.2.2.1 Physical Methods

The ability of DNA alone to transfect cells is very poor due to low cellular uptake and its susceptibility to nuclease degradation (Herweijer and Wolff 2003). Several physical methods have been developed to enhance DNA delivery efficiency including electroporation, pressurized intravascular delivery, sonoporation, laser irradiation, and magnetofection (Mehier-Humbert and Guy 2005). Electroporation increases the permeability of cell membranes to plasmid DNA by exposing the target cells to a series of electrical pulses. Pressurized intravascular delivery has been used to successfully transfect cells in a variety of tissue types including liver and skeletal muscle. Sonoporation, which usually involves the use of a low-dose ultrasound, is another physical method to enhance gene transfer efficiency. The application of ultrasound or laser irradiation leads to transient formation of small pores on the cell membrane and enhances permeability (Taniyama et al. 2002). Magnetofection involves the application of a magnetic field to enhance the uptake of plasmid DNA coupled with magnetic nanoparticles (Scherer et al. 2002). Most of these physical methods enhance the entry of DNA into cells by overcoming barriers posed by the cell membrane, but are often associated with significant cytotoxicity. Furthermore, the challenges associated with intracellular transport to the nucleus remain to be addressed (Geng et al. 2011; Guo and Huang 2011; Rychahou and Evers 2010).

20.2.2.2.2 Cationic Biomaterials for Nonviral Gene Delivery

Biomaterials, such as cationic polymers and lipids, are capable of condensing DNA into nanoparticles for enhanced delivery into target cells (Nguyen et al. 2009). Because of their lipophilic nature, cationic lipids can penetrate cell membranes and deliver DNA into cells, and have been widely used for nonviral-based gene delivery *in vitro*. However, clinical applications of cationic lipids for gene therapy are impeded by their unfavorable biodistribution, as they tend to accumulate in the lung and liver when delivered *in vivo*. Furthermore, they may cause high levels of cytotoxicity by disrupting the cell membrane.

Cationic polymers are promising gene delivery carriers due to their relative ease of manufacture and enhanced safety compared with viruses (Goldberg et al. 2008). However, polymer-mediated gene delivery is typically much less efficient than viral vectors due to the multistep barriers they need to overcome to achieve efficient expression (Green et al. 2008; Nguyen et al. 2009). The polymer must initially bind to, condense, or encapsulate the DNA in the form of a nanoparticle to protect it from degradation. Once the nanoparticles get trafficked into the cell via endocytosis, they must be capable of escaping the endosome into the cytoplasm before endosomal degradation. Finally, the nanoparticles need to travel through the nuclear membrane and release the DNA cargo for gene expression. Two cationic polymers that have been widely used for gene delivery are polyethylenimine (PEI) and poly(L-lysine) (PLL; Goldberg et al. 2008; Green et al. 2008). However, PEI is nondegradable and has high cytotoxicity effects and a low transfection efficiency in comparison to viral vectors and high cytotoxicity effects (Green et al. 2008; Moghimi et al. 2005). PLL is biodegradable and can be conjugated with bioactive ligands to enhance target-specific delivery (Luo and Saltzman 2000). However, like PEI, PLL suffers from low transfection efficiency and high cytotoxicity (Brazeau et al. 1998). Because of the complex multistep barriers of the gene delivery process, it is very difficult to develop a polymer with high efficiency using the conventional design-driven, iterative synthesis approach.

20.3 HIGH-THROUGHPUT APPROACH TO DISCOVERING NOVEL MATERIALS FOR GENE DELIVERY

To accelerate the discovery of novel biomaterials for effective gene delivery, high-throughput screening has emerged as a novel approach to help better understand the structure–function relationships that govern polymer-mediated gene delivery. In a high-throughput screening approach, a large library of polymers with diverse chemical structures can be synthesized in a combinatorial approach, and the polymer/DNA nanoparticles can then be screened for their transfection efficiency and cytotoxicity using high-throughput screening assays. The versatility of polymer chemistry, combined with computational tools for predicting structure–function relationships, can be used in combinatorial library synthesis to generate a broad range of cationic polymer vectors with specific properties (Goldberg et al. 2008). Compared with the conventional, iterative approach, such a high-throughput approach greatly accelerated the development of novel polymers with enhanced gene delivery efficiency (Akinc et al. 2003a,b; Anderson et al. 2003; Green et al. 2008; Thomas et al. 2007). The results of such screening studies yield valuable information on structure–function relationships, which can guide the design of the next generation of polymer libraries synthesis for further performance improvement. Poly(β-amino esters) (PBAEs) are an excellent example of biodegradable polymers for gene delivery that have arisen from such combinatorial synthesis and high-throughput screening approaches (Green et al. 2008).

20.3.1 The Development of PBAEs for DNA Delivery

PBAEs constitute a particularly diverse class of cationic polymers for gene delivery. PBAE has several advantages including its biodegradability via hydrolytically degradable ester groups, reduced toxicity, capacity for structural diversity, and ability to trigger endosomal escape (Lynn and Langer 2000). Most importantly, PBAEs can be synthesized via a one-step reaction with conjugate addition between diacrylate and amine monomers, which makes it feasible for facile combinatorial synthesis (Figure 20.5a and b; Green et al. 2008). The first high-throughput effort to probe the

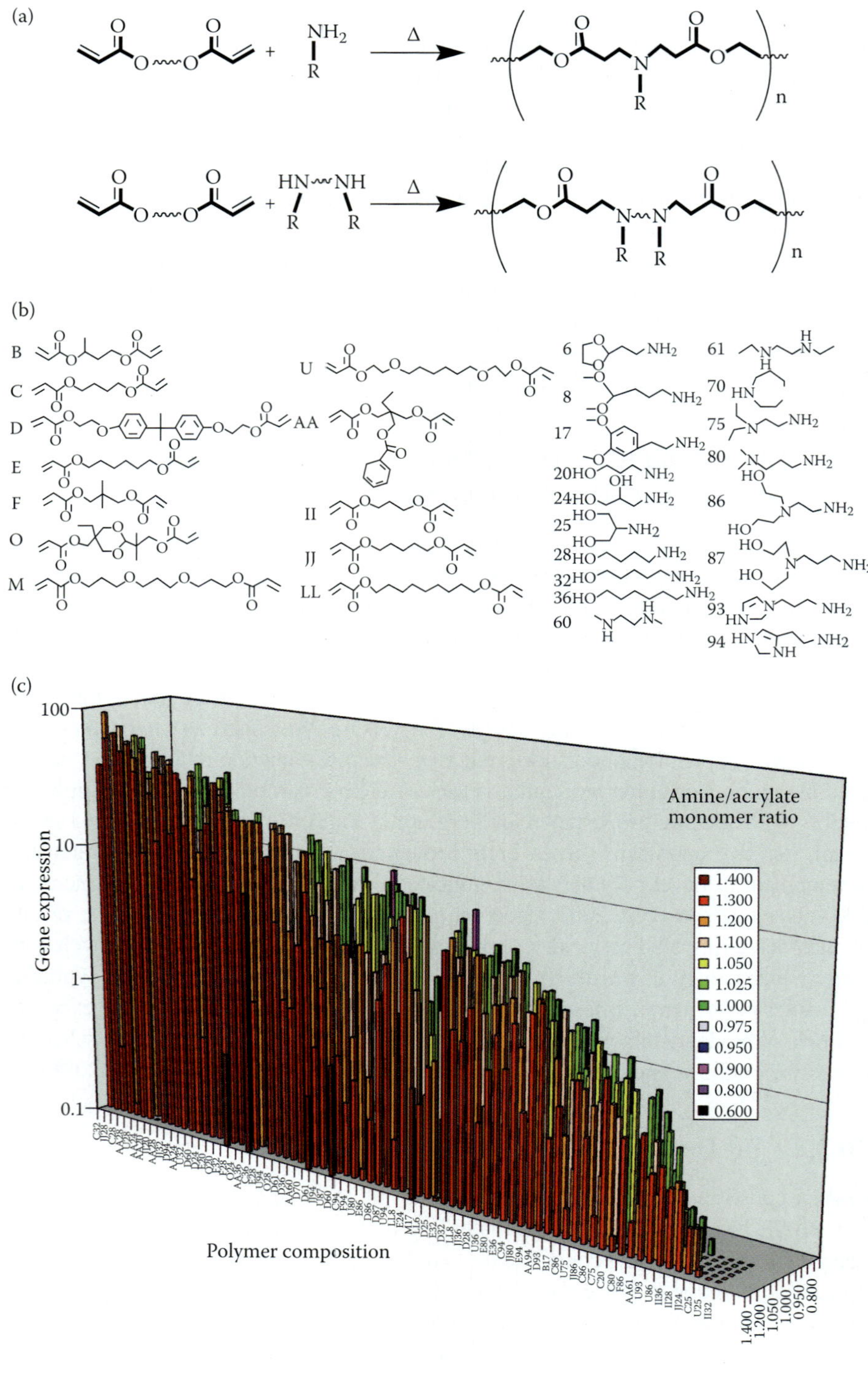
(a)
(b)
(c)
Gene expression
Polymer composition
Amine/acrylate monomer ratio
1.400
1.300
1.200
1.100
1.050
1.025
1.000
0.975
0.950
0.900
0.800
0.600

structure–function relationship of PBAEs synthesized a combinatorial library of 140 PBAEs composed of 20 amine monomers and 7 diacrylate monomers (Akinc et al. 2003a). Results from this screening showed that polymer structures can significantly influence the size and charge of the polymer/DNA nanoparticles, which has a significant impact on transfection efficiency. Lead polymers formed complexes smaller than 250 nm and had a positive ζ-potential, and demonstrated four to eight times higher transfection efficiency than PEI (Akinc et al. 2003b). Two leading PBAEs were further optimized by varying other parameters, including chemical structures of the polymer end groups, molecular weight, and polymer to DNA ratios (Akinc et al. 2003b). Using a semiautomated, parallel synthesis and screening process, Anderson et al. created a library of 2350 structurally unique, degradable PBAEs (Anderson et al. 2003). This combinatorial library was synthesized by diluting monomers in dimethyl sulfoxide, a low-viscosity medium, and using a fluid-handling robot and 12-channel micropipette to automate monomer mixing and thereby simultaneously setting up all 2350 reactions (Anderson et al. 2003). The library was later screened for its ability to bind DNA and transfect COS-7 cells, an easy to transfect cell line, under serum-free conditions. The screening identified a subset of 46 polymers with performance superior to the PEI control (Anderson et al. 2003; Green et al. 2008). These studies found that lead polymers can condense DNA into nanoparticles with smaller sizes and more positive surface charge, and such lead polymers also showed structural similarities (Figure 20.5c). Among the top nine polymers, all were formed from a conjugate addition of an amino alcohol and a hydrophobic diacrylate molecule (Anderson et al. 2005; Green et al. 2008). Moreover, the three top-performing polymers (i.e., C28, C32, and JJ28) all had converging structures (Anderson et al. 2005). These high-throughput studies highlight the importance of combinatorial synthesis and screening platforms in uncovering structure–function relationships and thereby optimizing polymeric gene delivery systems.

20.3.2 Combinatorial Approaches to siRNA Delivery Systems

Recent advances in RNA interference (RNAi) have provided another powerful tool for regulating cell behavior via gene silencing. RNAi is a gene-silencing mechanism that involves double-stranded RNA-mediated sequence-specific mRNA degradation and is a powerful mechanism for controlling cell behavior. However, the success of RNAi-based therapeutics requires an efficient delivery system to transport the 21–25 nucleotide double-stranded siRNAs into the target cells. Chemically synthesized siRNA often degrades rapidly *in vivo* and direct injection of naked siRNA often

FIGURE 20.5 (a) High-throughput synthesis and screening of PBAEs for gene delivery. PBAE synthesis by the conjugate addition of amines to diacrylate groups. (b) Acrylate and amino monomers used for the synthesis of the PBAE library. (c) Transfection proficiency of the PBAE polymer library. COS-7 cells were transfected with PBAE/DNA nanoparticles. PBAE polymers with different amine/acrylate monomer ratios were screened. (Adapted from Anderson D. G. et al. *Mol. Ther.* 11(3), 426–434, 2005 and Green J. J. et al. *Acc. Chem. Res.* 41(6), 749–759, 2008. With permission.)

leads to poor gene silencing (Urban-Klein et al. 2005). The low success rate of early studies highlighted the importance of developing delivery systems that would protect siRNA from degradation and facilitate its cellular uptake. Earlier trials showed that PEI can complex synthetic siRNAs into nanoparticles and help preserve the bioactivity of these molecules (Urban-Klein et al. 2005); however, PEI/siRNA complexes have significant toxicity effects (Akinc et al. 2008). In an effort to develop novel biomaterials for safe and efficient siRNA delivery, Akinc et al. synthesized a combinatorial library of lipid-like materials (lipidoids; Akinc et al. 2008). From a large library of more than 1200 structurally diverse lipidoids, Akinc and colleagues identified lipidoids that facilitate high levels of specific siRNA-mediated silencing of endogenous gene transcripts. Top-performing lipidoids also share common structures including amide linkages, more than two alkyl tails of eight to twelve carbons and a secondary amide (Akinc et al. 2008). Such information provides valuable guidelines for further optimizing siRNA delivery vectors.

20.4 A COMBINATORIAL APPROACH TO IDENTIFY SYNERGISTIC GENETIC SIGNALS

Gene therapy offers a promising approach for promoting desired cellular processes and tissue development by upregulating inductive genes via DNA delivery or downregulating inhibitory genes via RNAi delivery. Codelivery of multiple genetic signals may act synergistically to accelerate the desired cellular processes. Although extensive work has been performed on delivering single genetic signals for gene therapy, efforts using synergistic genetic signals are only beginning to emerge. The cellular microenvironment is highly dynamic and multifactorial. Many signaling pathways are interconnected, affecting cell fate in a synergistic or antagonistic manner. For example, bone morphogenic protein-6 (BMP-6) or insulin-like growth factor-1 (IGF-1) alone could not induce chondrogenesis, whereas on codelivery, such factors with TGF-β3 can synergistically promote chondrogenic differentiation (Indrawattana et al. 2004). Likewise, most tissue morphogenesis processes are tightly regulated by interactive signals that either turn on an activator gene, or turn off an inhibitor gene. Identifying synergistic genetic signals that promote the desired cellular processes would provide a powerful tool for directing tissue regeneration.

High-throughput screening has been explored to facilitate the discovery of potent gene targets in regulating stem cell differentiation. For example, Zhao et al. screened a synthetic siRNA library targeting 5000 human genes, which yielded 12 candidate suppressors for osteogenic specification in human MSCs (Zhao and Ding 2007). A recent study examined the effects of codelivering two genes, *BMP2* and core-binding factor α-1 (*Cbfa1*), on osteogenic differentiation of adipose-derived stem cells (ADSCs; Lee et al. 2010). *BMP2*- and *Cbfa1*-transduced ADSCs showed an upregulation of osteogenic markers and increased mineralization. Codelivery of *BMP2* and dexamethasone, a small molecule activator of osteogenesis, also led to increased expression of early-stage osteogenic markers in mouse embryonic stem cells (Blum et al. 2004). BMP-transduced, mouse muscle-derived stem cells also showed increased angiogenesis and mineralization when cotransduced with VEGF (Peng et al. 2002). These studies demonstrated that codelivering multiple genetic signals may synergistically promote lineage-specific differentiation. A recent study has also explored the potential benefits of codelivering multiple inductive and suppressive genes on osteogenic differentiation (Ramasubramanian et al. 2011). The gene expression of three target genes, *BMP2*, an osteogenic inducer, as well as *GNAS* and *Noggin*, osteogenic suppressors, were modulated in a combinatorial manner using biomaterials-mediated gene delivery. Compared with *BMP2* DNA delivery alone, codelivery of *BMP2* DNA and either siGNAS or siNoggin significantly accelerated osteogenic differentiation in human ADSCs, with enhanced osteogenic gene expression and mineralization. These results suggest that inductive or suppressive genetic switches interact in a complex, nonlinear manner. Given the complex interactions among various genetic signals, it is highly desirable to develop high-throughput screening assays to facilitate the identification of synergistic genetic signals for gene therapy (Nguyen et al. 2009; Ramasubramanian et al. 2011).

20.5 CONCLUSION

Advances in nanotechnology have provided powerful tools to control cell fate and tissue regeneration. These strategies can be broadly classified into two categories, the "outside–in" or "inside–out" approaches. The outside–in approach focuses on engineering the cellular microenvironment via recapitulating the sophisticated structure and biological functionalities found in the native ECM. Nanotechnology facilitates the incorporation of nanoscale features into scaffold design to better mimic native tissues, which helps elucidate the complex interactions between cells and their surrounding microenvironment. The inside–out approach influences cell behavior via directly regulating genetic signals using biomaterials-mediated gene delivery. Toward this end, the use of cationic lipids or polymers in gene delivery systems offers great versatility in chemical modification to circumvent barriers in gene delivery. High-throughput synthesis and screening approaches have demonstrated great promise in accelerating the rapid discovery of novel biomaterials that can condense nucleic acids into nanoparticles for efficient gene delivery. Identifying potential synergistic signals using high-throughput screening would also provide potentially more powerful gene targets for promoting desired tissue formation. Finally, validating the efficacy of these strategies in appropriate animal models will be crucial for the successful translation of their final applications in the clinical setting.

Although significant advances have been made in applying nanotechnology for advancing tissue regeneration, challenges remain before such therapies can be successfully translated from bench-to-bedside. For example, extensive efforts have been dedicated to understanding individual types of microenvironmental cues (e.g., nanotopography), but how complex interactive cues regulate cell fate and tissue development remains largely unknown. Moreover, many of the molecular mechanisms underlying these effects have yet to be discovered. As for the application of nanomaterials for gene delivery, there is an impending need to develop technology that facilitates efficient targeting to specific cell types or tissues *in situ*. Biomaterial-based vectors allow the possibility of conjugating biofunctional moieties such as antibodies that target specific cells or tissues. In sum, nanotechnology will continue to provide a powerful tool for both fundamental and applied research in tissue regeneration.

The contributions of nanotechnology to the tissue engineering field will also become even more prominent as the field continues to evolve and better integrate with multiple disciplines, such as molecular and cell biology, materials science, computational biology, and medicine.

REFERENCES

Akinc, A., D. G. Anderson, D. M. Lynn, and R. Langer. 2003a. Synthesis of poly(beta-amino ester)s optimized for highly effective gene delivery. *Bioconjugate Chemistry* 14(5), 979–988.

Akinc, A., D. M. Lynn, D. G. Anderson, and R. Langer. 2003b. Parallel synthesis and biophysical characterization of a degradable polymer library for gene delivery. *Journal of the American Chemical Society* 125(18), 5316–5323.

Akinc, A., A. Zumbuehl, M. Goldberg, E. S. Leshchiner, V. Busini, N. Hossain et al. 2008. A combinatorial library of lipid-like materials for delivery of RNAi therapeutics. *Nature Biotechnology* 26(5), 561–569.

Anderson, D. G., A. Akinc, N. Hossain, and R. Langer. 2005. Structure/property studies of polymeric gene delivery using a library of poly(beta-amino esters). *Molecular Therapy* 11(3), 426–434.

Anderson, D. G., D. M. Lynn, and R. Langer. 2003. Semi-automated synthesis and screening of a large library of degradable cationic polymers for gene delivery. *Angewandte Chemie (International ed. in English)* 42(27), 3153–3158.

Babensee, J. E., L. V. McIntire, and A. G. Mikos. 2000. Growth factor delivery for tissue engineering. *Pharmaceutical Research* 17(5), 497–504.

Banerjee, A., M. Arha, S. Choudhary, R. S. Ashton, S. R. Bhatia, D. V. Schaffer et al. 2009. The influence of hydrogel modulus on the proliferation and differentiation of encapsulated neural stem cells. *Biomaterials* 30(27), 4695–4699.

Blum, J. S., M. B. Parrott, A. G. Mikos, and M. A. Barry. 2004. Early osteoblastic differentiation induced by dexamethasone enhances adenoviral gene delivery to marrow stromal cells. *Journal of Orthopaedic Research* 22(2), 411–416.

Brazeau, G. A., S. Attia, S. Poxon, and J. A. Hughes. 1998. *In vitro* myotoxicity of selected cationic macromolecules used in non-viral gene delivery. *Pharmaceutical Research* 15(5), 680–684.

Brittberg, M., A. Lindahl, A. Nilsson, C. Ohlsson, O. Isaksson, and L. Peterson. 1994. Treatment of deep cartilage defects in the knee with autologous chondrocyte transplantation. *New England Journal of Medicine* 331(14), 889–895.

Buehler, M. J. 2006. Nature designs tough collagen: Explaining the nanostructure of collagen fibrils. *Proceedings of the National Academy of Sciences of the United States of America* 103(33), 12285–12290.

Check, E. 2005. Gene therapy put on hold as third child develops cancer. *Nature* 433, 561.

Dalby, M. J., N. Gadegaard, R. Tare, A. Andar, M. O. Riehle, P. Herzyk et al. 2007. The control of human mesenchymal cell differentiation using nanoscale symmetry and disorder. *Nature Materials* 6(12), 997–1003.

Engler, A. J., S. Sen, H. L. Sweeney, and D. E. Discher. 2006. Matrix elasticity directs stem cell lineage specification. *Cell* 126(4), 677–689.

Ferkol, T., J. C. Perales, F. Mularo, and R. W. Hanson. 1996. Receptor-mediated gene transfer into macrophages. *Proceedings of the National Academy of Sciences of the United States of America* 93(1), 101–105.

Ferrara, N. 2010. Binding to the extracellular matrix and proteolytic processing: Two key mechanisms regulating vascular endothelial growth factor action. *Molecular Biology of the Cell* 21(5), 687–690.

Geng, T., Y. Zhan, J. Wang, and C. Lu. 2011. Transfection of cells using flow-through electroporation based on constant voltage. *Nature Protocols* 6(8), 1192–1208.

Goldberg, M., R. Langer, and X. Jia. 2007. Nanostructured materials for applications in drug delivery and tissue engineering. *Journal of Biomaterials Science. Polymer Edition* 18(3), 241–268.

Goldberg, M., K. Mahon, and D. Anderson. 2008. Combinatorial and rational approaches to polymer synthesis for medicine. *Advanced Drug Delivery Reviews* 60(9), 971–978.

Green, J. J., R. Langer, and D. G. Anderson. 2008. A combinatorial polymer library approach yields insight into nonviral gene delivery. *Accounts of Chemical Research* 41(6), 749–759.

Guo, X., and Huang, L. 2011. Recent advances in nonviral vectors for gene delivery. *Accounts of Chemical Research*. DOI: 10.1021/ar200151m.

Hartgerink, J. D., E. Beniash, and S. I. Stupp. 2001. Self-assembly and mineralization of peptide–amphiphile nanofibers. *Science* 294(5547), 1684–1688.

Heilbronn, R., and S. Weger. 2010. Viral vectors for gene transfer: Current status of gene therapeutics. *Handbook of Experimental Pharmacology* 197, 143–170.

Herweijer, H., and J. A. Wolff. 2003. Progress and prospects: Naked DNA gene transfer and therapy. *Gene Therapy* 10(6), 453–458.

Hollon, T. 2000. Researchers and regulators reflect on first gene therapy death. *Nature Medicine* 6, 6.

Holmes, T. C., S. de Lacalle, X. Su, G. Liu, A. Rich, and S. Zhang. 2000. Extensive neurite outgrowth and active synapse formation on self-assembling peptide scaffolds. *Proceedings of the National Academy of Sciences of the United States of America* 97(12), 6728–6733.

Hosseinkhani, H., M. Hosseinkhani, F. Tian, H. Kobayashi, and Y. Tabata. 2006. Osteogenic differentiation of mesenchymal stem cells in self-assembled peptide–amphiphile nanofibers. *Biomaterials* 27(22), 4079–4086.

Huebsch, N., P. R. Arany, A. S. Mao, D. Shvartsman, O. A. Ali, S. A. Bencherif et al. 2010. Harnessing traction-mediated manipulation of the cell/matrix interface to control stem-cell fate. *Nature Materials* 9(6), 518–526.

Indrawattana, N., G. Chen, M. Tadokoro, L. H. Shann, H. Ohgushi, T. Tateishi et al. 2004. Growth factor combination for chondrogenic induction from human mesenchymal stem cell. *Biochemical and Biophysical Research Communications* 320(3), 914–919.

Kay, M. A., J. C. Glorioso, and L. Naldini. 2001. Viral vectors for gene therapy: The art of turning infectious agents into vehicles of therapeutics. *Nature Medicine* 7(1), 33–40.

Kisiday, J., M. Jin, B. Kurz, H. Hung, C. Semino, S. Zhang et al. 2002. Self-assembling peptide hydrogel fosters chondrocyte extracellular matrix production and cell division: Implications for cartilage tissue repair. *Proceedings of the National Academy of Sciences of the United States of America* 99(15), 9996–10001.

Kolácná, L., J. Bakesová, F. Varga, E. Kostáková, L. Plánka, A. Necas et al. 2007. Biochemical and biophysical aspects of collagen nanostructure in the extracellular matrix. *Physiological Research* 56, Suppl 1, S51–60.

Kootstra, N. A., and I. M. Verma. 2003. Gene therapy with viral vectors. *Annual Review of Pharmacology and Toxicology* 43, 413–439.

Kumbar, S. G., R. James, S. P. Nukavarapu, and C. T. Laurencin. 2008. Electrospun nanofiber scaffolds: Engineering soft tissues. *Biomedical Materials* 3(3), 034002.

Lamers, E., R. van Horssen, J. te Riet, F. C. van Delft, R. Luttge, X. F. Walboomers et al. 2010. The influence of nanoscale topographical cues on initial osteoblast morphology and migration. *European Cells* and *Materials* 20, 329–343.

Langer, R., and J. P. Vacanti. 1993. Tissue engineering. *Science* 260(5110), 920–926.

Lee, S. J., S. W. Kang, H. J. Do, I. Han, D. A. Shin, J. H. Kim et al. 2010. Enhancement of bone regeneration by gene delivery of BMP2/Runx2 bicistronic vector into adipose-derived stromal cells. *Biomaterials* 31(21), 5652–5659.

Luo, D., and W. M. Saltzman. 2000. Synthetic DNA delivery systems. *Nature Biotechnology* 18(1), 33–37.

Lutolf, M. P., and J. A. Hubbell. 2005. Synthetic biomaterials as instructive extracellular microenvironments for morphogenesis in tissue engineering. *Nature Biotechnology* 23(1), 47–55.

Lynn, D. M., and R. Langer. 2000. Degradable poly(beta-amino esters): Synthesis, characterization, and self-assembly with plasmid DNA. *Journal of the American Chemical Society* 122(44), 10761–10768.

Ma, Z., M. Kotaki, R. Inai, and S. Ramakrishna. 2005. Potential of nanofiber matrix as tissue-engineering scaffolds. *Tissue Engineering* 11(1–2), 101–109.

Matthews, J. A., G. E. Wnek, D. G. Simpson, and G. L. Bowlin. 2002. Electrospinning of collagen nanofibers. *Biomacromolecules* 3(2), 232–238.

Mehier-Humbert, S., and R. H. Guy. 2005. Physical methods for gene transfer: Improving the kinetics of gene delivery into cells. *Advanced Drug Delivery Reviews* 57(5), 733–753.

Moghimi, S. M., P. Symonds, J. C. Murray, A.C. Hunter, G. Debska, and A. Szewczyk. 2005. A two-stage poly(ethylenimine)-mediated cytotoxicity: Implications for gene transfer/therapy. *Molecular Therapy* 11, 990–995.

Nguyen, D. N., J. J. Green, J. M. Chan, R. Langer, and D. G. Anderson. 2009. Polymeric materials for gene delivery and DNA vacinnation. *Advanced Materials* 21(8), 847–867.

Oh, S., K. S. Brammer, Y. S. Li, D. Teng, A. J. Engler, S. Chien et al. 2009. Stem cell fate dictated solely by altered nanotube dimension. *Proceedings of the National Academy of Sciences of the United States of America* 106(7), 2130–2135.

Pack, D. W., A. S. Hoffman, S. Pun, and P. S. Stayton. 2005. Design and development of polymers for gene delivery. *Nature Reviews Drug Discovery* 4(7), 581–593.

Papachroni, K. K., D. N. Karatzas, K. A. Papavassiliou, E. K. Basdra, and A. G. Papavassiliou. 2009. Mechanotransduction in osteoblast regulation and bone disease. *Trends in Molecular Medicine* 15(5), 208–216.

Papisov, M. I. 1998. Theoretical considerations of RES-avoiding liposomes: Molecular mechanics and chemistry of liposome interactions. *Advanced Drug Delivery Reviews* 32(1–2), 119–138.

Pek, Y. S., A. C., Wan, and J. Y. Ying. 2010. The effect of matrix stiffness on mesenchymal stem cell differentiation in a 3D thixotropic gel. *Biomaterials* 31(3), 385–391.

Peng, H., V. Wright, A. Usas, B. Gearhart, H. C. Shen, J. Cummins et al. 2002. Synergistic enhancement of bone formation and healing by stem cell-expressed VEGF and bone morphogenetic protein-4. *Journal of Clinical Investigation* 110(6), 751–759.

Pollard, H., G. Toumaniantz, J. L. Amos, H. Avet-Loiseau, G. Guihard, J. P. Behr et al. 2001. Ca^{2+}-sensitive cytosolic nucleases prevent efficient delivery to the nucleus of injected plasmids. *Journal of Gene Medicine* 3(2), 153–164.

Ramasubramanian, A., S. Shiigi, G. K. Lee, and F. Yang. 2011. Non-viral delivery of inductive and suppressive genes to adipose-derived stem cells for osteogenic differentiation. *Pharmaceutical Research* 28(6), 1328–1337.

Rho, J. Y., L. Kuhn-Spearing, and P. Zioupos. 1998. Mechanical properties and the hierarchical structure of bone. *Medical Engineering and Physics* 20(2), 92–102.

Rho, K. S., L. Jeong, G. Lee, B. M. Seo, Y. J. Park, S. D. Hong et al. 2006. Electrospinning of collagen nanofibers: Effects on the behavior of normal human keratinocytes and early-stage wound healing. *Biomaterials* 27(8), 1452–1461.

Rubinson, D. A., C. P. Dillon, A. V. Kwiatkowski, C. Sievers, L. Yang, J. Kopinja et al. 2003. A lentivirus-based system to functionally silence genes in primary mammalian cells, stem cells and transgenic mice by RNA interference. *Nature Genetics* 33(3), 401–406.

Rychahou, P. G., and B. M. Evers. 2010. Hydrodynamic delivery protocols. *Methods in Molecular Biology* 623, 189–195.

Scherer, F., M. Anton, U. Schillinger, J. Henke, C. Bergemann, A. Krüger et al. 2002. Magnetofection: Enhancing and targeting gene delivery by magnetic force *in vitro* and *in vivo*. *Gene Therapy* 9(2), 102–109.

Schultz, G. S., and A. Wysocki. 2009. Interactions between extracellular matrix and growth factors in wound healing. *Wound Repair and Regeneration* 17(2), 153–162.

Shah, R. N., N. A. Shah, M. M. Del Rosario Lim, C. Hsieh, G. Nuber, and S. I. Stupp. 2010. Supramolecular design of self-assembling nanofibers for cartilage regeneration. *Proceedings of the National Academy of Sciences of the United States of America* 107(8), 3293–3298.

Shin, H., S. Jo, and A. G. Mikos. 2003. Biomimetic materials for tissue engineering. *Biomaterials* 24(24), 4353–4364.

Silver, F. H., J. W. Freeman, and G. P. Seehra. 2003. Collagen self-assembly and the development of tendon mechanical properties. *Journal of Biomechanics* 36(10), 1529–1553.

Taniyama, Y., K. Tachibana, K. Hiraoka, M. Aoki, S. Yamamoto, K. Matsumoto et al. 2002. Development of safe and efficient novel nonviral gene transfer using ultrasound: Enhancement of transfection efficiency of naked plasmid DNA in skeletal muscle. *Gene Therapy* 9(6), 372–380.

Tessmar, J. K., and A. M. Göpferich. 2007. Matrices and scaffolds for protein delivery in tissue engineering. *Advanced Drug Delivery Reviews* 59(4–5), 274–291.

Thomas, C. E., A. Ehrhardt, and M. A. Kay. 2003. Progress and problems with the use of viral vectors for gene therapy. *Nature Reviews Genetics* 4(5), 346–358.

Thomas, M., J. J. Lu, C. Zhang, J. Chen, and A. M. Klibanov. 2007. Identification of novel superior polycationic vectors for gene delivery by high-throughput synthesis and screening of a combinatorial library. *Pharmaceutical Research* 24(8), 1564–1571.

Urban-Klein, B., S. Werth, S. Abuharbeid, F. Czubayko, and A. Aigner. 2005. RNAi-mediated gene-targeting through systemic application of polyethylenimine (PEI)-complexed siRNA *in vivo*. *Gene Therapy* 12(5), 461–466.

Wu, G. Y., and C. H. Wu. 1988. Receptor-mediated gene delivery and expression *in vivo*. *Journal of Biological Chemistry* 263(29), 14621–14624.

Wu, G. Y., J. M. Wilson, F. Shalaby, M. Grossman, D. A. Shafritz, and C. H. Wu. 1991. Receptor-mediated gene delivery *in vivo*. Partial correction of genetic analbuminemia in Nagase rats. *Journal of Biological Chemistry* 266(22), 14338–14342.

Yim, E. K., R. M. Reano, S. W. Pang, A. F. Yee, C. S. Chen, and K. W. Leong. 2005. Nanopattern-induced changes in morphology and motility of smooth muscle cells. *Biomaterials* 26(26), 5405–5413.

Yoshimoto, H., Y. M. Shin, H. Terai, and J. P. Vacanti. 2003. A biodegradable nanofiber scaffold by electrospinning and its potential for bone tissue engineering. *Biomaterials* 24(12), 2077–2082.

Yu, J. Y., M. A. Vodyanik, K. Smuga-Otto, J. Antosiewicz-Bourget, J. L. Frane, S. Tian et al. 2007. Induced pluripotent stem cell lines derived from human somatic cells. *Science* 318(5858), 1917–1920.

Zhang, L., and T. J. Webster. 2009. Nanotechnology and nanomaterials: Promises for improved tissue regeneration. *Nano Today* 4(1).

Zhao, Y., and S. Ding. 2007. A high-throughput siRNA library screen identifies osteogenic suppressors in human mesenchymal stem cells. *Proceedings of the National Academy of Sciences of the United States of America* 104(23), 9673–9678.

21 Nanobiomaterial-Based Short Interfering RNA Delivery Systems

Yoon N. Cho, Murugan Ramalingam, and Ziyad S. Haidar

CONTENTS

21.1 INTRODUCTION

A range of organisms (*Drosophila*, *Caenorhabditis elegans*, *Neurospora*, plants, and mammals) undergo sequence-specific posttranscriptional gene silencing. The mechanism behind this process is a result of the normal defense against viruses and the mobilization of genetic elements (transposons; Tijsterman et al. 2002). This process is called "RNA interference" (RNAi), and results in gene silencing. On the basis of RNAi, the use of short interfering RNA (siRNA) opened a new era in the treatment of various diseases because of the following reasons. First, siRNA only interferes with translation, not transcription, in the nucleus, which has previously raised concerns about the possible adverse gene alterations after gene therapy using DNA manipulation. In other words, siRNA only interacts with mRNA, not proteins, so that siRNA inhibits unnecessary or harmful proteins before translation into protein (Oh and Park 2009). Another merit is that only a few siRNA molecules are needed for gene silencing. siRNA also shows the diversity of applicable targets for the treatment of various diseases because siRNA-based drugs can target any mRNA of interest (Bumcrot et al. 2006; Kim and Rossi 2007). Fire et al. (1998) first described double-stranded RNA (dsRNA) as a tool for genetic interference by using *C. elegans* and demonstrated that dsRNA was more efficient in inducing interference compared with single-stranded RNA (ssRNA). Likewise, the sequence-specific knockdown of mammalian cells by using synthetic siRNA was also successful. This experiment also specified the appropriate length of siRNA in the suppression of endogenous and heterologous genes as

TABLE 21.1
Current Clinical Trials with siRNA

Sponsor	Product	Target Gene	Diseases	Phase
Alnylam	ALN-VSP01	VEGF/KSP	Liver cancer	I
Calando	CALAA-01	RRM2	Solid tumor	I
Duke University	Proteasome siRNA and tumor antigen RNA-transfected dendritic cells	Immunoproteasome	Melanoma	I
Gradalis, Inc.	Autologous FANG vaccine	TGFβ1 and TGFβ2	Solid tumor	I
Quark	QPI-1007	Caspase-2	Glaucoma or acute eye injury	I
Quark	I5NP	Not found	Kidney injury or acute renal failure undergoing major cardiovascular surgery	I
Tekmira	Apo8 SNALP	Apo8	Hypercholesterolemia	I
TransDerm	TD101	Keratin 6a (N171K mutation)	Pachyonychia	Ib
Silence (Augen)	Atu-027	PKN3	Solid tumor	I
Sylentis	SYL-040012	ADRB2	Ocular hypertension	I
Quark	I5NP	Not found	Delayed graft function/kidney transplant	I/II
Alnylam	ALN-RSV01	RSV	Viral nucleocapsid	II
Quark, Pfizer	PF-04523655	RTP801	Wet AMD or diabetic macular edema	II
Quark	QP1-1002	p53	Acute renal failure	II
ZaBeCor	Excellair	SYK kinase	Asthma	II

Source: Watts, J. K., and D. R. Corey. *Bioorg. Med. Chem. Lett.* 20, 3203–3207, 2010; Chen, S.-H., and G. Zhaori. *Eur. J. Clin. Invest.* 41, 221–232, 2010.

21 bp (Elbashir et al. 2001a). Since those findings, there have been trials for the use of siRNA in the treatment of various diseases, especially in targeting cancer (Xia et al. 2007; Niu et al. 2006; Halder et al. 2006; Takeshita et al. 2005). In 2004, synthetic siRNA was first introduced to patients with blinding choroidal neovascularization using direct intraocular siRNA injections (Davis et al. 2010). This was followed by other human clinical trials including cancer trials (Table 21.1). Calando Pharmaceuticals Pasadena, CA (USA) is currently conducting phase I solid tumor treatments using CALAA-01, which contains siRNA with cyclodextrin-based polymers targeting ribonucleotide reductase (RRM2). Alnylam Pharmaceutics and Silence Therapeutics London, UK are also conducting phase I trials using lipid nanoparticles for liver cancer and solid tumors, respectively (Watts and Corey 2010). Nevertheless, only two reports about siRNA in human clinical trials have been published thus far (Vincenzoa et al. 2008; Leachman et al. 2010). Another study even suggests non-RNAi mechanisms on siRNA administration in humans (Davis et al. 2010). Accordingly, more investigations into the use of siRNAs are needed to show direct evidence of the RNAi mechanism in humans. Moreover, this may reflect the difficulty behind the systemic delivery of siRNA targeting human cancer. In this respect, this review describes not only the available siRNA delivery systems but also the recent progress in different types of cancers and suggests future directions of feasible methods.

21.2 siRNA AND ITS ENVIRONMENTAL EFFECT

The nature of siRNA is basically insufficient to represent drug-like properties because the phosphodiester backbone of siRNA has approximately 40 anionic charges, which makes diffusion through the cell membrane difficult. Moreover, siRNA shows high molecular weight (>30 kDa) and large size according to its DNA structure (two turns of a nucleic acid double helix). In particular, siRNA is extremely hydrophilic and is further hydrated by a sugar phosphate backbone in aqueous solution (Bumcrot et al. 2006). Apart from the nature of siRNA, there are more considerations in terms of the cell environment after the delivery of naked siRNA into the cytoplasm. First, siRNA shows short half-lives *in vivo* because it is not stable in serum after serum nuclease activity (Bumcrot et al. 2006). Second, siRNA is likely to face "off-target" effects because short motifs in the mRNA of other genes share partial homology with siRNA, which could lead to the inhibition of nontarget genes (Jackson et al. 2003). Finally, siRNA could activate the immune system by interferon (IFN) response (Sledz et al. 2003; Karikó et al. 2004). It is known that dsRNA activates the immune response through RNA-dependent protein kinase (PKR) and retinoic acid-induced gene 1 (RIG-1) in the cytoplasm, and toll-like receptor (TLR3) in the cell surface as well as in the endosomes. In particular, ssRNA containing specific sequences (UGUGU and GUCCUUCAA) activate IFN response by TLR7 and TLR8 in endosomes (Marques and Williams 2005). Furthermore, it has been noted that IFN response caused by siRNA is length dependent. That is, dsRNA (>30 bp) can activate the γ-IFN pathway in mammalian systems (Elbashir et al. 2001a). Therefore, the characteristics of siRNA and issues regarding the cell environmental cell's environment should be carefully considered before using siRNA as a biopharmaceutical drug.

21.3 siRNA DELIVERY SYSTEMS IN CANCER TREATMENT

Cancer therapy has taken up a large proportion of all siRNA-based drugs since 2007, and most of these trials have been performed through systemic administration. Compared with local administration, which has the advantage of being less complex, costing less, and having fewer risks with regard to side effects and uptake by other tissues (Watts and Corey 2010), the *in vivo* systemic route requires more careful approaches when used. Basically, the delivery system has to be biocompatible, biodegradable, and nonimmunogenic. In addition, protection of the active dsRNA against serum nuclease activity as well as renal clearance or target tissue-specific distribution avoiding rapid hepatic clearance are necessary for the stability and long half-lives of siRNA drugs. Lastly, an effective endosomal release system into the cytoplasm is crucial to allow siRNA interaction with RNA induced silencing complex (RISC) (Aigner 2007; Juliano et al. 2008). Currently, there are various strategies available to meet such requirements in *in vitro* and *in vivo* experiments.

21.4 CHEMICAL MODIFICATION

Chemical modifications of the nucleic acid backbone have been attempted to enhance the stability and alter the biodistribution of siRNA. Various chemical modifications of siRNA have been created. There are five different types of modifications according to sugar, backbone linkage, base, architecture modifications, and terminal conjugates (Deleavey et al. 2009).

21.4.1 Sugar Modifications

The most widely studied sugar modification is 2′-*O*-methyl modification. It was reported that siRNA modifications with 2′-*O*-Me in either strand decreased siRNA activity (Elbashir et al. 2001b; Braasch et al. 2003; Chiu and Rana 2003; Czauderna et al. 2003), and only fully modified 2′-*O*-Me in siRNA sense strands makes it functional (Choung et al. 2006; Kraynack and Baker 2006). siRNA modifications with locked nucleic acids (LNA) containing methylene linkage between 2′-oxygen and 4′-carbon also showed an increase of stability and reduced off-target gene regulation (Elmén et al. 2005; Mook et al. 2007). However, a recent study demonstrated that unlocked nucleic acid-modified siRNAs dramatically increased biostability in *in vivo* and *in vitro* cancer experiments without heavy modifications such as LNA modification (Laursen et al. 2010).

21.4.2 Backbone Modifications

A well-known backbone modification is the substitution of the phosphodiester (PO_4) group to phosphothioates (PS) at the 3′-end of the siRNA sequence (Chiu and Rana 2003); however, some cytotoxicity was reported with extensive PS modification (Geary et al. 2001) and even seemed irrelevant to biodistribution (Braasch et al. 2004). siRNA with boranophosphate linkage is another popular modification. The boranophosphate linkage showed a significant increase in stability against nuclease activity, avoiding modification at the central position of the antisense strand (Hall et al.

2004). One report described another effective modification using phosphonoacetate linkage, which led to esterification and produced neutral backbones, allowing uptake without transfection reagent. This modification also showed its ability for retention in the cell by cleavage of ester group through cellular esterases (Sheehan et al. 2003; Yamada et al. 2007). Furthermore, there is a combination type of backbone modification. siRNA modification of 4′-thioribose with 2′-*O*-methyl or 2′-*O*-methoxyethyl significantly improved siRNA activity and plasma stability (Dande et al. 2006).

21.4.3 Base Modifications

Although DNA base modification has more limitations compared with other methods, there are several cases of siRNA modifications by switching bases into 4-thiouridine, 5-bromo, 5-iodo, 5-unidine, or 5-inosine (Parrish et al. 2000). For example, adenine and uracil can be replaced with diaminopurine and 5-bromouracil or 5-iodouracil, respectively. However, these modifications were tolerated in siRNA duplexes and their activity was reduced (Chiu and Rana 2003). It was also reported that small base modifications with 5-Me-U and 5-Me-C could improve siRNA stability (Terrazas and Kool 2009).

21.4.4 Architecture Modifications

siRNA duplex architecture can be also altered and used as a useful tool for both research and drug development. For instance, asymmetric siRNAs require deletion at sense or antisense strands (or both), which leads to an efficacious silencing activity (Sun et al. 2008; Chu and Rana 2008). Small internally segmented interfering RNAs (sisiRNA) are another architectural modification that makes it highly functional. It is known that sisiRNA consists of a naked antisense strand and two shorter (10–12 nucleotides) complementary sense strands, and prevents unwanted off-target effects as only the antisense strand can be loaded by RISC (Bramsen et al. 2007). Moreover, other alternative designs are available: dgRNAs, ss-siRNA, short-hairpin siRNA, and dumbbell-shaped RNA (Gaglione and Messere 2010).

21.4.5 Terminal Conjugations

There are different groups conjugating to the termini of the sense strand in the siRNA duplex. The conjugates can be any foreign molecules that provide specific cell targets and enhance the nuclease stability, thermodynamics, biodistribution, and pharmacokinetics of siRNAs. Moreover, most of the conjugations use cleavage linkages including acid-labile and reducible bonds for the efficient release of intact siRNA in the cell. Lipid, for example, is a well-known material that may improve siRNAs' uptake using membrane permeability or receptor-mediated endocytosis (Gaglione and Messere 2010). Cholesterol is a type of lipid that can be covalently conjugated to siRNA by an aminocaproic acid pyrrolidine noncleavable linker or a reducible disulfide linker (Soutschek et al. 2004; Moschos et al. 2007), although it has been reported that this only improved the duration of siRNA without mRNA knockdown (Moschos et al. 2007). α-Tocopherol is an alternative steroid conjugate, covalently bound to the antisense strand of siRNA and designed to be cleaved by Dicer. Moreover, intravenous administration of the α-tocopherol–siRNA showed improved efficiency in the reduction of endogenous apoB mRNA in the liver without IFN response and side effects (Nishina et al. 2008).

Target delivery of siRNA is also available by ligand–receptor interaction using aptamers, peptides, hormones, and antibodies. Aptamer is a synthetically prepared nucleic acid and is known to inhibit the activity of many pathogenic proteins with more convenient modification, chemical versatility and stability, and reduced immunogenicity (Nimjee et al. 2005). One of the well-studied aptamers uses prostate-specific membrane antigen (PSMA) that can be directly conjugated to siRNA by biotin–streptavidin interaction (Chu et al. 2006). Recently, more optimized PSMA aptamers targeting prostate cancer-specific prosurvival genes (Plk1 and Bcl) were developed by changing into blunt or truncated siRNA terminal ends, and they indeed led to the regression of PSMA-expressing tumors after systemic administration in athymic mice (Dassie et al. 2009). Peptides are another option to enhance cellular uptake of siRNA delivery. Covalent conjugates with cell-penetrating peptides (CPPs), also called protein transduction domains, have been proposed for siRNA delivery. CPPs are characterized by a high content of basic amino acid sequences that are 10–30 residues in length (Gaglione and Messere 2010). Penetratin and transportan are the most widely studied CPPs conjugated to thiol-containing siRNA, and their conjugations are achieved by disulfide linkage that is labile in the cytoplasm (Muratovska and Eccles 2004). Although there are only a few reports available, the use of transactivator protein from HIV-1 (TAT) also demonstrated an improvement of siRNA delivery and a downregulation of the target protein. For the binding linkage, a disulfide bond can be used between TAT and the 3′ terminus of the antisense strand (Chiu et al. 2004). A disulfide bond is again applicable for the complex formation between siRNA and poly(ethylene glycol) (PEG). One research group concluded that siRNA–PEG conjugates improved the stability and prolonged the half-life, whereas reducing the IFN response of the conjugates (Jung et al. 2010). Kim et al. (2006) demonstrated another feasible molecule using a hydrophobically modified protein–protein transduction domain, cholesteryl oligo-D-arginine (Chol-R9). This noncovalent system targeted vascular endothelial growth factor (VEGF) mRNA and showed stability and improved knockdown of the VEGF. Additionally, an amphipathic peptide (MPG), derived from the fusion peptide of HIV-1 gp41 protein and the nuclear localization signal of SV40 large T-antigen is conjugated to siRNA by a noncovalent reaction (Simeoni et al. 2003). Furthermore, there are other peptide systems for noncovalent conjugations that have recently used endosomolytic CPP (EB1; Lundberg et al. 2007) and cationic arginine residues (CADY; Crombez et al. 2009). However, there are concerns about the penetrating efficacy of CPPs because the use of cationic peptide with anionic siRNA may lead to the neutralization of CPPs (Muratovska and Eccles 2004). Cytotoxicity is another issue with the use of CPPs because it could be triggered by cell membrane perturbation or immunogenicity. In most CPP-siRNA systems, cytotoxicity was not clearly evaluated; however, one study indeed demonstrated innate immune response by introducing siRNA conjugated by penetratin or TAT in mice (Moschos et al. 2007). In particular, there is a peptide applicable to the tumor vasculature.

RGD peptides, promising ligands, target αv-integrins which are highly expressed on the surface of endothelial cells and other cells in tumor. Once the RGD motif binds to αv-integrins, it will cleave the peptide by cell surface-associated proteases and expose the cryptic CendR element, RXXk/R, which renders itself translocated by neutropillin-1 (Sugahara et al. 2009). Thus, RGD peptides have also emerged as a useful tool for siRNA delivery.

21.5 POLYMERIC AND DENDRIMERIC NANOCARRIERS

Polymers are ideal siRNA carrier candidates because they provide a wide range of treatment options for efficient transfection and release of nucleic acids based on optimal parameters. The use of polymers essentially requires adjusting hydrophobicity/hydrophilicity, polymer molecular weight, and charge density. In particular, cationic polymers (one of the major carriers), must consider pH, ionic strength, and the ratio of protonatable polymer amine groups to nucleic acid phosphate groups as DNA and cationic polymers form a complex through electrostatic interaction (Cherng et al. 1999). In most nanocarriers, the size should be up to 150 nm for cellular uptake by nonspecific endocytosis, and the positive surface charge of nanocarriers is not always preferential *in vivo* as they could encounter several negatively charged molecules, such as albumin. Moreover, aggregation with other nanoparticles by repulsion should also be considered (Bishop 1997; Oupický et al. 2000).

Cationic polymer carriers were used because self-assembly of DNA or RNA was formulated with cationic polymers through electrostatic binding. For instance, polycationic polymers (Figure 21.1a), poly(L-lysine) (PLL) and polyethyleneimine (PEI), were synthesized and effectively formed complexes with DNA through electrostatic interactions in *in vitro* experiments (Abdallah et al. 1996). In particular, PEI has been extensively used as a cationic polymer that shows high transfection efficiency and pH-sensitive release of DNA (Boussif et al. 1996). However, its complexation with high molecular weight and increasing branching led to cytotoxicity (Boeckle et al. 2004; Hunter 2006; Thomas and Klibanov 2002; Fischer et al. 2002). To overcome these problematic issues, the "core–shell" structure was introduced by using block copolymers of PEI and PEG. In an aqueous solution, DNA condensed into a compact structure (called a polyplex) by electrostatic interaction with cationic segments of PEI and PEG moieties, forming an outer layer to protect the polyplex (Davis 2002). An important aspect of this complex is PEG chain length, which influences nuclease resistance and stability (Mao et al. 2006). Remarkable progress has been shown in this system using an antiangiogenic tumor therapeutic strategy. PEI-graft-PEG was branched to siRNA against VEGF receptor 1 (VEGFR1), and the RGD peptide was conjugated to a PEG spacer. As a result, the system showed improved stability in serum as well as in *in vivo* tumor suppression and in *in vitro* mRNA VEGFR1 inhibition (Kim et al. 2011). PLL-graft-PEG, another combination of copolymers, has also been indicated in prolonged circulation times of *in vivo* siRNA (Sato et al. 2007). Cyclodextrin is a synthetic polycation that leads to self-assembly with siRNA, and systemic delivery of siRNA against the *EWS-FLI1* gene was attempted by complexation with cyclodextrin-PEG-transferrin. Transferrin, a target ligand for transferring receptor expressed on tumor cells, can enhance cellular uptake, and

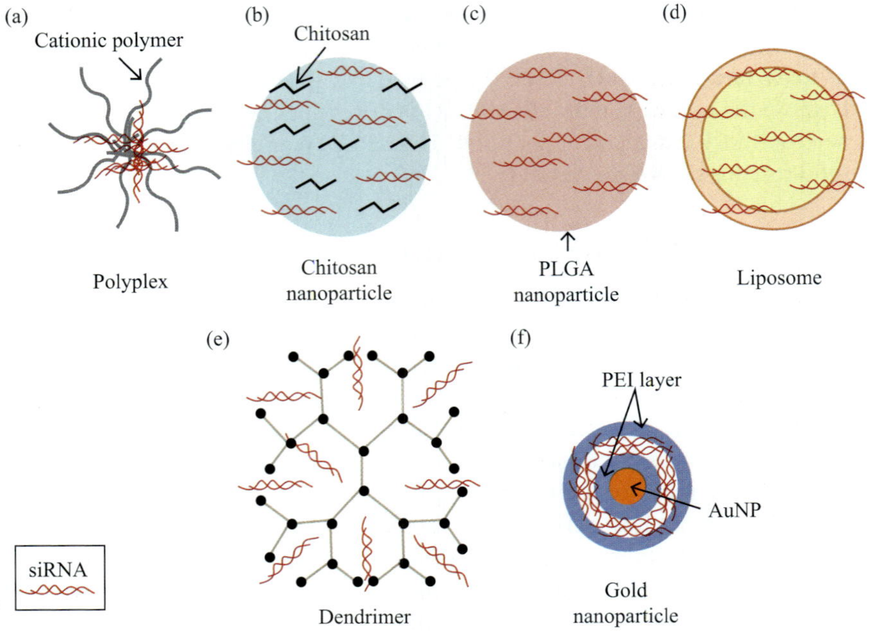

FIGURE 21.1 Examples of various siRNA delivery carriers, which includes the delivery through (a) cationic polymers such as polyplex, (b) natural polymers such as chitosan, (c) synthetic polymers such as PLGA, (d) lipid bilayers such as liposome, (e) dendritic macromolecules, and (f) charged particles such as gold nanoparticles.

this target delivery indeed resulted in a reduction and prevention of tumor growth in a mouse model (Hu-Lieskovan et al. 2005). Although no clinical sign of toxicity was reported in a nonhuman primate experiment (Heidel et al. 2007). Recently, the synthetic polycation poly-dimethylaminoethyl methacrylate (poly-DMAEMA), was used for siRNA delivery, particularly for designing pH-sensitive response polymeric micelles. This system was highlighted in effective endosomal escape after endocytosis and the system consisted of cationic poly-DMAEMA to mediate siRNA and a pH-responsive block that is composed of DMAEMA, anionic propylacrylic acid, and hydrophobic butyl methacrylate. Consequently, endosomal release of the siRNA was indicated as well as significant mRNA knockdown with low-cytotoxicity *in vitro* experiments (Convertine et al. 2010).

On the other hand, natural cationic polymers are more preferable as siRNA carriers as they are biocompatible, biodegradable, and nontoxic. The most commonly used polymers are chitosan nanocarriers (Figure 21.1b). In one study, 84% deacetylated chitosan (114 kDa)–siRNA polymer was designed and intranasally administered to enhanced green fluorescent protein (EGFP)-tagged mice for 5 consecutive days. These polymers were there through reduced bronchiolo–epithelial EGFP expression (Howard et al. 2006). Pillé et al. (2006) demonstrated intravenous delivery of anti-RhoA siRNA using chitosan-coated polyisohexylcyanoacrylate nanocarriers in xenografted breast cancer mouse model and thus indeed indicated nontoxic delivery and reduction in tumor growth and angiogenesis. Atelocollagen, a purified pepsin-treated type I collagen from calf dermis, is also a natural cationic polymer. Basically, atelocollagen is characterized by thermosensitive changeability because it changes its liquid property at 4°C into a gel at 37°C. Although telopeptide deficiency is known to trigger immunogenicity to a certain extent, atelocollagen showed increased cellular uptake as well as stability of the genes and oligoculeotides (Ochiya et al. 1999; Hanai et al. 2006). In previous research, 300 kDa atelocollegens, complexed with siRNA, were given intratumorally to mice with xenograft tumors and intravenously to mice with metastatic bone tumors, and resulted in a reduction and inhibition of tumor growth, respectively (Minakuchi et al. 2004; Takeshita et al. 2005).

The application of a biodegradable poly(DL-lactide-co-glycolide) (PLGA) polymer has recently emerged as a siRNA delivery tool (Figure 21.1c). In a previous study, siRNA encapsulation in the core of PLGA nanocarriers was achieved by double-emulsion solvent evaporation and produced <300 nm nanocarriers (Cun et al. 2010). For efficient endosomal escape, another preparation method was also suggested. PLGA was conjugated to the 3′-end of siRNA by a disulfide bond, and the conjugates were then self-assembled in a core–shell structure (~20 nm) in an aqueous condition (Lee et al. 2010). One problematic issue with PLGA nanocarriers is its negative charge. In this respect, cationic PEI can be complexed with siRNA to achieve a positive charge on the surface layer (Lee et al. 2010). To enhance biocompatibility and biodegradability, natural cationic polymers can also be used. Chitosan-modified PLGA indeed showed higher encapsulation efficiency and more prolonged half-life of siRNA (Yuan et al. 2010; Tahara et al. 2010).

The dendrimeric structure (Figure 21.1e) has shed light on siRNA delivery because its multivalent polycationic property creates strong and stable binding with nucleic acids. It has a tree-like structure including hyperbranched polymer chains (Lee et al. 2005). Poly(amidoamine) (PAMAM) is the most widely used dendrimeric structure and it has been demonstrated that the ability of siRNA adaptation depends on the generation of PAMAM produced, and on the pH due to backfolding (Pavan et al. 2010). For the efficient and safe delivery of siRNA, there are ongoing studies to improve the system using PAMAM. One study, for example, attempted to modify fifth-generation PAMAM with the addition of cyclic RGD peptide, and siRNA delivery in a multicellular spheroid model of malignant glioma indicated a potential role for penetration into malignant tumors and interference with tumor angiogenesis (Waite and Roth 2009).

21.6 BIOPARTICLE-LIKE BASED DELIVERY

Liposomes are one of the lipid-based systems developed for *in vivo* delivery of siRNA (Figure 21.1d). Cationic liposomes have been predominantly used for interfering RNA since they were demonstrated for nucleic acid delivery in *in vitro* studies (Felgner et al. 1987; Wu and Wu 1987). The advantages of using liposomes include the simple manipulation needed to form complexes with negatively charged siRNA, its high transfection efficiency, its low cytotoxicity, and its immunogenicity. The use of liposomes can also protect siRNA from enzymatic degradation and reduce renal clearance (Oh and Park 2009).

Cationic liposomes require efficient cationic lipid sources to form a nanoparticle. For instance, *N*-[1-(2,3-dioleyloxy)propyl]-*N*,*N*,*N*-trimethylammoium chloride was the first synthesized cationic lipid for plasmid DNA transfer into mammalian cells (Felgner and Ringold 1989). Since siRNA emerged as a potential drug, cationic lipid nanoparticles have become a routine delivery system for siRNA and therefore various types of cationic lipids have been generated. Dioleoyl-L-α-phosphatidylethanolamine (DOPE) and cholesterol are the most common cationic lipid sources that readily mix with nucleic acids for cationic liposome formation (Martin et al. 2005). Recently, *N′*,*N′*-dioctadecyl-*N*-4,8-diaza-10-aminodecanoylglycine amide (DODAG), a cationic lipid, was applied to anti-HBV siRNA delivery and the system indicated minimal toxicity and immune stimulation in *in vivo* conditions (Mével et al. 2010). In another study, siRNA was delivered by cationic liposomes, formulated by the cationic lipids, 1,2-dioleoyl-3-trimethylammoium propane (DOTAP) with DOPE, cholesterol, and poly-L-arginine-conjugated PEG (PLR-PEG). It was reported that this liposome showed high siRNA transfection efficiency with low cytotoxicity in an *in vitro* environment compared with conventional cationic liposomes without PEGylation (Kim et al. 2010).

Exosomes can also be used in siRNA delivery. Exosome, a natural intracellular molecule (50–100 nm), is endocytic in origin and is derived from invaginations and budding from the limiting membrane of the multivesicular body with plasma membrane. Its functions are not well understood but it is believed that they mediate immune response and affect the function of neighboring cells by intracellular transfer of mRNA, miRNA, receptors, and enzymes (Boorn et al. 2011). Exosomes may be good candidates as siRNA carriers for the following reasons. First, it should be less immunogenic because exosomes will be collected from the patients' own cells. Second, exosome is relatively stable because expression of CD55 and CD59 on its surface is likely to bypass opsonins, coagulation factors, and complement. In particular, the tetraspanin C9 on its surface facilitates membrane fusion in the target cells and releases its contents into the cytosol. Moreover, the small size of the endosome (<100 nm) does not induce phagocytosis (Boorn et al. 2011). One recent study demonstrated siRNA delivery using exosomes. This process was achieved by engineering the dendritic cells to express Lamp2b, an exosomal membrane protein transfected by neuron-specific RVG peptide plasmid. RVG exosomes, derived from dendritic cells, were then purified and GAPDH siRNA was loaded through electroporation. This exosomal delivery indicated strong mRNA and protein knockdowns of target mRNA after systemic injection in mouse (Alvarez-Erviti et al. 2011).

21.7 LAYER-BY-LAYER DELIVERY FOR THE PROTECTION OF CARRIERS

Recently, siRNA was conjugated to the surface of gold nanoparticles (AuNPs) by a thiol group (Figure 21.1f). Their nanocarriers showed prolonged half-lives and improved gene knockdown (Giljohann et al. 2009). There are several advantages with the use of AuNPs such as biocompatibility, straightforward synthesis with easy modification, and availability of diverse size distributions. However, problematic issues can also be shared with their use because AuNPs tend to aggregate even after interaction with nucleic acid, and its system also showed the low efficiency of nucleic acid activity after delivery. Moreover, the system showed dependence on an additional transfection reagent for the efficacy of the nucleic acid. Consequently, one study demonstrated layer-by-layer assembled AuNPs for siRNA delivery to overcome the problems of AuNP–siRNA conjugates. These polyelectrolyte layers are composed of oppositely charged layers of cationic PEI and anionic siRNA (Elbakry et al. 2009). There are other layer-by-layer coated gold nanoparticles using a combination of different cationic polyelectrolytes (PLL representing protease degradability). Their layers contained three PLL layers, incorporated siRNA layers, and therefore, the nanoparticles released siRNA gradually and showed extensive gene silencing effect (Lee et al. 2011).

21.8 siRNA DELIVERY IN CANCER TREATMENT

Every year, more than 10 million cancer cases are diagnosed but a better understanding of tumor biology and improved diagnostic and treatment technologies has reduced mortality in patients with cancer. Currently, cancer can be treated by using surgical intervention, radiation, and chemotherapeutic drugs; however, these methods have always exposed the patient to side effects that kills healthy cells and triggers toxicity (Peer et al. 2007). In this respect, RNAi could potentialy be used in cancer treatment as a less invasive tool. Because of the developments in diagnosis and treatment in molecular genetics, there have been many types of applicant genes and its related pathways identified (Vogelstein and Kinzler 2004). siRNA therapy in cancer is generally dependent on the manipulation of cell growth and survival or death signaling pathways, for example, 1) the inhibition of angiogenesis by inhibiting VEGF signaling; 2) the inhibition of tumor survival and induction of apoptosis by inhibiting the Wnt pathway, NF-kβ (Nuclear Factor-Kappaβ) pathway, EGFR (epidermal growth factor receptor), *Her-2/neu*, telomerase, MDM2, and p53 or Bcl-2; 3) the enhancement of radiosensitivity or chemosensitivity by inhibiting MDR1; and 4) combination siRNA therapy targeting multiple oncogenic pathways (Tseng et al. 2009). Using siRNA targeting a wide range of candidate genes, there are various types of siRNA delivery systems that have been introduced in many different types of cancers (Table 21.2).

21.8.1 Breast Cancer

All over the world, breast cancer is responsible for most cancer deaths in women (23% of all cancers; Parkin et al. 2005). In the development of potential approaches, various siRNA delivery systems have been attempted for breast cancer treatment. In one study, chitosan nanoparticles were designed by encapsulating quantum dots in the core and thus HER2/neu siRNA was coated in the outer layer through electrostatic interaction. It was reported that the system was effectively internalized into SKBR3 breast cancer cell lines with the silencing effect of HER2/neu siRNA (Tan et al. 2007). Cationic liposomes, composed of siRNA targeting cyclin D1, dioctadecylamidoglycylspermidine, and DOPE were also used to transfect MCF-7 breast cancer cells with low cytotoxicity and high uptake of siRNA (Lavigne and Thierry 2007). This group again demonstrated the successful inhibition of plasminogen activator inhibitor-1 (proto-oncogene) mRNA in the

TABLE 21.2
Current siRNA Strategy in Cancer Treatment

Delivery Systems	Target Gene	Types of Cancer	Reference
Liposome (SNALP)	HBV	Liver cancer	Morrissey et al. 2005b
Cationic liposome	Bcl-2	Liver metastasis	Yano et al. 2004
Cationic liposome	Integrin αv	Prostate cancer	Bisanz et al. 2005
Cationic liposome	CD31	Prostate cancer	Santel et al. 2006
Cationic liposome	Bcl-2	Prostate cancer	Yano et al. 2004
Cationic liposome	scL-siRNA targeting metastasis	Lung cancer	Pirollo et al. 2006
Cationic liposome	Cyclin D1	Breast cancer	Lavigne and Thierry 2007
Cationic cardiolipin liposome	Raf-1	Prostate cancer	Pal et al. 2005
Cationic cardiolipin analogue-based liposomes	c-raf	Breast cancer	Chien et al. 2005
Hyaluronic acid-cationic liposome	Antitelomerase	Lung cancer	Taetz et al. 2009
R8 (arginine peptide)-liposome	HDM2	Lung cancer	Zhang et al. 2006a
Neutral liposome (DOPC)	EphA2	Ovarian cancer	Landen et al. 2006
Neutral liposome (DOPC)	FAK	Ovarian cancer	Halder et al. 2006
Neutral liposome (DOPC)	ADRB2	Ovarian cancer	Thaker et al. 2006
Neutral liposome (DOPC)	IL-8	Ovarian cancer	Merritt et al. 2008
Liposome-polycation-DNA	EGFR	Lung cancer	Li et al. 2007
Cationic immunoliposome	EGFR	Lung cancer	Pirollo et al. 2006
Immunoliposome	Her-2	Breast cancer	Hogrefe et al. 2006
CaCO3 Nanoparticle	VEGF	Gastric cancer	He et al. 2008
Chitosan-coated nanoparticle	RhoA	Breast cancer	Pillé et al. 2006
Chitosan-coated nanoparticle	HER2/neu	Breast cancer	Tan et al. 2007
Chitosan-nanoparticle	POSTN, FAK, PLSDC1	Ovarian cancer	Han et al. 2010
Folated lipid nanoparticle	Her-2	Nasopharyngeal cancer	Yoshizawa et al. 2008
PEG-LPD nanoparticle	EGFR	Lung cancer	Chen et al. 2009
PEI polymer	Her-2	Ovarian cancer	Urban-Klein et al. 2005
PEI polymer	PTN	Orthotopic glioblastoma	Grzelinski et al. 2006
PEI-G-PEG Copolymer	[a]	Breast cancer	Sutton et al. 2006
Cationic polymer (chitosan-graft-PEI)	EGFP, Akt1	Lung cancer	Jere et al. 2009
Poly(ester amine) polymer	Akt1	Urethane-induced lung cancer	Xu et al. 2008
Polycationic polymer	VEGF	Prostate cancer	Kim et al. 2009
Amphiphilic copolymer	Bcl-2	Breast cancer	Wang et al. 2006

(continued)

TABLE 21.2 (Continued)
Current siRNA Strategy in Cancer Treatment

Delivery Systems	Target Gene	Types of Cancer	Reference
Cationic micelle	VEGF	Breast cancer	Zhu et al. 2010
Atelocollagen	HPV18 type E6 and E7	Cervical cancer	Fujii et al. 2006
Atelocollagen	Bcl-xL	Prostate cancer	Mu et al. 2009
Chemical modification	HBV	HBV vector	Morrissey et al. 2005a
Carbon nanotube	TERT	Lewis lung tumor	Zhang et al. 2006b
Cyclodextrin-containing polycation	EWS-FLI1	Metastatic Ewing's sarcoma	Hu-Lieskovan et al. 2005
Aptamer (PSMA)	Plk-1	Prostate cancer	Dassie et al. 2009
PAMAM dendrimer	Hsp27	Prostate cancer	Liu et al. 2009
Fusion protein (Protamine, HIV-1 envelop Ab)	c-myc, MDM2, VEGF	Subcutaneous B16 melanoma tumor	Song et al. 2005
Electroporation	EGFP	B16F10 melanoma tumor cells	Golzio et al. 2007
Bioparticle (HAS)	[b]	Breast cancer	Abbasi et al. 2011
Bioparticle (rHDL)	FAK, STAT3	Ovarian cancer	Shahzad et al. 2011
Polyelctrolytic complex (PEC)	VEGF	Ovarian cancer	Kim et al. 2008

Source: Oh, Y.-K., and T. G. Park. *Adv. Drug Del. Rev.*, 61, 850–862, 2009.

[a] Signaling peptide of secretory clusterin (sCLU).

[b] AllStars Cell Death siRNA (Qiagen) targeting ubiquitously expressed human genes essential for cell survival.

human breast cancer cell line MDA-MB-231 using the same cationic liposome system (Meryet-Figuières et al. 2007). In addition, a codelivery system for breast cancer treatment was introduced. The system, delivering paclitaxel together with siRNA targeting Bcl-2, was achieved by self-assembled amphiphillic copolymers and was shown to increase the sensitivity of MDA-MB-231 human breast cancer cells to paclitaxel (Wang et al. 2006). In a recent study, another codelivery system was designed by biodegradable cationic micelles that were prepared from PDMAEMA–PCL–PDMAEMA triblock copolymers. VEGF siRNAs were complexed with PDMAEMA and paclitaxel drug was loaded into the core of PCL. This system, transfected to MDA-MB-231, also improved gene knockdown efficiency with low cytotoxicity (Zhu et al. 2010). Moreover, novel biocompatible nanocarriers have also been developed lately. Human serum albumin was used as a scaffold core and coated with siRNA-loaded PEI. In this system, AllStars Cell Death siRNA (Qiagen, Valencia, CA) was delivered into MCF-7 breast cancer cells, which led to cell death (Abbasi et al. 2011). Furthermore, there are more siRNA carriers available for breast cancer, such as cationic immunoliposomes modified with antitransferrin receptor single-chain antibody fragment (Hogrefe et al. 2006) and copolymers of PEI and PEG (PEI-g-PEG; Sutton et al. 2006).

21.8.2 Prostate Cancer

Prostate cancer is the fifth most common cancer worldwide and is the second most common cancer among men (Parkin et al. 2005). Previously, most studies regarding prostate cancer treatment mentioned cationic liposomes with different modifications carrying siRNA targeting Raf-1, anti-integrin alpha V, or anti-CD31 *in vivo* studies (Oh and Park 2009). Recently, there have been many trials with different siRNA carrier systems. One study systemically delivered siRNA targeting Bcl-xL, anti-apoptotic factor into PC-3, and human prostate cancer cells via Bcl-xL-atelocollagen complexes; no side effects were observed with this system. They further suggested the use of this system along with cisplatin (CDDP), and showed a significant inhibition effect compared with CDDP alone (Mu et al. 2009). Triethanolamine core PAMAM dendrimers were also used for Hsp27 siRNA delivery into PC-3 cells. These dendrimer complexes showed efficient delivery of Hsp27 siRNA with a potent anticancer effect representing knockdown of Hsp27 mRNA (Liu et al. 2009). On the basis of the peptide delivery system, VEGF siRNA was also delivered into PC-3 cells using prostate cancer-binding peptide (PCP) conjugated with PEI via a PEG linker. Therefore, the polycationic carrier (~150 nm), VEGF siRNA/PEI-PEG-PCP, enhanced gene silencing of VEGF activity under serum conditions (Kim et al. 2009). In addition, an aptamer-based system was applied for siRNA delivery by one research group. PSMA binding aptamer was conjugated with Plk-1 siRNA in their first demonstration and the first generation of PSMA-Plk1 nanocarrier was recently modified to increase the efficiency of enzyme dicer recognition and favor the loading of the siRNA antistrand by RISC. Therefore, the aptamer portion of a A10-Plk1 chimera (first generation) was reduced to a 39-nucleotide base with a 2′-fluoropyrimidine modification at the 3′-end of the longer strand and the 3′-end of the siRNA portion was modified by a 2-nucleotide (UU) overhang at the 3′-end sense strand, or introducing wobble base pairs (C→U) in the sense strand. This system demonstrated potent antitumor activity after systemic administration of their optimized chimeras in an *in vivo* experiment (Dassie et al. 2009).

21.8.3 Lung Cancer

Because the incidence of lung cancer has increased globally, especially among men (Parkin et al. 2005), numerous approaches, pronominally lipid-based systems have been applied to siRNA delivery in lung cancer. One example is by the use of liposomes with arginine octamers (R8-liposome). Thus, human double-minute gene 2 (*HDM2*), loaded into R8-liposome demonstrated a significant stability in the blood serum with higher transfection efficiency in lung tumor cells, including lung squamous cell carcinoma (SK-MES-1), non–small cell lung carcinoma (A549), and small cell lung carcinoma (NCI-H446). Moreover, the system indicated effective gene silencing effect and reduction in the proliferation of cancer cells (Zhang et al. 2006a). Lipid polycation-DNA (LPD) was also developed using siRNA targeting human survivin in PEGylated LPD nanoparticles consisting of di-stearoyl-phosphatidyl-ethanolamine (DSPE)-PEG-anisamide (AA), which showed antitumor effects with improved stability (Li and Huang 2006). Recently, their group designed another LPD-PEG-AA-containing DOTAP or *N,N*-di-stearoyl-*N*-methyl-*N*-2[*N'*-(*N*6-guanidino-L-lysinyl)] aminoethyl ammonium chloride (DSGLA). LPD-PEG-AA with either combination efficiently delivering and effectively silencing EGFR siRNA in H460 cells; however, a more synergistic cell killing effect was observed after DSGLA-PEG-AA delivery (Chen et al. 2009). Also, cationic liposome assembly by DOTAP and DOPE has been used with targeted delivery of siRNA delivery for lung cancer. One study modified the cationic liposomes by conjugation with an antitransferrin receptor single-chain antibody fragment and delivered them into animal lung cancer models induced by MDA435/LCC6 (Pirollo et al. 2006). Another recent study used hyaluronic acid modification with cationic liposomes and delivered antitelomerase siRNA into CD44 receptor–expressing lung cancer cells with improved stability as well as reduced cytotoxicity (Taetz et al. 2009). The most recent cationic polymers in siRNA delivery in lung cancer are based on chitosan-graft-PEI (CHI-g-PEI) copolymers that displayed efficient delivery of EGFP and Akt1 siRNA and thereby silenced EGFP expression and oncoprotein Akt1, respectively (Jere et al. 2009).

21.8.4 Ovarian Cancers

Recently, VEGF siRNA was delivered into ovarian cancer cells (A2780 and SK-OV-3) using a polyelectrolytic complex (PEC). A luteinizing hormone-releasing hormone (LHRH), cancer-targeting ligand peptide was coupled with a PEG–siRNA conjugate to form PEC and PEC (siRNA-PEG-LHRL), self-assembled with the addition of PEI via ionic interaction. As a result, these PEC micelles displayed the silencing efficiency of VEGF expression by a LHRH receptor-specific manner (Kim et al. 2008). Another study developed RGD peptide-labeled chitosan nanoparticles (RGD-CH-NP) targeting multiple growth-promoting genes, POSTN, FAK, and PLSDC1, in orthotopic animal models of ovarian cancer (SKOV3ip1, HeyA8, and A2780 models) and demonstrated their silencing efficacy. They further investigated the system with PLXDC1 siRNA into the $\alpha_v\beta_3$ integrin-positive tumor endothelial cells in the A2780 tumor-bearing mice and showed significant inhibition of its tumor growth (Han et al. 2010). In tumor biology, for maintaining its growth, tumor cells are known to eliminate high-density lipoprotein (HDL) particles by overexpressing its receptors (scavenger receptor type B1). With this in mind, one study formulated reconstituted HDL (rHDL) nanoparticles incorporated into siRNA. They delivered siRNA targeting FAK or STAT3 involved in cancer growth and metastasis into orthotopic ovarian cancer mice models (SKOV3ip1 and HEYA8-MDR) and demonstrated efficient inhibition of these two protein expressions in ovarian cancers (Shahzad et al. 2011).

21.9 FUTURE PROSPECTS

Since the comprehensive understanding of RNAi, siRNA has been known as a promising agent to inhibit the expression of unwanted proteins, especially in cancer treatment, because a tumor is formed by the overexpression of specific proteins or by the suppression of a protein inducing apoptosis. Therefore, different approaches have been developed to deliver target siRNA into the tumor region. Many nonviral nanodelivery systems have been introduced using siRNA chemical modification, biodegradable polymeric nanoparticles, polyplexes, liposomes, lipid nanoparticles, lipoplexes, and dendrimers. Accordingly, most studies demonstrated promising results as efficient transfection and inhibition using targeting siRNA *in vitro* experiments and their systems indeed showed the silencing efficacy of siRNA when locally administered in *in vivo* mouse models. Nevertheless, only a few types of siRNA nanocarriers have entered clinical trials because most of the nanocarriers developed thus far are often faced with issues regarding stability hindered by serum nuclease activity, hepatic and renal clearance, aggregation with other biomolecules, or immune response. Because systemic delivery is necessary in cancer treatment, there must be a more promising manner to bypass such problems. The best way to overcome these issues is to use biocompatible and biodegradable biomolecules. One recently introduced example is the exosome. Derived from dendritic cells of recipient *in vitro* level, exosomes can be loaded with siRNA that can be delivered directly into the cytoplasm, that is, siRNA-loaded exosomes can bypass the immune response during internalization. rHDL is another example of such a biomolecule. However, liposomal delivery, which is already known for biocompatibility, also shows somewhat reduced stability because of its tendency toward degradation or aggregation with other serum molecules such as albumin. As a result, it sometimes leads to leakage of the content from the liposome (Takeuchi et al. 2001). Therefore, more promising results should be demonstrated in the use of biomolecule-based systems in *in vivo* experiments. Apart from biomolecules, enhanced stability of any sort of systemic delivery system could also be achieved using a layer-by-layer system. This delivery system concept was demonstrated by liposome and doxorubicin drug delivery using a natural polymer, anionic arginine and cationic chitosan, consisting of several layers. Accordingly, the system indeed showed improved stability (Zhao et al. 2007; Haidar et al. 2008). Likewise, any nanocarriers coupled with siRNA could be coated with several layers of polyelectrolytes for their protection during delivery. Thus, more investigations are needed to support the safety of siRNA delivery using the layer-by-layer system. Another crucial issue could be derived from siRNA processing in the RNAi mechanism. The length of siRNA is 21–23 bp to avoid the IFN response; however, siRNA has a potential problem that could induce off-target effects after its release from the endosome. This is mainly because siRNA is not optimally synthesized for the specific mRNA target, or else, there are some homologue sequences that share other mRNA sequences. Consequently, an optimal RISC substrate is necessary for target-specific silencing. In this respect, long dsRNAs can be used instead of siRNA. dsRNA is suspected to be internalized through systemic RNAi-deficient 1 (Duxbury et al. 2005) and they are cleaved into numerous siRNA (21–23 bp), which would be optimal for a RISC substrate as well as for target mRNA in the cytoplasm of mammalian cells. However, It has been understood that long dsRNAs (>30 bp) induce an IFN response (Strat et al. 2006). Therefore, before the use of long dsRNA delivery for therapeutic purposes, an innovative strategy would be required to bypass the initial immune response. To date, the significant problem in most systems is insufficient information about the evaluation of toxicity when the siRNA carrier is systemically administered in *in vivo* models. For clinical use, thorough verification at the internationally recognized level is essential. All in all, the achievement of effective and optimized delivery systems with thorough verification would be the key for systemic administration in cancer treatment.

ACKNOWLEDGMENTS

This work was supported by the South Korean Ministry of Knowledge and Education and the Incheon Free Economic Zone in the framework of several founding operating grants to the Utah–Inha Drug Delivery Systems and Advanced Therapeutics Research Centre. The authors thank T. G. Ramasamy for literature compilation and V. S. Joo for grammatical editing.

REFERENCES

Abbasi, S., A. Paul, and S. Prakash. 2011. Investigation of siRNA-loaded polyethylenimine-coated human serum albumin nanoparticle complexes for the treatment of breast cancer. *Cell Biochemistry and Biophysics* 61(2), 277–287.

Abdallah, B., A. Hassan, C. Benoist, D. Goula, J. P. Behr, and B. A. Demeneix. 1996. A powerful nonviral vector for *in vivo* gene transfer into the adult mammalian brain: Polyethylenimine. *Human Gene Therapy* 7, 1947–1954.

Aigner, A. 2007. Nonviral *in vivo* delivery of therapeutic small interfering RNAs. *Current Opinion in Molecular Therapeutics* 9, 345–352.

Alvarez-Erviti, L., Y. Seow, H. Yin, C. Betts, S. Lakhal, and M. J. A. Wood. 2011. Delivery of siRNA to the mouse brain by systemic injection of targeted exosomes. *Nature Biotechnology* 29, 341–345.

Bisanz, K., J. Yu, M. Edlund, B. Spohn, M.-C. Hung, L. W. K. Chung, and C.-L. Hsieh. 2005. Targeting ECM–integrin interaction with liposome-encapsulated small interfering RNAs inhibits the growth of human prostate cancer in a bone xenograft imaging model. *Molecular Therapy* 12, 634–643.

Bishop, N. E. 1997. An update on non-clathrin-coated endocytosis. *Reviews in Medical Virology,* 7, 199–209.

Boeckle, S., K. V. Gersdorff, S. V. D. Piepen, C. Culmsee, E. Wagner, and M. Ogris. 2004. Purification of polyethylenimine polyplexes highlights the role of free polycations in gene transfer. *The Journal of Gene Medicine* 6, 1102–1111.

Boorn, J. G. V. D., M. Schlee, C. Coch, and G. Hartmann. 2011. siRNA delivery with exosome nanoparticles. *Nature Biotechnology* 29, 325–326.

Boussif, O., M. A. Zanta, and J. P. Behr. 1996. Optimized galenics improve *in vitro* gene transfer with cationic molecules up to 1000-fold. *Gene Therapy,* 3, 1074–1080.

Braasch, D. A., S. Jensen, Y. Liu, K. Kaur, K. Arar, M. A. White, and D. R. Corey. 2003. RNA interference in mammalian cells by chemically-modified RNA. *Biochemistry,* 42, 7969–7975.

Braasch, D. A., Z. Paroo, A. Constantinescu, G. Ren, O. K. Öz, R. P. Mason, and D. R. Corey. 2004. Biodistribution of phosphodiester and phosphorothioate siRNA. *Bioorganic and Medicinal Chemistry Letters* 14, 1139–1143.

Bramsen, J. B., M. B. Laursen, C. K. Damgaard, S. W. Lena, B. R. Babu, J. Wengel, and J. Kjems. 2007. Improved silencing properties using small internally segmented interfering RNAs. *Nucleic Acids Research* 35, 5886–5897.

Bumcrot, D., M. Manoharan, V. Koteliansky, and D. W. Y. Sah. 2006. RNAi therapeutics: A potential new class of pharmaceutical drugs. *Nature Chemical Biology* 2, 711–719.

Chen, S.-H., and G. Zhaori. 2010. Potential clinical applications of siRNA technique: Benefits and limitations. *European Journal of Clinical Investigation* 41, 221–232.

Chen, Y., J. Sen, S. R. Bathula, Q. Yang, R. Fittipaldi, and L. Huang. 2009. Novel cationic lipid that delivers siRNA and enhances therapeutic effect in lung cancer cells. *Molecular Pharmaceutics* 6, 696–705.

Cherng, J.-Y., H. Talsma, R. Verrijk, D. J. A. Crommelin, and W. E. Hennink. 1999. The effect of formulation parameters on the size of poly((2-dimethylamino)ethyl methacrylate)-plasmid complexes. *European Journal of Pharmaceutics and Biopharmaceutics* 47, 215–224.

Chien, P.-Y., J. Wang, D. Carbonaro, S. Lei, B. Miller, S. Sheikh, S. M. Ali et al. 2005. Novel cationic cardiolipin analogue-based liposome for efficient DNA and small interfering RNA delivery *in vitro* and *in vivo*. *Cancer Gene Therapy* 12, 321–328.

Chiu, Y.-L., A. Ali, C.-Y. Chu, H. Cao, and T. M. Rana. 2004. Visualizing a correlation between siRNA localization, cellular uptake, and RNAi in living cells. *Chemistry and Biology* 11, 1165–1175.

Chiu, Y.-L., and T. M. Rana. 2003. siRNA function in RNAi: A chemical modification analysis. *RNA* 9, 1034–1048.

Choung, S., Y. J. Kim, S. Kim, H.-O. Park, and Y.-C. Choi. 2006. Chemical modification of siRNAs to improve serum stability without loss of efficacy. *Biochemical and Biophysical Research Communications* 342, 919–927.

Chu, C.-Y., and T. M. Rana. 2008. Potent RNAi by short RNA triggers. *RNA* 14, 1714–1719.

Chu, T. C., K. Y. Twu, A. D. Ellington, and M. Levy. 2006. Aptamer mediated siRNA delivery. *Nucleic Acids Research* 34, e73.

Convertine, A. J., C. Diab, M. Prieve, A. Paschal, A. S. Hoffman, P. H. Johnson, and P. S. Stayton. 2010. pH-responsive polymeric micelle carriers for siRNA drugs. *Biomacromolecules* 11, 2904–2911.

Crombez, L., G. Aldrian-Herrada, K. Konate, Q. N. Nguyen, G. K. McMaster, R. Brasseur, F. Heitz, and G. Divita. 2009. A new potent secondary amphipathic cell-penetrating peptide for siRNA delivery into mammalian cells. *Molecular Therapy* 17, 95–103.

Cun, D., C. Foged, M. Yang, S. Frøkjær, and H. M. Nielsen. 2010. Preparation and characterization of poly(dl-lactide-co-glycolide) nanoparticles for siRNA delivery *International Journal of Pharmaceutics* 390, 70–75.

Czauderna, F., M. Fechtner, S. Dames, H. Aygün, A. Klippel, G. J. Pronk, K. Giese, and J. Kaufmann. 2003. Structural variations and stabilising modifications of synthetic siRNAs in mammalian cells. *Nucleic Acids Research* 31, 2705–2716.

Dande, P., T. P. Prakash, N. Sioufi, H. Gaus, R. Jarres, A. Berdeja, E. E. Swayze et al. 2006. Improving RNA interference in mammalian cells by 4′-thio-modified small interfering RNA (siRNA): Effect on siRNA activity and nuclease stability when used in combination with 2′-*O*-alkyl modifications. *Journal of Medicinal Chemistry* 49, 1624–1634.

Dassie, J. P., X.-Y. Liu, G. S. Thomas, R. M. Whitaker, K. W. Thiel, K. R. Stockdale, D. K. Meyerholz et al. 2009. Systemic administration of optimized aptamer-siRNA chimeras promotes regression of PSMA-expressing tumors. *Nature Biotechnology* 27, 839–846.

Davis, M. E. 2002. Non-viral gene delivery systems. *Current Opinion in Biotechnology* 13, 128–131.

Davis, M. E., J. E. Zuckerman, C. H. J. Choi, D. Seligson, A. Tolcher, C. A. Alabi, Y. Yen et al. 2010. Evidence of RNAi in humans from systemically administered siRNA via targeted nanoparticles. *Nature* 464, 1067–1070.

Deleavey, G. F., J. K. Watts, and M. J. Damha. 2009. Chemical modification of siRNA. *Current Protocols in Nucleic Acid Chemistry* 39, 16.3.1–16.3.22.

Duxbury, M. S., S. W. Ashley, and E. E. Whang. 2005. RNA interference: A mammalian SID-1 homologue enhances siRNA uptake and gene silencing efficacy in human cells. *Biochemical and Biophysical Research Communications* 331, 459–463.

Elbakry, A., A. Zaky, R. Liebl, R. Rachel, A. Goepferich, and M. Breunig. 2009. Layer-by-layer assembled gold nanoparticles for siRNA delivery. *Nano Letters* 9, 2059–2064.

Elbashir, S. M., J. Harborth, W. Lendeckel, A. Yalcin, K. Weber, and T. Tuschl. 2001a. Duplexes of 21-nucleotide RNAs mediate RNA interference in cultured mammalian cells. *Nature* 411, 494–498.

Elbashir, S. M., J. Martinez, A. Patkaniowska, W. Lendeckel, and T. Tuschl. 2001b. Functional anatomy of siRNAs for mediating efficient RNAi in *Drosophila melanogaster* embryo lysate. *EMBO Journal* 20, 6877–6888.

Elmén, J., H. Thonberg, K. Ljungberg, M. Frieden, M. Westergaard, Y. Xu, B. Wahren et al. 2005. Locked nucleic acid (LNA) mediated improvements in siRNA stability and functionality. *Nucleic Acids Research* 33, 439–447.

Felgner, P. L., T. R. Gadek, M. Holm, R. Roman, H. W. Chan, M. Wenz, J. P. Northrop et al. 1987. Lipofection: A highly efficient, lipid-mediated DNA-transfection procedure. *Proceedings of the National Academy of Sciences of the United States of America* 84, 7413–7417.

Felgner, P. L., and G. M. Ringold. 1989. Cationic liposome-mediated transfection. *Nature* 337, 387–388.

Fire, A., S. Xu, M. K. Montgomery, S. A. Kostas, S. E. Driver, and C. C. Mello. 1998. Potent and specific genetic interference by double-stranded RNA in *Caenorhabditis elegans. Nature* 391, 806–811.

Fischer, D., A. V. Harpe, K. Kunath, H. Petersen, Y. Li, and T. Kissel. 2002. Copolymers of ethylene imine and *N*-(2-hydroxyethyl)-ethylene imine as tools to study effects of polymer structure on physicochemical and biological properties of DNA complexes. *Bioconjugate Chemistry* 13, 1124–1133.

Fujii, T., M. Saito, E. Iwasaki, T. Ochiya, Y. Takei, S. Hayashi, A. Ono et al. 2006. Intratumor injection of small interfering RNA-targeting human papillomavirus 18 E6 and E7 successfully inhibits the growth of cervical cancer. *International Journal of Oncology* 29, 541–548.

Gaglione, M., and A. Messere. 2010. Recent progress in chemically modified siRNAs. *Mini Reviews in Medicinal Chemistry,* 10, 578–595.

Geary, R. S., R. Z. Yu, and A. A. Levin. 2001. Pharmacokinetics of phosphorothioate antisense oligodeoxynucleotides. *Current Opinion in Investigational Drugs* 2, 562–571.

Giljohann, D. A., D. S. Seferos, A. E. Prigodich, P. C. Patel, and C. A. Mirkin. 2009. Gene regulation with polyvalent siRNA–nanoparticle conjugates. *Journal of the American Chemical Society* 131, 2072–2073.

Golzio, M., L. Mazzolini, A. Ledoux, A. Paganin, M. Izard, L. Hellaudais, A. Bieth et al. 2007. *In vivo* gene silencing in solid tumors by targeted electrically mediated siRNA delivery. Electrically mediated gene silencing in tumors. *Gene Therapy* 14, 752–759.

Grzelinski, M., B. Urban-Klein, T. Martens, K. Lamszus, U. Bakowsky, S. Höbel, F. Czubayko et al. 2006. RNA interference-mediated gene silencing of pleiotrophin through polyethylenimine-complexed small interfering RNAs *in vivo* exerts antitumoral effects in glioblastoma xenografts. *Human Gene Therapy* 17, 751–766.

Haidar, Z. S., R. C. Hamdy, and M. Tabriziana. 2008. Protein release kinetics for core–shell hybrid nanoparticles based on the layer-by-layer assembly of alginate and chitosan on liposomes. *Biomaterials* 29, 1207–1215.

Halder, J., A. A. Kamat, C. N. Landen, Jr. L. Y. Han, S. K. Lutgendorf, Y. G. Lin, W. M. Merritt et al. 2006. Focal adhesion kinase targeting using *in vivo* short interfering RNA delivery in neutral liposomes for ovarian carcinoma therapy. *Clinical Cancer Research* 12, 4916–4924.

Hall, A. H. S., J. Wan, E. E. Shaughnessy, B. R. Shaw, and K. A. Alexander. 2004. RNA interference using boranophosphate siRNAs: Structure–activity relationships *Nucleic Acids Research* 32, 5991–6000.

Han, H. D., L. S. Mangala, J. W. Lee, M. M. K. Shahzad, H. S. Kim, D. Shen, E. J. Nam et al. 2010. Targeted gene silencing using RGD-labeled chitosan nanoparticles. *Clinical Cancer Research* 16, 3910–3922.

Hanai, K., F. Takeshita, K. Honma, S. Nagahara, M. Maeda, Y. Minakuchi, A. Sano et al. 2006. Atelocollagen-mediated systemic DDS for nucleic acid medicines. *Annals of the New York Academy of Sciences* 1082, 9–17.

He, X.-W., T. Liu, Y.-X. Chen, D.-J. Cheng, X.-R. Li, Y. Xiao, and Y.-L. Feng. 2008. Calcium carbonate nanoparticle delivering vascular endothelial growth factor-C siRNA effectively inhibits lymphangiogenesis and growth of gastric cancer *in vivo*. *Cancer Gene Therapy* 15, 193–202.

Heidel, J. D., Z. Yu, J. Y.-C. Liu, S. M. Rele, Y. Liang, R. K. Zeidan, D. J. Kornbrust et al. 2007. Administration in non-human primates of escalating intravenous doses of targeted nanoparticles containing ribonucleotide reductase subunit M2 siRNA. *Proceedings of the National Academy of Sciences of the United States of America* 104, 5715–5721.

Hogrefe, R. I., A. V. Lebedev, G. Zon, K. F. Pirollo, A. Rait, Q. Zhou, W. Yu et al. 2006. Chemically modified short interfering hybrids (siHYBRIDS): Nanoimmunoliposome delivery *in vitro* and *in vivo* for RNAi of HER-2. *Nucleosides, Nucleotides and Nucleic Acids* 25, 889–907.

Howard, K. A., U. L. Rahbek, X. Liu, C. K. Damgaard, S. Z. Glud, M. Ø. Andersen, M. B. Hovgaard et al. 2006. RNA interference *in vitro* and *in vivo* using a chitosan/siRNA nanoparticle system. *Molecular Therapy* 14, 476–484.

Hu-Lieskovan, S. J. D. Heidel, D. W. Bartlett, M. E. Davis, and T. J. Triche. 2005. Sequence-specific knockdown of EWS-FLI1 by targeted, nonviral delivery of small interfering RNA inhibits tumor growth in a murine model of metastatic Ewing's sarcoma. *Cancer Research* 65, 8984–8992.

Hunter, A. C. 2006. Molecular hurdles in polyfectin design and mechanistic background to polycation induced cytotoxicity. *Advanced Drug Delivery Reviews* 58, 1523–1531.

Jackson, A. L., S. R. Bartz, J. Schelter, S. V. Kobayashi, J. Burchard, M. Mao, B. Li, G. Cavet, and P. S. Linsley. 2003. Expression profiling reveals off-target gene regulation by RNAi. *Nature Biotechnology* 21, 635–637.

Jere, D., H.-L. Jiang, Y.-K. Kim, R. Arote, Y.-J. Choi, C.-H. Yun, M.-H. Cho et al. 2009. Chitosan-graft-polyethylenimine for Akt1 siRNA delivery to lung cancer cells. *International Journal of Pharmaceutics* 378, 194–200.

Juliano, R., M. R. Alam, V. Dixit, and H. Kang. 2008. Mechanisms and strategies for effective delivery of antisense and siRNA oligonucleotides. *Nucleic Acids Research* 36, 4158–4171.

Jung, S., S. H. Lee, H. Mok, H. J. Chung, and T. G. Park. 2010. Gene silencing efficiency of siRNA-PEG conjugates: Effect of PEGylation site and PEG molecular weight. *Journal of Controlled Release* 144, 306–313.

Karikó, K., P. Bhuyan, J. Capodici, and D. Weissman. 2004. Small interfering RNAs mediate sequence-independent gene suppression and induce immune activation by signaling through Toll-like receptor 3. *Journal of Immunology* 172, 6645–6549.

Kim, D. H., and J. J. Rossi. 2007. Strategies for silencing human disease using RNA interference. *Nature Reviews Genetics* 8, 173–184.

Kim, H.-K., E. Davaa, C.-S. Myung, and J.-S. Park. 2010. Enhanced siRNA delivery using cationic liposomes with new polyarginine-conjugated PEG-lipid. *International Journal of Pharmaceutics* 392, 141–147.

Kim, J., S. W. Kim, and W. J. Kim. 2011. PEI-g-PEG-RGD/small interference RNA polyplex-mediated silencing of vascular endothelial growth factor receptor and its potential as an anti-angiogenic tumor therapeutic strategy. *Oligonucleotides* 21, 101–107.

Kim, S. H., J. H. Jeong, S. H. Lee, S. W. Kim, and T. G. Park. 2008. LHRH receptor-mediated delivery of siRNA using polyelectrolyte complex micelles self-assembled from siRNA-PEG-LHRH conjugate and PEI. *Bioconjugate Chemistry* 19, 2156–2162.

Kim, S. H., S. H. Lee, H. Tian, X. Chen, and T. G. Park. 2009. Prostate cancer cell-specific VEGF siRNA delivery system using cell targeting peptide conjugated polyplexes. *Journal of Drug Targeting* 17, 311–317.

Kim, W. J., L. V. Christensen, S. Jo, J. W. Yockman, J. H. Jeong, Y.-H. Kim, and S. W. Kim. 2006. Cholesteryl oligoarginine delivering vascular endothelial growth factor siRNA effectively inhibits tumor growth in colon adenocarcinoma. *Molecular Therapy* 14, 343–350.

Kraynack, B. A., and B. F. Baker. 2006. Small interfering RNAs containing full 2′-*O*-methylribonucleotide–modified sense strands display Argonaut e2/eIF2C2-dependent activity. *RNA* 12, 163–176.

Landen, C. N., W. M. Merritt, L. S. Mangala, A. M. Sanguino, C. Bucana, C. Lu, Y. G. Lin et al. 2006. Intraperitoneal delivery of liposomal siRNA for therapy of advanced ovarian cancer. *Cancer Biology and Therapy* 5, 1708–1713.

Laursen, M. B., M. M. Pakula, S. Gao, K. Fluiter, O. R. Mook, F. Baas, N. Langklaer et al. 2010. Utilization of unlocked nucleic acid (UNA) to enhance siRNA performance *in vitro* and *in vivo*. *Molecular BioSystems* 6, 862–870.

Lavigne, C., and A. R. Thierry. 2007. Specific subcellular localization of siRNAs delivered by lipoplex in MCF-7 breast cancer cells. *Biochimie* 89, 1245–1251.

Leachman, S. A., R. P. Hickerson, M. E. Schwartz, E. E. Bullough, S. L. Hutcherson, K. M. Boucher, C. D. Hansen et al. 2010. First-in-human mutation-targeted siRNA phase Ib trial of an inherited skin disorder. *Molecular Therapy* 18, 442–446.

Lee, C. C., J. A. MacKay, J. M. Fréchet, and F. C. Szoka. 2005. Designing dendrimers for biological applications. *Nature Biotechnology* 23, 1517–1526.

Lee, S. H., H. Mok, Y. Lee, and T. G. Park. 2010. Self-assembled siRNA–PLGA conjugate micelles for gene silencing. *Journal of Controlled Release*, 30, 152–158.

Lee, S. K., M. S. Han, S. Asokan, and C.-H. Tung. 2011. Effective gene silencing by multilayered siRNA-coated gold nanoparticles. *Small (Weinheim an der Bergstrasse, Germany)* 7, 364–370.

Li, S.-D., Y.-C. Chen, M. J. Hackett, and L. Huang. 2007. Tumor-targeted delivery of siRNA by self-assembled nanoparticles. *Molecular Therapy* 16, 163–169.

Li, S.-D., and L. Huang. 2006. Targeted delivery of antisense oligodeoxynucleotide and small interference RNA into lung cancer cells. *Molecular Pharmaceutics* 3, 579–588.

Liu, X.-X., P. Rocchi, F.-Q. Qu, S.-Q. Zheng, Z.-C. Liang, M. Gleave, J. Iovanna et al. 2009. PAMAM dendrimers mediate siRNA delivery to target Hsp27 and produce potent antiproliferative effects on prostate cancer cells. *ChemMedChem* 4, 1302–1310.

Lundberg, P., S. El-Andaloussi, T. Sütlü, H. Johansson, and Ü. Langel. 2007. Delivery of short interfering RNA using endosomolytic cell-penetrating peptides. *FASEB Journal* 21, 2664–2671.

Mao, S., M. Neu, O. Germershaus, O. Merkel, J. Sitterberg, U. Bakowsky, and T. Kissel. 2006. Influence of polyethylene glycol chain length on the physicochemical and biological properties of poly(ethylene imine)-graft-poly(ethylene glycol) block copolymer/siRNA polyplexes. *Bioconjugate Chemistry* 17, 1209–1218.

Marques, J. T., and B. R. G. Williams. 2005. Activation of the mammalian immune system by siRNAs. *Nature Biotechnology* 23, 1399–1405.

Martin, B., M. Sainlos, A. Aissaoui, N. Oudrhiri, M. Hauchecorne, J.-P. Vigneron, J.-M. Lehn et al. 2005. The design of cationic lipids for gene delivery. *Current Pharmaceutical Design* 11, 375–394.

Merritt, W. M., Y. G. Lin, W. A. Spannuth, M. S. Fletcher, A. A. Kamat, L. Y. Han, C. N. Landen et al. 2008. Effect of interleukin-8 gene silencing with liposome-encapsulated small interfering RNA on ovarian cancer cell growth. *Journal of the National Cancer Institute* 100, 359–372.

Meryet-Figuiéres, M., S. Resinaa, C. L. Bar:ovatz-Meimon, G. Lebleu, B., and Thierry, A. R. 2007. Inhibition of PAI-1 expression in breast cancer carcinoma cells by siRNA at nanomolar range, *Biochimie* 89, 1228–1233.

Mével, M., N. Kamaly, S. Carmona, M. H. Oliver, M. R. Jorgensen, C. Crowther, F. H. Salazar et al. 2010. DODAG; a versatile new cationic lipid that mediates efficient delivery of pDNA and siRNA. *Journal of Controlled Release* 143, 222–232.

Minakuchi, Y., F. Takeshita, N. Kosaka, H. Sasaki, Y. Yamamoto, M. Kouno, K. Honma et al. Atelocollagen-mediated synthetic small interfering RNA delivery for effective gene silencing *in vitro* and *in vivo*. *Nucleic Acids Research* 32, E109.

Mook, O. R., F. Baas, M. B. D. Wissel, and K. Fluiter. 2007. Evaluation of locked nucleic acid-modified small interfering RNA *in vitro* and *in vivo*. *Molecular Cancer Therapeutics* 6, 833–843.

Morrissey, D. V., K. Blanchard, L. Shaw, K. Jensen, J. A. Lockridge, B. Dickinson, J. A. McSwiggen et al. 2005a. Activity of stabilized short interfering RNA in a mouse model of hepatitis B virus replication. *Hepatology* 41, 1349–1356.

Morrissey, D. V., J. A. Lockridge, L. Shaw, K. Blanchard, K. Jensen, W. Breen, K. Hartsough et al. 2005b. Potent and persistent *in vivo* anti-HBV activity of chemically modified siRNAs. *Nature Biotechnology* 23, 1002–1007.

Moschos, S. A., S. W. Jones, M. M. Perry, A. E. Williams, J. S. Erjefalt, J. J. Turner, P. J. Barnes et al. 2007. Lung delivery studies using siRNA conjugated to TAT(48–60) and penetratin reveal peptide induced reduction in gene expression and induction of innate immunity. *Bioconjugate Chemistry* 18, 1450–1459.

Mu, P., S. Nagahara, N. Makita, Y. Tarumi, K. Kadomatsu, and Y. Takei. 2009. Systemic delivery of siRNA specific to tumor mediated by atelocollagen: Combined therapy using siRNA targeting Bcl-xL and cisplatin against prostate cancer. *International Journal of Cancer* 125, 2978–2990.

Muratovska, A., and M. R. Eccles. 2004. Conjugate for efficient delivery of short interfering RNA (siRNA) into mammalian cells. *FEBS Letters* 558, 63–68.

Nimjee, S. M., C. P. Rusconi, and B. A. Sullenger. 2005. Aptamers: An emerging class of therapeutics. *Annual Review of Medicine* 56, 555–583.

Nishina, K., T. Unno, Y. Uno, T. Kubodera, T. Kanouchi, H. Mizusawa, and T. Yokota. 2008. Efficient *in vivo* delivery of siRNA to the liver by conjugation of α-tocopherol. *Molecular Therapy* 16, 734–740.

Niu, X. Y., Z. L. Peng, W. Q. Duan, H. Wang, and P. Wang. 2006. Inhibition of HPV 16 E6 oncogene expression by RNA interference *in vitro* and *in vivo*. *International Journal of Gynecological Cancer* 16, 743–751.

Ochiya, T., Y. Takahama, S. Nagahara, Y. Sumita, A. Hisada, H. Itoh, Y. Nagai et al. 1999. New delivery system for plasmid DNA *in vivo* using atelocollagen as a carrier material: The minipellet. *Nature Medicine* 5, 707–710.

Oh, Y.-K., and T. G. Park. 2009. siRNA delivery systems for cancer treatment. *Advanced Drug Delivery Reviews* 61, 850–862.

Oupický, D., Č. Koňák, K. Ulbrich, M. A. Wolfert, and L. W. Seymour. 2000. DNA delivery systems based on complexes of DNA with synthetic polycations and their copolymers. *Journal of Controlled Release* 65, 149–171.

Pal, A., A. Ahmad, S. Khan, I. Sakabe, C. Zhang, U. N. Kasid, and I. Ahmad. 2005. Systemic delivery of Raf siRNA using cationic cardiolipin liposomes silences Raf-1 expression and inhibits tumor growth in xenograft model of human prostate cancer. *International Journal of Oncology* 26, 1087–1091.

Parkin, D. M., F. Bray, J. Ferlay, and P. Pisani. 2005. Global cancer statistics, 2002. *Cancer Journal for Clinicians* 55, 74–108.

Parrish, S., J. Fleenor, S. Xu, C. Mello, and A. Fire. 2000. Functional anatomy of a dsRNA trigger: Differential requirement for the two trigger strands in RNA interference. *Molecular Cell* 6, 1077–1087.

Pavan, G. M., L. Albertazzi, and A. Danani. 2010. Ability to adapt: Different generations of PAMAM dendrimers show different behaviors in binding siRNA. *Journal of Physical Chemistry B* 114, 2667–2675.

Peer, D., J. M. Karp, S. Hong, O. C. Farokhzad, R. Margalit, and R. Langer. 2007. Nanocarriers as an emerging platform for cancer therapy. *Nature Nanotechnology* 2, 751–760.

Pillé, J.-Y., H. Li, E. B. Bertrand, J.-R. Pritchard, L.-L. Opolon, P. Maksimenko, A. Lu et al. 2006. Intravenous delivery of anti-RhoA small interfering RNA loaded in nanoparticles of chitosan in mice: Safety and efficacy in xenografted aggressive breast cancer. *Human Gene Therapy* 17, 1019–1026.

Pirollo, K. F., G. Zon, A. Rait, Q. Zhou, W. Yu, R. Hogrefe, and E. H. Chang. 2006. Tumor-targeting nanoimmunoliposome complex for short interfering RNA delivery. *Human Gene Therapy* 17, 117–124.

Santel, A., M. Aleku, O. Keil, J. Endruschat, V. Esche, B. Durieux, K. Löffler et al. 2006. RNA interference in the mouse vascular endothelium by systemic administration of siRNA-lipoplexes for cancer therapy. *Gene Therapy* 13, 1360–1370.

Sato, A., S. W. Choi, M. Hirai, A. Yamayoshi, R. Moriyama, T. Yamano, M. Takagi et al. 2007. Polymer brush-stabilized polyplex for a siRNA carrier with long circulatory half-life. *Journal of Controlled Release* 122, 209–216.

Shahzad, M. M., L. S. Mangala, H. D. Han, C. Lu, J. Bottsford-Miller, M. Nishimura, E. M. Mora et al. 2011. Targeted delivery of small interfering RNA using reconstituted high-density lipoprotein nanoparticles. *Neoplasia* 13, 309–319.

Sheehan, D., B. Lunstad, C. M. Yamada, B. G. Stell, M. H. Caruthers, and D. J. Dellinger. 2003. Biochemical properties of phosphonoacetate and thiophosphonoacetate oligodeoxyribonucleotides. *Nucleic Acids Research* 31, 4109–4118.

Simeoni, F., M. C. Morris, F. Heitz, and G. Divita. 2003. Insight into the mechanism of the peptide-based gene delivery system MPG: Implications for delivery of siRNA into mammalian cells. *Nucleic Acids Research* 31, 2717–2724.

Sledz, C. A., M. Holko, M. J. D. Veer, R. H. Silverman, and B. R. G. Williams. 2003. Activation of the interferon system by short-interfering RNAs. *Nature Cell Biology* 5, 834–839.

Song, E., P. Zhu, S.-K. Lee, D. Chowdhury, S. Kussman, D. M. Dykxhoorn, Y. Feng et al. 2005. Antibody mediated *in vivo* delivery of small interfering RNAs via cell-surface receptors. *Nature Biotechnology* 23, 709–717.

Soutschek, J., A. Akinc, B. Bramlage, K. Charisse, R. Constien, M. Donoghue, S. Elbashir et al. 2004. Therapeutic silencing of an endogenous gene by systemic administration of modified siRNAs. *Nature* 173–178.

Strat, A., L. Gao, T. Utsuki, B. Cheng, S. Nuthalapaty, J. M. Mathis, Y. Odaka et al. 2006. Specific and nontoxic silencing in mammalian cells with expressed long dsRNAs. *Nucleic Acids Research* 34, 3803–3810.

Sugahara, K. N., T. Teesalu, P. P. Karmali, V. R. Kotamraju, L. Agemy, O. M. Girard, D. Hanahan et al. 2009. Tissue-penetrating delivery of compounds and nanoparticles into tumors. *Cancer Cell* 16, 510–520.

Sun, X., H. A. Rogoff, and C. J. Li. 2008. Asymmetric RNA duplexes mediate RNA interference in mammalian cells. *Nature Biotechnology* 26, 1379–1382.

Sutton, D., S. Kim, X. Shuai, K. Leskov, J. T. Marques, B. R. Williams, D. A. Boothman et al. 2006. Efficient suppression of secretory clusterin levels by polymer-siRNA nanocomplexes enhances ionizing radiation lethality in human MCF-7 breast cancer cells *in vitro*. *International Journal of Nanomedicine* 1, 155–162.

Taetz, S., A. Bochot, C. Surace, S. Arpicco, J.-M. Renoir, U. F. Schaefer, V. Marsaud et al. 2009. Hyaluronic acid-modified DOTAP/DOPE liposomes for the targeted delivery of anti-telomerase siRNA to CD44-expressing lung cancer cells. *Oligonucleotides* 19, 103–116.

Tahara, K., H. Yamamoto, N. Hirashima, and Y. Kawashima. 2010. Chitosan-modified poly(D,L-lactide-co-glycolide) nanospheres for improving siRNA delivery and gene-silencing effects. *European Journal of Pharmaceutics and Biopharmaceutics* 74, 421–426.

Takeshita, F., Y. Minakuchi, S. Nagahara, K. Honma, H. Sasaki, K. Hirai, T. Teratani et al. 2005. Efficient delivery of small interfering RNA to bone-metastatic tumors by using atelocollagen *in vivo*. *Proceedings of the National Academy of Sciences of the United States of America* 102, 12177–12182.

Takeuchi, H., H. Kojima, H. Yamamoto, and Y. Kawashima. 2001. Evaluation of circulation profiles of liposomes coated with hydrophilic polymers having different molecular weights in rats. *Journal of Controlled Release* 75, 83–91.

Tan, W. B., S. Jiang, and Y. Zhang. 2007. Quantum-dot based nanoparticles for targeted silencing of HER2/neu gene via RNA interference. *Biomaterials* 28, 1565–1571.

Terrazas, M., and E. T. Kool. 2009. RNA major groove modifications improve siRNA stability and biological activity. *Nucleic Acids Research* 37, 346–353.

Thaker, P. H., L. Y. Han, A. A. Kamat, J. M. Arevalo, R. Takahashi, C. Lu, N. B. Jennings et al. 2006. Chronic stress promotes tumor growth and angiogenesis in a mouse model of ovarian carcinoma. *Nature Medicine* 12, 939–944.

Thomas, M., and A. M. Klibanov. 2002. Enhancing polyethylenimine's delivery of plasmid DNA into mammalian cells. *Proceedings of the National Academy of Sciences of the United States of America* 99, 14640–14645.

Tijsterman, M., R. F. Ketting, and R. H. A. Plasterk. 2002. The genetics of RNA silencing. *Annual Review of Genetics* 36, 489–519.

Tseng, Y.-C., S. Mozumdar, and L. Huang. 2009. Lipid-based systemic delivery of siRNA. *Advanced Drug Delivery Reviews* 61, 721–731.

Urban-Klein, B., S. Werth, S. Abuharbeid, F. Czubayko, and A. Aigner. 2005. RNAi-mediated gene-targeting through systemic application of polyethylenimine (PEI)-complexed siRNA *in vivo*. *Gene Therapy* 12, 461–466.

Vincenzoa, J. D., J. E. Cehelsky, R. Alvarez, S. Elbashir, J. Harborth, I. Toudjarska, L. Nechev et al. 2008. Evaluation of the safety, tolerability and pharmacokinetics of ALN-RSV01, a novel RNAi antiviral therapeutic directed against respiratory syncytial virus (RSV). *Antiviral Research* 77, 225–231.

Vogelstein, B., and K. W. Kinzler. 2004. Cancer genes and the pathways they control. *Nature Medicine* 10, 789–799.

Waite, C. L., and C. M. Roth. 2009. PAMAM-RGD conjugates enhance siRNA delivery through a multicellular spheroid model of malignant glioma. *Bioconjugate Chemistry* 20, 1908–1916.

Wang, Y., S. Gao, W.-H. Ye, H. S. Yoon, and Y.-Y. Yang. 2006. Co-delivery of drugs and DNA from cationic core–shell nanoparticles self-assembled from a biodegradable copolymer. *Nature Materials* 5, 791–796.

Watts, J. K., and D. R. Corey. 2010. Clinical status of duplex RNA. *Bioorganic and Medicinal Chemistry Letters* 20, 3203–3207.

Wu, G. Y., and C. H. Wu. 1987. Receptor-mediated *in vitro* gene transformation by a soluble DNA carrier system. *Journal of Biological Chemistry* 262, 4429–4432.

Xia, C-F., Y. Zhang, Y. Zhang, R. J. Boado, and W. M. Pardridge. 2007. Intravenous siRNA of brain cancer with receptor targeting and avidin-biotin technology. *Pharmaceutical Research* 24, 2309–2316.

Xu, C.-X., D. Jere, H. Jin, S.-H. Chang, Y.-S. Chung, J.-Y. Shin, J-E. Kim et al. 2008. Poly(ester amine)-mediated, aerosol-delivered Akt1 small interfering RNA suppresses lung tumorigenesis. *American Journal of Respiratory and Critical Care Medicine* 178, 60–73.

Yamada, C. M., D. J. Dellinger, and M. H. Caruthers. 2007. Synthesis and biological activity of phosphonocarboxylate DNA. *Nucleosides, Nucleotides and Nucleic Acids* 26, 539–546.

Yano, J., K. Hirabayashi, S.-I. Nakagawa, T. Yamaguchi, M. Nogawa, I. Kashimori, H. Naito et al. 2004. Antitumor activity of small interfering RNA/cationic liposome complex in mouse models of cancer. *Clinical Cancer Research* 10, 7721–7726.

Yoshizawa, T., Y. Hattori, M. Hakoshima, K. Koga, and Y. Maitani. 2008. Folate-linked lipid-based nanoparticles for synthetic siRNA delivery in KB tumor xenografts *European Journal of Pharmaceutics and Biopharmaceutics* 70, 718–725.

Yuan, X., B. A. Shah, N. K. Kotadia, J. Li, H. Gu, and Z. Wu. 2010. The development and mechanism studies of cationic chitosan-modified biodegradable PLGA nanoparticles for efficient siRNA drug delivery. *Pharmaceutical Research* 27, 1285–1295.

Zhang, C., N. Tang, X. Liu, W. Liang, W. Xu, and V. P. Torchilin. 2006a. siRNA-containing liposomes modified with polyarginine effectively silence the targeted gene. *Journal of Controlled Release* 112, 229–239.

Zhang, Z., X. Yang, Y. Zhang, B. Zeng, S. Wang, T. Zhu, R. B. S. Roden et al. 2006b. Delivery of telomerase reverse transcriptase small interfering RNA in complex with positively charged single-walled carbon nanotubes suppresses tumor growth. *Clinical Cancer Research* 12, 4933–4939.

Zhao, Q., B. Han, Z. Wang, C. Gao, C. Peng, and J. Shen. 2007. Hollow chitosan-alginate multilayer microcapsules as drug delivery vehicle: Doxorubicin loading and *in vitro* and *in vivo* studies. *Nanomedicine: Nanotechnology, Biology and Medicine* 3, 63–74.

Zhu, C., S. Jung, S. Luo, F. Meng, X. Zhu, T. G. Park, and Z. Zhong. 2010. Co-delivery of siRNA and paclitaxel into cancer cells by biodegradable cationic micelles based on PDMAEMA–PCL–PDMAEMA triblock copolymers. *Biomaterials* 31, 2408–2416.

22 Sugar–Glass Nanoparticles for Protein Delivery

Jyotsnendu Giri and Marcus T. Cicerone

CONTENTS

22.1 INTRODUCTION

The proteins involved in recognition, signaling, and catalysis are often intrinsically unstable and may have short functional lifetimes in a living organism. This is acceptable because proteins are continuously cleared from the body, and high concentrations of damaged proteins are not likely to accumulate in a healthy organism. However, these same types of intrinsically unstable proteins are used as therapeutics in biopharmaceutical and drug delivery applications where long-term stability is critically important. Protein degradation in a product leads to uncontrolled dosing of the active drug, and introduction of degraded protein into the body can lead to adverse immune response.

Compounding the fact that these biomolecules are intrinsically unstable is the need to process them—incorporate them into a delivery vehicle or depot—as well as ship and store them. These steps frequently expose the bioactive components to process-related stresses that can lead to accelerated degradation (Putney and Burke 1998; Fu et al. 2000; Zhu et al. 2000; Perez et al. 2002). We have developed methods to encapsulate proteins and other bioactive molecules in nanoparticles that provides a tailored environment for the active molecule while protecting it from solvents and other process-related stresses. These nanoparticles disperse well in nonpolar solvents, yielding very good handing and delivery properties.

In this contribution, we first discuss the practice and physical chemistry of protein stabilization from a biopharmaceutical perspective. We show that the matrix in a freeze-dried protein formulation can be designed to provide optimal stability to the protein drug. We then show that these formulations can be prepared and freeze-dried in the form of nanoparticles, allowing them to be processed into polymeric carriers without losing biological function.

22.2 PHYSICAL CHEMISTRY OF PROTEIN STABILIZATION

Proteins are often dried without excipient or even mixed directly into polymeric materials in drug delivery and drug depot applications. As we will show in the following sections, these environments lead to very high losses in processing as well as high degradation rates even after the proteins have been processed. Nature has found a much more effective way to protect proteins and other functional biological components from dehydration stress. Many plants and animals possess the remarkable ability to survive long periods of time in extremely dry environments by producing high intracellular concentrations of sugars in response to dehydration stress. Delicate biological structures and labile macromolecules are protected by these sugars, even under complete desiccation, creating a suspended state of biological activity in the dry state from which they can recover from almost miraculously upon hydration. The role of sugars in protecting dried biological structures and molecules was first recognized by Crowe et al. (1984) for lipid membranes and, later, for proteins by Carpenter et al. (1987), who subsequently demonstrated that sugar-based glasses could be used to stabilize labile analytical proteins for biotechnological applications.

Starting from these pioneering initial studies, the practice of preserving biological agents in carbohydrate glasses has become widespread. Indeed, the maintenance of protein function after desiccation is essential in many fields: analytical enzymes, foods, and biopharmaceuticals all rely on sugar-based glassy matrices to stabilize proteins. Despite the importance of this protein preservation method and decades of practical application, the mechanisms by which sugar-based preservative matrices ("sugar–glasses") impart dehydration stability to proteins are simply not well understood. In fact, it is only now starting to become clear what factors most strongly influence the rates of these processes and how to best control or predict them.

Proteins degrade in glassy matrices by essentially the same modes as they do in aqueous solution. They can undergo chemical degradation, typically through deamidation, hydrolysis or oxidation in the glass, or aggregation upon rehydration (Lai and Topp 1999). The major chemical degradation mechanisms are known for proteins in aqueous solution (Manning et al. 2010) and typically involve protein conformational mobility as well as transport of small molecule reactants, such as oxygen or water. There are likely to be differences in the details of these degradation mechanisms in glass, but the requirement of local protein mobility and reactant transport is undoubtedly retained. Similarly, protein aggregation in glass is likely to require local protein mobility while in the glass. It is believed that, while in the glass, proteins may undergo small conformational changes that expose hydrophobic groups, generating an "aggregation competent" state, and that aggregation occurs subsequently when the protein–sugar–glass matrix is redissolved in buffer and the protein has significant translational mobility.

Two hypotheses for how sugar–glasses stabilize proteins have been widely discussed: "water replacement" and "vitrification." In the water replacement hypothesis, the sugar–glass matrix is proposed to substitute for water, thermodynamically stabilizing the protein against degradation by stabilizing its native structure (Prestrelski et al. 1993). There is much evidence for a correlation between protein stability and near-native conformation in the glass (Tanaka et al. 1991; Costantino et al. 1998; Cleland et al. 2001), but this relation has not been quantifiable and, as we have recently shown, it is caused by a simple correlation (Cicerone et al. 2011). The alternative vitrification hypothesis assumes rather that the rates of chemical and physical degradation processes are simply slowed down by the sluggish structural relaxation of the glass (Franks et al. 1991). This hypothesis is normally considered in terms of the α relaxation time, τ_α, larger τ_α being associated with slower degradation (Duddu et al. 1997; Guo et al. 2000; Pikal et al. 2008). Investigations into this relationship sometimes indicate a correlation and sometimes do not. When temperature dependence of α relaxation and degradation are compared, some correlation is observed (Duddu et al. 1997; Pikal et al. 2008), but when degradation rates in multiple glass formers are compared with α relaxation time, there is often no correlation found (Rossi et al. 1997; Buera et al. 1999; Davidson and Sun 2001).

We and others (Cicerone and Soles 2004; Moznine et al. 2003; Yoshioka et al. 2006; Yoshioka et al. 2007; Wang et al. 2009a,b) have reported that high-frequency "β relaxation" processes in sugar–glasses correlate strongly with protein or other drug stability; that is, with rates of physical and chemical degradation of processes in the glass. This correlation holds for changes in temperature as well as for formulation changes and is consistent with a transport-limited reaction rate hypothesis. It is generally true that reactions in glassy systems are transport-limited, and it has been shown that high-frequency relaxation processes in glass are strongly coupled to transport properties.

In general, glasses exhibit at least two distinct high-frequency relaxation or β processes, the so-called "fast β" (β_{fast}) and the slow, or Johari–Goldstein β (β_{JG}), processes (Johari and Goldstein 1971). The β_{fast} relaxation is associated with sub-ballistic exploration of the molecular cage of neighbors and occurs on a picosecond timescale (Allegrini et al. 1999), whereas the β_{JG} relaxation is associated with small amplitude intermolecular motions and typically occurs on a timescale of microseconds to milliseconds in sugar–glass (Obrzut et al. 2010; Kaminski et al. 2009). These two relaxation processes seem to have an underlying commonality, as they both respond similarly to antiplasticizing additives (Psurek et al. 2008; Riggleman et al. 2010; Simmons and Douglas 2011), and both have been associated with α relaxation through theory and phenomenological correlation (Hall and Wolynes 1987; Dyre et al. 1996; Ngai and Capaccioli 2004; Kessairi et al. 2008; Buchenau and Zorn 1992; Larini et al. 2008). Also, both processes are linked to transport in the glass. Small amplitude protein motions seem to be "slaved" to the β_{fast} dynamics of the fluid matrix (Caliskan et al. 2004; Dirama et al. 2005; Frauenfelder et al. 2009). Small molecule transport seems to be coupled to both β_{fast} and β_{JG} (Starr et al. 2002; Ngai et al. 2001; Gusev et al. 1993; Molinero and Goddard 2005). Furthermore, it has long been held that high-frequency motions in general can gate reactive and diffusive processes in highly viscous media (Grote et al. 1984; Sumi 1991).

Under conditions that prevail in these glasses (Lievonen and Roos 2002), one expects k_d to be directly proportional to D_T, the diffusion coefficient of the relevant species, consistent with predictions made by both the Smoluchowski theory of diffusion-limited reactions (Collins and Kimball 1949) and the classical Kramers theory under diffusion-limited reaction conditions (Kramers 1940). The fact that β relaxation correlates strongly with degradation rates in all the systems that we have investigated suggests that either the fast or slow β processes is controlling D_T of the relevant molecular species in each of these reactions.

In the following sections, we use the antiplasticization phenomenon to show that degradation processes are indeed coupled to β relaxation rather than α relaxation. Antiplasticizers are molecular additives that lower T_g and speed up α relaxation in the glass, while simultaneously slowing down the higher frequency β relaxation process. This is in contrast with ordinary plasticization in which both α and β relaxation are accelerated with the addition of a diluent (Anopchenko et al 2006; Psurek et al. 2008; Stukalin et al. 2010; Riggleman et al. 2010).

Figure 22.1a provides an illustration of the antiplasticization effect, as exhibited by the behavior of the β_{JG} relaxation for a series of melt-quenched trehalose

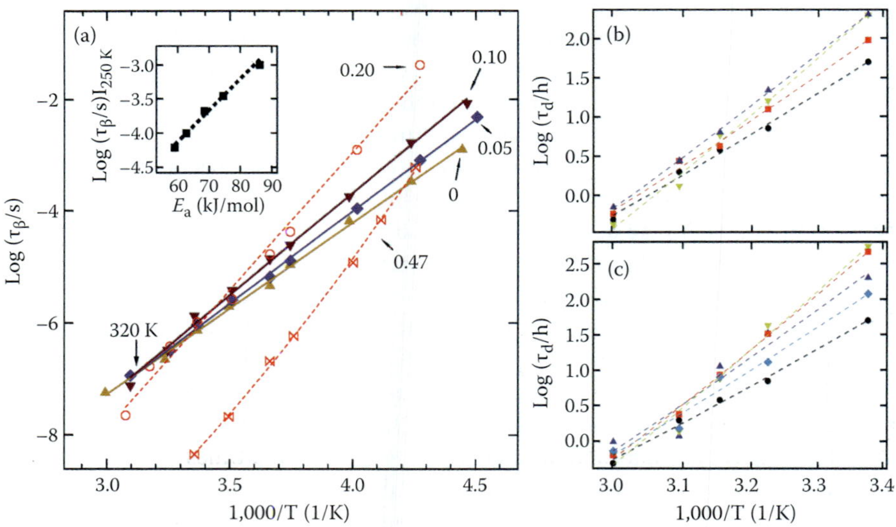

FIGURE 22.1 Antiplasticization effects in sugar/diluent glasses. (a) Dielectric relaxation times for melt-quenched trehalose glasses containing varying amounts of glycerol, indicated as a mass fraction. Relaxation times at 320 K are independent of the glycerol content for a glycerol mass fraction of ≲0.2. Inset: linear dependence of log(τ_β) on activation energy E_a for a glycerol mass fraction of ≲0.2. (b) and (c) Arrhenius plots of HRP degradation rates in antiplasticized maltitol glasses. (b,) Maltitol with the following mass fractions of glycerol: 0 (●), 0.05 (▼), 0.1 (▲), 0.2 (■). (c) Maltitol with the following mass fractions of DMSO: 0 (●), 0.05 (▲), 0.1 (▼), 0.15 (■), 0.2 (◆). (From Anopchenko A. et al. Dielectric study of the antiplasticization of trehalose by glycerol. *Phys. Rev. E* 74:031501, 2006. With permission.)

glasses with different amounts of glycerol (Anopchenko et al. 2006; Obrzut et al. 2010; Psurek et al. 2008). We see that as glycerol is added to trehalose, the activation energy for the β_{JG} relaxation process, $E_{a,\beta}$, increases and the logarithm of the Arrhenius prefactor, $\log(a_{\infty,\beta})$, decreases proportionally in such a way that there is a crossing temperature, the antiplasticization temperature T_a, below which the β_{JG} relaxation time slows down with the additive, and above which, it speeds up. This proportionality also leads to a linear relationship between $E_{a,\beta}$ and $\log(\tau_\beta)$ at all temperatures investigated, as shown explicitly for glycerol mass fractions of up to 0.2 at 23°C in the inset to Figure 22.1a. At high glycerol mass fractions (>0.2), the proportionality relation between $E_{a,\beta}$ and $\log(a_{\infty,\beta})$ breaks down, and τ_β decreases at all temperatures with any further addition of glycerol. We have recently shown that this antiplasticization behavior can be understood on the basis of simple analytic arguments related to the activation free energy governing the β_{JG} relaxation process (Psurek et al. 2008). The effect, often termed "entropy–enthalpy compensation," arises when the activation entropy and enthalpy terms have opposite signs and the leading (linear) concentration term in a Taylor expansion of the activation free energy dominates so that the activation energy and entropy are proportional and extensive in the amount of diluent added to the glass (Psurek et al. 2008). Because the leading term in a Taylor series expansion is, by definition, linear in the perturbation (such as the addition of an antiplasticizer), entropy–enthalpy compensation is fulfilled over a sufficiently small range for almost any variable that significantly affects the reaction rate. Of note here is that, for the dielectric data of Figure 22.1a (which introduces protein stability data), this linear response persists for a more than 20% change in the formulation concentration and holds for a change of more than six orders of magnitude in dynamic response!

Taking advantage of entropy–enthalpy compensation for τ_β, we differentially vary α and β_{JG} relaxation rates in a series of antiplasticized glasses, showing that β relaxation has a dominant effect on protein stability. Figure 22.1b and c show the degradation times (τ_d) of a model protein, horseradish peroxidase (HRP), in two such series of antiplasticized glasses. Values of τ_d were determined by measuring reconstituted HRP enzymatic activity after holding the freeze-dried protein–sugar samples under constant temperature for varying times. Error bars represent intervals of 1 standard deviation (SD) derived from at least three replicate determinations of τ_d at each temperature. The τ_d values vary precisely, as seen for τ_β in Figure 22.1a: τ_d behaves in an Arrhenius manner, having E_a values that are tuned by varying the glycerol or dimethyl sulfoxide (DMSO) concentrations. Also, the apparent antiplasticization temperature, T_a, for each of these plots is similar to the 320 K value observed in Figure 22.1a, a common finding in systems exhibiting an entropy–enthalpy compensation relationship (Liu and Guo 2001; Anopchenko et al. 2006; Riggleman et al. 2006, 2007, 2010; Psurek et al. 2008). The trends are also nonmonotonic with antiplasticizer concentration, having a reversal in the sign of the effect on τ_d somewhere in the range of a mass fraction DMSO or glycerol (0.05–0.1), which is a common pattern in entropy–enthalpy compensation with antiplasticizing additives. Each of these behaviors clearly shows that changing the β relaxation time of the glass significantly affects the behavior of HRP stability. We have measured the enzyme stability in more than 100 such antiplasticized glasses and found this same behavior in every case. The collected data are plotted in Figure 22.2.

Figure 22.2 shows log(τ_d) for HRP and equine alcohol dehydrogenase (ADH) at room temperature (T = 23°C) for 22 antiplasticized sugar–glass systems, comprising more than 100 glass formulations. The lower abscissa is the apparent activation energy, E_a, for the temperature dependence of τ_d in the various glasses. The solid lines are best fits to the data. The linear relationship here between log(τ_d) and E_a is a clear sign that β_{JG} relaxation plays an important role in the protein degradation processes for all of these sugar–glasses.

In addition to being linear, the relationship between log(τ_d) and E_a has the same slope as that for log(τ_β) and E_a in the inset to Figure 22.1a. We demonstrate this by using the linear relationship shown in the inset of Figure 22.1a to obtain log(τ_β) values for a process with E_a values from the bottom axis of Figure 22.2 and plot those values on the top axis of Figure 22.2. Obtained in this way, the slope of the fits to log(τ_d) against log(τ_β) is 1 for both ADH and HRP. Thus, we find the degradation rates scale linearly with τ_β in all of these glasses, even though the timescales of τ_β and τ_d are separated by more than 10 decades in time. This strengthens our conclusion that β relaxation of the sugar matrix is a predominant kinetic parameter governing protein degradation in these systems.

We contrast this direct relationship between protein stability and β_{JG} relaxation with a consideration of the relationship between protein stability and α relaxation.

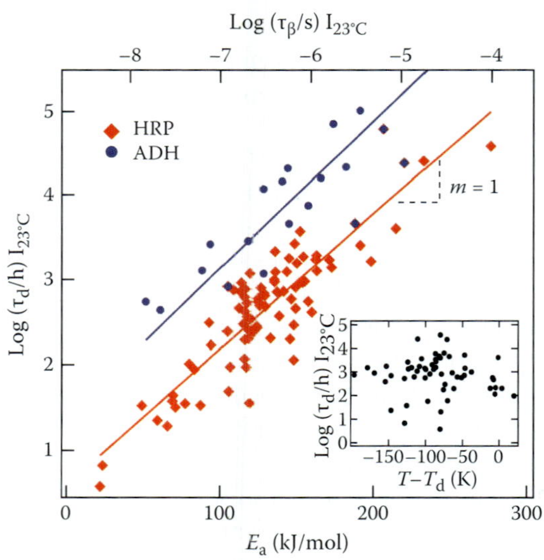

FIGURE 22.2 Enzyme degradation rates at 23°C in more than 100 plasticized and antiplasticized sugar–glasses. Top abscissa values are derived from activation energies for degradation and the relationship shown in the inset to Figure 22.1. The solid lines are best fits to the data and yield a slope near 1 for log(τ_d) vs. log(τ_β). Inset: T–T_g for most of the glasses shown in the main figure, plotted against the enzyme degradation rate. No correlation was found between the abscissa value (a surrogate for τ_α) and enzyme stability.

The inset to Figure 22.2 shows τ_d values of HRP plotted against values of (T–T_g) for the systems shown in the main figure. Here, (T–T_g) values are expected to correlate with τ_α; a larger negative value of (T–T_g) generally indicates a slower α relaxation. Although the precise relationship between (T–T_g) and log(τ_α) depends on the relative heat capacity difference between the glass and the supercooled liquid (Shamblin et al. 1999), which varies somewhat from glass to glass, a reasonably strong correlation between (T–T_g) and log(τ_α) is expected for these glasses. In contrast with this expectation, there is no evident correlation between τ_d and (T–T_g) in the measurements. The lack of any discernable trend in the inset to Figure 22.2 indicates that α relaxation does not have a direct and significant influence on protein stability, at least for formulations deep in the glassy regime. This observation is consistent with the data of Figure 22.1b and c, in which the protein stability in the glass state increases as we add DMSO or glycerol to maltitol, even though τ_α is reduced.

With the relevance of reactant diffusion rates in mind, we can rationalize the lack of correlation between α relaxation and degradation rate shown in the inset to Figure 22.2. One expects Stokes–Einstein relation, $D_T \sim \tau_\alpha^{-1}$, to hold in a dynamically homogeneous system. Glasses are dynamically heterogeneous, and this is sufficient to cause dramatic violation of the classical Stokes–Einstein behavior (Douglas and Leporini 1998). In these systems, a "fractional Stokes–Einstein" relationship ($D_T \sim \tau_\alpha^{-\delta}$) is observed, where $0 < \delta < 1$ (Pollack 1981). This altered scaling relationship between D_T and τ_α has significant implications for rates of reaction in glass-forming and other complex liquids in that it implies a corresponding fractional power–law relationship between the rate of reaction k_d and τ_α, that is, $k_d \sim \tau_\alpha^{-\delta}$. In glassy systems, D_T can be enhanced by as much as five orders of magnitude at T_g (Cicerone et al. 1997) over the value expected in the classical Stokes–Einstein picture. The magnitude of this Stokes–Einstein violation is strongly influenced by the properties of the most mobile domains in the glass, with the degree of enhancement in D_T being dependent on the breadth of the distribution of dynamics or, correspondingly, the fragility of the glass (Cicerone et al. 1997; Zwanzig 1989; Kumar et al. 2006). Thus, it is reasonable that we see a spread in degradation rates over a factor of 10^5 with constant α relaxation.

Being coupled to β relaxation, degradation rates of proteins can change from formulation to formulation in nonintuitive ways. For example, even though proteins dried in the presence of polymer or protein alone can have a high T_g (and thus slow α relaxation), degradation in these dried formulations can be quite rapid. As mentioned previously, many protein depot or delivery vehicles are formulated essentially as proteins mixed with polymer. The black circle in Figure 22.2 indicates the characteristic degradation rate of a protein dried either alone or in the presence of a polymer. Simply adding sugars to the protein formulation before drying can increase stability by more than 10-fold, even though the glass transition temperature of the material is reduced. Furthermore, optimizing the sugar–glass formulation can result in a 10^3-fold increase in stability.

Although optimizing freeze-drying formulations can yield proteins that are stable against degradation during storage, drying in the presence of sugars does not necessarily make the proteins impervious to other process stresses, nor does it necessarily yield a protein product that is easy to handle and that disperses well in polymer systems. For this, we turn to nanotechnology.

22.3 PROTEIN DELIVERY AND CHALLENGES

The development of a depot system for biomolecules such as proteins is more difficult than for small molecule pharmaceutical drugs mainly because of their larger size and structure-sensitive functional properties. An ideal drug depot system would provide high encapsulation efficiency; preserve proteins in an active state after processing, during storage, and in *in vivo* release; provide release profiles without burst phase; and allow for loading of multiple biomolecules. To date, no depot system has been able to meet this ideal.

Currently, emulsion-based methods are often used where protein in aqueous phase is emulsified into organic solvent to incorporate it into a polymer depot. Biomolecules undergo several process and storage-related stresses throughout the life of this type of product that can result in significant degradation, loss of bioactivity, reduced efficiency/efficacy, and increased safety concerns. The stresses include hydrophobic surface (polymer and organic solvents), mechanical agitation, and elevated temperature (Kokai et al. 2010; Fu et al. 2000; Castellanos et al. 2005; He et al. 2006; Giteau et al. 2008a,b). Moreover, this process also features limited encapsulation efficiency as well as burst release, and it is difficult to incorporate multiple biomolecules as individual entities. A number of approaches have been developed to ameliorate the effect of individual stresses in emulsion-based methods. These include the use of interface stabilizers (Kokai et al. 2010; He et al. 2006), protein crystallization (Giteau et al. 2008a,b), and covalent protein modification (Castellanos et al. 2005). Far from meeting the ideal of solving all processing issues, many approaches for improving one aspect of performance are neutral or deleterious to others. For example, solid-in-oil-in-water emulsions (Putney and Burke 1998) expose the protein to fewer solvent-based stresses compared with water-in-oil-in-water methods; however, solid-in-oil-in-water methods have thus far led to reduced encapsulation efficiency or to significant burst release. Approaches like spray-drying, spray freeze-drying, or freeze-drying with polyethylene glycol (PEG) have been used to generate smaller stabilizer-bearing protein particles for sustained release, but these are associated with low protein recovery (Wang et al. 2004) and can still result in burst release (Morita et al. 2000). In addition to the trade-offs in performance with current methods, many approaches developed are specific to the polymer used, or to particular proteins. This is problematic for drug delivery applications because new approaches may be needed for each new drug. Furthermore, fabrication of protein-encapsulated scaffolds for tissue engineering applications will likely require a range of polymer/solvent systems (Lee and Mooney 2003; Tayalia and Mooney 2009).

Finding a generic approach to protein stabilization and sustained delivery that will be compatible with a range of polymer and protein systems for protein drug depots is widely expected to be a formidable challenge (Tessmar and Gopferich 2007; Tayalia and Mooney 2009; Yang et al. 2009). In addition to being compatible with many polymer/solvent systems, a generic platform must facilitate the use of additives to protect the protein from various processing and storage stresses (Paillard-Giteau 2010; Putney and Burke 1998; Fu et al. 2000; Zhu et al. 2000). Such a platform also needs to produce small, preferably nano-sized, surface-functionalized particles to achieve good dispersion and sustained release (Perez et al. 2002; Giteau et al. 2008a,b).

22.4 ENGINEERING NANOTECHNOLOGY: NOVEL SUGAR–GLASS NANOPARTICLES

We have used a nanotechnology-based approach to address the problems associated with the development of a generic protein depot. We developed a novel all-in-one "sugar–glass nanoparticle" (SGnP) system for efficient incorporation and stabilization of proteins in a polymer depot. We encapsulated proteins along with sugars and other excipients as needed into nanoparticles (30–160 nm) that were functionalized with a surfactant to promote dispersion in organic solvent. A schematic drawing of these protein-bearing nanoparticles is shown in Figure 22.3. The generation of these particles combines practices from enzyme chemistry and biopharmaceutics. In the enzyme chemistry field, proteins have been protected from nonpolar solvent exposure by dispersing them in water pools of inverse micelles. This practice provides excellent dispersion of protein while protecting against water–oil interface and, to some extent, elevated temperatures (Savelli et al. 2000; Naoe et al. 2004; Mathew and Juang 2005; Anarbaev et al. 2009). Here, we incorporate sugars and other stabilizers along with the proteins in inverse micelles. In biopharmaceutics, it has long been the practice to freeze-dry proteins in the presence of sugars and other excipients to protect them from chemical and physical degradation during storage (Wang et al. 2009a,b). Here, we freeze-dry the nanoparticles by flash-freezing the inverse micelle suspension; first drying off the solvent, and then the water, leaving protein-laden, surfactant-coated SGnPs. We observed a loss of activity of only 3% to 5% of protein during the encapsulation process. The proteins remain active after

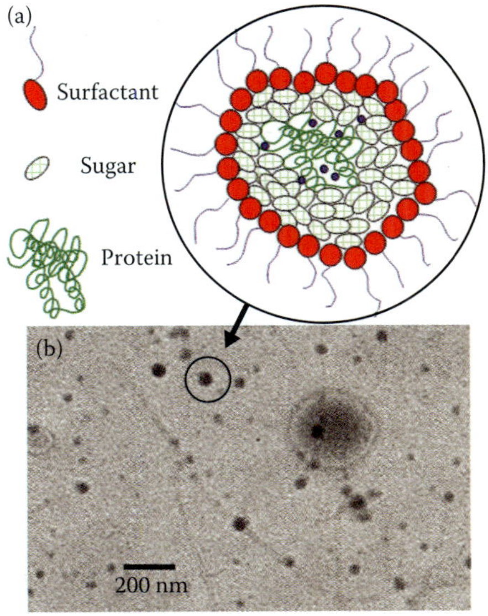

FIGURE 22.3 (a) Schematic of the biomolecule-encapsulated SGnP. The sugar–glass matrix stabilizes the biomolecules (e.g., protein) from process-related stresses and storage. The surfactant coating on the nanoparticle surface provides excellent dispersion properties in organic solvent–polymer solutions. (b) Representative typical TEM image of protein–SGnP.

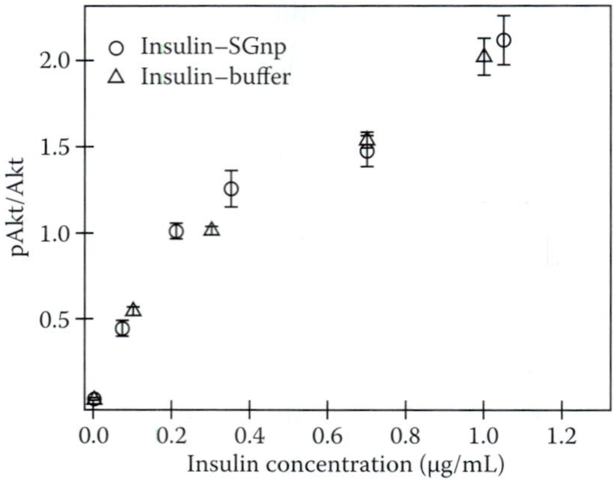

FIGURE 22.4 Insulin dose-dependent bioactivity (pAkt production) plot when L6 myoblast cells were treated with different concentrations of insulin (added into the medium) in the form of insulin–SGnP and insulin–buffer (before encapsulation). The total pAkt was normalized with the total Akt, which is proportional to the number of cells. The intensity ratio of pAkt/Akt was used to express the extent of induction of the phosphorylation of Akt. Values are the average ± 1 SD of $n = 4$ samples for each concentration.

encapsulation into the SGnP system and show bioactivity similar to its buffer form before encapsulation into sugar–glass. Figure 22.4 shows no loss of insulin activity after encapsulation into the SGnP system through a pAkt production assay using L6 myoblast cells. These cells produce pAkt at the same rate regardless of whether they are treated with insulin directly from the buffer or with the same dose of insulin that was first encapsulated in SGnPs.

22.5 ADVANTAGES OF SGnP PLATFORMS IN DELIVERY

22.5.1 Protection of Activity during Processing: Solvent, Polymer, and Processing Conditions

Once the protein is encapsulated into the SGnPs, it is extremely robust to subsequent processing stresses such as solvent exposure, mechanical agitation, etc. Nonaqueous solvent exposure is known to induce denaturation and degradation for many proteins in buffer (Putney and Burke 1998; Perez et al. 2002; Schwendeman 2002). We have characterized the ability of SGnPs to protect proteins against damage from a range of solvents. We exposed four proteins to nine solvents, either directly from buffer, mimicking the conventional emulsion method, or while encapsulated in SGnPs. Two of the proteins we tested (HRP and lipase) are model enzymes, commonly used for stability studies, whereas the other two proteins (insulin and bone morphogenetic protein 2; BMP-2) are clinically important signaling proteins (Chen et al. 2010). We challenged these proteins by exposure to organic solvents with a wide range of polarity indices (PI) ranging from 0 to 6.5. Water-immiscible solvents completely deactivated all four proteins when they were introduced from the buffer. In contrast, more than 80% of the initial enzyme activity was retained in all cases for protein in SGnPs (Figure 22.5). Encapsulation in SGnPs also conferred similar benefits for exposure to the water-miscible solvents tetrahydrofuran, acetone, and hexafluoroisopropanol ($p < .05$). No benefit was observed for exposure to methanol or dimethylformamide. However, unlike the results from HRP and lipase, two proteins (BMP-2 and insulin) showed a moderate level of activity in water-miscible solvent systems even when exposed directly to solvent from buffer.

Preservation of protein activity by SGnPs can be interpreted based on the reasonable assumption that trehalose glass absorbs polar solvents to a greater extent than nonpolar solvents, which is supported by the fact that trehalose glasses swelled considerably in high PI solvents such as methanol (Engasser et al. 2008). When proteins were presented directly to solvent from buffer, each of the solvents deactivated the proteins to some extent. Thus, the sugar–glass was effective at acting as a barrier between the protein and solvent with a PI of ≤5, but was less effective in a solvent with a PI of ≥5. The degree of degradation in a particular solvent varied slightly according to protein, likely because of differences in their specific structure and function (Doukyu and Ogino 2010). However, there was a uniformly high degree of protection from solvents with PI < 5 for all these proteins on encapsulation in the sugar–glass, suggesting that the SGnP platform may be widely effective for biomolecule protection against nonpolar solvents.

The nanoparticles also provide an environment of sugars that stabilizes the proteins against degradation during further drug depot device processing, such as

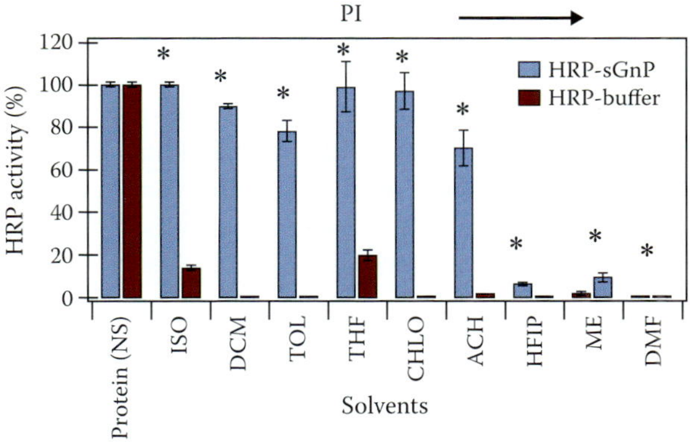

FIGURE 22.5 Presenting activity retention data for HRP as a model protein system exposed to nine solvents, either directly from the buffer or while encapsulated in SGnPs. Here, protein (NS) was represented as protein–SGnP and protein–buffer samples without any solvent treatment and considered 100% activity. The residual activity of protein in different solvents-treated protein samples was calculated relative to their corresponding "protein" (NS) samples. Values are average ± SDs of $n = 4$ samples for each groups. Statistically significant differences ($p < .05$) in protein activity between protein–buffer and protein–SGnP samples after incubation with a particular solvent are indicated by an asterisk (by t test). Protein (NS), protein–SGnP, and protein–buffer samples without any solvent treatment; ISO, isooctane; DCM, dichloromethane; TOL, toluene; THF, tetrahydrofuran; CHLO, chloroform; ACH, acetone; HFIP, hexafluoroisopropanol; ME, methanol; DMF, dimethylformamide.

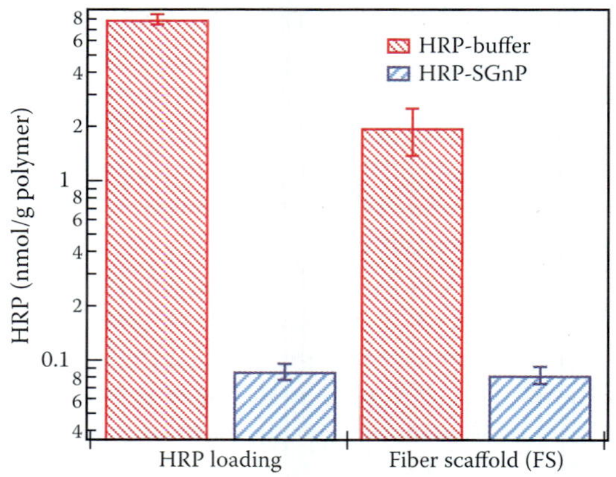

FIGURE 22.6 Process-related activity loss for HRP in nanofiber scaffolds electrospun from chloroform–PEO solutions in conventional and novel SGnP methods. Values are average ± SD of $n = 4$ samples for each group.

polymer scaffold formation. We obtained a 10,000-fold enhancement of residual protein activity when SGnP-protected protein was exposed to PCL/chloroform as compared with the introduction of protein from buffer. Figure 22.6 shows a snapshot of process-related activity loss for HRP in nanofibrous scaffolds electrospun from chloroform-poly(ethylene oxide) (PEO) solutions. In generating these scaffolds, HRP was added to PEO solutions as SGnPs or added directly from buffer. As with the PCL solutions, 95.5% of the protein activity was lost when protein was added to the PEO solution from buffer at 8 nmol HRP/g PCL. Seventy-five percent of the remaining protein activity was lost during scaffold electrospinning. Conversely, only 5% of the total activity was lost in all the steps through electrospinning for HRP protected in SGnPs. Additionally, the protein in these scaffolds was much more stable during storage as discussed in the next section.

22.5.2 Storage Stability

Storage stability is crucial for efficacy in long-term release and is important for manufacturing. The SGnP-HRP encapsulated PEO-nanofiber scaffolds were stored for 150 days at 23°C and 35°C. Residual enzyme activity was measured at 45 and 150 days. Figure 22.7 shows that protein sequestered in SGnPs was more than ten times more stable than unprotected protein. Stabilized protein had a stability lifetime of 210 days at 23°C, whereas the unprotected protein had a lifetime of only 15 days at this temperature. At 35°C, the benefit of SGnP encapsulation was similar. If unprotected protein were introduced into a tissue scaffold using conventional methods, it would likely degrade entirely in just a few days, making long-term delivery impossible. In contrast, proteins encapsulated in SGnPs are much more stable for storage. The encapsulated protein is also very stable as a suspension in isooctane and can be stored at room temperature for at least a month, or for several months at −20°C with a

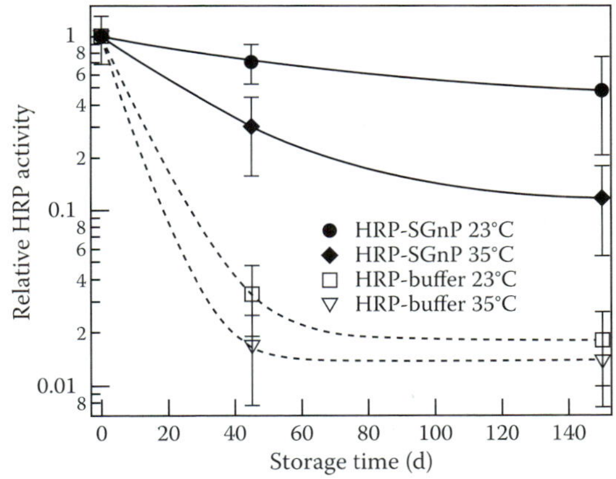

FIGURE 22.7 Comparative activity loss of HRP in the nanofiber scaffold where protein is encapsulated into the scaffold in the form of protein–SGnP or conventional protein–buffer, stored for 150 days at 23°C and 35°C. Values are average ± SD of $n = 4$ samples for each group.

less than 5% loss of protein activity. This is especially advantageous for manufacturing if a just-in-time approach is not desirable.

22.5.3 Encapsulation Efficiencies and Multiple Biomolecule Encapsulation

In addition to protection from solvent, the nanoencapsulation approach also provides excellent dispersion properties in solvents and homogeneous distribution of the protein throughout the carrier polymer (Figure 22.8). This property provides a platform to incorporate protein-bearing SGnPs within the depot at a loading level as high as

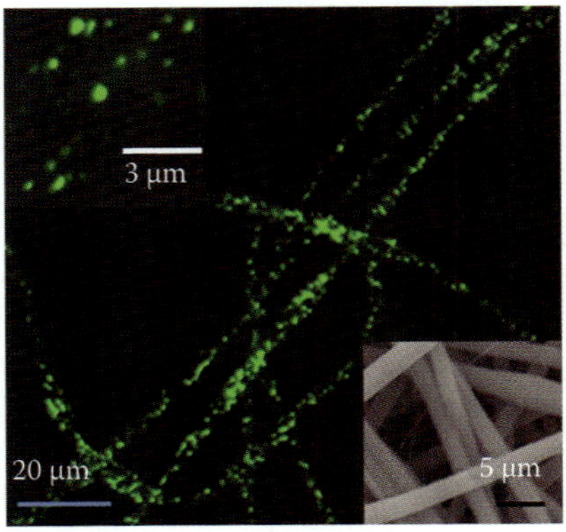

FIGURE 22.8 SGnPs distribution within the nanofiber matrix. Confocal image of 2% FITC–SGnP (mass fraction) loaded PCL nanofiber scaffold. Inset, higher magnification image of the same fiber showing distinct FITC–SGnP in the fiber matrix (top left) and SEM images of PCL nanofiber (bottom right).

40% of the mass fraction of the polymer matrix. These high loading levels do not affect scaffold preparation outcomes, at least for electrospinning and gas-foaming. Once packaged in SGnPs, distinct bioactive components present identically, thus this approach is applicable to the incorporation of multiple bioactive species in a single delivery system such as functional scaffold for tissue regeneration.

22.5.4 Slow-Released Profiles

Uniform dispersion is desirable to prevent burst release and provide uniform sustained release of the biomolecule. Burst release is often observed from carrier systems where separation occurs between the polymer and protein-bearing phases (Chen et al. 2010; Sy et al. 2009). Many approaches have been described to control drug release kinetics, including the tailoring of protein–polymer chemical interactions, the use of additional matrix materials, and the secondary incorporation of drug into microspheres (Chew et al. 2006; Chen et al. 2010). Using the SGnP system, HRP and insulin release profiles from gas-foamed and electrospun nanofibrous scaffolds made of PCL were essentially devoid of burst features. Figure 22.9 shows fraction of active protein released from PCL scaffolds as a function of time. In panel b, 11% and 8% initial release of HRP was seen from electrospun fibrous and gas-foamed scaffolds, respectively. This was followed by a slower release of approximately 0.85% and 0.33% per day, for up to 9 days, from the nanofibrous and gas-foamed scaffolds, respectively, and sustained release of HRP (0.1%/day) from the scaffold up to 33 days. It seems that a layer of surfactant that persists on the SGnPs facilitates the excellent dispersion. The SGnPs lose their ability to form stable suspensions when this surfactant layer is removed by excessive solvent washing.

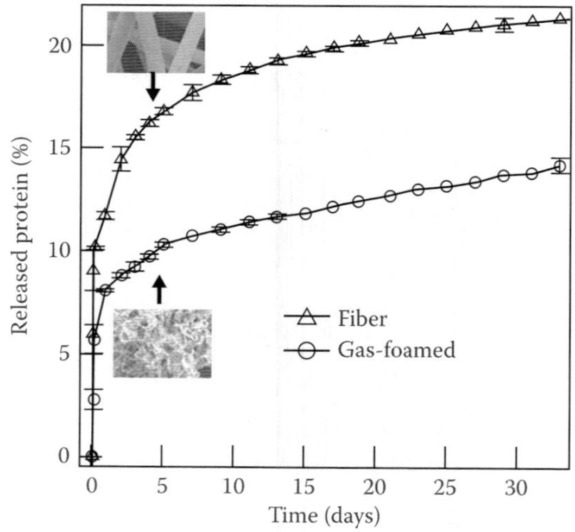

FIGURE 22.9 Released profile of HRP from the two types of PCL scaffold, nanofiber and gas-foamed. Insets, SEM micrographs of the corresponding scaffolds.

22.6 CONCLUSIONS AND FUTURE DIRECTION

The SGnP approach to stabilizing and delivery of proteins is nearly ideal in that a single approach yields excellent protection from process-related stresses, very good encapsulation efficiency and storage stability, as well as giving burst-free sustained release for essentially any protein and polymer system of interest. The approach borrows from a significant amount of work done in biopharmaceutics toward understanding and controlling protein degradation processes in sugar-based glasses. The required level of control is not possible with the methods currently used to incorporate biomacromolecules into polymeric carriers, but is readily met by the SGnP nanoencapsulation approach described here. Furthermore, because well-established biopharmaceutical glass technology is used, we expect that this approach will be amenable to other bioactive species such as viruses, DNA, or siRNA.

ACKNOWLEDGMENTS

J. Giri acknowledges postdoctoral support from the U.S. National Research Council. This work was funded in part under NIH/NIBIB grant R01 EB006398-01A1.

REFERENCES

Allegrini, P., J. F. Douglas, and S. C. Glotzer. 1999. Dynamic entropy as a measure of caging and persistent particle motion in supercooled liquids. *Physical Review E* 60: 5714–5724.

Anarbaev, R. O., A. L. Rogozina, and O. I. Lavrik. 2009. DNA polymerase β reveals enhanced stability in reverse microemulsions. *Journal of Molecular Catalysis B: Enzymatic* 59 (1–3): 64–69.

Anopchenko, A., T. Psurek, D. van der Hart, J. F. Douglas, and J. Obrzut. 2006. Dielectric study of the antiplasticization of trehalose by glycerol. *Physical Review E* 74: 031501.

Buchenau, U., and R. Zorn. 1992. A relation between fast and slow motions in glassy and liquid selenium. *Europhysics Letters* 18: 523–528.

Buera, M. P., S. Rossi, S. Moreno, and J. Chirife. 1999. DSC confirmation that vitrification is not necessary for stabilization of the restriction enzyme *Eco*RI dried with saccharides. *Biotechnology Progress* 15: 577–579.

Caliskan, G., D. Mechtani, J. H. Roh, A. Kisliuk, A. P. Sokolov, S. Azzam, M. T. Cicerone, S. Lin-Gibson, and I. Peral. 2004. Protein and solvent dynamics: How strongly are they coupled? *Journal of Chemical Physics* 121: 1978–1983.

Carpenter, J. F., L. M. Crowe, and J. H. Crowe. 1987. Stabilization of phosphofructokinase with sugars during freeze-drying—Characterization of enhanced protection in the presence of divalent-cations. *Biochimica et Biophysica Acta* 923: 109–115.

Castellanos, I. J., G. Flores, and K. Griebenow. 2005. Effect of the molecular weight of poly(ethylene glycol) used as emulsifier on alpha-chymotrypsin stability upon encapsulation in PLGA microspheres. *Journal of Pharmacy and Pharmacology* 57(10): 1261–1269.

Chen, F. M., M. Zhang, and Z. Wu. 2010. Toward delivery of multiple growth factors in tissue engineering. *Biomaterials* 31(24): 6279–6308.

Chew, S. Y., Y. Wen, Y. Dzenis, and K. W. Leong. 2006. The role of electrospinning in the emerging field of nanomedicine. *Current Pharmaceutical Design* 12(36): 4751–4770.

Chirico, G., M. Placidi, and S. Cannistraro. 1999. Fractional Stokes–Einstein relationship in biological colloids: Role of mixed stick–slip boundary conditions. *Journal of Physical Chemistry B* 103: 1746–1751.

Cicerone M. T., and C. L. Soles. 2004. Fast dynamics and stabilization of proteins: Binary glasses of trehalose and glycerol. *Biophysical Journal* 86: 3836–3845.

Cicerone, M. T., P. A. Wagner, and M. D. Ediger. 1997. Translational diffusion on heterogeneous lattices: A model for dynamics in glass forming materials. *Journal of Physical Chemistry B* 101: 8727–8734.

Cicerone, M. T., Q. Zhong, J. Johnson, K. A. Aamer, and M. Tyagi. 2011. Surrogate for Debye–Waller factors from dynamic Stokes shifts. *Journal of Physical Chemistry Letters* 1464–1468.

Cleland, J. L., X. Lam, B. Kendrick, J. Yang, T. H. Yang, D. Overcashier, D. Brooks, C. Hsu, and J. F. Carpenter. 2001. A specific molar ratio of stabilizer to protein is required for storage stability of a lyophilized monoclonal antibody. *Journal of Pharmaceutical Sciences* 90: 310–321.

Collins, F. C., and G. E. Kimball. 1949. Diffusion-controlled reaction rates. *Journal of Colloid Science* 4: 425–437.

Costantino, H. R., K. G. Carrasquillo, R. A. Cordero, M. Mumenthaler, C. C. Hsu, and K. Griebenow. 1998. Effect of excipients on the stability and structure of lyophilized recombinant human growth hormone. *Journal of Pharmaceutical Sciences* 87: 1412–1420.

Crowe, L. M., R. Mouradian J. H. Crowe, S. A. Jackson, and C. Womersley. 1984. Effects of carbohydrates on membrane stability at low water activities. *Biochimica et Biophysica Acta* 769: 141–150.

Davidson, P., and W. Q. Sun. 2001. Effect of sucrose/raffinose mass ratios on the stability of co-lyophilized protein during storage above the Tg. *Pharmaceutical Research* 18: 474–479.

Dirama, T. E., G. A. Carri, and A. P. Sokolov. 2005. Coupling between lysozyme and glycerol dynamics: Microscopic insights from molecular-dynamics simulations. *Journal of Chemical Physics* 122: 244910.

Douglas, J. F., and D. Leporini. 1998. Obstruction model of the fractional Stokes–Einstein relation in glass-forming liquids. *Journal of Non-Crystalline Solids* 235: 137–141.

Doukyu, N., and H. Ogino. 2010. Organic solvent-tolerant enzymes. *Biochemical Engineering Journal* 48(3): 270–282.

Duddu, S. P., G. Z. Zhang, and P. R. DalMonte. 1997. The relationship between protein aggregation and molecular mobility below the glass transition temperature of lyophilized formulations containing a monoclonal antibody. *Pharmaceutical Research* 14: 596–600.

Dyre, J. C., N. B. Olsen, and T. Christensen. 1996. Local elastic expansion model for viscous-flow activation energies of glass-forming molecular liquids. *Physical Review B* 53: 2171–2174.

Engasser, J. M., F. Chamouleau, L. Chebil, and M. Ghoul. 2008. Kinetic modeling of glucose and fructose dissolution in 2-methyl 2-butanol. *Biochemical Engineering Journal* 42(2): 159–165.

Franks, F., R. H. M. Hatley, and S. F. Mathias. 1991. Materials science and the production of shelf-stable biologicals. *Biopharm International* 4: 38–55.

Frauenfelder, H., G. Chen, J. Berendzen, P. W. Fenimore, H. Jansson, B. H. McMahon, I. R. Stroe, J. Swenson, and R. D. Young. 2009. A unified model of protein dynamics. *Proceedings of the National Academy of Sciences of the United States of America* 106: 5129–5134.

Freed, K. F. 2011. Entropy–enthalpy compensation in chemical reactions and adsorption: An exactly solvable model. *Journal of Physical Chemistry B* 115: 1689–1692.

Fu, K., A. M. Klibanov, and R. Langer. 2000. Protein stability in controlled-release systems. *Nature Biotechnology* 18(1): 24–25.

Giteau, A., M. C. Venier-Julienne, A. Aubert-Pouëssel, and J. P. Benoit. 2008a. How to achieve sustained and complete protein release from PLGA-based microparticles. *International Journal of Pharmaceutics* 350(1–2): 14–26.

Giteau, A., M. C. Venier-Julienne, S. Marchal, J. L. Courthaudon, M. Sergent, C. Montero-Menei, J. M. Verdier, and J. P. Benoit. 2008b. Reversible protein precipitation to ensure stability during encapsulation within PLGA microspheres. *European Journal of Pharmaceutics and Biopharmaceutics* 70(1): 127–136.

Grote, R. F., G. van der Zwan, and J. T. Hynes. 1984. Frequency-dependent friction and solution reaction rates. *Journal of Physical Chemistry* 88: 4676–4684.

Guo, Y. S., S. R. Bryn, and G. Zografi. 2000. Physical characteristics and chemical degradation of amorphous quinapril hydrochloride. *Journal of Pharmaceutical Sciences* 89: 128–143.

Gusev, A. A., S. Arizzi, U. W. Suter, and D. J. Moll. 1993. Dynamics of light gases in rigid matrices of dense polymers. *Journal of Chemical Physics* 99: 2221.

Hall, R. W., and P. G. Wolynes. 1987. The aperiodic crystal picture and free-energy barriers in glasses. *Journal of Chemical Physics* 86: 2943–2948.

He, J. T., H. B. Su, G. Li, X. Tao, W. Mo, and H. Song. 2006. Stabilization and encapsulation of a staphylokinase variant (K35R) into poly(lactic-co-glycolic acid) microspheres. *International Journal of Pharmaceutics* 309(1–2): 101–108.

Johari, G. P., and M. Goldstein. 1971. Viscous liquids and the glass transition. III. Secondary relaxations in aliphatic alcohols and other nonrigid molecules. *Journal of Chemical Physics* 55: 4245–4252.

Kaminski, K., E. Kaminska, K. L. Ngai, M. Paluch, P. Wlodarczyk, A. Kasprzycka, and W. Szeja. 2009. Identifying the origins of two secondary relaxations in polysaccharides. *Journal of Physical Chemistry. B* 113: 10088–10096.

Kessairi, K., S. Capaccioli, D. Prevosto, M. Lucchesi, S. Sharifi, and P. A. Rolla. 2008. Interdependence of primary and Johari–Goldstein secondary relaxations in glass-forming systems. *Journal of Physical Chemistry. B* 112: 4470–4473.

Kokai, L. E., H. P. Tan, S. Jhunjhunwala, S. R. Little, J. W. Frank, and K. G. Marra. 2010. Protein bioactivity and polymer orientation is affected by stabilizer incorporation for double-walled microspheres. *Journal of Controlled Release* 141(2): 168–176.

Kramers, H. A. 1940. Brownian motion in a field of force and the diffusion model of chemical reactions. *Physica* 7: 284–304.

Kumar, S. K., G. Szamel, and J. F. Douglas. 2006. Nature of the breakdown in the Stokes–Einstein relationship in a hard sphere fluid. *Journal of Chemical Physics* 124: 214501.

Lai, M. C., and E. M. Topp. 1999. Solid-state chemical stability of proteins and peptides. *Journal of Pharmaceutical Sciences* 88: 489–500.

Larini, L., A. Ottochian, C. De Michele, and D. Leporini. 2008. Universal scaling between structural relaxation and vibrational dynamics in glass-forming liquids and polymers. *Nature Physics* 4: 42–45.

Lee, K. Y., and D. J. Mooney. 2003. Controlled growth factor delivery for tissue engineering. *Advances in Controlled Drug Delivery: Science, Technology, and Products* 846: 73–83.

Lievonen, S. M., and Y. H. Roos. 2002. Nonenzymatic browning in amorphous food models: Effects of glass transition and water. *Journal of Food Science* 67: 2100–2106.

Liu, L., and Q.-X. Guo. 2001. Isokinetic relationship, isoequilibrium relationship, and enthalpy–entropy compensation. *Chemical Reviews* 101: 673–696. Available at: http://dx.doi.org/10.1021/cr990416z.

Manning, M. C., D. K. Chou, B. M. Murphy, R. W. Payne, and D. S. Katayama. 2010. Stability of protein pharmaceuticals: An update. *Pharmaceutical Research* 27: 544–575.

Mathew, D. S., and R. S. Juang. 2005. Improved back extraction of papain from AOT reverse micelles using alcohols and a counter-ionic surfactant. *Biochemical Engineering Journal* 25(3): 219–225.

Molinero, V., and W. A. Goddard. 2005. Microscopic mechanism of water diffusion in glucose glasses. *Physical Review Letters* 95: 045701.

Morita, T., Y. Sakamura, Y. Horikiri, T. Suzuki, and H. Yoshino. 2000. Protein encapsulation into biodegradable microspheres by a novel S/O/W emulsion method using poly(ethylene glycol) as a protein micronization adjuvant. *Journal of Controlled Release* 69(3): 435–444.

Moznine, R. E., G. Smith, E. Polygalov, P. M. Suherman, and J. Broadhead. 2003. Dielectric properties of residual water in amorphous lyophilized mixtures of sugar and drug. *Journal of Physics D: Applied Physics* 36: 330.

Naoe, K., K. Noda et al. 2004. Higher order structure of proteins solubilized in AOT reverse micelles. Colloids and Surfaces B-Biointerfaces 38(3–4): 179–185.

Ngai, K. L., L. R. Bao, A. F. Yee, and C. L. Soles. 2001. Correlation of positron annihilation and other dynamic properties in small molecule glass-forming substances. *Physical Review Letters* 87(21): 215901.

Ngai, K. L., and S. Capaccioli. 2004. Changes of the primary and secondary relaxation of sorbitol in mixtures with glycerol. *Journal of Physical Chemistry B* 108: 11118–11123.

Obrzut, J. A. Anopchenko, J. F. Douglas, and B. W. Rust. 2010. Relaxation and antiplasticization measurements in trehalose-glycerol mixtures—A model formulation for protein preservation. *Journal of Non-Crystalline Solids* 356: 777–781.

Paillard-Giteau, A., V. T. Tran, O. Thomas, X. Garric, J. Coudane, S. Marchal, I. Chourpa, J. P. Benoît, C. N. Montero-Menei, and M. C. Venier-Julienne. 2010. Effect of various additives and polymers on lysozyme release from PLGA microspheres prepared by an s/o/w emulsion technique. *European Journal of Pharmaceutics and Biopharmaceutics* 75(2): 128–136.

Perez, C., I. J. Castellanos, H. R. Constantino, W. Al-Azzam, and K. Griebenow. 2002. Recent trends in stabilizing protein structure upon encapsulation and release from bioerodible polymers. *Journal of Pharmacy and Pharmacology* 54(3): 301–313.

Pikal, M. J., D. Rigsbee, M. L. Roy, D. Galreath, K. J. Kovach, B. Wang, J. F. Carpenter, and M. T. Cicerone. 2008. Solid state chemistry of proteins—Part II: The correlation of storage stability of freeze-dried human growth hormone (hGH) with structure and dynamics in the glassy solid. *Journal of Pharmaceutical Sciences* 97: 5106–5121.

Pollack, G. L. 1981. Atomic test of the Stokes–Einstein law: Diffusion and solubility of Xe. *Physical Review. A* 23: 2660–2663.

Prestrelski, S. J., N. Tedeschi, T. Arakawa, and J. F. Carpenter. 1993. Dehydration-induced conformational transitions in proteins and their inhibition by stabilizers. *Biophysical Journal* 65: 661–671.

Psurek, T., C. L. Soles, K. A. Page, M. T. Cicerone, and J. F. Douglas. 2008. Quantifying changes in the high-frequency dynamics of mixtures by dielectric spectroscopy. *Journal of Physical Chemistry B* 112: 15980–15990.

Putney, S. D., and P. A. Burke. 1998. Improving protein therapeutics with sustained-release formulations. *Nature Biotechnology* 16(2): 153–157.

Riggleman, R. A., J. F. Douglas, and J. J. de Pablo. 2007. Tuning polymer melt fragility with antiplasticizer additives. *Journal of Chemical Physics* 126: 234903–234913.

Riggleman, R. A., J. F. Douglas, and J. J. de Pablo. 2010. Antiplasticization and the elastic properties of glass-forming polymer liquids. *Soft Matter* 6: 292–304.

Riggleman, R. A., K. Yoshimoto, J. F. Douglas, and J. J. de Pablo. 2006. Influence of confinement on the fragility of antiplasticized and pure polymer films. *Physical Review Letters* 97: 045502.

Rossi, S., M. P. Buera, S. Moreno, and J. Chirife. 1997. Stabilization of the restriction enzyme *Eco*RI dried with trehalose and other selected glass-forming solutes. *Biotechnology Progress* 13: 609–616.
Savelli, G., N. Spreti, and P. D. Profioa. 2000. Enzyme activity and stability control by amphiphilic self-organizing systems in aqueous solutions. *Current Opinion in Colloid and Interface Science* 5(1–2): 111–117.
Schwendeman, S. P. 2002. Recent advances in the stabilization of proteins encapsulated in injectable PLGA delivery systems. *Critical Reviews in Therapeutic Drug Carrier Systems* 19(1): 73–98.
Shamblin, S. L., X. Tang, L. Chang, B. C. Hancock, and M. J. Pikal. 1999. Characterization of the time scales of molecular motion in pharmaceutically important glasses. *Journal of Physical Chemistry B* 103: 4113–4121.
Simmons, D. S., and J. F. Douglas. 2011. Nature and interrelations of fast dynamic properties in a coarse-grained glass-forming polymer melt. *Soft Matter* 22(7): 11010–11020.
Starr, F. W., S. Sastry, J. F. Douglas, and S. C. Glotzer. 2002. What do we learn from the local geometry of glass-forming liquids? *Physical Review Letters* 89: 125501–125504.
Stukalin, E. B., J. F. Douglas, and K. F. Freed. 2010. Plasticization and antiplasticization of polymer melts diluted by low molar mass species. *Journal of Chemical Physics* 132: 084504.
Sumi, H. 1991. Theory on reaction rates in nonthermalized steady states during conformational fluctuations in viscous solvents. *Journal of Physical Chemistry* 95: 3334–3350.
Sun, W. Q., and P. Davidson. 2001. Effect of dextran molecular weight on protein stabilization during freeze-drying and storage. *Cryo Letters* 22: 285–292.
Sy, J. C., A. S. Klemm, and V. Prasad Shastri. 2009. Emulsion as a means of controlling electrospinning of polymers. *Advanced Materials* 21(18): 1814.
Tanaka, K., T. Takeda, and K. Miyajima. 1991. Cryoprotective effect of saccharides on denaturation of catalase by freeze-drying. *Chemical and Pharmaceutical Bulletin* 39: 1091–1094.
Tayalia, P., and D. J. Mooney. 2009. Controlled growth factor delivery for tissue engineering. *Advanced Materials* 21(32–33): 3269–3285.
Tessmar, J. K., and A. M. Gopferich. 2007. Matrices and scaffolds for protein delivery in tissue engineering. *Advanced Drug Delivery Reviews* 59(4–5): 274–291.
Wang, J. J., K. M. Chua, and C. Wang. 2004. Stabilization and encapsulation of human immunoglobulin G into biodegradable microspheres. *Journal of Colloid and Interface Science* 271(1): 92–101.
Wang, B. Q., S. Tchessalov, M. T. Cicerone, N. W. Warne, and M. J. Pikal. 2009a. Impact of sucrose level on storage stability of proteins in freeze-dried solids—Part II: Correlation of aggregation rate with protein structure and molecular mobility. *Journal of Pharmaceutical Sciences* 98: 3145–3166.
Wang, B. Q., S. Tchessalov, N. W. Warne, and M. J. Pikal. 2009b. Impact of sucrose level on storage stability of proteins in freeze-dried solids—Part I: Correlation of protein–sugar interaction with native structure preservation. *Journal of Pharmaceutical Sciences* 98(9): 3131–3144.
Yang, S. X., W. E Yuan, and T. Jin. 2009. Formulating protein therapeutics into particulate forms. *Expert Opinion on Drug Delivery* 6(10): 1123–1133.
Yoshioka, S., T. Miyazaki, and Y. Aso. 2006. Beta-relaxation of insulin molecule in lyophilized formulations containing trehalose or dextran as a determinant of chemical reactivity. *Pharmaceutical Research* 23: 961–966.
Yoshioka, S., T. Miyazaki, Y. Aso, and T. Kawanishi. 2007. Significance of local mobility in aggregation of beta-galactosidase lyophilized with trehalose, sucrose or stachyose. *Pharmaceutical Research* 24: 1660–1667.

Zhu, G. Z., S. R. Mallery, and S. P. Schwendeman. 2000. Stabilization of proteins encapsulated in injectable poly(lactide-co-glycolide). *Nature Biotechnology* 18(1): 52–57.

Zwanzig, R. 1989. Diffusion in a dynamically disordered continuum. *Chemical Physics Letters* 164: 639–642. Available at: http://www.sciencedirect.com/science/article/B6TFN-44XDT2M-48/2/1d2b75b19fa8b6d43e4ef595d57ab3e9.

23 Clinical Relevance of Tissue Engineering

Koteshwar N. Sridhar and Daniel G. Miller

CONTENTS

23.1 INTRODUCTION

Tissue engineering (TE) is an emerging multidisciplinary application that involves biology, medicine, and engineering and has the potential for restoring, maintaining, and enriching tissue and organ function. Medical or surgical procedures that have attempted to partially replace lost tissue account for a large part of healthcare resources, which exceeds $400 billion[1] per year.

The surgical reconstruction of defects or deformities by traditional procedures such as autologous tissue transfers and artificial mechanical devices have the inherent problem of scarcity, susceptibility to infection, and uncertain long-term outcomes. The basis of TE for therapeutic and diagnostic applications is the ability to exploit living cells in a variety of ways supported by research in the following areas:

1. Biomaterials and scaffolds: novel biomaterials help in the growth and differentiation of cells by providing physical and chemical cues and a framework for expansion in two and three dimensions.
2. Cell culture: acquiring cells for proliferation and differentiation depending on the specific needs.
3. Biomolecules including angiogenic and growth factors.

In human beings, the aspects of regenerative medicine such as solid organ transplant and cell therapy have been practiced for many years, for instance, kidney and bone marrow transplants. The various tissues studied for therapeutic regeneration include skin, cartilage, and cornea; with organs such as the liver, kidney, and heart having been studied as well. This chapter reviews the clinical relevance of the science of TE and its current status.

23.2 STRATEGIES IN TE

TE intends to restore the functional and structural characteristics of tissues by providing suitable living cells with or without biomaterials.

23.2.1 Cell Types

Various cell types have been used in laboratory studies. These include stem cells and mature cells. The stem cells include embryonic stem (ES) cells, adult stem cells (ASCs), hemopoietic stem cells (HSCs), mesenchymal stem cells (MSCs), and induced pluripotent stem (iPS) cells. Mature cells, such as keratinocytes, fibroblasts, and chondrocytes, have shown the capacity to dedifferentiate and grow.

23.2.1.1 ES Cells

Human ES cells were discovered in 1998 by Thomson et al.[2] These cells are capable of differentiating into lineages of all three germinal layers (ectoderm, endoderm, and mesoderm), and this ability has aroused interest in the use of ES cells for regenerative medicine. However, the use of ES cells has been limited for various reasons, including those related to ethical issues of destroying human embryos, their propensity to produce tumors when injected in mice,[3] and the possibility of viral transfer because a large number of approved cells are being cultured on mouse feeder lines. In view of these limitations, the alternative sources of stem cells, called ASCs, have been pursued.[4]

23.2.1.2 Adult Stem Cells

ASCs are present in approximately 1% to 2% of the total cell population within a specific tissue and are vital to the maintenance of local tissue homeostasis by continuously contributing to tissue regeneration, replacement of cells lost due to apoptosis, and direct injury. ASCs are held in an undifferentiated state within their niche until they receive a stimulus to differentiate. ASCs have limited differentiation potential compared with ES cells because they occur further down the development pathway. These problems, along with accessibility, low frequency, and poor growth, have limited their use in regenerative medicine. The best characterized and most widely understood ASCs are the HSCs, which are used in bone marrow transplant to replenish the marrow after irradiation.

23.2.1.3 Mesenchymal Stem Cells

The bone marrow also has progenitor cells called MSCs, which are noncirculating stromal cells that are capable of developing into mesenchymal lineages such as cardiac muscle, or transdifferentiating into ectodermal lineage such as neurons. MSCs also have limitations as they can only divide a finite number of times and, with time, may accumulate genetic changes.[5] A simple method has been described in collecting MSCs from the posterior maxilla and expanding them into cells of the required mesodermal lineages.[6] There are ongoing efforts aimed at obtaining MSCs from peripheral blood. A recent study compared the MSCs obtained from peripheral blood with the MSCs derived from bone marrow, and evaluated their ability to form a mesenchymal lineage. It was found that in a study of 20 patients, both cells were comparable.[7]

23.2.1.4 iPS Cells

In 2006, iPS cells were cultured from the skin cells of adult and embryonic mice by reprogramming using Yamanaka factors.[8] In comparison with ES, iPS cells are specific to the individual, have superior potential to differentiate into specific lineages, but frequently show signs of apoptosis and senescence.[9]

23.2.2 Cell Expansion

Cells are cultured either statically or dynamically. Static cultures are either two-dimensional (2-D) or three-dimensional (3-D). A routine culture in a Petri dish is static and is a 2-D culture. The cells are regularly monitored and the dishes are periodically changed to allow for growth and confluence. This method allows for the differentiation of a small number of cells, which is time consuming, labor intensive, and cumbersome.[10]

Three-dimensional cultures can produce an organized cohesive tissue. However, this requires the use of suitable biomaterial in a bioreactor. A bioreactor reproduces the physiological environment specific to the tissue that is to be regenerated, including oxygen and nutrients, removes catabolites, monitors pH, and applies mechanical stresses to stimulate the formation of extracellular matrix (ECM).[11] Cells can thus be cultured and encapsulated in a semipermeable membrane that isolates the cells from the body, allows nutrition to flow in and waste to flow out, but prevents macromolecules to access the cells. This method has been used in the production of bioartificial liver using xenogenic hepatocytes.[12] A dynamic 3-D culture can be kept for a long time using bioreactors.

23.2.3 Biomaterials

Biomaterials in TE act as an artificial ECM, supporting cell adhesion, growth, and differentiation. Although cell suspensions have been injected in areas like the knee, biomaterials are usually required for their localization because most mammalian cells are anchorage-dependent for survival.[13] A vast majority of biomaterials are used as carriers of cells or to deliver therapeutic agents to the transplanted area.[14] A scaffold should have the following requisites to be useful:

1. Material should be biocompatible and biodegradable.
2. Degradation products should be nontoxic.
3. Material should be porous enough to allow for the movement of nutrition and waste products.
4. A mathematical model to understand the effects of scaffold porosity and initial cell density has been studied in a theoretical framework.[15] In this study, it was concluded that total cell distribution and yield were directly related to the scaffold porosity and initial cell density, an observation to be kept in mind when designing the scaffolds.
5. Material must possess robust mechanical properties.

The scaffold aims to provide 3-D environments to bring cells in close proximity so that they can assemble to form tissues. As the scaffold degrades over time, the cells deposit their own ECM molecules and eventually form tissues. The most commonly used materials, as shown in Table 23.1, are synthetic polymers such as lactic-glycolic acid or polyacrylonitrile, polyvinyl chloride, or natural materials like collagen, hydroxyapatite, acellular tissue matrices (bladder mucosa and small intestinal mucosa), or alginates. Collagen and alginate have been approved by the Food and Drug Administration (FDA) as wound dressing material. Collagen is widely available in animal tissues and is purified by enzyme treatment and gets degraded by lysosomal enzymes. There is an increasing interest in developing scaffolds with the imprint of the cells on them. This is performed by using the ECM derived from the cultured cells to prepare the scaffolds. As there is a hidden imprint of the cell of origin on these scaffolds, growing the cells toward a required phenotype has been easier.[16] Nanotechnology has brought in new ideas in the field of biomaterials because nanomodified surfaces can induce good cellular response and sustained and robust cell differentiation.[17] Nanotechnology focuses on assembling simple elements to form complex structures. The cells are composed of many nanosized components, all working together harmoniously. These nanocomponents modulate the production,

TABLE 23.1
Commonly Used Biomaterials

	Natural Material	Synthetic Material	
		Biodegradable	Nondegradable
1	Collagen	Aliphatic polyesters	Polyethylene derivatives
2	Chitosan	PLLA = poly-L-lactic acid	Polyurethane
3	—	PLGA = poly(lactic-co-glycolic acid)	
4	—	Polyanhydrides	

composition, and architecture of the ECM, which, in turn, influence the behavior of other cells and tissues. The potential of nanotechnology is yet to be fully exploited.

23.2.4 Present Status of TE

23.2.4.1 Challenges for Clinical Transition

Clinical relevance stems from the fact that there is a high level of expectation from the patient and physician communities to improve the quality of life of patients immediately. However, in practice, TE is still not able to offer reliable solutions to these unmet medical needs primarily because of the following factors:

1. Safety and efficacy factors for the stem cell lineages
2. Improper understanding of the cell factors and their microenvironment
3. Nonavailability of good biomaterials for cell attachment proliferation and differentiation
4. Improper enhancement of angiogenesis
5. Reliable monitoring of the cell engraftment and survival
6. Lack of clear regulatory guidelines adds to the uncertainty
7. Unreliable clinical trial data
8. Fear of transmission of xenozoonotic diseases and rejection in the case of xenotransplants
9. Cost to the healthcare system as there is uncertainty about healthcare reimbursement
10. Safety concerns about the gene therapy

23.2.4.2 Clinical Applications of TE

Clinical trials in TE are represented by the soft tissues such as skin and cartilage and hard tissue such as bone. These will highlight the clinical challenges of TE. All of them are presented with the clinical need, and the three components of TE, that is, cells, scaffolds, and biomaterials.

23.2.4.2.1 Bone Regeneration by TE

A significant number of bone injuries require bone grafting to improve the healing process; most of them are autografts (from the same patient), allografts (outside donor), or synthetic materials. Autografts have the advantage of histocompatibility but are usually associated with uncertain morbidity and inadequate volume. In addition, because they are morcelized during harvesting, the structure is damaged. Allografts are either from cadavers or structural grafts and pose problems of uncertain osteogenic activity, therefore creating a need for synthetic grafts.

23.2.4.2.1.1 Principles of Bone Regeneration This follows the natural bone healing, that is, acute inflammation releasing bioactive molecules, which, in turn, attract the MSCs to the injured area. These MSCs adhere to the scaffolds in the area and differentiate into osteoblasts and secrete the osteoid, which becomes mineralized to form the bone.[18] Bioactive materials provide signals to stimulate bone regeneration such as catalysts. Bone growth-promoting proteins, which are called

TABLE 23.2
Proteins Approved in Bone Regeneration Therapy

	Protein Biological for Bone Regeneration
1	Transforming growth factor-β
2	Bone morphogenetic proteins
3	Insulin-like growth factors
4	Platelet-derived growth factors
5	Fibroblast growth factors

bone morphogenetic proteins (BMPs), activate MSCs toward bone formation.[19] They are a class of low molecular weight proteins. Recently, the U.S. FDA has approved recombinant BMP-2 and BMP-7 for specific and limited clinical indications. The biomolecules used to augment the bone cell reaction are given in Table 23.2. Instead of providing these bioactive materials as growth supplements, it would be a technological improvement if osteogenic cells are genetically modified to secrete the very same biological stimulants locally depending on the need, which would make bone growth closer to the natural situation. Gene therapy involves the transfer of genetic material with the help of a vector into the cell, which, in turn, secretes growth factors, locally stimulating the MSCs toward bone formation,[20] which have been successfully demonstrated in animal models.[21]

23.2.4.2.1.2 Scaffolds A bioactive nanoscaffold has been described composed of poly-caprolactone and nanosized hydroxyapatite or beta-tricalcium phosphate, which was able to support the growth of human bone marrow MSCs and guide their osteogenic differentiation at the same time without needing additional biomaterials.[22]

23.2.4.2.1.3 Sources of Cells Osteogenic cells are the precursors of osteoblasts, which form the bone MSCs and are the main source of osteogenic cells, and are abundant in the bone marrow. The bone marrow could be from the same patient (autologous) or aspirated from the iliac crest or allogeneic (donor) from universal cell banks. Although allogeneic bank cells provide the volume required and have very little immunogenicity, regulatory issues have prevented their widespread use. Therefore, alternative sources of MSCs from muscle and fat have been pursued, and these are reported[23] to have improved the osteogenicity of acellular bone grafts. Adipose MSCs were studied and found to be superior to bone marrow MSCs in a recent study.[24]

23.2.4.2.1.4 Preclinical Studies These studies face the challenges of variation in the cellular profile of the species, the limited bone marrow volumes, and the lack of a standard cell culture technique. Despite these variations, *in vivo* studies on bone marrow in animals have been promising.[25] When bone marrow rich in osteogenic cells was used over a carrier such as hydroxyapatite or inside ceramic tubing, bone defects in animal models have shown healing. Tripathi et al.[26] have shown that scaffolds loaded with osteoblasts demonstrated bone formation radiologically in rabbit radial defects. Interestingly, cell-enriched grafts with either allogeneic or xenograft MSCs have not shown any immunological reaction, which may be due to the lack of surface markers on the MSCs.

23.2.4.2.1.5 Clinical Studies Autologous bone marrow with suitable matrices have shown the clinical fusion of the defects in trauma[27] and nonunion cases.[28] Cell-enriched grafts using 1000 progenitor cells/cm with a total volume of approximately 30,000 progenitor cells have shown good healing.[29]

23.2.4.2.1.6 Role of Electrical Stimulation and Distraction Osteogenesis Mechanical loading and electromagnetic signals have shown biophysical effects, inducing bone formation in animals and in clinical studies.[30] The precise cellular response to the stimulation signals is not clearly understood; hence, this remains as an investigational tool at the moment. In distraction osteogenesis, separated fragments unite because of various mechanical and biochemical stimuli, and this has been a clinically useful method in craniofacial bone healing.[31] The future of bone regeneration is expected to be in exploring growth factors and their delivery, or in inducing bone formation *in situ* with percutaneous methods.

23.2.4.2.2 Cartilage

The demand for the TE of cartilage stems from the large number of traumatic injuries as well as for the treatment for degenerative joints in the aging population. The present modalities of treatment, like arthroplasty, have a finite clinical span and some morbidity. Novel TE have been attempted to eliminate the damaged cartilage.

23.2.4.2.2.1 Cells Ideally, a fully differentiated chondrocyte would be the cell of choice as it is already programmed toward the cartilage phenotype. In 1994, Brittberg et al.[13] described the technique of autologous chondrocyte implantation, in which autologous chondrocytes were injected under the periosteal flap, which is now marketed as Carticel by Genzyme and is the only FDA-approved cell-based therapy for cartilage repair. This procedure has not been universally used because of the expensive procedure, the limited availability of chondrocytes, the potential donor site morbidity, and the potential of chondrocyte to dedifferentiate into monolayer cultures. The other types of cells used are MSCs from the bone marrow, which have been induced to differentiate toward chondrocytes.[32] ASCs from the dermal layer of skin are an attractive alternative to primary cells for meniscus engineering because they may be easily obtained and used autologously. Recently, chondroinducible dermis cells from caprine skin have shown promising characteristics for cartilage TE. In a recent study, their multilineage differentiation capacity was determined, and the expansion and TE methods for these cells were investigated and were found to be a reliable source of mesenchymal cells.[33]

23.2.4.2.2.2 Scaffolds Hyaline cartilage serves as a low-friction and wear-resistant articulating surface in load-bearing, diarthrodial joints. Unfortunately, as the avascular, alymphatic nature of cartilage significantly impedes the body's natural ability to regenerate, damage resulting from trauma and osteoarthritis necessitates attempts at repair. Current clinical methods are generally limited in their ability to regenerate functional cartilage. A variety of synthetic and natural materials have been used as carriers of MSCs, most commonly in the form of hydrogels,[34] and these systems have been tuned for optimal nutrient diffusion, connectivity of deposited matrix, degradation, soluble factor delivery, and mechanical loading for enhanced matrix production and organization. As the effect of scaffolds has been variable on the cells, newer scaffolds with biomimetic properties have been attempted. Electrospun nanofibers with structural similarity to ECM have been attempted with promising results.[35] Even with these promising advances, the complex mechanical properties and biochemical composition of native cartilage have not been achieved, and engineering cartilage tissue still remains a significant challenge.

23.2.4.2.2.3 Biomolecules Various signaling molecules including members of transforming growth factor-β and BMPs have been used to promote cell proliferation, differentiation, and ECM production, but the optimum method of administration is yet to be standardized.[36] As the cartilage cells are weight-bearing cells, eventual function depends on their ability to manage mechanical stress. Many attempts have been made to incorporate mechanical stress in the bioreactor so that truly functional cartilage can be produced.[37]

23.2.4.2.2.4 Clinical Studies Various attempts to regenerate the cartilage have been the subject of a few clinical studies, such as the following:

1. Covering the cartilage defect with collagen membrane at the level of the periosteum.[38]
2. Cultured chondrocytes were seeded on hyaluronic scaffolds or collagen membrane or gels and then affixed to the cartilage defect area by fibrin glue.[39]
3. Chondrospheres, which are chondrocytes cultured in high density without scaffold material, were encapsulated into a spheroid and implanted arthroscopically.[40] These chondrospheres are a subject of a pilot study on more than 500 patients.[41]
4. The sole supplement to the culture medium is the patient's own serum, and the implantation technique is quite simple, performed through a keyhole using an arthroscope. These features make this an attractive proposition in clinical practice.

23.2.4.2.3 Skin

The largest organ in the body primarily serves as a protective barrier against the external environment, and the loss of a large portion of skin results in the evaporation of fluid and electrolytes and allows for infection. Adult skin consists of two layers, a superficial epidermis and a deeper supporting layer of dermis. The TE of the skin is mainly used for the management of burns and large wound defects such as diabetic ulcers. The wound can be superficial or deep. In the case of burns, immediate coverage with semisynthetic biomaterials like collagen can be performed, followed by expanded keratinocytes. The disadvantage of this method is the time delay required for the culture of keratinocytes, which is usually 3–4 weeks. The cells that are involved in the regeneration are multipotent skin stem cells, present in the hair follicle bulge, aided by growth factors. Despite this treatment, in severely burned patients, the skin coverage is suboptimal because there is serous fluid between the transplanted skin sheet and the wound. This interferes with fluid exchange and leads to the loss of grafted cells.[42] However, another problem in wound healing is the formation of scars due to the disorganized arrangement of collagen.

23.2.4.2.3.1 Cell Application to the Wound by Spraying Cell application by keratinocyte sheet application would require a large number of cells and time for expansion. In comparison with cell application in the form of sheets, spraying of cultivated subconfluent keratinocytes covers a larger area with the same number of cells, thus shortening application time. The application of primary cell suspensions, which are obtained from autologous cells and applied immediately, has the advantage of providing all types of cells including melanocytes. Because these cells do not have sufficient time to dedifferentiate in the cultures, they adhere better and help in keratinocyte expansion.[43] In Table 23.3, we provide the list of materials available for wound healing approved by the FDA.[44]

TABLE 23.3
FDA-Approved Commercially Available Skin Substitutes for TE

Product	FDA-Approved Indications (PMA and HDE)	Competitive Advantages
Alloderm	Burns/full-thickness wounds (allograft)	Not rejected; no cases of viral transmission after >100,000 product applications; 2-year shelf life
Apligraf	Venous/diabetic ulcers (PMA)	Mimics function of dermis; cryopreserved product
Dermagraft	Diabetic foot ulcers (PMA); ulcers secondary to epidermolysis bullosa (HDE)	Mimics function of dermis; cryopreserved product
Epicel	Deep partial- and full-thickness burns (HDE); congenital nevi (HDE)	Autologous cells; no rejection, high incidence of permanent take
Integra	Deep partial- and full-thickness burns (PMA)	Two layers; good barrier function; used in >10,000 patients; moderate shelf life
OrCel	Split-thickness donor sites (PMA); mitten hand deformity surgery of epidermolysis bullosa (HDE)	Mimics cytokine expression of healing skin; 9-month shelf life cryopreserved
TransCyte	Full- and partial-thickness burns (PMA)	1.5 year shelf life cryopreserved

Note: PMA, premarket approval; HDE, humanitarian device exemption.

23.2.5 FDA-Approved Products

The first FDA approval of an engineered tissue product, without a living cellular component, occurred with the premarket application (PMA) approval for Integra artificial skin on March 1, 1996. This was followed by Organogenesis's Apligraf, which contained a living component, and subsequently the Advanced Tissue Sciences application of Dermagraft with a proprietary matrix, and cryopreserved cellular component was approved in 2001. Together, Apligraf and Dermagraft have been used in more than 400,000 patient applications with few adverse events reported, suggesting that if used as indicated, these products are both safe and effective. Regulatory authorities in Europe and in the United States have held extensive public discussions of the issues emerging from TE, and developed consensus regulatory guidelines to provide researchers and commercial organization with the basic code of conduct. Currently, there are 168 tissue-engineered products and procedures undergoing clinical evaluation worldwide. The work is driven in large part by the perceived market for tissue-engineered products. The TE products shown in Table 23.4 have been approved by the FDA for use in human medicine. However, the path to market authorization has been rather long-drawn and limited by scientific advances, which affect tissue-engineered products. In Table 23.5, we provide a synopsis of the key regulatory milestones for various types of drugs, pharmaceuticals, and cell-based therapies.

23.2.6 Future

The opportunities for TE are enormous as we are facing an increasing aging population with different ailments. Scientific fervor in this field is demonstrated by the exciting publications of bladder reconstruction[45] and tracheal replacement.[46] Ultimately, these depend on the innovativeness of the researchers and the clinical challenges because TE is not only a new medical field but also a new way of thinking. Basic research and good clinical trials should proceed simultaneously to reach this goal. It is estimated that the number of patients needing TE therapy is very large, and some information on the procedures conducted in the United States alone is shown in Table 23.6.

The following is a list of innovative concepts on the horizon waiting to be clinically relevant:

1. Automation to scale-up the products with the help of robots that could provide consistency and reproducible quality.
2. Drug delivery systems that can monitor the milieu dissociation of molecules and products on a screen.
3. Preventive measures such as cell banks or biomolecules, which can be used before the anticipated injury so that the damage is minimized. Stents used in tubular structures such as cardiac stents may have cell sheaths that deposit the cells after the angioplasty.
4. Cultured cells can be used for studying the disease *in vitro* or to study the effect of drugs.
5. The field of biomaterials is intensely researched toward developing suitable models for cell growth and differentiation and proper scaffolds for tissue reconstruction.

TABLE 23.4
FDA-Approved TE Products and Therapies

	Therapy	Product Name	Company	Indication
1	Autologous cellular immunotherapy	Provenge (sipuleucel-T),	Dendreon Corporation	For the treatment of asymptomatic or minimally symptomatic metastatic castrate-resistant (hormone refractory) prostate cancer Center for Biologics Evaluation and Research, US FDA (CBER); April 29, 2010
2	Autologous cultured chondrocytes	Carticel	Genzyme Biosurgery	For repair of articular cartilage defects of the medial or lateral condyles, or trochlea having inadequate response to prior non-Carticel surgical treatment (CBER); June 21, 2007
3	Cultured epidermal autografts	Epicel	Genzyme Biosurgery	For use in patients who have deep dermal or full-thickness burns comprising a total body surface area of ≥30%. It may be used in conjunction with split-thickness autografts, or alone in patients for whom four split-thickness autografts may not be an option due to the severity and extent of their burns (CDRH, HDE); October 25, 2007
4	Azficel-T	LAVIV	Fibrocell Technologies, Inc.	Indicated for improvement of the appearance of moderate to severe nasolabial fold wrinkles in adults (CBER); June 20, 2011
5	Graftskin; cultured; allogeneic keratinocytes and fibroblasts	Apligraf	Organogenesis	For use with standard therapeutic compression for the treatment of noninfected partial and full-thickness skin ulcers due to venous insufficiency of >1-month duration and that have not adequately responded to conventional ulcer therapy. For use in the treatment of full-thickness neuropathic diabetic foot ulcers of >2 weeks duration that extend through the dermis but without tendon, muscle, capsule, or bone exposure (CDRH); May 22, 1998
6	Cryopreserved human fibroblast-derived dermal substitute with a bioabsorbable matrix	Dermagraft	Advanced Tissue Sciences, Inc.	For full-thickness diabetic foot ulcers that have been present for >6 weeks and for ulcers that extend deeper into the skin where the blood vessels are located, but do not involve tendon, muscle, joint capsule, or bone (CDRH); September 28, 2001

TABLE 23.5
Key Regulatory Milestones for Drugs, Pharmaceuticals, and Cell-Based Therapies

Year	Milestone
1973	Federal Regulation: General Biological Products (21 CFR 610, 38 FR 32056)
1975	Federal Regulation: Blood and Blood Components (21 CFR 606, 40 FR 53532)
1978	Federal Regulations: Finished Pharmaceuticals (21 CFR 211, 43 FR 45077)
1993	Application of current authorities to human somatic cell therapy products (58 FR53248). "The relevant function of the cells, if known, and/or relevant products biosynthesized by the cells should be defined and quantitated as a measure of potency"
1993	Human tissue intended for transplantation. FDA indicates that transplantation falls within their jurisdiction and that they may regulate the industry at some future time (58 FR 65514)
1993	Animal cells and tissues as somatic therapy (58 FR 5324 10/14/93)
1996	Guidance on applications for products comprised of living autologous cells manipulated *ex vivo* and intended for structural repair of reconstruction (FR 04/26/96)
1998	Establishment registration and listing for manufacturers of human cell and tissue-based products (FR 05/14/98)
1998	Human tissue intended for transplantation. Regulation effective January 26, 1998 (FR 06/29/98)
1999	Suitability determination for donors of human cellular and tissue-based products (FR 09/30/99)
1999	Guidance for industry: public health issues posed by the use of nonhuman primate xenografts in humans
2001	Current good tissue practice for manufacture of human cellular and tissue-based products; inspection and enforcement; proposed rule (66 FR 1508)
2001	PHS guideline on infectious disease issues in xenotransplantation
2003	Guidance for industry: source animal, product, preclinical, and clinical issues concerning the use of xenotransplantation products in humans
2010	Guidance for industry: characterization and qualification of cell substrates and other biological materials used in the production of viral vaccines for infectious disease indications

Source: Center for Biologicals Evaluation and Research, U.S. Food and Drug Administration.
Note: CFR, Code of Federal Regulations; FR, Federal Register; PHS, Public Health Service.

Cell therapy may be expensive in the beginning; efforts are underway to reduce this by using simpler cell growth and expansion methods.

23.2.7 Conclusion

Despite being an exciting science, TE has not translated to significant clinical use because of the many lacunae in available scientific data. Experimental studies oftentimes are not repeatable primarily owing to the nonuniformity of the regeneration process and assessment protocols. These, in addition to insufficient clinical trial data, make it difficult for us to scale-up TE-based solutions. Therefore, well-conducted

TABLE 23.6
Potential Tissue and Organ Transplant Market in the United States

	Structure	Procedures/Year
1	Skin	4,750,000
2	Cartilage	1,132,100
3	Blood vessels	1,100,000
4	Pancreas	728,000
5	Kidney	600,000
6	Breast	261,000
7	Liver	155,000
8	Tendon and ligament	123,000
9	Intestine	100,000
10	Ureter and urethra	81,900
11	Heart valves	65,000
12	Bladder	57,200

Source: Brockbank, K. G. M. 2002. Tissue engineering constructs and commercialization. In *Biomaterials in the Design and Reliability of Medical Devices*, edited by M. N. Helmus. Austin, TX: Landes Biosciences. ISBN: 0-306-47690-8.

clinical trials and a good understanding of cellular behavior is an urgent need for TE to become a mainstay solution for some of the unmet medical needs. Currently, TE is still a complementary armamentarium to a reconstructive surgeon. There is a good opportunity now for clinicians to join hands with scientists to build a focused program that develops TE solutions for the current day-to-day problems faced by clinicians. To achieve these goals, the clinician would have to become a student, a researcher, a strategist, and an entrepreneur.

REFERENCES

1. Langer, R., and J. R. Vacanti. 1995. Tissue engineering. *Science* 260: 920–926.
2. Thomson et al. 1998. Embryonic stem cell lines derived from human blastocysts. *Science* 282: 1145–1147.
3. Reubinoff, B. E. et al. 2000. Embryonic stem cell lines from human blastocysts: Somatic differentiation *in vitro*. *Nature Biotechnology* 18(4): 399–404.
4. Fuchs, E. et al. 2004. Socializing with neighbor: stem cells and their niche. *Cell* 116(6): 769–778.
5. Pittenger, M. F. et al. 1999. Multilineage potential of adult MSC. *Science* 284(5411): 143–147.
6. Lohberger, B., M. Payer, B. Rinner, C. Bartmann, E. Stadelmeyer, E. Traunwieser, T. Devaney, N. Jakse, A. Leithner, and R. Windhager. 2011. *Journal of Oral and Maxillofacial Surgery*. PMID: 219400092.
7. Chong, P. P., L. Selvaratnam, A. A. Abbas, and T. Kamarul. 2011. *Journal of Orthopaedic Research*.

8. Takahashi, K., and S. Yamanaka. 2006. Induction of pluripotent stem cells from mouse embryonic and adult fibroblast cultures by defined factors. *Cell* 126: 663–676.
9. Clarke, L., and D. Van der Kooy. 2009. A safer stem cell: Inducing pluripotency. *Nature Medicine* 15: 1001–1002.
10. Polak, J. M., and A. Mantalaris. 2008. Cell expansion, cell encapsulation, 3-D cultures. In *Advances in Tissue Engineering*, edited by J. M. Polak, S. Mantalaris and S. E. Harding, 503–515. London: Imperial College Press.
11. Dvir, T., and S. Cohen. 2008. Bioreactor engineering regenerate the dynamics cell microenvironment. In *Advances in Tissue Engineering*, edited by J. M. Polak, S. Mantalaris, and S. E. Harding, 517–535. London: Imperial College Press.
12. Strom, S., and R. Fisher. 2003. Hepatocyte transplantation: new possibilities for therapy. *Gastroenterology* 124: 568–571.
13. Brittberg, M. et al. 1994. Treatment of deep cartilage defects in the knee with autologous chondrocyte transplantation. *New England Journal of Medicine* 331: 889–895.
14. Kim, B. S., and D. J. Mooney. 1998. Development of biocompatible synthetic extracellular matrices for tissue engineering. *Trends in Biotechnology* 16: 224–230.
15. Shakeel, M. et al. 2011. A continuum model of cell proliferation and nutrient transport in a perfusion bioreactor. *Mathematical Medicine and Biology*. PMID: 21994793.
16. Lu, H., T. Hoshiba, N. Kawazoe, I. Koda, M. Song, and G. Chen. 2011. Cultured cell-derived extracellular matrix scaffolds for tissue engineering. *Biomaterials*. PMID: 21937104.
17. Gentleman, E. et al. 2009. Comparative material differences revealed in engineered bone as a function of cell-specific differentiation. *Nature Materials* 8: 763–770.
18. Whang, P. G., and J. R. Liebermann. 2003. Clinical issues in the development of cellular systems for use as a bone graft substitutes. In *Bone Graft Substitutes*, edited by C. T. Laurencin, 142–155. West Conshohocken, PA: ASTM International.
19. Urist, M. R. 1965. Bone formation by autoinduction. *Science* 150: 893–899.
20. Nussenbaum, B., and P. H. Krebsback. 2004. Practical matters in the application of tissue engineered products for skeletal regeneration in the head and neck region. In *Tissue Engineering in Musculoskeletal Clinical Practice*, edited by L. J. Sandell and J. Grodzinsky, 154. Rosmont, IL: AAOS.
21. Musgrave, D. S. et al. 2002. Gene therapy and tissue engineering in orthopedic surgery. *Journal of the American Academy of Orthopaedic Surgeons* 10: 6–15.
22. Polini, A. et al. 2011. Osteoinduction of human mesenchymal stem cells by bioactive composite scaffolds without supplemental osteogenic growth factors. *PLoS One* 6(10): e26211. PMID: 22022571.
23. Cowan, C. M. et al. 2004. Adipose derived adult stromal cells heal critical-size mouse calvarial defects. *Nature Biotechnology* 22 (5): 560–567.
24. Schubert, T. et al. 2011. The enhanced performance of bone allografts using osteogenic differentiated adipose derived mesenchymal stem cells. *Biomaterials*. PMID: 21872925.
25. Miniamide, A. et al. 2005. The use of cultured bone marrow cells in type I collagen gel and porous hydroxyapatite for posterolateral lumbar spine fusion. *Spine* 30(10): 1134–1138.
26. Tripathi, A. et al. 2011. Tissue-engineered osteogenesis in bone defects by homologous osteoblasts loaded on sterile bioresorbable coral scaffold in rabbits. *Surgical Science* 2: 369–375.
27. Chapman, M. W. et al. 1997. Treatment of acute fractures with a collagen–calcium phosphate graft material. *Journal of Bone and Joint Surgery* 79A(4): 494–502.
28. Sim, R. et al. 1993. Autologous marrow injection in the treatment of delayed and non-union in long bones. *Singapore Medical Journal* 34: 412–417.
29. Hernigou, P. et al. 2005. Percutaneous autologous bone marrow grafting of nonunions: Influence of the number and concentration of progenitor cells. *Journal of Bone and Joint Surgery, American Volume* 87: 1430–1437.

30. Chao, E. Y. et al. 2003. Biophysical stimulation of bone fracture repair, regeneration, and remodeling. *European Cells and Materials* 6: 72–84.
31. Choi, I. H. et al. 2000. Vascular proliferation and blood supply during distraction osteogenesis: A scanning EM observation. *Journal of Orthopaedic Research* 18: 698–705.
32. Wakitani, S. et al. 2002. Human autologous culture expanded bone marrow mesenchymal cell transplantation for repair of cartilage defects in osteoarthritic knees. *Osteoarthritis and Cartilage* 10: 199–206.
33. Sanchez-Adams, J., and K. A. Athanasiou. 2011. Dermis isolated adult stem cells for cartilage tissue engineering. *Biomaterials* 33(1): 109–19 PMID:21959004.
34. Park, Y. et al. 2004. Bovine primary chondrocyte culture in synthetic matrix metalloproteinase-sensitive poly (ethylene glycol)-based hydrogels as a scaffold for cartilage repair. *Tissue Engineering* 10: 515–522.
35. Li, W. J. et al. 2005. A three-dimensional nanofibrous scaffold for cartilage tissue engineering using human mesenchymal stem cells. *Biomaterials* 26: 599–609.
36. Chen, F. H. et al. 2006. Technology insight: adult stem cells in cartilage regeneration and tissue engineering. *Nature Clinical Practice. Rheumatology* 2: 373–382.
37. Miyanishi, K. et al. 2006. Dose- and time-dependent effects of cyclic hydrostatic pressure on transforming growth factor-beta-3 induced chondrogenesis by adult human mesenchymal cells *in vitro*. *Tissue Engineering* 12: 2253–2262.
38. Steinwachs, M., and P. C. Kreuz. 2007. Autologous chondrocyte implantation in chondral defects of the knee with a typeI/III collagen membrane: A prospective study with a 3-year follow-up. *Arthroscopy* 23(4): 381–387.
39. Peterson, L. et al. 2000. Two to 9 year outcome after autologous chondrocyte transplantation of the knee. *Clinical Orthopaedics and Related Research* 374: 212–234.
40. Anderer, U., and J. Libera. 2002. *In vitro* engineering of human autogenous cartilage. *Journal of Bone and Mineral Research* 17: 1420–1429.
41. De Codon. 2009. Company published results. In *Fundamentals of Tissue Engineering and Regenerative Medicine*, edited by U. Meyer, T. Meyer, J. Handschel, and H. P. Wiesmann, 237–241. Heidelberg, Germany: Springer.
42. Barret, J. P. et al. 2000. Cost-efficacy of cultured epidermal auto grafts in pediatric burns. *Annals of Surgery* 231(6): 869–876.
43. Stoner, M. L., and F. M. Wood. 2000. The treatment of hypopigmented lesions with cultured epithelial autograft. *Journal of Burn Care and Rehabilitation* 21(1): 41–46.
44. Center for Biologicals Evaluation and Research, U.S. Food and Drug Administration.
45. Atala, A. et al. 2006. Tissue engineered autologous bladders for patients needing cystoplasty. *Lancet* 367: 1241–1246.
46. Macchiarini, P. et al. 2008. Clinical transplantation of tissue engineered airway. *Lancet* 372: 2023–2030.
47. Brockbank, K. G. M. 2003. Tissue engineering constructs and commercialization. In *Biomaterials in the Design and Reliability of Medical Devices*, edited by M. N. Helmus, Chapter 7, 191–197. New York: Plenum Publishers.

24 Future Perspectives of Biomaterials and Stem Cells in Regenerative Medicine

Dilip Dey

CONTENTS

24.1 INTRODUCTION

Because the life expectancy is very high in the industrialized world, an increasing number of graying populations are being added every year. This requires healthcare support system specialties such as regenerative medicine for organ repair or replacement. In particular, war veterans returning from Iraq and Afghanistan, who have significant organ damage or amputations, need organ replacements. Therefore, older population and war veterans require the help of regenerative medicine, and there is an increasing pressure to develop procedures for their treatment and to improve their quality of life.

Regenerative medicine is a new branch of medicine, and the derivation of stem cells [both embryonic stem cells (ESCs) and induced pluripotent stem cells (iPS)] and the advancement of integration to biomedical engineering have given it a new impetus. In nature, some of the lower vertebrates are well known for their ability to regenerate lost organs, for example, starfish, lizards, and salamanders (Brockes and Kumar 2010). Singing birds generate neuronal cells that contribute to song during the breeding season. These cells later die and more cells regenerate every year (Nottebohm 1989). However, with very few exceptions, mammals irreversibly lose their ability to regenerate organs. It is assumed that mammals have gone through the trade-off of organ regeneration to prevent the development of cancer. However, if a part of the embryo is cut off during gestation, this embryo can regenerate the lost organs (Muneoka et al. 2008). Moreover, one organ may regenerate in lower vertebrates but fails to do so in their mammalian counterpart, specifically, in humans. For instance, zebra fish (*Danio rerio*) can regenerate part of a lost heart, but humans are unable to do this (Jopling et al. 2010). However, in the adult animal, wound healing is a universal phenomenon in which partial regeneration of tissues takes place and closes the wound.

Tissue engineering, which acts as an interface between biology and engineering, is an emerging field that offers the possibility of replacing tissues and organs such as bone and neural cells (Bono et al. 2010). Tissue engineering was first described in 1985 and is synonymous with regenerative medicine. However, there have been significant recent developments. The emergence of stem cell biology and the development of a suitable scaffold (biomaterial) on which the tissue repair and regeneration can take place has revolutionized this field. However, the development of complex tissues and scaling up their production is a challenging task. It is difficult to create an intact vasculature in tissue-engineered constructs, and recapitulate the size or complexity of the organ (Rustad et al. 2010). The discovery of stem cells has opened the vista of tissue regeneration *in vitro* and the repair of damaged organs *in vivo*. Organ damage and its severity can be diverse, ranging from peripheral nerve damage to brain tissue damage/death due to stroke, Alzheimer disease, Parkinson disease, kidney failure, ischemic heart muscle, liver, and other organ failures because of the extended life expectancy. Workplace accidents or battlefield injuries that result in brain trauma, amputation, or loss of a limb are debilitating and can decrease the quality of life. The cost of their treatment is staggering and burdens the healthcare system. In other cases, organ/tissue replacements become imperative in various organ failures such as in burn patients (skin tissue), liver replacement due to liver failure, kidney replacements, and heart muscle replacements. Naturally, we are constantly turning over various cells in the body such as red blood cells, skin epithelium, and gut epithelium. On the other hand, some tissues rarely regenerate once they reach the adult stage. Our heart muscle and brain cells are vital but do not regenerate once they are damaged. For neuronal tissue, until the early 1990s, people believed that brain cells (neurons) did not divide and had no possibility of regeneration. However, by later incorporating the proliferation marker BrdU and performing histochemistry on cadaver brains, it was found that in a few cases, neurons do divide (Gould et al. 1999).

Cells are generally quiescent in adult tissues. However, with the proper manipulation, quiescent cells can be dedifferentiated to pluripotent cells that can be turned into particular cell lineages different from its origin. Stem cells are pluripotent cells that can self-renew and give rise to various types of adult/differentiated cells. There are three types of stem cells—ESCs, organ-specific multipotent adult stem cells, and iPS cells, each having its advantages and disadvantages as well (Figure 24.1). The embryonic pluripotent stem cells can go through unlimited cell division and give rise to various differentiated tissues. First isolated from human blastocysts by Thomson et al. (1998), ESCs have broadened the scope of regenerative medicine. Although ESCs have tremendous potential to be used for therapeutic purposes, its use is stymied by ethical considerations. Moreover, ESCs are prone to immune rejection. In addition to ESCs, there are adult stem cells in various organs and tissues that are multipotent or progenitor cells and have the capacity to give rise to only limited types of tissues when necessary. It is difficult to isolate the rare adult stem cells for clinical use—for example, liver stem cells cannot be isolated noninvasively and probably will not give rise to myoepithelial or cardiomyocytic cells. This technical barrier was overcome by the landmark work of Takahashi and Yamanaka (2006). They pinpointed several genes that were sifted from the microarray data and transfected adult cells by viral delivery of four genes (*Oct3/4*, *Sox2*, *Nanog*, and *Klf4*)

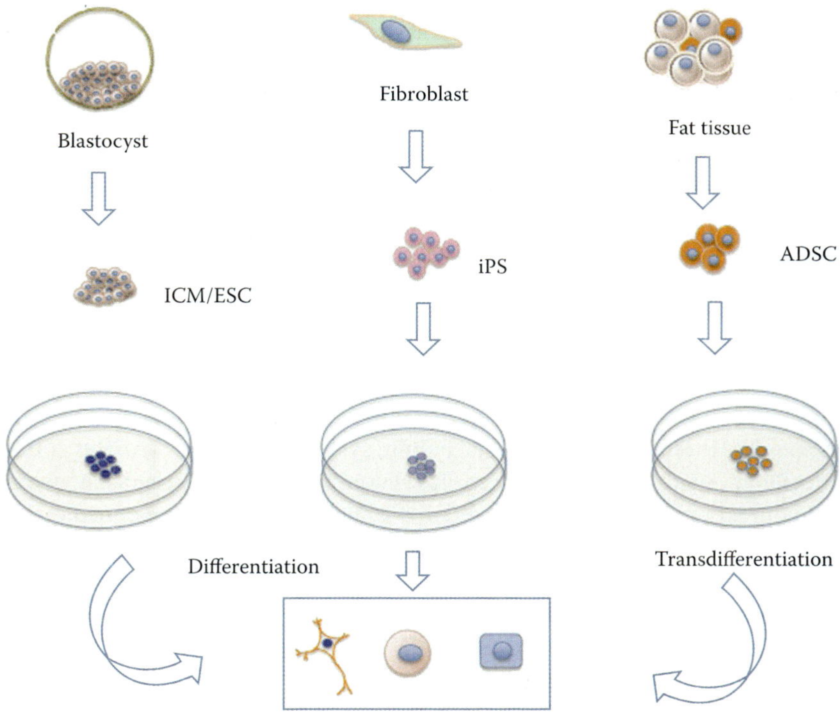

FIGURE 24.1 Various sources of stem cells and their derivation methods. Stem cells may be derived from various sources. ESCs are derived from the inner cell mass (ICM) of blastocysts. iPS is derived from any adult cells by four reprogramming factors. Adult stem cells are located in various tissues and are multipotent or progenitor cells. ES and iPS are pluripotent. However, iPS cells are not as good as ES cells due to incomplete epigenetic modifications and other unknown reasons.

and green flourescent protein marker. After culturing them in appropriate culture conditions, they found that adult cells were slowly being reprogrammed and forming pluripotent colonies. Although the generation of iPS is considerably simple, not all iPS are the same at the molecular level. This is due to incomplete epigenetic modification. During the reprogramming of adult cells, epigenetic modification (silencing of genes by DNA methylation or histone deacetylation) takes place (Mikkelsen et al. 2008). Epigenetics may be defined as an inheritable change in phenotype that is not caused by DNA alteration (Bird 2002). There are hundreds of methylation sites throughout the genome and very often, epigenetic modifications are not complete during the reprogramming process, resulting in heterogeneous populations of iPS cells. The iPS cells can be considered a hub from where the various final destinations of cells can be reached. The selected iPS clones need to be treated further to generate various lineage-specific differentiations. From a biochemical perspective, the reprogramming to iPS and differentiation to various tissues from iPS are complex changes that are not yet well understood. The iPS cells are grown on polystyrene flasks, which are coated with attachment factor (Invitrogen, Carlsbad, CA), Matrigel (BD Biosciences, San Diego, CA), collagen, laminin, and a feeder layer. Then, iPS cells are plated on mitotically inactive mouse embryonic fibroblast feeder layers, and the growth medium supplemented with knockout serum and basic fibroblast growth factor (bFGF) also known as FGF2. The knockout serum and bFGF maintain the human ES and iPS cells in a self-renewing, pluripotent state and prevent differentiation. In case of mouse ESC or iPS, they need leukemia-inhibitory factor (LIF) in the culture medium. LIF keeps mouse ESC/iPS in an undifferentiated, pluripotent state, whereas withdrawal of LIF initiates them for differentiation (Levenstein et al. 2006).

24.2 GROWING CELLS IN TWO DIMENSIONS VERSUS THREE DIMENSIONS

On a flat surface, cells grow in two dimensions (2-D), which hardly recapitulates/mimics the natural environment. Although 2-D cell cultures provide useful information about the biological process in tissue engineering, the analysis of cell growth dynamics in a three-dimensional (3-D) culture system is necessary (Abbott 2003). In multicellular organisms, cells grow in close coordination and communicate with the surrounding tissue. The cells are nested in a complex and not fully defined extracellular matrix (ECM). This ECM provides structural support as well as physiological and biochemical support to the cell (Lin and Bissell 1993). The gap of knowledge about the tissue-surrounding matrix *in vivo* stymies our effort for *in vitro* tissue generation or differentiation from stem cells.

The ECM is a very complex and dynamic structure surrounding the cells and tissues. Some of the components of ECM that support cells include fibronectin, laminin, and collagen. ECM is also a reservoir of various growth factors, hormones, cytokines, chemokines, and structural proteins (Ponting and Dexter 1988). They may present as immobilized or sequestered and are released by proteolytic actions to become an active molecule (Figure 24.2). Fibronectin is a ubiquitous ECM protein with an arginine–glycine–aspartic acid (RGD) signaling peptide motif (Pierschbacher and Ruoslahti 1984). Fibronectin remains as a globular unfolding

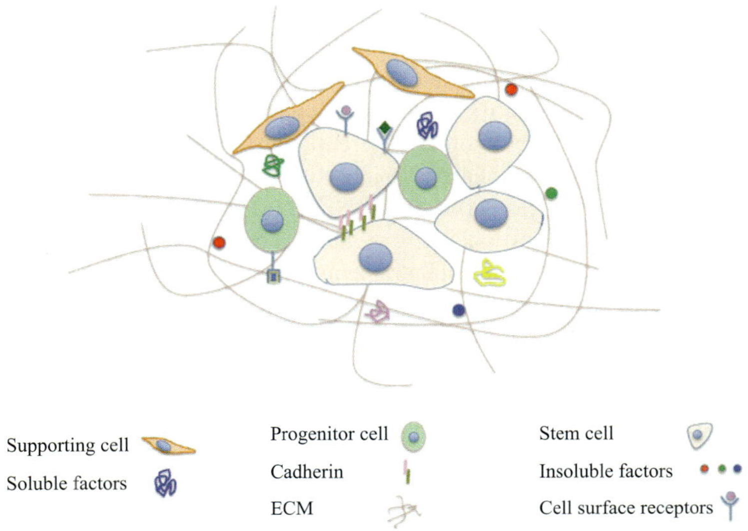

FIGURE 24.2 Role of ECM on stem cell maintenance, growth, and differentiation. Adult stem cells are surrounded by ECM, progenitor cells, and supporting cells. When required, they differentiate into cells of the corresponding tissues. ECM may contain both soluble and insoluble growth factors, as well as various structural proteins (such as fibronectin) that provide a unique environment for the maintenance of adult stem cells.

molecule and expresses cryptic domains that are exposed in cell membrane integrin-mediated processes (Salmeron-Sanchez et al. 2011). A variety of cells cross-talk, and supporting cells aid adult stem cell growth and maintenance. The cells are tethered with the matrix and these ECM signals are relayed/transmitted to the cytoplasm by cell membrane receptors. The receptors "sense" the signal and transmit inside the cell where biochemical activities are modulated depending on the signal message. With some exceptions, a typical cell membrane receptor has an extracellular domain, a transmembrane domain, and a cytoplasmic domain. This cytoplasmic domain is connected to different adapter/effecter molecules and an actin cytoskeleton (Hynes 1992). The receptor molecules are mediators between the extracellular and intracellular environment of a cell. The receptors may be members of integrins, ephrins, etc., and these families of proteins have profound effects (in both normal and pathological situations) on cell survival, growth, division, migration, etc. (Horowitz 1997).

24.2.1 Biomaterials in Stem Cell Differentiation

In their niche, stem cells are relatively quiet and divide only occasionally. On injury or tissue damage, stem cells come out from their niche and divide rapidly to compensate for the damaged area and thus help in wound healing (Gurtner et al. 2007). In their niche, stem cells are surrounded by a complex milieu of chemical and physical conditions that change temporally and spatially to favor its "stemness." Fibronectin in the ECM connects with cell surface integrin molecules. Any significant changes that occur in the ECM are relayed to the cytoplasm via integrins. These promote the expression of genes related to cell survival, proliferation, migration, etc. The role of ECM on cell lineage selection is beyond the scope of this article. Thus, we shall limit our discussion to the various biomaterials and their roles in stem cell lineage selection (differentiation). Matrix mechanical property and topography is important for retaining a stem cell or promoting differentiation. Biomaterials with a combination of microfluidics and microfabrication allow for the control of variables to stem cell fate determination (Dawson et al. 2008). This artificial environment mimics the stem cell niche but is never equal to it. Thus, there are ongoing advances in the area of (smart) biomaterials to design new smart molecules and deconstruct the stem cell niche. By using advanced biomaterials, stem cells can be grown on the mimicked host microenvironment. Stem cells cultured to preserve its stemness or, when needed, differentiation to pluripotent cells. Stem cells are grown on flat surfaces coated with laminin, collagen, or feeder cells. These simplistic cell culture conditions are different from the *in vivo* environment of stem cells. However, 2-D culture conditions have an advantage in that they elucidate the role of individual components in cell growth and cell fate. For high-throughput screening, an ECM array has been used to determine the factors that convert mammary progenitor cells to myoepithelial cells (Underhill and Bhatia 2007).

24.2.2 Biomaterials in Stem Cell Fate Determination

Biomaterials are defined as nonviable materials used in a medical device, intended to interact with biological systems (European Society of Biomaterials Conference 1986). During the course of time, the types of biomaterials and their applications have increased dramatically. Biomaterials are extensively used in tissue engineering as a scaffold in which inert, natural, or synthetic materials are used for the growth of biological samples *in vitro* or *in vivo* (Hubbell 1995). This biomaterial is mostly used as a scaffolding molecule that facilitates/enhances the development of various cell types to its end. The *in vivo* ECM environment can be partially achieved *in vitro* by the use of biomaterials and scaffolds. Biomaterial sources may be natural or synthetic, each with its own advantages and disadvantages. Synthetic scaffold is used as an alternative to natural scaffolds. Combining biomaterials, cells, ECM, scaffold, and microenvironmental cues that mimic the *in vivo* environment (Toh et al. 2011), cells can be directed to various lineages. The delivery of biomaterial in tissue engineering requires a suitable scaffold that supports tissue repair and regeneration.

An ideal biomaterial should have absorbable, nonimmunogenic, less toxic properties during application. Additionally, biomaterials should (a) allow cell attachment, (b) promote growth and migration, (c) deliver and stabilize biochemical factors, (d) enable diffusion and delivery of cell nutrients, and (e) in some cases, exert mechanical and biological influences to modify the behavior of the cell (Willerth and Sakiyama-Elbert 2008). Biocompatibility of the scaffold and immune response to the scaffold are also criteria to be considered.

The biomaterial and scaffold materials may be protein-based, polysaccharide-based, peptide-based, polymer-based, or ceramic-based. An updated list of some of the destination cells and their differentiation procedures may be found in recent articles.

24.2.3 Natural Biomaterials

24.2.3.1 Polysaccharide-Based Biomaterials

Polysaccharide is made of sugar monomers and has a role in biomaterial formation. The polysaccharide can be obtained from both plants and animals. Depending on the source and method of isolation, polysaccharides can sometimes elicit an immune response (Willerth and Sakiyama-Elbert 2008). Scaffolds made of polysaccharides can form gels and can be injected into an injury site. Some examples of polysaccharide scaffolds include agarose, alginate, hyaluronan, and chitosan. Stem cells use these scaffolds for differentiation into various tissues.

24.2.3.1.1 Agarose

Agarose is a prevalent biomaterial derived from the seaweed named agar agar. It is a polymer of sugars and can be readily prepared by heating in a suitable solution. The concentration of agarose gel depends on the amount of agar used (w/v) for the preparation of gel. It can have various desirable degrees of porosity, toughness/stiffness, and mechanical strength depending on the concentration of agar. Agarose scaffolds have been used for stem cell differentiation to dopaminergic neuron, chondrogenesis, cartilage, heart tissue, etc. (Ando et al. 2007; Chen et al. 2007; Finger et al. 2007).

24.2.3.1.2 Alginate

Alginate is a widely used natural scaffold material derived from algae, which forms by ionic linking. Alginates encapsulate cells and allow growth, proliferation, and secretion of ECM. Alginate scaffolds have been used in many laboratories especially for the development of cartilage (Wayne et al. 2005; Hannouche et al. 2007). Alginate has also been used for neural tissue engineering as well as for differentiation of adipose derived stem cells (ADSC) to chondrocytes (Hannouche et al. 2007). Various proteins or peptides covalently attached with alginate provide sites for cell attachment. Alginate-entrapped neurons (astrocytes, microglia, and neurons) display outgrowth property (Frampton et al. 2011).

24.2.3.1.3 Chitosan

Chitosan is a polysaccharide-based scaffold derived by the deacetylation of chitin. Chitin is an exoskeleton molecule present in crustacean shells and is a relatively inert material that does not elicit a chronic immune response. This material may be useful and thus considered for long-term application and spinal cord repair (Kim et al. 2011). Chitosan has been used in tissue engineering because it is structurally similar to glycoaminoglycans, which is an ECM component. Incorporation of both phosphonic and sulfonic function in chitosan enhances stem cell growth. Thus, phosphorylated polyelectrolyte complex is stem cell-compatible and may have tissue engineering applications (Yeh and Lin 2011). Chitosan has been used to regenerate skin, bone, and nerve tissue. The scaffold can be used for stem cell differentiation both *in vitro* and *in vivo*.

24.2.3.1.4 Hyaluronan (Hyaluronic Acid)

Hyaluronic acid is found in abundance in the ECM. Mesenchymal stem cells (MSC) have been differentiated into cartilage, bone, nerve, etc., inside hyaluronic acid scaffolds. Additionally, high molecular weight hyaluronic acid supplied to MSC for chondrogenesis in chitosan scaffold dose-dependent effect on matrix production improves cartilage production (Schwartz et al. 2011).

24.2.3.1.5 Hydrogel

Hydrogels are commonly used as biomaterials for cell encapsulation to mimic the matrix *in vivo* (Kachouie et al. 2010). ADSCs can grow in hydrogel containing peptides that mimic 3-D growth environments. Cells are seeded in peptide hydrogel polymer scaffolds (Xie et al. 2011) with the appropriate growth medium. After 10 days, cells converted to neurons and expressed both BrdU and nestin (cell proliferation and neuronal markers, respectively).

24.2.3.1.6 Biocoral

Biocoral scaffold is a porous natural hydroxyapatite. Human dental pulp stem cells (DPSC), and human osteoblasts were seeded on 3-D biocoral scaffolds both in normal and rotative (3-D) conditions. DPSCs differentiated into osteoblasts and osteoblast secreted the ECM that helped a diffuse bone formation in biocoral samples (Mangano et al. 2011). RT-PCR showed the expression of mineralization tissues such as osteocalcin and osteopontin.

24.2.3.1.7 Cardiogel

Cardiogel is an efficient 3-D ECM that supports the growth of bone marrow-derived MSCs and facilitates better growth, cell division, and enhanced survival and has the potential for cardiac tissue regeneration (Sreejit and Verma 2011). Rat ESC, in combination with insoluble collagenous bone matrix, promotes ectopic bone formation *in vivo* (Kahle et al. 2010).

24.2.3.1.8 Hydroxyapatite

Hydroxyapatite is the mineral component of human bones and teeth enamel and is used as a synthetic biomaterial and a bone graft extender. Composite hydroxyapatite and porous 3-D silk fibroin scaffolds (biodegradable and biocompatible) induce the formation of mineralized bone matrix by human MSCs (hMSCs) in the absence of osteogenic growth factor (Bhumiratana et al. 2011).

Stem cells have the potential for ischemic disease treatment, but they can be poorly grafted, which can be overcome by the use of cell transplantation scaffold or stem cell delivery with a combination of growth factors. Human ADSCs cultured and grafted in spheroids facilitate cell survival, angiogenic factor secretion, neovascularization, etc., in the hypoxic conditions of the transplanted area. Therefore, spheroid basal cell delivery could be a good strategy for stem cell therapy in which poor engraftment is a barrier (Bhang et al. 2011). Polymers used as scaffolds have the same properties as natural ECM and they are used as a substitute for ECM. By changing the composition of polymers, scaffold properties and biocompatibility can be altered. Abed et al. (2011) synthesized polysaccharide-based hydrogels that are made of dextran and pullulan. Bone regeneration solely by osteogenic factors often has poor success, as blood supply through angiogenesis is also essential for bone survival (Kujala et al. 2002). Delivery of osteogenic and angiogenic factors together in a biodegradable scaffold enhances bone regeneration. Xiao and others combined bone morphogenetic protein-2 (BMP-2) and vascular endothelial growth factor (VEGF) with coral scaffolds for repair of critical-sized bone defects in rabbits (Xiao et al. 2011). Human MSCs can be used for xenogenic transplantation because they have low immunogenicity, differentiation potential for several tissues, and high proliferation rates. Thus, MSCs are ideal cells for tissue engineering and Zong et al. (2010) used poly-lactic-co-glycolic acid scaffolds for osteoblast-like cell differentiation from hMSCs.

24.2.3.1.9 Decellularized Tissues

Decellularized tissues are chemically stripped of its cells, leaving behind the intricate structure of the ECM. Until now, decellularized tissues have been isolated from animal cadavers, which can elicit an immune response in the host. Also, it can potentially harbor pathogens (Datta et al. 2006). In contrast, biomaterial composed of ECM is safely derived from human fat (adipose tissue), which is easily extractable, abundantly available, and expandable. Choi et al. (2011) suggested that decellularized ECM containing biological and chemical cues of native human ECM could be used as a scaffold for autologous as well as allograft tissue engineering. In bone tissue engineering, a porous scaffold is necessary to support cells and tissue interaction as well as the formation of a vascular network for the repair of bone tissue.

24.2.3.2 Protein-Based Biomaterial

Protein is found in the ECM and in the cell membrane, in many cases, in glycosylated form. Protein-based biomaterials mimic the ECM and thus have the advantage of being biomaterial candidates that direct cell growth and migration during tissue regeneration (Romano et al. 2011). This helps in the stabilization of encapsulated cells. There are several protein-based biomaterials.

24.2.3.2.1 Collagen

Collagen is the major insoluble fibrous protein in the ECM and in connective tissue molecules that has been used in various tissue-engineering purposes. Collagen can be used either alone or in combination with other molecules. Monkey ESCs can differentiate into an embryoid body in a collagen scaffold (Michelini et al. 2006). Inside a collagen scaffold, human ESCs can differentiate to form blood vessels and hepatocytes (Baharvand et al. 2006). Ghanaati et al. (2011) used a bilayered collagen matrix to treat gingival tissue that promoted its ingrowth and reversed tissue recession. By the improvement of biodegradation elastomeric materials that use ECM proteins, a new kind of material is developed, one which has superior degradation kinetics, desirable biocompatibility, and soft tissue-like mechanical properties (Bettinger 2011). Dried myocardial tissue forms a necrotic tissue scar. Biomaterials that contain growth factors and cells provide mechanical support as well as growth factors for cell growth (Kolettis et al. 2011).

24.2.3.2.2 Fibrin

Fibrin has long been used clinically for hemostasis and sealing, use in other applications has been limited because of its relatively rapid resorption *in vivo*. Fibrin has been used for neural cell, vasculature, bone regeneration, etc. (Kania et al. 1998).

24.2.3.2.3 Spider Silk

Spider silk is stronger and tougher than native biological fibers such as collagen. It is also biocompatible and biodegradable—suitable for a biomaterial candidate. However, its application is limited due to the inability to mimic complex natural tissue ultrastructures (Brown et al. 2011b).

24.2.3.3 Synthetic Biomaterials

Synthetic biomaterials have physical properties that are comparable to or even superior over natural biomaterials. They have been developed with a defined chemical composition, and can be modified chemically. They show reproducibility and their degradation rates can be controlled by the mechanical property of the scaffold (Engler et al. 2006). The rate of degradation of the scaffold also affects the rate of drug release. However, they lack a site for cell adhesion and degradation of synthetic biomaterial product is toxic for cell culture (Willerth and Sakiyama-Elbert 2008). The cell adhesion site can be overcome by attaching different chemical groups on the surface. Besides, they show adverse physiological reactions such as infection and inflammation (Bazaka et al. 2010).

24.2.3.4 Polymer-Based Biomaterials

Polymer-based scaffolds have been used for stem cell differentiation and tissue engineering. Polymer-based biomaterials lacks sites for cell adhesion. Also, after degradation, they produce toxic material/chemicals. There are several polymer-based biomaterial scaffolds, for example, poly(lactic-co-glycolic acid) (PLGA) and poly(ethylene glycol), and pol(ε-caprolactone) (Li et al. 2003). PLGA consists of glycolic acid and lactic acid. PLGA is biocompatible, is approved by the Food and Drug Administration, and has the ability to modulate its rate of degradation. It is used in various kinds of tissue engineering. However, it lacks controllable drug delivery and changes its environment to an acidic pH during degradation.

Cells including MSC can sense substrate mechanical properties and respond accordingly. By using a polyacrylamide gel system, Lees et al. (2007) showed that loss of substrate modulus affected hMSC morphology, proliferation, and differentiation potential. A 3-D culture using alginate scaffold maintained a variety of neural cell types with their viability and function. It provides a higher degree of controlled cell growth environment, with a more physically relevant diffusion property.

24.2.3.5 Modifications of Biomaterials

24.2.3.5.1 Growth Factor Delivery: Free or Immobilized

Immobilization of growth factors by covalently attaching to biomaterial matrices has advantages over freely diffusible growth factors because covalently immobilized growth factors are more stable and promote sustained growth factor delivery (Masters 2011). On the other hand, the development and synthesis of new polymers has led to a dramatic improvement in medicine, tissue engineering, drug delivery, and cell therapy. To be considered a suitable biomaterial, it should have the desired microstructure, permeability, degradation rates, and suitable mechanical properties. By using chemical modification techniques, Lopez-Cebral et al. (2011) modified gelatin and the suitable characteristics of the modified protein to develop new nanomedicines—nanoparticles using the ionotropic gelation technique. Cells may perceive nanomaterials in bizarre ways. Therefore, each component of the cell, starting from the membrane to the nucleus, needs to be tested for its response to nanoparticles. MSCs isolated from human placenta, cord blood, fetal bone marrow, and adipose tissue cultured in the presence of nanomaterial were monitored for differentiation into the osteogenic phenotype (Sujatha et al. 2011). During this process, some cells die and others take up nanoparticles. The viable cells both divide and differentiate into osteogenic cell types.

In bone regeneration, scaffolds with controlled porosity are necessary to allow cell growth, diffusion of nutrients, and vasculogenesis (Grandi et al. 2011). Synovium-derived MSCs can be used for cartilage repair in chitin–alginate composite 3-D porous scaffolds supplemented with transforming growth factor β3 (TGF-β3) and BMP-2. Cells attach, proliferate, and secrete ECM. A percentage of cells in S and G_2/M phases increased, and those in G_0/G_1 phase decreased, indicating cell proliferation (Qi et al. 2011). Endothelial progenitor cells are used for the revascularization of tissues. Biocompatible polymers, such as fibrin, can induce cell profiles, differentiation, and functional activity of endothelial progenitor cells. Fibronectin is a very important molecule for cell and tissue regeneration, cell proliferation, migration, and collagen reorganization. Fibronectin is upregulated in tissue injury and a decreased level of fibronectin has been implicated in delayed wound healing. Living cells undergo fibronectin matrix assembly, converting the soluble form of inactive fibronectin to biologically active ECM fibrils. Engineering of fibronectin mimetics has bypassed cell-dependent fibronectin assembly and is used in tissue repair (Roy et al. 2011). ES or iPS cells are grown on a feeder layer or Matrigel that is extracted from mouse sarcoma. This Matrigel is rich in a basement membrane component and remains chemically undefined. The basement membrane is a thin sheet of fibers/ECM that underlies the epithelium in normal tissues. It is mainly composed of laminin, collagen, heparin sulfate proteoglycans, and growth factors such as FGF, TGF-β, EGF, etc. (Benton et al. 2011). Cells cultured on Matrigel show complex *in vivo* type cellular behavior that is not observed in a cell culture dish without Matrigel. Endothelial cells form a cobweb-like intricate network on a Matrigel-coated dish but not on the surface of a plastic dish.

24.2.3.6 Nanobiomaterials

Nanomaterials are nanosized material that may be nanoparticle (1–100 nm) or nanofibers (diameter 1–100 nm). The protein, lipid, and carbohydrate properties are determined by size, folding, orientation, and pattern at the nanoscale level (Christenson et al. 2007). All biomolecules interact or react with each other. For example, in a protein, antibody only recognizes and interacts with a small motif of proteins. Many individuals are allergic to certain kinds of pollens. The protein components that are responsible for compatibility factors determinants with same species plant female gamete causes immune responses (allergic reaction) in human—not the whole pollen. Fibronectin is a well-known protein molecule present in the cell ECM. Other biomolecules can interact with fibronectin's motif (R–G–D). Transcription factors bind with DNA at the nanoscale level. For all these reasonable phenomena, nanobiomaterial has become so prominent in modern-day technology. From carbon nanotubes to drug delivery, nanotechnology is becoming more widely accepted among biotechnologists. Among biomaterials, the nanometer-scale substrate is biologically readily acceptable (Li et al. 2006). Nanobiomaterials have higher surface area to volume ratios than microscale biomaterials. This ratio and their crystal grains change surface energy for protein adsorption (Christenson et al. 2007).

24.2.3.7 Future Perspectives of Biomaterial in Stem Cell Differentiation

The design of a physiologically relevant structure and the fabrication of biomaterials for *in vivo* use have facilitated the incorporation of microvascular networks, transport and filtration of fluids, and 3-D microenvironments resulting in the regulation of cell behavior (Inamdar and Borenstein 2011). During the last decade, a large amount of interdisciplinary studies combining tissue engineering, biomaterials, developmental biology, polymer science, bioreactors, and stem cells have been performed. Biomaterials in layer-by-layer organization creates a 3-D niche when a single marrow-derived stem cell population differentiates into zone-specific chondrocytes and organize into a complex tissue structure. Spatially varying biomaterial structures in a 3-D scaffold stimulate the regeneration of multilayered complex tissue from a single stem cell population (Roy et al. 2011). The delivery of soluble factors bound with synthetic biomaterial is now a reality, along with an increasing complexity that is comparable to native tissue. Thus, biomaterials received another boost for translational medicine (Kretlow and Mikos et al. 2011). Prestwich (2011) created a hyaluronic acid-based synthetic ECM that can be used in many applications such as 3-D culture of cells for *in vitro* and *in vivo* translational purposes. They developed injectable and biocompatible vehicles for delivery, growth, and differentiation of stem cells.

Raof and his group have synthesized an array of one-dimensional alginate gel microstrands and aqueous microstrands through an SU-8 filter device by capillary action. ES cells collected from one-dimensional aqueous microstrands favored the differentiation toward endoderm and mesoderm cell lineages, whereas gelled microstrands preferred to differentiate into ectoderm and mesoderm lineages (Raof et al. 2011). High-performance biomaterials and devices were developed by controlling the interfacial and physical properties of biomaterials to regulate stem cell behavior and differentiation (Sun et al. 2011). The combined effects of microstructured tissue culture polystyrene substrate (modulates cell shape and orientation) and laminar shear stress were applied in endothelial cells to observe cell behavior (Brown et al. 2011a). Surface topography in nanoscale and fluid sheer force can also direct cell lineage selection (Lim et al. 2011). Research is currently underway for the therapeutic application of substrate that is more defined, such as immobilized human plasma vitronectin. The vitronectin protein helps in the adhesion of cells onto a surface. Li et al. (2010) showed that adsorbed vitronectin was capable of supporting the expansion of human ESCs (>10 passages). Cells express the pluripotent marker Oct4 comparable to Matrigel, which is a highly complex substrate.

24.3 CONCLUSION

Tissues are grown and maintained in 3-D in their natural environment. This natural *in vivo* environment is very difficult to recreate in plastic dishes. The factors involved in stem cell lineage commitment and terminal differentiations are not well understood. Work from Lim et al. (2011) shows that the ECM provides physical and biochemical cues for stem cell lineage selection. The biochemical molecules may be chemokines, cytokines, or growth factors. Future development of smart biomaterials and external and internal molecules of stem cell differentiation will bring the bench work to clinical applications.

ACKNOWLEDGMENTS

The author likes to thank Shana Kalaria for critical reviewing of this chapter.

REFERENCES

Abed, A., N. Assoul, M. Ba, S.M. Derkaoui, P. Portes, L. Louedec, P. Flaud, I. Bataille, D. Letourneur, and A. Meddahi-Pellé. 2011. Influence of polysaccharide composition on the biocompatibility of pullulan/dextran-based hydrogels. *Journal of Biomedical Materials Research Part A* 96(3): 535–542.

Abbott, A. 2003. Cell culture: biology's new dimension. *Nature* 424, 870–872. doi:10.1038/424870a.

Ando, T., H. Yamazoe, K. Moriyasu, Y. Ueda, and H. Iwata. 2007. Induction of dopamine-releasing cells from primate embryonic stem cells enclosed in agarose microcapsules. *Tissue Engineering* 13(10): 2539–2547.

Baharvand, H., S.M. Hashemi, S. Kazemi Ashtiani, and A. Farrokhi. 2006. Differentiation of human embryonic stem cells into hepatocytes in 2-D and 3-D culture systems *in vitro*. *International Journal of Developmental Biology* 50, 645–652.

Bazaka, K., M.V. Jacob, R.J. Crawford, and E.P. Ivanova. 2011. Plasma-assisted surface modification of organic biopolymers to prevent bacterial attachment. *Acta Biomaterialia* 7(5): 2015–2028 [Epub 2010 Dec 29].

Benton, G., H.K. Kleinman, J. George, and I. Arnaoutova. 2011. Multiple uses of basement membrane-like matrix (BME/Matrigel) *in vitro* and *in vivo* with cancer cells. *International Journal of Cancer* 128(8): 1751–1757.

Bettinger, C.J. 2011. Biodegradable elastomers for tissue engineering and cell–biomaterial interactions. *Macromolecular Bioscience* 11(4): 467–482.

Bhang, S.H., S.W. Cho, W.G. La, T.J. Lee, H.S. Yang, A.Y. Sun, S.H. Baek, J.W. Rhie, and B.S. Kim. 2011. Angiogenesis in ischemic tissue produced by spheroid grafting of human adipose-derived stromal cells. *Biomaterials* 32(11): 2734–2747.

Bhumiratana, S., W.L. Grayson, A. Castaneda, D.N. Rockwood, E.S. Gil, D.L. Kaplan, and G. Vunjak-Novakovic. 2011. Nucleation and growth of mineralized bone matrix on silk-hydroxyapatite composite scaffolds. *Biomaterials* 32(11): 2812–2820.

Bird, A. 2002. DNA methylation patterns and epigenetic memory [review]. *Genes and Development* 16(1): 6–21.

Bono, E., S.H. Mathes, N. Franscini, and U. Graf-Hausner. 2010. Tissue engineering—The gateway to regenerative medicine. *Chimia (Aarau)* 64(11): 808–812.

Brockes, J., and A. Kumar. 2008. Comparative aspects of animal regeneration. *Annual Review of Cell and Developmental Biology* 24: 525–549.

Brown, A., G. Burke, and B.J. Meenan. 2011a. Modeling of shear stress experienced by endothelial cells cultured on microstructured polymer substrates in a parallel plate flow chamber. *Biotechnology and Bioengineering* 108(5): 1148–1158. doi: 10.1002/bit.23022 [Epub 2011 Jan 27].

Brown, C.P., F. Rosei, E. Traversa, and S. Licoccia. 2011b. Spider silk as a load bearing biomaterial: tailoring mechanical properties via structural modifications. *Nanoscale* 3(3): 870–876 [Epub 2011 Jan 7].

Chen, S.S., W. Fitzgerald, J. Zimmerberg, H.K. Kleinman, and L. Margolis. 2007. Cell–cell and cell–extracellular matrix interactions regulate embryonic stem cell differentiation. *Stem Cells* 25(3): 553–561.

Choi, J.S., B.S. Kim, J.Y. Kim, J.D. Kim, Y.C. Choi, H.J. Yang, K. Park, H.Y. Lee, and Y.W. Cho. 2011. Decellularized extracellular matrix derived from human adipose tissue as a potential scaffold for allograft tissue engineering. *Journal of Biomedical Materials Research Part A* 97(3): 292–299.

Christenson, E.M., K.S. Anseth, J.J. van den Beucken, C.K. Chan, B. Ercan, J.A. Jansen, C.T. Laurencin et al. 2007. Nanobiomaterial applications in orthopedics. *Journal of Orthopaedic Research* 25(1): 11–22.

Datta, N., Q.P. Pham, U. Sharma, V.I. Sikavitsas, J.A. Jansen, and A.G. Mikos. 2006. *In vitro* generated extracellular matrix and fluid shear stress synergistically enhance 3-D osteoblastic differentiation. *Proceedings of the National Academy of Sciences of the United States of America* 103: 2488–2493.

Dawson, E., G. Mapili, K. Erickson, S. Taqvi, and K. Roy. 2008. Biomaterials for stem cell differentiation. *Advanced Drug Delivery Reviews* 60(2): 215–228 [Epub 2007 Oct 11].

Engler, A.J., S. Sen, H.L. Sweeney, and D.E. Discher. 2006. Matrix elasticity directs stem cell lineage specification. *Cell* 126: 677–689.

European Society of Biomaterials Conference, Chester, England, March 3–5, 1986.

Finger, A.R., C.Y. Sargent, K.O. Dulaney, S.H. Bernacki, and E.G. Loboa. 2007. Differential effects on messenger ribonucleic acid expression by bone marrow-derived human mesenchymal stem cells seeded in agarose constructs due to ramped and steady applications of cyclic hydrostatic pressure. *Tissue Engineering* 13(6): 1151–1158.

Frampton, J.P., M.R. Hynd, M.L. Shuler, and W. Shain. 2011. Fabrication and optimization of alginate hydrogel constructs for use in 3-D neural cell culture. *Biomedical Materials* 6(1): 015002 [Epub 2011 Jan 5].

Ghanaati, S., M. Schlee, M.J. Webber, I. Willershausen, M. Barbeck, E. Balic, C. Görlach, S.I. Stupp, R.A. Sader, and C.J. Kirkpatrick. 2011. Evaluation of the tissue reaction to a new bilayered collagen matrix *in vivo* and its translation to the clinic. *Biomedical Materials* 6(1): 015010 [Epub 2011 Jan 17].

Gould, E., A.J. Reeves, M.S. Graziano, and C.G. Gross. 1999. Neurogenesis in the neocortex of adult primates. *Science* 286(5439): 548–552.

Grandi, C., R. Di Liddo, P. Paganin, S. Lora, D. Dalzoppo, G. Feltrin, C. Giraudo, M. Tommasini, M.T. Conconi, and P.P. Parnigotto. 2011. Porous alginate/poly(ε-caprolactone) scaffolds: Preparation, characterization and *in vitro* biological activity. *International Journal of Molecular Medicine* 27(3): 455–467.

Gurtner, G.C., M.J. Callaghan, and M.T. Longaker. 2007. Progress and potential for regenerative medicine. *Annual Review of Medicine* 58: 299–312.

Hannouche, D., H. Terai, J.R. Fuchs, S. Terada, S. Zand, B.A. Nasseri, H. Petite, L. Sedel, and J.P. Vacanti. 2007. Engineering of implantable cartilaginous structures from bone marrow-derived mesenchymal stem cells. *Tissue Engineering* 13: 87–99.

Horowitz, A.F. 1997. Integrins and health. *Scientific American* 68–75.

Hubbell, J.A. 1995. Biomaterials in tissue engineering. *Biotechnology (NY)* 13(6): 565–576.

Hynes, R.O. 1992. Integrins: Versatility, modulation, and signaling in cell adhesion [review]. *Cell* 69(1): 11–25.

Inamdar, N.K., and J.T. Borenstein. 2011. Microfluidic cell culture models for tissue engineering. *Current Opinion in Biotechnology* 22: 1–9.

Jopling, C., E. Sleep, M. Raya, M. Martí, A. Raya, J.C. Belmonte. 2010. Zebrafish heart regeneration occurs by cardiomyocyte dedifferentiation and proliferation. *Nature* 464(7288): 606–609.

Kachouie, N.N., Y. Du, H. Bae, M. Khabiry, A.F. Ahari, B. Zamanian, J. Fukuda, and A. Khademhosseini. 2010. Directed assembly of cell-laden hydrogels for engineering functional tissues. *Organogenesis* 6(4): 234–244.

Kahle, M., H.P. Wiesmann, K. Berr, R.A. Depprich, N.R. Kübler, C. Naujoks, M. Cohnen, M.A. Ommerborn, U. Meyer, and J. Handschel. 2010. Embryonic stem cells induce ectopic bone formation in rats. *Biomedical Materials Engineering* 20(6): 371–380.

Kania, R.E., A. Meunier, M. Hamadouche, L. Sedel, and H. Petite. 1998. Addition of fibrin sealant to ceramic promotes bone repair: long-term study in rabbit femoral defect model. *Journal of Biomedical Materials Research* 43(1): 38–45.

Kim, H., C.H. Tator, and M.S. Shoichet. 2011. Chitosan implants in the rat spinal cord: Biocompatibility and biodegradation. *Journal of Biomedical Materials Research Part A* 97(4): 395–404.

Kolettis, T.M., A. Vilaeti, K. Dimos, N. Tsitou, and S. Agathopoulos. 2011. Tissue engineering for post-myocardial infarction ventricular remodeling. *Mini Reviews in Medicinal Chemistry* 11(3): 263–270.

Kretlow, J.D., and A.G. Mikos. 2011. Founder's award to Antonios G. Mikos, Ph.D., 2011 Society for Biomaterials Annual Meeting and Exposition, Orlando, Florida, April 13–16, 2011. Bones to biomaterials and back again—20 years of taking cues from nature to engineer synthetic polymer scaffolds. *Journal of Biomedical Materials Research Part A* 1;98(3): 323–31.

Kujala, S., T. Raatikainen, J. Ryhänen, O. Kaarela, and P. Jalovaara. 2002. Composite implant of native bovine bone morphogenetic protein (BMP) and biocoral in the treatment of scaphoid nonunions—A preliminary study. *Scandinavian Journal of Surgery* 91(2): 186–190.

Lees, J.G., S.A. Lim, T. Croll, G. Williams, S. Lui, J. Cooper-White, L.R. McQuade, B. Mathiyalagan, and B.E. Tuch. 2007. Transplantation of 3-D scaffolds seeded with human embryonic stem cells: Biological features of surrogate tissue and teratoma-forming potential. *Regenerative Medicine* 2(3): 289–300.

Levenstein, M.E., T.E. Ludwig, R.H. Xu, R.A. Llanas, K. van den Heuvel-Kramer, D. Manning, and J.A. Thomson. 2006. Basic fibroblast growth factor support of human embryonic stem cell self-renewal. *Stem Cells* 24(3): 568–574.

Li, J., J. Bardy, L.Y. Yap, A. Chen, V. Nurcombe, S.M. Cool, S.K. Oh, and W.R. Birch. 2010. Impact of vitronectin concentration and surface properties on the stable propagation of human embryonic stem cells. *Biointerphases* 5(3): FA132–142.

Li, W.J., K.G. Danielson, P.G. Alexander, and R.S. Tuan. 2003. Biological response of chondrocytes cultured in three-dimensional nanofibrous poly(ecaprolactone) scaffolds. *Journal of Biomedical Materials Research Part A* 15;67(4): 1105–1114.

Li, W.J., Y.J. Jiang, and R.S. Tuan. 2006. Chondrocytes phenotype in engineered fibrous matrix is regulated by fiber size. *Tissue Engineering* 12: 1775–1785.

Lim, J.Y., A.E. Loiselle, J.S. Lee, Y. Zhang, J.D. Salvi, and H.J. Donahue. 2011. Optimizing the osteogenic potential of adult stem cells for skeletal regeneration. *Journal of Orthopaedic Research* doi: 10.1002/jor.21441 [Epub ahead of print].

Lin, C.Q., and M.J. Bissell. 1993. Multifaceted regulation of cell differentiation by extracellular matrix. *FASEB Journal* 7: 737–743.

Lopez-Cebral, R., M. Martín-Pastor, J.E. Parraga, G.K. Zorzi, B. Seijo, and A. Sanchez. 2011. Chemically modified gelatin as biomaterial in the design of new nanomedicines. *Medicinal Chemistry* 7(3): 145–154.

Mangano, C., F. Paino, R. d'Aquino, A. De Rosa, G. Iezzi, A. Piattelli, L. Laino et al. 2011. Human dental pulp stem cells hook into biocoral scaffold forming an engineered biocomplex. *PLoS One* 6(4): e18721.

Meissner A., T.S. Mikkelsen, H. Gu, M. Wernig, J. Hanna, A. Sivachenko, X. Zhang et al. 2008. Genome-scale DNA methylation maps of pluripotent and differentiated cells. *Nature* 454(7205): 766–70.

Masters, K.S. 2011. Covalent growth factor immobilization strategies for tissue repair and regeneration. *Macromolecular Bioscience* 11(9):1149–63. doi: 10.1002/mabi.201000505 [Epub 2011 Apr 20].

Michelini, M., V. Franceschini, S. Sihui Chen, S. Papini, A. Rosellini, F. Ciani, L. Margolis, and R.P. Revoltella. 2006. Primate embryonic stem cells create their own niche while differentiating in three-dimensional culture systems. *Cell Proliferation* 39: 217–229.

Muneoka, K., M. Han, and D.M. Gardiner. 2008. Regrowing limbs: Can people regenerate body parts? *Scientific American* 298(4): 56–63.

Nguyen, L.H., A.K. Kudva, N.S. Saxena, and K. Roy. 2011. Engineering articular cartilage with spatially varying matrix composition and mechanical properties from a single stem cell population using a multi-layered hydrogel. *Biomaterials* [Epub ahead of print].

Nottebohm, F. 1989. From bird song to neurogenesis [review]. *Scientific American* 260(2): 74–79.

Pierschbacher, M.D., and E. Ruoslahti. 1984. Cell attachment activity of fibronectin can be duplicated by small synthetic fragments of the molecule. *Nature* 309(5963): 30–33.

Ponting, I.L., and T.M. Dexter. 1988. The role of colony stimulating factors, interleukin 1 and stromal extracellular matrix in the regulation of stem cell development. *Behring Institute Mitteilungen* (83): 48–55.

Prestwich, G.D. 2011. Hyaluronic acid-based clinical biomaterials derived for cell and molecule delivery in regenerative medicine. *Journal of Controlled Release* 155(2): 193–199.

Qi, J., A. Chen, H. You, K. Li, D. Zhang, and F. Guo. 2011. Proliferation and chondrogenic differentiation of CD105-positive enriched rat synovium-derived mesenchymal stem cells in three-dimensional porous scaffolds. *Biomedical Materials* 6(1): 015006 [Epub 2011 Jan 5].

Raof, N.A., M.R. Padgen, A.R. Gracias, M. Bergkvist, and Y. Xie. 2011. One-dimensional self-assembly of mouse embryonic stem cells using an array of hydrogel microstrands. *Biomaterials* 32(20): 4498–4505 [Epub 2011 Apr 2].

Romano, N.H., D. Sengupta, C. Chung, and S.C. Heilshorn. 2011. Protein-engineered biomaterials: Nanoscale mimics of the extracellular matrix [review]. *Biochimica et Biophysica Acta* 1810(3): 339–349 [Epub 2010 Jul 18].

Roy, D.C., S.J. Wilke-Mounts, and D.C. Hocking. 2011. Chimeric fibronectin matrix mimetic as a functional growth- and migration-promoting adhesive substrate. *Biomaterials* 32(8): 2077–2087.

Rustad, K.C., M. Sorkin, B. Levi, M.T. Longaker, and G.C. Gurtner. 2010. Strategies for organ level tissue engineering. *Organogenesis* 6(3): 151–157.

Salmeron-Sanchez, M., P. Rico, D. Moratal, T.T. Lee, J.E. Schwarzbauer, and A.J. Garcia. 2011. Role of material-driven fibronectin fibrillogenesis in cell differentiation. *Biomaterials* 32(8): 2099–2105 [Epub 2010 Dec 24].

Schwartz, Z., D.J. Griffon, L.P. Fredericks, H.B. Lee, and H.Y. Weng. 2011. Hyaluronic acid and chondrogenesis of murine bone marrow mesenchymal stem cells in chitosan sponges. *American Journal of Veterinary Research* 72(1): 42–50.

Sreejit, P., and R.S. Verma. 2011. Cardiogel supports adhesion, proliferation and differentiation of stem cells with increased oxidative stress protection. *European Cells and Materials* 21: 107–121.

Sujatha, G., A. Sinha, and S. Singh. 2011. Cells behaviour in presence of nanoscaffolds. *Journal of Biomedical Nanotechnology* 7(1): 43–44.

Sun, T., G. Qing, B. Su, and L. Jiang. 2011. Functional biointerface materials inspired from nature. *Chemical Society Reviews* 40(5): 2909–2921.

Takahashi, K., and S. Yamanaka. 2006. Induction of pluripotent stem cells from mouse embryonic and adult fibroblast cultures by defined factors. *Cell* 126(4): 663–676.

Thomson, J.A., J. Itskovitz-Eldor, S.S. Shapiro, M.A. Waknitz, J.J. Swiergiel, V.S. Marshall, and J.M. Jones. 1998. Embryonic stem cell lines derived from human blastocysts. *Science* 282(5391) 1145–1147.

Toh, W.S., M. Spector, E.H. Lee, and T. Cao. 2011. Biomaterial-mediated delivery of microenvironmental cues for repair and regeneration of articular cartilage. *Molecular Pharmacology* 8(4): 994–1001.

Underhill, G. H., and S. N. Bhatia. 2007. High-throughput analysis of signals regulating stem cell fate and function. *Current Opinion in Chemical Biology* 11: 357–366.

Xiao, C., H. Zhou, G. Liu, P. Zhang, Y. Fu, P. Gu, H. Hou, T. Tang, and X. Fan. 2011. Bone marrow stromal cells with a combined expression of BMP-2 and VEGF-165 enhanced bone regeneration. *Biomedical Materials* 6(1): 015013 [Epub 2011 Jan 21].

Xie, X., Z. Tang, J. Chen, J. Yang, W. Zeng, N. Liu, and Y. Liu. 2011. Neurogenesis of adipose-derived stem cells in hydrogel. *Journal of Huazhong University of Science and Technology. Medical Sciences* 31(2): 174–177 [Epub 2011 Apr 20, 52].

Wayne, J.S., C.L. McDowell, K.J. Shields, and R.S. Tuan. 2005. *In vivo* response of polylactic acid–alginate scaffolds and bone marrow-derived cells for cartilage tissue engineering. *Tissue Engineering* 11, 953–963.

Willerth, S.M., and S. E. Sakiyama-Elbert. 2008. Combining stem cells and biomaterial scaffolds for constructing tissues and cell delivery. *StemBook* [Internet]. Cambridge, MA: Harvard Stem Cell Institute.

Yeh, H.Y., and J.C. Lin. 2012. Surface phosphorylation for polyelectrolyte complex of chitosan and its sulfonated derivative: Surface analysis, blood compatibility and adipose derived stem cell contact properties. *Journal of Biomaterials Science Polymer Edition* 23(1–4): 233–250.

Zong, C., D. Xue, W. Yuan, W. Wang, D. Shen, X. Tong, D. Shi, L. Liu et al. 2010. Reconstruction of rat calvarial defects with hMSCs and osteoblast-like cells in poly-lactic-co-glycolic acid scaffolds. *European Cells and Materials* 20: 109–120.

Index

Page numbers followed by *f* and *t* indicate figures and tables, respectively.

D

E

F

J

K

L

M

N

T

U

V

W

X

Y

Z